Intracellular Niches of Microbes

Edited by
Ulrich E. Schaible and Albert Haas

Related Titles

Carter, J., Saunders, V.

Virology

Principles and Applications

2007

ISBN: 978-0-470-02387-7

Jungblut, P. R., Hecker, M. (eds.)

Proteomics of Microbial Pathogens

2007

ISBN: 978-3-527-31759-2

zur Hausen, H.

Infections Causing Human Cancer

2006

ISBN: 978-3-527-31056-2

Deretic, V. (ed.)

Autophagy in Immunity and Infection

A Novel Immune Effector

2006

ISBN: 978-3-527-31450-8

Frosch, M., Maiden, M. C. J. (eds.)

Handbook of Meningococcal Disease

Infection Biology, Vaccination, Clinical Management

2006

ISBN: 978-3-527-31260-3

Schumann, W.

Dynamics of the Bacterial Chromosome

Structure and Function

2006

ISBN: 978-3-527-30496-7

Khardori, N. (ed.)

Bioterrorism Preparedness

Medicine - Public Health - Policy

2006

ISBN: 978-3-527-31235-1

Intracellular Niches of Microbes

A Pathogens Guide Through the Host Cell

Edited by
Ulrich E. Schaible and Albert Haas

WILEY-VCH

WILEY-VCH Verlag GmbH & Co. KGaA

The Editors

Prof. Ulrich E. Schaible
Research Center Borstel
Zentrum für Medizin und Biowissenschaften (FZB)
Department of Molecular Infection Research
Parkallee 1 – 40
23845 Borstel
Germany

Prof. Albert Haas
University of Bonn
Cell Biology Institute
Ulrich-Haberland-Straße 61 a
53121 Bonn
Germany

Library of Congress Card No.: applied for

British Library Cataloguing-in-Publication Data
A catalogue record for this book is available from the British Library.

Bibliographic information published by the Deutsche Nationalbibliothek
The Deutsche Nationalbibliothek lists this publication in the Deutsche Nationalbibliografie; detailed bibliographic data are available on the Internet at http://dnb.d-nb.de.

Typesetting Thomson Digital, Noida, India
Printing betz-druck GmbH, Darmstadt
Bookbinding Litges & Dopf GmbH, Heppenheim
Cover Design Adam-Design, Weinheim

Printed in the Federal Republic of Germany
Printed on acid-free paper

ISBN: 978-3-527-32207-7

Foreword

Over the last thirty years, the combined use of bacterial genetics, molecular and cell biology, and more recently genomics, have illuminated our understanding of the virulence of the major human and veterinary pathogens. Among those, intracellular bacterial pathogens have played a dominant role in this endeavor because they recapitulate most of the existing steps of microbe-host interaction. The progress made has been impressive, not only conceptually, but also technically. Imaging, for instance, has often prepared and preceded the discoveries in this area where *seeing is believing*, it has also tremendously benefited from these models in return. Adherence of pathogens to cells, diversity of mechanisms of entry, a variety of mechanisms for intracellular survival and growth, from vacuolar rupture followed by escape into the cytoplasm, to remodeling of vacuoles to avoid phagolysomal fusion – a combinatorial synthesis of these steps of interaction has led to an amazing breadth of diversified strategies representing the complex solutions "developed" by nature for pathogens to achieve survival as species, or pathovars among these species, in the ferocious struggle for life. These microorganisms are exposed to harsh environmental conditions such as fighting against protozoan predators (i.e. amoebas) for the so called environmental pathogens like *Legionella*, or to the immune system that has appeared early in the world of multicellular organisms. *Nothing makes sense in biology if not seen under the angle of evolution* said Theodosius Dobzansky. The world of intracellular microbes is a perfect illustration of this statement. With the possible exception of environmental pathogens, intracellular pathogens bear in their genomes the traces of their construction under selective pressure of their host. It results in a complex mixture of gene acquisition and gene deletion that strongly differentiate them from their closest commensal cousins. A permanent flux of genes, largely perpetrated by bacteriophages and plasmids permanently occurs, particularly in the gut lumen of animals, thereby permanently offering microbes options to improve their fit with the host. It is difficult to decipher the timing of the genetic events that have led to such complex combinations. However, one would like to believe that some key steps have occurred at some stage, like the acquisition of a large pathogenicity island encoding invasive capacities, thereby propelling the microorganisms in a new hostile environment to which it was not prepared. Only the acquisition and/or loss of genes

Intracellular Niches of Microbes. Edited by Ulrich E. Schaible and Albert Haas

ISBN: 978-3-527-32207-7

allowing the pathogen to cope with these new conditions was able to secure its survival. *It is not the strongest of the species that survives…nor the most intelligent that survives. It is the one that is the most adaptable to change* said Darwin in *On the origin of species by way of natural selection*. It is interesting to consider that our "contemporary" pathogens still have a large capacity to evolve, but that only the stochastic acquisition of new genes by horizontal transfer can achieve the quantum leap changes that move the evolutionary process. It is also interesting to observe that several genomes of intracellular pathogens, particularly those that have become obligate intracellular parasites (i.e. *Rickettsia spp.*, *Chlamydia spp.*, *Mycobacterium leprae*) show highly degraded genomes, with massive amount of gene loss. Whether this is a true reflection of the need for compensatory gene deletion to the acquisition of new intracellular pathogenic properties, or simply the loss in absence of selective pressure of genes (i.e. metabolic pathways) that are no longer useful to the pathogen since it benefits from the cell nutrients, is still an open and debated question. In any event, the progression towards complete intracellular parasitism that is often accompanied by restriction in species specificity is a dominant feature of intracellular microorganisms that certainly needs further investigation.

Last but not least, it seems that the ultimate option for these microorganisms is to become true symbionts. Indeed, some molecular systems such as secretory apparatus that are essential to deliver virulence effectors are conserved in true symbionts, probably with the aim to deliver symbiotic effectors. The ultimate example is the absolute harnessing by eukaryotic cells of *Rickettsia spp.* to generate mitochondria as O_2 appeared on earth. No need to reach this extreme, however, *Wolbachia spp.* do parasitize insects in a very "stealthy" manner, and only recently was it observed that this symbiosis, beside its well known effect on fecundity, was stimulating a background level of innate immunity allowing these insects to resist viral infection.

The world of intracellular pathogens, and symbionts that should now be part of this global concept, appears as a gold mine of strategies whose purpose often still needs to be understood.

In an amazing series of contributions by renowned world experts, this volume provides the first integrated, coherent and exhaustive review of the biology of intracellular pathogens and symbionts. It offers a fascinating vision of the evolutionary logics that have led microorganisms to venture into cells and to adapt to the environment of this niche in a way that led to genomic modifications which made them for ever different and highly adapted to this particular life style. For some of them it evolved to the ultimate point of becoming obligate intracellular pathogens, or even symbionts.

Ulrich Schaible and Albert Haas should be congratulated for their exceptional vision of the field, contagious enthusiasm that helped assemble this key stone volume, and, last but not least, exceptional service to our community.

Philippe J. Sansonetti
Professeur au Collège de France
Professeur à l'Institut Pasteur

Contents

Intracellular Niches of Microbes. Edited by Ulrich E. Schaible and Albert Haas

ISBN: 978-3-527-32207-7

List of Contributors

Lee-Ann H. Allen
Departments of Microbiology and
Internal Medicine
The Inflammation Program and the
VA Medical Center
Iowa City, IA 52242
USA

Damien Balestrino
Institut Pasteur
Unité des Interactions Bactéries
Cellules
75015 Paris
France
and
INSERM
U604
75015 Paris
France
and
INRA
USC2020
75015 Paris
France

Kristine von Bargen
Cell Biology Institute
University of Bonn
Ulrich-Haberland-Str. 61a
53121 Bonn
Germany

Martin Baumgartner
Division of Molecular Pathobiology
DCR-VPH
Vetsuisse Faculty Bern
University of Bern
Länggassstr. 122
3012 Bern
Switzerland

Ulrike Becken
Cell Biology Institute
University of Bonn
Ulrich-Haberland-Str. 61a
53121 Bonn
Germany

William J. Broughton
LBMPS
Sciences III
Université de Genève
30 Quai Ernest-Ansermet
1211 Geneva 4
Switzerland

Philip D. Butcher
Medical Microbiology
Centre for Infection
Division of Cellular & Molecular
Medicine
St. George's University of London
Cranmer Terrace, Tooting
London SW17 0RE
UK

Intracellular Niches of Microbes. Edited by Ulrich E. Schaible and Albert Haas

ISBN: 978-3-527-32207-7

Gabriela Cosío
Program in Cell Biology
Hospital for Sick Children
555 University Avenue
Toronto, Ontario M5G 1X8
Canada

Pascale Cossart
Institut Pasteur
Unité des Interactions Bactéries
Cellules
75015 Paris
France
and
INSERM
U604
75015 Paris
France
and
INRA
USC2020
75015 Paris
France

Deborah Dean
Center for Immunobiology and
Vaccine Development
Children's Hospital Oakland Research
Institute
Oakland, CA
USA
and
UCB and UCSF Joint Graduate Group
in Bioengineering
University of California at Berkeley
Berkeley, CA
USA
and
University of California at San Francisco
San Francisco, CA
USA
and
Department of Medicine
University of California at San Francisco
San Francisco, CA
USA

William J. Deakin
LBMPS
Sciences III
Université de Genève
30 Quai Ernest Ansermet
1211 Geneva 4
Switzerland

Albert Descoteaux
INRS-Institut Armand-Frappier
Université du Québec
Laval, Quebec H7V 1B7
Canada

Régis Dieckmann
Département de Biochimie
Faculté des Sciences
Université de Genève
Sciences II
30 quai Ernest Ansermet
1211 Geneva 4
Switzerland

Dirk Dobbelaere
Division of Molecular Pathobiology
DCR-VPH
Vetsuisse Faculty Bern
University of Bern
Länggassstr. 122
3012 Bern
Switzerland

Heike Feldhaar
Lehrstuhl für Verhaltensbiologie
Universität Osnabrück
Barbarastr. 11
49076 Osnabrück
Germany

Stacey D. Gilk
Coxiella Pathogenesis Section
Laboratory of Intracellular Parasites
Rocky Mountain Laboratories
National Institute of Allergy and
Infectious Diseases
National Institutes of Health
Hamilton, MT 59840
USA

Jean-Pierre Gorvel
Centre d'Immunologie de
Marseille-Luminy
case 906
13288 Marseille Cedex 9
France

David B. Guiliano
Infectious and Tropical Diseases
Division of Infection and Immunity
Windeyer Building
46 Cleveland Street
London W1T 4JF
UK

Gareth Griffiths
Electronmicroscopical Unit
for Biological Sciences
University of Oslo
P.O. Box 1062, Blindern
0316 Oslo
Norway

Sergio Grinstein
Program in Cell Biology
Hospital for Sick Children
555 University Avenue
Toronto, Ontario M5G 1X8
Canada

Roy Gross
Lehrstuhl für Mikrobiologie
Biozentrum
Universität Würzburg
Am Hubland
97074 Würzburg
Germany

Maximiliano G. Gutierrez
Helmholtz-Zentrum für
Infektionsforschung GmbH
Inhoffenstr. 7
38124 Braunschweig
Germany

Albert Haas
Cell Biology Institute
University of Bonn
Ulrich-Haberland-Str. 61a
53121 Bonn
Germany

Ted Hackstadt
Host–Parasite Interactions Section
Laboratory of Intracellular Parasites
Rocky Mountain Laboratories
National Institute of Allergy and
Infectious Diseases
National Institutes of Health
Hamilton, MT 59840
USA

Robert A. Heinzen
Coxiella Pathogenesis Section
Laboratory of Intracellular Parasites
Rocky Mountain Laboratories
National Institute of Allergy and
Infectious Diseases
National Institutes of Health
Hamilton, MT 59840
USA

Monica Hagedorn
Département de Biochimie
Faculté des Sciences
Université de Genève
Sciences II
30 quai Ernest Ansermet
1211 Geneva 4
Switzerland

Alyssa Ingmundson
Section of Microbial Pathogenesis
Yale University School of Medicine
Boyer Center for Molecular Medicine
295 Congress Avenue
New Haven, CT 06536
USA

Monika Kalde
Centre d'Immunologie de
Marseille-Luminy
case 906
13288 Marseille Cedex 9
France

Kumiko Kambara
LBMPS
Sciences III
Université de Genève
30 Quai Ernest Ansermet
1211 Geneva 4
Switzerland

Daniel S. Korbel
Centre for Digestive Diseases
The Blizard Institute
Barts and The London School of
Medicine and Dentistry
Queen Mary, University of London
4 Network Street
London E1 2AT
UK
and
London School of Hygiene and
Tropical Medicine
Department of Infectious and
Tropical Diseases
Immunology Unit
Keppel Street
London WC1 7HT
UK

Goran Kovacevic
Faculty of Science
University of Zagreb
Rooseveltov trg 6
10000 Zagreb
Croatia

Wolfgang Löffelhardt
Max F. Perutz Laboratories
University of Vienna
Department of Biochemistry
Dr. Bohrgasse 9
1030 Vienna
Austria

Klaus Lingelbach
FB Biology
Department of Parasitology
Philipps-University Marburg
Karl-von-Frisch-Str. 8
35043 Marburg
Germany

Julia Mallégol
INRS-Institut Armand-Frappier
Université du Québec
Laval, Quebec H7V 1B7
Canada

Christine Matte
INRS-Institut Armand-Frappier
Université du Québec
Laval, Quebec H7V 1B7
Canada

Edgardo Moreno
Programa de Investigación
en Enfermedades Tropicales
Escuela de Medicina Veterinaria
304–3000 Heredia
Costa Rica

Simon L. Newman
Division of Infectious Diseases
University of Cincinatti Col. Med.
P.O. Box 670560
Cincinatti, OH 45267-0560
USA

Yelena Oksov
Laboratory of Electron Microscopy
Lindsey F. Kimball Research Institute
New York Blood Center
New York, NY 10021
USA

Jude M. Przyborski
FB Biology
Department of Parasitology
Philipps-University Marburg
Karl-von-Frisch-Str. 8
35043 Marburg
Germany

Elena Rydkina
Departments of Microbiology/
Immunology and Medicine
University of Rochester Medical School
of Medicine and Dentistry
601 Elmwood Avenue
Rochester, NY 14642
USA

Yasuko Rikihisa
Department of Veterinary Biosciences
College of Veterinary Medicine
The Ohio State University
1925 Coffey Road
Columbus, OH 43210
USA

Craig R. Roy
Section of Microbial Pathogenesis
Yale University School of Medicine
Boyer Center for Molecular Medicine
295 Congress Avenue
New Haven, CT 06536
USA

Sanjeev K. Sahni
Departments of Microbiology/
Immunology and Medicine
University of Rochester Medical School
of Medicine and Dentistry
601 Elmwood, Avenue
Rochester NY 14642
USA

Philippe Sansonetti
Unité de Pathogénie Microbienne
Moléculaire
Institut Pasteur
Paris
France
and
Unité 786
Institut National de la Santé et
de la Recherche Médicale
Inserm U786
Paris
France

Ulrich E. Schaible
Research Center Borstel
Leibniz-Zentrum für Medizin und
Biowissenschaften
Department of Molecular Infection
Research
Parkallee 1-40
23845 Borstel
Germany

and
London School of Hygiene and
Tropical Medicine
Department of Infectious and
Tropical Diseases
Immunology Unit
Keppel Street
London WC1 7HT
UK

Bianca E. Schneider
Infectious and Tropical Diseases
Immunology Unit
London School of Hygiene and
Tropical Medicine
Keppel Street
London WC1 7HT
UK

Grant S. Schulert
Departments of Microbiology and
Internal Medicine
The Inflammation Program and the
VA Medical Center
Iowa City, IA 52242
USA

L. David Sibley
Department of Molecular Microbiology
Washington University School of
Medicine
St. Louis, MO 63110
USA

David J. Silverman
Department of Microbiology and
Immunology
University of Maryland School of
Medicine
Baltimore, MD
USA

Thierry Soldati
Département de Biochimie
Faculté des Sciences
Université de Genève
Sciences II
30 quai Ernest Ansermet
1211 Geneva 4
Switzerland

Naraporn Somboonna
Center for Immunobiology and
Vaccine Development
Children's Hospital Oakland Research
Institute
Oakland, CA
USA
and
UCB and UCSF Joint Graduate Group
in Bioengineering
University of California at Berkeley
Berkeley, CA
USA
and
University of California at San Francisco
San Francisco, CA
USA
and
National Center for Genetic
Engineering and Biotechnology
National Science and Technology
Development Agency
Pathumthani 12120
Thailand

Olivia Steele-Mortimer
Laboratory of Intracellular Parasites
National Institutes of Allergy and
Infectious Diseases
National Institutes of Health
Rocky Mountain Laboratories
Hamilton, MT 59840
USA

Jürgen M. Steiner
Max F. Perutz Laboratories
University of Vienna
Department of Biochemistry
Dr. Bohrgasse 9
1030 Vienna
Austria

Guy Tran Van Nhieu
Inserm U 971
Unité de Communications
Intercellulaires et Infections
Microbiennos
Collège de France
11, Place Marcelin Berthelot
75005 Paris Cedex
France

Michael Steinert
Institut für Mikrobiologie
Technische Universität Braunschweig
Spielmannstr. 7
38106 Braunschweig
Germany

Joanne M. Stevens
Division of Microbiology
Institute for Animal Health
Compton
Berkshire RG20 7NN
UK

Mark P. Stevens
Division of Microbiology
Institute for Animal Health
Compton
Berkshire RG20 7NN
UK

Martin C. Taylor
London School of Hygiene and
Tropical Medicine
Keppel Street
London WC1 7HT
UK

Daniel E. Voth
Coxiella Pathogenesis Section
Laboratory of Intracellular Parasites
Rocky Mountain Laboratories
National Institute of Allergy and
Infectious Diseases
National Institutes of Health
Hamilton, MT 59840
USA

Simon J. Waddell
Medical Microbiology
Centre for Infection
Division of Cellular & Molecular
Medicine
St. George's University of London
Cranmer Terrace, Tooting
London SW17 0RE
UK

Markus Winterberg
FB Biology
Department of Parasitology
Philipps-University Marburg
Karl-von-Frisch-Str. 8
35043 Marburg
Germany

Part I
General Aspects

Intracellular Niches of Microbes. Edited by Ulrich E. Schaible and Albert Haas

ISBN: 978-3-527-32207-7

1
Introduction: The Evolution of Intracellular Life Forms and their Niches

Ulrich E. Schaible and Albert Haas

> "As species are produced and exterminated by slowly acting and still existing causes, and not by miraculous acts of creation and by catastrophes; and as the most important of all causes of organic change is one which is almost independent of altered and perhaps suddenly altered physical conditions, namely, the mutual relation of organism to organism, – the improvement of one being entailing the improvement or extermination of others". Charles Darwin

1.1
A Short History of Theories and Discoveries

The complex mutual relationship between intracellular microbes and their host cells is a challenging field of research and requires the perspective of evolution biology. The individual host–microbe interactions covered in this book all raise the following questions: how do microbes enter, survive and proliferate in, and how do they exit host cells? And how can intracellular niches be characterized and what are the benefits of intracellular life for the microbes and its consequences for the host cell? The question, however, is how and under what selective pressure did these interactions evolve? The year 2009 marks the 200th birthday of Charles Darwin (1809–1882; 12th February 1809), and, more importantly, the 150th anniversary of the publication of his most important book *The Origin of Species by Means of Natural Selection* (24th November 1859) [1]. In this eminent and highly disputed and provocatively revolutionary work, Darwin outlined the concept of evolution by natural selection in the struggle of life. The concept of interspecies competition as the driving force for the evolution of all bacterial, animal and plant species laid the basis for modern day biology.

Louis Pasteur (1822–1895) and others proved that microbial life did not arise spontaneously and miraculously, but rather due to the omnipresence of microorganisms, an important fact for food preservation and the consequential establishment of sterilization techniques. The seminal work of the nineteenth-century microbiologists

Intracellular Niches of Microbes. Edited by Ulrich E. Schaible and Albert Haas

ISBN: 978-3-527-32207-7

set the path to study the novel complexity of interspecies interactions in natural science and medical research. Although infectious diseases were an important determinant for human history, causing migration, settlement and conflict behavior, it was not until the nineteenth century that infectious agents were identified as causative agents for certain diseases rather than the diseases being of mysterious origins. The time between the end of the nineteenth and the beginning of the twentieth century was the high season of bacteriology, during which a huge number of microbial species were identified using newly developed culture techniques. Many of these microbes were associated with humans, animals or plants, and they were either pathogens, beneficial symbionts or commensals. A number of those microbes had chosen other unicellular or multicellular organisms as their ecological niches. Finally, infectious diseases were recognized as the driving force for the evolution of the innate and, in vertebrates, the acquired immune systems (Chapter 12).

Robert Koch (1843–1910) and his colleagues identified the first intracellular pathogenic bacterium, the tubercle bacillus (*Mycobacterium tuberculosis*). In the late nineteenth century tuberculosis was the prime cause of death in the metropolitan areas of Europe and North America, stirring up intensive medical and scientific interest. At around the same time, an important virulence trait of the tubercle bacillus, that is, living in macrophages, was described by Elie Metchnikoff (1845–1916), the founder of phagocyte biology. This is still a prime topic in tuberculosis research today (see Chapter 19). Metchnikoff was the first to observe the phagocytosis of bacteria by phagocytes in 1883 during his time at the Viennese Institute of Zoology and he also pointed out the importance of these cells in host response and inflammation [2, 3]. The term macrophage was attributed to him and made him the founder of innate immunity. In 1908, he received the Nobel Prize for his achievements. Metchnikoff was also the first to observe tubercle bacilli thriving intracellularly in macrophages (Figure 1.1) [4]. However, it was not until the last quarter of the twentieth century that scientists started to study the virulence factors of pathogens, and that intracellular pathogens (and symbionts) were highlighted for their unique capabilities to survive within and manipulate their host cells.

The identification of intracellular survival mechanisms was made possible by novel techniques in cell biology and the arrival of modern molecular genetics. J. A. Armstrong and Philip D'Arcy Hart [5, 6] were the first to show inhibition of phagolysosome fusion by the tubercle bacillus. Similar peculiarities of *Toxoplasma gondii*- and *Chlamydia psittaci*-containing vacuoles were published in 1979 and 1981, respectively [7, 8]. In the last decade of the twentieth century, many virulence traits of intracellular microbes were elucidated. Genome analyses and molecular techniques, paired with novel model systems such as yeast two-hybrid screening technology, uncovered pathogenicity islands and plasmids, virulence factors, as well as host cell target structures. It was discovered that throughout evolution there must have been a tremendous horizontal gene transfer between different microbes as well as between bacteria and eukaryotes (Chapter 2). Many of those pathogens and their virulence traits will be covered in this book. Some important intracellular microbes, such as *M. leprae*, *Chlamydia* and *Rickettsia*, are not yet accessible to

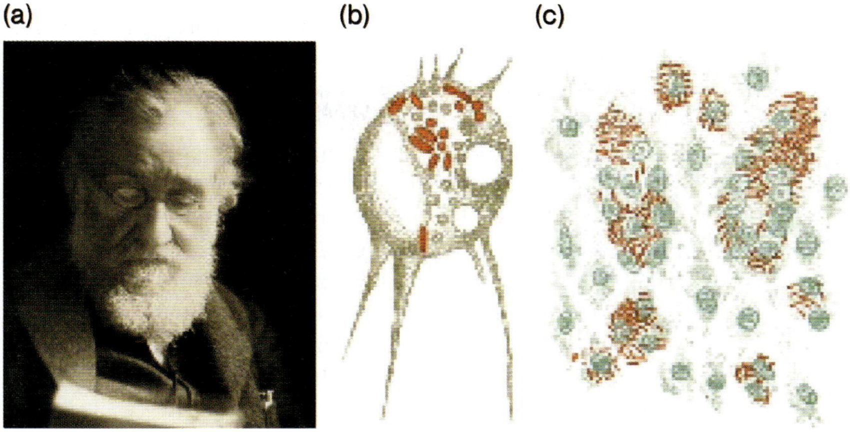

Figure 1.1 (a) Elie Metchnikoff (1845–1916) in his later life. He was the founder of phagocyte biology and coined the term "macrophage." Metchnikoff was also the first scientist to observe tubercle bacilli within macrophages and suggested they be able to survive within these cells, which otherwise are able to kill microbes. (b) Metchnikoff's depiction of a pigeon macrophage containing mycobacteria. (c) Macrophage culture infected with mycobacteria as observed by Elie Metchnikoff. These pictures were kindly provided by Stefan H. E. Kaufmann, Max-Planck-Institute of Infection Biology, Berlin, Germany.

manipulation by molecular genetics and future attempts will focus on generating targeted mutants in such organisms. A peek into the book of evolution of intracellular microbe genomes suggests that many of these virulence traits were established early in evolution, though probably not exactly for the purposes they are used for today.

1.2 A Look Through the Microscope of Evolution

> *Now, here, you see, it takes all the running you can do, just to keep in the same place.* The Red Queen in Lewis Carroll's *Alice in Wonderland*. (*The Origin of Species by the Means of Natural Selection*, 1859)

According to Darwin, life may have started in some "warm little pond." Rather than in warm ponds, it is believed today that primitive *Bacteria* and *Archaea* arose in the vicinity of hot vents in the ancient oceans some 4 billion years ago; and microbes still rule the earth today. It was only after the first half of life's history that microbes started to share the world with eukaryotes, and at this point they successfully explored these larger organisms as ecological niches. The *Cyanobacteria* laid the groundwork for the development of higher (aerobic) life forms by inventing photosynthesis and the production of oxygen, and α-proteobacteria contributed to the formation of eukaryotes by providing the ancestors of mitochondria. According to the (now widely accepted) endosymbiont hypothesis, mitochondria, plastids and hydrogenosomes originated from free-living bacteria. Without these essential endosymbiotic organelles, unicellular eukaryotes and subsequently larger multicellular organisms would probably never have seen the light of day.

Mitochondria and plastids probably represent the most intimate relationship between pro- and eukaryotic cells, and are the prime example of intracellular life. During the initial invasion step the host cell was likely just a membrane sac with a membrane compartment to concentrate the genetic information – an ancient nucleus precursor. Most likely it was still a fellow prokaryote. In 1967, Lynn Margulis (b.1938) reintroduced the endosymbiont theory to the field of evolution biology [9]. According to this hypothesis, mitochondria and plastids originate from ancient bacterial and cyanobacterial symbionts, respectively (Figure 1.2). This hypothesis was supported by the presence of two membranes surrounding these organelles,

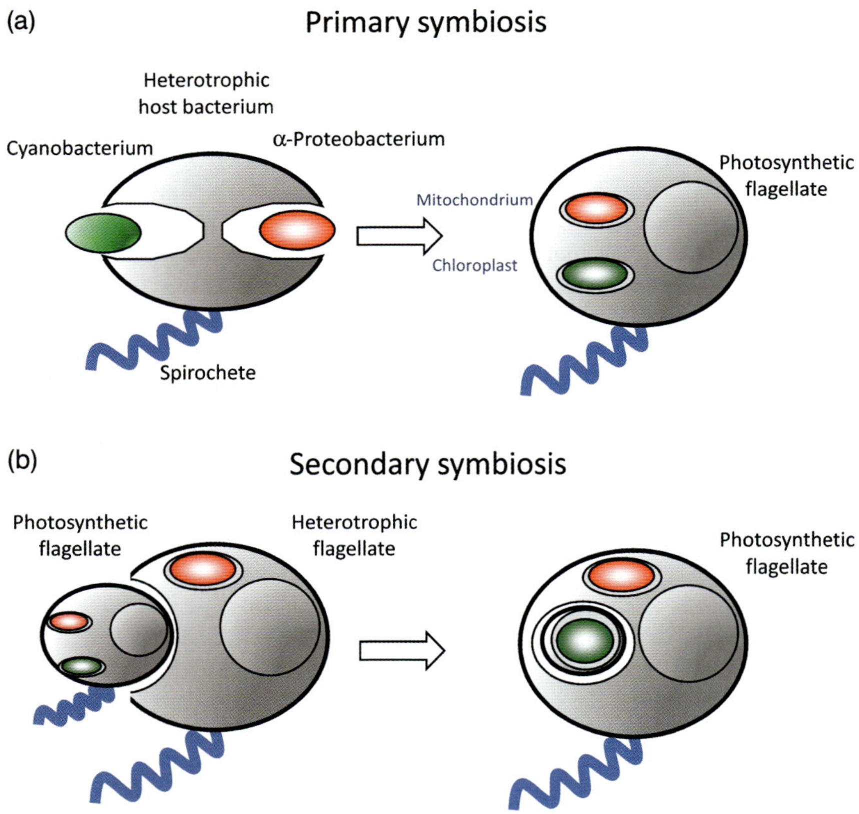

Figure 1.2 (a) Schematic drawing on the symbiont hypothesis how primary symbiosis between two (or three, or more) prokaryote microbes led to eukaryote cells with mitochondria, chloroplasts and flagellae. (b) Secondary symbiosis arose between heterotrophic and photosynthetic eukaryotes namely flagellate species. In both scenarios, the ultimate relationship lead to full dependency of both partners on each other and their loss of autonomy.

many structural similarities and the presence of bacterial DNA in these organelles. The theory dates back to 1883, when the German botanist Andreas Franz Wilhelm Schimper (1856–1901) postulated that chloroplasts are derived from photosynthetic bacteria, and was renewed by Konstatin Sergejewitsch Merschkowski (1855–1921) in 1905. Our current understanding is that more than 1.5 billion years ago an α-proteobacterium-like microbe invaded an ancient host cell, which was most likely another bacterium with compartmentalized chromosomes similar as found in *Gemmata obscuriglobus* and other δ-proteobacteria [10].

According to scenarios based on the argument that metabolic needs may have promoted formation of symbiosis between two prokaryote species, host cell and invader may have been a methanogenic *Archaea* and a methanotroph, respectively (reviewed in Dyall *et al.* [11]). In this scenario, the essential event of eukaryote evolution is set in the anoxic era, whereas others placed it in the aerobic age where an anaerobic archaeal host was protected from toxic oxygen by an aerobic symbiont.

Another version suggests that mitochondria are derived from photosynthetic bacteria due to the following arguments: (i) they release photosynthates such as glycollate for metabolic use by the heterotrophic partner (through peroxisomes as realized in higher plants); (ii) their morphological features resemble cristae of mitochondria; and (iii) 31 of the most conserved mitochondrial genes are closely related to genes in the phototrophic bacterium *Rhodospirillum rubrum* [12].

Whichever microbe the mitochondrial ancestor was, it is most likely that the process of mitochondrial endosymbiogenesis succeeded just once since all known current eukaryotes contain a number of original genes from the α-proteobacterial ancestor.

During a process starting some 3.5 billion years ago, atmospheric oxygen accumulated through the metabolic activity of photosynthetic bacteria. Shortly after eukaryotes with mitochondria started roaming the earth, another invasion event by cyanobacteria led to the emergence of the ancestors of green algae and higher plants (*Chlorophyta*) and subsequently of red (*Rhodophyta*) and brown algae (*Glaucophyta*) (see below) [13]. Amitochondrial amoebal, trichomonad, ciliat and anaerobic fungal species still exist today. In those organisms, ATP-producing organelles, the hydrogenosomes, play a role similar to mitochondria. Although hydrogenosomes do not contain a genome, proteomic analyses suggest their relationship with mitochondria, but this is controversially discussed [11]. In some other amitochondrial organisms, such as *Giardia*, *Entamoeba* and *Microsporidia*, in which mitochondria-like remnants have been found, it is not clear whether their loss is a secondary event. The emergence of eukaryotes from a get-together of different bacterial species probably represents the first type of intracellular life on earth.

It should, however, be mentioned that the bacterium *Bdellovibrio bacteriovorus* is a specialized parasite of other bacteria and invades their periplasmic space. A symbiotic α-proteobacterium, *Midichloria mitochondrii*, has recently been described residing in the mitochondria of tick ovary cells [14]. These facts may lead to the bold hypothesis that intrabacterial parasites/symbionts may have preceded the rise of eukaryotes, and that they were the first intracellular life forms.

The current view that interbacterial symbiosis formed the basis for the evolution of *bona fide* organelles is further corroborated by the identification of more recent "domestication" events of (cyano-)bacterial symbionts by eukaryotes. The filose amoeba *Paulinella chromatophora* harbors photosynthetic *Synechococcus*-type cyanobacteria as symbionts, which have totally lost their autonomy thereby forming a primary symbiont [13] (Figure 1.2).

Subsequently to endosymbionts becoming mitochondria and chloroplasts, evolution led to further examples of endosymbioses. After green, red and brown algae emerged, secondary endosymbiosis (Figure 1.2) was born when aplastid flagellates incorporated red algae cells, thus joining the photosynthetic community. The genera *Cryptophyta*, *Dinophyta*, *Heterokontaphyta* and *Haptophyta* were the results of these joint ventures. Also, the parasite phylum *Apicomplexa* originated from such an endosymbiosis, which explains why parasites such as *Plasmodium* and *Toxoplasma* are affected by herbicides that target plastid enzymes [15]. Moreover, unicellular eukaryotes, probably heterotrophic flagellates, incorporated green algal cells and gave rise to *Euglenophyta* and *Chlorarachniophyta*. A very recently evolved secondary endosymbiosis is the union between the colorless flagellate "Hatena" and a green algae of the genus *Nephroselmis* [16]. Upon engulfment of free-living flagellated *Nephroselmis* cells by Hatena, the symbiont loses flagellae, cytoskeleton and endomembranes but retains nucleus, plastides, mitochondria and eyespot. The complex feeding apparatus of the colorless host flagellates disappears after uptake of the symbiont. This event seems to coincide with the host cell's switch from heterotrophic predator to autotrophic algae. After cell division, the daughter cell lacking the symbiont becomes heterotrophic again and develops a feeding apparatus to catch a new symbiont. It has been suggested that "Hatena" could be a model for the early development of secondary symbiosis. This example shows that not only the symbiont but also the host cell may go through cellular changes upon formation of endosymbiosis. The latter notion is also corroborated by findings from the genome of the pathogenic filarial nematode *Brugia malayi*, which revealed adaptations, which had most probably evolved in response to the presence of *Wolbachia* symbionts [17].

On several occasions later in evolution, α-proteobacteria such as *Wolbachia*, *Rickettsia* and *Ehrlichia* species, as well as members of the *Chlamydiales* became settlers of eukaryotic cells as highly specialized obligate intracellular mutualists or pathogens. In free-living amoeba, more than 20 bacterial symbionts have been identified so far, belonging to the α-proteobacteria, β-proteobacteria, *Bacteroidetes* and *Chlamydiales* [18–20]. Interestingly, symbiotic *Chlamydia* species in amoebae have a biphasic lifecycle between metabolically active reticulate and inactive elementary bodies similar to that of pathogenic species in mammals, suggesting common ancestry between the groups. Among amoebae symbionts, differentiation between symbiosis and parasitism is difficult. In the case of *Parachlamydia*-related symbionts, their association with amoebal partners can also be detrimental to the host cell as they lyse their hosts at temperatures above ambient. In contrast, *Neohartmanella hartmanellae* is a *bona fide* mutualist since this bacterium promotes growth of its amoebal host [21, 22]. This suggests that there can be fine lines between mutualistic and parasitic companionships between bacteria and eukaryotes, depending on factors such as environmental conditions.

Some hypotheses of intermicrobial symbiosis go beyond metabolic mutualisms. The evolution of motility and cytoskeleton elements has been suggested to originate from a hypothetical spirochetal symbiont forming a consortium with an archea [10]. It should be noted that "living fossils" for such a scenario exist in the form of *Chlorochromatium aggregatum* and *Mixotricha paradoxa*. *C. aggregatum* evolved from a consortium comprising green sulfur bacterial epibionts surrounding a central motile β-proteobacterium. *M. paradoxa* is a flagellate in the gut of the termite *Mastotermes darwiniensis* and is coated with *Bacteroides* species and spirochetes for motility [23, 24].

Box 1.1
Definitions of interspecies relationships

Symbiosis: Partnership between two different species. Often used synonymously with mutualism, meaning that advantages usually outweigh the disadvantages for both partners.
Mutualism: Partnership between two different species with benefits for both partners.
Parasitism: Relationship between two different species, in which the smaller one, the parasite, lives in or on a larger host organism, gaining benefits and causing harm to the host.
Commensalism: Relationship between two different species benefiting one partner, the commensal, but without (known) benefits or disadvantages for the other one.
Pathogen: Microbe that induces one or more infectious diseases.

1.3
Continuous Exchange of Information

Information – Inspiration, Shuggy Otis, musician, 1971

Stable symbiosis with the newly formed eukaryotes caused a drastic reduction in the sizes of the mitochondrial and plastid genomes [25]. Genes missing from mitochondria and plastids were either totally lost or transferred to the host cell genome. The genome of free-living cyanobacteria is typically between 2000 and 4000 kb in size, whereas plastids have 200 kb coding for fewer than 200 genes. Horizontal gene transfer from the mitochondrial to the host cell genome is reflected by the presence of around 400 proteins in this organelle's proteome that are encoded by nuclear genes [26]. Consequently, the proper localization of mitochondrial or plastid proteins synthesized by the host cell requires mechanisms of transport into these organelles. Between 15 and 25 protein import translocases have been identified in mitochondria and plastids, and some of them have striking homology to bacterial protein exporters [11]. A significant reduction in genome size is also observed in obligate intracellular bacteria such as *Rickettsiae, Chlamydiae* and *M. leprae* (see Chapters 15, 19 and 26). Lateral gene transfer has occurred frequently between bacteria and eukaryotes. Genes of the intracellular symbiont *Wolbachia pipientis* have been found in the fruit fly (*Drosophila melanogaster*) genome but also in other host genomes, including those of wasps and nematodes [27]. *W. pipientis* is a maternally inherited endosymbiont in at least 20% of arthropod species. Relatives of this α-proteobacterium live in the germline of filarial nematodes (see Chapter 30). With respect to host–microbe interaction, one can envisage that bacterial genes inserted within host genomes may promote mutualism, but may also be employed by the host for its own purposes. Therefore, lateral gene transfer during microbe–host interactions may be a motor of evolution in both partner organisms.

1.4 Evolution of Intracellular Parasitism

> "Many microorganisms of diverse phylogenetic beginnings have adapted to intracellular life, each in its own unique way, and sometimes remarkable resemblances in behaviour among intracellular parasites are best ascribed to convergence in similar intracellular habits rather than to divergence from a common origin. Finally it should be remembered that adaptation to intracellular life, although by no means rare, is not easy. After all, most parasites still live extracellularly". James W. Moulder (1985) Comparative biology of intracellular parasitism, *Microbiological Reviews* 49, 298.

About 2 billion years ago, probably in a shallow laguna, bacteria encountered for the first time malicious shapeless little eukaryotes eating them. Some heterotrophic eukaryotic cells started a new business in life and became predators, whereas bacteria faced a new challenge, not to fall prey. The prototype of a phagocyte, an amoeba that feeds on bacteria, was probably the first to impose on bacteria the selection pressure to maintain or evolve new genes that facilitate survival within the predators (Figure 1.3). Today, still, amoebal creatures roam the world in search of bacterial prey. One of them, *Hartmanella*, is a notorious settler in cooling water systems and showerheads, an opportunistic human parasite itself, and one of the natural hosts for the human pathogen *Legionella pneumophila* (Chapter 18). The slime mold, *Dictyostelium discoideum*, has become a prime model organism for the study of some intracellular pathogens such as *Mycobacterium marinum, M. avium* and *L. pneumophila* (Chapter 4). It can be hypothesized that these free-living bacteria, which can survive and possibly even multiply in amoebae, carry pre-adaptations to divert phagosome trafficking, which may help to counter microbi-

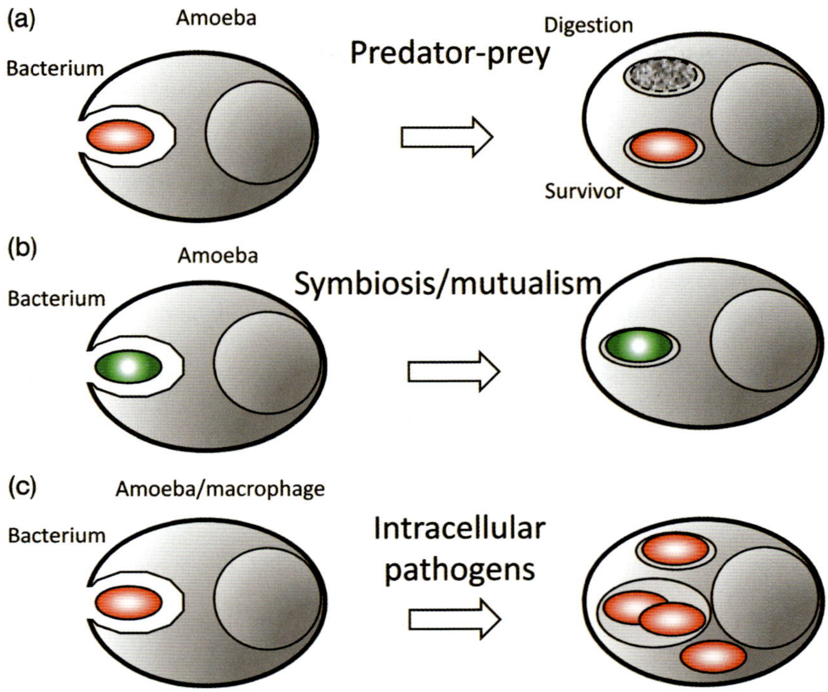

Figure 1.3 Drawings depicting the hypothetical transitions from bacteria falling prey to amoebal predators (a) to symbionts or parasites in amoeba (b). (c) The latter then give rise to intracellular pathogens of multicellular organisms including humans. This was due to preadaptations selected and sharpened during interactions with amoebal hosts.

cidal macrophages. These bacteria have the potential to become either pathogens of cold-blooded animals (e.g., *M. marinum*, a natural pathogen of fish and amphibians) or, when able to replicate at 37 °C, pathogens of mammals (e.g., *L. pneumophila*).

L. pneumophila is a prime example of how a prey of freshwater protozoa became a parasite within a unicellular host and thereby gained the prerequisites – or preadaptations to stay with the terminology of evolution biology – to survive inside mammalian phagocytes and become a pathogen (see Chapter 18). *Legionella* apparently benefited from the fact that mechanisms and factors of phagocytosis and intracellular trafficking are highly conserved throughout the animal kingdom. In principal, this allows intracellular bacteria to parasitize any phagocytic cell in the world. *L. pneumophila* lives in fresh water, especially artificial warm water distribution systems. When inhaled, it can cause a severe pneumonia called legionnaires' disease in immunocompromised and elderly patients. This pathogen seems to have a complex lifestyle which is regulated by nutrient availability and characterized by distinct gene expression patterns and switching from flagellated free forms to intracellular nonflagellated and spore-like stages [28]. Inside amoebae and macrophages it survives and proliferates in compartments made of endoplasmic reticulum (ER) membranes (see Chapter 18) [29]. It has been suggested that during coevolution between *Legionella* species and protozoa the bacteria acquired a significant number of eukaryotic-like genes that are not found in most other bacterial genomes and that are possibly involved in protein–protein interactions with host cell structures [30–32].

Interestingly, the *L. pneumophila* genome contains an unusual number of ankyrin-like domains as well as F-box and serine/threonine protein kinase genes. Eukaryotic ankyrin repeat proteins are typically involved in protein–protein and protein–cytoskeleton interactions. F-box proteins play a role during ubiquitination, and protein Ser/Thr kinases are eukaryotic signaling molecules. It is fascinating to note that other intracellular pathogens share these genomic features of *L. pneumophila*. *Wolbachia*, *Coxiella* and *Rickettsia* species also carry ankyrin-like domain genes, and *M. tuberculosis* and the extracellular *Yersinia pseudotuberculosis* have Ser/Thr protein kinase genes involved in their virulence [33]. Homologs of these genes are found in protozoa, suggesting that these virulence traits may have been gained through lateral transfer of protozoan genes, thereby preadapting *Legionella* to become a mammalian pathogen [34]. It will be interesting to find out whether other mammalian pathogens that dwell within protozoa, such as *M. marinum*, *M. avium*, *Listeria monocytogenes*, *Francisella* species or *B. pseudomallei*, have developed similar virulence features during coevolution with ancient predatory protozoa. Unlike specific pathogens, these ubiquitous and opportunistic organisms require a broad genetic flexibility to thrive in different environments.

Unlike other bacterial genomes, over 7% of the *L. monocytogenes* genome encodes regulatory proteins, whereas the similar-sized genome of the extracellular pathogen *S. aureus* has half as much regulatory DNA. This suggests that *L. monocytogenes* needs flexibility in its gene regulation in response to sudden environmental changes [35]. Development of the *L. monocytogenes* PrfA-regulated virulence gene cluster, which is essential for intracellular survival and spread from cell to cell, was an important prerequisite to becoming an intracellular pathogen in mammals (see Chapter 24). Interestingly, this gene cluster is also present in the nonpathogenic *L. seeligeri*, a close relative of *L. monocytogenes* but does not (yet) give *L. seeligeri* the license to become a mammalian pathogen. However, it has been suggested that it may allow freshwater listeria to survive in invertebrate hosts and, therefore, the PrfA system may represent an ancient preadaptation to becoming a pathogen [36].

It is also noteworthy that intracellular pathogens are rarely found in higher plants. This may be either because of the robust plant cell wall, which restricts entry or because of the inability of plant cells to phagocytose. The enigmatic protists of the order *Plasmodiophorida*, such as *Plasmodiophora brassicae* and *Spongospora* species, are intracellular plant pathogens of economical importance [37]. Although these microbes have recently became a focus of more detailed genetic studies, an understanding of the invasion pathways, intracellular survival and niches of these pathogens is still lacking.

1.5 Intracellular Symbionts: Tamed or Acclimatized Parasites?

Give it to me. Timbaland, *Timbaland Presents Shock Value* album, 2007.

Symbiosis is an important factor in the earth's ecosystem. Bacteria and cyanobacteria form symbiont–host relationships with protozoa, plants and invertebrates. Marine organisms, such as mussels, clams and tubeworms, within the dense animal communities at deep-sea hydrothermal vents often depend entirely on chemoautotrophic methano- or thiotrophic bacterial symbionts for carbon fixation [38]. The vesicomyid clams carry their symbionts in gill epithelial cells and transmit their endosymbionts to the next generation through their eggs. Vestimentiferan tubeworms, however, have to ingest their symbionts during the juvenile stage. The worm's sulfur-oxidizing bacteria are contained within bacteriocytes concentrated within a specialized organ, the trophosome. Another vent settler, the mussel *Bathymodiolus* seems to access its thiotrophic symbionts, which also live in gill epithelial cells, from the environment [39]. The entire hydrothermal vent community probably depends on these symbionts, which strongly influence the local marine ecology. However, very little is known about their intracellular lifestyle.

Unicellular algae such as zooxanthellae as well as green algae of the genus *Chlorella* also became intracellular symbionts in plants, protozoa and invertebrate animals. These symbioses are of prime ecologically importance. Cyanobacterial or algal symbionts provide photosynthates to the heterotrophic host and remove its carbon dioxide. *Chlorella* forms a symbiotic relationship with the freshwater polyp *Hydra viridis*, and is transmitted to its offspring through the eggs. *Chlorella* avoids digestion in the polyp's cells by inhibiting phagolysosome formation [40–42], as do several intracellular pathogens. Zooxanthellae of the genus *Gymnopedium* are intracellular symbionts of coral reef polyps, which have formed the biggest natural sculptures in the world. *Gymnopedia* are close relatives of dinoflagellates. They provide up to 90% of the polyp's energy and are therefore essential for the growth of the coral reefs in the otherwise oligotrophic oceans of the tropics.

The close relatives of some intracellular symbionts are important bacterial pathogens, which employ the intracellular niche as a survival mechanism, though with the outcome of disease. *Legionella-*, *Francisella-*, *Chlamydia-* and *Rickettsia*-like species have been found in free-living amoeba. Under experimental conditions, *L. pneumophila* can infect up to 14 amoeba species, including *Hartmanella* and *Acanthamoeba* spp., and two species of ciliated protozoa, whereas its close relative, the rarely pathogenic *L. micdadei*, only infects *Hartmanella* amoebae [33, 43]. Compared with *L. pneumophila*, *L. micdadei* inhabits a different intra-amoeba compartment and does not exit through pore formation and cytolysis [44]. *L. micdadei* may therefore represent an earlier stage during the evolution of *L. pneumophila* to become a successful survivor and ultimately a parasite of amoeba.

Establishment of a symbiosis is likely "a large step for the microbial species," yet "a small one for an individual microorganism." Transition from a pathogenic to a symbiotic microbe and vice versa could happen frequently, but selective pressure can stabilize the respective relationship. It has been demonstrated in the laboratory that transition from an extracellular microbe to an intracellular mutualist can be observed within several hundred generations time, which corresponds to a few years [45]. The partner organisms in this experiment were *Legionella*-like bacteria and *Amoeba proteus*. The resulting symbiosis was stable in that removal of the bacteria killed the host cells [45]. This feature of a mutualistic relationship often goes hand in hand with the fact that symbiotic microbes cannot be cultivated on their own in the laboratory, at least not with the currently known culture methods.

Relatives of pathogenic *Rickettsia*, *Coxiella* and *Francisella* spp. – causative agents of louse- or tick-borne zoonoses – have been found in arthropods, including ticks [18, 36]. Mutagenesis of *Wolbachia* can reprogam this fruit fly symbiont to become a parasite, suggesting that it can be a small step between both types of relationships [46]. *Wolbachia* spp. in nematodes are symbionts, whereas those dwelling in arthropods can cause disease, including sexual morphological alterations, cytoplasmic infertility and reduced reproductivity [47, 48]. An unexpected benefit for the insect host, however, is that *Wolbachia* promotes innate immunity in *Drosophila*, protecting against infection with the *Drosophila C virus* [49].

Rhizobium meliloti, a symbiont of legumes, depends on homologous genes, *bacA* and the *bvrR-bvrS* two-component regulatory system, for intracellular survival in root cells, both of which are also essential for another α-proteobacterium, the zoonotic *Brucella abortus* living inside macrophages [50]. This suggests that general adaptations are required for intracellular survival in organisms as distant as plants and mammals, and are independent of whether a pathogen or a mutualist is involved. Mutualism between symbiont and host is a fine balance and often the equilibrium is not far from parasitism. The distinction between symbiosis and parasitism is hard to make since both partners gain benefits as well as exploit each other.

1.6
An Ecological View of Intracellular Life

> *Pathogens lack malice; they are just trying to survive.*
> Arno Karen, in *Biography of a Germ*, 2000.

We have argued that bacteria developed ways to escape or survive the attack of predatory amoeba in the early ponds of evolution. However, successful intracellular survivors did not only gain the benefit of not being digested but also were rewarded with protection from environmental conditions and the possibility of accessing novel food sources. Exploiting the newly inhabited intracellular niche for survival and growth was a profitable result of novel adaptations to avoid or resist phagocyte digestion (Figure 1.3). The intracellular niche can provide microbes with nutrients, including essential micronutrients they otherwise have to compete for with fellow microbes in the extracellular environment and/or have to capture or synthesize by themselves:

Rickettsiae graze on the host cell's energy sources including ATP [51]. Mycobacteria inhibit phagosome maturation in macrophages and thus inhabit early phagosomes where they can scavenge iron from transferrin because this compartment intersects with the iron import pathway of the host cells (i.e., the transferrin–transferrin receptor uptake system into early endosomes) [52]. *Coxiella burnetti* and *Leishmania* amastigotes exploit the harsh lysosomal environment for growth and probably feast on hydrolytic degradation products such as amino acids [53–55]. *L. mexicana* amastigotes probably exploit autophagolysosomes to access purines [56]. *Leishmania* species are purine auxotrophs and require host cell-derived purine sources such as autophagosomes. Certain *C. trachomatis* strains are unable to synthesize the amino acid tryptophan.

Nutrient limitation – the bacteria's Achilles heel – is targeted by the host response in an interferon γ (IFNγ)-induced manner. Activated macrophages express the gene for indoleamine 2,3-dioxygenase (IDO), which depletes tryptophan by degradation to kynurein [57]. Sequestration of *C. trachomatis* from tryptophan drives the bacteria to differentiate into the nongrowing residual body form and causes latent infection. The genome of *C. psittaci* contains a more complete tryptophan synthesis machinery and resistance to IDO of this *Chlamydia* species is due to efficient recycling of the amino acid [57].

Genome reduction is a common consequence of colonization for pathogens or symbionts which became highly adapted to the intracellular lifestyle and entirely dependent on their host cells. This is seen in diverse obligate intracellular microbes such as the insect symbionts of the *Buchnera* genus, or *M. leprae, Rickettsia* or *Chlamydia* species, as well as in extracellular *Mycoplasma* species. The host provides a pretty constant environment as well as nutrients and metabolic resources. As a consequence of this close relationship, obligate intracellular microbes lost their ability to survive and proliferate outside of the host cell and become metabolically dependent. This often leads to loss of genes required for the synthesis of organic molecules such as amino acids and, ultimately, to the inability to generate ATP.

In an ecological view of interspecies relationships, interactions between two partners also determine interactions beyond this partnership. Thus, simple coevolution is unlikely, because the broad ecological context with its entire range of interacting factors, including food competition and predator–prey interactions, also influence host–parasite/symbiont interactions [58]. Intracellular microbes probably also influence each other. In the broader ecological context it can be hypothesized that intracellular symbionts such as *Rhizobium* or *Wolbachia* species enhance the fitness of legumes or parasitic nematodes, respectively. Studies on the evolution of virulence have found that the more virulent parasites are, the higher their transmission rates and the less they are controlled by immunity. It has been shown that immune pressure selects for more virulent parasites [59, 60]. These studies, however, never took into account that hosts with higher parasite loads may be an easier prey for predators and are therefore removed more quickly from the population [58]. This would increase the resistance to the parasite within the host population and would eventually lead to an equilibrium in an individual host–pathogen relationship (i.e., between defense and virulence). It may become disturbed, however, when the predator is removed, for example, or a new pathogen is introduced that affects herd immunity. A good example is the increased tuberculosis rate as boosted by the HIV pandemic.

1.7
The Immunologist's View

> *Thus spears and swords gave rise to shields and body armour, and radar defenses to the Stealth bomber.* Nesse and Williams, in *Why We Get Sick*, 1994.

Just as in warfare, acquisition of a new weapon by one party will cause development of a defense system by the opposition, allowing them to counter, a microbial challenge promotes the development of sophisticated defense mechanisms in the host organisms. Microbicidal effectors of amoebae comprise acidic pH, porins, bactericidal peptides and lysozymes. Multicellular invertebrates also primarily depend on defensin-like microbicidal peptides but additionally employ motile amoeboid phagocytes as eliminators of pathogenic invaders such as the haematocytes within the arthropod's hemolymph. Cytokines such as interleukin 1 and tumor necrosis factor (TNF) mediate phagocyte activation. Invertebrate cells sense microbial stimuli. *Drosophila* and other arthropods use Toll-like receptors (TLR) to distinguish between pathogenic types such as fungi, viruses or bacteria [61]. The immune deficiency (IMD) signaling pathway, homologous to the TNF receptor signaling pathway, is also essential for the fruit fly to survive infection [62].

In higher vertebrates, particularly in mammals, phagocytes have become diverse in function. Macrophages still fulfill their ancient functions of eliminating microbial invaders from tissue and clearing away dead cells. Interferons, in particular IFNγ, facilitate macrophage activation in order to better regulate expression of highly effective antimicrobial mechanisms, which can be destructive to normal tissue. Neutrophils – microphages as Elie Metchnikoff called them – became specialized in rapid recruitment to sites of microbial invasion and killing. These cells, however, also came with the danger of collateral tissue damage and therefore required control mechanisms, such as their short lifetime. Finally, dendritic cells took over the job of processing foreign antigens in order to present them in the context of self-molecules to the acquired immune system. In vertebrates, and particularly in mammals, the acquired immune system, comprising B cells, T cells, immune memory functions and a complex regulatory cytokine network, confronted pathogenic microbes with a totally novel challenge. Nevertheless, many pathogens thrive within the mammalian host. The diversity of tissues and entry ports opened up new niches, immune privileged sites and host cells within the multicellular organism such as Schwann cells for *M. leprae* or erythrocytes for *Bartonella*, *Plasmodium* and *Babesia* species. Active invasion at the epithelial interface and entry mediated by arthropod vectors provided new ways to access different host cells. Inflammation-mediated tissue damage opened up novel paths to exit the host in order to facilitate transmission to a new one. Lung lesions in people with active tuberculosis allow tubercle bacilli to be spread through coughing [63]. The development of a highly complex and versatile adaptive immune system by large multicellular organisms such as mammals is most likely the result of coevolution arising from continuous interactions with pathogenic microbes.

1.8 The Public Health View

> *Also, because of their rapid evolution and constantly changing circumstances of human life, they [pathogenic microorganisms] continue to present threats of future pestilence.*
> Cedric A. Mims in *The Pathogenesis of Infectious Disease*, 1988.

Protozoan or invertebrate hosts are likely to provide breeding grounds for newly arising human pathogens as they must have in the past. *L. pneumophila*, trained by predatory amoebae, became a relatively recent addition to the list of human pathogens due to the development by humans of extensive water distribution and air condition systems within large building complexes, including hospitals and pools. *Legionella*'s protozoan hosts thrive under these ecological conditions, making the bacterium a pathogen and public health issue. In the light of this, human infection with *L. pneumophila* can be thought of as an "accident," but it also shows how far preadaptations can get you as a pathogen, once you arrive in a permissive environment.

Preadaptations for survival within protozoan and invertebrate hosts can yield more infectious agents for humans in the future due to environmental changes and novel ecological niches generated by humankind. For pathogens such as *B. pseudomallei*, *Afipia felis* and the extracellular *M. ulcerans*, which recently gained more public health coverage, it is also hypothesized that reservoir hosts – freshwater amoeba or insects – exist, which may have been the training ground for those human pathogens [64–66]. Finally, mutualists, which are already equipped with genes for intracellular survival, could also become pathogens when the mutualist's behavior changes towards uncontrolled growth and pathology, or when it is transmitted to a new host. More importantly, mutualists living in ectoparasites of mammals such as blood-sucking insects, ticks and leeches can become future infectious burdens to humans and livestock.

It can also become a public health issue when free-living protozoa support pathogenic bacteria in surviving and persisting in the environment. Already established intracellular pathogens of mammals can "re-employ" their virulence factors to survive within free-living protozoa. The anaerobic pathogen and causative agent of vaginitis, *Mobiluncus curtisii*, does not normally replicate under aerobic conditions but does when amoeba are added to the cultures [67]. The facultative intracellular opportunist, *M. avium*, can even survive encystation of its host amoeba [68]. More important for hygiene measures are reports that residency within amoebae protects microbes from common water disinfectant procedures such as chlorination, as shown with *M. avium* ssp. *paratuberculosis* [2]. An experiment using an *Arabidopsis* model has due to their virulence-associated type III secretion system shown the unexpected result that *S. typhimurium* is able to invade and survive within plant cells. This makes lettuce cells a potential reservoir for typhoid fever when exposed to *Salmonella*-contaminated water [69].

1.9 The Book

> *If none of the microorganisms associated with man did any damage, and none was notably beneficial, they would be interesting but relatively unimportant objects.* Cedric A. Mims in *The Pathogenesis of Infectious Disease*, 1988.

Since Elie Metchnikoff's first observations of phagocytosed bacteria, intracellular pathogens have attracted the attention of microbiologists and attempts have been made to describe their biology in a systematic and comparative way [70]. In this book, specialists studying the different species, pro- and eukaryotes, parasites and symbionts give detailed insight into the intracellular lifestyle of these microorganisms, their ways of entering, surviving and proliferating within host cells, the diseases they cause and the benefits they have for partners in symbiosis. The main theme is discussion of the evolutionary and ecological aspects of this fascinating field of interspecies interactions.

Acknowledgments

The authors thank their individual lab members, present and past, for their enthusiasm and scientific contribution to the success of both groups. UES acknowledges the financial support from the Medical Research Council UK, the Wellcome Trust UK, the Bill and Melinda Gates Foundation and the Deutsche Forschungsgemeinschaft. AH thanks in particular the Deutsche Forschungsgemeinschaft and the Volkswagen Foundation. We thank Dr Susi Pätzold for critically reading the manuscript, and Dr Stefan H. E. Kaufmann for kindly sharing pictures of Elie Metchnikoff from his personal archive with us.

References

1 Darwin, C. (1859) *The Origin of Species by Means of Natural Selection*, Penguin Classics, Harmondsworth, Middlesex, UK.

2 Whan, L., Grant, I.R. and Rowe, M.T. (2006) *BMC Microbiol*, **6**, 63.

3 Metchnikoff, E. (1884) *Allg Wein med Ztg*, **27/29**, 307–332.

4 Metchnikoff, E. (1888) *Virchows Arch*, **113**, 63–94.

5 Armstrong, J.A. and Hart, P.D. (1971) *J Exp Med*, **134**, 713–740.

6 Armstrong, J.A. and Hart, P.D. (1975) *J Exp Med*, **142**, 1–16.

7 Bannister, L.H. (1979) *Proc R Soc Lond B Biol Sci*, **204**, 141–163.

8 Eissenberg, L.G. and Wyrick, P.B. (1981) *Infect Immun*, **32**, 889–896.

9 Margulis, L. (1971) *Sci Am*, **225**, 48–57.

10 Margulis, L., Chapman, M., Guerrero, R. and Hall, J. (2006) *Proc Natl Acad Sci USA*, **103**, 13080–13085.

11 Dyall, S.D., Brown, M.T. and Johnson, P.J. (2004) *Science*, **304**, 253–257.

12 Cavalier-Smith, T. (2006) *Proc Biol Sci*, **273**, 1943–1952.

13 Bhattacharya, D., Archibald, J.M., Weber, A.P. and Reyes-Prieto, A. (2007) *Bioessays*, **29**, 1239–1246.

14 Epis, S., Sassera, D., Beninati, T., Lo, N. *et al.* (2008) *Parasitology*, **135**, 485–494.

15 Bhattacharya, D., Yoon, H.S. and Hackett, J.D. (2004) *Bioessays*, **26**, 50–60.

16 Okamoto, N. and Inouye, I. (2005) *Science*, **310**, 287.

17 Ghedin, E., Wang, S., Spiro, D., Caler, E. *et al.* (2007) *Science*, **317**, 1756–1760.

18 Horn, M. and Wagner, M. (2004) *J Eukaryot Microbiol*, **51**, 509–514.

19 Greub, G., La Scola, B. and Raoult, D. (2004) *Emerg Infect Dis*, **10**, 470–477.

20 Greub, G. and Raoult, D. (2004) *Clin Microbiol Rev*, **17**, 413–433.

21 Greub, G., La Scola, B. and Raoult, D. (2003) *Ann N Y Acad Sci*, **990**, 628–634.

22 Horn, M. (2008) *Annu Rev Microbiol*, **62**, 113–131.

23 Wanner, G., Vogl, K. and Overmann, J. (2008) *J Bacteriol*, **190**, 3721–3730.

24 Radek, R. and Nitsch, G. (2007) *Eur J Protistol*, **43**, 281–294.

25 Muller, M. and Martin, W. (1999) *Bioessays*, **21**, 377–381.

26 Gray, M.W., Burger, G. and Lang, B.F. (1999) *Science*, **283**, 1476–1481.

27 Hotopp, J.C., Clark, M.E., Oliveira, D.C., Foster, J.M. *et al.* (2007) *Science*, **317**, 1753–1756.

28 Molofsky, A.B. and Swanson, M.S. (2004) *Mol Microbiol*, **53**, 29–40.

29 Abu Kwaik, Y. (1996) *Appl Environ Microbiol*, **62**, 2022–2028.

30 Cazalet, C., Rusniok, C., Bruggemann, H., Zidane, N. *et al.* (2004) *Nat Genet*, **36**, 1165–1173.

31 de Felipe, K.S., Pampou, S., Jovanovic, O.S., Pericone, C.D. *et al.* (2005) *J Bacteriol*, **187**, 7716–7726.

32 Al-Khodor, S., Price, C.T., Habyarimana, F., Kalia, A. and Abu Kwaik, Y. (2008) *Mol Microbiol*, **70**, 908–923.

33 Albert-Weissenberger, C., Cazalet, C. and Buchrieser, C. (2007) *Cell Mol Life Sci*, **64**, 432–448.

34 Habyarimana, F., Al-Khodor, S., Kalia, A., Graham, J.E. *et al.* (2008) *Environ Microbiol*, **10**, 1460–1474.

35 Buchrieser, C. (2007) *Microbes Infect*, **9**, 1147–1155.

36 Goebel, W. and Gross, R. (2001) *Trends Microbiol*, **9**, 267–273.

37 Bulman, S., Siemens, J., Ridgway, H.J., Eady, C. and Conner, A.J. (2006) *FEMS Microbiol Lett*, **264**, 198–204.

38 McMullin, E., Hourdez, S., Schaeffer, S.W. and Fisher, C.R. (2003) *Symbiosis*, **34**, 1–41.

39 Won, Y.J., Hallam, S.J., O'Mullan, G.D., Pan, I.L. *et al.* (2003) *Appl Environ Microbiol*, **69**, 6785–6792.

40 Hohman, T.C., McNeil, P.L. and Muscatine, L. (1982) *J Cell Biol*, **94**, 56–63.

41 Muscatine, L., Cook, C.B., Pardy, R.L. and Pool, R.R. (1975) *Symp Soc Exp Biol*, **29**, 175–203.

42 Muscatine, L. and McAuley, P.J. (1982) *Cytobios*, **33**, 111–124.

43 Hoffman, P., Friedman, H. and Bendinelli, M. (2007) *Infectious Agents and Pathogenesis*, Springer, New York, pp. 185–202.

44 Harb, O.S., Gao, L.Y. and Abu Kwaik, Y. (2000) *Environ Microbiol*, **2**, 251–265.

45 Jeon, K.W. (2004) *J Eukaryot Microbiol*, **51**, 502–508.

46 Min, K.T. and Benzer, S. (1997) *Proc Natl Acad Sci USA*, **94**, 10792–10796.

47 Stouthamer, R., Breeuwer, J.A. and Hurst, G.D. (1999) *Annu Rev Microbiol*, **53**, 71–102.

48 Werren, J.H., Baldo, L. and Clark, M.E. (2008) *Nat Rev Microbiol*, **6**, 741–751.

49 Hedges, L.M., Brownlie, J.C., O'Neill, S.L. and Johnson, K.N. (2008) *Science*, **322**, 702.

50 LeVier, K., Phillips, R.W., Grippe, V.K., Roop, R.M. 2nd and Walker, G.C. (2000) *Science*, **287**, 2492–2493.

51 Schaible, U.E. and Kaufmann, S.H. (2005) *Trends Microbiol*, **13**, 373–380.

52 Schaible, U.E. and Kaufmann, S.H. (2004) *Nat Rev Microbiol*, **2**, 946–953.

53 Voth, D.E. and Heinzen, R.A. (2007) *Cell Microbiol*, **9**, 829–840.

54 Lang, T., Hellio, R., Kaye, P.M. and Antoine, J.C. (1994) *J Cell Sci*, **107** (Pt 8), 2137–2150.

55 Antoine, J.C., Prina, E., Jouanne, C. and Bongrand, P. (1990) *Infect Immun*, **58**, 779–787.

56 Schaible, U.E., Schlesinger, P.H., Steinberg, T.H., Mangel, W.F. *et al.* (1999) *J Cell Sci*, **112** (Pt 5), 681–693.

57 Wood, H., Roshick, C. and McClarty, G. (2004) *Mol Microbiol*, **52**, 903–916.

58 Moller, A.P. (2008) *Ann N Y Acad Sci*, **1133**, 180–186.

59 Mackinnon, M.J. and Read, A.F. (2004) *PLoS Biol*, **2**, E230.

60 Fenner, F. and Fantini, B. (1999) *Biological Control of Vertebrate Hosts: The History of Myxomatosis – an Experiment in Evolution*, CABI Publishing, Wallingford, UK.

61 Medzhitov, R. and Janeway, C.A. Jr. (1997) *Cell*, **91**, 295–298.

62 Tanji, T., Hu, X., Weber, A.N. and Ip, Y.T. (2007) *Mol Cell Biol*, **27**, 4578–4588.

63 Hunter, R.L., Jagannath, C. and Actor, J.K. (2007) *Tuberculosis (Edinb)*, **87**, 267–278.

64 Inglis, T.J. and Sagripanti, J.L. (2006) *Appl Environ Microbiol*, **72**, 6865–6875.

65 La Scola, B. and Raoult, D. (1999) *Lancet*, **353**, 1330.

66 Mosi, L., Williamson, H., Wallace, J.R., Merritt, R.W. and Small, P.L. (2008) *Appl Environ Microbiol*, **74**, 7036–7042.

67 Tomov, A.T., Tsvetkova, E.D., Tomova, I.A., Michailova, L.I. and Kassovski, V.K. (1999) *Anaerobe*, **5**, 19–23.

68 Steinert, M., Birkness, K., White, E., Fields, B. and Quinn, F. (1998) *Appl Environ Microbiol*, **64**, 2256–2261.

69 Schikora, A., Carreri, A., Charpentier, E. and Hirt, H. (2008) *PLoS One*, **3**, e2279.

70 Moulder, J.W. (1985) *Microbiol Rev*, **49**, 298–337.

2
Limited Genomes and Gene Transfer in the Evolution of Intracellular Parasitism and Symbiosis

Naraporn Somboonna and Deborah Dean

2.1
Introduction to Gene Transfer and Evolution

Gene transfer can be categorized into horizontal (lateral) and vertical pathways. Horizontal gene transfer conveys genetic material between strains or species, while vertical gene transfer involves clonal reproduction and internal evolutionary forces that may involve mutations, recombination and rearrangement within the genome [1, 2]. In order for gene transfer to be successful, the genes must be stably integrated into the genome and fixed within the population [1–3].

The recent expansion in the number of genomes sequenced, while still limited, along with advances in genome analyses has provided new evidence among intracellular microorganisms on the scope, rate and consequences of horizontal gene transfer. This evolutionary mechanism was previously believed to be extremely limited for intracellular parasites and symbionts where genome reduction or differential gene loss due to segregated and less diverse niches was considered a primary mechanism for their host adaptation and evolution [3–11].

Historically, the concept of no sexual mode of reproduction in bacteria has been accepted [12]. Sherman and Wing [13] first attempted to detect recombination in mixed cultures of *Escherichia coli* and *Aerobacter aerogenes*. However, although a new combinatorial biochemical characteristic was determined, the authors could not conclude that gene exchange had occurred as each species was found equally in the culture mixture [12]. The possibility of horizontal gene transfer and genetic recombination was described in 1947 when Lederberg and Tatum discovered genetic conjugation in *E. coli* [12, 14]. For intracellular microorganisms, Akiba *et al.* [4] discovered in 1960 the first interspecies gene transfer involving mobile antibiotic resistance genes among *Shigella* strains and between *Shigella* and *E. coli* strains. Recently, Koonin *et al.* [3] identified several horizontally transferred genes in *Chlamydia trachomatis, Chlamydophila pneumoniae, Rickettsia prowazekii, Mycobacterium thermoautotrophicum* and *Mycobacterium tuberculosis* using bioinformatics analyses. These findings were consistent with phylogenetic analyses using several sequences from prokaryotes and eukaryotes [3]. In addition, Dean and colleagues [5, 15–18]

Intracellular Niches of Microbes. Edited by Ulrich E. Schaible and Albert Haas

ISBN: 978-3-527-32207-7

discovered widespread interstrain, interspecies and intergenera recombination for members of the *Chlamydiaceae*.

Horizontal gene transfer creates the potential for a bridge to form between unique microbial communities and host environments [1, 19]. Indeed, these transfer events provide an essential opportunity for host–microbe adaptive interactions and evolution, as evidenced by the emergence of microbial genetic and phenotypic diversity as well as new strains and species that are capable of exploiting new and unique niches [2, 20–22].

The known mechanisms for gene transfer are transformation, conjugation and transduction [23, 24]. Transformation refers to the stable uptake of a DNA fragment into the chromosome of competent recipients [24, 25]. Artificial transformation (e.g., electroporation, which uses electricity to induce short-lived pores in membranes) refers to a laboratory technique that allows microorganisms to enhance their DNA uptake [26]. Conjugation requires direct cell-to-cell contact for the exchange of mobile genetic elements, which comprise transposons and insertion sequence like elements and plasmids. For conjugative transposons, proteins encoding for excision of the donor DNA and formation of a conjugative bridge to transfer the DNA into the recipient must be present [27]. Conjugation can also be mediated by a plasmid that is transferable. For both transformation and conjugation, the recipient microorganisms must, therefore, be competent or contain the genes that support genome integration. Transduction requires a virus, known as a phage, as a vector to carry DNA from the donor to the recipient by infection [24]. In this case, the vector must be able to infect the recipient microorganism and protect itself from degradation by host endonucleases [27]. The frequency of gene transfer depends on the availability of foreign DNA along with the probability of successful integration and the degree of benefit to the recipient chromosome [1].

The transferred genes can be subdivided into new genes, existing genes (paralogs) and genes representing a different lineage (xenologs) with respect to homologous genes (if present) in the recipient genome [3]. From bioinformatics analyses, different intracellular parasites and symbionts have different rates of acquisition for the different genes. Although the nature of these acquisitions remains unclear, the persistence of certain transferred genes in the organism highlights the functional gain to the recipients [3]. Hence, their persistence in the genome likely underlies essential biologic and ancestral concepts for these organisms.

Some of the major indicators of gene transfer include incongruent phylogenetic tree topologies, unusual phyletic patterns, similarity with homolog(s) in a distant taxon, codon position-specific nucleotide composition, codon usage bias, G + C content dissimilarities compared across the genome, and gene order conservation in distantly related taxa [1, 3, 22, 23]. For instance, codon usage and G + C content suggest that up to 15% of the genes in prokaryotic genomes may be horizontally transferred [28, 29]. Further, due to the lack of conservation of gene order (synteny) in distantly related genomes, the synteny of three or more genes suggests gene transfer [3, 30]. Likewise, the presence of an operon among a few distant genomes is highly suggestive of gene transfer. Indeed, a number of operons appear to be horizontally transferred, which might be explained by the advantage of a new or replaced function that occurs in a single transfer event once integrated into the genome [1, 3]. The archeal H^+-ATPase operon that has replaced the bacterial ATPase is one example [1, 3]. However, the frequency of transferred operons and genes is reflected in the ability to be integrated into the genome and mechanisms that limit their uptake, such as mismatch repair or restriction modification systems [1]. Interestingly, in general, the prokaryotic genetic components required for replication, transcription and translation appear to be acquired from eukaryotes, whereas the components for metabolic enzymes, cell wall synthesis and uncharacterized proteins appear to be acquired from other prokaryotes [2, 31, 32]. Consequently, identification of horizontally transferred genes improves our understanding of host–parasite/symbiont, parasite–parasite or symbiont–symbiont interaction and evolution.

Unlike horizontal gene transfer, vertical gene transfer is typical. Inherited genes occur by sexual reproduction, which is common in higher eukaryotes and plants, or asexual reproduction, which includes cell division and budding. Asexual reproduction is a usual replication method for bacteria and fungi [4, 20].

The concept of horizontal gene transfer has become more significant in recent years as the mounting evidence for horizontal gene transfer among intracellular microorganisms fuels the hope of one day being able to genetically manipulate these organisms to discover the functions of a broad diversity of genes and improve our understanding of disease pathogenesis [5, 33, 34], not to mention the capabilities for discovering new drug targets and vaccines. Consequently, this chapter will focus mainly on horizontal gene transfer, and will interchangeably use horizontal gene transfer with gene transfer.

2.2 Gene Transfer in Intracellular Bacterial Parasites

Trafficking for intracellular bacteria can be categorized into intraphagosomal and intracytoplasmic pathways, in which different relationships with the host cell compartment may involve different modes of gene transfer and different types of acquired genes. Intraphagosomal pathogens include *Afipia, Brucella, Burkholderia, Chlamydia, Coxiella, Ehrlichia, Francisella, Legionella, Mycobacterium, Rhodococcus, Nocardia* and *Salmonella,* while intracytoplasmic pathogens include *Listeria, Rickettsia* and *Shigella* [6, 9, 35]. Specific genes are major targets for gene transfer [6]. These include those encoding proteins for type III and IV secretion systems, which are responsible for infection/virulence, and polymorphic surface proteins, which are responsible for attachment and/or antigenic variation. As expected, genes involved in housekeeping functions have been documented to undergo gene transfer less frequently [6].

Using sequences from various prokaryotes and eukaryotes in 2001, Koonin *et al.* [3] identified several horizontally transferred genes such as DNA gyrase A and B subunits among intracellular bacteria, including *C. trachomatis, C. pneumoniae, R. prowazekii, M. thermoautotrophicum* and *M. tuberculosis,* and *S*-adenosylmethionine (SAM)-dependent methyltransferase genes for *M. tuberculosis.* The authors calculated the number of domain transfer events among these different prokaryotic genomes and reported interdomain transfer to be less frequent in parasitic bacteria, except for the families *Chlamydiaceae* and *Rickettsiales,* compared with free-living bacteria, in which approximately 3% of the genomes were predicted to be involved.

Several transposons and foreign genes, such as the 14 *tra* genes that are usually involved in pilus formation and conjugal DNA transfer, have recently been discovered in the genomes of *R. prowazekii*, *Rickettsia massiliae* and *Rickettsia bellii*, representing both conjugation and transformation events in these organisms [6, 7]. Using quantitative assessments [3] and score-based identification of genomic islands [19], the numbers of horizontal gene transfer events, which include transfers from prokaryotes, archaea and eukaryotes, were highest in *R. prowazekii* (3.6% of the genome), *C. pneumoniae* (3.1%), *Ehrlichia ruminantium* (3.0%) and *M. tuberculosis* (3.0%). *Ehrlichia* is a member in the order *Rickettsiales* [19]. The high gene transfer rate in rickettsias may relate to the multiple copies of genes involved in facilitating transfer, including the *tra* genes. Whereas *R. prowazekii* (2.8% of the genome) and *C. pneumoniae* (2.2%) appear to have a predilection for paralogous and xenologous gene transfers, *M. tuberculosis* (1.7%) has an abundance of new gene transfers [3]. These findings support characteristic frequencies for gene transfer in different microorganisms. Nonetheless, the actual number of transferred genes may be substantially higher as some genes may not have sufficient similarity scores with prokaryotes, archaea or eukaryotic genes, especially for those that have not recently been transferred because they have become part of the core genes for those lineages and have undergone the same directional mutational pressures as the rest of the genome. Examples of core genes include those involved in antibiotic-resistance, type III secretion and antigenic surface proteins [3, 4].

Pathogenicity islands, also called genomic islands or virulence cassettes, refer to an area where there is excess genetic variation within the species. The islands often occur at tRNA and tRNA-like loci, which are frequent sites for integration of foreign sequences [23, 27]. The short direct target or inverted repeat sequences flanking the islands are hallmarks of insertion sequences [23]. Consequently, genomic islands generally serve as an indicator for horizontal gene transfer via conjugation. Pathogenicity islands tend to encode numerous important physiologic functions such as iron uptake and type III secretion, and are promiscuously transferred, facilitating a similar pathogenic lifestyle even for quite disparate bacterial taxa [1, 23, 27, 36, 37].

Genetic exchange among intracellular microorganisms is biologically feasible, as mixed infections do occur *in vivo*. For example, up to 57% of *C. trachomatis* sexually transmitted infections represent co-infections with different strains of *C. trachomatis* [38–43], or with *Neisseria gonorrhoeae*, *Trichomonas vaginalis* and human papillomavirus [42, 44–46]. In *in vitro* studies, different *C. trachomatis* strains can infect the same cell, leading to fusion of the phagosomes with the opportunity for transfer and recombination [47]. Cohabitation and vacuole fusion of *Coxiella* within the host cell with other intracellular pathogens, including *Mycobacteria*, *Salmonella* and *Leishmania*, has also been described [48], although evidence for horizontal transfer in *Coxiella* remains minimal.

Chlamydiaceae species contain DNA elements that support transformation, conjugation and transduction, including genes involved in DNA repair and recombination such as *recA* and *yqgF* [5, 36, 37, 47, 49–51]. Phylogenetic, statistical and modeling analyses using membrane-encoding genes such as *ompA* and *pmpA* to *pmpI*, and the tryptophan synthesis operon *trpRBA*, for example, support horizontal transfer between genera and among different chlamydial species and indicate phenotypic differences, including niche specificity [52], tissue tropism [15, 17], differential host immune responses [15] and persistence of the organism [15]. Intraspecies recombinations of these genes have also been detected, which also likely assist the bacteria in surviving under host immune pressure and in establishing infection in new cell types [5, 15–18]. Figure 2.1 shows examples of intraspecies recombination among recent clinical strains of *C. trachomatis* and examples of recombinants created by *in vitro* experimentation [53] that may represent transformation or conjugation [54]. Interestingly, consistent with the clinical findings of *C. trachomatis* strain Da, which infects both the eye and genital tract, there is demonstrated evidence for intraspecies recombination whereby this strain has acquired genetic components from both ocular and urogenital strains [17]. Furthermore, although genetic transfer between strains of the same species may be subtle compared with interspecies transfer, these variations may be critical, depending on the physiochemical type and structural location of the amino acid substitution [52, 55]. The evidence for conjugation includes the acquisition by *C. suis* of tetracycline-resistant genomic islands and the insertion element, IS605, which is homologous to another gut pathogen, *Helicobactor pylori*, inserted into the chlamydial invasion (*inv*)-like gene [37]. Transduction involving different phages is a possibility for certain species of *Chlamydiaceae*, including *Chlamydophila caviae*, *Chlamydophila abortus*, *Chlamydophila pneumoniae*, *Chlamydophila psittaci* and *Chlamydophila pecorum* [36, 52, 56–59].

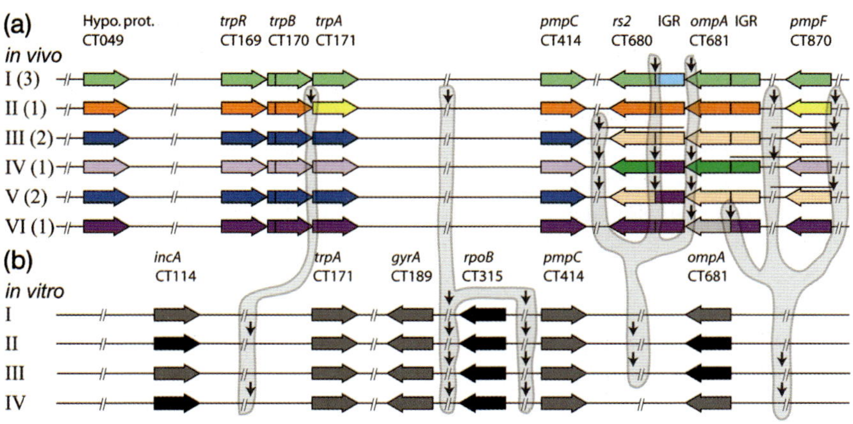

Figure 2.1 Diagram of *Chlamydia trachomatis* recombination types based on *in vivo* observation of genomic sequences and *in vitro* experimentation. (a) Six different recombination types were detected among 10 recent clinical isolates [5]. The number of clinical isolates for each type is shown in parentheses. The genes and intergenic regions (IGRs) that were sequenced are labeled on top of each arrow. The CT designation followed by numbers represents the gene number within the *C. trachomatis* genome. The arrows denote crossover regions. (b) Four main types were noted for *in vitro* recombinants of reference strains D and L1 [53]. The shaded gray groupings of crossovers represent regions that may be similar or identical to those observed *in vivo* or, vice versa, *in vitro*.

Horizontal transfer is also important in the evolution of food-borne pathogens including *Listeria* and *Salmonella* [60]. In *Listeria*, genomic differences between pathogenic *Listeria monocytogenes* and non-pathogenic *Listeria innocua* have resulted from transformation, conjugation and transduction [61]. Six virulence genes (*prfA, plcA, hly, mpl, actA* and *pclB*) are present in all *Listeria*, clustered in a region known as the LAPI-1 (*Listeria* pathogenicity island 1) [60] while the non-pathogenic *Listeria* strains have lost this cluster during evolution [60]. Interestingly, this genomic region appears to have evolved by vertical transfer from an ancestral strain and not by horizontal gene transfer as there is no evidence of insertion sequence-like elements or transposons [60, 63]. However, the plasmid pLM80 and transposons containing tetracycline-resistant determinants have been identified in *L. monocytogenes* [60]. Bertrand and colleagues [62] suggested that tetracycline resistance might be horizontally transferred from other Gram-positive bacteria, such as *Enterococcus faecalis*. While bacteriophages have been reported in *L. monocytogenes* and *L. innocua*, transduction has not been considered a predominant mechanism in the acquisition of virulence genes due to the narrow host range of its bacteriophages, resulting in low cross-species genetic changes [60, 61]. In contrast, the genomes of many species and strains of *Salmonella* exhibit several mobile genetic elements, virulence cassettes, plasmids and bacteriophages. Examples of transformation and conjugation in *Salmonella* are genes encoding for the type III secretion system and lipopolysaccharide, and the *pef* fimbrial operon located on the plasmid, respectively [60]. Gene transfer by all three methods has been authenticated as the main mechanism for pathogenic evolution in *Salmonella* [60, 64].

Evidence for gene transfer has also been discovered in other intracellular pathogens, including *Afipia, Brucella, Burkholderia, Francisella, Nocardia, Rhodococcus, Coxiella* and *Legionella* [65–69], which were previously thought to undergo minimal if any transformation or conjugation. *Rhodococcus* and *Mycobacterium*, although phylogenetically distinct based on 16S rRNA analyses, share a highly homologous haloalkane dehalogenase gene. This gene has recently evolved and functions to degrade an environmental pollutant known as synthetic haloalkane [69]. Synthetic haloalkanes are used as refrigerants, solvents and fire extinguishing agents. Extensive gene transfer events, including recombination and rearrangement involving the methionyl-tRNA synthetase and methyl-directed mismatch repair (*mutS*) genes, have also been described for *Brucella suis* and *Salmonella enterica* SARB strains [8, 68]. Interestingly, *mutS* also functions to block foreign DNA insertion, serving as a barrier for horizontal gene transfer [8]. Hence, mutations in *mutS* may affect its function by facilitating DNA insertion and gene shuffling. Horizontal gene transfer has also been responsible for genetic diversity among *Burkholderia*, especially within pathogenicity islands [70]. Several homologs between *Pseudomonas aeruginosa* and *Burkholderia cepacia* that function in the pathogenesis of cystic fibrosis suggest frequent genetic exchange [71]. *Legionella pneumophila*, a human pathogen causing legionnaires' disease (pneumoniae) and Pontiac fever (respiratory illness), is also competent in DNA transformation. *In vitro*, many *L. pneumophila* loci, especially those regions that are responsible for surface pili proteins, were shown to uptake extrachromosomal DNA, demonstrating pilus variants that correlated with the transformed DNA [72].

In contrast to transferred genes from prokaryotes that may often resemble paralogs, gene transfer from eukaryotes may result in new or xenologous genes that reveal key host–parasite associations and adaptive virulence mechanisms [23]. Due to the acquisition of several plant genes, such as ATP/ADP translocases, by members of the *Chlamydiaceae* family, these organisms have been hypothesized to have originally shared a symbiotic relationship with plants, later becoming parasites of humans and animals through evolution [3]. As human parasites, chlamydiae acquired eukaryotic domains and genes, such as the actin-dependent regulator of chromatin (SWIB) domain [3], aminoacyl-tRNA synthetase genes [30] and ATP/ADP translocase genes [3, 73], to facilitate their *in vivo* lifestyle.

Plasmids have also been identified in the intracellular bacteria listed above, and were expected to confer infectivity and/or virulence characteristics to the host pathogens. Examples of plasmids with these characteristics include the pFNL10 in *Francisella* [74] and the pXT107 in many *Nocardia* and *Rhodococcus* species [75]. In contrast, the 7.5 kb cryptic plasmid of *C. trachomatis* has no known function [76]. Supportive data for conjugation also comes from experimental studies by Voeykova *et al.* [77]. The frequency of conjugation for the *E. coli* plasmid pT01 into *Nocardia* and *Rhodococcus* was reported to range from 1×10^{-3} to 1×10^{-5}.

Bacteriophages have been identified in *Brucella* [78], *Burkholderia* [79], *C. caviae* [52], *C. abortus* [36, 56], *C. pneumoniae* [80], *C. psittaci* [59], *C. pecorum* [57], *Legionella* [81], *Mycobacterium* [82], *Rhodococcus* [83], *Salmonella* [60], *Listeria* [60] and *Shigella* [79]. Surprisingly, while phage have been documented in many *Chlamydiaceae* genera and species, none have been detected in *C. trachomatis* [36] nor among *Rickettsia* [84]. The absence of phages in *C. trachomatis* and *Rickettsia* suggest a preference for transformation and conjugation by these microorganisms. Many genes involved in pathogenicity, including niche-specific and virulence factors, and antibiotic resistance exhibit sequences reminiscent of bacteriophages, suggesting horizontal acquisition of virulence genes by the phages [27, 83]. For instance, transduction was believed to be a major gene acquisition method for *Burkholderia* evolution in order to survive in diverse niches [79]. Indeed, genetic exchange by bacteriophages has been a central mechanism in genetic communication for shiga-like toxins and antibiotic-resistant genes in *Shigella* and *Salmonella* species [85]. Furthermore, while vaccine and drug development for intracellular bacteria are difficult, understanding of the interrelationship between phage and host genomes may promote development of phage-directed therapy [82].

2.3 Gene Transfer in Intracellular Symbionts

Unlike intracellular bacterial parasites, intracellular symbionts have a mutual or commensal relationship with their hosts [6]. Mutualism describes a relationship where both species receive a fitness gain, and commensalism is where one gains but the other is not significantly harmed [86]. However, living together allows symbiotic species to genetically coevolve in a way that can benefit both [87]. Thus, symbiosis is regularly referred to as mutualism [88, 89].

Given that many organisms share symbiotic relationships, there is an opportunity for evolutionary change. Hence, researchers have begun to evaluate gene transfer among these species. A primitive example is the early eukaryotic (plant) symbiotic relationship with free-living photosynthetic bacteria. Algae were believed to engulf chloroplasts from these bacteria over a billion years ago, leaving the plants with a powerful photosynthesis ability and the bacteria with the possibility for reductive genome evolution towards an obligate intracellular lifestyle [90, 91].

Genetic exchanges in endosymbionts commonly affect host coevolution [92]. The genome of a *Buchnera* γ-proteobacterium symbiont of a pea aphid (also known as plant lice), although small in size and relying on the aphid for nutrients due to its intracellular compartmentalization, contains genes and pathways responsible for synthesizing essential amino acids that are missing in the aphid [93]. *Wolbachia*, the most common symbiont of arthropods, infecting an estimated 17–76% of all insect species, has attracted special attention because of their impact on host reproduction and evolutionary processes [92]. The main reproductive alterations found in *Wolbachia*-infected arthropods are cytoplasmic incompatibility (infected males become unable to reproduce with uninfected or infected females), parthenogenesis induction (reproduction of infected females without males), feminization of genetic males (infected males become fertile or infertile females), death of infected males and oogenesis completion [94]. A single *Wolbachia* strain can produce different phenotypes in different hosts [94, 95], suggesting rampant horizontal gene transfer and divergent evolution. Indeed, intergenic recombination has frequently been identified throughout the *Wolbachia* genomes, including the housekeeping genes [96]. Another example where gene transfer underlies phenotypic change comes from *Parachlamydia*, an amoeba symbiont and opportunistic pathogen of humans [6]. The family *Parachlamydiaceae* is in the order *Chlamydiales* because of its *Chlamydiaceae*-like developmental cycle. Like *Wolbachia*, strains of *Parachlamydia* that reside in dissimilar tissues and hosts exhibit dissimilar phenotypes [6], suggesting horizontal transfer and divergent evolution. For example, *Parachlamydia* have acquired numerous virulence factors similar to *Chlamydiaceae*, including type III secretion for communication/infection with host cells, ATP/ADP translocase for energy import and surface proteins for diversity [6, 97]. In addition, genomic islands have also been identified [98]. Surprisingly, *Parachlamydia acanthamoebae*, unlike *Chlamydiaceae*, contains the *tra* genes that are essential for the F-like conjugative transfer system [6, 98]. In this transfer system, the *tra* genes are responsible for sex pilus movement and mating pair stabilization, and thus likely facilitate DNA uptake [98]. Consequently, the similarity and dissimilarity between *Parachlamydiaceae* and *Chlamydiaceae* implies the shared common ancestor and evolutionary divergence of each family to suit a specific niche [97].

As with the intracellular bacterial parasites, gene transfer within species and with other intracellular species is also possible when multiple infections occur. An example of intraspecies recombination in *Wolbachia* occurs for the gene encoding surface proteins such as *wsp*, in which recombination results in different surface recognition or host interaction due to altered antigenicity [92, 96]. Based on various recombination detection programs such as MaxChi and Geneconv, intragenic recombination, although occurring less frequently compared to intergenic transfer, has been detected among various genes including the housekeeping gene *gltA* [96].

Phylogenetic clustering of different strains of *Wolbachia* that reside in distantly related host ranges also emphasizes horizontal transfer during some period of the evolutionarily process [99]. For cross-species gene transfer among intracellular symbionts, sequencing analysis identified DNA exchange between *Wolbachia pipientis* and *Trichogramma karkai* that exhibited multiple infections in the same host egg [100]. Indeed, the genetic data highlighted interspecies recombination and gene transfer for many genes, including *glt*, *wsp* and several other housekeeping and surface encoding genes [96]. Further, the data supported the phylogenetic tree topologies for *Wolbachia* and *Trichogramma* [92].

Evidence of gene transfer from the intracellular symbionts to their host organisms has also been described. Indeed, horizontal gene transfer from organelles, which can be occupied by intracellular microorganisms, to the eukaryotic nucleus is common [87]. The traditional example with regard to the presence of eubacterial genes in eukaryotes is the mitochondrial endosymbiont, which may have originated from *Rickettsia* [101]. As another example of gene transfer to the host, approximately 11 kb of DNA from *Wolbachia* was identified in the chromosome of the symbiotic bean beetle [102]. Genome-wide studies have also reported that an average of 18% of the *Arabidopsis* genome may have been derived from intracellular symbiotic bacteria [103]. Furthermore, genome-wide studies reported that more than 50% of the yeast *Saccharomyces cerevisiae* genome was transferred from a variety of bacteria, including a cyanobacterium (*Synechocystis* 6803), a proteobacterium (*E. coli*) and a methanogen (*Methanococcus jannaschii*) [87].

Phages and transposons have also been identified in *Wolbachia* and *Parachlamydia* [6, 93]. For instance, a lytic phage named WO was discovered in insect- and nematode-infecting *Wolbachia*. WO induces five reproductive alterations in the *Wolbachia*-infected arthropods [93]. Several repetitive DNA and mobile elements denoted in these microorganisms also support the possibility of high genome flux [104, 105]. Disrupted synteny also infers inversion and translocation that may be a result of genetic recombination [6]. These findings support the importance of horizontal transfer and transduction among symbionts.

2.4 Gene Transfer in Intracellular Parasites and Fungi

Parasitic microorganisms depend on and cost a burden to their hosts [6]. Parasitism is another type of symbiotic relationship. Gene transfer events include eukaryotic parasites such as fungi (*Histoplasma*) and protozoa, including *Leishmania*, *Plasmodia*, *Theileria*, *Toxoplasma* and *Trypanosoma*. *Histoplasma*, *Leishmania*, *Plasmodia*, *Toxoplasma* and *Trypanosoma* are human pathogens, responsible for serious respiratory infections, skin and visceral infections, malaria, congenital neurological birth defects and cardiac myopathy, respectively. *Theileria* are primarily cattle pathogens, causing fever and enlarged lymph nodes near the tick bite [106, 107]. In *Histoplasma*, transposable elements, such as cryptons, which encode tyrosine recombinases, have been identified [107]. While the functional domains of the cryptons are relatively conserved, the insertion sites have been shown to vary among different strains, suggesting recent mobility and divergent evolution [107]. Transformation and conjugation in *Histoplasma* are further supported by *in vitro* studies in which 80% of naked DNA was integrated into a single genome site of *Histoplasma capsulatum* [108] and construction of a telomeric plasmid-mediated protein was overexpressed in the organism [109].

For *Leishmania, Plasmodia, Theileria, Toxoplasma* and *Trypanosoma,* although direct evidence indicating transformation, conjugation and transduction is relatively limited, analyses of several genes revealed horizontal transfer either from plant or bacterial origins as one mechanism that causes divergent evolution and blurs their phylogenies [66–68, 110–112]. For example, methionyl-tRNA synthetase, phosphofructokinase (a central enzyme for glycolysis), and β-fructosidase (a cleavage enzyme for disaccharides) were identified in these pathogens, although the genes should be evolutionarily constrained as they have a core function in energy metabolism [67, 110]. The findings of these genes across different domains suggest horizontal transfer, and each species carries a single yet unique amino acid encodement in an active site of these enzymes, representing functional specificity and divergent evolution [67]. Genes encoding α-amylase, glutamine synthetase and the 1,4-α-glucan branching enzyme were likely transferred from bacteria such as *Burkholderia* and *Chlamydophila* given the high degree of homology between the bacterial and protozoan sequences [66]. The pressure of certain nucleotide changes at critical residues leading to structural and functional differences is consistent with a large body of literature [5, 55, 113, 114].

Interestingly, lateral gene transfer from *C. trachomatis* to *Leishmania* and *Trypanosoma* has been suggested for certain genes with unknown function, such as CT387 and CT043, due to their homologous sequences with unrelated species [111]. Gene transfer via transformation has also been suggested for *Trypanosoma brucei* based on antigenic diversity, and via conjugation for *Trypanosoma cruzi* based on the plasmid detection and discovery of various clones of drug-resistant organisms [114, 115]. Hall *et al.* [116] discovered a linear, extrachromosomal DNA in which the genes were expressed throughout the lifecycle in *Theileria annulata*. This extrachromosomal element likely served as a vector for the introduction of genes into protozoan cells [116]. Further, cohabitation of intracellular protozoa, such as *Leishmania,* with other intracellular pathogens, such as *Coxiella, Mycobacteria* and *Salmonella,* inside the host cells have also been documented and offer the possibility for genetic exchange [48].

2.5
Summary and Outlook

Intracellular microorganisms employ gene transfer for tremendous advantage, including *in vivo* host adaptation, antigenic variation, development of antimicrobial resistance and gain of metabolic traits [6, 22, 27, 117]. Host adaptation allows intracellular parasites and symbionts to successfully invade and somehow prevent the host cell from undergoing apoptosis or necrosis, and thrive in an adverse intracellular environment, avoiding the onslaught of lysosomes and other host cell factors. This also includes the capability to adapt to the presence of other intracellular pathogens. Acquisition of genes encoding antigenic proteins can provide profound diversity to membrane structures to avoid host immune pressure and continue their developmental cycle or abort development in favor of a persistent state, awaiting more favorable conditions for replication. Other successfully acquired genes provide resistance to antibiotics and facilitate adaptation to new tissues, occupation of diverse niches and expansion in host range. In addition, the identification of transferred genes can aid our understanding of phylogenetic tree topologies (which can be confounded by gene transfer as this distorts the molecular clock for the organism [118]), ancestral derivatives and the evolutionary relationship to closely related organisms [3]. Recently, the 16S rRNA gene, the most common gene used for phylogenetic reconstruction of prokaryotes because of its relative stability, was reported to be horizontally transferred [10]. This will provide a new challenge for evolutionary studies. Identification of gene transfer will also assist in distinguishing transferred genes from the signature genes that represent the core set of functional genes for that organism and that also serve to classify that particular organism [119–121].

The unique characteristics of the genomes of intracellular microorganisms, especially obligate intracellular microorganisms, include genome downsizing, specialization to specific enclosed host environments, and specific genomic dynamics compared with those of free-living organisms. The genome diversity of intracellular organisms appears to result from a combination of vertical transfer, in which mutations affect clonal diversity, and horizontal transfer, in which mounting evidence supports intra- and interspecies and genus-specific gene transfer. Here, we have described several examples of horizontal gene transfer that affect genotypic and phenotypic variation, virulence, pathogenesis and evolution. Interestingly, evidence for such transfers in addition to new data have led to speculation on the actual existence of a “universal ancestor.” According to Dutta and Pan [27], “There is no guarantee that a gene currently found in all three domains of life (Bacteria, Archea and Eukarya) was present in their common ancestor – it could have arisen in one domain and spread to others.” While this concept will take some time to work out, genome reduction and gene transfer represent ongoing evolutionary processes that occur among intracellular bacteria, symbionts, parasites and fungi to accomplish non-essential gene reduction/deletion and promote a selective advantage in their respective hosts. The next steps require genome sequencing of additional species and strains for all of these intracellular organisms to further accelerate our understanding of human and animal disease pathogenesis, evolution of each within their host and how we might harvest this knowledge to genetically manipulate these pathogens to promote drug discovery and the development of efficacious vaccines.

Acknowledgments

This chapter was supported in part by Public Health Service grants from the National Institutes of Health, R01 AI059647 and R01 AI039499 (to D. Dean), and the Founder Region Fellowship (to N. Somboonna).

References

1 Lawrence, J.G. (1999) *Curr Opin Microbiol*, **2**, 519–523.
2 Koonin, E.V. and Galperin, M.Y. (1997) *Curr Opin Genet Dev*, **7**, 757–763.
3 Koonin, E.V., Makarova, K.S. and Aravind, L. (2001) *Annu Rev Microbiol*, **55**, 709–742.
4 Akiba, T., Koyama, K., Ishiki, Y., Kimura, S. and Fukushima, T. (1960) *Jpn J Microbiol*, **4**, 219–227.
5 Gomes, J.P., Bruno, W.J., Nunes, A., Santos, N. *et al.* (2007) *Genome Res*, **17**, 50–60.
6 Wernegreen, J.J. (2005) *Curr Opin Genet Dev*, **15**, 572–583.
7 Blanc, G., Ogata, H., Robert, C., Audic, S. *et al.* (2007) *Genome Res*, **17**, 1657–1664.
8 Brown, E.W., Mammel, M.K., LeClerc, E. and Cebula, T.A. (2003) *Proc Natl Acad Sci USA*, **100**, 15676–15681.
9 Gross, R., Hacker, J. and Goebel, W. (2003) *Mol Microbiol*, **47**, 1749–1758.
10 Jain, R., Rivera, M.C., Moore, J.E. and Lake, J.A. (2002) *Theor Popul Biol*, **61**, 489–495.
11 Syvanen, M. (1985) *J Theor Biol*, **112**, 333–343.
12 Tatum, E.L. and Lederberg, J. (1947) *J Bacteriol*, **53**, 673–684.
13 Sherman, J.M. and Wing, H.U. (1937) *J Bacteriol*, **33**, 315–321.
14 Lederberg, J. (1951) *Science*, **114**, 68–69.
15 Millman, K.L., Tavare, S. and Dean, D. (2001) *J Bacteriol*, **183**, 5997–6008.
16 Gomes, J.P., Bruno, W.J., Borrego, M.J. and Dean, D. (2004) *J Bacteriol*, **186**, 4295–4306.
17 Gomes, J.P., Nunes, A., Bruno, W.J., Borrego, M.J. *et al.* (2006) *J Bacteriol*, **188**, 275–286.
18 Dean, D., Myers, G.S. and Read, T.D. (2006) in *Chlamydia: Genomics and Pathogenesis* (eds P.M. Bavoil and P.B. Wyrick), Horizon Bioscience, Wymondham, UK, pp. 1–24.
19 Collins, N.E., Liebenberg, J., de Villiers, E.P., Brayton, K.A. *et al.* (2005) *Microbiol*, **102**, 838–843.
20 Andersson, S.G., Zomorodipour, A., Andersson, J.O., Sicheritz-Pontén, T. *et al.* (1998) *Nature*, **396**, 133–140.
21 Brosch, R., Gordon, S.V., Eigimeier, K., Garnier, T. and Cole, S.T. (2000) *Res Microbiol*, **151**, 135–142.
22 Wren, B.W. (2000) *Nature Rev*, **1**, 30–39.
23 Ochman, H., Lawrence, J.G. and Groisman, E.A. (2000) *Nature*, **405**, 299–304.
24 Holmes, R.K. and Jobling, M.G. (1996) in *Medical Microbiology*, 4th edn (ed. S. Baron) University of Texas Medical Branch, Galveston, TX, section 5.
25 Chen, I. and Dubnau, D. (2004) *Nat Rev Microbiol*, **2**, 241–249.
26 Weaver, J.C. (1993) *J Cell Biochem*, **51**, 426–435.
27 Dutta, C. and Pan, A. (2002) *J Biosci*, **27**, 27–33.
28 Garcia-Vallve, M.Y., Romeu, A. and Palau, J. (2000) *Genome Res*, **10**, 1719–1725.
29 Grantham, R., Gautier, C., Gouy, M., Mercier, R. and Pave, A. (1980) *Nucleic Acids Res*, **8**, R49–R62.
30 Wolf, Y.I., Rogozin, I.B., Kondrashov, A.S. and Koonin, E.V. (2001) *Genome Res*, **11**, 356–372.
31 Olsen, G.L. and Woese, C.R. (1997) *Cell*, **89**, 991–994.
32 Edgell, D.R. and Doolittle, W.F. (1997) *Cell*, **89**, 990–995.
33 Keasling, J.D. (2008) *ACS Chem Biol*, **3**, 64–76.
34 Cohen, S., Chang, A., Boyer, H. and Helling, R. (1973) *Proc Natl Acad Sci USA*, **70**, 3240–3244.
35 Russell, D.G., Sturgill-Koszycki, S., Vanheyningen, T., Collins, H. and Schaible, U.E. (1997) *Phil Trans R Soc Lond B*, **352**, 1303–1310.
36 Read, T.D., Brunham, R.C., Shen, C., Gill, S.R. *et al.* (2000) *Nucleic Acids Res*, **28**, 1397–1406.
37 Dugan, J., Rockey, D.D., Jones, L. and Andersen, A.A. (2004) *Antimicrob Agents Chemother*, **48**, 3989–3995.

38 Dean, D., Oudens, E., Bolan, G., Padian, N. and Schachter, J. (1995) *J Infect Dis*, **172**, 1013–1022.

39 Brunham, R., Yang, C., Maclean, I., Kimani, J. *et al.* (1994) *J Clin Invest*, **94**, 458–463.

40 Dean, D., Suchland, R.J. and Stamm, W.E. (2000) *J Infect Dis*, **182**, 909–916.

41 Hayes, L.J., Yearsley, P., Treharne, J.D., Ballard, R.A. *et al.* (1994) *Infect Immun*, **62**, 5659–5663.

42 Lin, J.S., Donegan, S.P., Heeren, T.C., Greenberg, M. *et al.* (1998) *J Infect Dis*, **178**, 1707–1712.

43 Somboonna, N., Mead, S., Lui, J. and Dean, D. (2008) *Emerg Infect Dis*, **14**, 445–453.

44 Garland, S.M., Tabrizi, S.N., Chen, S., Byambaa, C. and Davaajav, K. (2001) *Infect Dis Obstet Gynecol*, **9**, 143–146.

45 Mgone, C.S., Lupiwa, T. and Yeka, W. (2002) *Sex Transm Dis*, **29**, 775–779.

46 Gomes, J.P., Tavira, L., Exposto, F., Prieto, E. and Catry, M.A. (2001) *Acta Trop*, **80**, 261–264.

47 Ridderhof, J.C. and Barnes, R.C. (1989) *Infect Immun*, **57**, 3189–3193.

48 Voth, D.E. and Heinzen, R.A. (2007) *Cell Microbiol*, **9**, 829–840.

49 Dean, D. and Millman, K. (2008) in *Evolutionary Biology of Bacterial and Fungal Pathogens* (eds F. Baquero, C. Nombela, G.H. Cassell and J.A. Gutiérrez), ASM Press, Washington, D.C., pp. 475–487.

50 Hsia, R.C. and Bavoil, P.M. (1996) *Gene*, **176**, 163–169.

51 Zhang, D.J., Fan, H., McClarty, G. and Brunham, R.C. (1995) *Infect Immun*, **63**, 676–680.

52 Read, T.D., Myers, G.S., Brunham, R.C., Nelson, W.C. *et al.* (2003) *Nucleic Acids Res*, **31**, 2134–2147.

53 Demars, R., Weinfurter, J., Guex, E., Lin, J. and Potucek, Y. (2007) *J Bacteriol*, **189**, 991–1003.

54 Demars, R. and Weinfurter, J. (2008) *J Bacteriol*, **190**, 1605–1614.

55 Betts, M.J. and Russell, R.B. (2003) in *Bioinformatics for Geneticists* (eds M.R. Barnes and I.C. Gray), John Wiley & Sons, Ltd, Chichester, pp. 289–316.

56 Skilton, R.J., Cutcliffe, L.T., Pickett, M.A., Lambden, P.R. *et al.* (2007) *J Bacteriol*, **189**, 4957–4959.

57 Garner, S.A., Everson, J.S., Lambden, P.R., Fane, B.A. and Clarke, I.N. (2004) *Virus Genes*, **28**, 207–214.

58 Everson, J.S., Garner, S.A., Lambden, P.R., Fane, B.A. and Clarke, I.N. (2003) *J Bacteriol*, **185**, 6490–6492.

59 Liu, B.L., Everson, J.S., Fane, B., Giannikopoulou, P. *et al.* (2000) *J Virol*, **74**, 3464–3469.

60 Kelly, B.G., Vespermann, A. and Bolton, D.J. (2009) *Food Chem Toxicol.*, **47**, 951–968.

61 Buchrieser, C., Rusniok, C., Kunst, F., Cossart, P. and Glaser, P. (2003) *FEMS Immun Med Microbiol*, **35**, 207–213.

62 Bertrand, S., Huys, G., Yde, M., D'Haene, K. *et al.* (2005) *J Med Microbiol*, **54**, 1151–1156.

63 Schmidt, H. and Hensel, M. (2004) *Clin Microbiol Rev*, **17**, 14–56.

64 Groisman, E.A. and Ochman, H. (1997) *Trends Microbiol*, **5**, 343–349.

65 Larsson, P., Oyston, P.C., Chain, P., Chu, M.C. *et al.* (2005) *Nat Genet*, **37**, 153–159.

66 Huang, J., Mullapudi, N., Lancto, C.A., Scott, M. *et al.* (2004) *Genome Biol*, **5**, R88.

67 Bapteste, E., Mareira, D. and Philippe, H. (2003) *Gene*, **318**, 185–191.

68 Brown, J.R. (2003) *Nat Rev*, **4**, 121–132.

69 Poelarends, G.J., Kulakov, L.A., Larkin, M.J., van Hylckama Vlieg, J.E.T. and Janssen, D.B. (2000) *J Bacteriol*, **182**, 2191–2199.

70 Tumapa, S., Holden, M.T., Vesaratchavest, M., Wuthiekanun, V. *et al.* (2008) *BMC Genomics*, **9**, 190.

71 Eberl, L. and Tũmmler, B. (2004) *Int J Med Microbiol*, **294**, 123–131.

72 Stone, B.J. and Kwaik, Y.A. (1999) *J Bacteriol*, **181**, 1395–1402.

73 Winkler, H.H. and Neuhaus, H.E. (1999) *Trends Biochem Sci*, **24**, 64–68.

74 Pomerantsev, A.P., Golovliov, I.R., Ohara, Y., Mokrievich, A.N. *et al.* (2001) *Plasmid*, **46**, 210–222.

75 Xia, H.Y., Tian, Y.Q., Zhang, R., Lin, K.C. and Qin, Z.J. (2006) *Acta Biochim Biophys Sin (Shanghai)*, **38**, 620–624.

76 Carlson, J.H., Whitmire, W.M., Crane, D.D., Wicke, L. *et al.* (2008) *Infect Immun.*, **76**, 2273–2283.

77 Voeykova, T., Emelvanova, L., Tabakov, V. and Mkrtumyan, N. (1998) *FEMS Microbiol Lett*, **162**, 47–52.

78 Riqby, C.E., Cerqueira-Campos, M.L., Kelly, H.A. and Surujballi, O.P. (1989) *Can J Vet Res*, **53**, 319–325.

79 Summer, E.J., Gill, J.J., Upton, C., Gonzalez, C.F. and Young, R. (2007) *Curr Opin Microbiol*, **10**, 410–417.

80 Rupp, J., Solbach, W. and Gieffers, J. (2007) *FEMS Microbiol Lett*, **273**, 45–49.

81 Lammertyn, E., Vande Voorde, J., Meyen, E., Maes, L. *et al.* (2007) *Microb Ecol.*, **56**, 191–197.

82 Papke, R.T. and Doolittle, W.F. (2003) *Curr Biol*, **13**, R606–R607.

83 Hondalus, M.K. (1997) *Vet Microbiol*, **56**, 257–268.

84 Wood, D.O. and Azad, A.F. (2000) *Infect Immun*, **68**, 6091–6093.

85 Brabban, A.D., Hite, E. and Callaway, T.R. (2005) *Foodborne Pathog Dis*, **2**, 287–303.

86 Dethlefsen, L., McFall-Ngai, M. and Relman, D.A. (2007) *Nature*, **449**, 811–818.

87 Hoffmeister, M. and Martin, W. (2003) *Environ Microbiol*, **5**, 641–646.

88 Wilkinson, D.M. (2001) *Nature*, **412**, 485.

89 Saffo, M.B. (1993) *Symbiosis*, **14**, 1–3.

90 Kneip, C., Voβ, C., Lockhart, P.J. and Maier, U.G. (2008) *BMC Evol Biol*, **8**, 30.

91 Raymond, J. and Blankenship, R.E. (2003) *Proc Natl Acad Sci USA*, **100**, 7419–7420.

92 Huigens, M.E., de Almeida, R.P., Booms, P.A., Luck, R.F. and Stouthamer, R. (2004) *Proc Biol Sci*, **271**, 509–515.

93 Shigenobu, S., Watanabe, H., Hattori, M., Sakaki, Y. and Ishikawa, H. (2000) *Nature*, **407**, 81–86.

94 Garvotte, L., Henri, H., Stouthamer, R., Charif, D. *et al.* (2007) *Mol Biol Evol*, **24**, 427–435.

95 Fujii, Y., Kageyama, D., Hoshizaki, S., Ishikawa, H. and Sasaki, T. (2001) *Proc R Soc Lond B*, **268**, 855–859.

96 Baldo, L., Bordenstein, S., Wernegreen, J.J. and Werren, J.H. (2006) *Mol Biol Evol*, **23**, 437–449.

97 Horn, M., Collingro, A., Schmitz-Esser, S., Beier, C.L. *et al.* (2004) *Science*, **304**, 728–730.

98 Greub, G., Collyn, F., Guy, L. and Roten, C. (2004) *BMC Microbiol*, **4**, 48.

99 Schulenburg, J.H., Hurst, G.D., Huigens, T.M., van Meer, M.M. *et al.* (2000) *Mol Biol Evol*, **17**, 584–600.

100 Huigens, M.E., Luck, R.F., Klaassen, R.H.G., Maas, M.F.P.M. *et al.* (2000) *Nature*, **405**, 178–179.

101 Emelyanov, W. (2003) *Eur J Biochem*, **270**, 1599–1618.

102 Kondo, N., Nikoh, N., Ijichi, N., Shimada, M. and Fukatsu, T. (2002) *Proc Natl Acad Sci USA*, **91**, 14280–14285.

103 Martin, W., Rujan, T., Richly, E., Hansen, A. *et al.* (2002) *Proc Natl Acad Sci USA*, **99**, 12246–12251.

104 Wu, M., Sun, L.V., Vamathevan, J., Riegler, M. *et al.* (2004) *PLoS Biol*, **2**, E69.

105 Salzberg, S.L., Hotopp, J.C., Delcher, A.L., Pop, M. *et al.* (2005) *Genome Biol*, **6**, R23.

106 Roos, D.S. (2005) *Science*, **309**, 72–73.

107 Goodwin, T.J., Butler, M.I. and Poulter, R.T. (2003) *Microbiology*, **149**, 3099–3109.

108 Sullivan, T.D., Rooney, P.J. and Klein, B.S. (2002) *Eukaryot Cell*, **1**, 895–905.

109 Fisher, K.L., Deepe, G.S. Jr and Woods, J.P. (1999) *Infect Immun*, **67**, 3312–3316.

110 Opperdoes, F.R. and Michels, P.A.M. (2007) *Trends Parasitol*, **23**, 470–476.

111 Gupta, R.S. and Griffiths, E. (2006) *Trends Microbiol*, **14**, 527–535.

112 Huang, J., Mullapudi, N., Sicheritz-Ponten, T. and Kissinger, J.C. (2004) *Int J Parasitol*, **34**, 265–274.

113 Karchin, R., Diekhans, M., Kelly, L., Thomas, D.J. *et al.* (2005) *Bioinformatics*, **21**, 2814–2820.

114 Stothard, J., Frame, I. and Miles, M. (1999) *Mem Inst Oswaldo Cruz*, **94**, 189–193.

115 Cross, G.A. (1996) *Bioessays*, **18**, 283–291.

116 Hall, R., Coggins, L., McKellar, S., Shiels, B. and Tait, A. (1990) *Mol Biochem Parasitol*, **38**, 253–260.

117 Stêpkowski, T. and Legocki, A.B. (2001) *Acta Biochim Pol*, **48**, 367–381.

118 Novichkov, P.S., Omelchenko, M.V., Gelfand, M.S., Mironov, A.A. *et al.* (2004) *J Bacteriol*, **186**, 6575–6585.

119 Klint, M., Fuxelius, H.H., Goldkuhl, R.R., Sharin, H. *et al.* (2007) *J Clin Microbiol*, **45**, 1410–1414.

120 Baldo, L. and Werren, J.H. (2007) *Curr Microbiol*, **55**, 81–87.

121 Baldo, L., Dunning Hotopp, J.C., Jolley, K.A., Bordenstein, S.R. *et al.* (2006) *Appl Environ Microbiol*, **72**, 7098–7110.

3
Phagocytosis: Early Events in Particle Recognition and Uptake

Gabriela Cosío and Sergio Grinstein

3.1
Phagocytosis: An Overview

The uptake of "large" (>0.5 μm) particles by cells, a process termed phagocytosis, has been recognized as an important mechanism for the internalization and subsequent destruction of pathogens by specialized cells of the immune system. In addition, phagocytosis plays a key role in the nutrition of unicellular organisms such as the free-living amoeba *Dictyostelium*, and in the clearance of apoptotic cells that is central to tissue remodeling and development. Macrophages and neutrophils have been acknowledged as professional phagocytes by virtue of their high efficiency of binding and uptake of targets, yet other types of cells are also able to internalize particles. In this respect, fibroblasts are able to take up apoptotic cells [1], thyroid and bladder epithelial cells can engulf erythrocytes [2] and retinal epithelial cells can internalize the effete ends of retinal rods [3]. The differences between professional and nonprofessional phagocytes may rely on the level of expression of a specialized group of distinct plasmalemmal proteins known as phagocytic receptors, responsible for binding the particles and triggering their uptake.

The phagocytic process is not simple; ligands on the surface of a given pathogen activate not one but many different receptors concomitantly, which collaborate in engulfing and ultimately destroying the microorganism. The activation of multiple receptors triggers several distinct signaling cascades which in all likelihood influence the different stages of phagocytosis. The complexity of phagocytosis and its outcome is exacerbated by the fact that pathogens are not always passive bystanders. Several types of bacteria and protozoa can subvert the phagocyte machinery and influence the uptake and/or maturation processes, thereby altering their own fate inside the phagocyte.

In a general way, phagocytosis can be subdivided into three main stages: particle binding, ingestion (i.e., phagosome formation) and phagosome maturation. Binding refers to the recognition of ligands on the surface of the particle by cognate receptors. Ingestion entails the generation of a separate intracellular membrane-bound vacuole

Intracellular Niches of Microbes. Edited by Ulrich E. Schaible and Albert Haas

ISBN: 978-3-527-32207-7

where the particle is sequestered. Lastly, maturation refers to the remodeling process that converts the nascent phagosome into a microbicidal phagolysosome.

In addition to the destruction of the pathogen, phagocytosis contributes in other ways to the immune response. Interaction with the pathogen can induce the production and release of chemical mediators, such as chemokines and cytokines. In addition, phagocytes are also key to the presentation of microbial antigens to T and B lymphocytes. Presentation of antigens by major histocompability complex (MHC) class I and II molecules fosters humoral antibody-mediated immunity and T cell-mediated cellular immunity. Thus, phagocytosis seems to be an important link between the innate and adaptive immune responses, which synergize to combat current and prevent future infections.

3.2
Recognition of Target Particles

Professional phagocytes express on their surface a sizable array of distinct receptors (Table 3.1) capable of recognizing a vast collection of self- and non-self ligands such as those present on apoptotic/necrotic cells as well as on infectious agents, respectively. As intimated above, these receptors rarely operate in isolation and costimulation of multiple receptor types is likely to be the norm [4]. It is noteworthy that different receptors activate distinct signaling pathways that interact in a cooperative manner to regulate the final response [4]. Phagocytic receptors can interact with their target by directly binding to molecular determinants on their surface. In some cases, however, the interaction is mediated by opsonins, serum components such as complement proteins or specific antibodies that coat the target surface [5].

3.2.1
Direct Recognition of Targets: Non-Opsonic Receptors

The interaction of macrophages with non-opsonized bacteria is based on the recognition of conserved microbial structures, known as pathogen-associated molecular patterns (PAMPs). PAMPs are unique to microbes and are invariant among a given class of microorganisms [25, 26]. They are recognized by one or more pattern recognition receptors (PRRs).

PAMPs vary greatly in their molecular nature. They include a complex array of polysaccharides and other glycoconjugates such as membrane-anchored lipopolysaccharides (LPS) of Gram-negative bacteria [27], lipoteichoic acid (LTA) on Gram-positive microbes, and peptidoglycans, the "structural scaffold" that maintains the shape and integrity of both Gram-positive and Gram-negative bacteria [28, 29]. Besides polysaccharides, various proteins and lipoproteins found on microbial pilli, flagella or capsule can act as PAMPs [5].

PRRs are grouped into different families according to their molecular structure. Some of the main PRRs are: scavenger receptors (SR), the glycosylphosphatidylinositol (GPI)-anchored glycoprotein CD14, C-type lectins, integrins and Toll-like

Table 3.1 Phagocytic receptors that bind to microbes.

Receptors	Ligands	Refs
Fc receptors:		
FcγRI	IgG-, CRP-opsonized targets	[6–9]
FcγRII	IgG-, CRP-opsonized targets	[6–9]
FcγRIII	IgG-, CRP-opsonized targets	[6–9]
FcεRI	IgE-opsonized targets	[10]
FcεRII	IgE-opsonized targets	[11]
FcαR	IgA-opsonized targets	[12]
Complement receptors:		
CR1	MBL-, C1q-, C3b-opsonized targets	[13]
CR3 ($\alpha_M\beta_2$, CD11b/CD18)	iC3b-opsonized targets	[14]
CR4 ($\alpha_x\beta_2$, CD11c/CD18)	iC3b-opsonized targets	[15]
Other integrins:		
$\alpha_5\beta_1$	Fibronectin/vitronectin-opsonized targets	[16]
$\alpha_4\beta_1$		
$\alpha_V\beta_3$		
Scavenger receptors:		
SRA	LPS, LTA, CpG DNA, *E. coli, S. aureus, N. meningitidis*	[17]
MARCO	LPS, *E. coli, S. aureus, N. meningitidis*	[18]
Mannose receptor	Mannan, *K. pneumoniae* LPS, *S. aureus* capsular polysaccharide, *C. perfringens* capsular polysaccharide	[19]
Dectin-1	β1,3-glucan	[20]
CD14	LPS, peptidoglycan	[21, 22]
C1qR(P)	C1q, MBL, SPA	[23]
CD44	Hyaluronan-coated prey	[24]

Adapted from [4].

receptors. SRs were initially defined by their ability to bind and internalize modified low-density lipoproteins [30]. However, new groups of proteins that share molecular features with SR have been added to the list, and SR have now been documented to recognize a variety of artificial and polyanionic ligands, including many microorganisms and apoptotic cells. SR-A and MARCO (macrophage receptor with a collagenous structure) are two members of class A SR important in microbial recognition. SR-A is able to recognize LPS [31], different LTA structures [32] and unmethylated bacterial CpG DNA [33]. MARCO has also been found to recognize intact Gram-positive and -negative bacteria, as well as isolated products such as LPS and CpG DNA [34, 35]. In addition to SR-A, other classes of SR bind entire bacteria or their products. These include SR-E LOX-1 (lectin-like OX-LDL receptor 1) [36], SR-G SR-PSOX (scavenger receptor for phosphatidylserine and OX-LDL) [37] and SR-H FEEL-1/2 (fascilin, EGF-like, laminin-type EGF-like, and link-domain-containing scavenger receptor) [38]. Likewise, members of the SR-B family, SRB-1 and CD36, recognize isolated bacterial products such as LPS and diacylated lipopeptide, respectively [39, 40].

CD14 is a GPI-anchored glycoprotein that lacks a membrane-spanning region and a cytoplasmic tail [41]. Soluble CD14 without the GPI anchor is also found in the plasma. CD14 recognizes peptidoglycan from Gram-positive microbes [41] and LPS from Gram-negative bacteria [42]. CD14 also binds LTA from Gram-positive bacteria, lipoproteins and lipoarabinomannan from mycobacteria, and mannuronic acid from Gram-negative bacteria [41, 43, 44].

C-type lectin receptors bind to carbohydrates from self and foreign sources, circulating proteins and pathogens [45]. C-type lectins may contain a single or multiple carbohydrate-recognition domains (CRD), which may or may not require calcium for ligand binding [5]. C-type lectins with a single CRD and that require calcium for ligand binding are type II membrane receptors and include DC-SIGN (dendritic cell-specific intercellular adhesion molecule 3-grabbing non-integrin) and Dectin-2 [5]. C-type lectins that contain various CRDs and that also require calcium for ligand binding, are type I transmembrane receptors and include the mannose receptor (MR), Endo180, DEC-205 and the phospholipase A_2 receptor [46]. The MR was one of the first receptors of this type to be cloned. It recognizes terminal mannose, fucose, *N*-acetyl glucosamine or glucose residues [47] and is also able to bind LPS [48]. Single CRD C-type lectins that do not require calcium for recognition include Dectin-1 and CD69. These molecules possess a single extracellular CRD, a transmembrane region and a cytoplasmic domain with or without signaling regions [49]. Dectin-1 has a tyrosine-based activation motif [20, 50], and its CRD recognizes β-1,3-linked and β-1,6-linked glucans, which are structures normally found in yeast [20, 51].

The main function of integrins is the attachment of cells to the extracellular matrix (ECM) and the transduction of signals from the ECM to the cell. However, they have also been found to be important for the recognition and phagocytosis of microorganisms. Complement receptors mainly interact with bacteria through complement proteins, though they have been found to recognize them directly in some instances. The complement receptors CR3 and CR4 are two β_2 integrins involved in microbe recognition. CR3 was found to be important for the non-opsonic phagocytosis of *Mycobacterium tuberculosis* [52] by binding to the bacterial antigen known as 85C [53] and to surface oligosaccharides [54]. CR4 also binds to *M. tuberculosis*, although the ligand is unknown [55]. CR3 mediates the non-opsonic phagocytosis of *Bordetella pertussis* through recognition of a bacterial filamentous hemagglutinin (FHA) [56]. Lastly, *Pseudomonas aeruginosa* is another pathogen that can be internalized via CR3-mediated recognition of LPS [57].

There are 10 members of the Toll-like receptor (TLR) family in humans. They have a diverse array of important functions [58], though their direct role as phagocytic receptors is currently the subject of debate. TLRs are type I transmembrane proteins with an extracellular leucine-rich domain and a cytoplasmic domain resembling that of the interleukin 1 (IL-1) receptor [59]. TLR1 has been shown to bind bacterial triacyl lipopeptides [60], while TLR2 recognizes Gram-positive peptidoglycan and LTA [58, 61]. TLR4 is able to bind LPS from Gram-negative bacteria [62, 63], whereas flagellin is a recognized ligand for TLR5 [64, 65]. TLR3, TLR7, TLR8 and TLR9 localize on endosomes and bind unmethylated CpG DNA [66]. TLRs may function as heterodimers, providing broader specificity. In this respect, the TLR2–TLR1 complex binds to diacylated lipopeptides, while TLR2–TLR6 recognizes triacylated lipopeptides [67].

3.2.2 Indirect Recognition of Targets: Opsonic Receptors

In order to increase the efficiency of recognition and clearance of pathogens, the immune system utilizes more generic receptors that recognize humoral components called opsonins. Opsonin is derived from the Greek word meaning "prepare for eating," and it refers to a serum substance able to coat foreign agents, making them vulnerable to phagocytosis.

The most widely recognized opsonins are complement-derived proteins and antibodies. However other opsonins include the lipopolysaccharide-binding protein (LBP) and the collectins: mannose-binding lectin (MBL), C1q and lung surfactant protein A (SP-A). The most important complement opsonins are the particles iC3b and C3b, which are derived from the complement protein C3 [68]. The C3 protein has a fundamental role in the activation of the classical and alternative complement cascades and C3b is one of the fragments resulting from its cleavage. After cleavage, C3b suffers a conformational change that exposes a highly reactive thioester which covalently attaches the peptide to the pathogen by interaction with free surface hydroxyl or amino groups. C3b may be further cleaved to produce the iC3b peptide.

The complement receptors with phagocytic properties are CR1 and CR3. CR1 can react with pathogens coated predominantly with C3b and can also bind particles coated with MBL, C4b, iC3b and C1q, though with lower affinity [68]. CR3 is a β_2 integrin composed of two subunits: CD11b or α_M and CD18 or β_2. CR3 binds mainly to iC3b but other ligands include fibronectin, factor X and ICAM-1 [68]. Complement receptors need an additional signal in order to elicit uptake. This second signal brings about the activation and clustering of the integrin molecules. Some of these signals are triggered by stimulation of the phagocyte by inflammatory cytokines, LPS, binding of the cell to extracellular matrix molecules such as fibronectin or laminin, or by the stimulation of protein kinase C by phorbol esters [69, 70].

Target particles can also be opsonized by antibodies. IgG is the best known opsonin, however IgA has also been shown to have opsonin activity [68]. The binding of antibodies neutralizes the pathogen and prepares them to be cleared by professional phagocytes. Fcγ receptors (FcγR) recognize the Fc portion of the pathogen-bound IgG and by these means initiate ingestion. FcγRs can be divided into two classes according to whether they activate or inhibit the phagocytic response [2]. FcγRI (a high-affinity receptor) and FcγRIIA and FcγRIIIA (low-affinity receptors) are activating receptors that signal through the phosphorylation of ITAM (immunoreceptor tyrosine-based activation motif) motifs that recruit kinases and activate phosphorylation cascades [6, 71]. By contrast, FcγRIIB contains an ITIM (immunoreceptor tyrosine-based inhibitory motif) motif that engages phosphatases such as SHIP, and inhibits the phagocytosis signal [6]. The balance between activating and inhibitory FcγRs determines the phagocytic and inflammatory outcome.

The MBL is a circulating opsonin with a C-terminal lectin domain that associates as multimers of trimers and binds to mannans [72, 73]. MBL is able to bind to a wide variety of organisms, including bacteria, yeasts, parasites and the glycoprotein envelope of viruses and apoptotic cells [74]. MBL-opsonized ligands can interact with the C1q receptor [72, 73], the complement receptor CR3 and collectin receptors [74].

Other opsonins include the SP-A, a C-type lectin that binds to carbohydrate structures on bacterial surfaces [72, 73], and pentraxins such as the C-reactive protein that binds to many bacterial surface ligands.

3.2.3 Cooperation Between Receptors

The broad spectrum of surface molecules present on pathogen surfaces can in principle be recognized by a variety of receptors on the phagocyte plasma membrane. The engagement of multiple receptors results in a cooperative effect between different signaling pathways that collectively contribute to ligand recognition and modulate the biological response of the host cell. In this respect, it has been shown that opsonization of pathogens with both complement and antibodies markedly augments the phagocytic efficiency [68]. In addition, binding of C1q and/or SP-A to their respective receptors can enhance immunoglobulin G (IgG)- or complement-mediated particle uptake [68]. CD14 and TLR4 receptors seem also to cooperate in the recognition and signaling of LPS. LPS and the LPS-binding protein (LBP), a component of serum, initiate signals through membrane-bound CD14, while TLR4 functions as a CD14 coreceptor that activates LPS signaling [75].

3.3 Signaling Events Upon Ligand Recognition

It is the combination of distinct signaling pathways elicited by receptors that defines the phagocyte response to a specific pathogen. However, to facilitate understanding of the phagocytic process, the modes of action of individual receptors are discussed separately below. The signaling events triggered by the Fcγ receptor and the CR3 receptor have been studied in greatest detail and will be used as paradigms in the following section.

3.3.1 FcγR-Mediated Phagocytosis

The binding of IgG-opsonized targets to their cognate receptors, FcγR, leads to their rapid internalization by a process driven by a remarkable actin-dependent rearrangement of the phagocyte plasma membrane. Actin remodeling is a biphasic event, with initial polymerization during pseudopod extension followed by rapid depolymerization during or immediately after sealing of the nascent vacuole [76]. The signals triggering this response will be described next.

Particle binding induces clustering of the receptors, which undergo tyrosine phosphorylation of their ITAM motifs. Src-family kinases are responsible for this phosphorylation [77]. The phosphorylated ITAM motifs function as docking sites for numerous proteins, including kinases and adapter proteins. One such protein is Syk, which binds to the dual phosphotyrosines of the ITAM through its suitably spaced tandem SH2 domains. The recruitment of Syk kinase to the phosphorylated ITAMs causes its autophosphorylation and activation [71], eliciting a variety of processes such as gene expression, actin cytoskeleton rearrangement and the production of inflammatory mediators [2]. The order of the subsequent events has not been fully defined, however many signaling molecules including phospholipase A_2 (PLA_2), phospholipase C gamma (PLCγ), protein kinase C (PKC), phospholipase D (PLCD), phosphatidylinositol 3-kinase (PI3K) and Rho-GTPases are all activated and seemingly important for phagocytosis.

Syk does not function in isolation to activate these effectors. Indeed, a variety of adapter proteins have been postulated to mediate effector activation. Most notably, the Cbl-Nck-Grb2 adapter complex has been implicated in phagocytosis. Gab2 is another adapter protein which has been shown to associate to p85, the regulatory subunit of PI3K. Gab2 is thought to be a key component of phagocytosis, amplifying PI3K signaling [78]. CrkII is another adapter that serves to recruit DOCK180, favoring the activation of Rac1, another key player in Fc-mediated particle uptake [79]. Another adapter, SLP-76, binds Fyb/SLAP which then recruits Ena/VASP proteins, profilin and Nck. The latter in turn recruits WASP (Wiscott-Aldrich syndrome protein) and Arp2/3 [80, 81].

Like proteins, lipids play key roles in phagosome formation and maturation. Cholesterol and sphingolipids are essential constituents of lipid rafts. These microdomains have been shown to recruit Fcγ receptors after they are crosslinked. Following stimulation, FcγRs colocalize with Lyn kinase in cholesterol-rich clusters of 200–300 nm in diameter, where phosphorylation likely occurs [82]. Accordingly, the disruption of rafts by cholesterol removal prevented the association of Lyn with the receptors [83].

Phosphoinositol lipids are major regulators of actin remodeling during several cellular phenomena, including phagocytosis [84]. Phosphoinositides are substrates for the phospholipases that generate second messengers such as diacylglycerol (DAG) and inositol 1,4,5-trisphosphate or (InsP3), and also play a role as membrane anchors for cytosolic proteins that have phosphoinositide-binding domains [85, 86]. Phosphatidylinositol 4,5-bisphosphate ($PI(4,5)P_2$) is important for actin remodeling since it recruits actin-capping, severing and monomer-binding proteins [87]. During phagocytosis $PI(4,5)P_2$ accumulates at the extending pseudopods and rapidly disappears upon phagosome sealing [88]. These dynamic changes suggest an essential function for actin assembly. Phosphatidylinositol 4-phosphate 5-kinase alpha (PIP5KIα) regulates the synthesis of $PI(4,5)P_2$ during phagocytosis and its inhibition blocks phagocytosis [89]. The disappearance of $PI(4,5)P_2$ could be due to either its hydrolysis by PLCγ, resulting in the production of DAG and InsP3, or its conversion to phosphatidylinositol 3,4,5-trisphosphate ($PI(3,4,5)P_3$) by class I PI3K. DAG in turn is able to recruit proteins with C1 domains such as PKCε that was shown to be recruited to the forming phagosome [90].

$PI(3,4,5)P_3$ production by class I PI3K was shown to be important for phagosome closure [91, 92]. $PI(3,4,5)P_3$ accumulates at the phagocytic cup after particle binding and persists on the sealed phagosome for about 1 min [93]. PI3K activity is thought to be important for the delivery of endomembrane vesicles necessary for the completion of ingestion, but not for actin polymerization [94]. Its inhibition by wortmannin or LY294002 prevented the uptake of large particles (>3 μm diameter) but not of smaller particles [91, 92]. $PI(3,4,5)P_3$ has also been shown to be necessary for the recruitment of myosin X, which has a role in propelling pseudopod extension and sealing the phagosome [95]. The 5′-phosphatase SHIP [96, 97] has been shown to terminate the $PI(3,4,5)P_3$ signal.

Phosphatidylinositol 3-phosphate (PI(3)P), the inositide monophosphorylated in position 3′, has also been shown to play a pivotal role in phagocytosis, particularly in phagosome maturation. PI(3)P originates mainly from the phosphorylation of PI by class III PI3K (Vps34), but can also be formed by dephosphorylation of $PI(3,5)P_2$ by 5-phosphatases or by sequential dephosphorylation of $PI(3,4,5)P_3$ by 4- and 5- phosphatases [98, 99]. PI3P was detected in the phagosomal membrane about 1 min after sealing and persisted for about 10 min [100]. This inositide is known to play a role in maturation because blocking its production using anti-class III PI3K antibodies or wortmannin prevented the fusion of phagosomes with late endosomes/lysosomes [93, 101]. PI3P is also important for the activation of the NADPH oxidase at the phagosome by recruiting the p40 subunit [102, 103].

Other glycerophospholipids and sphingolipids also contribute to signal phagocytosis. Some of these lipids are important to anchor specific signaling and effector proteins. In addition, several lipid species act in conjunction to provide negative charge to membranes and thereby promote the recruitment of proteins with polycationic motifs. Phosphatidic acid (PA) is a glycerophospholipid that is found at the phagosomal cup and on nascent phagosomes [104]. It may have a role during phagocytosis by activating PIP5K [105, 106] or by being dephosphorylated to DAG [107]. Phosphatidylserine (PS) is another glycerophospholipid with multiple roles in phagocytosis. When exposed extracellularly it identifies apoptotic cells as targets for ingestion [108]. On the inner leaflet of the membrane PS can recruit enzymes with C2 domains, such as classic and novel PKC isoforms [109] and appears also to be important for NADPH oxidase activity [110, 111]. Cytosolic phospholipase A_2 ($cPLA_2$) catalyzes the hydrolysis of glycerophospholipids to generate arachidonic acid [112], which can activate the NADPH oxidase [113, 114]. Sphingosine kinase phosphorylates sphingosine to yield sphingosine-1-phosphate. This kinase has been shown to localize to phagocytic sites and its inhibition blocks the acidification of lysosomes as well as the acquisition of lysosomal markers [115, 116].

Signaling pathways activated during FcγR-mediated uptake converge in the activation of monomeric GTPases of the Rho family. Rho-GTPases have been acknowledged to regulate cytoskeletal remodeling, transcriptional activation and superoxide production [117], impinging on multiple aspects of the phagocytic response. RhoA, Rac1 and Cdc42 have all been shown to accumulate at the nascent phagosome during FcγR-mediated ingestion [118] and their inhibition by *C. difficile* toxin B blocks phagocytosis [118, 119]. Rac1 and Cdc42 appear to be essential for phagocytosis since dominant inhibitory variants of both GTPases inhibit FcγR-mediated ingestion [118–120]. In addition, macrophages obtained from Rac1/2-null mice also had impaired FcγR-mediated uptake [121]. The role of RhoA during FcγR-mediated uptake has been controversial. Caron and Hall [118] showed that RhoA was dispensable for FcγR-mediated phagocytosis whereas a previous study showed that RhoA was required for FcγR-mediated uptake [122]. A recent study using primary bone marrow-derived macrophages showed that RhoA inactivation by TAT-C3 transferase, significantly reduced the number of particles ingested via FcγR [121]. These discrepancies may be due to the inherent differences between the primary cells and immortalized cell lines used in the studies.

The exact function for each one of the Rho-GTPases during FcγR-mediated phagocytosis is unclear. However, they have been suggested to mediate different processes during phagosome formation. In one study that examined the structural effects on phagocytosis mediated by FcεRI, a receptor with signaling properties similar to those of FcγRs [71], inhibition of Cdc42 function interfered with pseudopod extension [123]. On the other hand, Rac1 inhibition prevented pseudopod fusion and phagosome closure [119]. Both Rac1 and Cdc42 interact with effectors that control actin rearrangement, such as WASP that in turn binds and activates the actin-nucleating activity of the Arp2/3 complex [124]. In fact, the Arp2/3 complex localizes to FcγR-mediated phagosomes [125]. The specific role of RhoA is unknown, but is thought to participate in a step distinct from the actin polymerization needed for cup formation and closure [121].

Actin rearrangement plays a critical role during phagocytosis. Formation of new F-actin filaments, increase of filament length, as well as crosslinking and bundling of filaments are all thought to occur [126]. The major actin nucleator is the Arp2/3 complex, which, as stated above, has been shown to localize to phagosomes and to be required for particle uptake [125]. WASP proteins and the related Scar/WAVE proteins are able to bind and activate the Arp2/3 complex. Cdc42 and $PI(4,5)P_2$ synergize to activate the WASP homolog N-WASP, which triggers actin polymerization via Arp2/3 [127]. Acting bundling by α-actinin is also thought to occur during the course of phagosome formation [128]. Myosins are thought to provide the contractile force for particle internalization. Myosin II, myosin IC, myosin V, myosin IXb and myosin X have all been implicated in FcγR-mediated particle uptake [129, 130].

The coordination between the polymerization and the depolymerization of actin is crucial for the proper formation and remodeling of actin structures. Some of the proteins important for actin disassembly are gelsolin and cofilin. These two severing proteins localize to phagosomes [131, 132] and inhibition of cofilin causes severe defects in phagocytosis [133].

3.3.2
Complement Receptor-Mediated Phagocytosis

The CR3 complement receptor is the second best characterized phagocytic receptor. It is an $\alpha_M\beta_2$ integrin and it has been largely recognized to bind particles opsonized with the complement component C3bi. Several studies have suggested that the mechanisms of ingestion elicited by the FcγR and CR3 receptors are different. Electron microcopy studies showed that while during FcγR-mediated uptake thin, actin-enriched pseudopods extend around the target particle, complement-opsonized particles passively sink upon CR3 engagement [134]. Recent studies [121], however, yielded differing results, showing that CR3-induced uptake also triggered the formation of thin pseudopods around the target, which were similar in length to those found during FcγR-mediated ingestion. The reasons for this discrepancy are unclear.

To initiate phagocytosis CR3 receptors need an activating signal that can be provided by chemokines, tumor necrosis factor α (TNFα) or by cell adhesion to fibronectin-coated surfaces [2]. The activation elicited by these agents may involve stimulation of protein kinase C (PKC) and likely induces an inside-out change in the conformation of the integrin that facilitates ligand binding [135]. The signaling pathway activated after CR3 engagement is not precisely defined. CR3-mediated uptake has been shown to be sensitive to PKC inhibitors but, unlike FcγR-mediated ingestion, is not affected by inhibition of Src-family tyrosine kinases [128].

Reorganization of the actin cytoskeleton also plays a fundamental role during CR3-mediated uptake. The Arp2/3 complex [125] as well as the formin mammalian diaphanous (mDia) [136] are recruited to sites of CR3-mediated phagocytosis. The Rho-GTPases responsible for these effects have not been unambiguously identified. Caron and colleagues reported that the inhibition of RhoA by C3 transferase hindered CR3-mediated uptake and that the expression of dominant-negative mutants of Cdc42 and Rac1 was without effect [118]. They therefore concluded that RhoA, but not Rac1 or Cdc42, was essential for CR3-mediated phagocytosis. On the other hand, while confirming the role of RhoA, a more recent analysis found Rac1 to be equally important [121]. In this study Rac1/2-null macrophages were found to have marked defects in Arp2/3 recruitment and actin polymerization during CR3-mediated uptake, and a constitutively active form of Rac1 rescued these defects [121]. Clearly, additional work is required to validate the role of the individual Rho-family GTPases in complement-mediated phagocytosis.

3.4 Membrane Dynamics During Phagocytosis

Phagocytes are able to internalize multiple large particles. In extreme cases the total volume internalized can be comparable to the phagocyte's own size. In the process, considerable expanses of membrane become internalized and it is safe to speculate that this is compensated in some way to maintain the cell's features. Indeed, experiments measuring the surface area of macrophages with the fluorescent dye FM1-43 showed that the total surface area of the phagocyte in fact *increases* in the course of particle uptake during phagocytosis [137]. Likewise, electrophysiological measurements of membrane capacitance, an index of the plasma membrane area, showed gradual increases during phagocytosis [138]. Together these data are a strong indication that exocytic events take place to compensate for membrane loss during particle uptake.

In support of this concept, several lines of evidence showed that the fusogenic machinery is important for phagocytosis. SNARE proteins are essential components of this machinery and their inhibition by treatment with tetanus neurotoxin [137] or the transfection of a dominant negative form of the ATPase NSF, an essential regulator of membrane fusion [81], impaired phagocytosis. The source of endomembranes has been investigated monitoring the fate of specific organellar markers. VAMP3, a SNARE protein found in recycling endosomes, was shown to be delivered to sites of phagocytosis [139, 140], indicating that such endosomes are a likely source of membrane. In addition, VAMP7 [141] as well as LAMP1, which localize to late endosomes and lysosomes, were recruited to sites of phagocytosis [139]. Moreover, it was also found that VAMP7, which has an essential role in the exocytosis of late endocytic compartments, is required for pseudopod extension [139]. In accordance with these findings, synaptotagmin VII, a calcium-binding protein important for the fusion of VAMP7-containing vesicles [142], was reported to support some forms of phagocytosis. Macrophages derived from synaptotagmin VII-knockout mice showed markedly deficient phagocytosis when large phagocytic loads were used [143]. This may indicate that the exocytosis of late endocytic/lysosomal vesicles is only required when a large amount of membrane is internalized and needs to be replaced [144].

The traffic of vesicles, in particular the tethering and docking of endomembranes, is thought to be controlled by GTPases of the Rab and Arf families. Rab11 and Arf6 have been shown to regulate the exocytosis of recycling endosomes during phagocytosis. Interestingly, dominant-negative mutants of Rab11 that impair secretion also block phagocytosis [145]. Arf6 mutants were also found to reduce the phagocytic efficiency and to prevent the focal delivery of VAMP3 to the phagocytic cup [146, 147].

The endoplasmic reticulum (ER) has been suggested to be a source of endomembrane during phagocytosis. However, the experimental evidence has been the subject of debate. Direct fusion of the ER with nascent phagosomes was suggested, based on electron microscopy and immunocytochemical methods [148]. In addition, the uptake of large particles was depressed by targeting the ER-localized SNARE ERS-24/Sec22 [149]. Yet, the role of ERS-24/Sec22 in phagocytosis is controversial, since another study found that impairment of this SNARE was without effect on phagocytosis [150]. The same study showed that interference with syntaxin18 and D12, two other ER SNAREs, modestly inhibited phagocytosis [150]. More recently, the work of Touret and colleagues [151] reassessed the contribution of the ER to phagosome formation and maturation by different means, including biochemical and immunological assays, plus fluorescence imaging and electron microcopy. These authors failed to find any evidence that the ER contributes to the phagosomal membrane at any of the stage during phagocytosis. Instead they confirmed the canonical view that the phagosome is initially derived from the plasma membrane and undergoes fusion with endocytic organelles during its maturation.

The notion that the ER is a source of endomembranes is appealing since this largest of organelles could readily provide a sufficient amount of membrane for the engulfment of large and/or multiple particles. In addition, it may provide a path to account for the cross-presentation of antigens. However, even in this case the antigen would have to follow a complicated path. For cross-presentation the antigen needs to be proteolytically processed by cytoplasmic proteasomes. Therefore, antigen that is putatively internalized into the ER would have to make its way to the cytosol and then re-enter the ER via the TAP transporter for cross-presentation. In addition, the postulate that the phagosome is composed largely of ER is not readily compatible with the well-accepted fact that phagosomes have an acidic lumen. By contrast, fusion of phagosomes with the endocytic machinery readily accounts for this fact since, unlike the ER, endocytic organelles are rich in proton-pumping V-ATPases.

The complete internalization of targets requires the fission of the nascent phagosome from the plasma membrane. The actual process and proteins involved have been partially characterized. Myosins IC, II, V, IXb [152] and X [95] were shown to associate with nascent phagosomes [153]. More importantly, blocking myosin-derived contraction was found to inhibit phagosome formation [94, 130]. These data suggest that myosins provide the force required for scission of the nascent phagosome from the plasmalemma.

3.5 Phagosome Maturation

After a foreign particle is taken up, the recently formed phagosome undergoes a process of maturation. In essence, phagosomes mature through a series of fusion and fission events with the endocytic pathway. The result of this process is the gradual modification of the phagosomal membrane and contents, ultimately yielding a hybrid organelle known as the phagolysosome. The content of the phagolysosome is recognized by its very low pH, the presence of hydrolytic enzymes and bactericidal peptides and the ability to generate toxic oxidative compounds [154–156].

The initial steps in phagosome maturation involve interaction with early endosomes, followed by late endosomes and lysosomes. Early endosomes are characterized by their tubulovesicular structure and by being mildly acidic (pH ~6.0) and largely, though not entirely devoid of proteases [100, 157]. Recycling endosomes reside within a juxtanuclear location and are less acidic than early endosomes (pH ~6.5) [157–159]. Late endosomes are characterized by their multivesicular organization, are more acidic (pH 5.5) and also more enriched in hydrolytic enzymes [100, 157]. Lysosomes contain hydrolytic enzymes such as cathepsin D and possess a very low pH (4.5–5.5) [157]. The components of the endocytic pathway are identifiable biochemically by a distinct set of molecular markers. Early endosomes are characterized by the presence of the Rab-GTPase Rab5 [160] and the early endosome antigen 1 (EEA1) [161]; recycling endosomes can be identified by the presence of Rab11 [157, 162]; late endosomes contain Rab7, Rab9, lysobisphosphatidic acid and mannose-6-phosphate receptor [157, 163]; lysosomal-associated membrane proteins (LAMPs) as well as hydrolytic enzymes are common to lysosomes and late endosomes [163, 164].

During the course of sealing or shortly thereafter, the phagosome starts to fuse with subcompartments of the endocytic pathway. As a result, the phagosome progressively acquires the biochemical characteristics of each of the endocytic compartments that it interacts with. The kinetics of maturation differs somewhat depending on the target particle and type of phagocyte [165], but typically the phagosome exhibits the molecular features of early endosomes within 10 min of sealing and fuses with late endosomes after 10–30 min [93, 101]. The best accepted model of phagosome maturation is that proposed by Desjardins [166], in which phagosomes fuse transiently with the endocytic organelles by a "kiss and run" process. This refers to the exchange of components between phagosomes and endosomes by transient and partial contact, instead of a complete fusion of both organelles.

Two main families of proteins are crucial for phagosome maturation: Rab-GTPases and SNAREs. SNARE (soluble *N*-ethylmaleimide-sensitive factor-attachment protein receptor) proteins, the *N*-ethylmaleimide-sensitive factor (NSF) and NSF-attachment proteins (SNAPs) are essential components of the fusogenic machinery necessary for vesicular transport [167, 168]. SNARE proteins on vesicles (v-SNAREs) interact with their counterpart or target SNAREs (t-SNAREs) to form a stable protein complex that enables the interaction and fusion of two compartments [168]. The dissociation of the complex is promoted by an ATPase, the NSF factor [169]. Recent evidence shows that members of this fusogenic machinery are present in phagosomes and that their function is required for maturation. t-SNAREs (syntaxins 2, 3, 4, 8, 13), v-SNAREs (VAMP2, VAMP3) and the NSF factor have all been identified in phagosomes [101, 140, 170–172].

Rab-GTPases are important for the selective docking and tethering of vesicles and are recognized as key directors of membrane traffic [173]. They cycle between an inactive GDP-bound state and an active GTP-bound state. This cycle is controlled by GDP dissociation inhibitors (GDIs), GTP-exchange factors (GEFs) and GTPase-activating proteins (GAPs). Multiple Rabs have been found to associate to phagosomes, including Rab3, Rab4, Rab5, Rab7, Rab9, Rab10, Rab11 and Rab 14 [172], but their function is not always known. Rab5 is the best characterized of these Rab-GTPases. One of the recognized functions of active Rab5 (Rab5-GTP) is to recruit EEA1 [161], which has two separate Rab5-binding sites and has been proposed to tether two different membranes displaying Rab5 [174, 175]. EEA1 forms a complex with NSF and two other Rab5 effectors, Rabex-5 and Rabaptin-5. This complex interacts with syntaxin 13 (a t-SNARE) and by this means triggers membrane fusion [176]. Rab5 has also been demonstrated to interact with class I and class III PI3Ks [177], which are thought to control membrane traffic during endocytosis [178]. Class I PI3K is responsible for the synthesis of phosphatidylinositol 3,4-bisphosphate and PI(3,4,5)P_3. The precise function of class I PI3K in endocytic traffic is still unknown. In contrast, the mammalian version of class III PI3K, VPS34, is key to endosome progression. VPS34 generates PI(3)P on early endosomes [177] and different proteins with an active role in endocytic traffic are recruited to early endosomes by binding to this inositide. PI(3)P ligands contain FYVE or Phox homology (PX) domains and include proteins with fusogenic properties like EEA1 [179], Rabenosyn-5 [180] and Hrs [181–183], among others.

Rab5 is thought to play a role during the earlier stages of phagosome maturation. It has been shown to transiently associate to nascent phagosomes [184] and the expression of a dominant-negative form of Rab5 blocked the acquisition of EEA1 and LAMP1. The action of Rab5 is mediated in part by PI(3)P, which seems to be crucial for phagosome maturation. Inhibitory drugs or anti-VPS34 antibodies precluded phagosome maturation, arresting the process prior to the fusion with lysosomes [93]. Class III PI3K products seem to be important also for the production of reactive oxygen species [185]. Of note, the $p40^{phox}$ and $p47^{phox}$ subunits of the NADPH oxidase contain PX domains capable of binding PI(3)P, suggesting that this inositide is important for the recruitment of oxidase subunits to the phagosome membrane [186].

Rab7 has been shown to mediate the transition between early and late endosomes [164, 187], as well as the fusion of late endosomes with lysosomes [164]. RILP (Rab-interacting lysosomal protein) is a Rab7 effector and has been shown to target Rab7-containing vesicles to microtubules [188]. These two proteins associate also with phagosomes, where they appear to play a similar functional role. The specific roles of other Rab isoforms in phagosome maturation remain to be defined.

The removal of phagosome membrane and contents by fission of vesicles seems to be as important during maturation, as is vesicle fusion. The COPI complex and clathrin, known to be important for vesicle traffic between the ER and the Golgi apparatus [189, 190], have been shown to be present on endosomal membranes [191–193]. Brefeldin A treatment, which prevents the assembly of COPI complexes, disrupts traffic between early to late endosomes [194, 195]. Brefeldin also prevents recycling of some proteins from phagosomes [196]. However, phagolysosome formation and acidification are not affected by COPI inactivation, which suggests the existence of ancillary fission mechanisms [197]. Clathrin is similarly not essential for phagosome maturation.

The traffic of vesicles to and from phagosomes depends in part on microtubules and their associated motors. Motor proteins propel organelles along microtubules in a centripetal (directed to the minus end of microtubules) or centrifugal (towards the plus end of microtubules) manner. Dynein and kinesin typify these motors. Phagosomes display bidirectional motility, though they seem to be preferentially propelled in a centripetal direction [198]. Such movements are thought to be important in phagosome evolution, because disruption of microtubules impairs maturation [199, 200]. This was made apparent when cells were treated with the microtubule-disrupting agent nocodazole, which impaired phagosomal acquisition of the late/endosome marker LAMP2 [201, 202].

3.6
Inflammatory Signals Linked to Phagocytosis

The engulfment of particles could be accompanied by either anti-inflammatory or proinflammatory responses. The outcome depends largely on the phagocytic receptor(s) engaged. Thus, phagocytosis mediated by FcγR generates an inflammatory response, while CR3- or complement-mediated uptake does not often trigger proinflammatory mediators [203, 204]. Apoptotic cell uptake, which is important to avoid excessive tissue damage and to encourage cell regeneration, is also anti-inflammatory [205–207].

In addition to the phagocytic receptors themselves, inflammation can be regulated by additional factors, such as Toll-like receptors (TLRs). TLRs are type I transmembrane proteins that are evolutionarily conserved between insects and humans [208]. TLRs are related to IL-1 receptors in their cytoplasmic domain but their extracellular portions differ markedly. Eleven human TLRs and 13 mouse TLRs have been identified [75]. They are expressed differentially among immune cells [209] and recognize the molecular signatures of microbial pathogens known as pathogen-associated molecular patterns (PAMPs), which range from lipids, lipoproteins, proteins to nucleic acids derived from various microorganisms, including bacteria, viruses, protozoa and fungi [75]. Some of the PAMPs include mannans of the yeast cell wall, formylated peptides, bacterial cell wall components, such as lipopolysaccharide (LPS), lipopeptides, peptidoglycans and teichoic acids [210]. TLRs can be classified into two groups based on subcellular localization. TLR1, 2, 4, 5 and 6 are present on the plasma membrane, while the second group localize to intracellular compartments such as endosomes and includes TLR3, 7, 8 and 9 [75]. In addition to the classification based on location, TLRs can also be classified into several groups based on the types of PAMPs they recognize: TLR1, 2, 4 and 6 recognize lipids; TLR5 and 11 recognize protein ligands and TLR3, 7, 8 and 9 detect nucleic acids [211]. In order to discriminate among the vast collection of molecular patterns found in nature, TLRs function in cooperation with one another [212] and also with a number of other pattern-recognition and phagocytic receptors that establish a combinatorial repertoire of recognition specificity.

Though some debate exists, TLRs do not seem to be themselves phagocytic receptors, and for many TLRs ligand engagement occurs intracellularly. During phagocytosis initiated by other receptors, TLRs are recruited to the phagosomes, where they sample the contents, determine the nature of the pathogen and elicit a specific response [213, 214].

After recognizing PAMPs, TLRs activate intracellular signaling pathways that lead to the production of inflammatory cytokine genes such as TNFα, IL-6, IL-1β and IL-12. Moreover, TLRs also signal to upregulate costimulatory molecules and are capable of inducing type I interferon to elicit antiviral responses [215]. Ligation of TLRs induces their dimerization, which triggers the recruitment of the adapter proteins MyD88, TIRAP (or MAL), TRIF (or TICAM-1) and TRAM (or TICAM-2) [216]. MyD88 is a universal adapter shared by all TLRs except for TLR3. Its recruitment leads to the activation of MAP kinases (ERK, JNK, p38) and of the transcription factor NFκB, responsible for the expression of inflammatory cytokines. TIRAP is involved in the signaling pathway downstream of TLR2 and TLR4 and is MyD88-dependent. TRIF is recruited by TLR4 and TLR3 and activates an alternative pathway that results in activation of NFκB, MAPKs and of the transcription factor IRF-3. TRAM is involved in the signaling pathway activated by TRIF after TLR4 engagement. Together, TLRs trigger a specific combination of adapters in order to activate an appropriate and effective response [211].

Phagocytes such as macrophages and neutrophils produce reactive oxygen species as part of their pathogen killing mechanisms and these can contribute to inflammation. The production of these oxygen metabolites, primarily superoxide, is due to the activity of the NADPH oxidase on phagosomal membranes [217]. The NADPH oxidase activity can be induced by Fc-receptors during phagocytosis [218, 219]. However, not all phagocytic receptors activate the NADPH oxidase, since CR3-mediated uptake does not lead to superoxide production [204, 220]. Several stimuli that are not themselves phagocytic but frequently coexist with phagocytic signals can also activate the NADPH oxidase. These include chemoattractants (fMet-Leu-Phe, C5a, LTB4) and chemokines (IL-8) [221].

3.7 Concluding Remarks

This chapter presents an abbreviated and simplified summary of the events that occur during phagocytosis. While the host intends these responses to clear invading microorganisms, all too often these manage to evade or overcome the microbicidal responses and manage to survive and in some cases to proliferate within the phagocytic cell itself. Subsequent chapters in this volume will describe the individual, ingenious strategies that microbes have evolved to avoid and in some instances even co-opt the innate immune response. We hope that the information in this chapter serves the reader to better understand the mode of action of microbial effectors.

References

1 Fadok, V.A. and Henson, P.M. (1998) Apoptosis: getting rid of the bodies. *Curr Biol*, **8**, R693–R695.

2 Aderem, A. and Underhill, D.M. (1999) Mechanisms of phagocytosis in macrophages. *Annu Rev Immunol*, **17**, 593–623.

3 Rabinovitch, M. (1995) Professional and non-professional phagocytes: an introduction. *Trends Cell Biol*, **5**, 85–87.

4 Underhill, D.M. and Ozinsky, A. (2002) Phagocytosis of microbes: complexity in action. *Annu Rev Immunol*, **20**, 825–852.

5 Pluddemann, A., Mukhopadhyay, S. and Gordon, S. (2006) The interaction of macrophage receptors with bacterial ligands. *Expert Rev Mol Med*, **8**, 1–25.

6 Ravetch, J.V. and Bolland, S. (2001) IgG Fc receptors. *Annu Rev Immunol*, **19**, 275–290.

7 Bharadwaj, D., Stein, M.P., Volzer, M., Mold, C. and Du Clos, T.W. (1999) The major receptor for C-reactive protein on leukocytes is Fcgamma receptor II. *J Exp Med*, **190**, 585–590.

8 Bharadwaj, D., Mold, C., Markham, E. and Du Clos, T.W. (2001) Serum amyloid P component binds to Fc gamma receptors and opsonizes particles for phagocytosis. *J Immunol*, **166**, 6735–6741.

9 Mold, C., Gresham, H.D. and Du Clos, T.W. (2001) Serum amyloid P component and C-reactive protein mediate phagocytosis through murine Fc gamma Rs. *J Immunol*, **166**, 1200–1205.

10 Raghavan, M. and Bjorkman, P.J. (1996) Fc receptors and their interactions with immunoglobulins. *Annu Rev Cell Dev Biol*, **12**, 181–220.

11 Richards, M.L. and Katz, D.H. (1991) Biology and chemistry of low affinity IgE receptor (Fc epsilon RII/CD23). *Crit Rev Immunol*, **11**, 65–86.

12 van Spriel, A.B., van den Herik-Oudijk, I.E., van Sorge, N.M., Vile, H.A., van Strijp, J.A. and van de Winkel, J.G. (1999) Effective phagocytosis and killing of *Candida albicans* via targeting FcgammaRI (CD64) or FcalphaRI (CD89) on neutrophils. *J Infect Dis*, **179**, 661–669.

13 Ghiran, I., Barbashov, S.F., Klickstein, L.B., Tas, S.W., Jensenius, J.C. and Nicholson-Weller, A. (2000) Complement receptor 1/CD35 is a receptor for mannan-binding lectin. *J Exp Med*, **192**, 1797–1808.

14 Ehlers, M.R. (2000) CR3: a general purpose adhesion-recognition receptor essential for innate immunity. *Microbes Infect*, **2**, 289–294.

15 Ross, G.D., Reed, W., Dalzell, J.G., Becker, S.E. and Hogg, N. (1992) Macrophage cytoskeleton association with CR3 and CR4 regulates receptor mobility and phagocytosis of iC3b-opsonized erythrocytes. *J Leukoc Biol*, **51**, 109–117.

16 Blystone, S.D. and Brown, E.J. (1999) Integrin receptors of phagocytes, in *Phagocytosis* (ed. S. Gordon), JAI, Stamford, CT, pp. 103–147.

17 Platt, N. and Gordon, S. (2001) Is the class A macrophage scavenger receptor (SR-A) multifunctional? – The mouse's tale. *J Clin Invest*, **108**, 649–654.

18 van der Laan, L.J., Dopp, E.A., Haworth, R., Pikkarainen, T., Kangas, M., Elomaa, O., Dijkstra, C.D., Gordon, S., Tryggvason, K. and Kraal, G. (1999) Regulation and functional involvement of macrophage scavenger receptor MARCO in clearance of bacteria in vivo. *J Immunol*, **162**, 939–947.

19 Ezekowitz, R.A., Sastry, K., Bailly, P. and Warner, A. (1990) Molecular characterization of the human macrophage mannose receptor: demonstration of multiple carbohydrate recognition-like domains and phagocytosis of yeasts in Cos-1 cells. *J Exp Med*, **172**, 1785–1794.

20 Brown, G.D. and Gordon, S. (2001) Immune recognition. A new receptor for beta-glucans. *Nature*, **413**, 36–37.

21 Schiff, D.E., Kline, L., Soldau, K., Lee, J.D., Pugin, J., Tobias, P.S. and Ulevitch, R.J. (1997) Phagocytosis of gram-negative bacteria by a unique CD14-dependent mechanism. *J Leukoc Biol*, **62**, 786–794.

22 Dziarski, R., Tapping, R.I. and Tobias, P.S. (1998) Binding of bacterial peptidoglycan to CD14. *J Biol Chem*, **273**, 8680–8690.

23 Nepomuceno, R.R., Henschen-Edman, A.H., Burgess, W.H. and Tenner, A.J. (1997) cDNA cloning and primary structure analysis of C1qR(P), the human C1q/MBL/SPA receptor that mediates enhanced phagocytosis *in vitro*. *Immunity*, **6**, 119–129.

24 Vachon, E., Martin, R., Plumb, J., Kwok, V., Vandivier, R.W., Glogauer, M., Kapus, A., Wang, X., Chow, C.W., Grinstein, S. and Downey, G.P. (2006) CD44 is a phagocytic receptor. *Blood*, **107**, 4149–4158.

25 Janeway, C.A. Jr (1989) Approaching the asymptote? Evolution and revolution in immunology. *Cold Spring Harb Symp Quant Biol*, **54** (Pt 1), 1–13.

26 Medzhitov, R. and Janeway, C.A. Jr (1997) Innate immunity: the virtues of a nonclonal system of recognition. *Cell*, **91**, 295–298.

27 Raetz, C.R. (1990) Biochemistry of endotoxins. *Annu Rev Biochem*, **59**, 129–170.

28 Wicken, A.J. and Knox, K.W. (1975) Lipoteichoic acids: a new class of bacterial antigen. *Science*, **187**, 1161–1167.

29 Scheffers, D.J. and Pinho, M.G. (2005) Bacterial cell wall synthesis: new insights from localization studies. *Microbiol Mol Biol Rev*, **69**, 585–607.

30 Goldstein, J.L., Ho, Y.K., Basu, S.K. and Brown, M.S. (1979) Binding site on macrophages that mediates uptake and degradation of acetylated low density lipoprotein, producing massive cholesterol deposition. *Proc Natl Acad Sci USA*, **76**, 333–337.

31 Hampton, R.Y., Golenbock, D.T., Penman, M., Krieger, M. and Raetz, C.R. (1991) Recognition and plasma clearance of endotoxin by scavenger receptors. *Nature*, **352**, 342–344.

32 Greenberg, J.W., Fischer, W. and Joiner, K.A. (1996) Influence of lipoteichoic acid structure on recognition by the macrophage scavenger receptor. *Infect Immun*, **64**, 3318–3325.

33 Zhu, F.G., Reich, C.F. and Pisetsky, D.S. (2001) The role of the macrophage scavenger receptor in immune stimulation by bacterial DNA and synthetic oligonucleotides. *Immunology*, **103**, 226–234.

34 Elomaa, O., Kangas, M., Sahlberg, C., Tuukkanen, J., Sormunen, R., Liakka, A., Thesleff, I., Kraal, G. and Tryggvason, K. (1995) Cloning of a novel bacteria-binding receptor structurally related to scavenger receptors and expressed in a subset of macrophages. *Cell*, **80**, 603–609.

35 Sankala, M., Brannstrom, A., Schulthess, T., Bergmann, U., Morgunova, E., Engel, J., Tryggvason, K. and Pikkarainen, T. (2002) Characterization of recombinant soluble macrophage scavenger receptor MARCO. *J Biol Chem*, **277**, 33378–33385.

36 Shimaoka, T., Kume, N., Minami, M., Hayashida, K., Sawamura, T., Kita, T. and Yonehara, S. (2001) LOX-1 supports adhesion of Gram-positive and Gram-negative bacteria. *J Immunol*, **166**, 5108–5114.

37 Shimaoka, T., Nakayama, T., Kume, N., Takahashi, S., Yamaguchi, J., Minami, M., Hayashida, K., Kita, T., Ohsumi, J., Yoshie, O. and Yonehara, S. (2003) Cutting edge: SR-PSOX/CXC chemokine ligand 16 mediates bacterial phagocytosis by APCs through its chemokine domain. *J Immunol*, **171**, 1647–1651.

38 Adachi, H. and Tsujimoto, M. (2002) FEEL-1, a novel scavenger receptor with in vitro bacteria-binding and angiogenesis-modulating activities. *J Biol Chem*, **277**, 34264–34270.

39 Vishnyakova, T.G., Bocharov, A.V., Baranova, I.N., Chen, Z., Remaley, A.T., Csako, G., Eggerman, T.L. and Patterson, A.P. (2003) Binding and internalization of lipopolysaccharide by Cla-1, a human orthologue of rodent scavenger receptor B1. *J Biol Chem*, **278**, 22771–22780.

40 Hoebe, K., Georgel, P., Rutschmann, S., Du, X., Mudd, S., Crozat, K., Sovath, S., Shamel, L., Hartung, T., Zahringer, U. and Beutler, B. (2005) CD36 is a sensor of diacylglycerides. *Nature*, **433**, 523–527.

41 Dziarski, R. (2003) Recognition of bacterial peptidoglycan by the innate immune system. *Cell Mol Life Sci*, **60**, 1793–1804.

42 Fujihara, M., Muroi, M., Tanamoto, K., Suzuki, T., Azuma, H. and Ikeda, H. (2003) Molecular mechanisms of macrophage activation and deactivation by lipopolysaccharide: roles of the receptor complex. *Pharmacol Ther*, **100**, 171–194.

43 Bernardo, J., Billingslea, A.M., Blumenthal, R.L., Seetoo, K.F., Simons, E.R. and Fenton, M.J. (1998) Differential responses of human mononuclear phagocytes to mycobacterial lipoarabinomannans: role of CD14 and the mannose receptor. *Infect Immun*, **66**, 28–35.

44 Flo, T.H., Ryan, L., Kilaas, L., Skjak-Braek, G., Ingalls, R.R., Sundan, A., Golenbock, D.T. and Espevik, T. (2000) Involvement of CD14 and beta2-integrins in activating cells with soluble and particulate lipopolysaccharides and mannuronic acid polymers. *Infect Immun*, **68**, 6770–6776.

45 McGreal, E.P., Martinez-Pomares, L. and Gordon, S. (2004) Divergent roles for C-type lectins expressed by cells of the innate immune system. *Mol Immunol*, **41**, 1109–1121.

46 East, L. and Isacke, C.M. (2002) The mannose receptor family. *Biochim Biophys Acta*, **1572**, 364–386.

47 Mullin, N.P., Hall, K.T. and Taylor, M.E. (1994) Characterization of ligand binding to a carbohydrate-recognition domain of the macrophage mannose receptor. *J Biol Chem*, **269**, 28405–28413.

48 Zamze, S., Martinez-Pomares, L., Jones, H., Taylor, P.R., Stillion, R.J., Gordon, S. and Wong, S.Y. (2002) Recognition of bacterial capsular polysaccharides and lipopolysaccharides by the macrophage mannose receptor. *J Biol Chem*, **277**, 41613–41623.

49 Taylor, P.R., Martinez-Pomares, L., Stacey, M., Lin, H.H., Brown, G.D. and Gordon, S. (2005) Macrophage receptors and immune recognition. *Annu Rev Immunol*, **23**, 901–944.

50 Adachi, Y., Ishii, T., Ikeda, Y., Hoshino, A., Tamura, H., Aketagawa, J., Tanaka, S. and Ohno, N. (2004) Characterization of beta-glucan recognition site on C-type lectin, dectin 1. *Infect Immun*, **72**, 4159–4171.

51 Herre, J., Gordon, S. and Brown, G.D. (2004) Dectin-1 and its role in the recognition of beta-glucans by macrophages. *Mol Immunol*, **40**, 869–876.

52 Stokes, R.W., Haidl, I.D., Jefferies, W.A. and Speert, D.P. (1993) Mycobacteria-macrophage interactions. Macrophage phenotype determines the nonopsonic binding of Mycobacterium tuberculosis to murine macrophages. *J Immunol*, **151**, 7067–7076.

53 Hetland, G. and Wiker, H.G. (1994) Antigen 85C on Mycobacterium bovis. BCG and M. tuberculosis promotes monocyte-CR3-mediated uptake of microbeads coated with mycobacterial products. *Immunology*, **82**, 445–449.

54 Thornton, B.P., Vetvicka, V., Pitman, M., Goldman, R.C. and Ross, G.D. (1996) Analysis of the sugar specificity and molecular location of the beta-glucan-binding lectin site of complement receptor type 3 (CD11b/CD18). *J Immunol*, **156**, 1235–1246.

55 Zaffran, Y., Zhang, L. and Ellner, J.J. (1998) Role of CR4 in Mycobacterium tuberculosis-human macrophages binding and signal transduction in the absence of serum. *Infect Immun*, **66**, 4541–4544.

56 Ishibashi, Y., Claus, S. and Relman, D.A. (1994) Bordetella pertussis filamentous hemagglutinin interacts with a leukocyte signal transduction complex and stimulates bacterial adherence to monocyte CR3 (CD11b/CD18). *J Exp Med*, **180**, 1225–1233.

57 Pollard, A.J., Currie, A., Rosenberger, C.M., Heale, J.P., Finlay, B.B. and Speert, D.P. (2004) Differential post-transcriptional activation of human phagocytes by different Pseudomonas aeruginosa isolates. *Cell Microbiol*, **6**, 639–650.

58 Takeda, K., Kaisho, T. and Akira, S. (2003) Toll-like receptors. *Annu Rev Immunol*, **21**, 335–376.

59 Stenger, S. and Modlin, R.L. (2002) Control of *Mycobacterium tuberculosis* through mammalian Toll-like receptors. *Curr Opin Immunol*, **14**, 452–457.

60 Takeuchi, O., Sato, S., Horiuchi, T., Hoshino, K., Takeda, K., Dong, Z., Modlin, R.L. and Akira, S. (2002) Cutting edge: role of Toll-like receptor 1 in mediating immune response to microbial lipoproteins. *J Immunol*, **169**, 10–14.

61 Iwaki, D., Mitsuzawa, H., Murakami, S., Sano, H., Konishi, M., Akino, T. and Kuroki, Y. (2002) The extracellular toll-like receptor 2 domain directly binds peptidoglycan derived from *Staphylococcus aureus*. *J Biol Chem*, **277**, 24315–24320.

62 da Silva, C.J., Soldau, K., Christen, U., Tobias, P.S. and Ulevitch, R.J. (2001) Lipopolysaccharide is in close proximity to each of the proteins in its membrane receptor complex. transfer from CD14 to TLR4 and MD-2. *J Biol Chem*, **276**, 21129–21135.

63 Dziarski, R., Wang, Q., Miyake, K., Kirschning, C.J. and Gupta, D. (2001) MD-2 enables Toll-like receptor 2 (TLR2)-mediated responses to lipopolysaccharide and enhances TLR2-mediated responses to Gram-positive and Gram-negative bacteria and their cell wall components. *J Immunol*, **166**, 1938–1944.

64 Hayashi, F., Smith, K.D., Ozinsky, A., Hawn, T.R., Yi, E.C., Goodlett, D.R., Eng, J.K., Akira, S., Underhill, D.M. and Aderem, A. (2001) The innate immune response to bacterial flagellin is mediated by Toll-like receptor 5. *Nature*, **410**, 1099–1103.

65 Reichhart, J.M. (2003) TLR5 takes aim at bacterial propeller. *Nat Immunol*, **4**, 1159–1160.

66 Dalpke, A., Frank, J., Peter, M. and Heeg, K. (2006) Activation of toll-like receptor 9 by DNA from different bacterial species. *Infect Immun*, **74**, 940–946.

67 Ulevitch, R.J., Mathison, J.C. and da Silva, C.J. (2004) Innate immune responses during infection. *Vaccine*, **22** (Suppl 1), S25–S30.

68 Peiser, L. and Gordon, S. (2001) Phagocytosis: enhancement, in *Encyclopedia of Life Sciences*, John Wiley & Sons, Ltd, Chichester, pp. 1–8.

69 Pommier, C.G., Inada, S., Fries, L.F., Takahashi, T., Frank, M.M. and Brown, E.J. (1983) Plasma fibronectin enhances phagocytosis of opsonized particles by human peripheral blood monocytes. *J Exp Med*, **157**, 1844–1854.

70 Wright, S.D. and Griffin, F.M. Jr (1985) Activation of phagocytic cells' C3 receptors for phagocytosis. *J Leukoc Biol*, **38**, 327–339.

71 Daeron, M. (1997) Fc receptor biology. *Annu Rev Immunol*, **15**, 203–234.

72 Epstein, J., Eichbaum, Q., Sheriff, S. and Ezekowitz, R.A. (1996) The collectins in innate immunity. *Curr Opin Immunol*, **8**, 29–35.

73 Tenner, A.J., Robinson, S.L. and Ezekowitz, R.A. (1995) Mannose binding protein (MBP) enhances mononuclear phagocyte function via a receptor that contains the 126,000 M(r) component of the C1q receptor. *Immunity*, **3**, 485–493.

74 Stuart, L.M. and Ezekowitz, R.A. (2005) Phagocytosis: elegant complexity. *Immunity*, **22**, 539–550.

75 Akira, S., Uematsu, S. and Takeuchi, O. (2006) Pathogen recognition and innate immunity. *Cell*, **124**, 783–801.

76 Greenberg, S., El Khoury, J., di Virgilio, F., Kaplan, E.M. and Silverstein, S.C. (1991) Ca^{2+}-independent F-actin assembly and disassembly during Fc receptor-mediated phagocytosis in mouse macrophages. *J Cell Biol*, **113**, 757–767.

77 Ghazizadeh, S., Bolen, J.B. and Fleit, H.B. (1994) Physical and functional association of Src-related protein tyrosine kinases with Fc gamma RII in monocytic THP-1 cells. *J Biol Chem*, **269**, 8878–8884.

78 Gu, H., Botelho, R.J., Yu, M., Grinstein, S. and Neel, B.G. (2003) Critical role for scaffolding adapter Gab2 in Fc gamma R-mediated phagocytosis. *J Cell Biol*, **161**, 1151–1161.

79 Lee, W.L., Cosio, G., Ireton, K. and Grinstein, S. (2007) Role of CrkII in Fcgamma receptor-mediated phagocytosis. *J Biol Chem*, **282**, 11135–11143.

80 Krause, M., Sechi, A.S., Konradt, M., Monner, D., Gertler, F.B. and Wehland, J. (2000) Fyn-binding protein (Fyb)/SLP-76-associated protein (SLAP), Ena/vasodilator-stimulated phosphoprotein (VASP) proteins and the Arp2/3 complex link T cell receptor (TCR) signaling to the actin cytoskeleton. *J Cell Biol*, **149**, 181–194.

81 Coppolino, M.G., Kong, C., Mohtashami, M., Schreiber, A.D., Brumell, J.H., Finlay, B.B., Grinstein, S. and Trimble, W.S. (2001) Requirement for N-ethylmaleimide-sensitive factor activity at different stages of bacterial invasion and phagocytosis. *J Biol Chem*, **276**, 4772–4780.

82 Strzelecka-Kiliszek, A., Korzeniowski, M., Kwiatkowska, K., Mrozinska, K. and Sobota, A. (2004) Activated FcgammaRII and signalling molecules revealed in rafts by ultra-structural observations of plasma-membrane sheets. *Mol Membr Biol*, **21**, 101–108.

83 Kwiatkowska, K. and Sobota, A. (2001) The clustered Fcgamma receptor II is recruited to Lyn-containing membrane domains and undergoes phosphorylation in a cholesterol-dependent manner. *Eur J Immunol*, **31**, 989–998.

84 Martin, T.F. (1998) Phosphoinositide lipids as signaling molecules: common themes for signal transduction, cytoskeletal regulation, and membrane trafficking. *Annu Rev Cell Dev Biol*, **14**, 231–264.

85 Itoh, T. and Takenawa, T. (2002) Phosphoinositide-binding domains: Functional units for temporal and spatial regulation of intracellular signalling. *Cell Signal*, **14**, 733–743.

86 Lemmon, M.A. (2003) Phosphoinositide recognition domains. *Traffic*, **4**, 201–213.

87 Takenawa, T. and Itoh, T. (2001) Phosphoinositides, key molecules for regulation of actin cytoskeletal organization and membrane traffic from the plasma membrane. *Biochim Biophys Acta*, **1533**, 190–206.

88 Botelho, R.J., Teruel, M., Dierckman, R., Anderson, R., Wells, A., York, J.D., Meyer, T. and Grinstein, S. (2000) Localized biphasic changes in phosphatidylinositol-4,5-bisphosphate at sites of phagocytosis. *J Cell Biol*, **151**, 1353–1368.

89 Coppolino, M.G., Dierckman, R., Loijens, J., Collins, R.F., Pouladi, M., Jongstra-Bilen, J., Schreiber, A.D., Trimble, W.S., Anderson, R. and Grinstein, S. (2002) Inhibition of phosphatidylinositol-4-phosphate 5-kinase Ialpha impairs localized actin remodeling and suppresses phagocytosis. *J Biol Chem*, **277**, 43849–43857.

90 Larsen, E.C., Ueyama, T., Brannock, P.M., Shirai, Y., Saito, N., Larsson, C., Loegering, D., Weber, P.B. and Lennartz, M.R. (2002) A role for PKC-epsilon in Fc gammaR-mediated phagocytosis by RAW 264.7 cells. *J Cell Biol*, **159**, 939–944.

91 Araki, N., Johnson, M.T. and Swanson, J.A. (1996) A role for phosphoinositide 3-kinase in the completion of macropinocytosis and phagocytosis by macrophages. *J Cell Biol*, **135**, 1249–1260.

92 Cox, D., Tseng, C.C., Bjekic, G. and Greenberg, S. (1999) A requirement for phosphatidylinositol 3-kinase in pseudopod extension. *J Biol Chem*, **274**, 1240–1247.

93 Vieira, O.V., Botelho, R.J., Rameh, L., Brachmann, S.M., Matsuo, T., Davidson, H.W., Schreiber, A., Backer, J.M., Cantley, L.C. and Grinstein, S. (2001) Distinct roles of class I and class III phosphatidylinositol 3-kinases in phagosome formation and maturation. *J Cell Biol*, **155**, 19–25.

94 Araki, N., Hatae, T., Furukawa, A. and Swanson, J.A. (2003) Phosphoinositide-3-kinase-independent contractile activities associated with Fcgamma-receptor-mediated phagocytosis and macropinocytosis in macrophages. *J Cell Sci*, **116**, 247–257.

95 Cox, D., Berg, J.S., Cammer, M., Chinegwundoh, J.O., Dale, B.M., Cheney, R.E. and Greenberg, S. (2002) Myosin X is a downstream effector of PI(3)K during phagocytosis. *Nat Cell Biol*, **4**, 469–477.

96 Nakamura, K., Malykhin, A. and Coggeshall, K.M. (2002) The Src homology 2 domain-containing inositol 5-phosphatase negatively regulates Fcgamma receptor-mediated phagocytosis through immunoreceptor tyrosine-based activation motif-bearing phagocytic receptors. *Blood*, **100**, 3374–3382.

97 Cox, D., Dale, B.M., Kashiwada, M., Helgason, C.D. and Greenberg, S. (2001) A regulatory role for Src homology 2 domain-containing inositol 5′-phosphatase (SHIP) in phagocytosis mediated by Fc gamma receptors and complement receptor 3 (alpha(M)beta(2); CD11b/CD18). *J Exp Med*, **193**, 61–71.

98 Shin, H.W., Hayashi, M., Christoforidis, S., Lacas-Gervais, S., Hoepfner, S., Wenk, M.R., Modregger, J., Uttenweiler-Joseph, S., Wilm, M., Nystuen, A., Frankel, W.N., Solimena, M., De, C.P. and Zerial, M.

(2005) An enzymatic cascade of Rab5 effectors regulates phosphoinositide turnover in the endocytic pathway. *J Cell Biol*, **170**, 607–618.

99 Nicot, A.S., Fares, H., Payrastre, B., Chisholm, A.D., Labouesse, M. and Laporte, J. (2006) The phosphoinositide kinase PIKfyve/Fab1p regulates terminal lysosome maturation in *Caenorhabditis elegans*. *Mol Biol Cell*, **17**, 3062–3074.

100 Vieira, O.V., Botelho, R.J. and Grinstein, S. (2002) Phagosome maturation: aging gracefully. *Biochem J*, **366**, 689–704.

101 Fratti, R.A., Backer, J.M., Gruenberg, J., Corvera, S. and Deretic, V. (2001) Role of phosphatidylinositol 3-kinase and Rab5 effectors in phagosomal biogenesis and mycobacterial phagosome maturation arrest. *J Cell Biol*, **154**, 631–644.

102 Suh, C.I., Stull, N.D., Li, X.J., Tian, W., Price, M.O., Grinstein, S., Yaffe, M.B., Atkinson, S. and Dinauer, M.C. (2006) The phosphoinositide-binding protein $p40^{phox}$ activates the NADPH oxidase during FcgammaIIA receptor-induced phagocytosis. *J Exp Med*, **203**, 1915–1925.

103 Ellson, C.D., Davidson, K., Ferguson, G.J., O'Connor, R., Stephens, L.R. and Hawkins, P.T. (2006) Neutrophils from $p40^{phox-/-}$ mice exhibit severe defects in NADPH oxidase regulation and oxidant-dependent bacterial killing. *J Exp Med*, **203**, 1927–1937.

104 Corrotte, M., Chasserot-Golaz, S., Huang, P., Du, G., Ktistakis, N.T., Frohman, M.A., Vitale, N., Bader, M.F. and Grant, N.J. (2006) Dynamics and function of phospholipase D and phosphatidic acid during phagocytosis. *Traffic*, **7**, 365–377.

105 Ktistakis, N.T., Delon, C., Manifava, M., Wood, E., Ganley, I. and Sugars, J.M. (2003) Phospholipase D1 and potential targets of its hydrolysis product, phosphatidic acid. *Biochem Soc Trans*, **31**, 94–97.

106 Jones, D.R., Sanjuan, M.A. and Merida, I. (2000) Type Ialpha phosphatidylinositol 4-phosphate 5-kinase is a putative target for increased intracellular phosphatidic acid. *FEBS Lett*, **476**, 160–165.

107 Kent, C. (1995) Eukaryotic phospholipid biosynthesis. *Annu Rev Biochem*, **64**, 315–343.

108 Marguet, D., Luciani, M.F., Moynault, A., Williamson, P. and Chimini, G. (1999) Engulfment of apoptotic cells involves the redistribution of membrane phosphatidylserine on phagocyte and prey. *Nat Cell Biol*, **1**, 454–456.

109 Corbalan-Garcia, S. and Gomez-Fernandez, J.C. (2006) Protein kinase C regulatory domains: the art of decoding many different signals in membranes. *Biochim Biophys Acta*, **1761**, 633–654.

110 Bey, E.A., Xu, B., Bhattacharjee, A., Oldfield, C.M., Zhao, X., Li, Q., Subbulakshmi, V., Feldman, G.M., Wientjes, F.B. and Cathcart, M.K. (2004) Protein kinase C delta is required for $p47^{phox}$ phosphorylation and translocation in activated human monocytes. *J Immunol*, **173**, 5730–5738.

111 Babior, B.M., Lambeth, J.D. and Nauseef, W. (2002) The neutrophil NADPH oxidase. *Arch Biochem Biophys*, **397**, 342–344.

112 Diaz, B.L. and Arm, J.P. (2003) Phospholipase A_2. *Prostaglandins Leukot Essent Fatty Acids*, **69**, 87–97.

113 Li, Q. and Cathcart, M.K. (1997) Selective inhibition of cytosolic phospholipase A2 in activated human monocytes. Regulation of superoxide anion production and low density lipoprotein oxidation. *J Biol Chem*, **272**, 2404–2411.

114 Shmelzer, Z., Haddad, N., Admon, E., Pessach, I., Leto, T.L., Eitan-Hazan, Z., Hershfinkel, M. and Levy, R. (2003) Unique targeting of cytosolic phospholipase A2 to plasma membranes mediated by the NADPH oxidase in phagocytes. *J Cell Biol*, **162**, 683–692.

115 Thompson, C.R., Iyer, S.S., Melrose, N., VanOosten, R., Johnson, K., Pitson, S.M., Obeid, L.M. and Kusner, D.J. (2005) Sphingosine kinase 1 (SK1) is recruited to nascent phagosomes in human

macrophages: inhibition of SK1 translocation by Mycobacterium tuberculosis. *J Immunol*, **174**, 3551–3561.

116 Malik, Z.A., Thompson, C.R., Hashimi, S., Porter, B., Iyer, S.S. and Kusner, D.J. (2003) Cutting edge: Mycobacterium tuberculosis blocks Ca^{2+} signaling and phagosome maturation in human macrophages via specific inhibition of sphingosine kinase. *J Immunol*, **170**, 2811–2815.

117 Aspenstrom, P. (1999) Effectors for the Rho GTPases. *Curr Opin Cell Biol*, **11**, 95–102.

118 Caron, E. and Hall, A. (1998) Identification of two distinct mechanisms of phagocytosis controlled by different Rho GTPases. *Science*, **282**, 1717–1721.

119 Massol, P., Montcourrier, P., Guillemot, J.C. and Chavrier, P. (1998) Fc receptor-mediated phagocytosis requires CDC42 and Rac1. *EMBO J*, **17**, 6219–6229.

120 Cox, D., Chang, P., Zhang, Q., Reddy, P.G., Bokoch, G.M. and Greenberg, S. (1997) Requirements for both Rac1 and Cdc42 in membrane ruffling and phagocytosis in leukocytes. *J Exp Med*, **186**, 1487–1494.

121 Hall, A.B., Gakidis, M.A., Glogauer, M., Wilsbacher, J.L., Gao, S., Swat, W. and Brugge, J.S. (2006) Requirements for Vav guanine nucleotide exchange factors and Rho GTPases in FcgammaR- and complement-mediated phagocytosis. *Immunity*, **24**, 305–316.

122 Hackam, D.J., Rotstein, O.D., Schreiber, A., Zhang, W. and Grinstein, S. (1997) Rho is required for the initiation of calcium signaling and phagocytosis by Fcgamma receptors in macrophages. *J Exp Med*, **186**, 955–966.

123 Chimini, G. and Chavrier, P. (2000) Function of Rho family proteins in actin dynamics during phagocytosis and engulfment. *Nat Cell Biol*, **2**, E191–E196.

124 Machesky, L.M. and Insall, R.H. (1998) Scar1 and the related Wiskott-Aldrich syndrome protein, WASP, regulate the actin cytoskeleton through the Arp2/3 complex. *Curr Biol*, **8**, 1347–1356.

125 May, R.C., Caron, E., Hall, A. and Machesky, L.M. (2000) Involvement of the Arp2/3 complex in phagocytosis mediated by FcgammaR or CR3. *Nat Cell Biol*, **2**, 246–248.

126 Pollard, T.D., Blanchoin, L. and Mullins, R.D. (2000) Molecular mechanisms controlling actin filament dynamics in nonmuscle cells. *Annu Rev Biophys Biomol Struct*, **29**, 545–576.

127 Rohatgi, R., Ma, L., Miki, H., Lopez, M., Kirchhausen, T., Takenawa, T. and Kirschner, M.W. (1999) The interaction between N-WASP and the Arp2/3 complex links Cdc42-dependent signals to actin assembly. *Cell*, **97**, 221–231.

128 Allen, L.A. and Aderem, A. (1996) Molecular definition of distinct cytoskeletal structures involved in complement- and Fc receptor-mediated phagocytosis in macrophages. *J Exp Med*, **184**, 627–637.

129 Mansfield, P.J., Shayman, J.A. and Boxer, L.A. (2000) Regulation of polymorphonuclear leukocyte phagocytosis by myosin light chain kinase after activation of mitogen-activated protein kinase. *Blood*, **95**, 2407–2412.

130 Swanson, J.A., Johnson, M.T., Beningo, K., Post, P., Mooseker, M. and Araki, N. (1999) A contractile activity that closes phagosomes in macrophages. *J Cell Sci*, **112** (Pt 3), 307–316.

131 Yin, H.L., Albrecht, J.H. and Fattoum, A. (1981) Identification of gelsolin, a Ca2 + -dependent regulatory protein of actin gel-sol transformation, and its intracellular distribution in a variety of cells and tissues. *J Cell Biol*, **91**, 901–906.

132 Aizawa, H., Fukui, Y. and Yahara, I. (1997) Live dynamics of Dictyostelium cofilin suggests a role in remodeling actin latticework into bundles. *J Cell Sci*, **110** (Pt 19), 2333–2344.

133 Nagaishi, K., Adachi, R., Matsui, S., Yamaguchi, T., Kasahara, T. and Suzuki, K. (1999) Herbimycin A inhibits both dephosphorylation and translocation of cofilin induced by opsonized zymosan in macrophagelike U937 cells. *J Cell Physiol*, **180**, 345–354.

134 Kaplan, G. (1977) Differences in the mode of phagocytosis with Fc and C3 receptors in macrophages. *Scand J Immunol*, **6**, 797–807.

135 Berton, G. and Lowell, C.A. (1999) Integrin signalling in neutrophils and macrophages. *Cell Signal*, **11**, 621–635.

136 Colucci-Guyon, E., Niedergang, F., Wallar, B.J., Peng, J., Alberts, A.S. and Chavrier, P. (2005) A role for mammalian diaphanous-related formins in complement receptor (CR3)-mediated phagocytosis in macrophages. *Curr Biol*, **15**, 2007–2012.

137 Hackam, D.J., Rotstein, O.D., Sjolin, C., Schreiber, A.D., Trimble, W.S. and Grinstein, S. (1998) v-SNARE-dependent secretion is required for phagocytosis. *Proc Natl Acad Sci USA*, **95**, 11691–11696.

138 Holevinsky, K.O. and Nelson, D.J. (1998) Membrane capacitance changes associated with particle uptake during phagocytosis in macrophages. *Biophys J*, **75**, 2577–2586.

139 Braun, V., Fraisier, V., Raposo, G., Hurbain, I., Sibarita, J.B., Chavrier, P., Galli, T. and Niedergang, F. (2004) TI-VAMP/VAMP7 is required for optimal phagocytosis of opsonised particles in macrophages. *EMBO J*, **23**, 4166–4176.

140 Bajno, L., Peng, X.R., Schreiber, A.D., Moore, H.P., Trimble, W.S. and Grinstein, S. (2000) Focal exocytosis of VAMP3-containing vesicles at sites of phagosome formation. *J Cell Biol*, **149**, 697–706.

141 Hong, W. (2005) SNAREs and traffic. *Biochim Biophys Acta*, **1744**, 120–144.

142 Martinez, I., Chakrabarti, S., Hellevik, T., Morehead, J., Fowler, K. and Andrews, N.W. (2000) Synaptotagmin VII regulates Ca(2 +)-dependent exocytosis of lysosomes in fibroblasts. *J Cell Biol*, **148**, 1141–1149.

143 Czibener, C., Sherer, N.M., Becker, S.M., Pypaert, M., Hui, E., Chapman, E.R., Mothes, W. and Andrews, N.W. (2006) Ca^{2+} and synaptotagmin VII-dependent delivery of lysosomal membrane to nascent phagosomes. *J Cell Biol*, **174**, 997–1007.

144 Huynh, K.K., Kay, J.G., Stow, J.L. and Grinstein, S. (2007) Fusion, fission, and secretion during phagocytosis. *Physiology (Bethesda)*, **22**, 366–372.

145 Cox, D., Lee, D.J., Dale, B.M., Calafat, J. and Greenberg, S. (2000) A Rab11-containing rapidly recycling compartment in macrophages that promotes phagocytosis. *Proc Natl Acad Sci USA*, **97**, 680–685.

146 Zhang, Q., Cox, D., Tseng, C.C., Donaldson, J.G. and Greenberg, S. (1998) A requirement for ARF6 in Fcgamma receptor-mediated phagocytosis in macrophages. *J Biol Chem*, **273**, 19977–19981.

147 Niedergang, F., Colucci-Guyon, E., Dubois, T., Raposo, G. and Chavrier, P. (2003) ADP ribosylation factor 6 is activated and controls membrane delivery during phagocytosis in macrophages. *J Cell Biol*, **161**, 1143–1150.

148 Gagnon, E., Duclos, S., Rondeau, C., Chevet, E., Cameron, P.H., Steele-Mortimer, O., Paiement, J., Bergeron, J.J. and Desjardins, M. (2002) Endoplasmic reticulum-mediated phagocytosis is a mechanism of entry into macrophages. *Cell*, **110**, 119–131.

149 Becker, T., Volchuk, A. and Rothman, J.E. (2005) Differential use of endoplasmic reticulum membrane for phagocytosis in J774 macrophages. *Proc Natl Acad Sci USA*, **102**, 4022–4026.

150 Hatsuzawa, K., Tamura, T., Hashimoto, H., Hashimoto, H., Yokoya, S., Miura, M., Nagaya, H. and Wada, I. (2006) Involvement of syntaxin 18, an endoplasmic reticulum (ER)-localized SNARE protein, in ER-mediated phagocytosis. *Mol Biol Cell*, **17**, 3964–3977.

151 Touret, N., Paroutis, P., Terebiznik, M., Harrison, R.E., Trombetta, S., Pypaert, M., Chow, A., Jiang, A., Shaw, J., Yip, C., Moore, H.P., van der Wel, N., Houben, D., Peters, P.J., de Chastellier, C., Mellman, I. and Grinstein, S. (2005) Quantitative and dynamic assessment of the contribution of the ER to phagosome formation. *Cell*, **123**, 157–170.

152 Diakonova, M., Bokoch, G. and Swanson, J.A. (2002) Dynamics of cytoskeletal proteins during Fcgamma receptor-mediated phagocytosis in macrophages. *Mol Biol Cell*, **13**, 402–411.

153 Araki, N. (2006) Role of microtubules and myosins in Fc gamma receptor-mediated phagocytosis. *Front Biosci*, **11**, 1479–1490.

154 Tjelle, T.E., Lovdal, T. and Berg, T. (2000) Phagosome dynamics and function. *Bioessays*, **22**, 255–263.

155 Hampton, M.B., Kettle, A.J. and Winterbourn, C.C. (1998) Inside the neutrophil phagosome: oxidants, myeloperoxidase, and bacterial killing. *Blood*, **92**, 3007–3017.

156 Tapper, H. (1996) The secretion of preformed granules by macrophages and neutrophils. *J Leukoc Biol*, **59**, 613–622.

157 Mukherjee, S., Ghosh, R.N. and Maxfield, F.R. (1997) Endocytosis. *Physiol Rev*, **77**, 759–803.

158 Lemmon, S.K. and Traub, L.M. (2000) Sorting in the endosomal system in yeast and animal cells. *Curr Opin Cell Biol*, **12**, 457–466.

159 Riezman, H., Woodman, P.G., van Meer, G. and Marsh, M. (1997) Molecular mechanisms of endocytosis. *Cell*, **91**, 731–738.

160 Barbieri, M.A., Roberts, R.L., Mukhopadhyay, A. and Stahl, P.D. (1996) Rab5 regulates the dynamics of early endosome fusion. *Biocell*, **20**, 331–338.

161 Mu, F.T., Callaghan, J.M., Steele-Mortimer, O., Stenmark, H., Parton, R.G., Campbell, P.L., McCluskey, J., Yeo, J.P., Tock, E.P. and Toh, B.H. (1995) EEA1, an early endosome-associated protein. EEA1 is a conserved alpha-helical peripheral membrane protein flanked by cysteine "fingers" and contains a calmodulin-binding IQ motif. *J Biol Chem*, **270**, 13503–13511.

162 Mohrmann, K. and van der Sluijs, S.P. (1999) Regulation of membrane transport through the endocytic pathway by rabGTPases. *Mol Membr Biol*, **16**, 81–87.

163 Somsel, R.J. and Wandinger-Ness, A. (2000) Rab GTPases coordinate endocytosis. *J Cell Sci*, **113** (Pt 2), 183–192.

164 Bucci, C., Thomsen, P., Nicoziani, P., McCarthy, J. and van, D.B. (2000) Rab7: a key to lysosome biogenesis. *Mol Biol Cell*, **11**, 467–480.

165 de, C.C. and Thilo, L. (1997) Phagosome maturation and fusion with lysosomes in relation to surface property and size of the phagocytic particle. *Eur J Cell Biol*, **74**, 49–62.

166 Desjardins, M. (1995) Biogenesis of phagolysosomes: the 'kiss and run' hypothesis. *Trends Cell Biol*, **5**, 183–186.

167 Gotte, M. and von Mollard, G.F. (1998) A new beat for the SNARE drum. *Trends Cell Biol*, **8**, 215–218.

168 Chen, Y.A. and Scheller, R.H. (2001) SNARE-mediated membrane fusion. *Nat Rev Mol Cell Biol*, **2**, 98–106.

169 Sollner, T.H. (2002) Vesicle tethers promoting fusion machinery assembly. *Dev Cell*, **2**, 377–378.

170 Desjardins, M., Nzala, N.N., Corsini, R. and Rondeau, C. (1997) Maturation of phagosomes is accompanied by changes in their fusion properties and size-selective acquisition of solute materials from endosomes. *J Cell Sci*, **110** (Pt 18), 2303–2314.

171 Hackam, D.J., Rotstein, O.D., Bennett, M.K., Klip, A., Grinstein, S. and Manolson, M.F. (1996) Characterization and subcellular localization of target membrane soluble NSF attachment protein receptors (t-SNAREs) in macrophages. Syntaxins 2, 3, and 4 are present on phagosomal membranes. *J Immunol*, **156**, 4377–4383.

172 Garin, J., Diez, R., Kieffer, S., Dermine, J.F., Duclos, S., Gagnon, E., Sadoul, R., Rondeau, C. and Desjardins, M. (2001) The phagosome proteome: insight into phagosome functions. *J Cell Biol*, **152**, 165–180.

173 Zerial, M. and McBride, H. (2001) Rab proteins as membrane organizers. *Nat Rev Mol Cell Biol*, **2**, 107–117.

174 Simonsen, A., Lippe, R., Christoforidis, S., Gaullier, J.M., Brech, A., Callaghan, J., Toh, B.H., Murphy, C., Zerial, M. and Stenmark, H. (1998) EEA1 links PI(3)K function to Rab5 regulation of endosome fusion. *Nature*, **394**, 494–498.

175 Callaghan, J., Nixon, S., Bucci, C., Toh, B.H. and Stenmark, H. (1999) Direct interaction of EEA1 with Rab5b. *Eur J Biochem*, **265**, 361–366.

176 McBride, H.M., Rybin, V., Murphy, C., Giner, A., Teasdale, R. and Zerial, M. (1999) Oligomeric complexes link Rab5 effectors with NSF and drive membrane fusion via interactions between EEA1 and syntaxin 13. *Cell*, **98**, 377–386.

177 Christoforidis, S., Miaczynska, M., Ashman, K., Wilm, M., Zhao, L., Yip, S.C., Waterfield, M.D., Backer, J.M. and Zerial, M. (1999) Phosphatidylinositol-3-OH kinases are Rab5 effectors. *Nat Cell Biol*, **1**, 249–252.

178 Backer, J.M. (2000) Phosphoinositide 3-kinases and the regulation of vesicular trafficking. *Mol Cell Biol Res Commun*, **3**, 193–204.

179 Lawe, D.C., Patki, V., Heller-Harrison, R., Lambright, D. and Corvera, S. (2000) The FYVE domain of early endosome antigen 1 is required for both phosphatidylinositol 3-phosphate and Rab5 binding. Critical role of this dual interaction for endosomal localization. *J Biol Chem*, **275**, 3699–3705.

180 Nielsen, E., Christoforidis, S., Uttenweiler-Joseph, S., Miaczynska, M., Dewitte, F., Wilm, M., Hoflack, B. and Zerial, M. (2000) Rabenosyn-5, a novel Rab5 effector, is complexed with hVPS45 and recruited to endosomes through a FYVE finger domain. *J Cell Biol*, **151**, 601–612.

181 Raiborg, C., Bache, K.G., Mehlum, A., Stang, E. and Stenmark, H. (2001) Hrs recruits clathrin to early endosomes. *EMBO J*, **20**, 5008–5021.

182 Raiborg, C., Bache, K.G., Gillooly, D.J., Madshus, I.H., Stang, E. and Stenmark, H. (2002) Hrs sorts ubiquitinated proteins into clathrin-coated microdomains of early endosomes. *Nat Cell Biol*, **4**, 394–398.

183 Raiborg, C. and Stenmark, H. (2002) Hrs and endocytic sorting of ubiquitinated membrane proteins. *Cell Struct Funct*, **27**, 403–408.

184 Roberts, R.L., Barbieri, M.A., Ullrich, J. and Stahl, P.D. (2000) Dynamics of rab5 activation in endocytosis and phagocytosis. *J Leukoc Biol*, **68**, 627–632.

185 Hawkins, P.T., Davidson, K. and Stephens, L.R. (2007) The role of PI3Ks in the regulation of the neutrophil NADPH oxidase. *Biochem Soc Symp*, **74**, 59–67.

186 Kanai, F., Liu, H., Field, S.J., Akbary, H., Matsuo, T., Brown, G.E., Cantley, L.C. and Yaffe, M.B. (2001) The PX domains of $p47^{phox}$ and $p40^{phox}$ bind to lipid products of PI(3)K. *Nat Cell Biol*, **3**, 675–678.

187 Feng, Y., Press, B. and Wandinger-Ness, A. (1995) Rab 7: an important regulator of late endocytic membrane traffic. *J Cell Biol*, **131**, 1435–1452.

188 Cantalupo, G., Alifano, P., Roberti, V., Bruni, C.B. and Bucci, C. (2001) Rab-interacting lysosomal protein (RILP): the Rab7 effector required for transport to lysosomes. *EMBO J*, **20**, 683–693.

189 Cosson, P. and Letourneur, F. (1997) Coatomer (COPI)-coated vesicles: role in intracellular transport and protein sorting. *Curr Opin Cell Biol*, **9**, 484–487.

190 Schekman, R. and Orci, L. (1996) Coat proteins and vesicle budding. *Science*, **271**, 1526–1533.

191 Raiborg, C., Bache, K.G., Mehlum, A., Stang, E. and Stenmark, H. (2001) Hrs recruits clathrin to early endosomes. *EMBO J*, **20**, 5008–5021.

192 Stoorvogel, W., Oorschot, V. and Geuze, H.J. (1996) A novel class of clathrin-coated vesicles budding from endosomes. *J Cell Biol*, **132**, 21–33.

193 Aniento, F., Gu, F., Parton, R.G. and Gruenberg, J. (1996) An endosomal beta COP is involved in the pH-dependent formation of transport vesicles destined for late endosomes. *J Cell Biol*, **133**, 29–41.

194 Gu, F., Aniento, F., Parton, R.G. and Gruenberg, J. (1997) Functional dissection of COP-I subunits in the biogenesis of multivesicular endosomes. *J Cell Biol*, **139**, 1183–1195.

195 Daro, E., Sheff, D., Gomez, M., Kreis, T. and Mellman, I. (1997) Inhibition of endosome function in CHO cells bearing a temperature-sensitive defect in the coatomer (COPI) component epsilon-COP. *J Cell Biol*, **139**, 1747–1759.

196 Beron, W., Mayorga, L.S., Colombo, M.I. and Stahl, P.D. (2001) Recruitment of coat-protein-complex proteins on to phagosomal membranes is regulated by a brefeldin A-sensitive ADP-ribosylation factor. *Biochem J*, **355**, 409–415.

197 Botelho, R.J., Hackam, D.J., Schreiber, A.D. and Grinstein, S. (2000) Role of COPI in phagosome maturation. *J Biol Chem*, **275**, 15717–15727.

198 Blocker, A., Severin, F.F., Burkhardt, J.K., Bingham, J.B., Yu, H., Olivo, J.C., Schroer, T.A., Hyman, A.A. and Griffiths, G. (1997) Molecular requirements for bi-directional movement of phagosomes along microtubules. *J Cell Biol*, **137**, 113–129.

199 Matteoni, R. and Kreis, T.E. (1987) Translocation and clustering of endosomes and lysosomes depends on microtubules. *J Cell Biol*, **105**, 1253–1265.

200 Gruenberg, J., Griffiths, G. and Howell, K.E. (1989) Characterization of the early endosome and putative endocytic carrier vesicles in vivo and with an assay of vesicle fusion *in vitro*. *J Cell Biol*, **108**, 1301–1316.

201 Desjardins, M., Huber, L.A., Parton, R.G. and Griffiths, G. (1994) Biogenesis of phagolysosomes proceeds through a sequential series of interactions with the endocytic apparatus. *J Cell Biol*, **124**, 677–688.

202 Desjardins, M., Celis, J.E., van Meer, G., Dieplinger, H., Jahraus, A., Griffiths, G. and Huber, L.A. (1994) Molecular characterization of phagosomes. *J Biol Chem*, **269**, 32194–32200.

203 Aderem, A.A., Wright, S.D., Silverstein, S.C. and Cohn, Z.A. (1985) Ligated complement receptors do not activate the arachidonic acid cascade in resident peritoneal macrophages. *J Exp Med*, **161**, 617–622.

204 Wright, S.D. and Silverstein, S.C. (1983) Receptors for C3b and C3bi promote phagocytosis but not the release of toxic oxygen from human phagocytes. *J Exp Med*, **158**, 2016–2023.

205 Savill, J. (1997) Recognition and phagocytosis of cells undergoing apoptosis. *Br Med Bull*, **53**, 491–508.

206 Savill, J. (1998) Apoptosis. Phagocytic docking without shocking. *Nature*, **392**, 442–443.

207 Platt, N., da Silva, R.P. and Gordon, S. (1998) Recognizing death: the phagocytosis of apoptotic cells. *Trends Cell Biol*, **8**, 365–372.

208 Anderson, K.V. (2000) Toll signaling pathways in the innate immune response. *Curr Opin Immunol*, **12**, 13–19.

209 Muzio, M., Bosisio, D., Polentarutti, N., D'amico, G., Stoppacciaro, A., Mancinelli, R., van't, V.C., Penton-Rol, G., Ruco, L.P., Allavena, P. and Mantovani, A. (2000) Differential expression and regulation of toll-like receptors (TLR) in human leukocytes: selective expression of TLR3 in dendritic cells. *J Immunol*, **164**, 5998–6004.

210 Janeway, C.A. Jr and Medzhitov, R. (1998) Introduction: the role of innate immunity in the adaptive immune response. *Semin Immunol*, **10**, 349–350.

211 Kawai, T. and Akira, S. (2007) TLR signaling. *Semin Immunol*, **19**, 24–32.

212 Ozinsky, A., Underhill, D.M., Fontenot, J.D., Hajjar, A.M., Smith, K.D.,

Wilson, C.B., Schroeder, L. and Aderem, A. (2000) The repertoire for pattern recognition of pathogens by the innate immune system is defined by cooperation between toll-like receptors. *Proc Natl Acad Sci USA*, **97**, 13766–13771.

213 Underhill, D.M., Ozinsky, A., Hajjar, A.M., Stevens, A., Wilson, C.B., Bassetti, M. and Aderem, A. (1999) The Toll-like receptor 2 is recruited to macrophage phagosomes and discriminates between pathogens. *Nature*, **401**, 811–815.

214 Aderem, A. (2001) Role of Toll-like receptors in inflammatory response in macrophages. *Crit Care Med*, **29**, S16–S18.

215 Kawai, T. and Akira, S. (2006) Innate immune recognition of viral infection. *Nat Immunol*, **7**, 131–137.

216 Akira, S. and Takeda, K. (2004) Toll-like receptor signalling. *Nat Rev Immunol*, **4**, 499–511.

217 DeLeo, F.R., Allen, L.A., Apicella, M. and Nauseef, W.M. (1999) NADPH oxidase activation and assembly during phagocytosis. *J Immunol*, **163**, 6732–6740.

218 Gresham, H.D., McGarr, J.A., Shackelford, P.G. and Brown, E.J. (1988) Studies on the molecular mechanisms of human Fc receptor-mediated phagocytosis. Amplification of ingestion is dependent on the generation of reactive oxygen metabolites and is deficient in polymorphonuclear leukocytes from patients with chronic granulomatous disease. *J Clin Invest*, **82**, 1192–1201.

219 Sakata, A., Ida, E., Tominaga, M. and Onoue, K. (1987) Arachidonic acid acts as an intracellular activator of NADPH-oxidase in Fc gamma receptor-mediated superoxide generation in macrophages. *J Immunol*, **138**, 4353–4359.

220 Yamamoto, K. and Johnston, R.B. Jr (1984) Dissociation of phagocytosis from stimulation of the oxidative metabolic burst in macrophages. *J Exp Med*, **159**, 405–416.

221 Segal, A.W., Wientjes, F., Stockley, R.W. and Dekker, L.V. (1999) Components and organization of the NADPH oxidase of phagocytic cells, in *Phagocytosis* (ed. S. Gordon), JAI, Stamford, CT, pp. 441–483.

4
Cellular Model Systems Used to Study Phagosome Biogenesis

Michael Steinert

4.1
From Grazing to Host Defense

Phagocytosis is an ancient and evolutionary highly conserved process by which cells engulf particles. Protozoa use phagocytosis primarily to obtain nutrients. In metazoa, with a developed immune system, phagocytosis is a feature of specialized, "professional phagocytes" such as macrophages, neutrophils and dendritic cells, which are capable of ingesting and killing a large variety of microorganisms [1]. In the light of evolution, protozoa and nonvertebrates may be viewed as an "ancient virulence school" for microbes which later developed into human pathogens. In the course of this coevolution, numerous bacterial species developed defense mechanisms against phagocytic cells, including the production of large filaments, antiphagocytic capsules or the secretion of toxins [2, 3]. Some pathogens even evolved strategies to reprogram their host cells, thereby ensuring intracellular survival or replication.

A better understanding of the cross-talk between phagocytic host cells and microbes was catalyzed by the observation that so-called model organisms can be infected by certain pathogens. In fact, it has been shown that many of the virulence factors required for pathogenicity in mammals are also important for pathogen survival during interactions with unicellular organisms and nonvertebrate hosts [4]. Because of their lower level of complexity, host models such as *Dictyostelium discoideum* and *Drosophila melanogaster* often allow a better dissection of the innate immune system, undisturbed by superimposed effects of the acquired immune system [5]. The organisms are easy to cultivate and, more importantly amenable to genetic manipulation, which facilitates identification and analysis of genes regulating phagocytosis and modulating host resistance to pathogens. Recent genome projects have also offered a variety of new target genes for testing expression patterns during infection. In the this chapter we will focus on phagocytosis and the outcome of ingestion of pathogens in the *D. discoideum* and the *D. melanogaster* model.

Intracellular Niches of Microbes. Edited by Ulrich E. Schaible and Albert Haas

ISBN: 978-3-527-32207-7

4.2
Dictyostelium, a Professional Phagocyte at the Border of Multicellularity

Dictyostelium cells are so-called "social amoebae" which live in the forest soil, feed on bacteria by phagocytosis and proliferate by binary fission [6]. As the food supply diminishes, the haploid cells stop growing and enter the social phase of their life cycle by assembling into multicellular aggregates. In a complex developmental process, approximately 10^5 cells form a migrating slug and finally a fruiting body, consisting of dead stalk cells and spores.

The phagocytic potential of *Dictyostelium* cells gradually declines when cells enter the multicellular stage. However, starving preaggregating or aggregating cells fully revert to the phagocytic stage when exposed to bacteria [7]. At the migrating slug stage "sentinel cells" may represent a reservoir of immune-like cells that engulf bacteria and sequester toxins while circulating within the slug [8]. This apparent innate immune function active during multicellular development is further substantiated by the occurrence of potential homologs to signaling proteins involved in the innate immunity of multicellular animals or plants, including Toll-like interleukin receptor (TIR) domain-containing proteins, WRKY transcription factors and potential leucine-rich repeat (LRR) domain receptors [8].

Dictyostelium cells are considered "professional phagocytes," as they ingest a large variety of bacteria, yeast, apoptotic cells, as well as inert particles. The phagocytosis capacity of *Dictyostelium* in the single cell growth phase exceeds that of neutrophils several fold, each cell being able to digest about 300 bacteria per hour. Both fluid-phase endocytosis and phagocytosis in *Dictyostelium* cells have been reviewed in detail [9–11].

In *Dictyostelium*, the first step in phagosome biogenesis is initiated by binding of the particle to cell surface receptors, which leads to rearrangement of the actin cortex, progressive surrounding of the particle by the plasma membrane and ingestion of the newly produced vesicle. Particle binding and phagocytic cup formation are inhibited in the Phg2-null mutant defective in a serine/threonine kinase that is constitutively associated to the plasma membrane [12].

Actin assembly and reorganization are known to be controlled by signals originating at the site of particle attachment and transmitted to the cell interior by heterotrimeric and monomeric G proteins of the Ras and Rac families, membrane lipids generated by the activity of phospholipases, phosphatidylinositol kinases and phosphatases, and calcium ions [13]. Phosphatidylinositol 4,5-bisphosphate ($PI(4,5)P_2$) appears to be important in both activation and localization of actin nucleation factors and actin-binding proteins, such as profilin, coronin, gelsolin, DAPI1 or CAP. Concomitant with the disassembly of the actin coat from the internalized phagosome, a decrease of $PI(4,5)P_2$ has been observed.

Phagocytosis in *D. discoideum* is efficiently inhibited by phospholipase C (PLC) inhibitors and intracellular calcium chelators [14, 15]. It has been proposed that the endoplasmic reticulum (ER), which has been shown to come into close association with phagosomes during uptake, may be required for local release of calcium. This calcium increase in turn may stimulate PLC activity and activate proteins involved in phagosomal membrane tethering and fusion [16, 17].

During matuarion, the phagosome sequentially fuses with acidic vesicles of the endocytic pathway. Early phagocytic steps are characterized by recruitment of the V-H^+-ATPase to the phagosome, shedding of the actin coat and delivery of proteins regulating vesicle fusion, such as the small G protein Rab7, the soluble *N*-ethylmaleimide-sensitive-factor attachment receptor (SNARE) Vti1, syntaxin 7, syntaxin 8 and the lysosomal marker LmpB. A second step in the maturation of phagosomes, starting between 3 and 15 min post uptake, is characterized by recruitment of lysosomal enzymes, such as cathepsin D and cysteine proteinases CP-p34, together with lysosomal membrane markers, such as the CD36/LIMP family members LmpA and LmpC, and the SNARE VAMP7. This is followed by the acquisition of lysosomal glycosidases, such as α-mannosidase and β-glucosidase. The post-lysosomal marker protein vacuolin B appears 60–90 min post internalization and indicates the late phase of phagosome maturation [18].

The cycle of phagosome biogenesis is considered complete when undigested material is excreted, though retrieval and recycling of several membrane components has been shown to occur from the very early steps [19]. Since post-lysosomal vesicles are recoated with actin before excretion, a close link between heterotrimeric G protein and actin reorganization at different steps of the phagocytic process is suggested [20].

The sequence of events presented here is consistent with a half-life of 45 min for degradation of bacteria [21]. The killing and digestion of the internalized bacteria is achieved by production of toxic oxygen radicals (respiratory burst), acidification of the phagosome lumen, depletion of essential divalent metals, degradation of the bacterial cell wall and digestion of cytoplasmic compounds by hydrolytic enzymes [22, 23]. Since the *Dictyostelium* genome contains several genes encoding for lysozyme homologs and a large family of pore-forming peptides, homologous to amoebapores and naegleriapores, further bactericidal effectors may be active in these amoebae.

4.3
Dictyostelium, a Surrogate Host for *Legionella*, *Mycobacterium* and Other Pathogens

Different bacterial species are taken up with different efficiencies and not all phagocytosed bacteria are killed by *D. discoideum*. In recent years *D. discoideum* has been established as a host model for several pathogens, including *Pseudomonas aeruginosa* [24, 25], *Mycobacterium avium*, *M. marinum* [26, 27], *Vibrio cholerae* [28], *Klebsiella pneumoniae* [29] and *Legionella pneumophila* [30, 31]. The research areas presently pursued include the use of *Dictyostelium* wild-type cells as screening system for virulence, the use of *Dictyostelium* mutant cells to identify genetic host determinants of susceptibility and resistance to infection and the use of reporter systems in *Dictystelium* cells [4] which allow the dissection of the complex host–pathogen cross-talk (Figure 4.1). Moreover, it was possible to analyze the *Dictyostelium* transcriptional host cell response upon infection [32].

A disadvantage of *D. discoideum* as a pathogenesis model is that the amoebae do not survive temperatures above 27 °C. This is especially critical for pathogens which

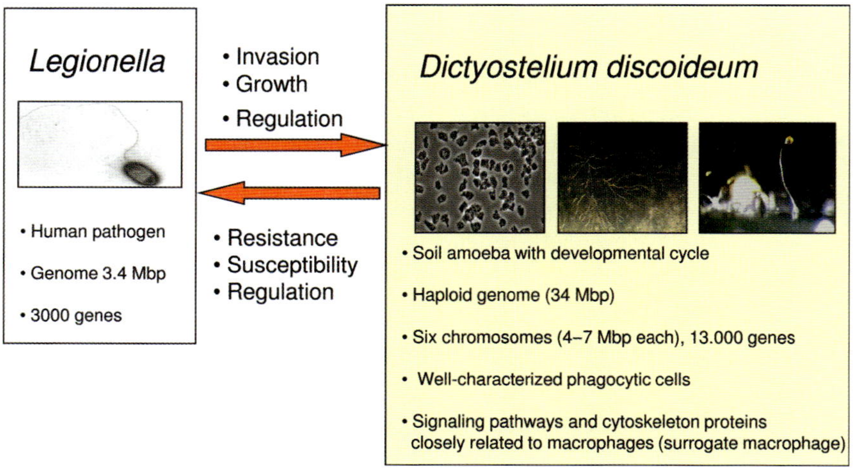

Figure 4.1 Key features of the haploid social amoeba *Dictyostelium* as a host model system for studying cellular aspects of *Legionella* pathogenicity. The deciphering of the genomes of both the pathogen and the host and the availability of an impressive molecular biological toolbox allow a detailed analysis of the host–pathogen cross-talk.

express their virulence traits at higher temperatures. The pathogens predominantly analyzed in *D. discoideum* are *L. pneumophila* and *Mycobacterium* spp. *L. pneumophila* is an environmental Gram-negative bacterium that often causes a severe and life-threatening pneumonia in humans. The transmission of *L. pneumophila* to humans occurs by inhalation of aerosols from contaminated water sources [33]. After uptake by alveolar macrophages the *Legionella*-containing phagosomes bypass endolysosomal degradation. The *Legionella*-harboring nascent phagosome interacts with the cytoskeleton, sequentially recruits smooth vesicles, mitochondria and rough endoplasmic reticulum (rER) and initially does not fuse with lysosomes (Figure 4.2). Furthermore, vacuole acidification is reduced and the rER-derived replicative niche undergoes a transition from tight to spacious vacuoles [34, 35]. Comparative experiments with several cellular infection models including *D. discoideum* have shown that the modulation of diverse host cell functions is a prerequisite for the biogenesis of the *Legionella*-specific replicative phagosome.

M. marinum is a pathogen of fish and amphibians that causes systemic tuberculosis-like diseases. This relative of *M. tuberculosis* provides a useful model to study the pathogenesis of tuberculosis in genetically tractable model organisms like *D. discoideum*. *Mycobacterium* ssp., similar to *Legionella*, proliferate intracellularly within *D. discoideum*, following a route somewhat comparable to what has been described for macrophages. *M. marinum*-containing phagosomes become less acidic, just like *L. pneumophila* vacuoles [36, 37]. At later stages of infection, *Mycobacterium* ssp. proliferate in neutral post-lysosomal, vacuolin-positive spacious phagosomes, which do not fuse with the plasma membrane, but rupture and release the bacteria into the cytosol [36]. The ultimate fate of the *Mycobacterium* phagosomes and the mechanisms

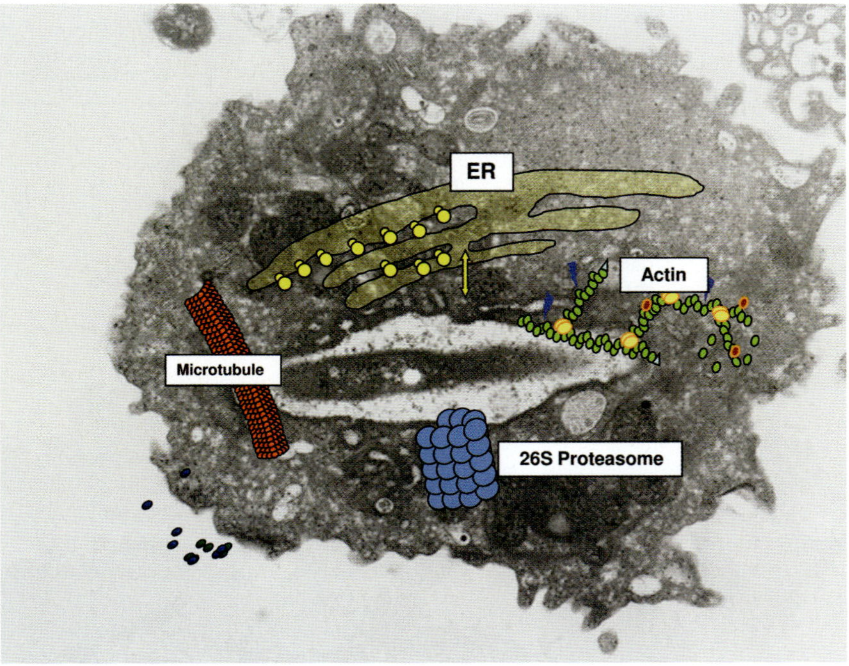

Figure 4.2 Intracellular interactions of the *Legionella*-containing phagosome with host cell structures. After actin-mediated uptake *L. pneumophila* avoids endosomal fusion and intercepts endoplasmic reticulum (ER)-derived vesicles that normally traffic to the Golgi. The newly internalized phagosome is rapidly transported about the cell on microtubules. The proteasome and ubiquitinated proteins are recruited to the *Legionella*-containing phagosome and are required for efficient bacterial replication.

of release are still largely unknown. Nevertheless, a number of host cell factors relevant for infection have been analyzed in detail (Table 4.1). One of them is Nramp1, which in mice (Bcg/Ity/Lsh locus) controls resistance to infection by the attenuated vaccine strain *M. bovis* BCG, *Salmonella* and *Leishmania* species. Intracellular growth, but not phagocytosis, was shown to be enhanced in a Nramp1-null *Dictyostelium* mutant for both *M. avium* and *L. pneumophila* [38]. In addition, polymorphic variants of the human homolog, which is encoded on chromosome 2, are faintly associated with susceptibility to tuberculosis and leprosy. Based on sequence conservation of the overall membrane structure and functional studies in different model systems, it is established that Nramp proteins are H^+-dependent divalent metal (e.g., iron) transporters (for a recent review see [39]). Nramp1 decorates the membrane of endolysosomal vesicles and the Golgi, and has been found to be recruited to macropinosomes and phagosomes containing non-pathogenic bacteria within 2 min following uptake [38]. Constitutive Nramp1 expression circumvents the *Legionella*-induced repression of the endogenous promoter and protects effectively against *Legionella* infection. Therefore it was concluded that control of iron homeostasis is important for establishing the *Legionella* intracellular niche for growth.

Table 4.1 Host cell factors that influence pathogen-directed phagosome biogenesis.

	Dictyostelium	*Drosophila*	Mammalian cells
Legionella	G protein (β subunit)		
	Actin		
	Coronin		
	Villidin		
	α-Actinin/filamin		
	Calreticulin		
	Calnexin		
	Microtubule		
	RatioA		
	Nramp1		Nramp1 (DMT2)
	PI3K		
	PI(4)P		PI(4)P
	Proteasome	Proteasome	Proteasome
		Sec22	Sec22
		TRAPP	TRAPP
		Arf1, Sar1, Rab5	Arf, Sar1, Rab1
		Cdc48/p97 (ERAD)	Cdc48/p97 (ERAD)
Mycobacterium	Nramp1		
	Coronin		Coronin (TACO)
		Scavenger receptor	Scavenger receptor
		CD36, Peste	

Intracellular growth of *M. marinum* is enhanced in a *Dictyostelium* RacH-null mutant. Conversely, the absence of one of the two *Dictyostelium* vacuolin isoforms renders the host more resistant to *M. marinum* [36]. Due to structural similarities with caveolin, it is assumed that lack of the vacuolin coat may facilitate fusion of the bacterium-containing phagosome with vesicles of the endocytic pathway, thus favoring recruitment of vacuolar ATPase and lysosomal enzymes, which would be counteractive for mycobacterial survival [36].

4.4 *Legionella*-directed Phagosome Biogenesis

Phagocytosis in *D. discoideum* and macrophages is known to be regulated by a heterotrimeric G protein-linked signal transduction (Table 4.1). By using pharmacological inhibitors and *Dictyostelium* mutants it has been demonstrated that *L. pneumophila* uptake occurs by conventional phagocytosis which requires heterotrimeric G proteins and the PLC pathway. Moreover, these experiments revealed that cytoplasmic calcium levels, the cytoskeleton proteins coronin, villidin and α-actinin/filamin as well as the calcium-binding ER proteins calreticulin and calnexin significantly influence this process [17, 40]. The newly internalized phagosome is rapidly transported through the cell on microtubules (Figure 4.1). The decreased intracellular replication of *L. pneumophila* in a *D. discoideum ratioA* (*rtoA*$^-$) mutant obviously results from a lowered efficiency of vesicle trafficking important for the integrity of the *Legionella*-containing compartment [35]. Of particular interest are the substrates of the *Legionella* Dot/Icm type IV secretion system [41]. SidM and LidA target the mammalian Rab1, a small GTPase regulating ER-to-Golgi traffic. RalF recruits and activates ADP-ribosylation factor 1 (Arf1), a small GTPase involved in retrograde vesicle transport from the Golgi apparatus to the ER [42, 43]. The recent analysis of the *D. discoideum* transcriptome upon infection with *L. pneumophila* revealed that by establishing its replicative niche *Legionella* not only interferes with bacterial degradation and intracellular vesicle transport and fusion but also profoundly influences and exploits the metabolism of its host. Furthermore, genes belonging to the *Dictyostelium* ubiquitination machinery are strongly upregulated after infection with *L. pneumophila* wild-type as compared to the *dotA* mutant [32]. Remarkably, the *L. pneumophila* genome analysis identified several homologs of eukaryotic genes and it has been speculated that the respective proteins may allow *Legionella* to communicate with eukaryotic cells [44].

Genetic and pharmacological evidence suggest that class I phosphatidylinositol 3-kinases (PI3Ks) are dispensable for phagocytosis of *L. pneumophila* but appear to play a major role for the establishment of the replicative vacuole [45]. *L. pneumophila* obviously subverts the host metabolism, favoring formation of specific phosphatidylinositol forms in the *Legionella*-containing vacuole (LCV). Among the different phosphatidylinositols, PI(4)P is enriched in the *Legionella*-containing phagosome and anchors specifically SidC, one of the secreted protein substrates of the *Legionella* Dot/Icm type IV secretion system. In the absence of functional PI3K, SidC recruitment to *Legionella*-containing phagosomes is increased, suggesting that PI (4)P sites are enriched [45].

Exciting progress has been made by proteome studies and the analysis of latex bead-containing *Dictyostelium* phagosomes which helped to define three maturation stages [19]. The first maturation stage is characterized by coronin and lysosomal glycoprotein (LmpB) acquisition. The second stage is characterized by the transfer of lysosomal enzymes. The third stage is characterized by quantitative retrieval of hydrolases from the phagolysosome and exocytosis of latex beads. So far, almost 200 phagosome proteins have been identified and ordered by their temporal appearance [20]. The phagosome proteins of *Legionella*-infected *Dictyostelium* cells also belong to a variety of functional categories, but specific alterations can be observed. Proteins involved in metabolism, protein biosynthesis as well as cytoskeleton and signal transduction make up the bulk of the *Legionella* phagosome proteome and agree well with the observed upregulation of their transcription [32]. The presence of elongation factors, ribosomal proteins, tRNA-synthetases as well as proteins typically associated with ER is consistent with the fact that the *Legionella*-containing phagosome intercepts secretory vesicles from ER exit sites and develops into a compartment that shares many features with the rER [46]. Unpublished results highlight the relevance of the ubiquitin-proteasome machinery (Shevchuk, Steinert *et al.*, unpublished). This view is further supported by the observation that *Legionella* translocates an E3 ubiquitin ligase that has two U-boxes. U-box 1 is critical to the ubiquitin ligation and U-box 2 mediates interaction with Clk1 (Cdc2-like kinase 1) [47].

Since actin assembled by phagosomes can provide tracks for lysosomes to move vectorially toward them, it is an interesting finding that infection with the pathogenic *L. pneumophila* Corby infection causes degradation of actin on phagosomes (Shevchuk, Steinert *et al.*, unpublished). The fact that phagosomes containing nonreplicative *L. hackeliae* do not display degraded actin indicates the importance of this phenomenon. Whether the cAMP-signaling system of *D. discoideum* plays a role in phagosome actin assembly and on phagosome maturation, as described for macrophages infected with mycobacteria, remains to be established [48].

The appearance of the actin-binding protein coronin on the phagosome raises another hypothesis. Coronin interacts with cytosolic NADPH oxidase components $p40^{phox}$ and $p67^{phox}$ and is known to be involved in their translocation to the phagosome [49]. Previous observations that *Legionella* replicates well in *Dictyostelium* mutants lacking coronin may thus be due to improper superoxide generation at the phagosome [30, 40]. Moreover, in neutrophils, phosphorylated $p47^{phox}$ protein has been shown to allow appropriate localization of PKC to coronin, leading to coronin phosphorylation. The phosphorylated coronin solubilizes, and the subsequent removal of the coronin coat facilitates phagolysosome fusion [49, 50, 51]. Since coronin as well as PKC inhibitors were observed on *Legionella* phagosomes we may have a first indication of how the *Legionella*-specific inhibition of the phagolysosome fusion is regulated.

4.5
The Whole Animal Model *Drosophila melanogaster*

D. melanogaster mount both cellular and humoral immune responses to pathogens. The innate immune response of the fruit fly *D. melanogaster* is characterized by a cellular immune response which depends on circulating phagocytic cells, a melanization response which produces reactive oxygen species at sites of infection, and the production of antimicrobial peptides in the fat body. *D. melanogaster* lacks an antibody-mediated adaptive immune response [52], but can react to different kinds of infections caused by Gram-positive or Gram-negative bacteria, fungi or parasitic protozoa. The sequencing of the entire genome revealed that a set of around 14 000 genes is sufficient for the generation of a multicellular organism of this kind that is able to perform complex immunogenic reactions [53]. Several thousand mutant fly strains with defects in one defined gene each are available for the genetic dissection of traits (http://flybase.net).

Most studies on the host–pathogen interaction with whole animals use injection methods to directly challenge the immune system, since only few pathogens are capable of naturally infecting *D. melanogaster*. Upon infection, a signaling cascade is activated which leads to the production of antimicrobial peptides in the fat body of the fly. It turned out that the activation of peptide production depends mainly on two distinct signal transduction pathways, the Toll and IMD (immune deficiency) pathways. Both pathways share striking similarities with the innate immune response pathways of mammals [52]. This innate immune system is activated differently depending on the nature of the attacker, thus discriminating between different classes of pathogens: fungi and Gram-positive bacteria induce the Toll pathway, whereas Gram-negative bacteria are sensed by the IMD pathway. Other signaling pathways, such as the JAK-STAT and the JNK pathways, may also be involved in the immune response, although the exact contribution of these pathways is not yet clear [54].

The most prominent pathogen studied in *D. melanogaster* is *P. aeruginosa.* Its virulence factors required for full infectivity in *D. melanogaster* are the same as for mammals. Moreover, the fact that the type III protein secretion system is activated during infection of both mammals and *Drosophila* recommends the fruit fly as a real *in vivo* model for the molecular dissection of virulence factor functions. Further virulence factors relevant in *D. melanogaster* are the GacAS two-component system, phenazine pigments, exotoxin A, as well as type IV pili [55]. Other pathogens currently under investigation in the fruit fly model are *Serratia marcescens, L. monocytogenes* and *M. marinum.* In adult flies, *M. marinum* initially infects hemocytes, where it blocks vacuolar acidification and replicates intracellularly [56]. *M. marinum* lacking the *mag24* gene is less virulent for *D. melanogaster,* similar to what has been observed for infections of *D. discoideum* [57]. Remarkably, *phag1* mutants, both in *D. melanogaster* and *D. discoideum,* also exhibit a specific susceptibility to *Klebsiella pneumoniae.*

Infections with fungi can also be monitored in *D. melanogaster.* In one study the authors infected immune-deficient flies with different *Candida albicans* mutants. Virulence patterns against *D. melanogaster* in these strains reproduced those in a murine model. Importantly, using this insect model they found additional virulence properties undetectable in the murine system [58]. Taken together, these studies support the hypothesis that evolutionary distant eukaryotic organisms share conserved strategies for resistance to infections.

4.6 Infection of *Drosophila* Phagocytes

Drosophila melanogaster cells such as primary macrophage-like phagocytes (hemocytes, differentiated plasmatocytes) and phagocytic cell lines can also be used to analyze the host–pathogen interaction. A major advantage of using *Drosophila* cells is the huge number of mutants that are defective in different aspects of the immune response and the relative ease with which single gene activities can be repressed by inhibitory RNA. The *D. melanogaster*-derived cell line S2, which was originally isolated from a 24-h-old embryo, has been used to study intracellular pathogens including *L. monocytogenes, M. marinum, M. fortuitum, Chlamydia trachomatis* and *Ehrlichia chaffeensis* [59]. S2 cells are classified as phagocytic hemocytes and known to be responsible for the production of antimicrobial peptides. Recent reports using RNA interference in *Drosophila* S2 cells helped to identify hundreds of host factors that affect *L. monocytogenes* entry, vacuole escape, intracellular growth and LLO (listeriolysin O) regulation [60, 61]. A genome-wide RNA interference screen in S2 cells has also been critical in identifying factors required for general phagocytosis, as well as those needed specifically for mycobacterial infection (Table 4.1). One specific factor, Peste (Pes), is a CD36 family member required for uptake of *M. fortuitum* [62]. Similar to the aforementioned proteome study in *D. discoideum*, an analysis of latex bead-containing phagosomes in cultured *D. melanogaster* S2 cells has been performed. This approach provided more than 600 phagosome-associated proteins for further molecular dissection [63].

Another *D. melanogaster* hemocyte-derived cell line utilized for infection studies is Kc167. RNA interference systems in this cell line allow the analysis of knockdowns of multiple genes simultaneously by adding double-stranded RNA against more than one target in a single culture. In particular, simultaneous depletion of the intermediate compartment and Golgi-tethering factor transport protein particle together with the ER SNARE protein Sec22 reduced *L. pneumophila* replication efficiently [64]. On the host side, additional genes encoding GTPases Arf1, Sar1, Rab5c as well as eukaryotic proteasome and cytosolic components of the ER-associated protein degradation (ERAD) pathway such as the Cdc48/p97 complex were obviously also required for intracellular replication. The same study revealed that membrane traffic funnels into the replication vacuole from multiple sources. On the bacterial side, intracellular replication of *L. pneumophila* was found to depend on the bacterial type IV secretion system Dot/Icm as in mammalian cells. A conclusion from these finding is that *L. pneumophila* exploits functionally redundant host pathways to create and maintain its intracellular niche.

4.7
Conclusions

In general, the outcome of host–pathogen interactions ranges from disease to asymptomatic carriage or even microbe elimination. So far the carrier state remains poorly understood, however, the realization of this state paved the way for defining host–pathogen interactions as regulated relationships. In recent years *Dictyostelium discoideum* and *Drosophila melanogaster* have proven useful as pathogenesis host models. Due to their ease in maintenance, their short generation times and genetic accessibility, they allow genomic screening for susceptibility – experiments that are much more complex to perform in mammals. A common criticism of these model systems is that they do not represent a real human infection. Therefore conclusions drawn from these models should take into account that certain properties are not represented in the model and some model properties cannot be found in the real system. In consequence, the selection of an appropriate model with an appropriate level of complexity is essential to elaborate valid results.

In this review we have highlighted several new proteins and genes that show that knowledge acquired through the study of model systems can be extended to mammalian phagosome biogenesis. Moreover, it becomes increasingly evident that many microbial products precisely mimic host cell determinants. Therefore, with deciphering of the genomes of both pathogen and host models, and an impressive toolbox including microarrays and proteomic analyses a comprehensive understanding of the respective host–pathogen cross-talk should be at hand. The expected gain of information will not only illuminate the mechanism by which pathogens subvert host cellular processes but also provide unique insight into the cellular processes themselves.

Acknowledgments

The work was supported by grants from the Deutsche Forschungsgemeinschaft (STE 838/3-4, SFB630 TP B1).

References

1. Stuart, L.M. and Ezekowitz, R.A. (2005) *Immunity*, **22**, 539–550.
2. Pieters, J. (2001) *Microbes Infect*, **3**, 249–255.
3. Celli, J. and Finlay, B.B. (2002) *Trends Microbiol*, **10**, 232–237.
4. Steinert, M. and Heuner, K. (2005) *Cell Microbiol*, **7**, 307–314.
5. Steinert, M., Leippe, M. and Roeder, T. (2003) *Int J Med Microbiol*, **293**, 321–332.
6. Kessin, R.H. (1981) *Cell*, **27**, 241–243.
7. Gambino, M., Kay, R.R. and Bozzaro, S. (1992) *Differentiation*, **49**, 133–141.
8. Chen, G., Zhuchenko, O. and Kuspa, A. (2007) *Science*, **317**, 678–681.
9. Maniak, M. (2001) *Biochim Biophys Acta*, **1525**, 197–204.
10. Maniak, M. (2002) *Int Rev Cytol*, **221**, 257–287.
11. Duhon, D. and Cardelli, J. (2002) *J Muscle Res Cell Motil*, **23**, 803–808.
12. Blanc, C., Charette, S., Cherix, N., Lefkir, Y. *et al.* (2005) *Eur J Cell Biol*, **84**, 951–960.
13. Van Haastert, P.J. and Devreotes, P.N. (2004) *Nat Rev Mol Cell Biol*, **5**, 626–634.
14. Peracino, B., Borleis, J., Jin, T., Westphal, M. *et al.* (1998) *J Cell Biol*, **141**, 1529–1537.
15. Seastone, D.J., Zhang, L., Buczynski, B., Rebstein, P. *et al.* (1999) *Mol Biol Cell*, **10**, 393–406.
16. Müller-Taubenberger, A., Lupas, A.N., Li, H., Ecke, M. *et al.* (2001) *EMBO*, **20**, 6772–6782.
17. Lu, M. and Clarke, M. (2005) *Cell Microbiol*, **7**, 995–1007.
18. Rauchenberger, R., Hacker, U., Murphy, J., Niewöhner, J. *et al.* (1997) *Curr Biol*, **7**, 215–218.
19. Gotthardt, D., Waratz, H.J., Henschel, O., Brückert, F. *et al.* (2002) *Mol Biol Cell*, **13**, 3508–3520.
20. Gotthardt, D., Blancheteau, V., Bosserhoff, A., Ruppert, T. *et al.* (2006) *Mol Cell Proteomics*, **5**, 2228–2243.
21. Maselli, A., Laevsky, G. and Knecht, D.A. (2002) *Microbiol*, **148**, 413–420.
22. Souza, G.M., Metha, D.P., Lammertz, M., Rodriguez-Paris, J. *et al.* (1997) *J Cell Sci*, **110**, 2239–2248.
23. Clarke, M., Köhler, J., Arana, Q., Liu, T. *et al.* (2002) *J Cell Sci*, **115**, 2893–2905.
24. Pukatzki, S., Kessin, R.H. and Mekalanos, J.J. (2002) *Proc Natl Acad Sci USA*, **99**, 3159–3164.
25. Cosson, P., Zulianello, L., Join-Lambert, O., Faurisson, F. *et al.* (2002) *J Bacteriol*, **184**, 3027–3033.
26. Skriwan, C., Fajardo, M., Hägele, S., Horn, M. *et al.* (2002) *Int J Med Microbiol*, **291**, 615–624.

27 Solomon, J.M., Leung, G.S. and Isber, R.R. (2003) *Infect Immun*, **71**, 3578–3586.
28 Pukatzki, S., Ma, A.T., Sturtevant, D., Krastins, B. *et al.* (2006) *Proc Natl Acad Sci USA*, **103**, 1528–1533.
29 Benghezal, M., Fauvarque, M.O., Tournebize, R., Froquet, R. *et al.* (2006) *Cell Microbiol*, **8**, 139–148.
30 Solomon, J.M., Rupper, A., Cardelli, J.A. and Isberg, R.R. (2000) *Infect Immun*, **68**, 2939–2947.
31 Hägele, S., Köhler, R., Merkert, H., Schleicher, M. *et al.* (2000) *Cell Microbiol*, **2**, 165–171.
32 Farbrother, P., Wagner, C., Na, J., Tunggal, B. *et al.* (2006) *Cell Microbiol*, **8**, 438–456.
33 Steinert, M., Heuner, K., Buchrieser, C., Albert-Weissenberger, C. *et al.* (2007) *Int J Med Microbiol*, **297**, 577–587.
34 Molmeret, M., Horn, M., Wagner, M., Santic, M. *et al.* (2005) *Appl Environ Microbiol*, **71**, 20–28.
35 Li, Z., Solomon, J.M. and Isberg, R.R. (2005) *Cell Microbiol*, **7**, 431–442.
36 Hagedorn, M. and Soldati, T. (2007) *Cell Microbiol*, **9**, 2716–2733.
37 Sturgill-Koszycki, S., Schaible, U.E. and Russell, D.G. (1996) *EMBO*, **15**, 6960–6968.
38 Peracino, B., Wagner, C., Balest, A., Balbo, A. *et al.* (2006) *Traffic*, **7**, 22–38.
39 Courville, P., Chaloupka, R. and Cellier, M.F. (2006) *Biochem Cell Biol*, **84**, 960–978.
40 Fajardo, M., Schleicher, M., Noegel, A. and Bozzaro, S. (2004) *Microbiology*, **150**, 2825–2835.
41 Hilbi, H., Segal, G. and Shuman, H.A. (2001) *Mol Microbiol*, **42**, 603–617.
42 Nagai, H., Kagan, J.C., Zhu, X. and Kahn, R.A. (2002) *Science*, **295**, 679–682.
43 Derre, I. and Isberg, R.R. (2005) *Infect Immun*, **73**, 4370–4380.
44 Jules, M. and Buchrieser, C. (2007) *FEBS Lett*, **581**, 2829–2838.
45 Weber, S.S., Ragaz, C., Reus, K., Nyfeler, Y. and Hilbi, H. (2006) *PLoS Pathog*, **2**, e46.
46 Machner, M.P. and Isberg, R.R. (2006) *Dev Cell*, **11**, 47–56.
47 Kubori, T., Hyakutake, A. and Nagai, H. (2008) *Mol Microbiol*, **67**, 1307–1319.
48 Kalamidas, S.A., Kuehnel, M.P., Peyron, P., Rybin, V. *et al.* (2006) *J Cell Sci*, **119**, 3686–3694.
49 Kaul, D. (2008) *FEMS Microbiol*, **278**, 10–14.
50 Grogan, A., Reeves, E., Keep, N., Wientjes, F.J. *et al.* (1997) *Cell Sci*, **110**, 3071–3081.
51 Harada, T., Miyake, M. and Imai, Y. (2007) *Microbiol Immunol*, **51**, 1161–1170.
52 Hoffmann, J.A. and Reichhart, J.M. (2002) *Nat Immunol*, **3**, 121–126.
53 Adams, M.D., Celniker, S.E., Holt, R.A., Evans, C.A. *et al.* (2000) *Science*, **287**, 2185–2195.
54 De Gregorio, E., Spellman, P.T., Tzou, P., Rubin, G.M. and Lemaitre, B. (2002) *EMBO*, **21**, 2568–2579.
55 Vodovar, N., Acota, C., Lemaitre, B. and Boccard, F. (2004) *Trends Microbiol*, **12**, 235–242.
56 Dionne, M.S., Ghori, N. and Schneider, D.S. (2003) *Infect Immun*, **71**, 3540–3550.
57 Solomon, J.M., Leung, G.S. and Isberg, R.R. (2003) *Infect Immun*, **71**, 3578–3586.
58 Alarco, A.M., Marcil, A., Chen, J., Suter, B. *et al.* (2004) *J Immunol*, **172**, 5622–5628.
59 Luce-Fedrow, A., Von Ohlen, T., Boyle, D., Ganta, R.R. and Chapes, S.K. (2008) *Appl Environ Microbiol*, **74**, 1886–1891.
60 Agaisse, H., Burrack, L.S., Philips, J.A., Rubin, E.J. *et al.* (2005) *Science*, **309**, 1248–1251.
61 Cheng, L.W., Viala, J.P., Stuurman, N., Wiedemann, U. *et al.* (2005) *Proc Natl Acad Sci USA*, **102**, 1346–1351.
62 Philips, J.A., Rubin, E.J. and Perrimon, N. (2005) *Science*, **309**, 1251–1253.
63 Stuart, L.M., Boulais, J., Charriere, G.M., Hennessy, E.J. *et al.* (2007) *Nature*, **445**, 95–101.
64 Dorer, M.S., Kirton, D., Bader, J.S. and Isberg, R.R. (2006) *PLoS Pathog*, **2**, e34.

5
Methods Used to Study Phagosome Biogenesis

Albert Haas

Phagosome biogenesis is a temporally and spatially ordered process which, in many aspects, parallels endosome maturation (see Chapters 3 and 5). Interest in microbial phagosome biogenesis has increased considerably in the past decade as cell biologists have recognized the usefulness of pathogen–host cell systems in the study of vesicle trafficking and the cytoskeleton and as microbiologists started to incorporate the more complex biology of eukaryotic cells into their views of infection. While many phagosome studies are sensibly performed using phagocytosed inert particles such as latex or paramagnetic beads, the study of the alterations of normal maturation by microbial pathogens has its own challenges and rewards and requires the authentic microorganism. This research comes with a higher degree of complexity, that is, a second living organism in place of a inanimate polymer bead. Similar points can be made about the study of symbiosis or of symbiotic compartments, although many compartments containing primary symbionts cannot be observed from the time of their creation, as the host cannot exist without them. An interesting case is the *Rhizobium*–legume symbiosis: The formation and development of a phagosome ("bacteriome") can be experimentally and reproducibly induced and followed using relatively simple tools. Many of the methods presented here for phagosome studies should be similarly applicable to the study of symbiosomes.

This chapter will present some of the methods that are currently used in the study of phagosome biogenesis (Figure 5.1). As these techniques are multifaceted and some of them quite sophisticated, this review will be limited to introducing the methods briefly and to giving some application examples. More substantial sections dedicated to methodological aspects of phagosome biogenesis can be found in Chapters 6–8.

5.1
Describing the World Within – Whole-Cell Assays for Phagosome Contents

Until recently, research into phagosome biogenesis was largely limited to analyzing which compartmental marker molecules are present on phagosomes at which stage

Intracellular Niches of Microbes. Edited by Ulrich E. Schaible and Albert Hass

ISBN: 978-3-527-32207-7

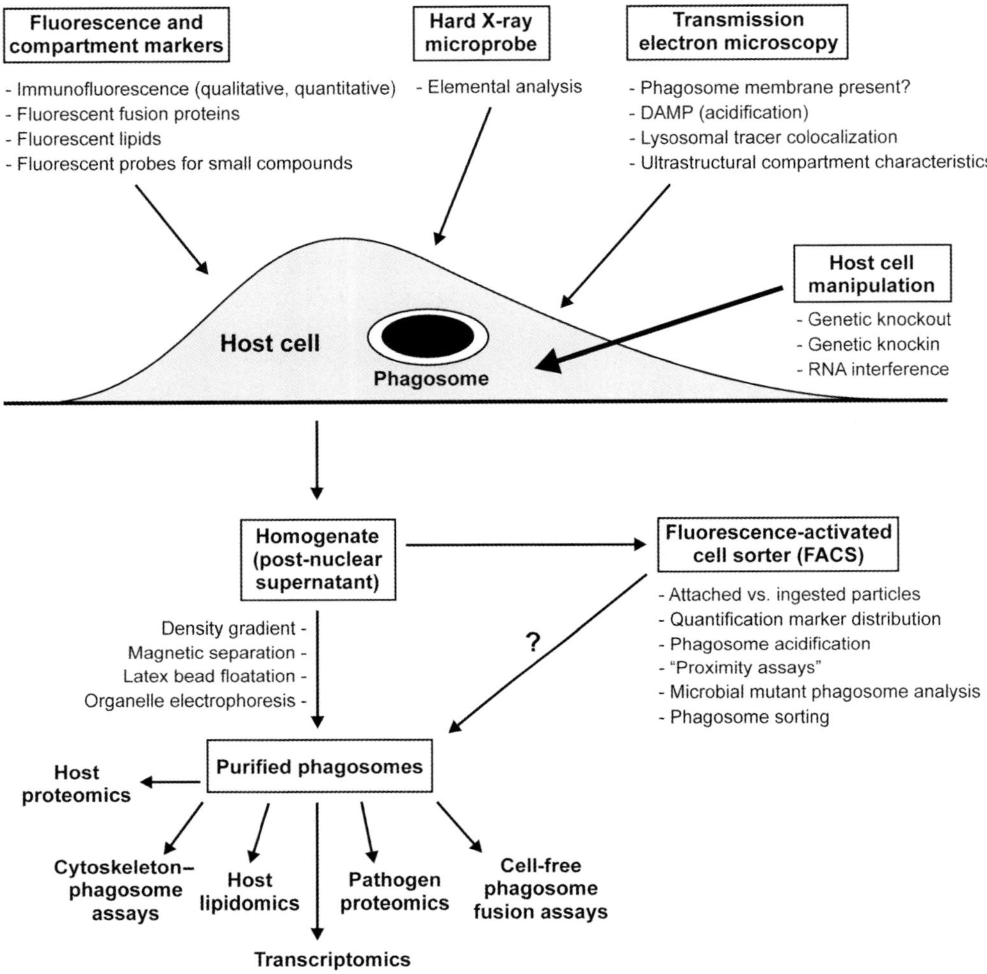

Figure 5.1 Various methods can be employed to study phagosome biogenesis, including microscopic (top), biochemical (bottom) and FACS-driven (right). Genetic host cell manipulation adds to the arsenal of methods that can be used to manipulate and understand phagosome biogenesis.

of maturation. In these studies the pathogen-containing phagosomes ("vacuoles") would be assigned certain compartment characteristics such as those of a sorting or late endosome. Knowing a pathogen's compartmentation allows the microbiologists to focus their virulence research on certain metabolic, environmental, ecological and physiological aspects of pathogen biology (e.g., a microorganism multiplying in the endoplasmic reticulum (ER) does not have to develop long-lasting acid stress resistance). Compartment identification also helps to focus on certain host aspects of infection (e.g., a pathogen living in a vacuole that resembles the early endosome will likely be able to access and probably requires some nutrients funneled into the endocytic system).

In the long run, however, in order to really understand the environment which a particle, pathogenic or not, encounters in the phagocyte, methods have to be developed that allow the quantification of intraphagosome parameters such as pH, oxygen pressure, concentrations of ions and micronutrients, presence of hydrolytic enzymes, or osmolarity. An added advantage would be the ability to monitor these parameters in nonfixed, that is, live cells in real time. Knowledge of the kinetics of the processes can then be compared with data obtained with microorganisms mutated in relevant metabolic or environmental response pathways, for example, oxygen resistance, glucose metabolism or ion scavenging. In the end, this information will provide important pieces in the reconstruction of the complete puzzle of "host cell–pathogen interplay."

Unfortunately, until recently few assays have been available that allow quantification of these variables. Applying some new methods, however, marker molecule distribution, intraphagosomal pH, lysosome content incorporation and hydrolytic activities in phagosomes can be quantified and rough estimates about their metal ion contents can be made.

5.1.1 Quantification of Lysosome Components in Phagosomes

Fusion of phagosomes with lysosomes is central to the destruction and killing of ingested materials and microorganisms (see Chapter 10). Phagosome–lysosome fusion is likely to occur in two ways, both of which may act on the same phagosome: (i) complete fusion of preexisting lysosomes with the maturing phagosome and (ii) transient fusion events between lysosomes and phagosomes, during which only some lysosome contents are transferred into phagosomes (and possibly vice versa). The latter has been termed "kiss-and-run fusion" [1].

5.1.1.1 Marker Protein Analysis

Lysosomes are defined by their low pH of ~4.0–5.0, their high concentrations in lysosomal hydrolases (such as cathepsin D and β-glucuronidase) and highly glycosylated transmembrane proteins (such as lysosome membrane-associated protein 1 (LAMP1)). It is very difficult, however, to distinguish between a late endosome and a lysosome based on these characteristics, as their differences are more quantitative than qualitative. This becomes even more complicated in professionally phagocytic cells, where the late endosomes and lysosomes do not seem to be completely separated. According to Griffiths [2], late endosomes can be regarded more as a cell stomach, while lysosomes have more the function of glands that contain digestive juice, meaning that lysosomes probably have "more of everything" than a late endosome. Marker molecules that are definitely found on late endosomes, but not lysosomes, are the mannose 6-phosphate receptors which deliver newly produced lysosomal hydrolases into the endocytic system and which are recycled from late endosomes before fusion with lysosomes.

There is, however, no single marker to positively and unambigously identify a lysosome.

Quantification of phagosome formation is usually done by microscopy, preferably laser-scanning confocal fluorescence microscopy. The microbe is stained with one fluor and the marker molecule with another (usually using a fluorescent antibody). Appearance of a "ring"-like fluorescent structure around a labeled particle in the channel of the second fluor or complete concurrence of the two fluors indicates "colocalization," that is, presence of the marker molecule on the phagosome membrane or in the observed phagosome and is counted as "positive." Analyzing many phagosomes in a sample (typically 50–100) in several independent experiments will help to identify the compartment. This method can be modified by quantifying the actual intensity of the fluorescent marker signal for each phagosome and can be used to test colocalization with any subcellular structure.

5.1.1.2 Endocytic Tracer Analysis

For the purpose of quantifying phagosome–lysosome fusion, the operational, albeit antiquated [3] definition of lysosomes as "terminally endocytic compartments" is often used. Tracer substances (fluorescent proteins or carbohydrates, ferritin, nanometer gold particles) are fed to phagocytes and are directed into the lysosomes by long chase times followed by feeding of the particle whose phagolysosome formation is to be investigated. A typical pulse-chase protocol would be 2–24 h of cultivation in presence of a fluorescent tracer, but absence of the phagocytic particle, followed by removal of uningested tracer, chase for 2–12 h and addition of the particle to be tested for 30 min. This is followed by a 1–2 h chase to allow the particle's phagosome to fuse with lysosomes and by sample fixation and preparation for (confocal) fluorescence microscopy. Alternatively, prelabeling of lysosomes with electron-dense markers such as 10 nm gold particles allows the exact determination in transmission electron microscopy (TEM) experiments of the average numbers of gold particles associated with control phagosomes and the comparison with test phagosomes [4]. As an internal control, test particles, such as pathogenic microorganisms, can be fed alongside non-interfering particles to quantify phagolysosome fusion for each of these in the same cell [4]. Rhodamine-labeled ~10 nm gold particles have been used as tracer material for lysosomes in fluorescence microscopy as well [5].

5.1.1.3 Fluorescence Resonance Energy Transfer

The fluorescence (or "Förster") resonance energy transfer (FRET) assay can be used to quantify the transfer of energy from a "donor" fluors to a suitable second, "acceptor" fluors. This interaction requires a very close contact between the two fluors (10 nm or less, 50% transmission at approximately 6–10 nm [6]) and provides an alternative approach to counting individual phagosomes microscopically when they appear at the same place ("colocalization"). While the latter indicates "how many phagosomes have fused with lysosomes," FRET results integrate over all phagosomes in a sample and show "how much" lysosome content has been transferred to phagosomes. The required close proximity of the two fluors can only be established by genuine vesicle fusion but not by an attachment of vesicles, as the width of a cellular membrane is typically 7–10 nm and the minimal possible distance between the lumina of two vesicles would therefore be 14 nm. In practice, the surfaces of inert beads [6] or bacteria [7] are covalently labeled with a donor fluor such as Alexa Fluor 488-SE and these particles are fed to macrophages whose lysosomes have been prelabeled with an acceptor fluor, for example, Alexa Fluor 594 [6]. FRET can be determined in a cuvette fluorometer [6] or microplate reader [7] format and is a measure for the average phagosome–lysosome fusion frequency in a large cohort of cells. The method can be applied to various microbial mutants [7] or to test the effect of drugs on phagolysosome formation [6]. It can also be used in fluorescence microscopy on single phagosomes [6]. As a cautionary comment, care must be taken to ensure that the dense labeling of the particle surface with a reactive dye does not in itself change trafficking of the particle.

5.1.2
Quantification of Phagosome pH

The first probe developed to explore the intraphagosome environment was a pH sensor. Taking advantage of an assay developed to measure endosome pH [8, 9], particles were surface labeled with a pH-sensitive fluor, usually fluorescein, whose emission at 520 nm quantifies the intracellular pH due to its exquisite sensitivity when excited at 490 nm [10, 11]. Fluorescein's maximum utility is in the pH range from 5.0 to 7.5, but pH-sensitive fluors with different pH ranges (e.g., Oregon Green) are also available [12]. To control and correct for the total particle volume ingested by phagocytes in each sample, a second, pH-insensitive fluorophore, for example, rhodamine, is also coupled to the particle surface. As an alternative, but less recommended choice, fluorescence emission of fluorescein can be determined at the fluor's isobestic point at which its fluorescence is little pH-dependent [12]. Comparison of pH-dependent and pH-independent signals would be used to determine intraphagosomal pH, saving the need for a second, calibrating fluor. In phagosome pH determination it is crucial that calibration of the system is done meticulously in each experiment. As the intraphagosome environment may change the fluorophore's emission relative to when it is in extracellular buffer solution, all calibrations must be done with parallel samples of infected cells, and the media are replaced by pH-calibrated buffer solutions in the presence of H^+-versus-metal$^+$ ionophores, e.g. nigericin, which equilibrate the pH between extracellular and intracellular world and between outside and inside phagosomes as well. Under these conditions, intraphagosomal pH equals cytoplasmic pH. Polynomal regression to a standard curve can then be used to determine pH in noncalibrated samples. In a modified protocol, calibration and pH determination can also be done microscopically on single phagosomes [13, 14].

An alternative, rarely used approach involving transmission electron microscopy (TEM), is labeling of phagosomes with DAMP (3-(2,4-dinitroanilino)-3′-amino-*N*-methyl dipropylamine). DAMP is a weak base which is membrane permeant but, once protonated in an acidic compartment, becomes membrane impermeant. The relative concentrations of DAMP in the phagosome are then determined in immuno-TEM after DAMP visualization with antibodies to the hapten dinitrophenol (cross-reacting with DAMP) [15]. In summary, this assay allows determination of the relative acidicity of two compartments. It is difficult to calibrate this assay, however, using extracellular pH variation and nigericin to pin down real phagosome pH. In some studies, the system was "calibrated" by counting immunogold particles not only in phagosomes but also in the nucleus, which was assumed to be at neutral pH, and the formula $pH = 7.0 - \log(D/N)$ to calculate phagosome pH was applied, with D being the gold density in the phagosome and N the density in the nucleus [16, 17].

A more readily applicable version of the DAMP assay is the "LysoTracker" assay (a brandmark of Invitrogen Inc.). Like DAMP, LysoTracker is a weak base which accumulates in acidic compartments, but LysoTracker is fluorescent. As with DAMP, determination of acidification of a compartment is possible by counting the number of LysoTracker-positive phagosomes [18, 19]. Again, calibration is extremely difficult because LysoTracker cannot be added together with a pH-calibrated acidic buffer. This would, by definition, lead to extracellular fluor protonation and lack of diffusion into the cell. One drawback of this otherwise very useful tool is that the pH at which a phagosome changes from positive to negative readout and vice versa is not known. While most publications simply state that a compartment is "acidic" when positive for LysoTracker, it is likely that organelles of slightly acidic pH do not appear positive. From many experiments, it is estimated that the pH required to turn a LysoTracker-negative compartment into a LysoTracker-positive compartment is around 5.8–6.0 (unpublished data). This estimate is also supported by data showing that the *Mycobacterium*-containing phagosome has a pH of 6.3 while it is negative for LysoTracker [19, 20].

A more specific example of an acidification sensor is the quantification of HCl-production in neutrophil phagosomes using a custom-tailored tetramethylrhodamine derivative whose fluorescence is dependent on HCl-induced de-spirocyclization [21].

5.1.3 Quantification of Phagosomes' Hydrolytic Activities

Lysosomes are the cell's dominant repository for large quantities of many species of hydrolytic enzymes [22]. Fusion of lysosomes with phagosomes denotes the end of phagosome maturation (Chapter 10) [23]. A thorough study into the phagosome hydrolase equipment, which also dissected the relative enzyme activities in early and late endosomes and lysosomes, found most hydrolase activities to be lysosomal [22]. Notable exceptions were cathepsin H (largely early endosomal) and cathepsin S (equally distributed between late endosomes and lysosomes). Although most of these hydrolase activities are lysosomal, probing the phagosome environment for hydrolase activities is not the same as quantifying phagolysosome formation. For example, newly formed immature (pro-)cathepsin D is trafficked from the *trans*-Golgi apparatus to early and late endosomes, being eventually activated in late endosomes and lysosomes to the mature enzyme. As the pathways of maturing phagosome and incoming proform hydrolases intersect [24], microorganisms can encounter some hydrolases early in phagosome maturation. In addition, pathogens may influence the intraphagosome activities of lysosome hydrolases [25] and, hence, measurement of these activities *in situ* is the only way of really knowing which hydrolases are active at which time.

5.1.3.1 **Protease Activities**

Phagosome protease activities can be measured using bead-coupled albumin which has been fluorescently labeled at self-quenching high density with DQ Green Bodipy fluor. In addition, the beads themselves were labeled with a second fluor (Alexa594), whose emission is protease-independent, to calibrate for bead number. Proteolytic cleavage leads to spatial separation of the fluor molecules, a decrease in FRET, and therefore to a fluorescence peak as a measure of proteolysis [6, 26]. A second approach to determine intraphagosomal protease activities is to couple small, synthetic fluorogenic protease substrates, for example, (biotin-LC-Phe-Arg)$_2$-rhodamine 110, a fluorogenic substrate for synthetic cysteine protease activity, to a carrier particle [6].

5.1.3.2 **Lipase Activities**

Triglyceride lipase can be determined by covering a phagocytic particle (here: silica beads) with a fluorogenic substrate, 1-trinitrophenyl-amino-dodecanoyl-2-pyrenedecanoyl-3-*O*-hexadecyl-*sn*-glycerol. Octadecyl rhodamine B is used as a hydrophobic red fluorescent calibration compound, again to normalize the observed lipase activities for the number of beads ingested [6]. Hydrolysis unquenches the pyrenedecaoic acid and yields increased fluorescence at 400 nm. The precise identity of the phagosomal lipase(s) and its localization are unknown.

5.1.3.3 **β-Galactosidase Activities**

β-Galactosidase activities were determined using methods similar to those described for lipase activities above by coating a particle with a lipid monolayer containing the fluorogenic substrate, here 5-dodecanoylaminofluorescein di-β-D-galactopyranoside (C_{12}-FDG). The substrate was used together with a pH-insensitive fluor (octadecyl rhodamine B) to correct for the number of ingested beads [26, 27]. This method was later adapted to study β-galactosidase activities in phagosomes containing *Mycobacterium marinum* wild-type vs. mutant cells and also proved useful for the study of phagosomes containing other microorganisms [28].

5.1.4
Intraphagosome Ion Concentrations

Measurement of intraphagosome ion concentrations comes with problems similar to those encountered in real-time detection of enzymatic phagosome activities. While quite a few very useful probes have been developed in recent years [29, 30], most of them cannot be firmly attached to microbial surfaces as they are. Such surface labeling would allow the quantification of intraphagosome ion concentrations while avoiding overlaying signals from other compartments as the signal is strictly phagosome-bound.

Use of software-aided microscopic tools that allow focusing onto one single phagosome makes it possible to load the infected host cells with membrane-permeant reporters and to quantify the intraphagosome environment. To do this, important requirements ought to be fulfilled: (i) the concentration of the compound to be detected should not be much higher in the cytoplasm surrounding the phagosome, which would very much limit the significance of data by surpassing the actual signal; (ii) there should be a protocol that allows for equilibration of the compound between phagosome interior, cytosol and outside world in order to standardize the system; (iii) the probes should be ratiometric, so that they can be excited at two different wavelengths, one yielding a ligand-independent signal to calibrate the probe concentration in the sample, and one being sensitive to ligand binding and indicating the concentration of ligand present in the phagosome.

A pioneering study that fulfilled at least some of these criteria [31] investigated the concentrations of calcium, potassium and sodium ions and pH, respectively, in *Chlamydia trachomatis*-containing vacuoles. All these concentrations were found to be the same in both cytosol and vacuole. It was concluded, in context with earlier studies, that substances smaller than some 500 Da could cross the vacuole membrane.

Another type of tool has recently been developed in which a protein with a tetracysteine tag is expressed in or on the cell of choice. A small (~1 kDa) molecule which serves as a biarsenical Ca^{2+} sensor (Calcium Green FlAsH) is added and binds tightly to the tetracystine motif. In this combination, Calcium Green FlAsH becomes ten times more fluorescent when calcium ions are present. This technique has not yet been used to probe intraphagosome calcium levels, but may become a very useful tool to do so [32].

Another case in which the lack of probes that can be firmly attached to particles is unproblematic is when particles are ingested by macropinocytosis or by other pathways into "spacious phagosomes." Then soluble and membrane-impermeable probes can be ingested together with the particle: This was examplified when during *Listeria monocytogenes* micropinocytosis, a soluble calcium sensor (fura dextran) was ingested together with the prey by the phagocyte and used to follow calcium concentrations within the phagosome [33]. For applications that do not require firm attachment, many fluorescent metal ion indicators are, in principle, available [29, 30], although not all may be useful for they may lack membrane permeability or because they respond to their target compounds only in a biologically irrelevant concentration range.

A novel approach to the assessment of intraphagosomal ion concentrations (i.e., chlorine, calcium, potassium, manganese, copper, phosphorus, sulfur, and zinc) was published by Wagner *et al.* [34] using a hard X-ray microprobe directly on biological samples. These authors studied differences in metal concentrations between vacuoles containing different mycobacteria at 1 and 24 h of infection by positioning the hard X-ray microprobe on the surface of the infected cell at a position corresponding to the phagosome below. Beyond producing an inventory of ion concentrations within the phagosome, this study provided experimental evidence for the validity of the approach by using a mutant of *M. tuberculosis* that cannot produce an extracellular iron-capturing siderophore. The authors showed that a siderophore mutant failed to accumulate iron in its phagosome over a 24 h period, while iron concentrations were 10-fold the concentration in vacuoles with wild-type bacteria. The method was further validated by independently feeding radioactive iron to the infected macrophages and demonstrating the accumulation of such iron in the mycobacterial phagosome. It can be expected that similar more physical approaches (using mass spectrometry and nuclear magnetic resonance) will, in the future, provide additional means to quantify intraphagosomal contents.

Finally, a method to measure intraphagosomal oxygen concentration was reported, based on cofeeding of yeast cell wall particles and a membrane-impermeant small oxygen probe, d-Cat_1 [35].

5.1.5
Problems with Intraphagosomal Enzyme Activity Assays

Most of the assays presented above have been developed and used with inert particles, such as carboxylated latex or silica beads. These industrially produced beads do not only provide uniform size, shape and quality, but also surfaces that can be used to couple reporter molecules, for example, by cross-linking protein amino groups to carboxylated latex beads or by adsorbing lipids on hydrophobic bead surfaces. While these investigations are important and valuable, they do not tell us what happens when *biological* particles in general and pathogenic microorganisms in particular are contained in a phagosome. Unfortunately, most microorganisms will not adsorb and tightly maintain hydrophobic substrates for lipolytic assays. Covalent coupling of reporter proteins to their surfaces, on the other hand, is often neither easy nor amenable to the viability of microbes. Therefore, new probes need to be developed. As many microorganisms can be surface-labeled to a reasonable degree of saturation with amino reactive groups (such as succinimidyl esters), such chemistry would allow many different reporter compounds to be coupled readily. It then still remains to be tested whether surface presentation of such compounds will leave the pathologically changed phagosome trafficking unaltered, because phagocytic receptor choice plays an important role in the fate of many phagosomes [36]. Also, virulence-relevant surface-located systems, such as protein secretion systems, may be inactivated by chemical cross-linking. Coupling of, for example, densely fluorescent-labeled mannosylated serum albumin as a protease reporter to a bacterial surface may lead to its increased uptake via macrophage mannose receptor and, hence, possibly altered trafficking.

5.1.6
Phagosome Content in and Accessibility for Lipids

This is a relatively little studied field. The *Chlamydia*-containing vacuole is the paradigm for labeling live, infected cells with lipid derivatives to probe their accessibility to a pathogen-containing phagosome. This vacuole intersects with the trafficking of certain sphingomyelin- and cholesterol-containing vesicles from the *trans*-Golgi network to the cell surface. Fluorescent lipid derivatives, such as NBD-C(6)-ceramide,

NBD-glucosylceramide or NBD-sphingomyelin were used to probe the accessibility of these lipids to the *Chlamydia* vacuole [37]. The experiments showed that while the glucosylceramide compound was not retained in the vacuole, sphingomyelin was.

While this was an assay for live cells, fixed cells can be analyzed with some lipid probes directly, although fixation of cells before fluorescence microscopy is usually little suited to retain lipids. This becomes even more critical when the preparation has to be permeabilized to allow staining agents access to the cell interior.

The small compound filipin is often used to label cholesterol [38], "lipid raft" assemblies can be visualized by the expression of a raft-localizing fluorescent protein [39] and ganglioside GM1 can be detected using fluorescent cholera toxin [40].

5.1.7 Genetic Probes and Real-Time Microscopic Imaging of Phagosomes

Another method used to analyze the rules of phagosome biogenesis is "life cell imaging", in which fusion proteins between a naturally fluorescent protein (such as green or yellow fluorescent protein) and a protein whose distribution and dynamics are to be analyzed, are expressed in the host cell [41]. Combinations of several such labels are possible, for example, a green fluorescent fusion protein for late phagocytic LAMP1 protein and a red fluorescent protein that labels bacteria [42]. This can, in principle, be extended to a couple of further marker molecules with different emission wavelengths that could be analyzed simultaneously. While the answers obtained in such experiments are normally not qualitatively different from those obtained with series of immunofluorescence samples, they can provide a lot of kinetic data on a single phagosome. As immunofluorescence microscopy requires sample fixation and therefore killing, only one short look is possible into the development of one particular phagosome and many parallel samples collected at various times are required to dissect the kinetics of the reaction.

Unfortunately, fusion protein experiments have the disadvantage that the attachment of the fluorescent tag can change the functionality, half-life and the distribution of the fusion protein in the cell. In addition, expression of the genetically altered construct is often at levels higher than those of endogenous proteins, which may lead to increased, "unspecific" binding to non-targets and, hence, to misinterpretation.

But with all its limitations, life cell imaging has started to make its way into phagosome research, for example, into the analysis of the interaction of phosphoinositides with phagosomes. As recombinant protein domains are available which very specifically bind to only one type of phosphoinositide (e.g., the tandem-FYVE domains binding to phosphatidylinositol 3-phosphate, PI(3)P), production and disappearance of phosphoinositides on phagosomes can be visualized in real time [43–46]. Similarly, recruitment to phagosomes of a green fluorescent protein coupled to the PI3P-binding domain of $p47^{phox}$, a component of NADPH oxidase, was studied in microglia [46].

While all of the above studies were done using transfected cell lines, Lippuner *et al.* [47] have expressed GFP-Rab5, GFP-Rab7, or GFP-transferrin receptor fusion proteins in transgenic mice. Primary bone marrow-derived macrophages from these

mice were successfully used for life cell microscopy with *Leishmania* parasites as phagocytic particles.

5.2 Biochemical Analysis and Assays with Purified Phagosomes

5.2.1 Proteomics

With the advent of proteomics methods in the last few years, phagosome analysis has also started to change from a biased technique ("Let's look for these 10 prime candidate proteins by Western blotting") to a largely non-biased approach ("Let's analyze all proteins within a phagosome sample"). This has not only yielded inventories of novel and sometimes unexpected proteins, but has also led to new discussions about the mechanism of phagosome biogenesis in general [48, 49]. All available investigations into the phagosome proteome were based on phagosomes containing inert paramagnetic or latex beads, because they can be easily purified either by magnetic [28, 50] or flotation gradient [23] techniques (for details, see Chapter 7). However, there is very little information on proteomes of phagosomes containing any microorganism which is largely due to the limitations of phagosome purification techniques [51–53]. These problems originate from, for example, the similar density of phagosomes and other organelles, such as mitochondria. Classical density gradient centrifugation analysis therefore hardly separates these organelles. In some cases, for example, with very spacious and bulky *Salmonella*-containing vacuoles from epithelial cells, gentle lysis of infected cells at the beginning of the purification scheme can already pose a big technical hurdle. Interestingly, a recent publication presented the proteome of the *Legionella*-containing vacuole in *Dictyostelium* [54]. The isolation procedure included affinity purification of vacuoles via the translocated *Legionella* effector protein SidC. This way, the authors analyzed more than 500 *Dictyostelium* proteins of the *Legionella* vacuole including all known suspects (see Chapter 4) such as the small GTPases Arf1, Rab1 and Rab7. Rab8 and Rab14 were examples of newly identified, unexpected proteins.

Density step gradients are mostly used in microbe-containing phagosome purification instead of discontinuous density gradients, because they speed up the purification process and they take into account the heterogeneity of phagosome compartments in a cell homogenate (some phagosomes having more early characteristics, others already having fused with lysosomes, some being single-bacterium phagosomes, others being communal, etc.). Efforts are being made to overcome such limitations, for example, by chemically cross-linking microorganisms to magnetic particles followed by magnetic separation [55]. Another alternative is separation of organelles by free flow electrophoresis [56]. Phagosomes obtained with either of these methods can be used in Western blotting [51–53, 55, 56], in enzyme activity determination [22, 51] or in cell-free organelle fusion assays (see Chapter 6). In addition, Trivedi *et al.* [57] described a method used to quantify very

close approximation of phagosomes and lysosomes in a cell-free system: Isolated phagosomes containing scintillant latex beads were mixed with lysosomes that contained tritiated marker. Only upon very close contact does scintillation occur and it can be quantified. This “tethering” reaction required ATP and cytosol, just as cell-free membrane fusion does (Chapter 6).

5.2.2 Lipidomics

Phagosomes do not only consist of proteins. Membrane materials comprise just as much lipids as proteins, but because of the lack of suitable detection methods, lipids have been neglected for many years. Recently, the group of Bernd Helms (Utrecht University), in cooperation with the group of Gareth Griffiths (University of Oslo, Norway), have identified hundreds of lipid species in phagosomes and have investigated changes in phagosome lipid composition over time (unpublished, cited after [36]).

5.3 Use of Fluorescence-Activated Cell Sorting (FACS) in Phagosome Research

FACS, although a well-established technique in immunological and hematological research, has only been used in a few phagosome studies. There are a number of reports on cell homogenates or partially purified phagosomes which were immunostained and analyzed by FACS for the presence and quantities of the respective markers on phagosomes [58–61]. In a modified protocol, pH was quantified in latex bead-containing phagosomes using a similar labeling technology as described above [62] and actin polymerization on phagosomes has also been investigated [63]. To the best of my knowledge, there is no published study on purification of phagosomes by FACS, largely due to the very high numbers of phagosomes required for proteomics and due to the mechanical stress exerted on these fragile organelles. In addition, safety considerations make it hard for most laboratories to sort phagosomes containing live biohazard level 2 (e.g., *Mycobacterium avium, Salmonella typhimurium*) or 3 (*Brucella melitensis, Mycobacteriun tuberculosis*) microorganisms, although some permissive instrumentation and arrangements became available recently [64]. In some cases, it may be possible to kill all microbes before FACS analysis without affecting the phagosomes too much. However, standard formaldehyde fixation and killing makes phagosomes worthless in many respects. Working with safety level 1 microbes, on the other hand, is much less complicated. The vaccination strain *Mycobacterium bovis* strain BCG (safety level 1) localizes to a pH ~6.5 early endocytic compartment in macrophages [65]. A mutant library has been established and phagosomes containing mutants in acidified (LysoTracker-positive) phagosomes have been enriched using FACS [66].

Another problem, beyond the biological safety considerations, is the small size of most phagosomes, which is close to the FACS detection limit [67], and, particularly when they are larger, their fragility.

5.4 Knockdown, Knockout and Phagosome Research

RNA interference presents us with a method to "silence" defined genes indirectly by inducing the destruction of their mRNA [68]. This "knockdown" method is based on the introduction of suitable small interfering RNAs (siRNAs), for example, gene-specific 21–28 nucleotide RNAs, into cells [69]. In a successful experiment, the decrease in steady state concentration of the affected proteins is decreased within some 12–24 h and the protein is eliminated within 48–72 h after transfection. These cells can then be infected. The knockdown efficiency depends largely on the effectiveness of transformation, on the nature of the used constructs, and on the half-life of the protein to be depleted.

Unfortunately, most professionally phagocytic cells are hard to transfect and, therefore, it is almost impossible to introduce an siRNA homogeneously into all cells of a culture. Nevertheless, microbial multiplication can be studied microscopically, as can intracellular marker protein distribution, and microscopic methods for content analysis (above) can still be applied. Such studies are limited, however, to those cells that have been transfected and cannot be extended to a biochemical analysis of the whole host cell population. Alternatively, instead of introducing siRNAs by transfection, they can be directly expressed in the host cell from a plasmid [70]. This plasmid can also be used to create cell lines that stably express siRNA and therefore constitutively repress expression of a defined protein [71]. Although the production of such cell lines is labour- and time-intensive, they have the general advantage that they represent a relatively homogeneous consortium of cells, all of which should have a low, if not, completely knocked down expression of the protein in question. In this way, RNA interference was used to investigate the role of the coronin-1 protein in the development of *Mycobacterium*-containing phagosomes in murine macrophages [71] and the contribution of clathrin vesicle-associated proteins in the uptake of *Listeria monocytogenes* [72].

Other tools which help not only to describe phagosomes, but to also gain functional, molecular insights into phagosome biogenesis, are knockout mice. For example, cells from mice deficient in integrin β1 synthesis have reduced phagolysosome formation [73] and cells from mice deficient in the small GTPase Rab27a show increased phagosome acidification and antigen presentation [74].

Non-mammalian phagocytes may be much more accessible to transformation and inhibitory RNA-mediated knockdown and therefore are, in some cases, excellent models for the study of host genetics in infection. *Drosophila* Schneider (S2) cells, for example, are easily transfectable and lend themselves to genome-wide RNA interference library screens. Cheng *et al.* [75] screened genome-wide for genes involved in the uptake of *Listeria monocytogenes* by S2 cells, in the escape from the phagosome and in intracellular multiplication. Similarly, using a selection of 86 siRNAs in S2 cells, several factors were identified which modulate the biogenesis of *Mycobacterium*-containing phagosomes [76]. In a proteomics approach, Stuart *et al.* [77] identified 617 proteins in preparations of latex bead-containing phagosomes from S2 cells.

Subsequently, siRNA to each of the corresponding genes were tested for effects on phagocytosis of bacteria. Approximately every fourth siRNA led to changes in phagocytosis rates.

These examples show that phagosome trafficking can be experimentally manipulated in non-mammalian phagocytes and, therefore, can be used to molecularly understand the host side of phagosome biogenesis (see Chapter 4). Limitations apply, in particular with pathogens that require temperatures around 37 °C to develop virulence characteristics, a temperature far too high for undisturbed growth of *Dictyostelium*, *Drosophila* and many other organisms.

5.5 Transcriptomics as a Phagosome Environmental Sensor

With the advent of transcriptome analyses in microorganisms, labs also became interested in analyzing the transcriptional response of microorganisms to host cells [73, 78] and, of course, also in the analysis of the host cell response to infection [79]. Transcriptomic analysis is not only interesting in the molecular analysis of how either of the two organisms involved reacts to exposure to the other, it can also, to some extent, deliver a picture of the phagosome microenvironment (see Chapter 9).

As shown by a recent study on *M. tuberculosis* in macrophages [80], these bacteria respond characteristically to a moderate, but characteristic, drop in the pH of the phagosome just after its creation (from a pH of ~7.2 to ~6.3). One of the outcomes of this study was the observation that acidification was an important environmental clue, stimulating pathogens to quickly adapt their physiology to intracellular persistence. Of course, the converse is also true: The transcriptional acid response can also be used to probe the approximate proton concentration in the phagosome. Such an approach could, with all its biological variability, be extended to other relevant factors, such as iron, exposure to which triggers a characteristic transcriptional response, or even exposure to certain host lipids (for a review, see [78]). In the further future, environmental sensor plasmids might become available which can be used, as GFP plasmids are used today, to visualize the intraphagosomal environmental conditions in real time. Some of these sensor proteins may well work in concert with small membrane-permeable fluorescent compounds that specifically bind to these probes and provide a kinetic view of phagosomes as they are created and eventually disappear before the experimenter's eyes.

Acknowledgments

Editorial help by Tobias Dykstra and comments by Kristine von Bargen and Ulrike Becken were highly appreciated. The author's own work was supported by grants from the Deutsche Forschungsgemeinschaft.

References

1 Desjardins, M. (1995) *Trends Cell Biol*, **5**, 183–186.

2 Griffiths, G. (1996) *Protoplasma*, **195**, 37–58.

3 Holt, O.J., Gallo, F. and Griffiths, G.M. (2006) *J Biochem*, **140**, 7–12.

4 Desjardins, M. and Descoteaux, A. (1997) *J Exp Med*, **185**, 2061–2068.

5 Kalamidas, S.A., Kuehnel, M.P., Peyron, P., Rybin, V. *et al.* (2006) *J Cell Sci*, **119**, 3686–3694.

6 Yates, R.M., Hermetter, A. and Russell, D.G. (2005) *Traffic*, **6**, 413–420.

7 Sydor, T., von Bargen, K., Becken, U., Spuerck, S. *et al.* (2008) *Vet Microbiol*, **128**, 327–341.

8 Ohkuma, S. and Poole, B. (1978) *Proc Natl Acad Sci USA*, **75**, 3327–3331.

9 Galloway, C.J., Dean, G.E., Marsh, M., Rudnick, G. and Mellman, I. (1983) *Proc Natl Acad Sci USA*, **80**, 3334–3338.

10 Lukacs, G.L., Rotstein, O.D. and Grinstein, S. (1990) *J Biol Chem*, **265**, 21099–21107.

11 Lukacs, G.L., Rotstein, O.D. and Grinstein, S. (1991) *J Biol Chem*, **266**, 24540–24548.

12 Schneider, B., Gross, R. and Haas, A. (2000) *Infect Immun*, **68**, 7039–7048.

13 Oh, Y.K. and Straubinger, R.M. (1996) *Infect Immun*, **64**, 319–325.

14 Schramm, N., Bagnell, C.R. and Wyrick, P.B. (1996) *Infect Immun*, **64**, 1208–1214.

15 Crowle, A.J., Dahl, R., Ross, E. and May, M.H. (1991) *Infect Immun*, **59**, 1823–1831.

16 Clemens, D.L., Lee, B.-Y. and Horwitz, M.A. (2004) *Infect Immun*, **72**, 3204–3217.

17 Newman, S.L., Gootee, L., Hilty, J. and Morris, R.E. (2006) *J Immunol*, **176**, 1806–1813.

18 Fernandez-Mora, E., Polidori, M., Lührmann, A., Schaible, U.E. and Haas, A. (2005) *Traffic*, **6**, 635–653.

19 Via, L.E., Fratti, R.A., McFalone, M., Pagan-Ramos, E. *et al.* (1998) *J Cell Sci*, **111** (Pt 7), 897–905.

20 Schaible, U.E., Sturgill-Koszycki, S., Schlesinger, P.H. and Russell, D.G. (1998) *J Immunol*, **160**, 1290–1296.

21 Kenmoku, S., Urano, Y., Kojima, H. and Nagano, T. (2007) *J Am Chem Soc*, **129**, 7313–7318.

22 Claus, V., Jahraus, A., Tjelle, T., Berg, T. *et al.* (1998) *J Biol Chem*, **273**, 9842–9851.

23 Garin, J., Diez, R., Kieffer, S., Dermine, J.F. *et al.* (2001) *J Cell Biol*, **152**, 165–180.

24 Ullrich, H.J., Beatty, W.L. and Russell, D.G. (1999) *Eur J Cell Biol*, **78**, 739–748.

25 Beaman, L., Paliescheskey, M. and Beaman, B.L. (1988) *Infect Immun*, **56**, 1652–1654.

26 Rohde, K., Yates, R.M., Purdy, G.E. and Russell, D.G. (2007) *Immunol Rev*, **219**, 37–54.

27 Yates, R.M., Hermetter, A., Taylor, G.A. and Russell, D.G. (2007) *Traffic*, **8**, 241–250.

28 Robinson, N., Kolter, T., Wolke, M., Rybniker, J. *et al.* (2008) *Traffic*, **9**, 1936–1947.

29 Terai, T. and Nagano, T. (2008) *Curr Opin Chem Biol*, **12**, 515–521.

30 Domaille, D.W., Que, E.L. and Chang, C.J. (2008) *Nat Chem Biol*, **4**, 168–175.

31 Grieshaber, S., Swanson, J.A. and Hackstadt, T. (2002) *Cell Microbiol*, **4**, 273–283.

32 Tour, O., Adams, S.R., Kerr, R.A., Meijer, R.M. *et al.* (2007) *Nat Chem Biol*, **3**, 423–431.

33 Shaughnessy, L.M., Hoppe, A.D., Christensen, K.A. and Swanson, J.A. (2006) *Cell Microbiol*, **8**, 781–792.

34 Wagner, D., Maser, J., Lai, B., Cai, Z. *et al.* (2005) *J Immunol*, **174**, 1491–1500.

35 James, P.E., Grinberg, O.Y., Michaels, G. and Swartz, H.M. (1995) *J Cell Physiol*, **163**, 241–247.

36 Haas, A. (2007) *Traffic*, **8**, 311–330.

37 Moore, E.R., Fischer, E.R., Mead, D.J. and Hackstadt, T. (2008) *Traffic*, **9**, 2130–2140.

38 Carabeo, R.A., Mead, D.J. and Hackstadt, T. (2003) *Proc Natl Acad Sci USA*, **100**, 6771–6776.

39 Zacharias, D.A., Violin, J.D., Newton, A.C. and Tsien, R.Y. (2002) *Science*, **296**, 913–916.

40 Watarai, M., Makino, S.-I., Fujii, Y., Okamoto, K. and Shirahata, T. (2002) *Cell Microbiol*, **4**, 341–355.

41 Giepmans, B.N.G., Adams, S.R., Ellisman, M.H. and Tsien, R.Y. (2006) *Science*, **312**, 217–224.

42 Drecktrah, D., Knodler, L.A., Howe, D. and Steele-Mortimer, O. (2007) *Traffic*, **8**, 212–225.

43 Cosío, G. and Grinstein, S. (2008) *Methods Mol Biol*, **445**, 287–300.

44 Ellson, C.D., Anderson, K.E., Morgan, G., Chilvers, E.R. *et al.* (2001) *Curr Biol*, **11**, 1631–1635.

45 Di, A., Brown, M.E., Deriy, L.V., Li, C. *et al.* (2006) *Nat Cell Biol*, **8**, 933–944.

46 Ueyama, T., Lennartz, M.R., Noda, Y., Kobayashi, T. *et al.* (2004) *J Immunol*, **173**, 4582–4589.

47 Lippuner, C., Paape, D., Paterou, A., Brand, J. *et al.* (2008) *FASEB J*, **23**, 483–491.

48 Gagnon, E., Duclos, S., Rondeau, C., Chevet, E. *et al.* (2002) *Cell*, **110**, 119–131.

49 Touret, N., Paroutis, P., Terebiznik, M., Harrison, R.E. *et al.* (2005) *Cell*, **123**, 157–170.

50 Lutz, D.A., Chen, X.M. and McLaughlin, B.J. (1993) *Anal Biochem*, **214**, 205–211.

51 Lührmann, A. and Haas, A. (2000) *Methods Cell Sci*, **22**, 329–341.

52 Howe, D. and Heinzen, R.A. (2008) *Methods Mol Biol*, **445**, 389–406.

53 Sturgill-Koszycki, S., Schaible, U.E. and Russell, D.G. (1996) *EMBO J*, **15**, 6960–6968.

54 Urwyler, S., Nyfeler, Y., Ragaz, C., Lee, H. *et al.* (2008) *Traffic*, **10**, 76–87.

55 Lönnbro, P., Nordenfelt, P. and Tapper, H. (2008) *BMC Cell Biol*, **9**, 35.

56 Hasan, Z., Schlax, C., Kuhn, L., Lefkovits, I. *et al.* (1997) *Mol Microbiol*, **24**, 545–553.

57 Trivedi, V., Zhang, S.C., Stockinger, W. and Nohturfft, A. (2007) *Sci STKE*, **392**, 3.

58 Steele-Mortimer, O., Méresse, S., Gorvel, J.P., Toh, B.H. and Finlay, B.B. (1999) *Cell Microbiol*, **1**, 33–49.

59 Scianimanico, S., Desrosiers, M., Dermine, J.F., Méresse, S. *et al.* (1999) *Cell Microbiol*, **1**, 19–32.

60 Ramachandra, L., Sramkoski, R.M., Canaday, D.H., Boom, W.H. and Harding, C.V. (1998) *J Immunol Methods*, **213**, 53–71.

61 Guermonprez, P., Saveanu, L., Kleijmeer, M., Davoust, J. *et al.* (2003) *Nature*, **425**, 397–402.

62 Savina, A., Jancic, C., Hugues, S., Guermonprez, P. *et al.* (2006) *Cell*, **126**, 205–218.

63 Defacque, H., Egeberg, M., Habermann, A., Diakonova, M. *et al.* (2000) *EMBO J*, **19**, 199–212.

64 Lennartz, K., Lu, M., Flasshove, M., Moritz, T. and Kirstein, U. (2005) *Cytometry A*, **66**, 119–127.

65 Hackam, D.J., Rotstein, O.D., Zhang, W.J., Demaurex, N. *et al.* (1997) *J Biol Chem*, **272**, 29810–29820.

66 Stewart, G.R., Patel, J., Robertson, B.D., Rae, A. and Young, D.B. (2005) *PLoS Pathog*, **1**, 269–278.

67 Jose, J., Betscheider, D. and Zangen, D. (2005) *Anal Biochem*, **346**, 258–267.

68 Bonetta, L. (2004) *Nat Methods*, **1**, 79–84.

69 Meister, G. and Tuschl, T. (2004) *Nature*, **431**, 343–349.

70 Tuschl, T. (2002) *Nat Biotechnol*, **20**, 446–448.

71 Jayachandran, R., Gatfield, J., Massner, J., Albrecht, I. *et al.* (2008) *Mol Biol Cell*, **19**, 1241–1251.

72 Veiga, E. and Cossart, P. (2005) *Trends Cell Biol*, **15**, 2–5.

73 Wang, Q.-Q., Li, H., Oliver, T., Glogauer, M. *et al.* (2008) *J Immunol*, **180**, 2419–2428.

74 Jancic, C., Savina, A., Wasmeier, C., Tolmachova, T. *et al.* (2007) *Nat Cell Biol*, **9**, 367–378.

75 Cheng, L.W., Viala, J.P.M., Stuurman, N., Wiedemann, U. *et al.* (2005) *Proc Natl Acad Sci USA*, **102**, 13646–13651.

76 Philips, J.A., Porto, M.C., Wang, H., Rubin, E.J. and Perrimon, N. (2008) *Proc Natl Acad Sci USA*, **105**, 3070–3075.

77 Stuart, L.M., Boulais, J., Charriere, G.M., Hennessy, E.J. *et al.* (2007) *Nature*, **445**, 95–101.

78 Kendall, S.L., Rison, S.C.G., Movahedzadeh, F., Frita, R. and Stoker, N.G. (2004) *Trends Microbiol*, **12**, 537–544.

79 Schnappinger, D., Ehrt, S., Voskuil, M.I., Liu, Y. *et al.* (2003) *J Exp Med*, **198**, 693–704.

80 Rohde, K.H., Abramovitch, R.B. and Russell, D.G. (2007) *Cell Host Microbe*, **2**, 352–364.

6
In Vitro Fusion Assays with Phagosomes

Ulrike Becken and Albert Haas

6.1
Introduction

Numerous basic cellular functions, such as reformation of organelles after cell division, protein sorting and interaction of the cell with the outside in the broadest sense, are closely linked with membrane fusion. During mitosis some membrane structures, such as the Golgi apparatus, disintegrate and membrane fusion leads to reformation of the organelle after cell division. Furthermore, many proteins are transported from one compartment to another by vesicles budding from one membrane and fusing with another. Essentials factors such as cholesterol or iron are taken up by receptor-mediated endocytosis and are eventually delivered to the lysosome, a process that involves several successive fusion and fission processes of endocytic organelles. The abundance of certain receptors on the cell surface is regulated by exo- and endocytosis of vesicles, leading to either the exposure of receptors at the cell surface or to receptor degradation in the lysosome. In addition to these general fusion events, some have evolved only in certain cells and for specialized functions. For example, professional phagocytes recognize different kind of particles such as microbes, cell debris or apoptotic cells via a legion of different phagocytic receptors whose binding and clustering mediates engulfment of the bound particle and formation of a phagosome. Phagosomes mature by fusion with organelles of the endocytic pathway, first with early endosomes and subsequently with late endosomes and lysosomes, acquiring antimicrobial properties such as an acidic pH and the presence of diverse hydrolytic enzymes. The phagocytosed particle is eventually killed and degraded and, in the case of a microbe, further fusion and fission events result in antigen presentation on the cell surface.

Membrane fusion is strictly regulated by a complex protein machinery, ensuring fusion of designated partners and preventing incorrect fusion processes. This machinery involves directed transport of organelles along cytoskeletal filaments, followed by a tight association of prospective fusion partners. Central to this machinery are the highly compartment-specific Rab-GTPases, which are recruited

Intracellular Niches of Microbes. Edited by Ulrich E. Schaible and Albert Haas

ISBN: 978-3-527-32207-7

to organelle membranes and in turn interact with various other cytosolic or membrane-bound proteins, so-called tethering factors. This establishes a connection between two organelles which necessarily precedes membrane fusion. The relatively loose attachment of membranes by Rab-GTPases and tethering factors is followed by a very tight binding of cognate SNARE (soluble *N*-ethylmaleimide-sensitive-factor attachment receptor) proteins present on both fusion partners, reducing the gap between the two membranes to just a couple of nanometers.

Much of the knowledge about membrane fusion mechanisms comes from cell-free systems. In such systems, isolated organelles are mixed and incubated under fusion-promoting conditions (appropriate buffer system, physiological temperature and presence of essential factors such as ATP or cytosolic proteins). Fusion efficiency is then quantified using a suitable assay.

The particular advantages of the *in vitro* study of membrane fusion are that (i) fusion reactions can be kinetically dissected (see below), (ii) reaction conditions can be exactly defined and (iii) reaction conditions can be easily manipulated. For instance, the function of a specific protein can be blocked by adding antibodies against it to the *in vitro* fusion reaction, an experimental approach that confirmed the necessity of an *N*-ethylmaleimide-sensitive fusion protein (NSF) for different membrane fusion events [1]. Cell-free fusion reactions can also be run with (genetically or biochemically) depleted components. For example, depletion of Rab5 from organelle membranes and cytosol abrogates homotypic fusion of early endosomes, showing this protein to be essential for endosome fusion [2].

Over the past 30 years many fusion events have been reconstituted *in vitro*, including trafficking between Golgi stacks [3], fusion of endoplasmic reticulum (ER) and Golgi vesicles with their target membrane [4], homotypic fusion of yeast vacuoles [5], or fusion between different compartments of the endocytic pathway [6–8] or with phagosomes [9]. In fact, the first *in vitro* membrane fusion assay ever, published by Oates and Touster [10], reconstituted homotypic fusion of phagosomes derived from *Acanthamoeba*, a protozoan which engulfs particles for nutrition. Fifteen years later, the group of Stahl at Washington University was the first to reconstitute fusion of early phagosomes and early endosomes from the mouse macrophage-like cell line J774E [9]. They found that phagosome–endosome fusion is similar to homotypic endosome fusion in that it is dependent on essential factors such as ATP, cytosolic proteins and NSF. They also showed that, like endosome–endosome fusion, the fusion competence of phagosomes and endosomes depends on the maturation stage of the organelles.

6.2
A Variety of Assays

Several *in vitro* membrane fusion assays have been developed to analyze phagosome maturation. In general all the approaches must fulfill two conditions: Firstly, the designated fusion partners are differentially labeled in a way that mixture of both labels results in formation of a specific, quantifiable product. Secondly, the assay must exclude that formation of this product is due to membrane leakiness instead of membrane fusion. The assays can be categorized in the following way:

6.2.1 Microscopic versus Biochemical Assays

Two alternatives to observe organelle fusion in a cell-free system are microscopical or biochemical detection.

For microscopic assays each of the fusion partners is tagged with a microscopically detectable label and organelle fusion presents itself as colocalization of both labels. All published microscopic *in vitro* phagosome–endosome fusion assays are based on electron microscopy. Endocytic compartments are labeled with electron-dense gold nanoparticles via endocytic uptake, while phagosomes can be readily discriminated in transmission electron microscopy without a specific label because of the size and unique structure of the enclosed particle. Fusion of the organelles results in a membrane-surrounded structure containing both the phagocytosed particle and endosomal gold (Figure 6.1d). To determine fusion efficiency, a sufficiently high number of phagosomes is analyzed and frequency of colocalization of phagocytosed particles and endosomal gold is quantified [9, 11, 12]. Biochemical assays for organelle fusion, on the other hand, are based on the formation of a specific fusion product that can be quantified enzymatically or fluorometrically. Some examples are listed in detail below.

Both microscopic and biochemical assays have certain advantages over the other. Microscopic assays are normally more sensitive than biochemical ones. Theroretically, a content of few hundred phagosomes in one microscopical sample is sufficient to calculate fusion efficiency, a number usually too low to overcome the detection limit in a biochemical assay containing many thousands of phagosomes. On the other hand, enzymatic or fluorometric quantification is typically much less laborious than the preparation and examination of electron microscopical samples and so allows a higher throughput. Furthermore, different aspects of membrane fusion are quantified in the two types of assay. While microscopic assays indicate how many phagosomes in a sample have turned into fusion products (e.g., phagolysosomes), biochemical assays integrate over the whole phagosome population and can state how much material has been transferred from endocytic to phagocytic organelles.

Our lab has recently established a fluorescence microscopy-based *in vitro* assay for fusion of lysosomes with bacteria-containing phagosomes (Becken and Haas, unpublished, similar to an assay described by Brandhorst *et al.* [13] for homotypic early endosome fusion. In our assay, bacteria are covalently surface-labeled with a green fluorescent dye before phagocytosis and lysosomes are preloaded with a red fluorophor. After vesicle fusion both dyes are present in one compartment. This can be visualized as a spatial overlap of both signals using conventional fluorescence microscopy (Figure 6.1e). Using this method there is little danger of false-positive signals caused by fluorescent dye leaking from ruptured lysosomes. If that did happen, it would be strongly diluted in the reaction buffer and would not be able to

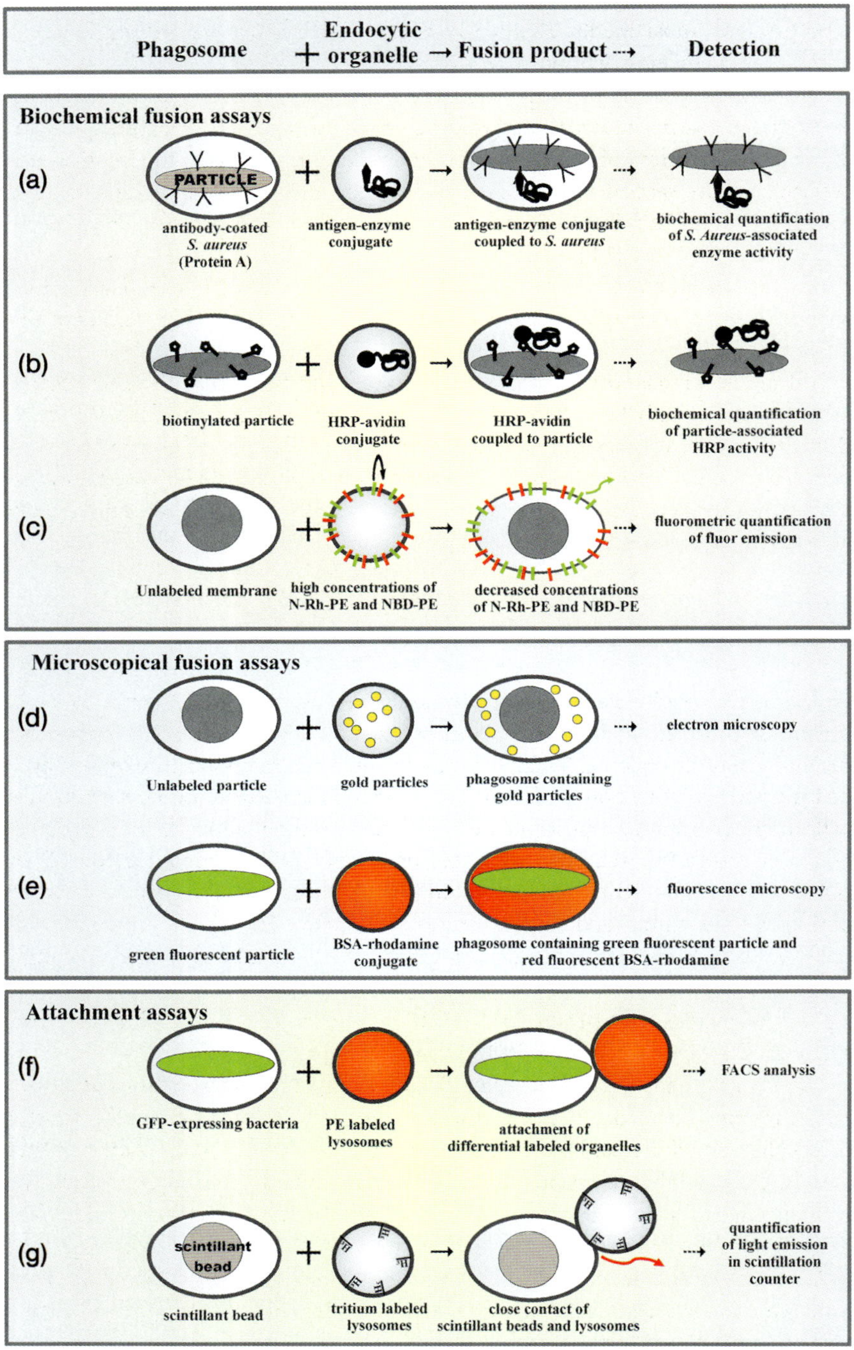

Figure 6.1 Assays reconstituting fusion or attachment of phagosomes and endocytic compartments *in vitro*. HRP, horseradish peroxidase; N-Rh-PE, lissamine rhodamine phosphatidylethanolamine; NBD-PE, benzoxadiazole phosphatidylethanolamine; BSA, bovine serum albumin; GFP, green fluorescent protein; PE, phycoerythrin; FACS, fluorescence-activated cell sorting. See text for details.

label further compartments. This means there is no need for special precautions to avoid a fusion-independent positive readout.

This assay system can be further validated by choosing suitable fluorophores with an overlap between the emission spectrum of the one and the excitation spectrum of the second fluorophore. Excitation of the first dye thus leads to the fluorescence emission of the second one in fused organelles (fluorescence resonance energy transfer, FRET). This interaction requires very close contact between the fluorophores (10 nm or less [14]), which can only be established by organelle fusion but not by an attachment of vesicles. Therefore, using FRET to quantify phagosome–lysosome fusion is an appropriate and even more stringent tool than colocalization studies.

Software tools to automate the taking and analyzing of fluorescence microscopic pictures have already been used in cell-free endosome–endosome fusion systems [13]. Use of such systems will hopefully also allow a time-efficient interpretation of microscopic data deriving from the described *in vitro* phagosome–lysosome fusion assay and so help to combine the high sensitivity of microscopy with the high capacity of biochemical approaches in one assay.

6.2.2
Content Mixing versus Membrane Mixing Assays

Fusion of two organelles includes two distinct events: mixing of membranes and mixing of contents, both of which can be used to detect phagosome–endosome fusion in cell-free systems. So far, the only example of a membrane mixing assay is that introduced by Peyron *et al.* [15]. These workers labeled lysosome membranes with the fluorescent phospholipid analogs lissamine rhodamine phosphatidylethanolamine (N-Rh-PE) and benzoxadiazole phosphatidylethanolamine (NBD-PE) in concentrations sufficiently high to repress fluorescence emission of the latter probe due to self-quenching and resonance energy transfer. Fusion of lysosomes with unlabeled phagosomes resulted in mixing of organelle membranes, decreased fluor density and, as a result, increased fluorescence intensity of NBD (Figure 6.1c). In this assay, a 1% increase in NBD fluorescence after membrane fusion was interpreted as 1% of the lysosomes having fused with phagosomes.

There are potential hazards of using dequenching of membrane fluors as a readout for membrane fusion. First, the assay must be carefully controlled for the transfer of label between donor and acceptor compartments via lipid-binding proteins rather than by true membrane fusion [16]. Second, membrane mixing assays are limited in that only fusion events actually involving membrane mixing are detected and they do not detect fusion after the "kiss and run" principle, where two organelles do not completely fuse but contents exchange only via a transient fusion pore. Both limitations are overcome in content mixing assays, in which the mixture of luminal labels from two fusion partners is detected.

The most commonly used cell-free phagosome–endosome fusion assays are based on an avidin–biotin binding system. In an approach developed by Alvarez-Dominguez *et al.* [17], phagosomes are labeled via surface biotinylation of the internalized bacteria and endosomes are tagged by endocytosed avidin–horseradish peroxidase (HRP) conjugate. The avidin–HRP conjugate binds to the biotinylated bacterial surface after phagosome–endosome fusion. The presence of unlabeled avidin as a scavenger outside the organelles prevents the binding of HRP–avidin due to membrane rupture and therefore false-positive signals. After separation of bacteria and soluble molecules, bacteria-associated peroxidase activity and therefore fusion rate is quantified in an enzymatic assay (Figure 6.1b).

6.2.3
Semi-intact Cells versus Cell-Free Assays

An alternative to a fully biochemically reconstituted system is the use of "semi-intact cells." In cell-free assays, cell lysates that contain the compartments of interest or organelles isolated from such lysates are used, while in approaches with semi-intact cells only the plasma membrane is permeabilized while the morphological integrity of the cell is retained. Pores, usually introduced through bacterial pore-forming toxins, allow fast exchange of soluble contents between inside and outside of the cells while most vesicles retain their position so that the inner architecture of the cell, including the association of vesicles and the cytoskeleton, is more or less conserved. This may be a reason for the generally higher fusion rate in semi-intact cells compared with that in cell-free systems. In fact, fusion of phagosomes with lysosomes, which is a low-frequency event even *in vivo* [18], was reconstituted in semi-intact cells for the first time [11]. In this system, phagosome–lysosome fusion is quantified using an avidin–biotin system: Lysosomes are preloaded with an avidin–HRP conjugate, cells are then allowed to engulf biotinylated particles for few minutes and subsequently are permeabilized. Almost the entire phagosome maturation sequence can proceed in these permeabilized cells as only phagosome–lysosome fusion as its last step is detected. This means that specific experimental conditions, for example, addition of drugs, can also affect other maturation steps and it is therefore difficult to assign their effect to one given maturation step. Nevertheless, organelle fusion in semi-intact cells is indeed more efficient than that in cell-free systems (64% [11] vs. 12% [12] of phagosomes matured).

6.2.4
Organelle Fusion versus Organelle Targeting

All the assays described above quantify fusion of phagocytic with endocytic compartments. Further assays have been established to detect attachment of phagosomes to endosomes. In 2001, Naori *et al.* determined the attachment rate of phagosomes containing GFP-expressing bacteria and fluorescently labeled lysosomes in a cell-free system by FACS (fluorescence-associated cell sorting) [19] (Figure 6.1f). Five years later, Stockinger *et al.* introduced a proximity scintillation assay for the same purpose, in which close contact (<600 nm) of phagosomes containing scintillant latex beads and tritium-labeled lysosomes results in light emission [20] (Figure 6.1g).

6.3
Studying Normal Maturation ...

Professional phagocytic cells are central to innate immunity. They recognize and subsequently internalize microbes that enter the sterile portions of the human body. The newly formed phagosome then matures by fusion with endocytic organelles into a phagolysosome. In almost all cases this maturation process is not actively influenced by the engulfed microorganism and leads to efficient killing and degradation of the microbe. Microorganisms that are not killed by the phagocytes but rather multiply in them have in most cases evolved strategies to divert normal phagolysosome formation (see Chapter 10). To understand how these pathogens alter phagosome maturation it is essential to understand the molecular machinery involved in the undisturbed process. As a first biochemical step in this direction, fusion between endocytic compartments and phagosomes containing particles that do not interfere with normal maturation has been reconstituted. Not many apathogenic bacteria have been used in such experiments but, as decribed below, three model systems have become popular for technical reasons.

Early *in vitro* experiments used formaldehyde-fixed *Staphylococcus aureus* as phagocytic probe [9]. These bacteria bind the constant part of IgG antibodies via their surface Protein A and therefore expose free variable (Fab) parts. This feature makes them particularly suitable for the creation of an *in vitro* fusion assay as they can be used as affinity reagents for any molecule to which a specific antibody is available. Phagosomes containing antibody-coated *S. aureus* are fused with endosomes enclosing a corresponding antigen–enzyme complex (e.g., anti-dinitrophenol antibody/dinitrophenol-derivatized β-glucuronidase). Fusion of both kinds of organelles can then be quantified as bacteria-associated enzyme activity (Figure 6.1a). An antigen–bovine serum albumin (BSA) conjugate present in the reaction buffer functions as a scavenger and impedes mutual binding of antibody-coated bacteria and antigen–enzyme complexes from broken organelles. In more recent approaches bacteria were substituted with synthetic particles of a similar size, either latex beads or paramagnetic beads [12, 21]. Their respective unique physical properties allow easy purification of bead phagosomes from cell lysates. Due to their relatively low density, latex bead phagosomes can be obtained in high purity in just one discontinuous sucrose density gradient centrifugation, while phagosomes containing magnetic particles can be isolated from cell lysates using a magnet. Furthermore, because of their uniform structure, size and shape beads can be easily identified in electron microscopy. Beads are commercially available with a variety of surface modifications, allowing covalent coupling of molecules such as avidin or biotin, making them excellent probes for *in vitro* fusion assays. However, inert beads do come with a price. As they are not at all degradable, their maturation features may be different from those of degradable materials. In addition, protocols established for *in vitro* reactions containing such handy particles are not directly useful in studies with microbial particles, as purification of the phagosome is vastly different and comes with different yields and subcellular contaminations.

Experiments in cell-free systems have great potential in the analysis of the molecular machinery involved in the fusion of phagosomes with endocytic compartments. First, such systems can be manipulated in many ways. While many experimental approaches to trafficking exist for intact cells, reaction conditions in cell-free systems or semi-intact cells can be changed readily in ways that are not possible with intact cells. For example, membrane-impermeant drugs or antibodies against relevant proteins can be added *in vitro*, and even ion concentrations can be adjusted. The even greater advantage of cell-free vesicle fusion systems is that they can be used to monitor an individual type of fusion step and that they can dissect the order of events, eventually leading to membrane fusion.

By purification of phagosome and endosome populations of defined ages, fusion between phagosomes and endosomes of different maturation stages can be investigated [12]. Such experiments, carried out by the group of Griffiths, revealed preference for fusion of organelles of the same maturation stage ("homotypic fusion") and indicated decreasing fusion competence with increasing organelle age. The Griffiths group also assessed the distinct involvement of actin in different phagosome maturation steps [22, 23]. They showed that membranes of late but not early phagosomes and endosomes are able to catalyze actin nucelation *in vitro* and that, accordingly, a correlation of actin nucleation and fusion efficiency is restricted to late endocytic and phagocytic compartments (see Chapter 8).

Membrane fusion is a multistep process, which involves directed transport of compartments along cytoskeletal filaments (targeting), a loose connection of prospective fusion partners via Rab proteins and tethering factors (tethering), followed by an extremely close connection of membranes mediated by SNARE proteins (docking) and subsequent membrane fusion. Essential to understanding the role of a molecular factor, whether it is a protein part of the eukaryotic membrane fusion machinery or a bacterial factor interfering with phagosome maturation, is identification of the subprocess in which it is involved. As a paradigm, yeast vacuoles can be used in the analysis of *in vitro* fusion studies, allowing the staging of the action of molecular factors [24]. Experimentally, this is achieved by reversible inhibition of the overall fusion reaction at a defined stage. For example, organelle docking is blocked using an anti-SNARE antibody while the fusion reaction is allowed to proceed up to the point where the block has been placed (here membrane docking). Subsequently, inhibition is reversed by addition of an excess of the corresponding antigen and concurrently an inhibitor of a functionally unknown second factor is added. If the latter drug inhibits overall fusion but does not inhibit the fusion reaction that remains after relieving SNARE inhibition, then it can be concluded that the substep of the fusion reaction inhibited by the compound had been passed while the antibody was inhibiting. Therefore, the inhibited step would be upstream of SNARE action (docking), that is, a priming or tethering reaction.

Recently, such staging experiments have been performed for phagosome–lysosome interaction *in vitro*. In a cell-free assay, Stockinger *et al.* demonstrated the differential requirements for actin, calmodulin and calcium in phagosome–lysosome targeting. Actin and calmodulin are needed to establish a complex of both organelles, then vesicle attachment becomes resistant to inhibitors of actin and calmodulin and a high calcium concentration is specifically required for stabilization of that complex [20].

A third important advantage of *in vitro* reconstituted membrane fusion is that most observed effects are a consequence of direct interference with the system. In living cells, knockdown of gene expression by RNA interference or gene knockout often leads to changed expression, activity and/or degradation of other factors as well. In an *in vitro* membrane fusion reaction, there is no transcription or translation and, therefore, effects after drug addition are much less prone to artefacts.

6.4
... and the Pathogenic Case

Various serious diseases, like tuberculosis or legionnaires' disease (this book gives many other examples), are caused by intracellular pathogens that are not killed by macrophages. In many cases this malfunction is thought to originate from bacteria-induced alteration of phagosome maturation, but the exact mechanism is poorly understood. Analysis of traffic alterations by pathogenic bacteria demands examination of two components: the vesicle fusion machinery of the phagocyte, which enables phagosome maturation and degradation of engulfed particles, and the pathogen interfering with this machinery.

To dissect which part of the fusion machinery is being manipulated by a pathogen, cell-free fusion assays can be used because they allow mixing of components from different cells. Using this modular principle Naroeni *et al.* [19] analyzed maturation of phagosomes containing *Brucella suis*, a pathogen that blocks fusion of its phagosomes with lysosomes. *Brucella* phagosomes and lysosomes do not bind to each other, even if both lysosomes and cytosolic proteins are from non-infected cells. This suggests that it is not fusion but upstream targeting of the phagosome that is disturbed and that inhibition of phagosome–lysosome targeting is caused by modifications of the phagosome membrane. Moreover, this *in vitro* experiment suggested that no secreted bacterial effectors are involved when seen together with data from coinfection experiments *in vivo*, in which the phagosome maturation block was restricted to *Brucella* phagosomes and maturation of latex bead phagosomes in the same cell was undisturbed.

Experiments analyzing traffic alterations caused by *Salmonella enterica* or *Trypanosoma cruzi* revealed a very different picture. Cytosol derived from cells infected with either pathogen did not support cell-free homotypic fusion of early endosomes, suggesting that it was not the phagosomes themselves, but bacterial factors secreted into the cytosol that were responsible for inhibited vesicle fusion [25, 26]. In the case of *Salmonella*, the type III secretion system effector protein spiC was reported to manipulate trafficking.

There are also some hints that a soluble bacterial factor is at least in part responsible for arresting maturation of *Mycobacterium*-containing phagosomes at an early stage in a neutrophil model. Use of cytosol derived from cells infected with *M. smegmatis* or simply cytosol preincubated with these bacteria in membrane fusion experiments resulted in decreased fusion of lysosomes with phagosomes containing inert latex beads [15]. In a similar experimental setting, Vergne *et al.* demonstrated an inhibitory effect of the purified mycobacterial phosphatase SapM on phagosome–late endosome fusion. Based on this finding the authors hypothesize that this enzyme alters the phosphorylation status of certain lipids on the phagosomal membrane, which in turn retards recruitment of eukaryotic proteins essential for membrane fusion [27].

Cell-free systems were not only used to analyze the involvement of certain bacterial factors in trafficking alterations, but also to investigate their mechanism of action: The mycobacterial lipid PIM (phosphatidylinositol-mannoside) is thought to be essential for maintaining fusion competence of *Mycobacterium*-containing phagosomes with early endosomes. Using *in vitro* fusion systems, Vergne *et al.* [21] investigated the effects caused by PIM in two respects. First, they demonstrated that PIM action stimulates homotypic fusion of early but not late endosomes. Second, they narrowed down the site of action of PIM, showing that this stimulatory effect was present only under conditions in which Rab proteins were limiting for organelle fusion.

In vitro experiments according to the modular principle have been used not only to investigate the action of pathogenicity factors, but also to study the immune activation of phagocytes in response to infection: Peyron *et al.* [15], again in a neutrophil model, demonstrated an increased fusion rate of phagosomes and lysosomes in the presence of cytosol from activated macrophages. This is particularly interesting as recent work has suggested that overall phagosome–lysosome fusion is, if anything, slower in activated than in resting macrophages [28]. Mukherjee *et al.* [29] recreated *in vitro* the altered trafficking of *Salmonella*-containing phagosomes after immune acitivation, and found decreased fusion competence of these phagosomes with early endosomes but increased fusion competence with lysosomes. Experiments combining phagosomes and cytosol coming either from activated or resting cells revealed that either component is influenced by immune activation: Fusion of early phagosomes and early endosomes is decreased, even if only one element is derived from activated cells; this effect is enhanced when both phagosomes and cytosol are from activated cells.

Cell-free fusion assays have been a useful tool in the investigation of both the mechanism of phagosome maturation in general and traffic alterations by pathogenic bacteria. But up till now, no systematic reconstitution of fusion of pathogen-containing phagosomes with some endocytic organelles but not with others has been done. Such an approach will be very valuable to further illuminate the microbial toolbox that allows intracellular pathogens to circumvent delivery to lysosomes. Such an *in vitro* approach will be particularly valuable in revealing the mechanism and details of interference of bacterial effector proteins with defined factors of the eukaryotic vesicle fusion machinery.

Acknowledgments

We thank the Volkswagen-Stiftung (I/77767) and the Deutsche Forschungsgemeinschaft (HA 1929/7-1) very much for their support of our studies on phagosome–lysosome fusion.

References

1 Beckers, C.J., Block, M.R., Glick, B.S., Rothman, J.E. and Balch, W.E. (1989) *Nature*, **339**, 397–398.

2 Barbieri, M.A., Hoffenberg, S., Roberts, R., Mukhopadhyay, A., Pomrehn, A. *et al.* (1998) *J Biol Chem*, **273**, 25850–25855.

3 Balch, W.E., Dunphy, W.G., Braell, W.A. and Rothman, J.E. (1984) *Cell*, **39**, 405–416.

4 Balch, W.E., Wagner, K.R. and Keller, D.S. (1987) *J Cell Biol*, **104**, 749–760.

5 Haas, A., Conradt, B. and Wickner, W. (1994) *J Cell Biol*, **126**, 87–97.

6 Diaz, R., Mayorga, L. and Stahl, P. (1988) *J Biol Chem*, **263**, 6093–6100.

7 Aniento, F., Emans, N., Griffiths, G. and Gruenberg, J. (1993) *J Cell Biol*, **123**, 1373–1387.

8 Mullock, B.M., Perez, J.H., Kuwana, T., Gray, S.R. and Luzio, J.P. (1994) *J Cell Biol*, **126**, 1173–1182.

9 Mayorga, L.S., Bertini, F. and Stahl, P.D. (1991) *J Biol Chem*, **266**, 6511–6517.

10 Oates, P.J. and Touster, O. (1976) *J Cell Biol*, **68**, 319–338.

11 Funato, K., Beron, W., Yang, C.Z., Mukhopadhyay, A. and Stahl, P.D. (1997) *J Biol Chem*, **272**, 16147–16151.

12 Jahraus, A., Tjelle, T.E., Berg, T., Habermann, A., Storrie, B. *et al.* (1998) *J Biol Chem*, **273**, 30379–30390.

13 Brandhorst, D., Zwilling, D., Rizzoli, S.O., Lippert, U., Lang, T. *et al.* (2006) *Proc Natl Acad Sci USA*, **103**, 2701–2706.

14 Wu, P. and Brand, L. (1994) *Anal Biochem*, **218**, 1–13.

15 Peyron, P., Maridonneau-Parini, I. and Stegmann, T. (2001) *J Biol Chem*, **276**, 35512–35517.

16 Kuwana, T., Mullock, B.M. and Luzio, J.P. (1995) *Biochem J*, **308** (Pt 3), 937–946.

17 Alvarez-Dominguez, C., Barbieri, A.M., Beron, W., Wandinger-Ness, A. and Stahl, P.D. (1996) *J Biol Chem*, **271**, 13834–13843.

18 Desjardins, M., Nzala, N.N., Corsini, R. and Rondeau, C. (1997) *J Cell Sci*, **110** (Pt 18), 2303–2314.

19 Naroeni, A., Jouy, N., Ouahrani-Bettache, S., Liautard, J.P. and Porte, F. (2001) *Infect Immun*, **69**, 486–493.

20 Stockinger, W., Zhang, S.C., Trivedi, V., Jarzylo, L.A., Shieh, E.C. *et al.* (2006) *Mol Biol Cell*, **17**, 1697–1710.

21 Vergne, I., Fratti, R.A., Hill, P.J., Chua, J., Belisle, J. *et al.* (2004) *Mol Biol Cell*, **15**, 751–760.

22 Jahraus, A., Egeberg, M., Hinner, B., Habermann, A., Sackman, E. *et al.* (2001) *Mol Biol Cell*, **12**, 155–170.

23 Kjeken, R., Egeberg, M., Habermann, A., Kuehnel, M., Peyron, P. *et al.* (2004) *Mol Biol Cell*, **15**, 345–358.

24 Wickner, W. and Haas, A. (2000) *Annu Rev Biochem*, **69**, 247–275.

25 Uchiya, K., Barbieri, M.A., Funato, K., Shah, A.H., Stahl, P.D. *et al.* (1999) *EMBO J*, **18**, 3924–3933.

26 Ochatt, C.M., Mayorga, L.S., Isola, E.L., Wilkowsky, S., Torres, H.N. *et al.* (1997) *J Eukaryot Microbiol*, **44**, 497–502.

27 Vergne, I., Chua, J., Lee, H.H., Lucas, M., Belisle, J. *et al.* (2005) *Proc Natl Acad Sci USA*, **102**, 4033–4038.

28 Yates, R.M., Hermetter, A., Taylor, G.A. and Russell, D.G. (2007) *Traffic*, **8**, 241–250.

29 Mukherjee, K., Parashuraman, S., Krishnamurthy, G., Majumdar, J., Yadav, A. *et al.* (2002) *J Cell Sci*, **115**, 3693–3701.

7
Phagosome Proteomes Unite! A Virtual Model of Maturation as a Tool to Study Pathogen-Induced Changes

Régis Dieckmann and Thierry Soldati

7.1
Background

Proteomics is the science that studies the complete set of proteins expressed in a given biological system at a given time. One of its key aims is to measure dynamic changes in protein quantities and posttranslational modifications as a function of physiological, pathological or experimental perturbations. Together with ultrastructural studies, comparative analysis of the protein composition of the organelles defines their identity and their relationships within the cell. Therefore, the exact identity of the phagosome is crucial to understanding its biology. The phagosome is a cytoplasmic organelle composed of a particle of extracellular origin enclosed by a membrane. It is built *de novo* upon particle uptake and its composition changes over time due to maturation. As early as the first published purification, the question of the phagosome identity was addressed and its lipid composition and enzymatic activities were compared with those of the plasma membrane [1, 2].

The proteomic characterization of phagosomes led to the stepwise identification of over 600 proteins in *Drosophila* and 800 in the mouse, among a total estimated number of about 1000 proteins [3, 4]. From these efforts have emerged the concept of a virtual model of the phagosome [5]. Groups of proteins can be recognized as functional entities, allowing phylogenetic reconstruction of their acquisition during evolution. The model has been used as a reference for comparisons with physiological [5] or pathogen-induced alterations [6] as well as a predictive tool for genes affecting uptake of bacteria [3].

The predictive power and biological validation of this proteomic-derived model depends both on the mastering of biological variation and on the robustness of the methodology used. On the one hand, the origin of the phagocytic cell, the type of particle and the time of maturation are the main sources of biological variations. Phagocytic cells can be phylogenetically diverse, from protozoan to higher vertebrates. In addition, the particle can be synthetic or natural. Its physical and biological properties impact on the receptors triggered, rate of uptake and subsequent

Intracellular Niches of Microbes. Edited by Ulrich E. Schaible and Albert Haas

ISBN: 978-3-527-32207-7

maturation. After engulfment, the process of maturation, which is naturally dedicated to microbe killing and degradation, will progressively change the environment around the particle, making the phagosome a transient and very dynamic organelle. On the other hand, because phagosome isolation from the cell is an obvious prerequisite of proteomic studies, organelle isolation as well as protein separation and identification techniques are known to introduce biases. In view of such complexity, we will discuss how the knowledge of the protein content of the phagosome has evolved and what are the advantages and drawbacks of a virtual phagosome as a comparative tool to study microbe-containing phagosomes or vacuoles (MCVs).

7.2
A Methods Guide to Phagosome Proteomics

For subsequent comparative proteomic analyses, an isolation method must allow easy access to relatively large amounts of biological material, typically above 100 μg of proteins. The method should be highly reproducible and relatively quick, so that it can be repeated with ease. To this end, three main strategies have been used: magnetic isolation, ultracentrifugation on sucrose gradients and organelle electrophoresis. The first two mainly make use of synthetic particles, and the last two have been applied to MCVs. Their respective principles, ease of use and quality control issues are discussed here.

7.2.1
Bead-Containing Phagosomes

Inert beads are chosen such that their intrinsic properties ensure separation specificity and greater purity. Their surface can be coated or opsonized with the ligand fitting the host-recognition systems and experimental purposes. Magnetic isolation is based on the fact that, after cell lysis, ingested magnetic beads can be isolated using a permanent magnet. It can be performed on the bench without heavy equipment, and it is relatively easy to scale up or down. Magnetic isolation has been used in *Tetrahymena thermophila* [7, 8], J774 mouse macrophages [9, 10] and *Entamoeba histolytica* [11–13]. In experiments in *E. histolytica*, the cytoskeleton surrounding the phagosomal membrane was copurified, making it well-suited for the study of actin-related changes induced by overexpression of a myosin [11].

The isolation of latex bead-containing phagosomes (LBPs) takes advantage of the low density of latex. After cell lysis, LBP isolation is performed via flotation on a sucrose step gradient in one separative centrifugation step. The method was first developed in *Acanthamoeba castellanii* [2], but it has now been used for proteomics in all the common phagocytic models, including *Dictyostelium discoideum* [14, 15], *Entamoeba histolytica* [16, 17], *Tetrahymena thermophila* [18], *Drosophila melanogaster* S2 cells [3], mouse and human macrophage cell lines [5, 19–23] and human neutrophils [24]. A range of detailed protocols have been published [22, 25, 26].

As the potential contaminations of LBP preparations have been much debated, many control experiments have been performed to assess LBP purity. This exhaustive list of experiments exemplifies which approaches can be used to check and improve the purity of a phagosome preparation. Flotation of LBPs on sucrose gradients yields up to 50-fold enrichment and results in one of the purest organelle preparations [22, 25]. Lipid droplets are the only potential contaminating organelle because their buoyancy is close to that of latex in the sucrose gradients [1, 27]. However, after dilution of the sucrose derived from the gradient, a further centrifugation step pellets the LBPs away from the lighter lipid droplets. As shown by electron microscopy, few contaminating membranes are present in the preparations [3, 14, 22]. Contamination has been assessed by immunoblot with an exhaustive panel of antibodies on successive dilutions of both total cell lysates and phagosome samples [3]. Contamination was estimated at less than 5% by using metabolic labeling. ^{35}S-Met-labeled phagosomes were compared by two-dimensional gel electrophoresis (2DE) with re-isolated phagosomes that had been mixed with ^{35}S-Met-labeled S2 cell lysates [3]. These data confirmed similar results obtained with magnetic beads [10]. Contamination can be reduced by incubating the cell lysates with physiological concentration of ATP in the cold prior to flotation [14]. This treatment releases the rigor mortis actin–myosin interactions, thereby reducing copurification of enmeshed organelles. As shown by blot immunodetection, ATP treatment drastically reduces the abundance of mitochondrial porin, and the endoplasmic reticulum (ER) proteins PDI or calreticulin in isolated phagosome fractions [14]. Progressive enzymatic digestion with pronase followed by 2DE separation can also distinguish between cytosolic components such as Rab7, Alix or V-ATPase A and B subunits and luminal components such as cathepsin D or PDI [20].

Mass spectrometry (MS) has been used to compare the relative abundance of proteins by differential peptide isotopic labeling after a carbonate wash [28]. Metabolic labeling as performed by Russell and colleagues [10] can also be performed using MS quantitation methods based on SILAC (discussed in Section 3.4). Such an experiment would, in principle, measure the level of contamination for each individual protein. In addition, correlation of the phagosome proteome with the first drafts of subcellular proteomic maps could help to quantify the contribution of proteins from plasma membrane, endosome or ER origin to the phagosome proteome [29, 30].

Furthermore, because the phagosome undergoes an orderly program of maturation, it can be assumed that a reproducible time-dependent variation in the maturation profile is indicative of a genuine component of the phagosome. Indeed, intuitively, the profile of a contaminant would rather be constant or vary stochastically. As mentioned, despite ATP treatment, some mitochondrial and ER proteins nevertheless copurify with phagosomes [14]. But, strikingly, these proteins show a relevant time-dependent profile (Dieckmann *et al.*, in preparation) and together with biotinylated surface proteins, actin-binding proteins and lysosomal proteases show clear successive waves of delivery and recycling [14]. Phagosome maturation can also be monitored by immunofluorescence on isolated LBPs, confirming little signal from contaminating organelles [26]. Immunoblot maturation profiles and immunofluorescence microscopy generally correlate well, as was shown for cathepsin D [14]. These profiles can also be compared to the *in vivo* localization of green fluorescence protein (GFP)-tagged proteins [31]. As discussed in other chapters, the quality and integrity of isolated LBPs are also confirmed by their use in *in vitro* functional assays testing their ability to bind to and move along microtubules [32], to bind actin filaments and promote actin nucleation and assembly ([33], reviewed in [34]) and to fuse with late endocytic compartments [21]. However, despite the many control experiments performed to test the purity of LBPs, the assignment of a function to a protein or a group of protein in phagocytosis is the ultimate proof that it is not a contaminant.

7.2.2
Microbe-Containing Vacuoles

The separation of MCVs by ultracentrifugation normally requires two centrifugation steps to achieve sufficient homogeneity. The first step is usually a sucrose gradient and the second a Ficoll or sucrose cushion. Such procedures have been applied to Gram-positive as well as Gram-negative bacteria [35]. Using variations on this common theme, vacuoles containing mycobacteria (*M. tuberculosis* and *M. avium*) [10, 36], *Chlamydia psittaci* [37] and *Francisella tularensis* [38] have been purified and used for comparative electrophoretic gel analysis. The mixing of ^{35}S-Met metabolically labeled macrophages with nonlabeled *M. avium*-containing macrophages before lysis did not reveal specific selection of contaminating proteins in purified MCVs, with a global contamination level estimated to be a low 6–8% [10].

Organelle electrophoresis is another suitable method for the separation of MCVs. Total cell lysates are loaded on Ficoll gradients and subjected to an electric field. Negatively charged organelles then migrate upwards to the anode. This method has been applied mainly to mycobacteria MCVs and led to some controversial results [39, 40]. Indeed, intracellular live mycobacteria in activated blood monocytes show distinct mobility shifts whether surrounded by a phagosomal membrane or not. Moreover, membrane-stripping might be due to the procedure itself [40]. Heterogeneity of the results indicates that additional controls are required to certify the origin of the bacterial population present in purified MCVs before their study by proteomics. As probably not all intracellular mycobacteria survive [41], the potential to separate vacuoles containing live versus dead bacteria by organelle electrophoresis [42] might become again an interesting issue.

Magnetic separation of MCVs can also be performed. For example, magnetic particles are pulsed to endosomes that will fuse with MCVs, allowing purification of mixed organelles [43]. Bacteria can also be chemically crosslinked to magnetic nanobeads without drastic alteration of the bacterial surface before being ingested by neutrophils [44]. Resulting MCVs are then isolated. Magnetic immuno-separation has been recently used to isolate *Legionella*-containing vacuoles from *D. discoideum* [45]. After cell homogenization, MCVs are affinity-purified on magnetic particles coupled to an antibody to the *Legionella pneumophila* effector protein SidC, which is a unique surface marker of *Legionella*-containing vacuoles.

7.2.3
Phagosomal Protein Separation and Identification

Bias can also be introduced by protein separation and identification techniques. In this section we outline a brief guide to the more accessible techniques used. Protocols and more extensive literature can be found on websites such as www.expasy.org, www.bio.cam.ac.uk/proteomics/ and in [29, 46]. The main tools for proteomic studies so far are a combination of 2DE and/or high-performance liquid chromatography (HPLC) for protein separation and MS-based protein identification.

Proteomics started with 2DE. This technique provides an immediate and comparative visualization of the protein content. It displays membrane-associated proteins but not proteins with multiple transmembrane domains. Around 50% of the proteins of a crude phagosomal extract separated by 2DE were shown to be membrane-associated by comparing samples after carbonate wash, a commonly used technique to strip off peripheral proteins from the membrane [19]. Identification performed after carbonate wash revealed an enrichment in prenylated and glycosylphosphatidylinositol (GPI)-anchored proteins [15]. 2DE has been used to follow close to 1000 spots during phagosome maturation in *D. discoideum* [15]. Differential in-gel electrophoresis (DIGE) is more recent. It is based on fluorescent labeling of the proteins with charge-matched dyes before 2DE separation and allows comigration of up to three samples in the same gel. After fluorescent scanning, gel images are perfectly superimposable. Migration discrepancies are reduced and quantitation is enhanced due to the linearity of the fluorescent signal. DIGE is quite convenient and accessible for comparative analysis, and spots can be identified by MS (Dieckmann *et al.*, in preparation).

For whole-proteome identification, phagosome isolates are digested with trypsin, resulting peptides are separated by liquid chromatography on a reverse-phase column and peptides are sequenced by tandem MS. Peptide sequences are screened against a protein database for protein identification, and thus a fully sequenced and thoroughly annotated genome is a requirement. If not, proteomic data can be used to further ascertain expressed sequence tag (EST) data and correct gene sequences but this is time consuming. To increase the number of identifications, an additional protein separation step is added. This step is commonly either protein separation in one-dimensional (1D) gel electrophoresis before enzymatic digestion or peptide separation on a cation-exchange column before the reverse phase [5, 24]. These last techniques are very sensitive. Therefore, the main problem then is getting a clean sample preparation, allowing discrimination between biologically relevant and irrelevant identifications.

An MS-based comparative approach, SILAC (stable isotopes labeling by amino acids in cell culture), has recently been applied to the study of phagosome maturation [28, 47]. The relative quantification is done by comparing the heavy and light isotopic peaks of the coeluted peptides in the MS spectrum, thus providing multiple simultaneous quantitative pairing.

7.3
Particle-Induced Impact on Phagosome Isolation

The phagocytic cell encounters a large diversity of particles that vary in size, shape, stiffness and surface properties. For example, red blood cells are round, deformable and between 6 and 40 μm in size; bacteria are rather stiff, round, spiral- or rod-shaped, 1 μm in size but can often be found in grapes or strings, like staphylococci or streptococci respectively. Dead cells and most bacteria are negatively charged while mycobacteria are hydrophobic [48]. Accordingly, particles are recognized by a large variety of specific host cell receptors and coreceptors that are considered to be key determinants for particle recognition and degradation [49, 50]. Receptor–substrate interactions and resulting signaling cascades have mostly been studied with immunoglobulin- and/or complement-opsonized particles in macrophages. Accordingly, IgG- or C3bi-coated particles induce different morphologies of phagocytosis and different signaling cascades [51–53]. Because mainly inert particles have been used in determining phagosomal protein composition, the physical properties of the beads and their influence on uptake and maturation are not to be underestimated. Indeed, particle shape, stiffness, size and surface properties influence the yield, rate and mechanism of uptake as well as the process of maturation.

Experimentally, the yield of a preparation is defined by the maximum number of ingested beads at steady-state and is determined mainly by cell volume [54]. Uptake rate and mechanism are more sensitive to particle properties. Indeed, the local shape of the particle in the zone of uptake often determines phagocytic rate. The smaller the radius of curvature, the quicker is particle uptake [55]. The uptake of soft versus stiff particles is dependent on Rac1 and triggers massive enrichment of myosin IE at the phagocytic cup [56, 57]. Uptake is also very sensitive to particle size. Below a diameter of 0.25 μm, the size limit of clathrin-coated pit endocytosis, particles are taken up by a less efficient process, which is probably receptor-mediated endocytosis [34, 58]. A lag in lysosome delivery to the bead-containing vacuole is observed for particle diameters between 0.2 μm and 0.75 μm [59, 60]. In addition, intermediate sizes of beads (1–4 μm) are engulfed with maximum phagocytosis rates [61]. Blocking the function of phosphatidylinositol 3-kinase (PI3K) [62], the ER SNARE ERS24/Sec22b [62, 63] or the adapter complex-1 [64] also has a size-dependent impact on uptake efficiency.

Surface properties or receptor-specific interactions influence uptake as well as phagosome maturation. Charge-dependent effects are seen in human cells and amoebae [65, 66]. For example, carboxylated latex beads instead of positively charged or hydrophobic beads were used in *E. histolytica* in order to increase yield for their proteomic analysis [16]. Interestingly, the bead surface properties may affect phagosome maturation. Indeed, after having fused with lysosomes, phagosomes containing IgG-opsonized sheep erythrocytes undergo more homotypic fusion than LBPs and their protein content is more rapidly degraded [67]. However, in *D. discoideum*, phagosomes containing either bacteria or latex beads showed no differences in maturation apart from a partial discrepancy in the distribution of the phosphoinositide probe CRAC-GFP [68].

Proteomic analysis of isolated phagosome is, therefore, sensitive to particle properties. While changes in the uptake mechanism and maturation pathway obviously alter the protein composition of the organelle, the rate of uptake influences directly the synchronicity and homogeneity of the isolated phagosome population. Accordingly, particles should be homogeneous in size and shape and be fed as a monodispersed solution to be engulfed as single events. When analyzing pathogens, microbe populations should be either standardized or varied enough to average out these effects. To date, no comparative proteomic analysis has been performed on phagosomes containing beads with different surface properties or different sizes or isolated after FcR- and complement receptor (CR)-mediated uptake. However, maturation profiles of LBPs containing 0.8 μm and 3 μm beads isolated from *D. discoideum* revealed major differences when analyzed by 1D gel electrophoresis [26].

7.4 Comparative Proteomics of Microbe-Containing Vacuoles

7.4.1 Building Up a Model Phagosome

Whole-proteome identification studies are progressively leading to the identification of about 1000 proteins [3, 4]. This number is close to the total number of proteins identified in mitochondria [69] and represents two-thirds of the total number of proteins identified in the secretory pathway in mammals [70]. Although 1000 expressed gene products represent, for example, close to 10% of the total *Dictyostelium* proteome, the total protein amount of the isolated phagosomes represents less than 1% of the total cellular protein mass when amoebae are filled with LBPs at steady-state of phagocytosis [16, 18, 26]. While phagosome protein composition analyses have been performed in many organisms, analysis of phagosome lipid composition is still being developed (B. Helms and B. Brügger, personnal communications). Recent studies on the phagosome proteome are listed in Table 7.1, together with the number of identified proteins, identification technique and type of beads used. These approaches have led to the discovery of a number of new proteins. Between 7% and 17% of the proteins identified in the *Drosophila*, *Dictyostelium* and neutrophil datasets have no known function [3, 15, 24]. Although 8% of the proteins have an unknown function in the *Dictyostelium* phagosome proteome, 29% have an unknown localization [15]. As a result, uncharacterized proteins are then assigned a phagosomal localization in the databases. Although these proteins might well have a function that is not mainly associated with phagosome formation, their function might still be tightly linked to metabolism, host defense or the maintenance of the endocytic pathway. Indeed, organellar proteomic studies have reinforced the

Table 7.1 Compilation of phagosome proteome studies.

Ref.	Particles Synthetic particle[a]	Comparison organism	Fraction analyzed, Phagosome maturation stage (pulse/chase)[b]	Identified proteins[c]	Method of analysis[d]
[21]	LB, 0.8 μm	Mouse MΦ J774, Human MΦ U937	PNS, total cell membrane, LBP (30/0, 60/60, 60/6 h)	0	2DE (^{35}S), GTPase overlay
[22]	LB, 0.8 μm	Mouse MΦ J774, BHK and NRK fibroblasts	(60/60)	≈20	2DE (^{35}S), 2DE map comparison, immunodetection
[19]	LB, 0.8 μm	Mouse MΦ J774, Human MΦ U937	LBP (60/60) before/after Triton X-114 extraction, carbonate wash, protease digestion	0	2DE (^{35}S)
[23]	LB, 0.8 μm	Mouse MΦ J774	Early and late endosomes (B cell lymphoma 2A4B) and LBP (60/0, 60/6 h)	0	2DE (^{35}S, SS)
[20]	LB, 0.8 μm	Mouse MΦ J774	Pronase digestion and (30/0, 60/60, 60/6 h, 60/24 h)	142	1DE, 2DE, LC-MS/MS
[5]	LB, 0.8 μm	Mouse MΦ RAW 264.7	Cells treated for 24 h with or without IFNγ, (60/60) and (15/0, 30/0, 60/60, 60/4 h)	≈170	2DE (SS), LC-MS/MS and immunodetection
[28]	LB, 0.8 μm, IgG	Mouse MΦ RAW 264.7	(10/10, 10/20, 10/35, 10/50, 10/80, 10/110)	382	SILAC, LC-MS/MS
[24]	LB, 2 μm, IgG-C3bi[e]	Isolated human neutrophils	CF 8 min then (30/0)	198	1DE, 2DE, LC-MS/MS
[3]	LB, 0.8 mm	*D. melanogaster* S2	(30/0) and (30/60)	617	LC-MS/MS and 2DE (^{35}S and SS)
[14]	LB, 0.8 mm	*D. discoideum* AX-2	(5/0, 15/0, 15/15, 15/45, 15/105, 15/165)	10	Immunodetection
[15]	LB, 0.8 μm	*D. discoideum* AX-2	(5/0, 15/0, 15/15, 15/45, 15/105, 15/165)	185	2DE (SS), LC-MS
[11]	MB, 2.8 μm, Human serum proteins	*E. histolytica* HM-1:IMSS WT and MyoIB overexpressor[f]	(3/0, 5/0, 15/0)	≈500	2DE (SS), LC-MS/MS
[13]	MB, 2.7 μm, GalNac-BSA	E. histolytica	(10/0)	20	LC-MS/MS

[12]	MB, 2.7 μm, carboxylated	*E. histolytica* HM-1:IMSS	CF 5 min then (0/0, 5/0, 10/0, 10/20)	>27	LC-MS/MS
[16]	LB, 2.8 μm, carboxylated	*E. histolytica* HM-1:IMSS	CF 5 min	85	LC-MS/MS
[17]	LB, 2.8 μm, carboxylated	*E. histolytica* HM-1:IMSS	CF 5 min then (0/0, 0/30, 0/60, 0/120)	243	LC-MS/MS
[7]	MB, 1 μm	*T. thermophila* SB281[g]	(0.5/5, 0.5/15, 0.5/90)	0	2DE, GTPase overlay
[8]	MB, 1 μm	*T. thermophila* SB281	(0.5, 0.5/5)	2	2DE (SS), Edman sequencing
[18]	LB, 2 μm	*T. thermophila* MN173, Gr11 Ex4.1A	(15/0, 30/0, 60/0)	73	LC-MS/MS
[92]	LB, 0.8 μm, *Neisseria gonorrhoeae* porin B	Isolated human MΦ	Porin +/− LBPs (60/0, 60/90, 60/7 h)	1 (7)	2DE map comparison, immunodetection
[91]	LB, 0.8 μm, *Listeria monocytogenes* internalin A or B	Human colon cancer cell line LoVo	Endosomal compartments, InlA$^+$ or InlB$^+$ LBP (30/0)	1 (12)	2DE, MS
[10]	*Mycobacterium avium*	Mouse MΦ J774, mouse bone marrow derived MΦ (BMDM)	LBP vs. *M. avium* MCVs (30/90)	1 (2)	2-DE (^{35}S)
			LBP vs. *M. avium* MCVs (30/60, 30/75, 30/165) including a 15 min ^{35}S metabolic labeling pulse	0 (6)	
			M. avium MCVs isolated from J774 MΦ (30/90) or BMDM MΦ (5 d, 12 d)	0 (2)	
[42]	*Mycobacterium bovis* BCG	Mouse MΦ J774	MCVs with live mycobacteria (60/12 h) or dead mycobacteria (60/120), late endosomes, lysosomes	1 (1)	2DE, MS
[38]	*Francisella tularensis*	Immortalized mouse bone marrow MΦ	MCVs isolated from ΔBcg or ΔNRAMP1 macrophages (2 h/48 h)	4 (≈10)	2DE (SS), MS

[a] Beads are normally washed in medium before use, and are, therefore, considered as medium- or serum-coated.
[b] Pulse/chase indication is only used for phagosome samples. Pulse/chase is expressed in minutes (min) where otherwise not indicated (h, hours; d, days). Centrifugal feeding (CF) is performed in addition before pulse/chase.
[c] Number in parenthesis specifies the number of spots differential in pathogen-related studies.
[d] ^{35}S indicates ^{35}S metabolic labeling, SS indicates silver staining.
[e] IgG cross-linking and then C3bi from adsorption of fresh human serum.
[f] *E. histolytica* are in trophozoite stage.
[g] *T. thermophila* cell lines are defective in dense core granules or mucocysts discharge.LB, latex beads; MB, magnetic beads; MΦ, macrophages; BHK, baby hamster kidney; NRK, normal rat kidney.

fact that proteins have a main static pool but might be dynamically and transiently relocated to other organelles.

Phagocytic cells are phylogenetically diverse and different phagocytic cell types even exist side by side in mammals. Building up a model phagosome helps to determine what is conserved or different between various cell types and organisms. In this modeling process, the first step is to classify the identified proteins in a functional pie chart [15, 17, 24]. Because the chart is tailored to the dataset obtained, direct comparison between different organisms is difficult. For example, the categories of function and localization are sometimes merged [24] and sometimes treated separately [15]. Illustrating the difficulty in building up unambiguous pie charts, a small subset of ER and mitochondrial proteins with unknown phagosomal function has been identified in all the various datasets. Nevertheless, some functional classes are found throughout phylogeny and include adhesion, signaling, cytoskeleton dynamics, vesicle and membrane trafficking and lysosomal enzymes. These classes are involved in the main steps of maturation (i.e., particle recognition, mechanics of engulfment and degradation).

Particle recognition is cell type-specific as confirmed by proteomic studies. For example, Gal/GalNAc lectin in *E. histolytica* or complement receptor 3 (CD11/CD18b) in neutrophils have been identified [11, 17, 24]. Surprisingly, Fcγ receptors could not be detected in any study. Signaling proteins such as Rac1, RhoGEF and the trimeric G proteins are found in the phagosome proteome from all organisms except from human neutrophils [24]. The mechanics of engulfment are very well conserved across organisms, notably the roles of talin and copine at the interface between plasma membrane and cytoskeleton, as well as the Arp2/3 complex, profilin and capping proteins to initiate and stop actin polymerization. Membrane trafficking proteins include the small GTPases Rab5, Rab7, Rab1, the membrane coats clathrin, COPI/II and NSF and various SNARE proteins to promote vesicle fusion [3, 5, 11, 12, 15, 17, 18, 20]. The raft protein flotillin and various enzymes involved in lipid metabolism are also found in a majority of phagosome proteomes [3, 15, 18, 20], suggesting active remodeling of the phagosomal membrane lipid composition. Degradative and metabolic enzymes are often pooled in the functional classification [15, 17, 24]. Metabolic enzymes encompass protein processing, sugar metabolism or protein synthesis. They are rarely discussed in proteomic papers because literature on their function in phagocytosis provides little information. However, their characterization might provide clues about pathogen susceptibility and potential host–pathogen coevolution mechanisms.

Turning now to degradative enzymes, these are present in macrophages as well as in non-mammalian models. Lysosomal cathepsins and other cysteine proteases are enriched in mouse macrophages, amoebae and *T. thermophila* [11, 15, 17, 20]. In contrast, neutrophils show a clear enrichment in specific antimicrobial molecules and display comparatively few cathepsins and lysosomal proteases in their destruction toolbox [24]. This distinct degradative capacity is accompanied by a relative enrichment in ER proteins compared with macrophage-like cells. Indeed, neutrophils show a clear enrichment of ER markers at the phagosomes by immunofluorescence [24]. It is truly intriguing that luminal ER proteins such as calreticulin, PDI, GRP78 and hsp70 are found in all phagosome preparations, independently of the organism used. Though the contribution of the ER to the phagosome is certainly modest [28, 71], its function might be of greater importance. In the innate immune system of mammals, the delivery of ER proteins is proposed to play a crucial role in antigen cross-presentation [72]. However, the presence of ER proteins in the phagosomes of more primitive cells, such as amoebae, which engulf particles as a food source, is less obvious to explain. We suggest that these ER proteins reflect an ancestral and crucial involvement of the ER-associated degradation (ERAD) machinery in phagocytosis. We imagine that some proteins, complex lipids, glycoconjugates or other yeast- or bacteria-specific components cannot be digested by the phagolysosomal system and necessitate ERAD, namely the participation of luminal and cytosolic chaperones to channel these components to the proteasomal degradation machinery. Indeed, essential components of ERAD are found in the *Dictyostelium* phagosome proteome (Boulais, Dieckmann *et al.*, in preparation).

The next analytical step in phagosome modeling is to build up a phagosomal protein–protein interaction network as illustrated in [3]. Proteins directly identified by proteomics are used to build up the core of the phagosomal network. Predicted interactions from interologs are then used to extend the network. Interologs are high confidence interaction networks between orthologous proteins from different species resulting from data mining. One of the strengths of this approach is to measure the confidence with which a functional group belongs to the phagosome in comparison to their chance appearance. Functional groups with a high confidence index are, for example, the ubiquitous vacuolar H^+-ATPase complex, but new candidates have emerged such as the coatomer complex, the chaperonin-containing T complex and the exocyst. For this last complex, a role in phagocytosis was demonstrated experimentally in the same study [3]. This approach also predicts any proteins, potentially belonging to the phagosome proteome, that might not be identified for technical or biological reasons, such as the formin-like diaphanous protein [3]. Indeed, orthologs of this protein are found in the *E. histolytica* phagosome proteome [11, 17] and mouse orthologs have an implication in CR3-mediated phagocytosis [73].

Matching orthologs of other species to the *D. melanogaster* phagosomal network can detect species-specific missing links. For example, no lysozyme is identified in *D. discoideum* phagosome proteome although the *alyA* strain with a lysozyme knockout has reduced enzymatic activity and grows poorly on bacteria [74]. Lysozyme is found, however, in the mammalian and *E. histolytica* phagosome proteome. Filling the gap might be really important if we consider that the phagosome proteome of three different strains of *E. histolytica*, one lab strain and two clinical isolates, might overlap by only 40% [17].

Until recently, comparative proteomics only highlighted qualitative differences; that is, presence or absence of single proteins in the phagosome. However, measuring changes in abundance of functional groups by quantitative proteomics should monitor exactly how much the organelle identity changes. A method that combines high-throughput protein sequencing and quantitation was used in a recent study [5] in which phagosomes isolated from macrophages treated or not with interferon γ (IFNγ) were analyzed. The whole phagosome proteome was digested into peptides and the peptides were compared and quantified according to their intensity in liquid chromatography/mass spectrometry (LC-MS) using validation criteria based on chromatographic retention time and peptide mass and charge. As a proof of principle, interferon-induced GTPases are increased on the phagosome of IFNγ-treated cells. In total, the abundances of 135 proteins were upregulated and 12 were downregulated. These results were further confirmed by comparison of silver-stained 2D gels.

7.4.2
The Phagosome is a Dynamic Organelle

Because organelle identity changes over the course of maturation, not all proteins identified in the proteome are present at all times at the phagosome. By analyzing phagosomal fractions, proteins have been shown to be acquired sequentially in mouse phagosomes [21]. Moreover, the acquisition of some proteins is a prerequisite for the next step to proceed. Proteins from early endosomes (e.g., Rab5 [75]), late endosome/lysosomes (e.g., Rab7 [76]) and lysosomes (e.g., LAMPs [77]) are sequentially acquired and this sequence is conserved in various organisms. Therefore, it is assumed that functionally related proteins show the same time profile. 2DE has been instrumental in showing that LBPs isolated at different times or stages during maturation have a rapidly changing protein pattern [21]. The sequence of delivery of classes of proteins such as the GTPases and hydrolases has been followed during maturation [7, 20, 21]. Semi-quantitative profiles have been drawn by 2DE for the delivery of hydrolases to the mouse phagosome and by MS for all proteins in *E. histolytica* phagosomes [17, 20].

The time-dependent acquisition of proteins during maturation has been particularly extensively studied in *D. discoideum*, notably because the maturation process is completed within 3 h. Using a precise pulse/chase protocol, temporal profiling by immunoblot defined six successive maturation steps: first, acquisition of trimeric G proteins, then acquisition of actin-remodeling proteins, retrieval of plasma membrane with concomitant acquisition of early endosomes and then delivery of lysosomal enzymes in two successive steps [14]. Indeed, the maturation sequence of several functional groups seems phylogenetically conserved although uptake rates, maturation times or even the purpose of phagocytosis are clearly diverging. Proteomics also strengthened experimentally the concept that functional protein modules are delivered to the phagosome to be retrieved at a later stage. In a 2DE study, quantitative protein profiles from about 1000 spots were clustered in a time-dependent manner [25]. Five clusters were characterized corresponding to successive functional steps involving actin remodeling, membrane trafficking and degradation. This study also highlights the profound plasticity and the dynamics of protein exchanges at the phagosome, where 40–50% of the spots appear and disappear between two separate stages of maturation. While some temporal profiles had only one maximum, others had two discrete maxima over the time-course. The discontinuity of the respective profiles reveals that a protein can be required at different steps of maturation and can be retrieved in-between.

This view was strengthened and extended in analyzing the quantitative maturation profiles of more than 380 proteins by SILAC [28]. Highly dynamic profiles suggest that the strictly linear acquisition of the markers is an oversimplification. Indeed, an explanation for these highly dynamic profiles could be that the acquisition of a Rab5 identity or the delivery of different types of lysosomal enzymes originates from successive emptying of the same vesicle and/or from the "kiss and run" of different types of vesicles present in the cell. Support for the proposed mechanism comes from the following studies. Remodeling of endosomes involves "kiss and run" fusion events regulated by Rab5 [78] and different types of Rab5-positive compartments exist in a cell [79]. Fluid phase delivery occurs at a different rate than delivery of membrane-associated proteins. Dextrans of different sizes taken up at the same time and colocalizing after uptake are segregated internally into different vesicles and delivered to the phagosome in discrete steps [80–82]. The cathepsins have different maturation profiles in isolated mouse macrophages, suggesting sequential delivery [20]. A similar sequential delivery sequence of lysosomal enzymes has also been observed in *D. discoideum* [83].

It should also be considered that the analysis of isolated phagosomes by proteomics obviously averages over a population of phagosomes. Although two markers may be found at the same time-point by proteomics, their presence on phagosomes might be mutually exclusive. In *D. discoideum*, biotinylated plasma membrane proteins and endocytosed glycoproteins labeled with a monoclonal antibody (PM4C4) have overlapping profiles in immunoblot but are segregated to different LBPs when analyzed individually by immunofluorescence [26]. These experiments illustrate the general concept that proteomics is looking at an average phagosome and not at the individual fate of each phagosome, which might follow different vesicle fusion and fission routes to get to the same destination or maturation state [84].

7.4.3
Proteomic Analysis of Pathogen-Induced Changes to Phagosome Identity

Successful isolation of MCVs is dependent on the efficient uptake of the microbe and its persistence into the cell. Pathogens can force their entry into the cell or trigger their uptake. Therefore, the morphology of pathogen uptake and following signaling cascades do not always correspond to conventional CR- and FcR-mediated phagocytosis. The physical properties of the pathogen, in particular shape, apparently have little impact on bacterial uptake compared with the importance of the type of recognition receptors engaged by the host. Indeed, rod-shaped bacteria such as *Legionella*, *Francisella* and enteropathogenic bacteria are taken up by coiling phagocytosis [85], asymmetric pseudopods [86] and "macropinocytic ruffling" [87], respectively. In addition, various spirochetes induce coiling phagocytosis but not at the same frequency, despite their common spiral shape [88]. These data illustrate that receptor-mediated recognition of the microbe is the dominant trait. However, microbe recognition and even engulfment is often mediated by more than one receptor, complicating the observation of a direct causal relationship between uptake morphology and microbe recognition [50].

Once inside the cell, the pathogen can rupture the vacuole or modify its biophysical characteristics. These changes might hinder the isolation of MCVs with a standard protocol. For example, *Listeria monocytogenes* permeabilizes and finally breaks apart the vacuole while *Mycobacterium tuberculosis* resides in a vacuole not easily accessible to cytosolic constituents [89]. Virulent strains of *Helicobacter pylori* stimulate phagosome fusion next to the cell surface, resulting in "megasomes" containing multiple viable organisms [90]. Each pathogen has its own strategies of survival that might vary even within the same species.

Proteomic studies of MCVs must compare samples that are relatively homogeneous and similar. Variations of MCV protein composition due to particle physical properties are avoided using a synchronized and homogeneous microbe culture or by randomizing it to average out the differences. Killed or attenuated/avirulent strains should be compared with the pathogenic microbe to ensure that the particle properties and the host receptors triggered are similar. Consequently, at least the general uptake mechanism and maturation pathway should be sufficiently similar to detect specific pathogen-induced changes. A straightforward comparison of MCVs with model LBPs is possible by coating the latex beads directly with the pathogen effectors, as has been done for internalin A/B of *Listeria monocytogenes* [91] or for the porin of *Neisseria gonorrhoeae* [92] but this is by no means possible for all pathogen effectors. Indeed, many effectors do not act from the phagosomal lumen but are injected into the host cytosol and thus cannot act if anchored on a bead surface.

So far, comparative proteomic studies of MCVs have been performed only by 2DE and mainly with mycobacteria-containing vacuoles. Proteomics studies have shown that internalized mycobacteria do not fundamentally change phagosome protein composition. Indeed, comparing endolysosomes to live and dead *M. bovis* BCG-containing vacuoles isolated by organelle electrophoresis from the mouse macrophage cell line J774, only few marked differences are observed after 1 h of infection [39, 42], the protein coronin being one of the characteristics of live MCVs. The differences are no larger when comparing 2-h-old *M. avium*-containing vacuoles isolated on gradients from J774 macrophages with broth-cultured bacteria and LBPs isolated after the same maturation time [10]. Proteomics studies also highlighted that the dissection of pathogen-induced changes in MCV protein patterns is also dependent on the host. The pattern of *M. avium*-containing vacuoles isolated after 5 days of infection in bone marrow-derived macrophages (BMDMs) displays a reduced set of spots compared with the pattern at 12 days of infection. Nevertheless, the MCV protein patterns at 12 days of infection in BMDMs or after 2 h infection in J774 macrophages are comparable. A further kinetic analysis combining a pulse/chase of metabolic labeling and 2DE demonstrated that *M. avium*-containing vacuoles are more accessible to newly synthesized proteins than LBPs of the same maturation time, and that a different set of proteins is delivered [10]. The only published protein profiles of MCVs containing microbes other than mycobacteria were obtained from *Chlamydia psittaci* [37] and *Francisella tularensis* [38] and compared by 1D or 2DE, respectively. Filling the gap, the full proteome of *Legionella*-containing vacuoles is now available, providing an extensive comparison of pathogen-induced changes to the model phagosome [45].

To date, differential analysis of MCVs has only concentrated on proteins that are present or absent upon infection. However, pathogen-induced changes might well impact only on the amount of given proteins delivered to the MCV during maturation. Unfortunately, no quantitative differential analysis of MCVs versus phagosomes has been performed by proteomics, mainly because of the lack of appropriate tools. Recently, quantitative proteomic tools have become available but require highly purified organelles to avoid mixed protein identification and quantitation [36].

7.5 Conclusions

The identification of all the proteins present on LBPs is coming to an end, with an estimated number of proteins of around 1000. Proteomics has fulfilled its role as a tool of discovery because 10% of the proteins identified had previously uncharacterized function and location [15, 17, 24]. Proteomic studies have further strengthened the knowledge that actin cytoskeleton remodeling, membrane trafficking and degradative activities represent the major part of the phagosome's lifestyle. These three steps reflect the major proteins fluxes on the cytosolic and luminal sides of the organelle.

Classification of the phagosomal proteins into functional groups results in the building of a model phagosome. Functional grouping showed a large degree of conservation between macrophages and non-mammalian models, suggesting that, in terms of phagocytosis at least, macrophages are less specialized cells than neutrophils and dendritic cells [24]. Functional grouping also highlighted the conserved presence of the coat proteins clathrin, COP I and II as well as luminal ER components in all the species. Coat formation by clathrin on the plasma membrane and COP I/II between the ER and Golgi is being unraveled, but the role and mechanisms of these coat constituents in phagocytosis is not clear. ER luminal proteins in phagocytosis have been proposed to play a role in antigen cross-presentation, providing a link between adaptive and innate immunity [72]. However, their presence in amoeba phagosomes cannot be explained straightforwardly.

Proteomics has also been instrumental in showing the progressive overall changes in protein composition over time. Systematic immunoblot analysis and 2DE have confirmed the hypothesis that maturation is sequential, which has also been demonstrated by more targeted studies [14, 20, 21]. However, more detailed and sensitive quantitative proteomic analysis suggests that the acquisition of maturation markers is not necessarily linear [25, 28]. The identity of LBPs seems to be acquired through different very dynamic vesicle fusion and fission events until it reaches a defined maturation checkpoint.

Proteomics has made a major contribution to our understanding of the modifications induced by pathogens on MCVs, mainly in mycobacteria. Indeed, it contributed to confirmation that mycobacteria reside in readily accessible vacuoles that do not diverge too much from the endolysosomal system and retain coronin, an early marker of maturation [10, 42]. Quantitative studies of highly purified MCVs should now be able to highlight the up- or downregulation of functional groups, providing more precise working hypotheses.

The LBP model also provides a list of potential candidates and functions that a microbe can try to manipulate in order to survive. Systematic RNAi or more targeted experiments based on this list have proven to be successful in identifying molecules responsible for pathogen susceptibility [3]. In order combine a powerful discovery tool with the necessary *in vivo* complementary data, quantitative comparative proteomics can now be coupled to functional assays by flow cytometry or automatic fluorescence imaging. This approach will allow us to build up comprehensive maps of the proteins affected by a pathogen.

References

1 Ulsamer, A.G., Wright, P.L., Wetzel, M.G. and Korn, E.D. (1971) *J Cell Biol*, **51**, 193–215.

2 Wetzel, M.G. and Korn, E.D. (1969) *J Cell Biol*, **43**, 90–104.

3 Stuart, L.M., Boulais, J., Charriere, G.M., Hennessy, E.J. *et al.* (2007) *Nature*, **445**, 95–101.

4 Griffiths, G. and Mayorga, L. (2007) *Genome Biol*, **8**, 207.

5 Jutras, I., Houde, M., Currier, N., Boulais, J. *et al.* (2008) *Mol Cell Proteomics*, **7**, 697–715.

6 Fratti, R.A., Vergne, I., Chua, J., Skidmore, J. and Deretic, V. (2000) *Electrophoresis*, **21**, 3378–3385.

7 Meyer, M., Mayer, T. and Tiedtke, A. (1998) *Electrophoresis*, **19**, 2528–2535.

8 Maicher, M.T. and Tiedtke, A. (1999) *Electrophoresis*, **20**, 1011–1016.

9 Lutz, D.A., Chen, X.M. and McLaughlin, B.J. (1993) *Anal Biochem*, **214**, 205–211.

10 Sturgill-Koszycki, S., Haddix, P.L. and Russell, D.G. (1997) *Electrophoresis*, **18**, 2558–2565.

11 Marion, S., Laurent, C. and Guillen, N. (2005) *Cell Microbiol*, **7**, 1504–1518.

12 Boettner, D.R., Huston, C.D., Linford, A.S., Buss, S.N. *et al.* (2008) *PLoS Pathog*, **4**, e8.

13 McCoy, J.J. and Mann, B.J. (2005) *Exp Parasitol*, **110**, 220–225.

14 Gotthardt, D., Warnatz, H.J., Henschel, O., Bruckert, F. *et al.* (2002) *Mol Biol Cell*, **13**, 3508–3520.

15 Gotthardt, D., Blancheteau, V., Bosserhoff, A., Ruppert, T. *et al.* (2006) *Mol Cell Proteomics*, **5**, 2228–2243.

16 Okada, M., Huston, C.D., Mann, B.J., Petri, W.A. Jr *et al.* (2005) *Eukaryot Cell*, **4**, 827–831.

17 Okada, M., Huston, C.D., Oue, M., Mann, B.J. *et al.* (2006) *Mol Biochem Parasitol*, **145**, 171–183.

18 Jacobs, M.E., DeSouza, L.V., Samaranayake, H., Pearlman, R.E. *et al.* (2006) *Eukaryot Cell*, **5**, 1990–2000.

19 Burkhardt, J., Huber, L.A., Dieplinger, H., Blocker, A. *et al.* (1995) *Electrophoresis*, **16**, 2249–2257.

20 Garin, J., Diez, R., Kieffer, S., Dermine, J.F. *et al.* (2001) *J Cell Biol*, **152**, 165–180.

21 Desjardins, M., Huber, L.A., Parton, R.G. and Griffiths, G. (1994) *J Cell Biol*, **124**, 677–688.

22 Desjardins, M., Celis, J.E., van Meer, G., Dieplinger, H. *et al.* (1994) *J Biol Chem*, **269**, 32194–32200.

23 Scianimanico, S., Pasquali, C., Lavoie, J., Huber, L.A. *et al.* (1997) *Electrophoresis*, **18**, 2566–2572.

24 Burlak, C., Whitney, A.R., Mead, D.J., Hackstadt, T. and Deleo, F.R. (2006) *Mol Cell Proteomics*, **5**, 620–634.

25 Gotthardt, D., Dieckmann, R., Blancheteau, V., Kistler, C. *et al.* (2006) *Methods Mol Biol*, **346**, 439–448.

26 Dieckmann, R., Gopaldass, N., Escalera, C. and Soldati, T. (2008) Methods in molecular biology, in *Autophagosomes and Phagosomes*, Vol. 445 (ed. V. Deretic), Humana Press, Totowa, NJ, pp. 317–328.

27 Yu, W., Cassara, J. and Weller, P.F. (2000) *Blood*, **95**, 1078–1085.

28 Rogers, L.D. and Foster, L.J. (2007) *Proc Natl Acad Sci USA*, **104**, 18520–18525.

29 Au, C.E., Bell, A.W., Gilchrist, A., Hiding, J. *et al.* (2007) *Curr Opin Cell Biol*, **19**, 376–385.

30 Foster, L.J., de Hoog, C.L., Zhang, Y., Xie, X. *et al.* (2006) *Cell*, **125**, 187–199.

31 Maniak, M., Rauchenberger, R., Albrecht, R., Murphy, J. and Gerisch, G. (1995) *Cell*, **83**, 915–924.

32 Blocker, A., Severin, F.F., Burkhardt, J.K., Bingham, J.B. *et al.* (1997) *J Cell Biol*, **137**, 113–129.

33 Al-Haddad, A., Shonn, M.A., Redlich, B., Blocker, A. *et al.* (2001) *Mol Biol Cell*, **12**, 2742–2755.

34 Desjardins, M. and Griffiths, G. (2003) *Curr Opin Cell Biol*, **15**, 498–503.

35 Luhrmann, A. and Haas, A. (2000) *Methods Cell Sci*, **22**, 329–341.

36 Mattow, J., Siejak, F., Hagens, K., Becher, D. *et al.* (2006) *Proteomics*, **6**, 2485–2494.

37 Zeichner, S.L. (1983) *Infect Immun*, **40**, 956–966.

38 Kovarova, H., Halada, P., Man, P., Golovliov, I. *et al.* (2002) *Proteomics*, **2**, 85–93.

39 Hasan, Z., Schlax, C., Kuhn, L., Lefkovits, I. *et al.* (1997) *Mol Microbiol*, **24**, 545–553.

40 Schuller, S., Neefjes, J., Ottenhoff, T., Thole, J. and Young, D. (2001) *Cell Microbiol*, **3**, 785–793.

41 Jordao, L., Bleck, C.K., Mayorga, L., Griffiths, G. and Anes, E. (2008) *Cell Microbiol*, **10**, 529–548.

42 Ferrari, G., Langen, H., Naito, M. and Pieters, J. (1999) *Cell*, **97**, 435–447.

43 Sturgill-Koszycki, S., Schlesinger, P.H., Chakraborty, P., Haddix, P.L. *et al.* (1994) *Science*, **263**, 678–681.

44 Lonnbro, P., Nordenfelt, P. and Tapper, H. (2008) *BMC Cell Biol*, **9**, 35.

45 Urwyler, S., Nyfeler, Y., Ragaz, G., Lee, H. *et al.* (2009) *Traffic*, **10**, 76–87.

46 Lau, K.W., Jones, A.R., Swainston, N., Siepen, J.A. and Hubbard, S.J. (2007) *Proteomics*, **7**, 2787–2799.

47 Ong, S.E., Blagoev, B., Kratchmarova, I., Kristensen, D.B. *et al.* (2002) *Mol Cell Proteomics*, **1**, 376–386.

48 Alsteens, D., Dague, E., Rouxhet, P.G., Baulard, A.R. and Dufrene, Y.F. (2007) *Langmuir*, **23**, 11977–11979.

49 Aderem, A. and Underhill, D.M. (1999) *Annu Rev Immunol*, **17**, 593–623.

50 Groves, E., Dart, A.E., Covarelli, V. and Caron, E. (2008) *Cell Mol Life Sci*, **65**, 1957–1976.

51 Allen, L.A. and Aderem, A. (1996) *J Exp Med*, **184**, 627–637.

52 Olazabal, I.M., Caron, E., May, R.C., Schilling, K. *et al.* (2002) *Curr Biol*, **12**, 1413–1418.

53 Chimini, G. and Chavrier, P. (2000) *Nat Cell Biol*, **2**, E191–196.

54 Weisman, R.A. and Korn, E.D. (1967) *Biochemistry*, **6**, 485–497.

55 Champion, J.A. and Mitragotri, S. (2006) *Proc Natl Acad Sci USA*, **103**, 4930–4934.

56 Beningo, K.A. and Wang, Y.L. (2002) *J Cell Sci*, **115**, 849–856.

57 Swanson, J.A., Johnson, M.T., Beningo, K., Post, P. *et al.* (1999) *J Cell Sci*, **112** (Pt 3), 307–316.

58 Roberts, J. and Quastel, J.H. (1963) *Biochem J*, **89**, 150–156.

59 de Chastellier, C. and Thilo, L. (1997) *Eur J Cell Biol*, **74**, 49–62.

60 Koval, M., Preiter, K., Adles, C., Stahl, P.D. and Steinberg, T.H. (1998) *Exp Cell Res*, **242**, 265–273.

61 Champion, J.A., Walker, A. and Mitragotri, S. (2008) *Pharm Res*, **25**, 1815–1821.

62 Cox, D., Tseng, C.C., Bjekic, G. and Greenberg, S. (1999) *J Biol Chem*, **274**, 1240–1247.

63 Becker, T., Volchuk, A. and Rothman, J.E. (2005) *Proc Natl Acad Sci USA*, **102**, 4022–4026.

64 Lefkir, Y., Malbouyres, M., Gotthardt, D., Ozinsky, A. *et al.* (2004) *Mol Biol Cell*, **15**, 861–869.

65 Thiele, L., Merkle, H.P. and Walter, E. (2003) *Pharm Res*, **20**, 221–228.

66 Benghezal, M., Cornillon, S., Gebbie, L., Alibaud, L. *et al.* (2003) *Mol Biol Cell*, **14**, 2890–2899.
67 Oh, Y.K. and Swanson, J.A. (1996) *J Cell Biol*, **132**, 585–593.
68 Giorgione, J. and Clarke, M. (2008) *Cell Motil Cytoskeleton*, **65**, 721–733.
69 Pagliarini, D.J., Calvo, S.E., Chang, B., Sheth, S.A. *et al.* (2008) *Cell*, **134**, 112–123.
70 Gilchrist, A., Au, C.E., Hiding, J., Bell, A.W. *et al.* (2006) *Cell*, **127**, 1265–1281.
71 Touret, N., Paroutis, P. and Grinstein, S. (2005) *J Leukoc Biol*, **77**, 878–885.
72 Houde, M., Bertholet, S., Gagnon, E., Brunet, S. *et al.* (2003) *Nature*, **425**, 402–406.
73 Colucci-Guyon, E., Niedergang, F., Wallar, B.J., Peng, J. *et al.* (2005) *Curr Biol*, **15**, 2007–2012.
74 Muller, I., Subert, N., Otto, H., Herbst, R. *et al.* (2005) *J Biol Chem*, **280**, 10435–10443.
75 Duclos, S., Diez, R., Garin, J., Papadopoulou, B. *et al.* (2000) *J Cell Sci*, **113** (Pt 19), 3531–3541.
76 Harrison, R.E., Bucci, C., Vieira, O.V., Schroer, T.A. and Grinstein, S. (2003) *Mol Cell Biol*, **23**, 6494–6506.
77 Huynh, K.K., Eskelinen, E.L., Scott, C.C., Malevanets, A. *et al.* (2007) *EMBO J*, **26**, 313–324.
78 Duclos, S., Corsini, R. and Desjardins, M. (2003) *J Cell Sci*, **116**, 907–918.
79 Lakadamyali, M., Rust, M.J. and Zhuang, X. (2006) *Cell*, **124**, 997–1009.
80 Desjardins, M. (1995) *Trends Cell Biol*, **5**, 183–186.
81 Berthiaume, E.P., Medina, C. and Swanson, J.A. (1995) *J Cell Biol*, **129**, 989–998.
82 Wang, Y.L. and Goren, M.B. (1987) *J Cell Biol*, **104**, 1749–1754.
83 Souza, G.M., Mehta, D.P., Lammertz, M., Rodriguez-Paris, J. *et al.* (1997) *J Cell Sci*, **110** (Pt 18), 2239–2248.
84 Griffiths, G. (2004) *Trends Cell Biol*, **14**, 343–351.
85 Horwitz, M.A. (1984) *Cell*, **36**, 27–33.
86 Clemens, D.L., Lee, B.Y. and Horwitz, M.A. (2005) *Infect Immun*, **73**, 5892–5902.
87 Cossart, P. and Sansonetti, P.J. (2004) *Science*, **304**, 242–248.
88 Rittig, M.G., Jagoda, J.C., Wilske, B., Murgia, R. *et al.* (1998) *Infect Immun*, **66**, 627–635.
89 Clemens, D.L., Lee, B.Y. and Horwitz, M.A. (2002) *Infect Immun*, **70**, 5800–5807.
90 Allen, L.A., Schlesinger, L.S. and Kang, B. (2000) *J Exp Med*, **191**, 115–128.
91 Pizarro-Cerda, J., Jonquieres, R., Gouin, E., Vandekerckhove, J. *et al.* (2002) *Cell Microbiol*, **4**, 101–115.
92 Mosleh, I.M., Huber, L.A., Steinlein, P., Pasquali, C. *et al.* (1998) *J Biol Chem*, **273**, 35332–35338.

8
Phagosome–Cytoskeleton Interactions

Maximiliano G. Gutierrez and Gareth Griffiths

8.1
Introduction

Phagocytosis is an evolutionarily ancient process that is believed to have evolved around the dawn of eukaryotic cell history to enable amoeboid cells to take up bacteria, which still provide a food source for the modern free-living amoeboid cells. In higher eukaryotes, this process is predominantly the function of three specialized "professional" phagocytes: macrophages, neutrophils and dendritic cells. The process of phagocytosis has been extensively covered by many excellent reviews [1–8]. Figure 8.1 summarizes the main stages involved in the life of a "permissive" latex bead phagosome (LBP) undergoing the full program of "maturation" [9]. This term describes how phagosomes fuse sequentially with different intracellular organelles, culminating in the end-stage, the phagolysosome. In parallel, lipids and proteins are lost from phagosomes by recycling of vesicles [10]. As seen in Figure 8.1, this process is accompanied by a progressive lowering of the organelle pH, under the control of the proton ATPase [11].

There are three major filament systems in eukaryotic cells, comprising actin, microtubules (MTs) and intermediate filaments (IFs). Since a role of IFs in phagocytosis has not been described, we will focus here on the actin and MT systems. Together, these two filament systems are important for the phagocytic uptake process, for recycling from the phagosome and for phagosome intracellular transport and fusion with endocytic organelles.

Here, we will address the interactions of the actin and MT systems with assembled phagosomes. Since most of the studies on these interactions have been carried out using the latex bead phagosome macrophage model system, most of our discussion will be centered on that system. We will not focus in detail on the role of the cytoskeleton in the phagocytic uptake process, a topic that has been extensively covered in a number of good reviews [12–14]. The ability of different pathogens to alter different aspects of phagosome maturation, including the cytoskeleton, has also been a major interest in the field. Many comprehensive reviews are available that cover most aspects of the cytoskeleton–pathogen interaction [12–19].

Intracellular Niches of Microbes. Edited by Ulrich E. Schaible and Albert Haas

ISBN: 978-3-527-32207-7

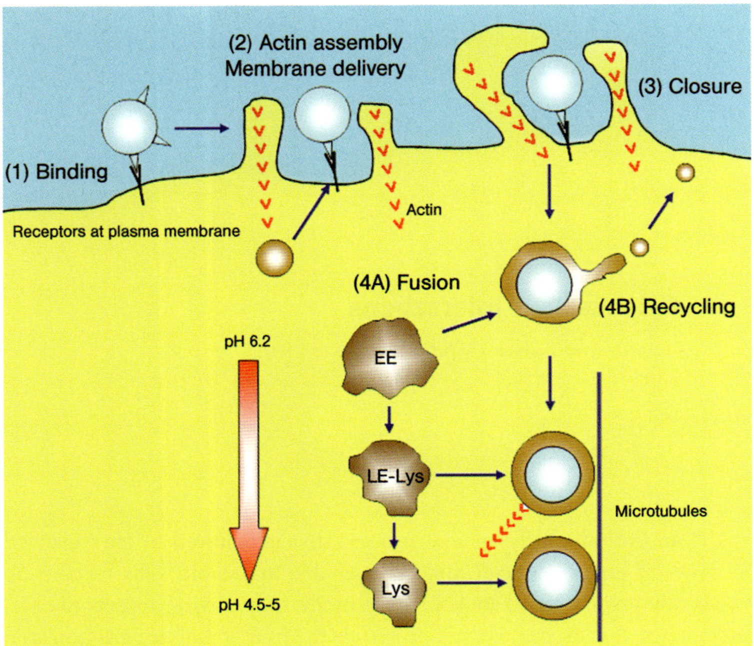

Figure 8.1 Scheme of phagosome maturation. Latex beads coated with ligands are recognized by receptors at the plasma membrane (1). Then the bead is taken up by pseudopodia generated in an actin-dependent way; membrane is provided by intracellular vesicles from different sources (2). After closure (3) the bead is located in phagosomes that follow a process of maturation fusing with components of the endocytic pathway (4A) or recycling its content (4B). The whole process is accompanied by a progressive acidification of the phagosome. This process of maturation is dependent on microtubules and actin. EE, early endosome; LE/Lys, late endosome–lysosome; Lys, lysosome.

8.2 The Actin Cytoskeleton. General Background

In Figure 8.2 we provide a simplified "global" picture of the main interactions between actin and microtubules with latex bead phagosomes that are important in both phagosome formation and maturation.

Actin filaments, like microtubules (see below) assemble asymmetrically, unlike intermediate filaments which are symmetrically assembled. Actin filaments are very dynamic and undergo permanent assembly and disassembly. The filaments grow faster at one end (the plus end) than at the other (minus end). The actin monomer (globular or G-actin) binds ATP and normally in cells it assembles into filaments in this form. After polymerization, ATP is hydrolyzed to ADP and free phosphate (Pi) that initially stays bound to the filament. The ADP-Pi-bound monomers in the filament lose Pi and the ADP form then depolymerizes at the minus end [20, 21].

The regulation of the actin system is complex since there are well over a hundred actin-binding proteins (ABPs) that mediate different effects on the actin cytoskeleton [22].

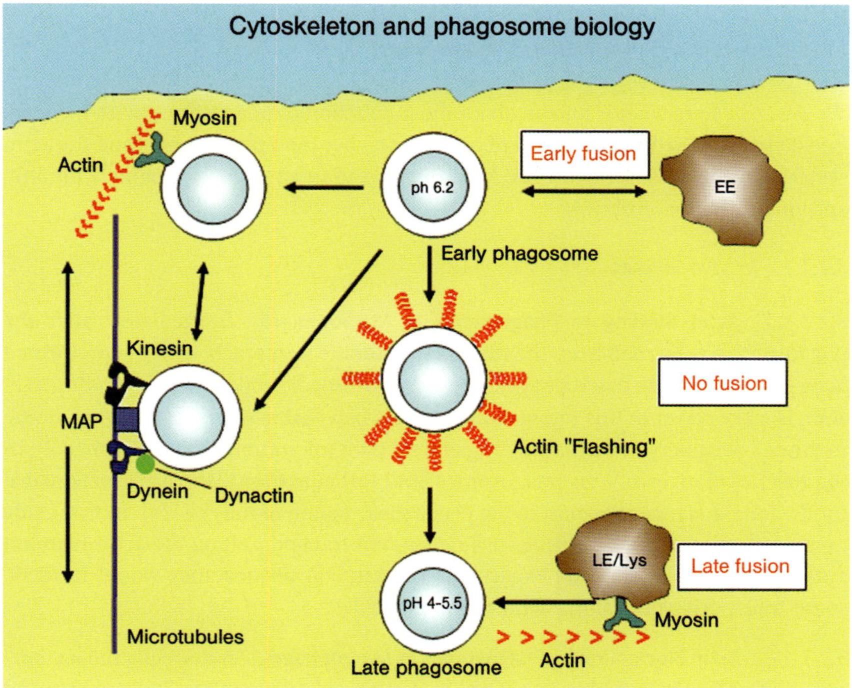

Figure 8.2 Interaction of phagosomes with actin and microtubule cytoskeleton. Early phagosomes interact with mysosin Va at the periphery. This process competes with the association of the phagosome with microtubules and subsequent transport to the perinuclear region mediated by dynein. Fusion events controlled by actin during maturation are depicted. Actin can mediate fusion with late endocytic vesicles or block it depending on the age of the phagosome. EE, early endosome; LE/Lys, late endosome–lysosome; MAP, microtubule-associated protein.

Most of these ABPs have been extensively analyzed *in vitro* but their interactions in cells are extremely complex, especially when they interact with membranes, a poorly understood set of processes. Among the different classes of ABP are those that (i) nucleate new filament assembly (e.g., ARP2/3, formins), (ii) block the actin plus (barbed) ends (e.g., capping proteins such as CapZ and gelsolin) or the pointed end (emerin), (iii) crosslink filaments (e.g., α-actinin tropomyosins, fimbrin, villin), (iv) bind lateral F-actin to membranes (e.g., myosin I, calmodulin), (v) bind (sequester or buffer) G-actin (e.g., profilin, thymosins) and (vi) facilitate depolymerization (e.g., cofilin or actin depolymerization factor (ADF)). One of the most difficult aspects of actin biology to address is the nucleation/polymerization of filaments by proteins present on biological membranes. In every known example the filament grows by insertion of actin monomers at the plus ends that are adjacent to the membrane [23]. This is generally a poorly understood process. The latex bead system offers a powerful system to elucidate the mechanisms of this process, as described below.

8.2.1 Actin Interactions with Assembled Phagosomes

We describe here what is known about the interaction of phagosomes with the actin cytoskeleton. Two different kinds of interactions between phagosomes and the actin system have been described from both *in vivo* and *in vitro* studies, namely binding and nucleation/assembly.

8.2.1.1 *In Vivo* Studies

8.2.1.1.1 Actin Binding to Phagosomes in Macrophages Immediately after the phagosome has formed from the plasma membrane it interacts with the preformed actin filaments in the macrophage cortex. Phagosome-bound myosin V binding to F-actin is a key player in this interaction, which is believed to maintain the phagosome in the cell periphery in a dynamic "tug of war" with microtubules [24] (Figure 8.2). In the absence of myosin V the phagosomes bind earlier to the MTs that are responsible for the long-distance transport to the perinuclear region of the cell [24]. In that study no evidence was found for myosin V-dependent transport along the actin filament system, although one would expect that myosins on phagosomes would facilitate some transport events along actin filaments [25].

8.2.1.1.2 Actin Nucleation by Phagosomes in Macrophages A dramatic nucleation/assembly of F-actin has been described by Zhang and collaborators [26] and Yam and Theriot [27]. Apparently random "flashing" of F-actin, a massive polymerization of actin, can be observed on some phagosomes in live macrophages expressing green fluorescent protein (GFP)-tagged actin. In most phagosomes this phenomenon is transient, with polymerization proceeding for 30 s to a few minutes, followed by an equally rapid depolymerization. Interestingly, in the analysis by Yam and Theriot [27], the authors showed that a surprising variety of ligand-coated beads displayed apparently random actin flashing, including even transferrin-conjugated beads taken up by MDCK epithelial cells.

Observations in our lab suggest that the initial uniform coat of phagosomal F-actin breaks symmetry and can be polarized to form a dynamic actin comet. This structure generates a propulsive force and moves the phagosome through the macrophage (Liebl and Griffiths, unpublished results). Such comets have been earlier described and characterized in detail on the surface of cytoplasmic pathogens (that have escaped from the phagosomal membrane), such as *Listeria*, *Shigella* or *Rickettsia* [28]. Comet motility under the control of phosphatidylinositol 4,5-bisphosphate ($PI(4,5)P_2$) and N-WASP has also been shown for late endosomes [29, 30] and for early endosomes [29, 31–33].

We have recently carried out an extensive analysis of this phenomenon using live cell imaging of macrophages labeled with GFP-tagged actin. In parallel the late endocytic organelles were labeled with rhodamine–dextran after pulse-chase endocytosis. This analysis revealed that the thick coat of actin around phagosomes is a steric barrier for phagolysosome fusion (Liebl and Griffiths, manuscript submitted). What is clear from this analysis is that the phagosomal actin nucleation machinery can be temporarily activated but only before phagolysosome fusion. Once the terminal phagolysosome is formed, and for at least 24 h, this compartment did not induce actin nucleation/assembly again. The above description suggests that the thick F-actin coat that polymerizes on phagosomes can prevent phagolysosome fusion. In addition, an excess of F-actin can also completely block phagolysosome fusion *in vitro* [34]. However, from *in vitro* studies (see below) we also postulated that F-actin can also facilitate phagolysosome fusion (Figure 8.2). We have no direct evidence for this model. It is based on three observations: (i) Phagosomes (and late endosomes/lysosomes) nucleate actin *de novo*. (ii) As in all known actin polymerization processes on membranes, the polarity of F-actin on phagosomes is such that the insertion of ATP-loaded monomers occurs (counterintuitively) close to the membrane (at the barbed or plus end of actin) [35]. Then, extrapolating from other studies [21], the monomers treadmill away from the membrane as the G-actin hydrolyzes ATP and fluxes outwards before the ADP-actin depolymerizes at the pointed or minus end of the filament [21]. (iii) A consequence of this filament polarity is that any organelle having a bound myosin that contacts the actin filament protruding, hair-like, from a phagosome will move its cargo towards the membrane. All of the 18 known myosins [36], except myosin VI [37], always move unidirectionally towards the barbed end of actin [38]. In other words, the actin filaments would allow a membrane vesicle such as a lysosome to be attracted vectorially towards the phagosome, thereby facilitating docking prior to fusion. This mechanism could in principle also facilitate phagosome–phagosome fusion.

Alternative models in which the lysosomes as well as the phagosomes nucleate actin filaments have also been considered; again homotypic lysosome fusion could also be driven by such a mechanism [39]. Finally, the assembly of actin by the plasma membrane may also facilitate transport of vesicles towards the plasma membrane during exocytosis. Evidence for this has been provided from many systems [40, 41]. A number of myosins have been detected on fully formed phagosomes; these include myosin I, II, V and X [42, 43]. Many myosins have also been described on late endosomes and lysosomes, including the myosin 1 family, myosin II, V, VI and VII [44, 45].

The difficulty in directly proving this model *in vitro* or in cells is the likelihood that a single actin filament could successfully carry out this mission. If so, it will be almost impossible with the present technology to visualize single actin filaments in cells, and an extremely demanding task even *in vitro*. Arguing in favor of this model are observations showing that beads coated with myosins indeed move along F-actin towards the barbed ends [46]. Further evidence for this "actin-track" model comes from the analysis of mycobacterial phagosomes (see below).

8.2.1.2 *In Vitro* Studies

Latex bead phagosomes can be purified by flotation in a sucrose gradient under different conditions, for example, different time-points. This allows the study of the influence of proteins and lipids present on the phagosomal membrane in actin binding or assembly (Figure 8.3). In addition, LBPs can be incubated with different

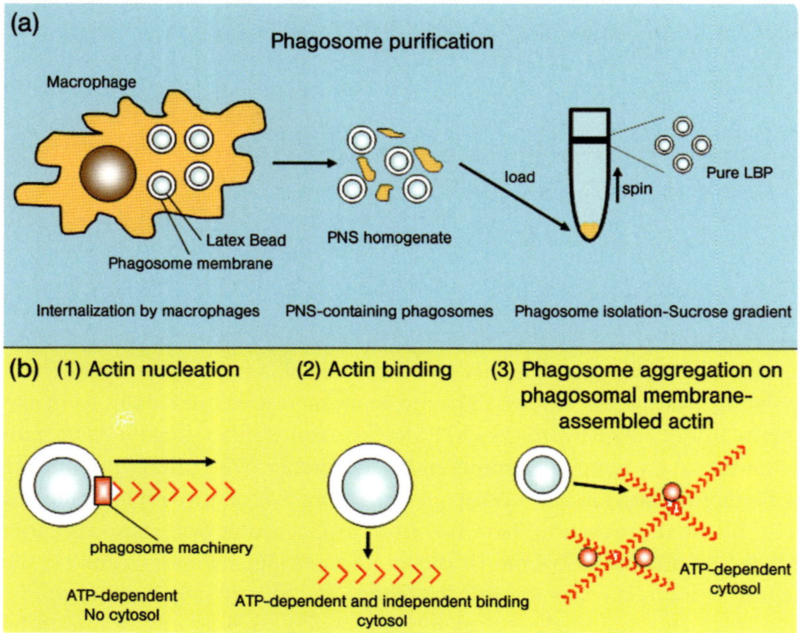

Figure 8.3 *In vitro* assays to monitor actin in latex bead phagosomes (LBPs). (a) Scheme showing the purification of LBPs from macrophages for biochemical analysis. Briefly, latex beads are internalized and then isolated using its property of flotation in a sucrose gradient. Then isolated LBPs are used for the assays described below. (b) Assays to assess the role of actin in phagosome biology. (1) Actin nucleation assay, with this approach LBPs are used in presence of ATP and the nucleation of actin filaments around the phagosome measured [35]. (2) Actin binding, actin is polymerized and then phagosomes incubated with this actin "network." Filamentous actin stabilized and labeled with rhodamine–phalloidin is perfused into the chamber and then a solution containing purified phagosomes are also perfused into the chamber and incubated for 20 min. Binding of phagosomes is analyzed by fluorescence microscopy. The bound phagosomes are counted by eye under the different conditions [24]. (3) Phagosome aggregation on phagosomal membrane-assembled actin [39]. Phagosomes are mixed with macrophage cytosol and rhodamine–G-actin and incubated in the presence of an ATP-regenerating system at 37 °C in a sealed chamber. Individual phagosomes and clusters are then quantified at different time-points using confocal microscopy.

components, for example, cytosol, ATP, allowing the outcome to be evaluated in different processes (Figure 8.3).

8.2.1.2.1 **LBP Actin-Binding Assay** The LBP actin-binding assay monitors the binding of LBPs with fluorescent beads to rhodamine–F-actin bound to glass [24]. In contrast to actin assembly on phagosome membranes, this binding requires cytosolic factors. At least three factors involved in phagosome binding to F-actin have been detected:

(i) a factor with ATPase activity, which has been identified as myosin Va, (ii) a high molecular weight factor (about 600 kDa), but without ATPase activity, possibly the actin-binding protein filamin, (iii) an unknown inhibitory factor with low molecular weight (about 30 kDa), which blocks binding activities of both myosin Va and filamin. This suggests that inhibitory activity may be involved in the regulation of the association of the filamin and myosin Va, which is responsible for the stimulation of F-actin–phagosome interaction with the phagosome surface. Recently the Kuznetsov group found that GTP and phosphatidylinositol bisphosphate (PIP_2) are also involved in the regulation of phagosome binding to F-actin. In agreement with these data many GTPases and PIP_2-regulated proteins have been implicated in the modulation of F-actin (see above). An intriguing observation in the comparison of the LBP actin-binding assay with the actin assembly assay is that the factors that stimulate binding tend to inhibit assembly and vice versa. For example, arachidonic acid (AA), PIP_2, sphingomyelin (SM) and sphingosine-1-phosphate (S1P) stimulated actin nucleation but inhibited binding of phagosomes to F-actin *in vitro* in the absence of cytosolic factors and ATP. Furthermore, binding assays in the presence of functional antibodies against different actin-associated proteins revealed that proteins known to activate actin nucleation on LBP phagosomal membranes in macrophages (ezrin, N-WASP, see below) downregulate the F-actin-binding capability (Hoffmann and Kuznetsov, unpublished observations).

8.2.1.2.2 ***In Vitro* Assembly of Actin by Phagosomes** The LBP actin assembly assay monitors the polymerization of rhodamine–actin by LBP using light microscopy. Key players involved in this process are ezrin and/or its close homolog moesin from the ezrin radixin moesin (ERM) protein family [35]. Recent data by our group argue that ezrin on phagosomes can bind and activate N-WASP, which may in turn switch on the actin nucleating complex Arp2/3. *In vitro* experiments suggest that ezrin oligomers in solution stimulate the ability of N-WASP to activate actin assembly via the Arp2/3 complex (Marion and Griffiths, unpublished observations). The synthesis of the phosphoinositides phosphatidylinositol-4-phosphate (PIP) and PIP_2 by LBP-bound phosphatidylinositol kinases are also essential for actin assembly and part of the role of PIP_2 is to bind to the N-terminal ERMAD domain of ezrin [47]. Both the ERM proteins and N-WASP bind PIP_2 and this interaction facilitates activation of both proteins. Activation of N-WASP exposes the VCA domain (verprolin, homology, cofilin homology and the acidic region), which in turn activates the Arp2/3 molecule to enhance the nucleation of actin [48]. As for all membrane systems, it is still not clear how the ERM protein, N-WASP, PIP_2 and other molecules interact in the LBP actin assembly.

Surprisingly, actin assembly occurs on LBPs in the absence of GTP since this molecule has no effect on this process [35]. This argues that the initial nucleation of actin assembly is upstream of GTPases that are expected to play a key role in regulating the signaling processes on phagosomes. Indeed many studies have shown that isolated phagosomes from many species contain many GTPases, including the actin regulatory switches Rho, Rac, Cdc42 (inhibitors of which have no effect on LBP actin assembly (unpublished data)). In most studies on membrane-associated actin assembly the Rho family proteins are considered to be upstream of membrane-dependent actin polymerization. However, evidence supports the notion that the ERM proteins can be upstream of the Rho family GTPases in a complex signaling network [49]. For example, ezrin binds and activates Dbl, the GEF for Cdc42 [50]. Indeed, most of these membrane-bound regulators are found on phagosomes. In addition, most of these proteins have multiple interaction partners, which makes it increasingly difficult to understand how the different machineries operate at the systems level.

With respect to the assembly of actin by phagosomes we have identified a number of key players using the *in vitro* assay. Gelsolin was also found to stimulate this process [51] while an antibody against profilin inhibited it (Defacque and Griffiths, unpublished data). LBPs have adenylate cyclase that can synthesize cyclic AMP (cAMP) from ATP. cAMP inhibits actin assembly by phagosomes, at least in part by activating phagosomal PKA [52]. Protein kinase A (PKA) is known to phosphorylate serine 66 on ezrin in some cell types and serine 10 of its close homolog merlin [53]. In fact, PKA phosphorylation of ezrin is associated with inhibition of actin-dependent processes such as cell motility [54], although it has not been directly shown whether the phosphorylation of ezrin by the activated PKA is part of the mechanism whereby cAMP inhibits phagosome actin assembly.

The concentration of ATP used in the LBP actin assembly assay plays an important role in the final outcome. The standard assay works best at low ATP (0.2 mM), which allows around 10–30% of LBPs to assemble actin. In contrast, physiological ATP (5 mM) blocks assembly. Since LBPs make significant cAMP with 5 mM but not with 0.2 mM this provides one mechanism by which cAMP via PKA can inhibit the system at high ATP. The addition of 1 mM ATP to LBPs from J774 macrophages leads to the phosphorylation of a large number of phagosomal proteins by phagosome-associated kinases [55], a multitude of which are found in the different phagosome proteomes [56]. With respect to the actin assembly process it is also clear that an important role of the ATP is to be used by kinases, such as phosphatidylinositol (PI), sphingosine (Sph) and ceramide (Cer) to their phosphorylated counterparts PIP, PIP_2, S1P and ceramide-1-phosphate (Cer-1P), respectively.

The identification of a key role for PIP and PIP_2 opened the door for a detailed analysis of the lipids that interconnect with these phosphoinositides in the LBP actin assembly process, as well as sphingolipids and fatty acids [57]. When PIP or PIP_2 are incorporated into the LBP membrane they stimulate actin assembly at low and high ATP. In contrast, other lipids such as phosphatidylcholine (PC), the polyunsaturated fatty acids eicosapentanoic acid (EPA) and docosahexanoic acid (DHA) inhibit the system in both conditions. AA is highly stimulatory under both ATP regimes. More intriguing is the effect of the sphingolipids, especially Sph and its downstream product S1P. Whereas Sph inhibits the system at high ATP and stimulates at low ATP, the opposite pattern is observed when S1P is added to LBPs. This lipid stimulates actin assembly at high ATP and inhibits it at low ATP [57]. The same pattern is seen with Cer and Cer-1P.

Our recent studies led us to an unexpected role of the phosphorylated lipids PIP, PIP_2, S1P, Cer-1P and phosphatidic acid (PA), which are all stimulatory in the LBP actin assay at high ATP. The incorporation of these lipids into the phagosomal membrane in the presence of ATP and ADP led to the translocation of ADP, but not ATP into the LBP lumen, as shown using radioactive nucleotides and HPLC. In the lumen the ADP becomes converted to ATP by an adenylate kinase (AK). An inhibitor of AK (Ap2P5) prevented the luminal ADP to ATP conversion. Although the AK remains to be determined, one of the five known AKs, AK2, has been detected in two LBP proteome analyses (M. Desjardins, personal communication; [58]).

A number of experiments suggested that the luminal ATP was able to stimulate the actin assembly on the cytoplasmic side of the LBP. For example LBPs prepared using beads conjugated with apyrase (that degrades ATP and ADP) on their surface were deficient in actin assembly.

The identification of a well-characterized ATP-binding transmembrane receptor, P2X7R, in the J774 macrophage LBP proteome (M. Desjardins, personal communication) led us to focus on this receptor. P2X7R is a cationic channel that is able to bind via its cytoplasmic domain to a complex of 12 proteins, including actin, α-actinin, supervillin and a phosphatidylinositol 4-kinase (PI4K) and 10 other proteins [59]. Whereas LBP prepared from wild-type P2X7R-positive bone marrow macrophages could assemble actin and was stimulated by S1P or PIP at high ATP, the equivalent LBP from P2X7R-knockout mice macrophages failed to be stimulated by these lipids. This argues for a model in which PIP and S1P induce ATP accumulation in the LBP lumen. This ATP then activates the P2X7R that signals downstream of the phagosome actin assembly machinery. Since PI4K interacts with the P2X7R, this kinase could also be an integral player of the system (Kuehnel *et al.*, 2009). Consistent with this idea, the synthesis of PIP and PIP_2 is essential for actin assembly in LBPs [47].

These recent experiments define new mechanisms but open even more questions to address with respect to the actin assembly process on LBPs. These include: (i) the identification of the ADP transporter and the mechanism of regulation by phosphorylated lipids, (ii) the identification of the AK and the mechanism that allows this protein to cross the phagosomal membrane and (iii) the molecular connection between the P2X7R and the PIP_2–ezrin–N-WASP machinery that assembles actin. Each player in this process has its own complexity. For example, PIP_2 can bind to a myriad of actin-binding proteins, including many that are established to be bone fide phagosome constituents such as ezrin, N-WASP, gelsolin, profilin, α-actinin, cofilin and some capping proteins such as CapZ [60]. Many of these proteins have multiple binding partners. For example, as mentioned above, the ERM proteins have multiple interactions both upstream and downstream of Rho, Rac and Cdc42 [49].

It is important to note that these complex processes of actin assembly and actin binding to phagosomes must depend on proteins and lipids that are found in the phagosome proteomics and lipidomics. The most recent data for the J774 macrophage phagosome proteome using more sensitive MS analysis has identified up to 2000 different proteins in the system (M. Desjardins, personal communication). This is complicated by the dynamic changes in phagosomes, since as they mature in cells their protein pattern, and their ability to assemble actin, changes dramatically [9, 35, 61, 62]. On the other hand, an ongoing analysis of the lipids present in LBPs from J774 macrophages by our collaborators, Jos Brouwers and Bernd Helms (University of Utrecht), and in LBPs from *Dictyostelium* (Soldati and

Brugger, personal communication) makes it clear that the lipids are also complex. Moreover, just like the proteins, the phagosome lipid pattern also changes significantly during phagosome maturation (Brouwers and Helms, personal communication). In both analyses many hundreds of different lipid species are present. The ongoing proteomic and Western blotting analyses show that many of the enzymes that interconvert lipids are detected on phagosomes and our studies show that many of these enzymes are active on phagosomes under different conditions. So in one way the lipids are even more complex than proteins. While proteins can be posttranslationally phosphorylated by kinases, the lipids can be newly synthesized and degraded or interconverted into other lipids by enzymatic machinery in the phagosome itself.

From all these studies described above, it is now becoming clear that phagosomes are not passive vesicles that only acquire proteins and lipids via vesicular transport, or (for proteins) directly from the cytoplasm, and lose these components by recycling or dissociation. These organelles are also themselves capable of a plethora of biochemical activities, for example the enzymatic synthesis of different lipids. Even in a single cell, functionally different populations of phagosomes are usually observed. This led to the concept of phagosome individuality [63]. Recent innovative analyses towards this goal are now beginning to draw the maps of molecular interactions that are possible for the many hundreds of proteins that have been detected in phagosome proteomic studies [56, 61, 64–67]. A dynamic description, including a predictive model of the interactions of lipids linked to PIP_2 has recently been initiated by our group. This includes a systems level analysis of these lipids and other molecules on phagosomal membranes (Kuehnel *et al.*, 2008).

8.2.1.2.3 **Phagosome Aggregation on Phagosomal Membrane-Assembled Actin** Initial studies showed that a number of actin-interfering reagents, such as cytochalasin D and latrunculin A, inhibited *in vitro* fusion of LBPs with early endosomes (EEs) and late endocytic organelles (LEOs). However, two actin-binding proteins stimulate LBP fusion with EEs/LEOs (Jahraus *et al.* [34]). The first is a gelsolin fragment that can sever actin and bind to the free barbed ends of actin. The second positive effector of fusion is thymosin β4 (Tβ4). To study in more detail the *in vitro* organization of LBPs, EEs and LEOs and F-actin, a confocal-based study using rhodamine–actin was developed to simultaneously visualize the interactions between the actin cytoskeleton assembled from cytosol and selectively labeled LBPs and/or EEs/LEOs (Figure 8.3). When postnuclear supernatants (PNS) containing fluorescent LBPs or fluorescent endosomes are added to the cytosol, a small number of these organelles were associated with distinct dots of labeled actin and these organelles were clustered with a concomitant formation of a visible actin network. Both LBPs and endosomes aligned along these actin fibers and formed large clusters at the junction points where many actin fibers crossed. This suggests that cellular membrane organelles can directly control an organized assembly of an actin network. This clustering of organelles along actin fibers is Tβ4-dependent and under ATP depletion actin polymerizes but there is no clustering. Interestingly, phagosome clustering is actin-dependent but MT-independent [39].

Alternative models have been developed to explain how actin filaments nucleated from the surface of a membrane organelle could "attract" other bound organelles toward it. First, because the barbed ends of actin are close to the membrane surface, the polarity of the filaments is such that most myosins bound to the organelles would carry their cargo toward the nucleating organelle in the presence of ATP. In a second model, both fusion partners nucleate actin and the double-headed myosin II is proposed to crosslink and slide the filament bundles in opposite directions. This might also facilitate membrane organelle aggregation, leading to docking and fusion [39].

8.2.2
Disruption of the Actin Cytoskeleton by Intraphagosomal Pathogens

There are many intracellular pathogens, including viruses, parasites and bacteria, that target eukaryotic components of the actin cytoskeleton [19, 28, 68–70]. This highlights the importance of the actin cytoskeleton in the cellular response against pathogens. The strategies rely mainly on the intracellular lifestyle and mechanisms of pathogenesis. Disruption or alteration of the actin cytoskeleton is a phenomenon that starts immediately upon contact of the microorganisms with the eukaryotic phagocytic cells. At the risk of oversimplification, the roles of actin in microorganism–phagocyte interactions can be used to categorize pathogens into four groups: (i) some use actin to enter into cells; (ii) some use it to avoid uptake by phagocytic cells; (iii) some use it to promote attachment to the host plasma membrane surface, forming specialized actin structures; and (iv) some escape from the phagosome and use the actin machinery to move within cells and travel from cell to cell.

Some pathogens included in group ii, such as *Yersinia*, *Helicobacter* and enteropathogenic *Escherichia coli* (EPEC), inject virulence factors across the plasma membrane of the phagocyte in order to prevent phagocytic uptake [16]. One of the most prominently studied pathogens with respect to the actin cytoskeleton is *Listeria monocytogenes*. This pathogen, which can be classified in group iv, targets the Arp2/3 complex via the *Listeria* surface protein ActA, which, together with other factors, allows actin-based motility [17]. Similarly, another pathogen that escapes from the phagosome, *Shigella*, produces a protein called IcsA that target N-WASP from the host cell to induce actin-based motility as well [71]. Moreover, *Rickettsia* spp. also target the Arp2/3 complex to polymerize actin and move within cells [16]. However, the mechanism seems to be different from the one used by *Listeria*.

In addition to the pathogens that have an "extraphagosomal" way of life, it is clear that some pathogens can modify the actin cytoskeleton from within the phagosome. We applied the actin assembly assay developed for LBPs (see above) to monitor the assembly of actin by mycobacteria, focusing on the nonpathogen *Mycobacterium smegmatis*, whose phagosomes mature fully and consequently the bacteria are killed, and the pathogens *M. avium* and *M. tuberculosis*, whose phagosomes are arrested at an early stage, blocking phagolysosomal fusion and therefore the bacteria survive in macrophages. In general, both heat-killed pathogens, as well as live nonpathogen-containing phagosomes assemble actin similarly to LBPs. However, under a wide range of conditions, phagosomes enclosing the live mycobacterial pathogens were unable to nucleate actin *in vitro* or *in vivo* [57]. We have accumulated a large set of correlative observations such that all conditions we have identified as being stimulating in the *in vitro* actin assembly assay (for beads or nonpathogenic *M. smegmatis* phagosomes) also stimulate (i) actin accumulation, seen as dots around phagosomes in cells labeled with phalloidin binding, (ii) phagosome fusion with late endosomes and lysosomes and (iii) killing of mycobacteria, a consequence of phagolysosome fusion. Conversely, compounds that inhibit *in vitro* actin assembly (i) do not increase F-actin on phagosomes, (ii) fail to promote phagolysosome fusion and (iii) favor growth of nonpathogenic and pathogenic mycobacteria [57].

Among the stimulatory factors in the above experiments are AA, S1P and low levels of intracellular cAMP, whose activities include the activation of phagosome-associated PKA [52, 57]. The inhibitors include the omega-3 fatty acids and eicosapentaenoic acid (EPA) and docosahexaenoic acid (DHA) [57], PC [72] and low levels of cAMP [52].

8.3
The Microtubule Cytoskeleton – General Background

Microtubules are made of subunits of tubulin that exist as an alpha–beta dimer. Like actin filaments, MTs are asymmetric with a preferred (plus) end for subunit addition during polymerization. In contrast to actin that binds ATP, tubulin binds GTP. Many microtubules in eukaryotic cells assemble by nucleating in the centrosome region or in other less-defined nucleation sites [73–75]; there MTs grow out by addition of GTP-tubulin to the fast-growing ends. After addition, the GTP is hydrolyzed into GDP by the intrinsic GTPase activity of tubulin. The growing MT always has a "cap" of GTP-bound monomers. When the cap is lost the MT spontaneously depolymerizes. Growth and shrinkage of MT is highly dynamic in cells and it has been referred to as "dynamic instability" [76].

Many proteins bind MTs, with the main classes being MT-associated proteins, (MAPs), the MT motors and the plus-end and minus-end binding proteins. Vesicles and other organelles can be transported by MTs by at least two different mechanisms. First, by motors; there are two major sets of MT motors: kinesins that drive motility in cells outwards towards the plus ends, and dynein, that, in conjunction with its cofactor dynactin, drives inward-directed motility towards the MT minus ends that are associated with the centrosome region or with noncentrosomal nucleation sites [74]. The second mechanism is a counterintuitive process referred to as MT "dynamics"; as MTs depolymerize at their plus ends this process facilitates the transport towards the minus ends of structures such as the kinetochore [77]. This process is perhaps akin to surfing a wave.

8.3.1
Microtubule Interactions with Assembled Phagosomes

After formation, phagosomes interact dynamically with actin filaments in the periphery of the cell in a process dependent on myosin Va [24]. Then phagosomes move centripetally to the perinuclear region in an MT-dependent way [25]. The current hypothesis established by the group of Sergei Kuznetzov is that myosin Va mediates phagosome binding to F-actin and this binding results in a delay in MT-dependent retrograde phagosome movement toward the cell center. In other words, this interaction appears to be a kind of tug-of-war between MTs and actin filaments and decides how long the phagosome is retained in the periphery [24]. In addition, MTs are not only participating in the movement of phagosomes, it seems they are also important for recycling components from the phagosome to the plasma membrane [10].

8.3.1.1 *In Vivo* Studies

8.3.1.1.1 *In Vivo* Movement of Phagosomes Along MTs Motility and positioning of phagosomes within the cells rely in part on its association with MTs. For a long time it has been known that MTs facilitate fusion events between phagosomes and endocytic organelles [78]. More recent studies using video analysis showed that in macrophages, phagosomes and late endosomes encounter one another repeatedly as both sets of organelles move along linear tracks consistent with MT transport [9]. Moreover, the acquisition by phagosomes of membrane protein- and fluid phase-markers of late endocytic organelles is inhibited by MT depolymerization [62, 79]. Other studies have addressed the mechanism by which endocytic organelles interact with MTs [80–82].

From time-lapse observations in J774 macrophages, phagosomes can be observed to follow two different kinds of movements within cells: a "slow" movement (around $0.1\,\mu m\,s^{-1}$), occurring immediately after their formation in the cell periphery, and a "fast" movement (above $0.1\,\mu m\,s^{-1}$) more frequently seen when these organelles moved towards the perinuclear region [83]. Interestingly, altering MT (but not actin) assembly inhibited only slow phagosome movements. Moreover, analysis of phagosome positioning within the cell showed that low pH in the medium induced a switch of the motors to the plus end predominantly [83, 84]. The use of this simple procedure revealed that even late phagosomes (24 h) mostly concentrated near the nucleus can still show plus-end MT motility and move out to the tips of the cell [85].

Immunocytochemistry was also used to analyze the transfer of a lysosomal marker LAMP2 to LBPs under normal conditions and in macrophages treated with the MT-depolymerizing drug nocodazole. Quantitative electron microscopy analysis of LAMP2 labeling showed that the amount of this marker was significantly lower on LBPs formed after a chase in the presence of nocodazole, compared with control cells. These experiments demonstrated a role for MTs in the transformation of phagosomes into phagolysosomes [9].

MT motors such as dynein and kinesin associate with phagosomes [86] and the interplay between these motors defines the position of phagosomes within the cell [12]. Following their formation at the cell periphery, phagosomes travel toward the centriole in an MT- and dynein-dependent manner, and this displacement is required for phagolysosome fusion [87]. Kinectin functions as a membrane anchor protein that attaches different organelles to the motor protein kinesin. The only known membrane receptor for kinesin is kinectin [88] and functional evidence argues that this is essential for phagosome motility *in vitro* [85]. However there is evidence arguing that kinectin is not essential for trafficking of phagosomes *in vivo* [89].

One protein shown to be involved in the association of phagosomes with MT is RILP, an effector of Rab7 [87]. It has been suggested that three proteins, namely Rab7, RILP and LAMP, are required for optimal association of late endosomes/lysosomes with dynein/dynactin. In fact, RILP expression induces the recruitment of functional dynein–dynactin motor complexes to Rab7-containing late endosomes and lysosomes [90]. Alternatively, phagosomes and late endosomes/lysosomes may interact with dynein/dynactin by more than one mechanism: the conventional Rab7/RILP association and alternatively by a second set of proteins involving LAMPs and either RILP, or an unknown adapter [91]. However, it is still unclear how LAMPs can be linked to the microtubular motor complex.

8.3.1.2 *In Vitro* Assays

8.3.1.2.1 Binding of Phagosomes to Microtubules The principle of this assay is to incubate isolated phagosomes with MTs that are attached to a glass surface. Counting phagosome–microtubule binding events using this assay showed that as with actin, the binding of phagosomes to MTs required factors present in the cytosol [79]. In addition, isolated early phagosomes (20 min after uptake) could bind MT while 1 h or 24 h phagosomes bound randomly. Therefore the ability of phagosomes to interact with MTs seems to be regulated during phagosome maturation and is an early requirement in the life of the phagosome [79]. Moreover, the binding of phagosomes to MTs was found to be likely mediated by a single cytosolic factor, with a molecular weight of approximately 200 kDa determined by gel filtration and displaying properties of a still to be identified microtubule-associated protein (MAP) [79]. MAP-depleted cytosol was not able to support phagosome binding to MTs. Interestingly, re-addition of a fraction enriched in MAPs to MAP-depleted cytosol restored the ability to phagosomes to bind MT plus ends, but the MAP fraction was not by itself able to support MT binding. These data argue that MAPs are necessary, but not sufficient, for early phagosomes to bind preferentially to MT plus ends.

8.3.1.2.2 *In Vitro* Motility of Phagosomes Along Microtubules The movement of purified phagosomes along polarity-marked MTs was reconstituted *in vitro* using a fluorescence video microscopy assay. A uniform layer of fluorescent microtubules marked at their minus ends by labeled MT seeds [92] was laid down on the coverglass of a perfusion chamber. A mixture of purified, salt-stripped phagosomes (containing weakly fluorescent latex beads), J774 macrophage cytosol, and an ATP-regenerating system were then added. In this system, 10–30% of the phagosomes that became bound to MTs moved along them at any time.

Using this *in vitro* assay to test movement of purified phagosomes along polarity-marked MTs it was found that phagosomes could move bidirectionally.

However, most of the phagosomes moved predominantly toward MT minus ends (70%) while around 30% moved to the plus ends, similar to their behavior *in vivo*. Motility required membrane proteins, as well as kinesin, cytoplasmic dynein, dynactin, and at least one other component that could be a dynactin-associated kinase modulating dynein-mediated organelle motility [83]. The fascinating question of how the system switches from one direction to other remains to be addressed.

8.3.2 Disruption of the Microtubule Cytoskeleton by Intraphagosomal Pathogens

It is clear that manipulation of actin's fate by intracellular pathogens within the host cytoplasm is a key event that determines bacterial survival. MTs are also targeted by microbes. Although many viruses are known to hijack MT motors, less is known about the use of MT-associated motors by intracellular bacteria. Evidence indicates that in order to manipulate the MT cytoskeleton, bacteria target directly or indirectly two classes of proteins: dyneins and kinesins. MT-dependent entry into host cells has been described for *Campylobacter jejuni*, *Citrobacter freundii* and *Serratia marcescens* [93, 94]. Also, inhibition of dynein decreases entry of *C. jejuni* in host cells, suggesting the involvement of this molecular motor in the MT-dependent entry of this pathogen into host cells [95].

Upon entry, movement towards the microtubule organizing center (MTOC) in a dynein-dependent way has been observed for several intracellular bacteria such as *Chlamydia trachomatis* [96, 97] and *C. jejuni* [95]. Chlamydia phagosomes for example, move towards the minus end of microtubules and aggregate at the MTOC. Trafficking towards MTOC is dynein-dependent although the activating function of the dynactin complex is replaced by still unidentified chlamydial protein(s). Importantly, chlamydial transcription and translation are required for intracellular trafficking [97].

Salmonella is one of the best-characterized intracellular pathogens that target the MT cytoskeleton once within the cell. This pathogen uses the MT cytoskeleton to generate the *Salmonella*-induced filaments (SIFs). These tubular structures associated with *Salmonella* phagosomes allow bacterial multiplication in HeLa cells [98, 99]. Evidence shows that the MT network and the associated motors play a critical role in the maturation of the *Salmonella*-containing vacuole (SCV), consistent with one of the first studies that showed the requirement of an intact MT network for the establishment of the *S. typhimurium* replication niche [100].

LAMP1 and LAMP2 double-deficient fibroblasts have reduced ability of phagosomes to move toward the MTOC. Moreover, those double knockout cells failed to kill engulfed *Neisseria gonorrhoeae*. In those cells, maturation is arrested prior to the acquisition of Rab7 and, as a result, the Rab7 effector RILP is not recruited to the phagosomes that were thus unable to undergo dynein/dynactin-mediated centripetal movement along MTs and remain in the cell periphery. The inability of such phagosomes to migrate towards lysosomes likely contributes to their incomplete maturation and inability to eliminate bacteria [101].

8.4 Concluding Remarks

The latex bead phagosome is now one of the best-characterized membrane organelles. Although the simplest imaginable organelle, in that it consists of a membrane and luminal content, there is no denying that its protein and lipid composition, its biochemical properties and its functional capacity, even after isolation, are remarkably complex, as are those of all biological membranes. Interactions with the cytoskeleton is only one aspect, albeit an important one, in the cellular functions of phagosomes. The challenge now is to use a systems biology type of global analysis to understand more holistically how these organelles function. This understanding will then provide the foundation for us to tackle the even more complex situation when a live pathogen such as *M. tuberculosis*, which has around 4000 genes, resides in the phagosome.

Acknowledgments

We are grateful to Sergei Kuznetsov, Luis Mayorga, David Liebl, Eik Hoffman and Sabrina Marion for critical reading of this manuscript and useful comments. The work discussed here was supported in part by grants from the German DFG to Gareth Griffiths. MGG was supported by a Research Fellowship from Alexander von Humboldt Foundation.

References

1 Aderem, A. and Underhill, D.M. (1999) *Annu Rev Immunol*, **17**, 593–623.
2 Swanson, J.A. and Hoppe, A.D. (2004) *J Leukoc Biol*, **76**, 1093–1103.
3 Booth, J.W., Trimble, W.S. and Grinstein, S. (2001) *Semin Immunol*, **13**, 357–364.
4 Vieira, O.V., Botelho, R.J. and Grinstein, S. (2002) *Biochem J*, **366**, 689–704.
5 Yeung, T., Ozdamar, B., Paroutis, P. and Grinstein, S. (2006) *Curr Opin Cell Biol*, **18**, 429–437.
6 Yeung, T., Terebiznik, M., Yu, L., Silvius, J. *et al.* (2006) *Science*, **313**, 347–351.
7 Haas, A. (2007) *Traffic*, **8**, 311–330.
8 Vergne, I., Chua, J., Singh, S.B. and Deretic, V. (2004) *Annu Rev Cell Dev Biol*, **20**, 367–394.
9 Desjardins, M., Huber, L.A., Parton, R.G. and Griffiths, G. (1994) *J Cell Biol*, **124**, 677–688.
10 Damiani, M.T. and Colombo, M.I. (2003) *Exp Cell Res*, **289**, 152–161.
11 Huynh, K.K. and Grinstein, S. (2007) *Microbiol Mol Biol Rev*, **71**, 452–462.
12 Harrison, R.E. and Grinstein, S. (2002) *Biochem Cell Biol*, **80**, 509–515.
13 Araki, N. (2006) *Front Biosci*, **11**, 1479–1490.
14 Castellano, F., Chavrier, P. and Caron, E. (2001) *Semin Immunol*, **13**, 347–355.
15 Bhavsar, A.P., Guttman, J.A. and Finlay, B.B. (2007) *Nature*, **449**, 827–834.
16 Gouin, E., Welch, M.D. and Cossart, P. (2005) *Curr Opin Microbiol*, **8**, 35–45.

17 Gruenheid, S. and Finlay, B.B. (2003) *Nature*, **422**, 775–781.

18 Henry, T., Gorvel, J.P. and Meresse, S. (2006) *Cell Microbiol*, **8**, 23–32.

19 Rottner, K., Stradal, T.E. and Wehland, J. (2005) *Dev Cell*, **9**, 3–17.

20 Carlier, M.F., Le Clainche, C., Wiesner, S. and Pantaloni, D. (2003) *Bioessays*, **25**, 336–345.

21 Pollard, T.D., Blanchoin, L. and Mullins, R.D. (2000) *Annu Rev Biophys Biomol Struct*, **29**, 545–576.

22 Winder, S.J. and Ayscough, K.R. (2005) *J Cell Sci*, **118**, 651–654.

23 Tilney, L.G. (1975) *Soc Gen Physiol Ser*, **30**, 339–388.

24 Al-Haddad, A., Shonn, M.A., Redlich, B., Blocker, A. *et al.* (2001) *Mol Biol Cell*, **12**, 2742–2755.

25 Toyohara, A. and Inaba, K. (1989) *J Cell Sci*, **94** (Pt 1), 143–153.

26 Zhang, F., Southwick, F.S. and Purich, D.L. (2002) *Cell Motil Cytoskeleton*, **53**, 81–88.

27 Yam, P.T. and Theriot, J.A. (2004) *Mol Biol Cell*, **15**, 5647–5658.

28 Stevens, J.M., Galyov, E.E. and Stevens, M.P. (2006) *Nat Rev*, **4**, 91–101.

29 Kanzaki, M., Furukawa, M., Raab, W. and Pessin, J.E. (2004) *J Biol Chem*, **279**, 30622–30633.

30 Taunton, J. (2001) *Curr Opin Cell Biol*, **13**, 85–91.

31 Chang, F.S., Han, G.S., Carman, G.M. and Blumer, K.J. (2005) *J Cell Biol*, **171**, 133–142.

32 Chang, F.S., Stefan, C.J. and Blumer, K.J. (2003) *Curr Biol*, **13**, 455–463.

33 Southwick, F.S., Li, W., Zhang, F., Zeile, W.L. and Purich, D.L. (2003) *Cell Motil Cytoskeleton*, **54**, 41–55.

34 Jahraus, A., Egeberg, M., Hinner, B., Habermann, A. *et al.* (2001) *Mol Biol Cell*, **12**, 155–170.

35 Defacque, H., Egeberg, M., Habermann, A., Diakonova, M. *et al.* (2000) *EMBO J*, **19**, 199–212.

36 Berg, J.S., Powell, B.C. and Cheney, R.E. (2001) *Mol Biol Cell*, **12**, 780–794.

37 Wells, A.L., Lin, A.W., Chen, L.Q., Safer, D. *et al.* (1999) *Nature*, **401**, 505–508.

38 Spudich, J.A. (1994) *Nature*, **372**, 515–518.

39 Kjeken, R., Egeberg, M., Habermann, A., Kuehnel, M. *et al.* (2004) *Mol Biol Cell*, **15**, 345–358.

40 Malacombe, M., Bader, M.F. and Gasman, S. (2006) *Biochim Biophys Acta*, **1763**, 1175–1183.

41 Sokac, A.M. and Bement, W.M. (2006) *Mol Biol Cell*, **17**, 1495–1502.

42 Allen, L.H. and Aderem, A. (1995) *J Exp Med*, **182**, 829–840.

43 Neuhaus, E.M. and Soldati, T. (2000) *J Cell Biol*, **150**, 1013–1026.

44 DePina, A.S. and Langford, G.M. (1999) *Microsc Res Tech*, **47**, 93–106.

45 Soni, L.E., Warren, C.M., Bucci, C., Orten, D.J. and Hasson, T. (2005) *Cell Motil Cytoskeleton*, **62**, 13–26.

46 Sheetz, M.P. and Spudich, J.A. (1983) *Nature*, **303**, 31–35.

47 Defacque, H., Bos, E., Garvalov, B., Barret, C. *et al.* (2002) *Mol Biol Cell*, **13**, 1190–1202.

48 Welch, M.D. and Mullins, R.D. (2002) *Annu Rev Cell Dev Biol*, **18**, 247–288.

49 Ivetic, A. and Ridley, A.J. (2004) *Immunology*, **112**, 165–176.

50 Prag, S., Parsons, M., Keppler, M.D., Ameer-Beg, S.M. *et al.* (2007) *Mol Biol Cell*, **18**, 2935–2948.

51 Defacque, H., Egeberg, M., Antzberger, A., Ansorge, W. *et al.* (2000) *Cytometry*, **41**, 46–54.

52 Kalamidas, S.A., Kuehnel, M.P., Peyron, P., Rybin, V. *et al.* (2006) *J Cell Sci*, **119**, 3686–3694.

53 Zhou, R., Cao, X., Watson, C., Miao, Y. *et al.* (2003) *J Biol Chem*, **278**, 35651–35659.

54 Laulajainen, M., Muranen, T., Carpen, O. and Gronholm, M. (2007) *Oncogene*, **27**, 3233–3243.

55 Emans, N., Nzala, N.N. and Desjardins, M. (1996) *FEBS Lett*, **398**, 37–42.

56 Griffiths, G. and Mayorga, L. (2007) *Genome Biol*, **8**, 207.

57 Anes, E., Kuhnel, M.P., Bos, E., Moniz-Pereira, J. *et al.* (2003) *Nat Cell Biol*, **5**, 793–802.

58 Burlak, C., Whitney, A.R., Mead, D.J., Hackstadt, T. and Deleo, F.R. (2006) *Mol Cell Proteomics*, **5**, 620–634.

59 Kim, M., Jiang, L.H., Wilson, H.L., North, R.A. and Surprenant, A. (2001) *EMBO J*, **20**, 6347–6358.

60 Yin, H.L. and Janmey, P.A. (2003) *Annu Rev Physiol*, **65**, 761–789.

61 Gotthardt, D., Warnatz, H.J., Henschel, O., Bruckert, F. *et al.* (2002) *Mol Biol Cell*, **13**, 3508–3520.

62 Desjardins, M., Celis, J.E., van Meer, G., Dieplinger, H. *et al.* (1994) *J Biol Chem*, **269**, 32194–32200.

63 Griffiths, G. (2004) *Trends Cell Biol*, **14**, 343–351.

64 Marion, S., Laurent, C. and Guillen, N. (2005) *Cell Microbiol*, **7**, 1504–1518.

65 Stuart, L.M., Boulais, J., Charriere, G.M., Hennessy, E.J. *et al.* (2007) *Nature*, **445**, 95–101.

66 Jutras, I., Houde, M., Currier, N., Boulais, J. *et al.* (2008) *Mol Cell Proteomics*, **7**, 697–715.

67 Gotthardt, D., Blancheteau, V., Bosserhoff, A., Ruppert, T. *et al.* (2006) *Mol Cell Proteomics*, **5**, 2228–2243.

68 Alonso, A. and Garcia-del Portillo, F. (2004) *Int Microbiol*, **7**, 181–191.

69 Hamon, M., Bierne, H. and Cossart, P. (2006) *Nat Rev*, **4**, 423–434.

70 Sibley, L.D. (2004) *Science*, **304**, 248–253.

71 Goldberg, M.B. (2001) *Microbiol Mol Biol Rev*, **65**, 595–626, table of contents.

72 Treede, I., Braun, A., Sparla, R., Kuhnel, M. *et al.* (2007) *J Biol Chem*, **282**, 27155–27164.

73 Bartolini, F. and Gundersen, G.G. (2006) *J Cell Sci*, **119**, 4155–4163.

74 Luders, J. and Stearns, T. (2007) *Nat Rev Mol Cell Biol*, **8**, 161–167.

75 Schiebel, E. (2000) *Curr Opin Cell Biol*, **12**, 113–118.

76 Mitchison, T.J. and Kirschner, M.W. (1987) *Cell Biophys*, **11**, 35–55.

77 Lombillo, V.A., Stewart, R.J. and McIntosh, J.R. (1995) *Nature*, **373**, 161–164.

78 Goldstein, I., Hoffstein, S., Gallin, J. and Weissmann, G. (1973) *Proc Natl Acad Sci USA*, **70**, 2916–2920.

79 Blocker, A., Severin, F.F., Habermann, A., Hyman, A.A. *et al.* (1996) *J Biol Chem*, **271**, 3803–3811.

80 Gruenberg, J., Griffiths, G. and Howell, K.E. (1989) *J Cell Biol*, **108**, 1301–1316.

81 Pierre, P., Scheel, J., Rickard, J.E. and Kreis, T.E. (1992) *Cell*, **70**, 887–900.

82 Scheel, J. and Kreis, T.E. (1998) *Methods Enzymol*, **298**, 381–389.

83 Blocker, A., Griffiths, G., Olivo, J.C., Hyman, A.A. and Severin, F.F. (1998) *J Cell Sci*, **111** (Pt 3), 303–312.

84 Heuser, J. (1989) *J Cell Biol*, **108**, 855–864.

85 Blocker, A., Severin, F.F., Burkhardt, J.K., Bingham, J.B. *et al.* (1997) *J Cell Biol*, **137**, 113–129.

86 Habermann, A., Schroer, T.A., Griffiths, G. and Burkhardt, J.K. (2001) *J Cell Sci*, **114**, 229–240.

87 Harrison, R.E., Bucci, C., Vieira, O.V., Schroer, T.A. and Grinstein, S. (2003) *Mol Cell Biol*, **23**, 6494–6506.

88 Sheetz, M.P. (1999) *Eur J Biochem*, **262**, 19–25.

89 Plitz, T. and Pfeffer, K. (2001) *Mol Cell Biol*, **21**, 6044–6055.

90 Jordens, I., Fernandez-Borja, M., Marsman, M., Dusseljee, S. *et al.* (2001) *Curr Biol*, **11**, 1680–1685.

91 Huynh, K.K., Eskelinen, E.L., Scott, C.C., Malevanets, A. *et al.* (2007) *EMBO J*, **26**, 313–324.

92 Howard, J. and Hyman, A.A. (1993) *Methods Cell Biol*, **39**, 105–113.

93 Hertle, R. and Schwarz, H. (2004) *BMC Infect Dis*, **4**, 16.

94 Oelschlaeger, T.A., Guerry, P. and Kopecko, D.J. (1993) *Proc Natl Acad Sci USA*, **90**, 6884–6888.

95 Hu, L. and Kopecko, D.J. (1999) *Infect Immun*, **67**, 4171–4182.

96 Clausen, J.D., Christiansen, G., Holst, H.U. and Birkelund, S. (1997) *Mol Microbiol*, **25**, 441–449.

97 Grieshaber, S.S., Grieshaber, N.A. and Hackstadt, T. (2003) *J Cell Sci*, **116**, 3793–3802.

98 Brumell, J.H., Goosney, D.L. and Finlay, B.B. (2002) *Traffic*, **3**, 407–415.

99 Brumell, J.H., Perrin, A.J., Goosney, D.L. and Finlay, B.B. (2002) *Curr Biol*, **12**, R15–R17.

100 Garcia-del Portillo, F., Zwick, M.B., Leung, K.Y. and Finlay, B.B. (1993) *Proc Natl Acad Sci USA*, **90**, 10544–10548.

101 Binker, M.G., Cosen-Binker, L.I., Terebiznik, M.R., Mallo, G.V. *et al.* (2007) *Cell Microbiol*, **9**, 2153–2166.

9
Intracellular Microbe Whole-Genome Expression Profiling: Methodological Considerations and Biological Inferences

Simon J. Waddell and Philip D. Butcher

9.1
Introduction

Understanding the mechanisms by which microbes survive in their host intracellular environments is an essential precursor to developing novel intervention strategies that might exploit the interplay between host and microbe. Transcriptional profiling allows us a glimpse of these molecular interactions and promises to reveal much of this dialogue in both pathogenic and symbiotic settings. Such studies will underpin the development of antivirulence strategies for new therapeutics. This chapter aims to describe the challenges of studying microbial RNA expression from intracellular and *in vivo* scenarios, and to detail a fraction of the insights acquired from this rapidly expanding field, driven by the recent availability of microbe whole-genome sequence information.

Microbial intracellular gene expression patterns may be used not only to recognize mechanisms vital for survival during infection, but also as a bioprobe to define the host microenvironments encountered. Furthermore, intracellular RNA profiling may decipher how these interactions change over time or with different genetic backgrounds of both host and microbe. This nonreductionist approach of profiling host–microbe crosstalk at the RNA, protein and metabolic levels will contribute to a systems biology view of microbial colonization. The major methodological bottlenecks and applications of these expression datasets are summarized in Figure 9.1, with each step broadly translating into a section of this chapter. We present examples of the use of bacterial and protozoan mRNA expression studies in commensal, mutualistic and parasitic scenarios, with a particular focus on the pathogen *Mycobacterium tuberculosis*, most familiar to the authors.

Intracellular Niches of Microbes. Edited by Ulrich E. Schaible and Albert Haas

ISBN: 978-3-527-32207-7

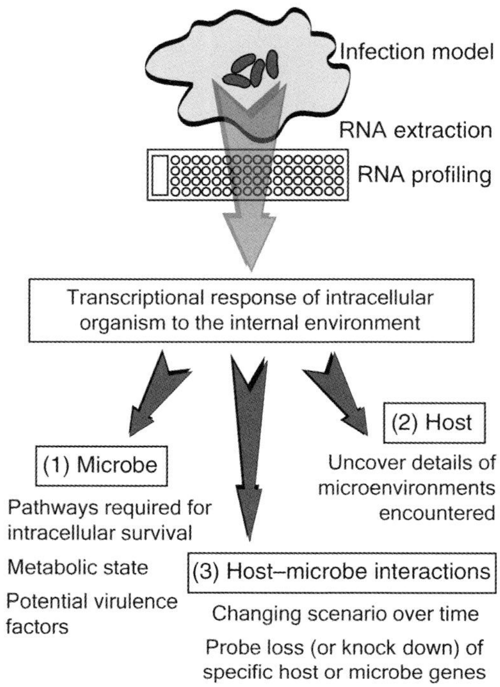

Figure 9.1 A cartoon illustrating the bottlenecks and applications of RNA profiling from intracellular microbes. The gene expression pattern of microbes intracellularly or *in vivo* may be interrogated (1) to identify gene products or metabolic states necessary for successful microbial infection; (2) as a bioprobe, to define the conditions confronting microbes during infection; or (3) to investigate the role of individual or multiple genes in the changing relationship between microbe and host.

9.2 Methodological Considerations

9.2.1 Expression Profiling Techniques

The development of methods over the last 20 years to identify microbial genes or proteins important in intracellular survival has led to significant advances in the understanding of host–microbe interactions in both disease and symbiotic infection states. More specifically, the determination of genes induced by facultative intracellular microbes from within host cells has been key to discovering novel virulence mechanisms. A range of techniques have been designed to sample the mRNA or protein from microbes in both direct and indirect assays. The availability of microbial genome sequences has had a massive impact on both the profiling techniques available, and the hypothesis-driven application of the resulting expression data. Although not a prerequisite for gene expression profiling [1], the ability to design unique gene-specific probes for reverse transcription polymerase chain reaction (RT-PCR) or microarray analysis has resulted in increased coverage and definition to microbial transcriptomes. Access to sequenced reference strains has also allowed expression datasets to be compared across experimental conditions on a whole-genome scale.

In proteomics, methodologies such as two-dimensional gel electrophoresis and mass spectroscopy (MS) remain the standards for the detection and identification of intracellularly expressed proteins [2–4]. However the profiling of intracellular proteomes is particularly challenging due to the low sensitivity and the limited resolution of protein moieties using these methods. Alternatively, antigen discovery techniques such as immunoblotting/autodisplay [5] and antigen microarrays [6] have identified microbial proteins that are recognized by the host after infection. At the transcriptional level, differential fluorescence induction [7], *in vivo* expression technology (IVET) [8] and recombinase-based *in vivo* expression technology (RIVET) [9], applied so successfully to determine *Salmonella* virulence factors, make use of reporter gene technology to reveal genes overexpressed intracellularly. Techniques such as serial analysis of gene expression (SAGE) [10] and massively parallel signature sequencing (MPSS) [11] have also been used to determine intracellularly expressed transcripts, and have contributed to the genome annotation of protozoan parasites. Other methods of measuring changes in RNA abundance such as RT-PCR, subtractive hybridization and microarray analysis rely on an input of representative total microbial RNA extracted from intracellular models of infection or *in vivo* scenarios. This chapter details the challenges involved in recovering this microbial RNA (or protein) population, and the insights resulting from, or reflected in, these microbial intracellular expression patterns.

9.2.2 Microbial RNA Extraction

9.2.2.1 RNA Stabilization

There are a number of obstacles that must be overcome to successfully profile microbial mRNA from intracellular models of infection. The most fundamental of these is the extraction of a representative mRNA population. The mRNA content of a microbe will change as differences are sensed in the environment in which the microbe finds itself, and this response is likely to affect the RNA profile within minutes, if not seconds [12]. The extraction process must therefore attempt to limit potential changes in gene expression introduced during RNA purification. The short half-life of microbial RNA, especially bacterial RNA (average of <8 min) [13], and the sensitivity of RNA to degradation when exposed to eukaryotic cellular fractions, increases the necessity for RNA preservation. RNA purification techniques must therefore ensure that microbial transcription is halted immediately, and that this RNA population is protected from further degradation during the extraction process. The potential bias introduced by RNA purification methods may be assessed using RT-PCR on a panel of sentinel genes or by microarray analysis; the bias will be revealed by comparing the expression pattern of genes in RNA extracted using different techniques from the same microbial source [14]. The development of guanidinium thiocyanate or sodium dodecyl sulfate (SDS)-based solutions, the extraction of RNA directly into lysing agents, or the use of commercially available solutions that penetrate cells and reduce degradation, have all been used successfully to extract an RNA population representative of the intracellular state of the microbe.

9.2.2.2 Separation of Microbe from Host Material

A second challenge to intracellular and *in vivo* expression profiling is the obligation to separate or distinguish between microbial and host RNA/protein. Moreover in many infection models host-derived RNA/protein will outnumber microbial RNA/protein by many factors. For some bacteria, such as *Mycobacterium, Salmonella, Shigella* and *Listeria,* it has been possible to develop differential lysis methods of RNA extraction using guanidinium thiocyanate [15, 16] or phenol/ethanol-based solutions [17], which penetrate bacterial cells to halt transcription and protect the RNA from degradation. These solutions are also capable of lysing eukaryotic cells in tissue culture models of infection while leaving bacterial cells intact; infecting bacteria can then simply be separated from host cell lysate by centrifugation.

The separation of microbe and host RNA populations may also be achieved using both negative and positive selection techniques. Graham and Clark-Curtiss [18] developed a positive system using excess genomic DNA to selectively capture transcribed sequences (SCOTS) to isolate *M. tuberculosis* genes overexpressed after phagocytosis against a background of host and ribosomal RNA. Subtractive hybridization methods have been used in conjunction with microarrays to pick out genes enriched in *M. tuberculosis* after infection of human macrophage-like cells [19]. Alternatively, oligo-dT labeled beads/columns have been exploited to remove or substantially reduce poly(A)-tailed eukaryotic RNA from mixed RNA populations containing limited bacterial RNA (which is not polyadenylated) [20]. Microbial gene specificity may also be introduced using gene-specific primers, as is the case of RT-PCR, or by using genome-directed primers (based on biasing primers towards the microbial genome sequence) that selectively label microbial RNA for microarray hybridization [21]. Alternatively, if microbial RNA yields account for a significant proportion of the total RNA population, microarray hybridization (to gene-specific elements) alone may be sufficient to quantify microbial RNA abundance against a background of copurified host RNA [22].

Techniques developed to study the protein profile of microbes in intracellular environments apply a range of differential lysis and gradient centrifugation steps to extract the microbe proteome without contaminating host proteins. The more stable nature of protein compared to nucleic acid has allowed longer and more stringent microbe-purification steps to be employed than would be applicable for extracting RNA [2].

9.2.2.3 Low RNA Yield

A third factor that has limited the application of intracellular and *in vivo* microbial transcriptomics is the low yield of microbial RNA recoverable from many infection models. Variation in the multiplicity of infection or timepoint assayed, development of microarray hybridization methods requiring less input RNA and gene-specific RT-PCR assays have all expanded the number of host–microbe scenarios amenable to RNA profiling. Likewise, the emerging field of bacterial RNA amplification [14], together with more established eukaryotic RNA amplification techniques (targeting eukaryotic poly(A)-tailed RNA) [23] have also allowed previously intractable interplay between host and infecting microbe to be explored.

9.2.3 Suitable Infection Model

An appropriate model of infection is required for intracellular microbial expression profiling. Often models are modified to obtain workable yields of bacteria, yet attention to the host–pathogen interaction should be the imperative, with method development (such as microbial RNA extraction and amplification techniques) required to retain the biological relevance of the model to reflect faithful host–pathogen crosstalk. Furthermore, the suitability of microbial RNA and protein extraction methods will depend both on the microbe studied and the host model utilized, as will the conclusions drawn. For example, mechanisms preventing phagolysosomal fusion may be more easily identified in the interaction between a microbial population and a macrophage cell line [24]. However, RNA profiling from whole organs of an animal model containing complex mixtures of host cell types and microbial subpopulations may better expose the global expression pattern of virulence factors or microbial metabolic states *in vivo* [25]. Alternatively, noninvasive human-derived samples may reveal antigenic factors or microbial transmission phenotypes [26]. The combination of biological inferences acquired from these diverse models (tissue culture to animal model to natural disease) enables different aspects of the multifactorial process of microbial infection to be interrogated. All these methods suffer, however, from the necessity that the microbial profiles generated represent the average of a mixed microbial population; that is, they are derived from several microbial subpopulations of potentially differing replicative and metabolic states, rather than from a synchronized population or single microbial cells.

It is important to clarify that the transcriptome of intracellular microbes (in most analyses) is the identification of gene products differentially expressed by microbes in a particular host microenvironment in comparison to a control condition (usually an *in vitro* condition such as mid-log phase growth). The comprehension of gene expression data, therefore, is dependent on both the use of a relevant infection model, and the choice of comparator conditions. Intracellular RNA has been compared to RNA derived from *in vitro* growth conditions, tissue culture controls or initial/final timepoints in an infection. Alternatively, analyses of relative RNA abundance have been applied to measure the microbial RNA profile in a single condition without recourse to additional comparisons [27].

9.3 Biological Inferences from Expression Profiling

The remainder of this chapter provides examples of investigations into host–microbe interactions, categorized into three sections by the biological inference gained from the intracellular or *in vivo* microbial expression profiles.

(1) What has been discovered about microbes in intracellular or *in vivo* scenarios?
(2) What can the intracellular profile of a microbe tell us about the host environment?
(3) What may we uncover about the changing crosstalk between host and microbe from comparing the microbial responses to infection with genetically distinct microbes or host cells?

9.3.1 Insights into the Microbe

The expression pattern of microbes from intracellular or *in vivo* settings enables metabolic pathways and virulence mechanisms required during successful infection to be recognized. This is of particular importance in the development of treatment strategies against obligate and facultative intracellular pathogens where all or part of the infectious cycle of the microbe is within host cells. A large number of genes have been identified as differentially regulated by microbes intracellularly or *in vivo*, including genes coding for products involved in microbial stress responses, DNA damage and the maintenance of the microbial cell wall/membrane. We have focused the following paragraphs, however, on aspects of microbial metabolism, effector molecule secretion and transcriptional regulation as paradigms for biological inference. For organism-specific mechanisms please refer to Chapters 10–33.

9.3.1.1 Metabolic State

The changing metabolic state of *M. tuberculosis* within murine or human macrophages has been described by several laboratories [16, 28, 29] with the induction of genes involved in the β-oxidation of fatty acids and the citric/glyoxylate cycles, as the carbon source switches from glucose or glycerol to fatty acids after macrophage infection. An equivalent metabolic shift has been observed in *Burkholderia pseudomallei*, in a hamster model of infection, with the downregulation of genes involved in the glycolytic/tricarboxylic acid (TCA) cycles, and the induction of genes involved in the metabolism of D-glucosamine and inositol, perhaps indicating a change in carbohydrate source [30]. The induction of genes coding for glycerol kinases and dehydrogenases suggests that *Listeria monocytogenes* growing in the cytosol of host cells may utilize glycerol as a source of carbon during infection, potentially derived from host phospholipids [31]. Adaptation of *L. monocytogenes* to the intracellular environment may also require a switch in nitrogen source, with ethanolamine a possible alternative. Perhaps the most striking descriptions of changing carbon and nitrogen metabolism come from the symbiotic bacteria. The differentiation of *Sinorhzobium meliloti* into bacteroids inside the legume nodule involves the induction of *nif*, *fix* and *fdx* gene families involved in nitrogen fixation, as the bacteria reduce nitrogen to ammonia in exchange for a carbon/energy source from the plant as part of the symbiotic relationship [32].

These metabolic changes may also be accompanied by variation in respiratory state; for example the adoption of a lipolytic lifestyle by *M. tuberculosis* coincides with modifications in respiratory state from aerobic to microaerophilic to anaerobic as murine infection continues [33]. The changing metabolic and respiratory state of intracellular bacteria is best exemplified, however, in bacteria that cycle between vegetative and replicative stages, such as *Chlamydia trachomatis* and *Legionella pneumophila*. The chlamydial elementary body (EB), which survives extracellularly and is responsible for cell invasion, differentiates into the reticulate body (RB), which replicates intracellularly, before redifferentiation back into EB for release and reinfection. Shaw *et al.* [34] followed the gene expression profile of *C. trachomatis* after epithelial cell infection, and identified sets of genes regulated at early/mid and late stages of infection. The early genes encode functions involved in the establishment of the chlamydial inclusion body and the acquisition of nutrients; the mid-stage genes for the rapid multiplication of bacteria; and late stage genes, the differentiation of RB back into EB [5]. Similarly, Bruggemann *et al.* [35] profiled the shift from replicative phase (RP) to transmissive phase (TP) of *L. pneumophila* during infection of its natural protozoan host. Genes involved in the cell division process and the metabolism of amino acids/carbohydrates were induced during RP, while genes coding for motility machinery and known virulence factors were activated during the transition from RP to TP, as the bacterium prepares for another cycle of infection.

The changing pattern of gene expression has also been profiled across the morphologically and physiologically distinct growth phases of the protozoan parasites. This is exemplified by descriptions of the transcriptome and proteome through the life cycle of the malaria parasite, *Plasmodium falciparum*, from mosquito to both asexual and sexual erythrocytic developmental forms [4, 36]. Le Roch *et al.* [36] demonstrated that 43% of *P. falciparum* expressed genes were cell-cycle regulated, identifying clusters of genes differentially regulated in one or more of nine time-points through the life cycle. This gene expression-derived metabolic map of pathogens across different environments may highlight drugable pathways, such as the induction of fatty acid biosynthetic genes by *Plasmodium yoelii* in a rat model of liver stage infection [37].

Finally, the transcriptomes of pathogens extracted directly from human samples promises to bridge the gap between models and natural infection. Rachman *et al.* [38] have described a unique *M. tuberculosis* gene expression profile extracted from human lung tissue following surgery, identifying, amongst others, induction of lipid biosynthetic and immunogenic PE/PPE genes by *M. tuberculosis in vivo*. Additionally, the transcriptional responses of *P. falciparum* and *M. tuberculosis* extracted from patient blood and sputum respectively, have led to hypothesized distinct physiological states for these important pathogens during natural infection [26, 39]. The induction of *M. tuberculosis* genes involved in cholesterol catabolism and anaerobic respiration, together with a slow/non-growing gene expression signature, indicates that the bacilli responsible for the transmission of *M. tuberculosis* are not actively replicating [26]. Further characterization of the metabolic state of infecting microbes may distinguish novel biosynthetic pathways and influence future drug development programs.

9.3.1.2 Secretion Apparatus

Analyzing the RNA or protein profile of intracellular microbes also reveals potential virulence mechanisms used by invading microorganisms to subvert the host immune response. Of these, perhaps the most recognizable are the secretion systems used by microbes to release effector molecules during infection. These immunomodulatory factors are carefully regulated and expressed in a time- and cell-specific manner, such as the type three-secretion systems (T3SS) of *Salmonella*. The T3SS located in SPI1 (*Salmonella* pathogenicity island 1), which is required for cell invasion, is induced on contact with human macrophages, and then repressed after cell entry [40]. However, SPI1 T3SS continues to be expressed after invasion of human epithelial cells. A second *Salmonella* T3SS located in SPI2, necessary for successful intracellular survival, is induced throughout infection in both epithelial and macrophage cell types [41]. Conversely the *ipa-mxi-spa* T3SS in *Shigella flexneri*, essential for cell entry, was determined to be repressed after infection of macrophage and epithelial cells [42], illustrating that the intracellular bacterial phenotype is dependent both on the induction and repression of genes under tight transcriptional control. The genes encoding the Dot/Icm T4SS of *L. pneumophila*, which are required for intracellular replication, have been shown to be upregulated in the protozoan host *Acanthamoeba castellanii*. Furthermore, genes coding for the substrates of this T4SS were preferentially induced in the transmissive phase, preparing the bacterium for subsequent rounds of infection [35].

Delivery systems analogous to bacterial type 3/4 secretion systems have also been investigated in *Apicomplexa* protozoa by proteome profiling of intracellular *Toxoplasma gondii* [43]. On *T. gondii* infection effector molecules are released from specialized secretory apical organelles, rhoptries. These proteins (ROPs and RONs) have been shown to play important roles in cell invasion and the immunomodulation of the host response [44]. The presence of genes coding for virulence factors involved in cell entry and survival are not restricted to pathogenic microbes. For example, both parasitic and mutualistic species of the endosymbiont *Wolbachia* contain a type 4 secretion apparatus, emphasizing the analogous mechanisms employed by bacteria to survive intracellularly irrespective of the very different infection outcomes [45]. The continued use of RNA and protein profiling techniques to map the expression of microbial secretion mechanisms and secreted factors within and on contact with host cells, together with the definition of specific receptor–ligand interactions, promises to be a productive approach for interpreting host–pathogen interplay.

9.3.1.3 Control of Gene Expression

A particularly conducive strategy for delineating microbial components that are functionally significant intracellularly and *in vivo* is to identify global regulators of gene expression, and the genes over which they have some effect. There are many illustrations of transcriptional regulators and their regulons that have been found to be necessary for successful infection, for example: the PrfA regulon containing genes required for cell entry and cytosolic replication of *L. monocytogenes* [46]; the role of the Fis transcriptional regulator in *Salmonella* colonization [47]; and the induction of the DosR regulon intracellularly by *M. tuberculosis* [16]. Moreover, the transcriptional profiling of sigma factor mutants has enabled the regulons of many bacterial sigma factors to be elucidated [48]. An additional advantage of whole-genome transcriptomics is the information generated about genes predicted to contain DNA-binding motifs. The differential expression of these genes, alongside others that may belong to a putative regulon, may define novel biosynthetic or regulatory features important for microbial survival. Recognition of these regulons will contribute to an understanding of the overlapping cascades of transcription that likely exist as part of the microbial response to multiple environmental signals *in vivo*.

The RNA profile of intracellular microbes may also uncover something of the global nature of the control of gene expression. For example, Bozdech *et al.* [49] described the intra-erythrocytic asexual developmental cycle of *P. falciparum* every hour over a 48 h period, and found that >80% of the transcriptome was differentially regulated in a rigid stage-specific manner. This highly structured cascade of gene expression may signify that, in this environment at least, transcription is controlled by a limited number of regulatory factors, and as such the parasite has a reduced transcriptional plasticity. This may be a feature of apicomplexan parasites where epigenetic mechanisms are especially important in the regulation of gene expression [50]. In stark contrast to the high percentage of genes differentially regulated within red blood cells by *P. falciparum*; <5% of *Leishmania* genes were up- or downregulated between morphologically distinct promastigote and amastigote forms. The constitutive expression of most genes in this kinetoplastid parasite, between sandfly and human macrophage stages of the life cycle, highlights the large part that post-transcriptional and post-translational regulation may play in the control of *Leishmania* gene expression [51].

9.3.2
Insights into the Host

Gene expression profiles of intracellular microbes may be used as a bioprobe, to yield information on the microenvironments experienced by infecting microbes. Contrasting profiles derived from intracellular or *in vivo* settings with gene expression patterns determined under defined *in vitro* conditions (such as low iron, low oxygen or carbon starvation) allow the microbial response to specific environmental cues to be defined. If such transcriptional signatures are revealed in intracellular or *in vivo* profiles, it is likely that similar environmental signals are transduced during infection. A good example of this comparative analysis is provided by Schnappinger *et al.* [16], with the comparison of gene expression profiles from intracellular *M. tuberculosis* with *in vitro*-derived low iron and low oxygen conditions. These intracellular datasets may also be used in conjunction with the transcriptional patterns derived from host immune cells after microbial infection or stimulation [52, 53] to create a molecular snapshot of parasitism [29].

The induction of fatty acid metabolism and iron sequestration encoding genes in *M. tuberculosis* after phagocytosis by murine macrophages reveals that this environment is likely to be low in carbohydrate and sugar, high in lipid and low in iron. On macrophage activation, the intracellular environment becomes increasingly hypoxic, although the DosR-mediated *M. tuberculosis* response may also be attributed to a rise in nitric oxide [16]. The low availability of iron is a feature of many but not all host microenvironments. Iron sequestration genes were demonstrated to be induced after *M. tuberculosis* infection of murine lungs [25], but were not found to be upregulated by *M. tuberculosis* inside human macrophage-like THP-1 cells [54]. *S. enterica* genes coding for magnesium and phosphate transportation were demonstrated to be induced after infection of murine macrophage-like cells, indicating that the phagocytic setting in which *Salmonella* finds itself is likely to be low in both magnesium and phosphate [17].

A fascinating example of the link between the constituents of the host environment and microbial gene expression comes from the bacterial endosymbiont *Buchnera aphidicola*. This obligate intracellular symbiont has retained the capacity to provide nine of the ten essential amino acids required by aphids that cannot be assimilated from, the carbohydrate-rich but amino acid-poor plant sap upon which the aphid host feeds. Furthermore, the selection of these amino acid biosynthetic genes (around 10% of the bacterial genome) may depend upon the nutritional requirements of the host organism, and may not include the full amino acid biosynthetic pathways depending on the precursors available from the aphid host [55]. In addition, *B. aphidicola* has lost the ability to regulate the expression of the majority of these biosynthetic genes. Further investigation into these bacterially encoded amino acid biosynthetic pathways requisitioned by the host may shed more light on the coevolution of such symbiotic relationships [56].

An essential element in the interpretation of intracellular or *in vivo*-derived gene expression patterns is the comparison with microbial transcriptional signatures derived from defined conditions *in vitro*; for example, the importance of the DosR-mediated hypoxic response in intracellular *M. tuberculosis* [16, 57]. Chemostats are one of the few methods of defining the transcriptional effect of a single variable while other experimental parameters remain constant [58], and it is only in comparison with such *in vitro* studies that the multiple overlapping signals may be dissected from the complex microbial response to the intracellular environment. The mapping of differentially regulated transcriptional factors, amino acid biosynthetic or transporter-encoding genes may also be exploited to elucidate the stimuli (and signal transduction pathways) that affect microbial gene expression *in vivo*. Predicted stimuli for microbe differentiation or adaptation to intracellular conditions include limited magnesium or phosphate [17], lack of amino acid moieties [20], nitric oxide [16] or a change in pH [24]. In many cases changing oxygen tension is likely to be an important stimulus [16, 32, 35]. Distinguishing which stimuli intracellular microbes respond to throughout infection will help to decipher the dialogue between host and parasite, and offer opportunities to influence the outcome of these interactions.

9.3.3 Insights into Host–Microbe Interactions

This section details some examples of how the microbial gene expression pattern may be used to explore the way in which changing microbe or host genetic backgrounds may affect the host-microbe relationship. A comparison of the transcriptional responses of intracellular bacteria with known deletions, or with host cells of different lineage for example, may illuminate functionally significant microbial or host immune response pathways. Rohde *et al.* [24] modified phagosome acidification of murine macrophages, and compared the intracellular RNA profiles of *M. tuberculosis* and the less virulent *M. bovis* BCG (bacillus Calmette-Guérin) to identify factors induced by decreased pH or constrained intracellular growth. Similarly, the shift from tachyzoite to bradyzoite stage in *T. gondii* was explored by contrasting the intracellular RNA derived from wild-type and differentiation-deficient mutants [59].

An alternative strategy is to vary the host microenvironment encountered by the microbe during infection; Belland *et al.* [20] used interferon gamma (IFNγ) treatment of an epithelial cell line to create a model of *C. trachomatis* persistence and reactivation. Similarly, Schnappinger *et al.* [16] compared the transcriptional response of *M. tuberculosis* to naive and IFNγ-activated macrophages derived from wild-type and $NOS2^{-/-}$ (nitric oxide synthase 2) knockout mice. The *M. tuberculosis* transcriptional profile has also been compared in human immune cells permissive and non permissive for mycobacterial growth, identifying factors differentially expressed by *M. tuberculosis* after invasion of macrophages and dendritic cells respectively [29]. Genes involved in the response to hypoxia (or nitric oxide), regulated by DosR, and genes associated with nutrient starvation and slow growth were induced 18 h after phagocytosis by dendritic cells compared to the intracellular profile of *M. tuberculosis* after 18 h macrophage infection. Definition of the *M. tuberculosis* transcriptional responses to both permissive and nonpermissive intracellular environments should enable both host and mycobacterial factors that influence the ability of mycobacteria to replicate successfully intracellularly to be recognized [29]. Tailleux *et al.* [29] also measured the transcriptional responses of dendritic cells and macrophages to *M. tuberculosis* infection, enabling a transcriptional snapshot of the dialogue between bacilli and phagocytic cell types to be created. The introduction of microarray platforms containing both host and microbial genomes should expedite these kinds of analysis further [60]. The integration of the transcriptomes from both infecting and infected cells, extracted simultaneously, and the identification of changing patterns of functionally related host and microbial genes should generate new hypotheses regarding the roles of both host and microbe throughout infection.

9.4 Concluding Remarks

Expression profiling has revealed much about the mechanisms employed by microbes to survive intracellularly and *in vivo*, and the different host microenvironments confronting microbes. The continued development of bioinformatic approaches to couple these genome, transcriptome and proteome datasets together will provide new insights into the molecular interactions during infection [61], and suggest novel treatment interventions [62]. Integrating these whole-genome datasets forms the bedrock of systems biology, a nonreductionist approach where experimental data informs models that can then be tested. The global nature of transcriptome (and proteome) approaches has created a huge resource of data, which may be especially useful in delineating the role of genes of unknown function. The comparison of these datasets, derived from complex *in vivo* settings as well as tractable intracellular models and (in a reductionist approach) defined *in vitro* conditions, may be exploited to build increasingly complex layers of experimentally determined information into our understanding of the multifactorial process of microbial infection.

Finally, sampling methods like microdissection and techniques such as bacterial RNA amplification will allow more distinct (and relevant) populations to be profiled during infection. Furthermore the application of new technologies such as microfluidics for single cell analysis [63] may enable the gene expression patterns associated with stochastic (clonal differentiation/persister cell generation) as well as genetic mechanisms (phase switching/hypermutation) to be explored. The continued utilization of gene expression datasets, in combination with expression patterns from novel models of infection, in intra- or interspecies comparisons, and in the definition of gene function from the profiling of genetically distinct organisms, will continue to shape our understanding of mutualistic and parasitic infections.

Acknowledgments

SJW is currently funded by a European Union 6th Framework Programme "New Medicines for Tuberculosis," NM4TB (project number LSHP-CT-2005-018923), and acknowledges funding from the European Union 6th Framework Programme Priority grant ("Molecular Markers of *M. tuberculosis* Early Interactions with Host Phagocytes," MM-TB, number LSHP-CT-2004-012187). PDB would like to acknowledge the Wellcome Trust and its Functional Genomics Resources Initiative for funding the multi-collaborative microbial pathogen microarray facility at St. George's (BμG@S).

References

1 Hayward, R.E., Derisi, J.L., Alfadhli, S., Kaslow, D.C. *et al.* (2000) *Mol Microbiol*, **35**, 6–14.

2 Monahan, I.M., Betts, J., Banerjee, D.K. and Butcher, P.D. (2001) *Microbiology*, **147**, 459–471.

3 Molestina, R.E., Klein, J.B., Miller, R.D., Pierce, W.H. *et al.* (2002) *Infect Immun*, **70**, 2976–2981.

4 Florens, L., Washburn, M.P., Raine, J.D., Anthony, R.M. *et al.* (2002) *Nature*, **419**, 520–526.

5 Heuer, D., Kneip, C., Maurer, A.P. and Meyer, T.F. (2007) *Int J Med Microbiol*, **297**, 569–576.

6 Eyles, J.E., Unal, B., Hartley, M.G., Newstead, S.L. *et al.* (2007) *Proteomics*, **7**, 2172–2183.

7 Valdivia, R.H. and Falkow, S. (1997) *Science*, **277**, 2007–2011.

8 Mahan, M.J., Slauch, J.M. and Mekalanos, J.J. (1993) *Science*, **259**, 686–688.

9 Brown, N.F., Vallance, B.A., Coombes, B.K., Valdez, Y. *et al.* (2005) *PLoS Pathog*, **1**, e32.

10 Radke, J.R., Behnke, M.S., Mackey, A.J., Radke, J.B. *et al.* (2005) *BMC Biol*, **3**, 26.

11 Bishop, R., Shah, T., Pelle, R., Hoyle, D. *et al.* (2005) *Nucl Acids Res*, **33**, 5503–5511.

12 Kendall, S.L., Movahedzadeh, F., Rison, S.C., Wernisch, L. *et al.* (2004) *Tuberculosis (Edinb)*, **84**, 247–255.

13 Bernstein, J.A., Khodursky, A.B., Lin, P.H., Lin-Chao, S. and Cohen, S.N. (2002) *Proc Natl Acad Sci USA*, **99**, 9697–9702.

14 Waddell, S.J., Laing, K., Senner, C. and Butcher, P.D. (2008) *BMC Genomics*, **9**, 94.

15 Mangan, J.A., Monahan, I.M. and Butcher, P.D. (2002) in *Functional Microbial Genomics* (eds B.W. Wren and N. Dorrell), Academic Press, London, pp. 137–151.

16 Schnappinger, D., Ehrt, S., Voskuil, M.I., Liu, Y. *et al.* (2003) *J Exp Med*, **198**, 693–704.

17 Eriksson, S., Lucchini, S., Thompson, A., Rhen, M. and Hinton, J.C. (2003) *Mol Microbiol*, **47**, 103–118.

18 Graham, J.E. and Clark-Curtiss, J.E. (1999) *Proc Natl Acad Sci USA*, **96**, 11554–11559.

19 Li, M.S., Monahan, I.M., Waddell, S.J., Mangan, J.A. *et al.* (2001) *Microbiology*, **147**, 2293–2305.

20 Belland, R.J., Nelson, D.E., Virok, D., Crane, D.D. *et al.* (2003) *Proc Natl Acad Sci USA*, **100**, 15971–15976.

21 Talaat, A.M., Hunter, P. and Johnston, S.A. (2000) *Nature Biotechnol*, **18**, 679–682.

22 Motley, S.T., Morrow, B.J., Liu, X., Dodge, I.L. *et al.* (2004) *Cell Microbiol*, **6**, 849–865.

23 Van Gelder, R.N., von Zastrow, M.E., Yool, A., Dement, W.C. *et al.* (1990) *Proc Natl Acad Sci USA*, **87**, 1663–1667.

24 Rohde, K.H., Abramovitch, R.B. and Russell, D.G. (2007) *Cell Host Microbe*, **2**, 352–364.

25 Talaat, A.M., Lyons, R., Howard, S.T. and Johnston, S.A. (2004) *Proc Natl Acad Sci USA*, **101**, 4602–4607.

26 Garton, N.J., Waddell, S.J., Sherratt, A.L., Lee, S.M. *et al.* (2008) *PLoS Med*, **5**, e75.

27 Sidders, B., Withers, M., Kendall, S.L., Bacon, J. *et al.* (2007) *Genome Biol*, **8**, R265.

28 McKinney, J.D., Honer zu Bentrup, K., Munoz-Elias, E.J., Miczak, A. *et al.* (2000) *Nature*, **406**, 735–738.

29 Tailleux, L., Waddell, S.J., Pelizzola, M., Mortellaro, A. *et al.* (2008) *PLoS ONE*, **3**, e1403.

30 Tuanyok, A., Tom, M., Dunbar, J. and Woods, D.E. (2006) *Infect Immun*, **74**, 5465–5476.

31 Joseph, B., Przybilla, K., Stuhler, C., Schauer, K. *et al.* (2006) *J Bacteriol*, **188**, 556–568.

32 Becker, A., Berges, H., Krol, E., Bruand, C. *et al.* (2004) *Mol Plant Microbe Interact*, **17**, 292–303.

33 Shi, L., Sohaskey, C.D., Kana, B.D., Dawes, S. *et al.* (2005) *Proc Natl Acad Sci USA*, **102**, 15629–15634.

34 Shaw, E.I., Dooley, C.A., Fischer, E.R., Scidmore, M.A. *et al.* (2000) *Mol Microbiol*, **37**, 913–925.

35 Bruggemann, H., Hagman, A., Jules, M., Sismeiro, O. *et al.* (2006) *Cell Microbiol*, **8**, 1228–1240.

36 Le Roch, K.G., Zhou, Y., Blair, P.L., Grainger, M. *et al.* (2003) *Science*, **301**, 1503–1508.

37 Tarun, A.S., Peng, X., Dumpit, R.F., Ogata, Y. *et al.* (2008) *Proc Natl Acad Sci USA*, **105**, 305–310.

38 Rachman, H., Strong, M., Ulrichs, T., Grode, L. *et al.* (2006) *Infect Immun*, **74**, 1233–1242.

39 Daily, J.P., Scanfeld, D., Pochet, N., Le, R.K. *et al.* (2007) *Nature*, **450**, 1091–1095.

40 Faucher, S.P., Porwollik, S., Dozois, C.M., McClelland, M. and Daigle, F. (2006) *Proc Natl Acad Sci USA*, **103**, 1906–1911.

41 Hautefort, I., Thompson, A., Eriksson-Ygberg, S., Parker, M.L. *et al.* (2008) *Cell Microbiol*, **10**, 958–984.

42 Lucchini, S., Liu, H., Jin, Q., Hinton, J.C. and Yu, J. (2005) *Infect Immun*, **73**, 88–102.

43 Bradley, P.J., Ward, C., Cheng, S.J., Alexander, D.L. *et al.* (2005) *J Biol Chem*, **280**, 34245–34258.

44 Boothroyd, J.C. and Dubremetz, J.F. (2008) *Nat Rev Microbiol*, **6**, 79–88.

45 Wernegreen, J.J. (2007) in *Bacterial Pathogenomics* (eds M.J. Pallen, K.E.

Nelson and G.M. Preston), ASM Press, Washington, D.C, pp. 196–212.

46 Chakraborty, T., Leimeister-Wachter, M., Domann, E., Hartl, M. *et al.* (1992) *J Bacteriol*, **174**, 568–574.

47 Thompson, A., Rowley, G., Alston, M., Danino, V. and Hinton, J.C. (2006) *Curr Opin Microbiol*, **9**, 109–116.

48 Rodrigue, S., Provvedi, R., Jacques, P.E., Gaudreau, L. and Manganelli, R. (2006) *FEMS Microbiol Rev*, **30**, 926–941.

49 Bozdech, Z., Llinas, M., Pulliam, B.L., Wong, E.D. *et al.* (2003) *PLoS Biol*, **1**, E5.

50 Hakimi, M.A. and Deitsch, K.W. (2007) *Curr Opin Microbiol*, **10**, 357–362.

51 Cohen-Freue, G., Holzer, T.R., Forney, J.D. and McMaster, W.R. (2007) *Int J Parasitol*, **37**, 1077–1086.

52 Jenner, R.G. and Young, R.A. (2005) *Nat Rev Microbiol*, **3**, 281–294.

53 Ehrt, S., Schnappinger, D., Bekiranov, S., Drenkow, J. *et al.* (2001) *J Exp Med*, **194**, 1123–1140.

54 Fontan, P., Aris, V., Ghanny, S., Soteropoulos, P. and Smith, I. (2008) *Infect Immun*, **76**, 717–725.

55 Zientz, E., Dandekar, T. and Gross, R. (2004) *Microbiol Mol Biol Rev*, **68**, 745–770.

56 Moran, N.A., Dunbar, H.E. and Wilcox, J.L. (2005) *J Bacteriol*, **187**, 4229–4237.

57 Voskuil, M.I., Schnappinger, D., Visconti, K.C., Harrell, M.I. *et al.* (2003) *J Exp Med*, **198**, 705–713.

58 Bacon, J. and Marsh, P.D. (2007) *Curr Mol Med*, **7**, 277–286.

59 Singh, U., Brewer, J.L. and Boothroyd, J.C. (2002) *Mol Microbiol*, **44**, 721–733.

60 Barnett, M.J., Toman, C.J., Fisher, R.F. and Long, S.R. (2004) *Proc Natl Acad Sci USA*, **101**, 16636–16641.

61 Hall, N., Karras, M., Raine, J.D., Carlton, J.M. *et al.* (2005) *Science*, **307**, 82–86.

62 Becker, D., Selbach, M., Rollenhagen, C., Ballmaier, M. *et al.* (2006) *Nature*, **440**, 303–307.

63 Gefen, O., Gabay, C., Mumcuoglu, M., Engel, G. and Balaban, N.Q. (2008) *Proc Natl Acad Sci USA*, **105**, 6145–6149.

10
Everybody Has A Home of Their Own – "The Phagosome Zoo"

Albert Haas

Most facultative and obligate intracellular microorganisms have specialized in multiplication within a membrane-bound compartment in their host cells. Their phagosomes either (i) do not mature into phagolysosomes, but share some characteristics with endocytic compartments, or (ii) are compartments that have no obvious connection with the endocytic continuum at all, or (iii) mature into phagolysosomes, but maturation is slowed by the pathogens to allow them to adapt to this environment. To add to this variety, some pathogens escape from the vacuole into the host cell cytoplasm. This stunning spectrum of possibilities will be introduced in this chapter and it will be argued that every intracellular pathogen has a compartment of its own. Furthermore, some emphasis will be placed on how professional vs. nonprofessional phagocytes are entered by microbes. Many more details on each of the discussed pathogens are found in the corresponding chapters of this book while this is only an introduction and overview. Phagosomes that are not transient compartments on the way to the phagolysosome will be termed "vacuoles" to express that they are noncanonical, stable "pathogenic" compartments.

10.1
Come On In, Get Killed! Death in the Shredder

Phagocytosis is defined as the ingestion of a particle by a cell (see Chapter 3). Phagocytosis in vertebrates is typically carried out by professional phagocytic cells, that is, polymorphonuclear cells (particularly neutrophils), dendritic cells, monocytes and macrophages, but is not unique to them. Most ingested particles are apoptotic cells, microorganisms and denatured protein complexes larger than some 0.4 μm in diameter. Their surface molecules are recognized by a variety of phagocyte surface receptors. Spatial accumulation of such receptors ("clustering") by the local crowding of their microbial partner molecules ("ligands") can then trigger signaling into the cell followed by membrane deformation, cytoskeleton rearrangement and a wrapping of the particle in host cell membrane. Eventually, the particle is ingested by

Intracellular Niches of Microbes. Edited by Ulrich E. Schaible and Albert Haas

ISBN: 978-3-527-32207-7

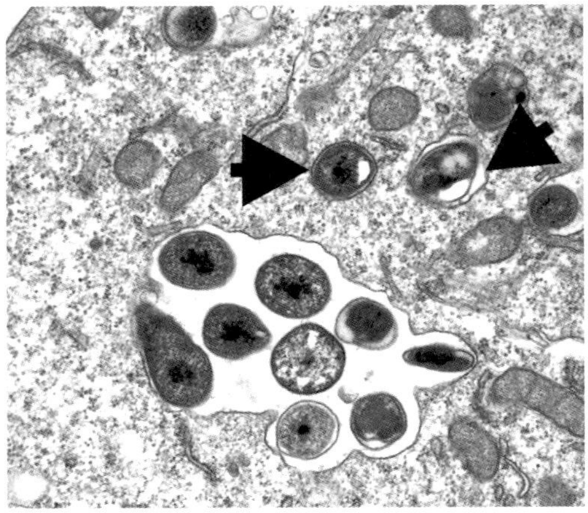

Figure 10.1 Phagosomes in macrophages. Murine J774E macrophage-like cells have been infected with *Afipia felis* for 24 h. A communal vacuole harboring several bacteria is found next to vacuoles containing single bacteria (arrowheads). Reprinted with permission from [2].

fusion at the tip of the nascent membrane bag and a phagosome is formed (Figure 10.1; see also Chapter 3).

In terms of cell biology, the newly created phagosome is an endosome specialized in microbial killing and, particularly in dendritic cells, in supporting antigen presentation. Ligation of only certain phagocyte receptors triggers the flushing of the phagosome lumen with antimicrobial compounds. As a central event in phagosome biogenesis, the phagosome lumen is strongly acidified (to pH ~4.5) by a membrane-embedded proton-pumping ATPase (Figure 10.2). An NADPH oxidase complex produces large quantities of superoxide radicals from molecular oxygen, and, in activated macrophages of some species, a nitric oxide synthase produces NO radicals from arginine. Through the cooperation of reactive oxygen and nitric oxide species, highly reactive peroxynitrite can be formed within the phagosome. Neutrophils are especially aggressive. They possess myeloperoxidase which, together with the above enzyme activities, leads to the formation of hyperchlorous acid and chloramines, hydroxyl radicals, and ozone, all of which are biocidal. As non-oxidative tools, some 50 different lysosomal hydrolases (proteases, DNases, lipases, etc.) compromise microbial viability and integrity, and defensin peptides and cationic proteins add to the killing potential [1, 2].

The development of the phagosome into a fully degradative phagolysosome is a temporally and spatially ordered process and parallels endosome maturation [3] (see Chapter 3). The newly formed "early" phagosome develops vectorially into a "late" phagosome after fusion with late endosomes and will finally become a phagolysosome by fusion with lysosomes (the normally terminal organelles of the endocytic pathway). Recent data suggests that more complex fusion patterns may exist [4, 5], but most features seen can be well explained by operational definition of

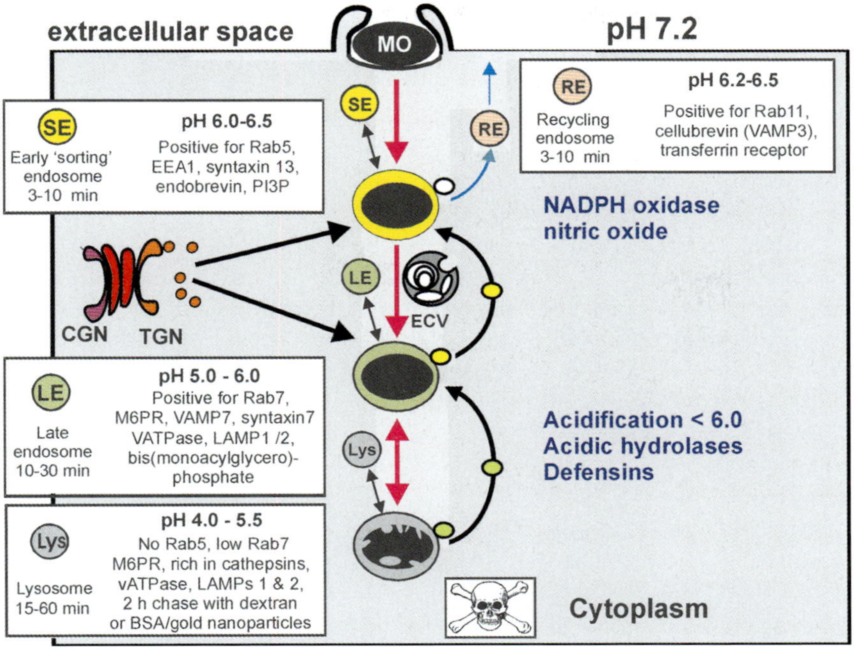

Figure 10.2 Outline of the phagocytic pathway in mammalian cells. When a harmless (non-interfering) microorganism (MO) is ingested by a phagocytic cell, it is wrapped in part of the host cell plasma membrane and a phagosome is formed. This will fuse successively with early (sorting) endosomes (SE), late endosomes (LE), and lysosomes (Lys). Some recycling of host cell surface components may occur via the recycling endosome (RE). The phagosome maturation steps are operationally defined as early and late phagosome and phagolysosome stages. Each of these phagocytic compartments is characterized by a set of marker proteins and lipids (as indicated) with a typical drop in pH along the continuum. NADPH oxidase, acidification, and acidic hydrolases plus defensins act at various steps of phagosome development. Killing of ingested microorganisms is likely most pronounced once the phagolysosome stage has been reached. Comprehensive recycling from the phagosomes to the previous stage of maturation is indicated by backwards arrows; this kind of recycling has not been experimentally addressed but likely happens as it is required to ensure compartment identity. ECV, endosomal carrier vesicle, a vesicle population which has been described to transport membrane proteins to be degraded between sorting and late endosomes. The roles of ECVs in phagosome maturation require further analysis. Blue lettering: killing mechanisms. Marker molecule abbreviations: Early endosome antigen-1 (EEA1), lysosome-associated membrane proteins (LAMP) 1 and 2, phosphatidylinositol 3-phosphate (PI3P), mannose 6-phosphate receptor (M6PR), Ras-like protein from rat brain (Rab) 5, 7, and 11, vesicle-associated membrane protein (VAMP) 3 and 7, proton-pumping vacuolar adenosine triphosphatase (vATPase), Scheme modified from [2].

these three phagocytic maturation stages. These stages can be experimentally discriminated by the presence or absence of stage-specific lipids and proteins or by chasing endocytic tracers into the endocytic continuum for different periods of time (Figure 10.2). Phagosomes containing the same cargo in the same cell type acquire and loose their maturation markers with almost identical kinetics, though exceptions to this rule may apply [5–7]. While a limited fusion competence between endocytic and phagocytic organelles of different maturation stages may actually be a prerequisite for phagosome maturation to occur, early phagosomes fuse mostly "homotypically" with early endosomes and, with much lower frequency, "heterotypically" with late endosomes [8, 9]. Late phagosomes are likely to fuse relatively avidly with late endosomes and less frequently with lysosomes, which again fuse predominately homotypically. Overall, fusion competence decreases as a phagocytic (or endocytic) compartment matures [9–11].

Rab proteins (mainly Rab 5, 7, and 11) are key regulators of phagosome fusion as they are in other fusion events of the endocytic and exocytic pathways (Figure 10.2). Rab proteins identify and, together with tethering factors, specifically bridge membranes to be fused [12]. SNARE proteins (e.g., syntaxins 7, 8 and 13, VAMPs 4, 7 and 8) on both partner membranes, as Rab downstream effectors, form extremely stable four-helix bundle complexes between the membranes and pull them together to a distance of a few nanometers [13]. This event may, in itself, be sufficient to catalyze membrane fusion and is likely supported by additional factors such as calcium-dependent proteins (e.g., calmodulin), ion channels (e.g., proton-pumping ATPase), and phospholipases. In addition to these key players, phagosome acidification can also contribute to maturation to some degree [14], as has been observed with early endosomes [15, 16]. While the contributions of cytoskeletal elements to phagocytosis, and phagosome maturation in particular, are still poorly defined, recent studies shed some light on the crucial role in phagosome maturation of the polymerization of tubulin [17] and actin [18, 19] (see also Chapter 8). Finally, protein phosphorylation [10] and lipid composition also change with and probably regulate phagosome maturation [10]. Some 1000 different lipid species have been identified on latex bead-containing phagosomes (LBP) from macrophages, adding to the amazing complexity of this organelle (cited after [3]). While the above describes the normal progression of phagosome maturation, some pathogenic, "intracellular" pathogens have specialized in diverting phagosome development. These will be presented below and in many other chapters in this book.

As models for host factors required for phagosome biogenesis both in physiological and pathological states, non-mammalian organisms have started to be used (for details, see Chapter 4). *Dictyostelium* (a slime-mold), in particular, has developed into an important model, as it is strongly phagocytic and a readily manipulable unicellular amoeba. Trafficking is a bit different from that in macrophages: Microorganisms, in particular bacteria, are taken up as food and this process seems to be more a particle-indiscriminating macropinocytosis than a (receptor-mediated) phagocytosis, although phagocytosis may be important as well [20]. Soon after ingestion, the macropinosome or phagosome acidifies and phagolysosome formation is initiated. After digestion of food or if the ingested materials have been unsuccessfully dealt with for some time, lysosomal hydrolases are recaptured and recycled by binding to specific receptors followed by alkalinization of the phagosome lumen. Nondigestible remainders are expelled. This is not normally so for macrophages, yet it should be mentioned that macrophages and likely neutrophils can exocytose a major proportion of their lysosomes when cytosolic calcium levels increase. Cells use this capability to release the contents of lysosome-like melanosomes or lytic granules from toxic T cells [21]. While release of indigestible material to "get rid of it" would work for *Dictyostelium*, as debris is expelled into the bulk extracellular world, it does not work well for macrophages, where another phagocyte would likely ingest the particle and then have the same problems.

10.2 Professionals and Laypersons

Professional phagocytes are not the only cells that intracellular pathogens have evolved to colonize. In fact, some pathogens, such as *Salmonella*, are better known for their abuse of epithelial cells than that of macrophages, although both cell types are targets and the pathogen may thrive in either environment. Even more impressive, some pathogens have developed mechanisms to invade strictly nonphagocytic cells, the paradigm of which is red blood cell (RBC) colonization by *Plasmodium*, the malaria parasite. The question is: How can eukaryotic cells be either professionally phagocytic, occasionally phagocytic or not phagocytic at all?

10.2.1 Professional Phagocytes

The professional phagocytes of vertebrates have been designed by nature to sniff out, bind to and ingest microbial foreign or nonphysiological particles, to decompose them and, where appropriate, present their antigenic structures to cells of the acquired immune system. Therefore, an arsenal of phagocytic receptors has evolved that takes care of the ingestion of a wide variety of particulate matter (see Chapter 3). These receptors either recognize particle structures directly or they identify "opsonizing" host proteins that have attached to the particles and branded them as material destined to be destroyed. While some receptors are more specific for certain ligands (e.g., complement receptor, immunoglobulin Fc receptor), others are not as discriminating (e.g., scavenger receptors). The various receptors are parts of different signaling pathways which regulate the reactions to the ingested particles. For example, a particle that has been covered with immunoglobulin G (IgG) can be interpreted as an enemy whose identity is known to the infected organism and that has previously activated an immune response and, therefore, is likely dangerous. Not surprisingly therefore, the macrophage's response to an IgG-covered particle is quite strong and mounts a strong production of oxygen radicals. On the other hand, an apoptotic cell which does not pose an infection threat to the organism is removed in an "immunologically silent" way after recognition by apoptosis-dedicated receptors such as phosphatidylserine receptors [22].

10.2.2 Nonprofessional Phagocytes

Nonprofessional phagocytes were discovered when formaldehyde-fixed RBCs were fed to macrophages and fibroblasts [23]. These particles were surprisingly taken up by either cell type. Coating RBCs with specific antibodies made them a particularly good meal for macrophages, while they were less avidly ingested by fibroblasts which do not possess immunoglobulin receptors. A recent paper by Gratton *et al.* [24] demonstrated that synthetic particles of up to 5 μm diameter can be ingested by epithelial (HeLa) cells, albeit with varying efficiency. Of 1 μm particles, a surprising 70% were taken up by these "nonprofessionally" phagocytic cells. However, the above studies used cell *lines* for analysis and it is possible that immortalization affects phagocytosis, although some cases of phagocytosis with primary "nonphagocytic" cells are clearly documented [25]. In general, it is probably fair to say that professional phagocytes have a much wider array of surface receptors at higher density connected to internalization signaling pathways and therefore ingest more types of particles more avidly than nonprofessionals do.

Nonprofessionally phagocytic cells, on the other hand, can be turned into professional phagocytes, at least with respect to some particles, when they are transfected with suitable receptor genes such as the immunoglobulin receptor [26, 27] or complement receptor 3 [27]. Such experimental model cells are termed "engineered phagocytes."

As particulate matter can, in principle, be ingested by nonprofessional phagocytes, experimental uptake of microorganisms by nonprofessional phagocytes, in particular low-level uptake, does not necessarily constitute evidence for a microbial strategy to force its uptake into these cells. For example, fibronectin is a very "sticky" protein of serum and can opsonize particles such as bacteria [28] and mediate phagocytosis [29], likely via host cell integrins. Most of the fibronectin is a constituent of the extracellular matrix, where it promotes the adhesion of cells such as fibroblasts. Therefore, phagocytosis of fibronectin-coated latex beads [30] by nonprofessional phagocytes maybe a mimic of colonization of substratum which, more by chance, results in phagocytosis when two different portions of the plasma membrane that have surround a fibronectin substratum fuse with each other. As with other virulence pathways, some pathogens, such as *Staphylococcus aureus* (a more extracellular pathogen), have "learned" to abuse this normality by expressing on their surfaces fibronectin-binding proteins [31].

Another important pathway of forced entry into nonprofessionals is macropinocytosis which is triggered by *Salmonella*. *Salmonella* uses a "type III protein secretion system" (T3SS) to inject effector proteins into the host cell that induce the formation of large lamellipodial structures that will eventually fuse with each other and form a large vesicle, the macropinosome [32]. Macropinocytosis can be normally induced in many cells [33], usually by exposure to growth hormones, such as epithelial growth factor. Therefore, *Salmonella* opens a host front door by using a fake key. A microbe's fate can be vastly different depending on which port of entry has been used into which type of host cell.

What happens when a microbe is finally inside, say, an epithelial cell? Some microorganisms will be killed [34] and others might thrive [35]. This means that the capability to enter a nonprofessional phagocyte does not necessarily ensure pathogen multiplication. In fact, the resulting host cell vacuoles are not automatically hospitable places, just because the cell around them is not specialized in killing microbes [36, 37].

10.2.3
"Not Phagocytic Cells at All"

Colonization of RBCs is the supreme discipline when it comes to colonizing vertebrate cells and only few pathogens, such as the malaria parasite *Plasmodium falciparum* (see Chapter 33), the apicomplexan *Babesia* spp. (see Chapter 33), or the pathogenic bacterium *Bartonella* spp. [38], have mastered it. The reason for this is that RBCs do not endocytose and, hence, this pathway cannot be hijacked by pathogens. They have to enter via mechanisms that involve major contributions from themselves (like the actin-propelled and gliding motilities of apicomplexans). The entry process of *Plasmodium* merozoites into RBCs has been studied in detail and involves secretion of parasite granules and shedding of a protein surface coat [39] (see Chapter 33). Important features of the RBC as a home are the lack of lysosomes and antimicrobicidal processes such as production of reactive oxygen metabolites. RBCs further provide a relatively long-lived (60–120 days life span) membrane-surrounded and protected place to live with no antigen-presentation machinery in the immediate neighborhood. RBCs can also be taken up in great numbers by blood-sucking insects and, hence, provide an appealing mode of distribution.

Although the above shows that, in principle, pathogens can thrive in any of these host cell types, macrophages are still the preferred host cells for many intracellular pathogens because they

- are professionally phagocytic and will, in contrast to nonprofessional phagocytes or to "phagocytosis-inert" cells such as RBC, actively ingest most particles so that entry of the phagocyte does not necessarily require evolution of invasion or "forced entry" mechanisms;
- can have a considerably longer lifetime (months to years) than neutrophils (12–72 h), a cell type which would be another obvious choice. This feature is particularly important for slow-growing pathogens such as *Mycobacterium tuberculosis* which would possibly not even start multiplying when the host cell was already dying by causes unrelated to the infection;
- are less microbicidal than neutrophils;
- are likely better food suppliers than RBCs, which contain plenty of hemoglobin but little else and which cannot supply a microbe-containing phagosome with nutrients via endocytosis.

10.3
Come On In, Have Fun! Life in a Golden Cage

In the following, intracellular pathogenic microbes will be categorized according to the prevailing characteristics of their vacuoles (Figure 10.3; Table 10.1). Some vacuoles share features of different compartments (Table 10.2), but for the sake of simplicity, this review

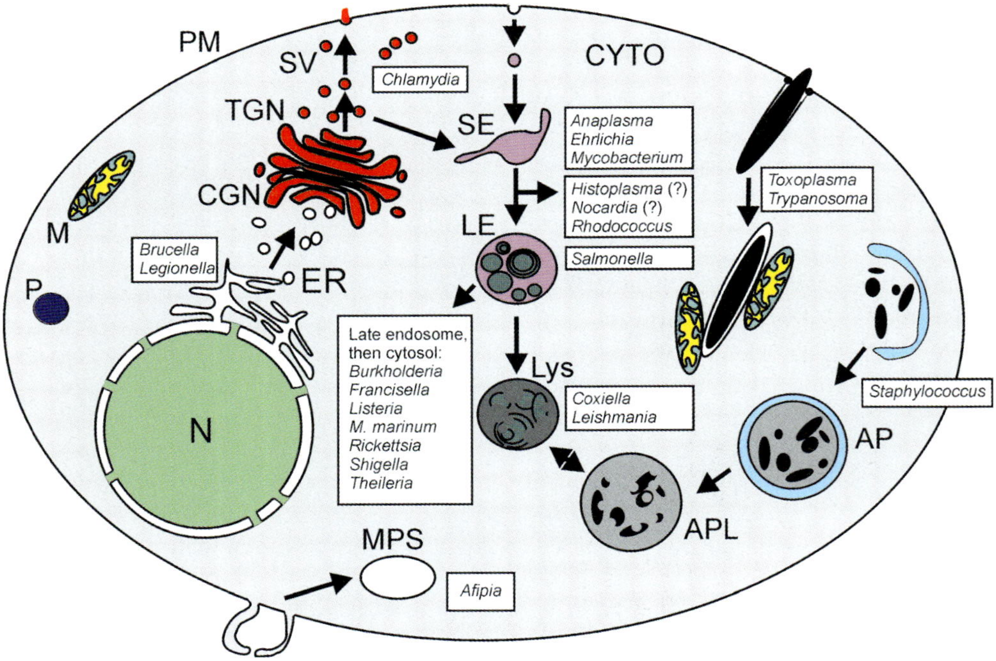

Figure 10.3 The subcellular compartments of a mammalian cell and the "phagosome zoo" within it. Shown is a scheme of a mammalian cell. Note that red blood cells do not have elaborate subcellular compartments but that, nevertheless, RBC-infecting eukaryotes have been included. The secretory pathway for the production and location of proteins to the endoplasmic reticulum (ER), the Golgi apparatus (*cis*-Golgi, CGN and *trans*-Golgi network, TGN), secretory vesicles (SV), and to the plasma membrane (PM) are shown on the left. Secretory proteins are synthesized on the (rough) ER, encased in transport vesicles that are shuttled to the *cis*-Golgi apparatus, further modified in the Golgi apparatus and finally secretory vesicles are released from the *trans*-Golgi network for transport to the plasma membrane. Some specific vesicles from the TGN fuse with sorting or late endosomes (as defined in Figure 10.2) and bring new proteins for endosomes and lysosomes. The second major pathway, the continuum of endocytic vesicles which includes the formation of small vesicles on the plasma membrane followed by fusion with early sorting lysosomes (SE) is shown on the right. These early endosomes mature into late endosomes (LE) and finally into lysosomes (Lys). Phagocytic compartments of different maturation stages possess a similar protein and lipid composition to their corresponding endosomal maturation stages. Macropinosomes (MPs) bring large quantities of liquid into the cell and can be hijacked as well. Some pathogens (*Toxoplasma*) enter the cell by true invasion and create a special parasitophorous vacuole which, in some cases, can associate with mitochondria (M). Autophagosomes (AP) are formed by mammalian cells when nutrients are scarce or the phagocyte is suffering from other forms of distress. These autophagosomes enclose cytoplasm (CYTO), possibly containing organelles, in a process which is only little understood. These autophagosomes then fuse with lysosomes to form autophagolysosomes (APL) in which the contents are degraded and from which the degradation products are made available to the cell for biosynthesis. *Staphylococcus aureus* is normally an extracellular pathogen but, in small quantities, can also be incorporated into autophagosomes, concomitant with inhibition of autophagolysosome formation. Subsequent escape into the cytoplasm allows for limited multiplication [125]. Pathogens are placed in this scheme according to which compartment resembles most their protein composition once they have established a stable compartment. Further compartments: nucleus (N), peroxisomes (P). For details, see text and corresponding chapters in this book.

Table 10.1 Intracellular vacuolar pathogens.

Pathogen (only genus) and illness caused by it	Host cells	Vacuole identity	Marker molecules of mature vacuoles	Vacuole pH	Average vacuole size	Average number of microorganisms/vacuole	Vacuole localization in cell	Time spent in host cell	How does the microbe exit?
Afipia (cat scratch disease)	Macrophages, likely freshwater amoebae	Not canonically endocytic compartment	No marker enriched, compartment negative for, e.g., LAMP1/2, proton-pumping ATPase, TfR, Rab5, Rab7, cathepsin D, EEA1. Ganglioside GM1 is present at least early in infection.	Not strongly acidic (4)	10 min after uptake: 1.7 × 2 m, 30 min: 0.3 × 1 μm	1–5	Perinuclear	ND	ND
Anaplasma (3) (anaplasmosis)	Granulocytes, endothelial cells	Autophagosome	LC3, beclin-1	Not acidic	0.3–8 μm	1–50	ND	4–7 ds	Host cell rupture, exocytosis
Brucella (brucellosis)	Macrophages, dendritic cells, epithelial cells	Endoplasmic reticulum	Calnexin, calreticulin, Rab1, Rab2, BIP, Sec61β	7.4	2 μm	1	Perinuclear	3 ds *in vitro*, several dd *in vivo*	Unknown
Chlamydia (3) (urogenital infections, conjunctivitis)	Epithelial cells	Post-Golgi compartment	Multiple chlamydial proteins (Incs), host lipids sphingomyelin and cholesterol	7.4	1–40 μm	1–1000	Perinuclear	2–3 ds	Lysis or exocytosis

(Continued)

Table 10.1 (*Continued*)

Pathogen (only genus) and illness caused by it	Host cells	Vacuole identity	Marker molecules of mature vacuoles	Vacuole pH	Average vacuole size	Average number of microorganisms/vacuole	Vacuole localization in cell	Time spent in host cell	How does the microbe exit?
Coxiella (3) (Q fever)	Macrophages	Lysosomal, autophago-lysosomal	LAMP1–3, cathepsin D, LC3, Rab7, Rab24, proton-pumping ATPase, cholesterol	4.5–5.3	12.5–25 μm (2)	200–1000 (2)	Perinuclear	>1 wk	Necrosis
Ehrlichia (3) (ehrlichiosis)	Macrophages monocytes	Early endosome	Rab5, EEA1, TfR	Slightly acidic	0.3–8 μm	1–50	ND	4–7 ds	Host cell rupture, exocytosis
Francisella (tularemia)	Macrophages	Unusual late endosome, then cytosolic	LAMP1, LAMP2, CD63	6.7 (at 4 h)	0.5 μm	1	Throughout cytoplasm	2–3 ds	Apoptosis: pyroptosis
Histoplasma (histoplasmosis)	Human macrophage	ND	Proton-pumping ATPase is excluded	6.5	2–5 μm	1	Throughout cytoplasm	2–3 d	Cell rupture by dividing yeasts
Legionella (legionnaire's disease)	Freshwater protozoa and macrophages	Endoplasmic reticulum	ARF1, Rab1, Calnexin, KDEL motif proteins, BIP, PDI, PI(4)P	7.0	1–30 μm	100	Peripheral–perinuclear	24 h	Respirable vesicles and cell necrosis
Leishmania (3) (leishmaniosis)	Macrophages	Lysosomal	Promastigote stage: poorly fusogenic with endosomes and lysosomes; rare	Promastigote stage: ?	5–10 μm	1 (8)	Throughout cytoplasm	weeks	ND

			acquisition of TfR, EEA1 or Rab7, slow acquisition of LAMP1						
			Amastigote stage: Rab7, LAMP1/2, proton-pumping ATPase, cathepsins B, D, H, L, annexins I and VI, lysosomal pre-loaded tracer; membrane made permeable for small substances by parasite pore proteins	Amastigote stage: 4.7–5.2		many (9)			
M. marinum (swimming pool granuloma)	*Dictyostelium*, macrophage	Amoebal post-lysosome, later cytoplasmic	Predicted copper transporter p80, flotillin-like vacuolin, NRAMP1	6.0–6.5 (estim.)	1–20 μm	1–30	Throughout cytoplasm	∼30 h	Nonlytic release
M. tuberculosis (tuberculosis)	Macrophages	Recycling endosome, later possibly cytoplasmic	Rab5, TfR, procathepsin D, exclusion of proton-pumping ATPase	6.3	1 μm	1	Perinuclear	2–6 d	Apoptosis, host cell lysis
Nocardia (nocardiosis)	Macrophages, endothelial cells	ND	ND	∼7.2	ND	Mostly singular	Throughout cytoplasm	ND	Possibly induction of host cell apoptosis

(Continued)

Table 10.1 *(Continued)*

Pathogen (only genus) and illness caused by it	Host cells	Vacuole identity	Marker molecules of mature vacuoles	Vacuole pH	Average vacuole size	Average number of microorganisms/vacuole	Vacuole localization in cell	Time spent in host cell	How does the microbe exit?
Plasmodium (3) (malaria)	Erythrocyte, hepatocyte, reticulocyte	Parasite-induced parasitophorous vacuole	Parasite proteins (5) PfEXP1, PfPV1, PfCG2			1	Throughout cytoplasm	48–72 h	Lysis (mechanism unknown
Rhodococcus (foal bronchio-pneumonia)	Macrophages	Stable compartment between early and late endosome	Rab7, LAMP1, LBPA. No Rab5, TfR, cathepsin D, β-galactosidase or proton-pumping ATPase	7.2	2–20 μm	1–40	Throughout cytoplasm	few days	Necrosis
Salmonella (typhoid fever)	Macrophages, epithelial cells, M cells, dendritic cells, neutrophils	Late endosome or lysosome	Proton-pumping ATPase, Rab7, Rab11, LAMPs, BMP bis(monoacylglycero)phosphate (LBPA), MHC class II, no CI-M6PR, syntaxin 13, PI(3)P, cholesterol	<5.0	Highly variable	up to 1000 (6)	Juxtanuclear	ND	ND
Toxoplasma (3) (toxoplasmosis)	All nucleated cells, virtually all warm-blooded animals	Parasite-induced parasitophorous vacuole	Host lipid raft proteins from the plasma membrane; absence of	6.8	5 to more than 30 μm	1–64	Perinuclear	12–72 h	Active exit; requires calcium signaling

			endocytic marker proteins and proton-pumping APTase; numerous secreted parasite proteins						
Trichinella (3) (trichinosis)	Skeletal muscle fiber	Early cytoplasmic, later membrane around cuticle	ND	ND		1–6 (per cell)	Central	20 d (7)	
Trypanosoma (3) (trypanosomiasis)	Cardiomyocytes, macrophages, neurons, dendritic cells, fibroblasts, others	Lysosome	LAMP1	5.5	variable	1 (vacuole), 400 (cytoplasmic)	Periphery	7–10 d	Lysis (mechanism unknown)

ND, not determined.
(1) *In vitro* also epithelial cells and fibroblasts.
(2) Stationary phase.
(3) Obligately intracellular.
(4) Negative for LysoTracker.
(5) Proteins produced by parasite and secreted into the parasitophorous vacuole.
(6) In human cervix epithelial HeLa cell line. Relevance for *in vivo* infections unclear.
(7) For development into muscle fiber, then persisting for years.
(8) For *L. donovani* and *L. major.*
(9) For *L. mexicana.*
ARF-1, ADP ribosylation factor-1; BIP, binding protein; (major chaperone of ER); CD63, cluster of differentiation protein 63; CI-M6PR, cation-independent mannose 6-phosphate receptor; EEA1, early endosome antigen-1; LAMP (1/2), lysosome-associated membrane protein (1 or 2); LC3, microtubule-associated protein 1 light chain 3; MHC class II, major histocompatibility complex class II; NRAMP-1, natural resistance-associated macrophage protein-1; PDI, protein disulfide isomerase; PI(4)P, phosphatidylinositol 4-phosphate; Rab (1/2/5/7/24)Ras-like protein from rat brain (1 or 2 or 5 or 7 or 24); Sec61β, secretion mutant 61 protein beta (part of ER protein translocase); TfR, transferrin receptor.

Table 10.2 Membrane-bound compartments of eukaryotic cells.

Compartment	Function	% of cell volume[a]	% of total cell membrane[b]	Size	Number/cell	pH	Protein content, mg ml^{-1}	Marker molecules
Cytosol	Gelatinous matrix for all organelles and cytoskeletal elements	54	0	N.A.	N.A.	7.0–7.5	200–300	Enzymes of glycolysis (e.g., glucose 6-phosphate isomerase)
Mitochondria	Round or oval double membrane organelles, often as long tubules, surrounded by two membranes; make ATP from food available as the universal energy ressource for the cell; participates in lipid synthesis	22	39	0.5–1.0 μm diameter 2–8 μm length	100–2000	Matrix 7.5–8.0 Inter-membrane space 8.0	600–700	Mitochondrial protein import proteins (TIM, TOM), porins, cardiolipin, ATP synthase (in plant cells also in chloroplasts). proteins of citric acid cycle
Rough ER	An extensive tubular network; originates from the nuclear envelope; rough ER is covered with ribosomes and spezializes in the production and initial modification of secretory proteins	9	35	~100 nm depth	N.A.	7.2	100[c]	Protein translocation channel proteins (TRAM, e.g., Sec61β) = typical rough ER, calnexin, calreticulin, protein disulfide isomerase (PDI), *N*-glycosyltransferase
Smooth ER	As rough ER, yet specializes in lipid production and detoxifica-	6[d]	16	Tubes of ~150 nm diameter	N.A.	7.2	100[c]	Glucose 6-phosphatase = typical smooth ER,

	tion; rough and smooth ER are interwoven; calcium store							PDI, calnexin, calreticulin
Nucleus	Large, round–oval organelle; stores, multiplies and makes available genetic information; surrounded by a double-membrane envelope with large transport pores (nucleopores)	6	0.2[e]	5–10 µm	1	7.7		Transcription factors, histones
Lysosome	Degradative compartment with very low pH (4.0–5.0), used to digest degradable substances from outside the cell, inactivated proteins from the resident membranes of the secretory and endocytic pathways; digests any material collected in the cytoplasma by autophagosomes	1	0.4	0.1–0.8 µm	100–1000	4.0–5.5	300	Acidic hydrolases (e.g., cathepsins, acid β-glucuronidase, acid sphingomyelinase), LAMPs)
Endosome	Intermediate compartments between plasma membrane and lyso-	1	0.4	0.05–0.4 µm	50–200	5.3–6.2		Rab5, transferrin receptor, EEA1

(Continued)

Table 10.2 (*Continued*)

Compartment	Function	% of cell volume[a]	% of total cell membrane[b]	Size	Number/cell	pH	Protein content, $mg\,ml^{-1}$	Marker molecules
	some; most plasma membrane receptor–ligand interactions are broken up here and the receptors are recycled to the plasma membrane							
Peroxisomes	Metabolize organic substances, in particular long-chain lipids; produce and inactivate reactive oxygen compounds; single membrane; sometimes crystalline-like interior; biosynthesis of nerve lipids	1	0.4	0.5–0.8 μm	5–10[f] 50–100[g]	6.9		Catalase, uric acid oxidase, D-amino acid oxidase
Golgi apparatus	Stack of membraneous sacs which are connected by transport vesicles; modification of secretory proteins, in particular by glycosylation	3	7	2.5 μm	1 (4–5 cisternae) in mammalian cell, up to several 100 in secretory plant cells	6.4		Mammals: Golgin-84, GM-130, α-2,6-sialyltransferase, α-1,2-mannosidase. Plants: RGP (reversibly glycosylated protein), α-1,2-mannosidase I (*cis*-Golgi), xyloglucan fucosyltransferase (*trans*-Golgi)
Plasma membrane	Separates the cell interior from the environment; signaling and communication platform of the cell	N.A.	4	6–7 nm in depth	1	N.A.	N.K.	Integrins, ion channels, G proteins, receptors (e.g. epidermal growth factor receptor)

Autophagosome	Sequestration of cytosolic components, including organelles, into a membrane-surrounded compartment leads to (macro-) autophagosome formation; constitutes an often stress-induced "self-digesting" activity of the cell; eventually fusion with lysosomes to form an autophagolysosome	variable[h]		0.3–0.9 μm (yeast) 0.5–1.5 μm (mammalian cells)	0–100	7.3–4.5 (drops during maturation)	approx. like cytosol	LC-3 (= Atg8), Atg5, Atg12, Atg16, at phagolysosome stage: see also lysosome marker molecules
Chloroplast (*)	Double membrane organelle; absorption of light energy and conversion into chemical energy such as ATP and production of carbohydrates from carbon dioxide and water	~10%, can be more than 50% (lower eukaryotes)		5–10 μm, up 1–2 μm diameter	1–10	Stroma 7.5–7.9		Proteins of the outer membrane (TOC) and inner membrane (TIC) protein import complexes; ribulose-1,5-bisphosphate carboxylase/oxygenase, RB60
Vacuole (*)	Storage organelle (e.g., pigments, sugars, organic acids, toxic metals), pH and ion homeostasis,	50 to >90		up to 10 μm	1–many	5.5–6.2		Tonoplast intrinsic proteins (TIPs), proton-pumping ATPase

(Continued)

Table 10.2 (*Continued*)

Compartment	Function	% of cell volume[a]	% of total cell membrane[b]	Size	Number/cell	pH	Protein content, mg ml^{-1}	Marker molecules
	maintenance of turgor pressure							
Cell wall (*)	Surrounds and mechanically strengthens plant cells	N.A.	N.A.	1 µm diameter		~5.5	N.A.	Lignin, cellulose

Animal cell compartments were ordered according to their relative contribution to the cell volume. Plant cell compartments were added.
N.A., not applicable.
*Is found exclusively in plant cells and eukaryotic photosynthesis-conducting algae.
[a]Some of this information was taken from [118].
[b]Percentages may vary considerably between cell types.
[c]Not discriminated between rough and smooth ER.
[d]Includes Golgi apparatus.
[e]Inner nuclear membrane only.
[f]Uninduced (sugar media) yeast cell.
[g]Mammalian liver cell.
[h]Autophagosome creation is an induced event; depending on environment from not detectable to comprehensive.

will list each of them in only one category. It is important to note that compartmentation is not necessarily the same for all isolates of a pathogen, nor is it for all hosts or even all cells of one host with the same pathogen. There can be considerable variability and the statements below refer to the best-investigated cases for each system. Additionally, mostly the genus names of the pathogens are indicated, although different species of a genus may behave differently.

10.3.1
Pathogen Vacuoles that Get Stuck in the Endocytic Continuum (*Histoplasma*, *Mycobacterium*, *Nocardia*, *Rhodococcus*, *Salmonella*)

One straightforward strategy that evolved in the interplay between pathogens and macrophages was to arrest maturation of the phagosome on its way to becoming a phagolysosome. This interference could, in principle, be caused by several mechanisms, for example, by injection of microbial effector proteins that manipulate the host cell's maturation factors or by altering phagosome membrane biophysical properties so that membranes lose their fusion competence. Both of these mechanisms have been realized.

The known vacuoles in this category do not have a general fusion block but rather selectively do not fuse with lysosomes and possibly some other defined compartments. *Mycobacterium* vacuoles still fuse with upstream transferrin receptor-containing vesicles [40] but they do not fuse with late endocytic organelles (see Chapters 19 and 25). *Salmonella*-containing vacuoles (see Chapter 21) in epithelial cells likely fuse with vesicles that are positive for Rab7 and LAMP1 but which are neither late endosomes nor lysosomes because they lack hydrolases such as cathepsin D [41]. Although there seemed to be an agreement that fusion of *Salmonella*-containing vacuoles with lysosomes would not occur, recent evidence suggests that it can [42]. This still leaves the possibility that this interaction is more dynamic than with control phagosomes but that a generally lower level of phagolysosome formation exists and all evidence points in this direction. *Rhodococcus equi* arrests phagosome maturation at a step intermediate between early and late endosomes – the vacuoles do not acidify, they are still accessible to externally added tracer molecules, but they do not fuse with lysosomes [43] (von Bargen and Haas, unpublished data) (see Chapter 20).

How microorganisms in this group arrest phagosome development at different maturation levels is not known in detail. *Mycobacterium* is arguably the best investigated case and lipids and proteins have been implicated as microbial effector molecules: While defined lipids can change the maturation of bead-containing phagosomes early in infection and therefore can be considered sufficient for early diversion of trafficking [44–46], there is little doubt that some proteins also specifically contribute to altered phagosome behavior [47]. *Salmonella*, on the other hand, may influence trafficking through the injection of effector proteins, such as SopE [48] into the host cell's cytoplasm which, for example, mimic eukaryotic GTPase exchange factors, and also by bacterial surface factors. While it had been assumed for some years that type III secretein system(s) would contribute to changes in trafficking, recent research suggests that they do not [42]. *Histoplasma capsulatum* (see Chapter 31) likely also belongs in this group of pathogens with arrested phagosome development. It resides in a membrane-surrounded pH 6.5 vacuole which excludes the proton-pumping ATPase [49], which does not fuse with lysosomes [49] and which is negative for late endocytic LAMP1 and LAMP2 [49, 50].

Phagosomes containing the symbiotic alga *Chlorella* in the freshwater cnidarian *Hydra viridis* also do not fuse with lysosomes [51]. Whether these vacuoles belong to the category of "stuck-in-the-endocytic continuum" microbes, is not clear.

10.3.2 The Tough Guys: Life in Hell (*Coxiella*, *Leishmania*)

There are at least two notable exceptions from the above rule that "intracellular pathogens evolved avoidance of delivery to phagolysosomes": The protozoan *Leishmania* and the intracellular bacterium *Coxiella* delay, but do not inhibit phagosome maturation into late phagosomes or phagolysosomes.

10.3.2.1 *Coxiella*

The pH in a mature *Coxiella* vacuole is around 4.7, matching lysosomal pH [52]. Adding a potent inhibitor of proton-pumping ATPase and, hence, of phagosome acidification, inhibits *Coxiella* multiplication while not that of *Chlamydia* (an inhabitant of a neutral pH vacuole) [52], showing that the normally hostile environment of the phagolysosome has become crucial for multiplication of this pathogen. Interestingly, phagolysosome formation is still delayed, giving the pathogen time to adapt, and autophagosome characteristics are important for maturation diversion [53]. As with other pathogens, a (type IV) protein secretion system plays a crucial role in altered phagosome biogenesis [54] (see Chapter 16).

10.3.2.2 *Leishmania*

Leishmania is a protozoan that exists in two developmental stages: the promastigote (with flagellum) and the amastigote (reduced flagellum) stages. The promastigote form lives first in the sandfly alimentary tract, then in its esophagus. This latter form is transmitted to warm-blooded animals, for example, humans, by an insect bite. The promastigote form has an amphipathic lipophosphoglycan on its surface that can stall phagosome maturation at an early endocytic stage [55]. It is assumed that, as the parasite matures and as it develops into an amastigote form, the lipophosphoglycan ceases to be produced and is successively lost from the parasite surface; the *Leishmania* vacuole then progresses towards a late phagosome or phagolysosome stage (pH 5.0) [56]. Eventually, infected macrophages are lysed and release the parasites for uptake by other phagocytes (see Chapter 32).

10.3.3 Life in the Secretory System (*Brucella*, *Chlamydia*, *Legionella*)

The secretory pathway of eukaryotes consists of the endoplasmic reticulum (ER) and the Golgi apparatus, including all their subcompartments and the vesicles that shuttle cargo between them and between Golgi and the plasma membrane (for an overview of organelle chacteristics, see Table 10.2). The secretory pathway serves the production and processing of proteins that are either (i) part of the secretory compartments themselves (e.g., Golgi enzymes), or that (ii) will be released into the extracellular space (e.g., protein hormones), or that (iii) are renewal constituents of the plasma membrane (e.g., transferrin receptor), or that (iv) are destined to replenish the lysosome protein pool. A well-described intersection between endocytic or phagocytic compartments and the secretory pathway is seen in vesicles that bring newly produced proteins (such as pro-cathepsin D) from the *trans*-Golgi network into the endocytic continuum. Therefore, the finding that intracellular pathogens can inhabit the secretory pathway was surprising.

Today, it is clear that endocytic and secretory pathways are connected with each other. For example, some enveloped viruses can enter via caveolin-containing plasma membrane domains and be transported in a backward ("retrograde") manner, first to a "caveosome" and then to the ER [57]. Similarly, some plant and bacterial toxins, such as ricin and shiga toxins, enter through different endocytic ports of entry, followed by "retrograde" transport to the Golgi, to the ER, and they are eventually translocated into the cytoplasm [58]. Other toxins, like diphtheria or anthrax toxin, are taken up by clathrin-mediated endocytosis, followed by crossing of the endosome membrane and entry into the cytoplasm [59].

Not only viruses, but also pathogenic microorganisms have found ways to manipulate the secretory (exocytic) pathway of their host cells, albeit in different ways. These pathogens include *Brucella*, *Chlamydia*, *Legionella* and, to some extent, *Toxoplasma*, which is, however, included below in the more appropriate category "vacuole off the beaten path" [60]. The pathways these pathogens take from the host cell surface to the secretory compartments differ as described in the following. Interestingly, transient tight association of phagosomes with mitochondria has been described most frequently for members of this group.

10.3.3.1 *Brucella*

Brucella trafficks normally through an early and into a late endosome-like compartment, before acquiring characteristics of the ER (see Chapter 14). This early endocytic episode is required for later conversion of the vacuole into an ER-like organelle [61]. Only 10–30% of the ingested bacteria divert phagosome maturation, but these will multiply quickly. At 4–8 h of infection, marker proteins of the ER (calreticulin, calnexin, Sec61β) appear on the vacuole. Similar to *Legionella*, *Brucella* intercepts secretory vesicles from ER exit sites to remodel its vacuole. Genetically, the possession of a type IV protein secretion system is crucial for the diversion of phagosome maturation [62].

10.3.3.2 *Legionella*

Legionella-containing vacuoles are never part of the normal endocytic pathway not even at an early step of their biogenesis [63] (see Chapter 18). Within minutes after their formation, *Legionella*-containing phagosomes bind to secretory vesicles from the ER on their way to the Golgi apparatus [64]. These vesicles likely fuse with the

Legionella phagosomes resulting in *Legionella*-containing vacuoles which more and more resemble the ER. Eventually, *Legionella* resides in what is morphologically clearly an ER, most impressively reflected by the presence of ribosomes on the vacuole's cytoplasmic surface after several hours of infection [65]. Such "ribosome-studded vacuoles" are not only found in infected macrophages, but also in infected amoeba, the likely authentic hosts and virulence-training ground for *Legionella* [66]. As with *Brucella*, a type IV protein secretion system is crucial in the establishment and maintenance of the unusual vacuole [64].

10.3.3.3 *Chlamydia*

Like *Legionella*, *Chlamydia* establishes its vacuole without entering the canonical endocytic pathway (see Chapter 15). Even early in infection, contact with endosomes is minimal [67], likely due to entry via specialized "lipid rafts" [68]. The resulting neutral-pH vacuole intercepts exocytic vesicles on their way from the Golgi appartus to the plasma membrane and these vesicles supply the bacteria with sphingomyelin and cholesterol [69]. Major bacterial factors that divert phagosome trafficking are chlamydial inclusion membrane proteins (Incs) some of which are found on the cytoplasmic face of the inclusion membrane and which are substrates for the chlamydial T3SS [70].

10.3.4 Pathogen Vacuoles off the Beaten Path (*Afipia*, *Plasmodium*, *Toxoplasma*)

The vacuoles containing some pathogens colocalize with neither endocytic nor secretory marker molecules. Although it seems a very "clever" strategy to "mask your vacuole" from the degradative endosomal system by having it acquire an identity that does not look like any stable compartment in the cell, this strategy can backfire. Isolation from killing mechanisms may come with isolation from nutrient supplies. To thrive in such compartments, pathogens must develop special food acquisition strategies [71, 72].

10.3.4.1 *Toxoplasma*

In fibroblasts, the *Toxoplasma* vacuole is stripped of transmembrane proteins during its formation [73] by a tight, moving junction located just beneath the host cell plasma membrane which physically restricts entry of host transmembrane proteins into the nascent membrane bag. Hence, the resulting parasitophorous vacuole is, at all times, void of marker molecules for early and late endosomes and lysosomes [74]. It does, however, contain ganglioside GM1, glycosylphosphatidylinositol (GPI)-anchored proteins, and cholesterol, all hallmarks of "lipid microdomains" which may be accumulated in this vacuole [73] (see Chapter 35). Nutrient supply of the parasite in its secluded home is ensured by parasite-derived pores in the parasitophorous vacuole which allow small (<1500 MW) substances to pass the membrane freely. Little is known about the pores' origin and incorporation mode, though. The *Toxoplasma* vacuole associates closely with ER and mitochondria early in infection which may help the parasite in securing host cell lipids [75–77]. Intriguingly, mitochondria are captured by the *Toxoplasma* vacuole through a parasite protein (ROP2), which presents a mitochondrial protein import sequence on the cytoplasmic surface of the vacuole [76]. The mitochondria pull themselves onto the vacuole surface in what can be considered a frustrated import reaction.

10.3.4.2 *Plasmodium*

Plasmodium is a peculiar member of this group in that its invasion of RBCs precludes its delivery into the endocytic system in the first place as RBCs do not endocytose. *Plasmodium* is localized to a vacuole that is free of most RBC membrane proteins but which incorporates secreted *Plasmodium* proteins (see Chapter 33). *Plasmodium* secretes pore proteins that are inserted into the vacuolar and plasma membranes and which cause the formation of an extensive tubular network [78, 79]. Up to 70% of the host cell hemoglobin, the most abundant protein in a RBC, can be degraded as a major source of food.

10.3.4.3 *Afipia*

Studies on *Afipia*-containing compartments suggested that the bacteria are taken up into macrophages via lipid microdomain-dependent macropinocytosis [80] followed by fast shrinking of the macropinosome, a typical process for this compartment (see Chapter 13). However, as a bacterial particle is contained, the macropinosome cannot shrink to naught and, hence, *Afipia* seems to "freeze" the macropinosome in a late stage of its existence, that is, after comprehensive recycling of membranes and ingested liquid. Of some two dozen marker molecules of the canonical endocytic pathway, such as EEA1, LAMP1, VAMP7, VAMP8, transferrin receptor, lysosomal β-glucosidase and others, not a single one is enriched in the *Afipia* phagosome, suggesting that the bacteria are in a rather peculiar vacuole [80].

10.3.5
Vacuole-Free Pathogens (*Babesia*, *Francisella*, *Listeria*, *Mycobacterium*, *Rickettsia*, *Shigella*, *Theileria*, *Trypanosoma*)

At first glance, the macrophage's cytoplasm seems to be less hostile than the degradative endocytic pathway and one wonders why not all intracellular pathogens have evolved mechanisms to disrupt their vacuole membranes to enter a world full of nutrients and energy sources. To get there should not be too difficult from an evolutionary point of view, as many microorganisms secrete membrane-damaging enzymes (e.g., phospholipases, sphingomyelinases, cholesterol oxidases, proteases), likely to combat preying organisms in the environment or to make food available by lysing neighbouring organisms' membranes. These enzymes could secondarily become the pathogens' key into the cytoplasm.

Unfortunately for most microorganisms, the cytoplasm is not a hospitable place: It contains microbicidal peptides (e.g., ubiquicidin [81] and calcium-binding calprotectin [82]), and recent work suggests that the cytoplasmic ubiquitin/proteasome protein degradation machinery and autophagy [83] can be involved in eliminating this group of pathogens from infected cells as well. Important microbial survival factors, such as cytoplasmic iron, are bound to high-affinity proteins (e.g., transferrin, ferritin), and the pathogens must provide capture and uptake machineries (e.g., iron siderophores) to acquire specific nutrients (see below). In addition, presence in the cytoplasm of a mammalian cell endangers the pathogen through the cytotoxic T-cell response when bacterial antigens are presented on the surface of the host cell, leading to killing of the infected cell altogether. Similarly, nucleotide-binding oligomerization domain (NOD)-like receptors (NLRs) play an important role in recognizing and combatting cytosolic pathogens by driving innate and adapted immune responses [84]. These complications may be the reason why only a few bacteria have succeeded in evolving in this direction, most notably, *Listeria monocytogenes, Shigella flexneri* and *Rickettsia prowazekii. Francisella tularensis* belongs to this group but is different because it is released from the phagosome only after several hours of enclosure in a compartment which acidifies only weakly, which lacks the lysosome hydrolase cathepsin D and which does not receive lysosome contents [85, 86] (see Chapter 23).

Theileria spp., *Trypanosoma cruzi* and *Babesia* spp. are eukaryotic members of this group. Similar to *Francisella, Trypanosoma* first resides in a membrane-bound vacuole that is positive for LAMP1 at 60 min of infection (see Chapter 36). *Trypanosoma* escapes into the cytoplasm where the trypomastigote form is rapidly converted into an amastigote form. *Babesia,* which parasitizes RBCs, lyses the parasitophorous vacuole soon after entry into the cell, possibly by the secretion of parasite proteins [87] (see Chapter 33) and so do transforming *Theileria* spp. [88] (see Chapter 34).

In an intriguing study, Götz *et al.* [89] microinjected different bacteria into epithelial Caco-2 cell cytoplasm to test whether they could multiply after being directly transferred into the host cytoplasm without the need to lyse the phagosome membrane. They found that *L. monocytogenes, S. flexneri* and, to some extent, enteroinvasive *Escherichia coli* could do so, while *S. enterica* var. *Typhimurium, Yersinia enterocolitica* or apathogenic *Bacillus subtilis* could not. This suggested that only bacteria which would naturally find their way into the cytoplasm and multiply there would do so after being microinjected. In *L. monocytogenes,* regulation of gene transcription via the virulence transcription regulator PrfA was necessary to enable these bacteria to multiply intracellularly after microinjection [89]. One target gene for the virulence regulator is the listerial *hpt* gene which likely codes for a glucose 1-phosphate uptake protein [89–91]. Glucose 1-phosphate is a typical host compound which is needed during gluconeogenesis. The bacterial knockout strain was strongly compromised in intracellular multiplication, further demonstrating that specific metabolic adaptations to the eukaryote are required for intracytoplasmic multiplication and that food may not be as readily available in the cytoplasm as it may seem. Similar to this study, it has been shown that export of membrane-disrupting listeriolysin O by pathogenic *Brucella* leads to release of some bacteria into the host cytoplasm, yet these bacteria could not multiply [92]. It must be noted, though, that two studies using either a harmless *Bacillus* [93] or an avirulent *E. coli* strain [94] presenting listeriolysin O on their surfaces (giving them access to the cytoplasm), came to different conclusions and state that, in principle, these bacteria can multiply intracellularly, although not having evolved to specifically do so.

Recent evidence has suggested that some mycobacteria may also rupture the vacuole membrane. *Mycobacterium marinum* has been described as cytoplasmic in late stages of the infection in two biological systems, *Dictyostelium discoideum* [95] and murine macrophages [96] (see Chapter 25). Similar to *L. monocytogenes*, *M. marinum* can induce actin tail formation [96], albeit less strongly. The relationship of such a cytoplasmic stage to infection remains to be investigated. In addition, recent studies have described that *M. tuberculosis* and *M. leprae* can also be found in the cytoplasm, but only at late times of infection, that is, beyond 48 h. The avirulent *M. bovis* vaccination strain BCG did not leave its vacuole [97]. These are surprising findings which will deserve further study.

10.3.6
From War to Peace Talks and Disarmament (*Rhizobium*, *Sodalis*, *Buchnera*)

Not all vacuoles contain pathogens. Some contain symbionts, either "primary" symbionts which exist in "bacteriomes" (specialized compartments) or "secondary" symbionts which are not restricted to life in a defined specialized compartment. Although the concepts of parasitism and symbiosis seem to be quite different, they are not (see Chapter 1). The way in which *Rhizobium* induces formation of nodules in legume plant is similar in many ways to infection by a pathogen. Similarly, a T3SS is involved in this symbiosome establishment. Such T3SS are also known from insect symbionts such as the tse tse symbiont *Sodalis glossinidius*. Here, invasion of host cells is dependent on T3SS effectors and this suggests that *Sodalis* may have evolved from an ancestor with a parasitic lifestyle [98].

A surge of interest in the cell biology of bacteriomes has started only a few years ago and corresponding research is still in its infancy. The best investigated bacteriomes in the plant world are the legume symbiosomes containing *Rhizobium* spp. (see Chapter 28). These symbiosomes are formed when suitable strains reach the inner plant cortex and are taken up by cortical cells in a process resembling endocytosis–phagocytosis [33]. Symbiosomes are the correlates to phagosomes in mammalian cells and resemble mammalian late endosomes, as a proteome analysis showed possession of Rab7, subunits of the proton-pumping ATPase and lysosome-associated integral membrane protein 1 (LIMP1) [99, 100]. This fits well with the finding that the soybean symbiosome contains several host enzymes with acidic pH optima [101]. Rhizobial symbiosome membranes also contain proteins involved in compound transport, metabolism, protein folding and signaling [102] and they possess characteristics of both plasma and vacuolar membranes. For example, symbiosomes contain nodulin-26 which is a tonoplast protein, and α-mannosidase, a plant vacuole marker enzyme, has been identified inside the symbiosome [103].

Rhizobia can, in incompatible associations, be delivered to and killed within a lysosome-like compartment which allows the host cell to control the bacterial influence [103]. This stunning similarity between (activated) mammalian professional phagocytes and plant root cells may reflect the evolutionarily relatively old necessity to remove microorganisms from the tissue whenever they turn out not to be beneficial (any more).

As for the precise compartmentation of insect bacteriomes, really nothing is known, likely due to a lack of tools, a scarcity in symbiont researchers and the high level of sophistication of experiments required to work with these little characterized, but amazing systems.

10.4 Immune Activation and the Loss of Home

If pathogenic microorganisms are so good at their job, why have higher organisms been able to evolve in their constant presence and not been eradicated a long time ago? One of the answers is that resident macrophages are not all that there is to counteract pathogens. For example, macrophages are the more "harmless" archaetype of a phagocytic cell when compared to neutrophils, which are much shorter lived but much more microbicidal. Why then should we have macrophages at all? Probably because macrophages help with antigen presentation to other immune cells and this is much more sensibly achieved when the ingested particle is not immediately trashed [104]. Macrophages can also be induced to be more killer cells, in that other immune cells, such as natural killer cells and a subset of T lymphocytes secrete interferon γ (IFNγ), a proinflammatory cytokine, during an infection (see Chapter 12). Activation of macrophages by IFNγ and possibly also by microbial substances such as lipopolysaccharide, which induces formation of tumor necrosis factor α (TNFα), leads to dramatically changed macrophage gene transcription patterns and pronounced alterations in the (latex bead-containing) phagosome proteome [105]. In parallel, the killing capacity toward intracellular pathogens is strongly upregulated. Relevant to this review, this increased killing is often paralleled by increased phagolysosome formation, for example, with *Mycobacterium* [106], *Salmonella* [107, 108] or *Francisella* [109]. This finding suggested that delivery of these microorganisms to a phagolysosome comes with increased killing – or that, of course, killing of a microbe comes with increased phagosome maturation. These two possibilities have historically not always been clearly separated and are very difficult to address experimentally. Hence, it was usually just assumed that "the more maturation, the more killing ensuing" was the better candidate. Some evidence, however, suggests this may not always be the case and that late endocytic conditions are not necessarily microbicidal even for pathogens that normally live in an early endosome-like compartment [110]. Furthermore, microbes may either adapt to the intralysosomal environment [111] or they can cease growing and enter a stage of latency.

Antimicrobial mechanisms in phagosomes that reach beyond exposure to oxidative compounds, acid, defensins and hydrolases, are nutrient starvation in activated macrophages by the use of ion transporters or depletion of nutrients such as the amino acid tryptophan by macrophage cytoplasmic depletion systems [91, 112] Finally, autophagy has been implicated as an important process in the eradication of intracellular pathogens, particularly in activated macrophages. It does so possibly by encasing pathogen phagosomes in a membrane which is destined to fuse with lysosomes.

10.5
Why Not Pathogens in Peroxisomal Vacuoles?

Having said the above, it seems that most intracellular niches have been occupied by microbial pathogens. But how about peroxisomes and nuclei? Are there any intracellular pathogens known that would localize to these organelles? While viral particles have been seen intranuclearly (e.g., Simian virus 40 [113]), reports of intranuclear bacterial pathogens are more anecdotical and seem to reflect single laboratory observations [114]. There is a notable exception for the α-proteobacterium *Nonospora macronucleata*, which multiplies in the somatic nucleus of *Paramecium caudatum* [115], although in a more symbiontic relationship.

Peroxisomal pathogens have, to the best of the author's knowledge, never been reported. While this may be due to nobody having looked thoroughly, there really might be none. What then makes a peroxisome so uninteresting from a microbial point of view? First, there has been no report of vesicular transport between the endocytic system and peroxisomes, making it hard for particles to reach this compartment directly from the outside world. Recent research has established that the ER is a major membrane and protein source for growth of peroxisomes [116], but this connection is likely characterized by vesicular traffic between ER and peroxisomes via vesicles too small for a microbe to squeeze into or the microbe would have to be taken up directly from the cytoplasm into the peroxisome. As for nutrients, the peroxisome might actually be a quite hospitable environment providing long-chain fatty acids. Microorganisms that could withstand or inactivate the oxidative stress in this compartment could possibly live quite happily.

This is different for the ER. To start with, it is also hard to understand how a pathogen can reach and then abuse this compartment, but *Legionella pneumophila* does. Its vacuole refuses to acquire typical early endocytic marker molecules, such as Rab5 or Rab7, immediately after uptake [117]. How *Legionella* "achieves" such exclusion from the start is not known. Absence of acquisition of degradative pathway markers is paralleled by the recruitment of small vesicles destined to bring cargo from the ER exit sites to the *cis*-Golgi network. This camouflages the *Legionella*-containing phagosomes and they turn themselves more and more into an ER, possibly by fusing first with these vesicles, then with ER cisternae. The ER is a great place for legionellae, as it is very extended, amounting to about half the internal membrane surface [118]. It contains neither aggressive hydrolases nor oxidative stress molecules, and it has a neutral pH. Finally, the rough ER is the place into which the cell stuffs newly produced secretory proteins through tight translocation channels. *Legionella*, on the other hand, has a predilection for amino acids as nutrient source and an extracellular protease that could make those available [119] – a perfect choice!

So, how about peroxisomes? The general rule goes that if there is an easy way to enter a compartment, it will be used. Peroxisomes seem to be too hard to reach and are possibly too small (~0.4 μm). This makes it all the more surprising that even a double-membraned organelle, such as the mitochondrion (a former bacterium itself), can be inhabited by an α-proteobacterium, *Midichloria mitochondrii* (a tick symbiont) [120]. Now that this is possible, it is likely only a question of time before an intra-peroxisomal pathogen will be found, possibly gaining access via a transport vesicle like the one *Legionella* uses to steal itself into the ER.

10.6 Looking for a New Home

Once a pathogen has managed to outsmart its host cell and can multiply within it, it needs to get out of it or it will eventually be eradicated. In fact, many facultative intracellular microbes proliferate extracellularly to some extent during active disease, although this does not happen during the latent phases of some chronic diseases such as tuberculosis or when a pathogen is handled from one host cell to another directly (e.g., *L. monocytogenes* between epithelial cells). Therefore, the eventual death of the exploited host cell is in the best interest of the pathogen. Phagocytic cells recruited to the place of inflammation may take up the released microorganisms and provide new homes. Egress of the pathogen from the host cell can be either without cell damage, by programmed cell death (apoptosis), necrosis, or by some form of cell death intermediate to apoptosis and necrosis.

10.7 Summary

Originally, formation of a phagolysosome was regarded as a single fusion step between a phagosome and a lysosome (see early seminal contributions such as that by Armstrong [121]). Our knowledge of the complexity of the endocytic continuum with its subcompartments has considerably increased in the past decade, and scientists in the field became increasingly aware of the similarities between endocytic and phagocytic compartments. This led to the realization that phagolysosome formation could be prevented not only by inhibiton of the proper fusion between a phagosome and a lysosome, but also by interference with phagosome maturation at an earlier stage. Later, partly inspired by the finding that *L. pneumophila*-containing vacuoles had the characteristics of the ER [122], it became clear that pathogen-containing phagosomes could be completely disconnected from the endocytic continuum. Therefore, inhibition of phagolysosome formation could also be achieved by complete diversion of trafficking from the endo-lysosomal continuum. Today, we can say that each type of pathogen-containing vacuole is different from the others. In some cases, these compartments may have a similar or identical composition to normally existing stable compartments (e.g., *Coxiella*-containing vacuoles and late endosomes and lysosomes), in other cases pathogens inhabit compartments that may not be normally found (*Chlamydia*-containing vacuoles). Future research will continue to describe the variety of these compartments and elucidate their cell biology relevance. These findings will also help to target antimicrobials to these vacuoles and they will increase our understanding of how, surprisingly, immune presentation may occur out of these non-endocytic compartments [123]. Future research will also look into what the precise intracompartmental conditions are, using the microbes as probes to study some otherwise hard-to-study compartments (see Chapter 5). Just as *Listeria* research strongly promoted cytoskeleton research [124], this kind of research will lead to more insights into cell pathology and it will also tremendously enrich our knowledge of how the eukaryotic cell works.

Acknowledgments

I thank the past and current members of the Haas laboratory for their contributions to the study of phagosome biogenesis, their dedication and their enthusiasm and comments on this manuscript. I thank Tobias Dykstra and Kristine von Bargen for editorial help and many of the authors of chapters on individual pathogens for their support in compiling Table 10.1. Work in the author's laboratory is supported by the Deutsche Forschungsgemeinschaft.

References

1 Brogden, K.A., Ackermann, M., McCray, P.B. and Tack, B.F. (2003) *Int J Antimicrob Agents*, **22**, 465–478.
2 Haas, A. (2007) *Traffic*, **8**, 311–330.
3 Rink, J., Ghigo, E., Kalaidzidis, Y. and Zerial, M. (2005) *Cell*, **122**, 735–749.
4 Anes, E., Peyron, P., Staali, L., Jordao, L. *et al.* (2006) *Cell Microbiol*, **8**, 939–960.
5 Rogers, L.D. and Foster, L.J. (2007) *Proc Natl Acad Sci USA*, **104**, 18520–18525.
6 Henry, R.M., Hoppe, A.D., Joshi, N. and Swanson, J.A. (2004) *J Cell Biol*, **164**, 185–194.
7 Griffiths, G. (2004) *Trends Cell Biol*, **14**, 343–351.
8 Aniento, F., Emans, N., Griffiths, G. and Gruenberg, J. (1993) *J Cell Biol*, **123**, 1373–1387.
9 Ward, D.M., Pevsner, J., Scullion, M.A., Vaughn, M. and Kaplan, J. (2000) *Mol Biol Cell*, **11**, 2327–2333.
10 Desjardins, M., Celis, J.E., van Meer, G., Dieplinger, H. *et al.* (1994) *J Biol Chem*, **269**, 32194–32200.
11 Jahraus, A., Tjelle, T.E., Berg, T., Habermann, A. *et al.* (1998) *J Biol Chem*, **273**, 30379–30390.
12 Zerial, M. and McBride, H. (2001) *Nat Rev Mol Cell Biol*, **2**, 107–117.
13 Jahn, R. and Scheller, R.H. (2006) *Nat Rev Mol Cell Biol*, **7**, 631–643.
14 Yates, R.M., Hermetter, A. and Russell, D.G. (2005) *Traffic*, **6**, 413–420.
15 van Weert, A.W., Dunn, K.W., Gueze, H.J., Maxfield, F.R. and Stoorvogel, W. (1995) *J Cell Biol*, **130**, 821–834.
16 Hurtado-Lorenzo, A., Skinner, M., Annan, J.E., Futai, M. *et al.* (2006) *Nat Cell Biol*, **8**, 124–136.
17 Funato, K., Beron, W., Yang, C.Z., Mukhopadhyay, A. and Stahl, P.D. (1997) *J Biol Chem*, **272**, 16147–16151.
18 Jahraus, A., Egeberg, M., Hinner, B., Habermann, A. *et al.* (2001) *Mol Biol Cell*, **12**, 155–170.
19 Kjeken, R., Egeberg, M., Habermann, A., Kuehnel, M. *et al.* (2004) *Mol Biol Cell*, **15**, 345–358.
20 Cornillon, S., Pech, E., Benghezal, M., Ravanel, K. *et al.* (2000) *J Biol Chem*, **275**, 34287–34292.
21 Holt, O.J., Gallo, F. and Griffiths, G.M. (2006) *J Biochem Tokyo*, **140**, 7–12.
22 Zhou, Z. (2007) *Dev Cell*, **13**, 759–760.

23 Rabinovitch, M. (1967) *J Cell Biol*, **35**, 108A–109.

24 Gratton, S.E.A., Ropp, P.A., Pohlhaus, P.D., Luft, J.C. *et al.* (2008) *Proc Natl Acad Sci USA*, **105**, 11613–11618.

25 Cardinali, G., Bolasco, G., Aspite, N., Lucania, G. *et al.* (2008) *J Invest Dermatol*, **128**, 558–567.

26 Joiner, K.A., Fuhrman, S.A., Miettinen, H.M., Kasper, L.H. and Mellman, I. (1990) *Science*, **249**, 641–646.

27 Caron, E. and Hall, A. (1998) *Science*, **282**, 1717–1721.

28 Hörmann, H. (1985) *Blut*, **51**, 307–314.

29 Mishima, H., Yasumoto, K., Nishida, T. and Otori, T. (1987) *Invest Ophthalmol Vis Sci*, **28**, 1521–1526.

30 McAbee, D.D. and Grinnell, F. (1983) *J Cell Biol*, **97**, 1515–1523.

31 Dziewanowska, K., Patti, J.M., Deobald, C.F., Bayles, K.W. *et al.* (1999) *Infect Immun*, **67**, 4673–4678.

32 Hardt, W.D., Chen, L.M., Schuebel, K.E., Bustelo, X.R. and Galán, J.E. (1998) *Cell*, **93**, 815–826.

33 Jones, A.T. (2007) *J Cell Mol Med*, **11**, 670–684.

34 Schueller, C., Schneider, B., Kempf, V.A.J. and Haas, A. (2007) *Microbes Infect*, **9**, 355–363.

35 Finlay, B.B. and Falkow, S. (1989) *Mol Microbiol*, **3**, 1833–1841.

36 Naderer, T. and McConville, M.J. (2008) *Cell Microbiol*, **10**, 301–308.

37 Muñoz-Elías, E.J. and McKinney, J.D. (2006) *Cell Microbiol*, **8**, 10–22.

38 Minnick, M.F., Mitchell, S.J. and McAllister, S.J. (1996) *Trends Microbiol*, **4**, 343–347.

39 Cowman, A.F. and Crabb, B.S. (2006) *Cell*, **124**, 755–766.

40 Sturgill-Koszycki, S., Schaible, U.E. and Russell, D.G. (1996) *EMBO J*, **15**, 6960–6968.

41 Méresse, S., Steele-Mortimer, O., Finlay, B.B. and Gorvel, J.P. (1999) *EMBO J*, **18**, 4394–4403.

42 Drecktrah, D., Knodler, L.A., Howe, D. and Steele-Mortimer, O. (2007) *Traffic*, **8**, 212–225.

43 Fernandez-Mora, E., Polidori, M., Lührmann, A., Schaible, U.E. and Haas, A. (2005) *Traffic*, **6**, 635–653.

44 Axelrod, S., Oschkinat, H., Enders, J., Schlegel, B. *et al.* (2008) *Cell Microbiol*, **10**, 1530–1545.

45 Kang, P.B., Azad, A.K., Torrelles, J.B., Kaufman, T.M. *et al.* (2005) *J Exp Med*, **202**, 987–999.

46 Vergne, I., Fratti, R.A., Hill, P.J., Chua, J. *et al.* (2004) *Mol Biol Cell*, **15**, 751–760.

47 Vergne, I., Chua, J., Singh, S.B. and Deretic, V. (2004) *Annu Rev Cell Dev Biol*, **20**, 367–394.

48 Schlumberger, M.C. and Hardt, W.D. (2005) *Curr Top Microbiol Immunol*, **291**, 29–42.

49 Strasser, J.E., Newman, S.L., Ciraolo, G.M., Morris, R.E. *et al.* (1999) *J Immunol*, **162**, 6148–6154.

50 Shi, L., Albuquerque, P.C., Lazar-Molnar, E., Wang, X. *et al.* (2008) *Eukaryot Cell*, **7**, 1109–1117.

51 Hohman, T.C., McNeil, P.L. and Muscatine, L. (1982) *J Cell Biol*, **94**, 56–63.

52 Heinzen, R.A., Scidmore, M.A., Rockey, D.D. and Hackstadt, T. (1996) *Infect Immun*, **64**, 796–809.

53 Romano, P.S., Gutierrez, M.G., Berón, W., Rabinovitch, M. and Colombo, M.I. (2007) *Cell Microbiol*, **9**, 891–909.

54 Pan, X., Lührmann, A., Satoh, A., Laskowski-Arce, M.A. and Roy, C.R. (2008) *Science*, **320**, 1651–1654.

55 Desjardins, M. and Descoteaux, A. (1997) *J Exp Med*, **185**, 2061–2068.

56 Antoine, J.C., Prina, E., Jouanne, C. and Bongrand, P. (1990) *Infect Immun*, **58**, 779–787.

57 Damm, E.-M., Pelkmans, L., Kartenbeck, J., Mezzacasa, A. *et al.* (2005) *J Cell Biol*, **168**, 477–488.

58 Wesche, J. (2002) *Int J Med Microbiol*, **291**, 517–521.

59 Sandvig, K. and van Deurs, B. (2005) *Gene Ther*, **12**, 865–872.

60 Roy, C.R., Salcedo, S.P. and Gorvel, J.-P.E. (2006) *Nat Rev Immunol*, **6**, 136–147.

61 Starr, T., Ng, T.W., Wehrly, T.D., Knodler, L.A. and Celli, J. (2008) *Traffic*, **9**, 678–694.

62 Arellano-Reynoso, B., Lapaque, N., Salcedo, S., Briones, G. *et al.* (2005) *Nat Immunol*, **6**, 618–625.

63 Joshi, A.D., Sturgill-Koszycki, S. and Swanson, M.S. (2001) *Cell Microbiol*, **3**, 99–114.

64 Robinson, C.G. and Roy, C.R. (2006) *Cell Microbiol*, **8**, 793–805.

65 Horwitz, M.A. (1987) *J Exp Med*, **166**, 1310–1328.

66 Brüggemann, H., Cazalet, C. and Buchrieser, C. (2006) *Curr Opin Microbiol*, **9**, 86–94.

67 Scidmore, M.A., Fischer, E.R. and Hackstadt, T. (2003) *Infect Immun*, **71**, 973–984.

68 Dautry-Varsat, A., Subtil, A. and Hackstadt, T. (2005) *Cell Microbiol*, **7**, 1714–1722.

69 Hackstadt, T., Rockey, D.D., Heinzen, R.A. and Scidmore, M.A. (1996) *EMBO J*, **15**, 964–977.

70 Shaw, E.I., Dooley, C.A., Fischer, E.R., Scidmore, M.A. *et al.* (2000) *Mol Microbiol*, **37**, 913–925.

71 Keeney, K.M., Stuckey, J.A. and O'Riordan, M.X.D. (2007) *Mol Microbiol*, **66**, 758–770.

72 Sinai, A.P. and Joiner, K.A. (1997) *Annu Rev Microbiol*, **51**, 415–462.

73 Charron, A.J. and Sibley, L.D. (2004) *Traffic*, **5**, 855–867.

74 Mordue, D.G. and Sibley, L.D. (1997) *J Immunol*, **159**, 4452–4459.

75 Sinai, A.P. and Joiner, K.A. (2001) *J Cell Biol*, **154**, 95–108.

76 Nakaar, V., Ngô, H.M., Aaronson, E.P., Coppens, I. *et al.* (2003) *J Cell Sci*, **116**, 2311–2320.

77 Crawford, M.J., Thomsen-Zieger, N., Ray, M., Schachtner, J. *et al.* (2006) *EMBO J*, **25**, 3214–3222.

78 Bano, N., Romano, J.D., Jayabalasingham, B. and Coppens, I. (2007) *Int J Parasitol*, **37**, 1329–1341.

79 Wickert, H., Wissing, F., Andrews, K.T., Stich, A. *et al.* (2003) *Eur J Cell Biol*, **82**, 271–284.

80 Lührmann, A., Streket, K., Schüttfort, A., Daniels, J.J. and Haas, A. (2001) *Proc. Natl. Acad. Sci. USA*, **98**, 7271–7276.

81 Hiemstra, P.S., van den Barselaar, M.T., Roest, M., Nibbering, P.H. and van Furth, R. (1999) *J Leukoc Biol*, **66**, 423–428.

82 Ross, K.F. and Herzberg, M.C. (2001) *Infect Immun*, **69**, 3248–3254.

83 Perrin, A.J., Jiang, X., Birmingham, C.L., So, N.S.Y. and Brumell, J.H. (2004) *Curr Biol*, **14**, 806–811.

84 Sirard, J.-C., Vignal, C., Dessein, R. and Chamaillard, M. (2007) *PLoS Pathog*, **3**, e152.

85 Clemens, D.L., Lee, B.-Y. and Horwitz, M.A. (2004) *Infect Immun*, **72**, 3204–3217.

86 Checroun, C., Wehrly, T.D., Fischer, E.R., Hayes, S.F. and Celli, J. (2006) *Proc Natl Acad Sci USA*, **103**, 14578–14583.

87 Yokoyama, N., Okamura, M. and Igarashi, I. (2006) *Vet Parasitol*, **138**, 22–32.

88 Dobbelaere, D.A., Fernandez, P.C. and Heussler, V.T. (2000) *Cell Microbiol*, **2**, 91–99.

89 Götz, M., Bubert, A., Wang, G., Chico-Calero, I. *et al.* (2001) *Proc Natl Acad Sci USA*, **98**, 12221–12226.

90 Chico-Calero, I., Suárez, M., González-Zorn, B. Scortti, M. *et al.* (2002) *Proc Natl Acad Sci USA*, **99**, 431–436.

91 Brown, S.A., Palmer, K.L. and Whiteley, M. (2008) *Nat Rev Microbiol*, **6**, 657–666.

92 Köhler, S., Layssac, M., Naroeni, A., Gentschev, I. *et al.* (2001) *Infect Immun*, **69**, 2753–2756.

93 Bielecki, J., Youngman, P., Connelly, P. and Portnoy, D.A. (1990) *Nature*, **345**, 175–176.

94 Monack, D.M. and Theriot, J.A. (2001) *Cell Microbiol*, **3**, 633–647.

95 Hagedorn, M. and Soldati, T. (2007) *Cell Microbiol*, **9**, 2716–2733.

96 Stamm, L.M., Morisaki, J.H., Gao, L.-Y., Jeng, R.L. *et al.* (2003) *J Exp Med*, **198**, 1361–1368.

97 van der Wel, N., Hava, D., Houben, D., Fluitsma, D. *et al.* (2007) *Cell*, **129**, 1287–1298.

98 Dale, C., Young, S.A., Haydon, D.T. and Welburn, S.C. (2001) *Proc Natl Acad Sci USA*, **98**, 1883–1888.

99 Cheon, C.I., Lee, N.G., Siddique, A.B., Bal, A.K. and Verma, D.P. (1993) *EMBO J*, **12**, 4125–4135.

100 Wienkoop, S. and Saalbach, G. (2003) *Plant Physiol*, **131**, 1080–1090.

101 Mellor, R.B. (1989) *J Exp Bot*, **40**, 831–839.

102 Catalano, C.M., Lane, W.S. and Sherrier, D.J. (2004) *Electrophoresis*, **25**, 519–531.

103 Verma, D.P.S. (1992) *Plant Cell*, **4**, 373–382.

104 Yates, R.M., Hermetter, A., Taylor, G.A. and Russell, D.G. (2007) *Traffic*, **8**, 241–250.

105 Jutras, I., Houde, M., Currier, N., Boulais, J. *et al.* (2008) *Mol Cell Proteomics*, **7**, 697–715.

106 Schaible, U.E., Sturgill-Koszycki, S., Schlesinger, P.H. and Russell, D.G. (1998) *J Immunol*, **160**, 1290–1296.

107 Kagaya, K., Watanabe, K. and Fukazawa, Y. (1989) *Infect Immun*, **57**, 609–615.

108 McCollister, B.D., Bourret, T.J., Gill, R., Jones-Carson, J. and Vázquez-Torres, A. (2005) *J Exp Med*, **202**, 625–635.

109 Santic, M., Molmeret, M. and Kwaik, Y.A. (2005) *Cell Microbiol*, **7**, 957–967.

110 Gomes, M.S., Paul, S., Moreira, A.L., Appelberg, R. *et al.* (1999) *Infect Immun*, **67**, 3199–3206.

111 Schröder, A., Kland, R., Peschel, A., von Eiff, C. and Aepfelbacher, M. (2006) *Med Microbiol Immunol*, **195**, 185–194.

112 Appelberg, R. (2006) *J Leukoc Biol*, **79**, 1117–1128.

113 Granboulan, N., Tournier, P., Wicker, R. and Bernhard, W. (1963) *J Cell Biol*, **17**, 423–441.

114 Tenner, K., Kaiserling, E., Lennert, K., Racz, P. and Wuthe, H.H. (1973) *Virchows Arch B Cell Pathol*, **14**, 35–45.

115 Kusch, J. and Görtz, H.-D. (2006) *Prog Mol Subcell Biol*, **41**, 61–76.

116 Hoepfner, D., Schildknegt, D., Braakman, I., Philippsen, P. and Tabak, H.F. (2005) *Cell*, **122**, 85–95.

117 Clemens, D.L., Lee, B.Y. and Horwitz, M.A. (2000) *Infect Immun*, **68**, 5154–5166.

118 Alberts, B., Johnson, A., Walter, P., Lewis, J. *et al.* (2008) *Molecular Biology of the Cell*, 5th edn, Taylor & Francis, New York.

119 Wieland, H., Ullrich, S., Lang, F. and Neumeister, B. (2005) *Mol Microbiol*, **55**, 1528–1537.

120 Sassera, D., Beninati, T., Bandi, C., Bouman, E.A.P. *et al.* (2006) *Int J Syst Evol Microbiol*, **56**, 2535–2540.

121 Armstrong, J.A. and Hart, P.D. (1971) *J Exp Med*, **134**, 713–740.

122 Katz, S.M. and Hashemi, S. (1982) *Lab Invest*, **46**, 24–32.

123 Brodsky, F.M., Lem, L., Solache, A. and Bennett, E.M. (1999) *Immunol Rev*, **168**, 199–215.

124 Hamon, M., Bierne, H. and Cossart, P. (2006) *Nat Rev Microbiol*, **4**, 423–434.

125 Schnaith, A., Kashkar, H., Leggio, S.A., Addicks, K. *et al.* (2007) *J Biol Chem*, **282**, 2695–2706.

11
Consequences of Pathogen Compartmentation for Therapeutic Intervention

Albert Haas and Ulrich Schaible

11.1
The Problem

As described in Chapter 10, each intracellular pathogen inhabits a different compartment in its host cell with characteristics ranging from the endoplasmic reticulum (ER) to an intermediate Golgi-to-plasma membrane compartment, or lysosomes or cytoplasm. These characteristic localizations directly affect the accessibility of the microbes to antimicrobial drugs and can be partly responsible for the differential susceptibilities of intracellular pathogens to therapeutic agents. The localization also affects immune effector mechanisms such as oxidative radicals, whose availability can be modulated by compartmentation. In addition, some facultative intracellular pathogens can exist or even proliferate in the extracellular space, during which time they would be expected to be yet again differently susceptible. The precise tissue infected and the remodeling of tissues due to pathological events or inflammation (e.g., granuloma formation) as a response to infection can additionally affect drug efficacy. These issues need all be taken into account when rational drug design is discussed. Here we focus on how the intracellular niches of pathogens affect drug efficacy and also how this knowledge can be exploited to promote drug delivery.

11.2
Consequences of Intracellular Localization for Treatment with Antibiotics

To be applicable as an effective drug against infections with intracellular pathogens, a compound must not only fulfill the obvious requirements such as high antimicrobial efficacy combined with low toxicity for the patient and easy applicability but also the following criteria:

(1) It must reach the infected cells within the tissue or cavity.
(2) It must efficiently enter infected cells and the pathogen-containing compartment.

Intracellular Niches of Microbes. Edited by Ulrich E. Schaible and Albert Haas

ISBN: 978-3-527-32207-7

(3) It must enter the microbe.
(4) It must work under the intracompartmental conditions, that is, at the given pH and oxygen tension and in the presence of enzymes which might be able to modify the drug.
(5) It should be active against both metabolically active, growing and nongrowing, quiescent forms of a given microbe (e.g., *Mycobacterium tuberculosis*).
(6) It should be active against a pathogen at different life stages, if applicable (i.e., promastigote and amastigote parasites or blood and liver stages).

Each of these steps can be rate-limiting in a given case and might complicate the treatment to a point where treatment is not at all possible.

Several complications can be envisaged. On a cellular level, a drug has to cross only one host membrane to reach cytoplasmically located *Listeria* while two membranes (plasma membrane plus ER membrane) have to be crossed to reach *Legionella*. Soluble, in particular synthetic, compounds enter infected cells by nonselective pino- or macropinocytosis and will be exported following different kinetics depending on their biophysical properties such as hydrophobicity and charge, all of which determine intracellular accumulation. Entering the endosomal system does not guarantee, however, that the drug can access a specified compartment, because some pathogens are cut off from the endosomal continuum. An additional factor when studying the efficacy of a drug against a given microorganism in a given cell type is that an antibiotic may be actively transported out of the mammalian cell [1] (Table 11.1). For example, macrophages express multidrug resistance proteins (MRPs) including organic anion-transporters in plasma membrane and endosomes, which can deplete the cytoplasm of drug molecules [1]. Consequently, the MRP inhibitor gemfibrozil increases the intracellular activity of macrolides against *Listeria* by retaining them in the cell [2]. Additionally, the type of infected host cell is crucial for antibiotic action, as azithromycin, a macrolide, is concentrated 150-fold, 50-fold or 10-fold in granulocytes, macrophages and parenchymal cells, respectively [3], impressively showing that not all cells are evenly "flushed" with antibiotics and reflecting the differences between different cell types in the net balance of endocytic activity and efflux. Finally, the intracellular milieu such as the intraphagosomal pH can be crucial for activity of antibiotics; for example gentamicin is inactive at the pH of lysosomes (4.0–5.0), the final destination for the endocytosed drug [4].

Treatment complications can also stem from the microbial side in that, for example, the wax-like cell walls of mycobacteria preclude the uptake of various antibiotics by these bacteria [5]. These pathogens further complicate successful therapy by falling into a "dormant" state in which many antimicrobial drugs have little effectiveness compared to their good activity against metabolically active and proliferating microbes. Consequently, drugs are inefficient when it comes to the eradication of latent tuberculosis – "Microorganisms which grow slowly, also die slowly" – which explains why most antibiotics are targeted against the metabolism of metabolically active microorganisms (such as cell wall biosynthesis or protein biosynthesis).

Table 11.1 Intracellular localization and accumulation levels of major antibiotic classes.

Pharmacochemical class	Antibiotic	Influx[a]	Efflux[a]	Accumulation level[b]	Activity at various pH[d]	Subcellular[c]
Beta-lactams	All	Fast	Variable	<1	−/++/+++	C
Macrolides	Erythromycin	Fast	Variable	4–10	−/+/+++	L/C
	Clarithromycin	Fast	Fast	10–20		
	Azithromycin	Fast	Slow	40–300		
	Telithromycin	Fast	Fast-slow	15–50		
Fluoroquinolones	All	Fast	Very fast	4–10	−/+/+++	C
Aminoglycosides	All	Very slow	Very slow	2–4 (after days)	−/+/+++	L
Lincosaminides	Clindamycin	Fast	Fast	5–20		U
	Lincomycin	Fast	Fast	1–4		
Tetracyclins	All	Fast	?	1–4	+/+++/++	U
Rifamycins	Rifampin	Fast	?	2–10	+++/++/++	C
Glycopeptides	Vancomycin	Slow	?	8 (24 h)		L
	Teicoplanin	Fast	?	60		C (not L)
	Oritavancin	Slow	?	150–300 (24 h)		L
Ozazolidinones	Linezolid	Fast	Fast	about 1		C

[a]Very fast: less than 3 min to equilibrium; fast: 3–15 min to equilibrium; slow: 15 min–3 h to equilibrium; very slow: more than 3 h to equilibrium.
[b]Accumulation ratio = intracellular divided by extracellular concentration.
[c]Suspected or demonstrated predominant subcellular localization; C = cytoplasm, L = lysosomes, U = unknown.
[d]Activities compared at pH <5/6/7; −, no activity; +, low activity; + +, moderate activity; + + +, maximum activity.
Information taken from [3, 6, 10].

Having said this, it is worth considering which of the major antibiotic classes (aminoglycosides, β-lactam antibiotics, macrolides, fluoroquinolones) are most useful in fighting an intracellular pathogen.

11.2.1
Aminoglycosides

Aminoglycosides (e.g., gentamicin, streptomycin, kanamycin, neomycin) act by inhibiting bacterial protein biosynthesis on ribosomes. They cannot cross the hydrophobic core of biological membranes due to their positive charge, but can be taken up by bulk phase endocytosis, eventually accumulating in the lysosomes where their efficacy is much reduced (Table 11.1). This is the reason why aminoglycosides are largely inactive against obligate and have little active against most facultative intracellular bacteria [6]. Neverthless, aminoglycoside antibiotics still are widely used, and it was the isolation of streptomycin, an aminoglycoside, which in 1943 constituted a turning point in the treatment of tuberculosis. This drug is still a second-line antibiotic in antituberculosis therapy. The efficacy of streptomycin against tuberculosis can be explained by the unusual compartmentation of the mycobacterial phagosomes in an early endocytic (barely acidic) stage, allowing newly ingested material, including antiobiotics, to reach it.

A recent study has also provided more direct evidence that aminoglycosides can act when *Salmonella* is located in a not fully acidic phagosome [7]. Similarly, Maurin and Raoult [6] showed that raising intralysosomal pH by providing lysosomotropic reagents together with amikacin, leads to enhanced intracellular killing of *Staphylococcus aureus*, although this finding probably cannot be readily translated into clinical treatment.

Gentamicin is widely used as a reagent to kill non-ingested bacteria in experimental phagocytosis assays ("gentamicin protection assay"). A surprising finding was that gentamicin can also traffic retrogradely from the plasma membrane, possibly via the Golgi, into the ER [8]. This feature could be important for drug efficacy against multiplying *L. pneumophila* or *C. trachomatis*, although with current regimens, gentamicin is not very effective against intracellular *Legionella* [9], which casts doubt over the *in vivo* relevance of this unusual uptake route.

11.2.2
Beta-lactam Antibiotics

Beta-lactam antibiotics (e.g., penicillin, ampicillin, cephalosporins, amoxicillin) mimic D-alanyl-D-alanine and inhibit the synthesis of the bacterial peptidoglycan layer of bacterial cell walls. As Gram-positive bacteria possess a much more comprehensive peptidoglycan than Gram-negatives, β-lactam antibiotics are particularly suited to combat diseases caused by Gram-positive bacteria. In general, these antibiotics can cross plasma membranes reasonably well but they do not accumulate in cells (Table 11.1). Due to one or more carboxylic acid functions, they tend to accumulate in the cytoplasm rather than acidic endolysosomal compartments of treated cells [10].

11.2.3
Macrolides

Macrolides (e.g., erythromycin, azithromycin clarithromycin) also block bacterial protein synthesis on ribosomes. Macrolides accumulate very well inside cells [11] although with quite some variability (Table 11.1). Low pH in the endosome–lysosome continuum creates a motive force which can lead to 100-fold accumulation of a single base drug and an even 10 000-fold accumulation of dibasic drugs [12]. Efflux transporters of the MRP family have been found in macrophages.

11.2.4
Fluoroquinolones

Fluoroquinolones (e.g., ofloxacin, pefloxacin, enrofloxacin) inhibit DNA gyrase or topoisomerase. They accumulate well within mammalian cells, and evidence points to a major concentration in the cytoplasm [13], even though active efflux – probably by MRP – is fast (Table 11.1).

Intracellular pathogens do not exist exclusively intracellularly. Particularly fast-growing pathogens will lyse their host cells, be released into the extracellular space and invade other host cells. During this time period, these microbes are accessible to drugs even if those do not enter the inside of an infected mammalian cell. A drug combination effective against both stages of the pathogen will therefore be most effective. Nevertheless, the definition of the intracellular localization of a given pathogen is crucial to enable the rational design of new drugs that penetrate the appropriate compartment to unfold their microbicidal activity, albeit only at the intracellular level.

11.3
Many Roads Lead Into the Host Cell

There are several ways in which antibiotics can be taken up by mammalian cells (Figure 11.1). Lipid-soluble antibiotics, such as rifampicin, chloramphenicol or penicillin, have the disadvantage of being less soluble in body fluids, they diffuse through the plasma membrane with little restriction, while most charged and hydrophilic β-lactam antibiotics and aminoglycosides cannot. But there are alternative routes into the cell, such as the use of small molecule transporters, e.g., nucleoside transporters, which can also quickly transport antibiotics such as clindamycin or macrolides [3]. The expression levels of small molecule transporters, however, depend strongly on cell type, cell location and the physiological state of the cell. Another route of entry is clathrin-mediated pinocytosis, which brings antibiotics into the endosome–lysosome continuum (reviewed in [14]).

Drug concentrations in transport vesicles are similar to those in the extracellular space, yet compounds can accumulate in lysosomes over time, particularly if there is no lysosomal drug exporter. Macropinocytosis is normally a growth

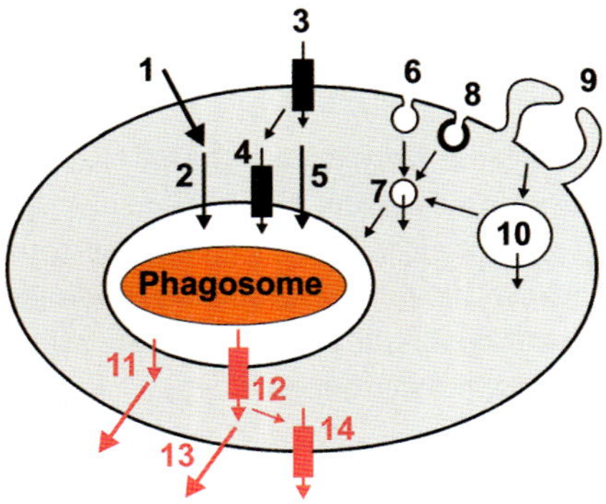

Figure 11.1 Uptake and loss of drugs by mammalian cells.Drugs can enter and leave mammalian cells via various routes. Shown is a cell with a microorganism (orange) contained in a phagosome. Entry (in black) is either by diffusion across the membrane (1) or by specific transporter proteins and channels (3), by clathrin-mediated endocytosis (6), by receptor-mediated endocytosis (8), or by macropinocytosis (9), leading to large fluid-filled vesicles (up to 1 μm in diameter; 10). Drugs that have entered by direct diffusion are usually expected to further diffuse into the phagosome space (2) and possibly out of it (11), depending on such factors as pH in the compartment and chemical modification of the drug in the phagosome. Substances taken up through receptors (3) might then diffuse into the phagosome lumen (particularly if modified in the cytosol so that charges are removed), or the may enter via a second transporter protein (4). Most endocytic pathways (6, 8, 9) lead into the early endosome and, finally, lysosome (7) from which substances may diffuse or be transported into the cytosol, while the drug-containing endocytic organelles themselves (7) may fuse with the phagosome and deliver their cargo. Pathways like those for entry (1, 3) can conversely also lead to drug loss from the cell (11, 13, 14). For more details, see text.

hormone-triggered uptake process feeding into the endosomal system and leading to vesicles sized ~1 μm. This is the cell's method of choice for ingesting large quantities and is likely the only endocytic process in which the soluble contents of the formed vesicle faithfully represent the cell environment; clathrin-dependent vesicles are too small to accommodate very large molecules and are, therefore, selective.

A "sophisticated" method of drug entry is receptor-mediated endocytosis, in which the drug is taken up with high specificity and in a highly regulated manner "on demand," as happens for epithelial growth factor (EGF) or transferrin [14]. Indeed, the antimycotic drug amphotericin B can use the low-density lipoprotein (LDL) receptor to enter mammalian cells [15]. Use of a collecting surface receptor can yield many-fold higher drug concentrations in transport vesicles than in the extracellular environment. Receptor-mediated uptake is an excellent candidate pathway to concentrate microbicidal drugs in the endosome–lysosome continuum for which there is proof of principle: the antimycobacterial agent para-aminosalicyclic acid has been coupled to maleylated bovine serum albumin (BSA). As BSA is highly glycosylated, it can be ingested by macrophages via the surface mannose receptor, bringing the substance into the endosome system in large quantities. Within the host cell, the pharmacologically active form of the drug is released [16].

11.4
And Off You Go . . .

While there are multiple ways in which a drug can enter a cell, there also exist sites where drugs can leave without having done their deed (Figure 11.1). Diffusion through the plasma membrane is a both-ways process. Exceptions to this rule are intracellular processes in which the drug is modified in the host cell cytoplasm after uptake, for example, by cleavage of an ester bond by esterases, creating charged carboxyl groups. Some drugs, such as erythromycin, can be rapidly exported by a carrier mechanism [17]. Finally, MRPs, such as small organic anion transporters, remove charged compounds from the cytoplasm into both late endosomes for exocytosis and directly across the plasma membrane [18]. As detailed above, inhibiting these transporters by specific compounds such as gemfibrozil can increase the retention period considerably. Interestingly, some intracellular pathogens intersect with the MRP pathway of transport; *L. mexicana*, for example, probably uses them to acquire small organic nutrients from the cytoplasm [19]. It can therefore be speculated that pathogens occupying similar niches as *Leishmania*, such as *Salmonella* and *Coxiella*, can be targeted by drugs pumped through these transporters. To further complicate matters, microorganisms themselves have efflux systems (e.g., the mycobacterial fluoroquinolone efflux system) [20].

11.5
"Small Molecules" Pave the Road to the Future

A sophisticated and rational approach to novel drug design against (intracellular) pathogens is represented by the generation of inhibitors to pathogen-specific virulence factors and/or physiological pathways active during infection. Small compound libraries can be screened for activity against specific microbial functions. Two recent publications [21, 22] have presented us with this very timely way of drug development against infectious diseases. These studies used an inhibitor designed for protein export from the *Yersinia pseudotuberculosis* type III protein secretion system, "compound 1" (*N*-(3,5-dibromo-2.-hydroxybenzylidene)-4-nitrobenzohydrazide), which reduces the growth of *Yersinia* in an HeLa cell infectivity model. Type III secretion systems are common protein export devices among pathogenic Gram-negative bacteria and can directly transfer specific virulence-associated proteins from the bacterial into the mammalian cell cytoplasm ("injection needles"; [23]). Clearly, these systems are important virulence factors. Studies like this one suggest that small molecule libraries represent a treasure trove for new drugs, with the possibility that compounds pulled out during specific screens can be further designed to enhance their specific efficacy, selectivity, targeting efficiency (and to reduce putative toxicity).

In a continuation of the above studies, the same inhibitor was also found to act against *Salmonella* and obligate intracellular *Chlamydia* as well, both pathogens with a type III secretion system [24, 25]. Interestingly, intracellular chlamydiae were not killed but were unable to multiply, an effect that was reversible upon drug removal.

Similarly, Alvesalo *et al.* [26] showed proof-of-principle for similarity-based virtual screening being useful in designing new antibacterial drugs: They identified small molecules against *Chlamydia pneumoniae* dimethyladenosine transferase. Pinkner and colleagues [27] rationally designed a small 2-pyridone to inhibit pilus biogenesis in uropathogenic bacteria, thereby inhibiting bacterial adherence and biofilm formation. Employing small molecule compounds against infections requires our knowledge of physiological pathways switched on during the intracellular life phase, such as the isocitrate shunt pathway in *M. tuberculosis* or the methyl citrate cycle in *Toxoplasma* [28–30], partly explaining the recently much increased interest in the physiology and the metabolism of intracellular pathogens.

In yet another variation of the small molecule theme, the identification of small avirulence molecules produced by the pathogens themselves [31] might open up the door to large-scale production of novel drugs that did not evolve through the algorithms of bioinformaticians, but through millions of years of biological evolution.

Small molecule compounds are also interesting drug candidates as toxin inactivators. Several low molecular weight compounds have been isolated which inactivate *Bacillus anthracis* lethal factor, a zinc-dependent metalloprotease; this is important because of the damage done by this potential warfare pathogen. Drug chemistry reaches from more classical protease inhibitors such as metal chelators to substances from random small molecule screens [32]. *Leishmania* and *Trypanosoma* produce virulence-associated membrane proteases, and specific inhibitors have been shown to be effective inhibitors of parasite growth in macrophages [33]. The investigation of small compounds may thus also benefit novel drug development against parasitic infections.

11.6
How Do You Come Up With A Successful "Small Molecule?"

Although screens for putative drugs against facultative intracellular pathogens can be carried out in broth cultures, this approach has the downside that the physiological state may be distinct from the more relevant intracellular state of life. Direct, intracellular assessment requires more sophisticated screening assays, which are obviously absolutely required when screening for drugs against obligate intracellular microbes such as *Chlamydia* or amastigote *Leishmania* parasites. More importantly, the assay read-out should not only be based on the numbers of pathogens, that is, colony forming units, but should also monitor the drug effects on the microbe's intracellular activities, such as toxin production.

For small compound screening a robust, fast, sensitive assay is needed to determine from a large number of samples whether the process of interest, for example, phagolysosome trafficking of a pathogen, is promoted by a certain drug. Screening requires a large-scale platform such as the 386-well format and an easily assessable assay, such as a fluorescence resonance energy transfer (FRET) assay for phagolysosome formation, or interaction with fluorescent proteins or fluorescent substrate conversion [34]. Using such screening procedures, a number of promising compounds, "hits," will be obtained. These hit compounds will be tested in an independent second assay for the process in question, such as microscopic analysis of phagolysosome formation. The compounds that are still promising after this second round of screening are carried to the next level of analysis. Toxicity, bioavailability and pharmacokinetics will be assessed and, again, all the points listed above for conventional antibiotics (entry into cells, extrusion from cell, degradation in the host organism, etc.) will have to be considered here, too. There is also hope that developing strongly expressed reporter systems using fluorescence or bioluminescence proteins in combination with sensitive cameras will lead to robust *in vivo* imaging systems to accelerate drug screening in proper animal models for the different infections.

11.7
A Few More, Maybe Even Practicable, Ideas For the Future ...

Small molecule biology will probably make a significant contribution to the development of novel antibiotics in years to come. Yet the complexity of the problem allows for many more novel approaches. We can also learn more from nature by using naturally occurring antibiotics such as the mammalian defensins, or small antimicrobial peptides derived from invertebrates or plants or immunostimulatory lipids derived from arachidonic acid metabolism. Some other suggestions are outlined below.

11.7.1
Iron Depletion

Iron availability is a limiting factor for pathogen survival in host cells [35, 36]. However, iron is sequestered from the pathogens by iron-binding host proteins such as transferrin and lactoferrin. Pathogens, on the other hand, employ iron-scavenging small molecules (siderophores) which compete with host factors for iron binding [35]. A strategy can be envisaged in which drugs specifically target microbial iron-scavenging factors and keep them from collecting or delivering iron. Lipocalin 2 expressed by neutrophils and macrophages can scavenge bacterial and fungal siderophores with high affinity and seems to be able to recycle looted iron to host cells [37]. Similarly, an oversupply of the host with host iron-binding factors such as lactoferrin could help to sequester iron away from the microorganisms as shown for tuberculosis and *Pseudomonas* infections [38, 39]. In the latter case, lactoferrin inhibited biofilm formation, which is a hallmark of infection in cystic fibrosis patients. Similar sequestration strategies can be envisaged for other nutrients essential for intracellular pathogens.

11.7.2 Transport into Phagosomes by Receptor-Mediated Endocytosis

Another future strategy based on the above experiments with para-aminosalicylic acid/BSA application could be one in which the specific host cell compartment is targeted by drugs using a host carrier molecule that delivers them to the place of action. For example, the human serum protein transferrin could be coupled to antimycobacterial drugs. As transferrin travels through the mycobacterial phagosome [40, 41], intravenously administered transferrin/drug complexes could unload a drug at moderately low pH in this early phagosome (pH around 6.3) thus locally increasing the drug concentration within the mycobacterial niche. This would yield a relatively specific drug for tubercle bacilli and, maybe, *Ehrlichia*, as these are the only pathogens known to reside in transferrin-positive phagosomes [42]. Furthermore, the strategy of specific host cell binding and drug accumulation can be combined with liposome technology, for example, a ligand that specifically targets a cell surface structure (receptor–ligand or specific antibody) will deliver a drug-containing vesicle to the cell of interest.

11.7.3 Retrograde Transport of Antibiotics

Similar to the above strategy of coupling drugs to cell surface receptor ligands, one could envisage a strategy in which the nontoxin moieties of toxins such as cholera toxin [43], ricin or shiga-like toxin [44] could promote "retrograde" transport into the ER where the drug is released due to its more oxidizing conditions. *Legionella* is a putative target pathogens for such a strategy.

11.7.4 Messing with Membrane Integrity

Lipophilic compounds such as the alkylphosphocholines miltefosine, edelfosine and ilmofosine are not only antitumor drugs, but also active against intracellular parasites such as *Leishmania* and *Trypanosoma*. These compounds act as synthetic ether–lipid analogs and alter the vesicle trafficking within the host cell and may therefore disturb access to nutrients and parasite growth [45] and they also induce parasite apoptosis and necrosis [46].

References

1 Cao, C.X., Silverstein, S.C., Neu, H.C. and Steinberg, T.H., (1974) *J Infect Dis*, **165**, 322–328.

2 Rudin, D.E., Gao, P.X., Cao, C.X., Neu, H.C. and Silverstein, S.C. (1992) *J Exp Med*, **176**, 1439–1447.

3 Hof, H. (1999) in *Opportunistic Intracellular Bacteria and Immunity* (eds L.J. Paradise *et al.*), Plenum Press, New York, pp. 281–293.

4 Eagle, H., Levy, M. and Fleishman, R. (1952) *Antibiot Chemother*, **2**, 563–575.

5 Lambert, P.A. (2002) *Symp Ser Soc Appl Microbiol*, **31**, 46S–54.

6 Maurin, M. and Raoult, D. (1994) *J Infect Dis*, **169**, 330–336.

7 Menashe, O., Kaganskaya, E., Baasov, T. and Yaron, S. (2008) *Antimicrob Agents Chemother*, **52**, 920–926.

8 Sandoval, R.M. and Molitoris, B.A. (2004) *Am J Physiol Renal Physiol*, **286**, F617–F624.

9 Walz, A., Nichterlein, T. and Hof, H. (1997) *Zentralbl Bakteriol*, **285**, 431–439.

10 Carryn, S., Chanteux, H., Seral, C., Mingeot-Leclercq, M.-P., van Bambeke, F. and Tulkens, P.M. (2003) *Infect Dis Clin N Am*, **17**, 615–634.

11 Carlier, M.B., Garcia-Luque, I., Montenez, J.P., Tulkens, P.M. and Piret, J. (1994) *Int J Tissue React*, **16**, 211–220.

12 de Duve, C., de Barsky, T., Poole, B., Trouet, A., Tulkens, P. and Van Hoof, F. (1974) *Biochem Pharmacol*, **23**, 2495–2531.

13 Carlier, M.B., Scorneaux, B., Zenebergh, A., Desnottes, J.F. and Tulkens, P.M. (1990) *J Antimicrob Chemother*, **26** (Suppl. B), 27–39.

14 Mayor, S. and Pagano, S.R. (2007) *Nat Rev Mol Cell Biol*, **8**, 603–612.

15 Brajtburg, J. and Bolard, J. (1996) *Clin Microbiol Rev*, **9**, 512–531.

16 Majumdar, S. and Basu, S.K. (1991) *Antimicrob Agents Chemother*, **35**, 135–140.

17 Laufen, H. and Wildfeuer, A. (1989) *Drug Res*, **39**, 233–235.

18 Steinberg, T.H., Swanson, J.A. and Silverstein, S.C. (1988) *J Cell Biol*, **107**, 887–896.

19 Schaible, U.E., Schlesinger, P.H., Steinberg, T.H., Mangel, W.F., Kobayashi, T. and Russell, D.G. (1999) *J Cell Sci*, **112**, 681–693.

20 De Rossi, E., Ainsa, J.A. and a Riccardi, G. (2006) *FEMS Microbiol Rev*, **30**, 36–52.

21 Kauppi, A.M., Nordfelth, R., Uvell, H., Wolf-Watz, H. and Elofsson, M. (2003) *Chem Biol*, **10**, 241–249.

22 Nordfelth, R., Kauppi, A.M., Norberg, H.A., Wolf-Watz, H. and Elofsson, M. (2005) *Infect Immun*, **73**, 3104–3114.

23 Galan, J.E. and Wolf-Watz, H. (2006) *Nature*, **444**, 567–573.

24 Wolf, K., Betts, H.J., Chellas-Géry, B., Hower, S., Linton, C.N. and Fields, K.A. (2006) *Mol Microbiol*, **61**, 1543–1555.

25 Hudson, D.L., Layton, A.N., Field, T.R., Bowen, A.J. and Wolf-Watz, H. (2007) *Antimicrob Agents Chemother*, **51**, 2631–2635.

26 Alvesalo, J.K., Siiskonen, A., Vainio, M.J., Tammela, P.S. and Vuorela, P.M. (2006) *J Med Chem*, **49**, 2353–2356.

27 Pinkner, J.S., Remaut, H., Buelens, F., Miller, E., Aberg, V. *et al.* (2006) *Proc Natl Acad Sci USA*, **103**, 17897–17902.

28 McKinney, J.D., Höner zu Bentrup, K., Muñoz-Elías, E.J., Miczak, A., Chen, B. *et al.* (2000) *Nature*, **406**, 735–738.

29 Muñoz-Elías, E.J., Upton, A.M., Cherian, J. and McKinney, J.D. (2006) *Mol Microbiol*, **60**, 1109–1122.

30 Gould, T.A., van de Langemheen, H., Muñoz-Elías, E.J., McKinney, J.D. and Sacchettini, J.C. (2006) *Mol Microbiol*, **61**, 940–947.

31 Prunier, A.L., Schuch, R., Fernández, R.E., Mumy, K.L., Kohler, H. *et al.* (2007) *Microbiology*, **153**, 2363–2372.

32 Turk, B.E. (2008) *Curr Pharm Biotechnol*, **9**, 24–33.

33 McKerrow, J.H., Engel, J.C. and Caffrey, C.R. (1999) *Bioorg Med Chem*, **7**, 639–644.

34 Yates, R.M., Hermetter, A. and Russell, D.G. (2005) *Traffic*, **6**, 413–420.

35 Radtke, A.L. and O'Riordan, M.X. (2006) *Cell Microbiol*, **8**, 1720–1729.

36 Schaible, U.E. and Kaufmann, S.H. (2004) *Nat Rev Microbiol*, **2**, 946–953.

37 Flo, T.H., Smith, K.D., Sato, S., Rodriguez, D.J., Holmes, M.A. *et al.* (2004) *Nature*, **432**, 917–921.

38 Schaible, U.E., Collins, H.L., Priem, F. and Kaufmann, S.H. (2002) *J Exp Med*, **196**, 1507–1513.

39 Singh, P.K., Parsek, M.R., Greenberg, E.P. and Welsh, M.J. (2002) *Nature*, **417**, 552–555.

40 Clemens, D.L. and Horwitz, M.A. (1996) *J Exp Med*, **184**, 1349–1355.

41 Sturgill-Koszycki, S., Schaible, U.E. and Russell, D.G. (1996) *EMBO J*, **15**, 6960–6968.

42 Barnewall, R.E., Rikihisa, Y. and Lee, E.H. (1997) *Infect Immun*, **65**, 1455–1461.

43 Chinnapen, D.J., Chinnapen, H., Saslowsky, D. and Lencer, W.I. (2007) *FEMS Microbiol Lett*, **266**, 129–137.

44 Sandvig, K., Grimmer, S., Lauvrak, S., Torgersen, M., Skretting, G. *et al.* (2002) *Histochem Cell Biol*, **117**, 131–141.

45 Urbina, J.A. (1997) *Parasitology*, **114** (Suppl.), S91–S99.

46 Azzouz, S., Maache, M., Dos Santos, M.F., Sarciron, M.E., Petavy, A.F. and Osuna, A. (2006) *J Parasitol*, **92**, 877–883.

12
The Immune Response to Intracellular Pathogens

Daniel S. Korbel and Ulrich E. Schaible

12.1
General Introduction

Mammals have evolved elaborate defense strategies to tackle the constant stream of pathogenic organisms seeking to invade and exploit the host organism for their own benefit. Pathogenic microbes invade through epithelia (e.g., in the gut or lungs) or exploit vectors, such as mosquitoes or ticks, to access the host. After invasion, pathogens establish infection in specific niches within the host. Colonization of such sites is sometimes preceded or followed by systemic spread of the pathogen. Intracellular pathogens inhabit host cells temporarily or for their whole life cycle. Facultative intracellular pathogens, on the one hand, do not necessarily require the intracellular lifestyle for proliferation, though it may protect them from humoral defense mechanisms such as antibodies and complement or may facilitate access to rare nutrients. Obligate intracellular pathogens, on the other hand, require the host cell for provision of essential nutrients and/or energy, and can usually not be grown in artificial media (Table 12.1).

A continuous physical barrier – the skin and the linings of mucosal surfaces (e.g., in the respiratory, intestinal and urogenital tracts) – directly protects areas of the body that are exposed to the environment. This line of defense is supported by the secretion of molecules with potent antimicrobial activity. Many potential invaders fail to overcome this first hurdle; however, those which succeed face a powerful defense strategy known as the immune system.

Mammals possess an innate and an adaptive (acquired) immune system. The innate arm is responsible for rapid recognition of, and timely response to a microbial attack, thereby limiting the initial pathogen burden. Innate immunity also instructs the adaptive immune system and, after the onset of adaptive immunity, will continue to contribute to the host response. At later stages, innate immunity is controlled by the adaptive immune system. Compared with innate immunity, initiation of an acquired immune response requires a longer time period. However, when re-encountering the microbe, the acquired immune system mounts a much

Intracellular Niches of Microbes. Edited by Ulrich E. Schaible and Albert Haas

ISBN: 978-3-527-32207-7

Table 12.1 Overview of intracellular pathogens and their intracellular niches.

Pathogen	Disease	Major host cells	Intracellular localization	Host defense in protection
Intracellular bacteria				
Brucella	Brucellosis	Macrophages	ER-derived compartment (BCV[a])	T_H1 cells, $CD8^+$ T cells, IFNγ, TNFα, activated macrophages and dendritic cells
Burkholderia	Melioidosis	Phagocytes, other cells	Cytoplasm	Neutrophils, macrophages, T_H1 cells, IFNγ, TNFα
Chlamydia	Conjunctivitis, urogential infection, lymphogranuloma venerum, pneumonia	Epithelial and endothelial cells, macrophages, neutrophils	Inclusion bodies	T_H1 cells, IFNγ, $CD8^+$ T cells, antibodies?
Coxiella burnetii	Q fever	Lung parenchyma cells, macrophages	Phagolysosomal compartment (RCV[b])	T_H1 cells, IFNγ, antibodies
Ehrlichia	Ehrlichiosis	Macrophages, neutrophils	Early endosome/phagosome	$CD4^+$ Tcells, $CD8^+$ Tcells, IFNγ, activated macrophages, IgG2a antibodies
Francisella tularensis	Tularemia	Macrophages, neutrophils	Phagosome, cytoplasm	IFNγ, TNFα, T_H1 cells, $CD8^+$ T cells, antibodies?
Legionella	Legionnaires' disease	Alveolar macrophages	ER-derived compartment (LCV[c])	IFNγ, other type 1 cytokines, NK cells
Listeria	Listeriosis	Macrophages, hepatocytes	Cytoplasm	Neutrophils, NK cells, $CD8^+$ T cells, T_H1 cells, antibodies
Mycobacterium avium	Disseminated infection in the immunocompromised	Macrophages	Early endosome/phagosome	T_H1 cells, activated macrophages
Mycobacterium bovis	Bovine tuberculosis	Macrophages	Early endosome/phagosome	T_H1 cells, activated macrophages
Mycobacterium leprae	Leprosy	Macrophages, Schwann cells	Early endosome/phagosome; cytoplasm?	T_H1 cells, activated macrophages

Mycobacterium marinum	Disseminated infection in the immunocompromised	Macrophages	Early endosome/phagosome cytoplasm	T_H1 cells, activated macrophages
Mycobacterium tuberculosis	Tuberculosis	Macrophages, dendritic cells	Early endosome/phagosome; cytoplasm?	T_H1 cells, activated macrophages, cathelicidin, $TNF\alpha_1$, T_H17
Rickettsia	Typhus, Rocky Mountain spotted fever	Vascular endothelium, smooth muscle cells	Cytoplasm	$CD8^+$ T cells
Rhodococcus, Nocardia	Nocardiosis	Macrophages	Early endosome/phagosome	T_H1 cells, IFNγ, IgM antibodies?
Salmonella	Typhoid fever	Macrophages, fibroblasts, intestinal epithelial cells	Late endosome/phagosome (SCV[d])	Antibodies, T_H1 cells, activated macrophages, neutrophils
Shigella	Shigellosis	Intestinal epithelial cells	Cytoplasm	Antibodies (IgA), neutrophils
Protozoan parasites and fungi				
Babesia	Babesiosis	Erythrocytes	Cytoplasmic parasitophorous vacuole (PV)	Activated macrophages, T_H1 cells
Cryptosporidium	Cryptosporidiosis	Intestinal epithelial cells, cholangiocytes	Extracytoplasmic PV	Neutrophils, NK cells
Histoplasma capsulatum	Histoplasmosis	Macrophages, dendritic cells	Late endosome, phagolysosome	TNFα, $CD4^+$ T cells, $CD8^+$ T cells?
Leishmania	Leishmaniosis (visceral and cutaneous)	Macrophages, neutrophils	Late endosome, phagolysosome	T_H1 cells, activated macrophages
Plasmodium	Malaria	Erythrocytes, hepatocytes	Cytoplasmic PV	Antibodies, $CD4^+$ T cells, $CD8^+$ T cells, IFNγ, TNFα
Theileria	Theileriosis, East Coast fever	Macrophages, lymphocytes	Cytoplasmic PV, cytoplasm	$CD8^+$ T cells, other immune cells?
Toxoplasma gondii	Toxoplasmosis	Wide range of host cells	Cytoplasmic PV	Antibodies, T_H1 cells, NK cells

[a] BCV, *Brucella*-containing vacuole.
[b] RCV, replicative *Coxiella* vacuole.
[c] LCV, *Legionella*-containing vacuole.
[d] SCV, *Salmonella*-containing vacuole.

stronger and highly antigen-specific response, delivering vigorous antimicrobial activities and mediating inflammatory reactions. Such responses can cause damage to host tissues and, therefore, fine-tuned regulatory mechanisms keep the balance between effective immune protection and immunopathology. Many principles of anti-infection immunity are similar between infections with extra- and intracellular pathogens. This is not surprising since only very few pathogens (such as *Chlamydia* and *Rickettsia*) spend all their metabolically active life exclusively within host cells. Many intracellular pathogens, including *Toxoplasma gondii*, *Leishmania* and *Plasmodium* parasites, disrupt their host cells and become free-living before entering new cells. Pathogens such as *Mycobacterium tuberculosis*, *Listeria* and *Salmonella* species live and even grow predominantly in the extracellular space during the active disease stage.

12.2 Innate Immunity

> *The presence of the cat's smell may signal the imminent arrival of its claws and teeth, and avoiding these is far more important than the stress of an unpleasant odour.* Randolph M. Nesse and George C. Williams in *Why we get Sick*, 1994.

12.2.1 Recognition of Microbes by Innate Receptors

To mount an efficient and rapid response to novel invaders, the host organism has to identify them quickly. Cells of the innate immune system express an array of molecular sensors to distinguish between "self" (or "safe") and "foreign" (or "potentially dangerous") structures. The detection of "foreign" relies on invariant structural characteristics unique to microbe-derived molecules termed microbe-associated molecular patterns (MAMPs). Prominent examples of MAMPs are bacterial flagellins and lipopolysaccharides (LPS) of Gram-negative bacteria, neither of which is produced by the host organism. Pattern recognition receptors (PRRs) are germline-encoded host molecules (either membrane-bound or cytosolic) involved in the sensing of microbial ligands. PRRs are principally involved in the stimulation of phagocytic cells such as macrophages, polymorphonuclear neutrophils (PMNs) and dendritic cells (DCs) but are also expressed on other immune cells such as natural killer (NK) cells, B cells and eosinophils, as well as non-immune cells such as epithelial cells. Engagement of PRRs typically results in the induction of proinflammatory cytokines such as type I interferons (IFNα, IFNβ), interleukin 1 (IL-1), IL-6, IL-12, IL-18, interferon γ (IFNγ), tumor necrosis factor α (TNFα) and chemokines. Figure 12.1 provides an overview of important PRRs, their ligands and their cellular localization.

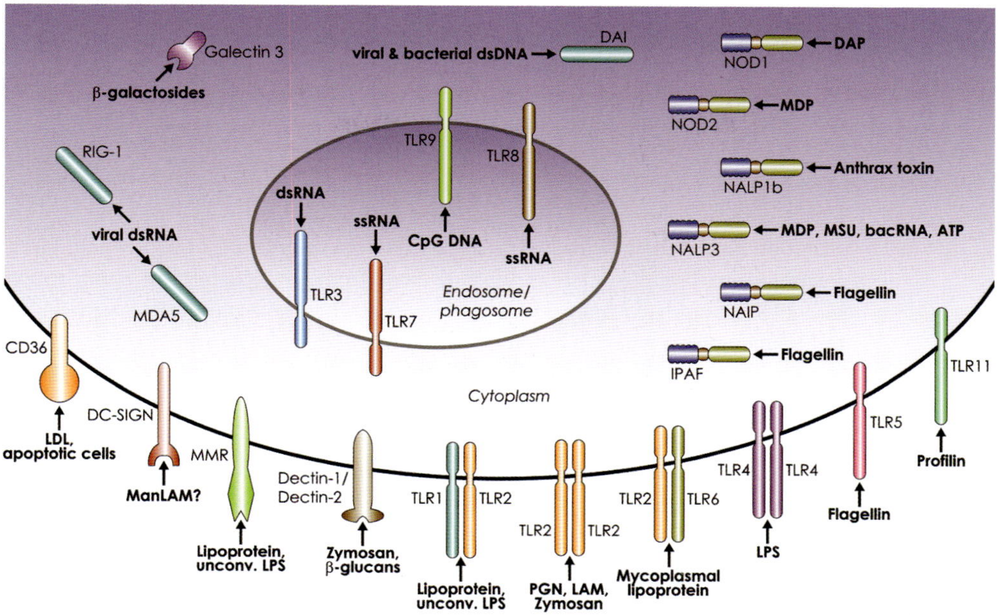

Figure 12.1 Innate recognition of pathogens. Germline-encoded pattern recognition receptors sense microbes by recognizing invariant structural characteristics called microbe-associated molecular patterns (MAMPs = **bold**). PRRs are found on the plasma membrane, in the cytoplasm and in endosomal/phagosomal compartments of phagocytes, dendritic cells, NK cells, B cells, eosinophils and other, non-immune cells. Depending on the nature of the ligand, engagement of PRRs can result in the induction of chemokines, type I interferons, IL-1, IL-6, IL-12, interferon γ (IFNγ), tumor necrosis factor α (TNFα) and other proinflammatory cytokines.ATP, adenosine triphosphate; CpG, unmethylated CpG-containing DNA motifs; DAP, diaminopimelic acid; dsDNA/RNA, double-stranded DNA/RNA; LPS, lipopolysaccharide; LAM, lipoarabinomannan; LDL, low-density lipoprotein; manLAM, mannosylated LAM; MDP, muramyl dipeptide; MMR, macrophage mannose receptor; MSU, monosodium urate; PfEMP1, *Plasmodium falciparum* erythrocyte membrane protein 1; PIMs, phosphatidylinositol mannosides; PGN, peptidoglycan; ssRNA, single-stranded RNA.

12.2.1.1 **Toll-Like Receptors**

The archetypal PRRs are the Toll-like receptors (TLRs). They have been identified in *Nematoda*, *Arthropoda*, *Echinodermata* and *Chordata* [1]. TLR-mediated recognition of microbes is a conserved pathway in innate immunity from invertebrates to mammals. In arthropods, which lack an acquired immune system, TLRs are essentially involved in the innate recognition of microbial infection. Shortly after the Toll protein was identified in the fruit fly as a central player in invertebrate immunity, the significance of the related TLR family in vertebrates was recognized ([2, 3]; and reviewed in [4]). In mammals, 11 TLRs have been characterized but more TLRs may join the family. TLRs are membrane-spanning molecules that share a common tripartite structure of a ligand-sensing domain, a transmembrane domain and a Toll/IL-1 receptor (TIR) domain that, upon ligation of the receptor, allows binding of soluble cytosolic signaling adapter molecules. While TLR1, 2, 4, 5, 6, 10 and 11 are expressed on the cell surface and recognize microbial proteins and glycolipids, other members of the TLR family sense nucleic acids and are localized within endosome and phagosome compartments (TLR7, TLR8, TLR9).

Microbial ligands recognized by TLR4, TLR5 and TLR9 include LPS, flagellin and unmethylated CpG DNA, respectively [5]. Most TLRs function as homodimers (e.g., TLR4) whereas some form heterodimers with other TLRs and thereby acquire different ligand specificities. The best-studied example is TLR2, which binds diacylated lipoproteins such as those derived from mycobacteria in combination with TLR1, triacylated *Mycoplasma* lipoproteins when heterodimerized with TLR6, and peptidoglycan when in a homodimer. TLR-mediated pattern recognition is not restricted to bacterial MAMPs [6]. The intracellular TLR3 binds double-stranded viral RNA. *Plasmodium* parasite DNA and, more recently, DNA derived from *Trypanosoma cruzi* have been described as ligands of TLR9. Profilin-like molecules from other apicomplexan parasites such as *T. gondii* have been associated with murine TLR11, a receptor that is nonfunctional in humans. So far, only few interactions between TLRs and fungal ligands have been described, such as TLR2 binding of zymosan. It should be noted that definite identification of novel TLR ligands relies on the availability of ultrapure ligand preparations, unequivocal test systems and meticulous analysis of receptor–ligand interactions.

The adapter molecule most commonly involved in the TLR signaling cascade is myeloid-differentiation 88 (MyD88). Binding of MyD88 to the TIR domain triggers activation and nucleus import of the transcription factor nuclear factor κB (NFκB) thereby inducing expression of a broad arsenal of antimicrobial agents and cytokines characteristic of the proinflammatory response. Some TLRs, such as TLR3 and TLR4, also recruit alternative adapter molecules which can trigger type I interferon-related responses [7, 8].

12.2.1.2 **NOD-Like Receptors**

Recently a novel family of intracellular PRRs, the NOD-like receptors (NLRs) were identified (reviewed in [9]). In humans, the NLR family has at least 20 members. Structurally, NLRs are characterized by their eponymous nucleotide-binding oligomerization domain (NOD), a leucine-rich domain that recognizes MAMPs, and a signaling domain that either triggers the proinflammatory NFκB and type I interferon pathways (e.g., NOD1 and NOD2) or induces inflammasome activation, that is, the proapoptotic caspases 1 and 3, and subsequent release of IL-1β and IL-18 (e.g., NALP3, IPAF) (reviewed in [10, 11]). The first members of the NLR family to be thoroughly characterized were NOD1 and NOD2 [12]. NOD2 and NALP3 sense muramyl dipeptides (MDPs) derived from bacterial peptidoglycans; the amino acid diaminopimelic acid specific for bacteria is recognized by NOD1. Recent studies suggest that anthrax toxin from *Bacillus anthracis* is recognized by NALP1, whereas flagellin can signal through NALP5 and IPAF. So far no interactions between NLRs and protozoa or fungi have been described; however, since the ligands for many of these receptors remain elusive, it is possible that some members of this family do recognize eukaryotic organisms.

12.2.1.3 Other Pattern Recognition Receptors

A number of PRRs are, in addition to signaling the recognition of MAMPs, also involved in phagocytosing microbes. Mammalian cells express a range of non-Toll-like PRRs on their surface which bind microbial structures, primarily carbohydrate residues. Scavenger receptors such as CD36 and SR1 and SR2 recognize a broad range of ligands including modified low-density lipoproteins (LDLs) and apoptotic cells but also bacteria and their products. CD36 facilitates phagocytosis of *P. falciparum*-infected erythrocytes by macrophages through binding of the *P. falciparum* erythrocyte membrane protein 1 (PfEMP1). LPS is bound by the soluble LPS-binding protein (LBP) which, in cooperation with the glycoprotein I (GPI)-anchored membrane molecule CD14, mediates engagement of and signaling through TLR4. Some PRRs are specifically expressed on phagocytes (e.g., macrophage mannose receptor (MMR) and DC-specific intercellular adhesion molecule 3-grabbing non-integrin (DC-SIGN)). MMR binds microbial glycans with terminal mannose, *N*-acetylglucosamine or fucose moieties, and it has been suggested that DC-SIGN, which is exclusively found on DCs and alveolar macrophages recognizes mycobacteria-derived mannosylated lipoarabinomannan (manLAM). It has been proposed that DC-SIGN and MMR engagement by mycobacteria may promote anti-inflammatory responses characterized by IL-10 secretion, which could counteract IFNγ function and therefore promote bacterial immune evasion. However, some reports suggest that DC-SIGN functions more as an adhesion receptor than as a signaling PRR.

Not all cytoplasmic PRRs are members of the NLR family. The evolutionarily unrelated CARD helicases (MDA5 and RIG1), DNA-dependent activator of IRFs (DAI) and galectin 3, for example, are also localized cytosolically and recognize double-stranded viral RNA, viral and bacterial DNA and β-galactosides, respectively.

Many microbes express multiple PRR ligands and therefore trigger a complex and characteristic signaling pattern. In some cases, mRNA microarray analyses or multi-cytokine/chemokine analysis of macrophages and/or DCs have revealed an innate response of surprising specificity upon encountering individual pathogens. It is now emerging that synergistic PRR signaling by multiple ligands leads to tailor-made immune responses specific to a given "molecular signature" [8, 13].

12.2.2 Phagocytic Cells

Engulfment of pathogens by phagocytes represents the earliest means of host defense in evolution. Phagocytosis was inherited from amoebal ancestors who used it to feed on bacterial prey. This strategy is still in use in primitive multicellular invertebrates such as *Porifera* and *Coelenterata*. Phagocytes in these organisms play a tripartite role: nutrition, tissue remodeling and antimicrobial defense. In higher multicellular organisms, the latter two have become the sole functions of phagocytes. In arthropods, hemocytes are the guarding phagocytes of the cellular innate defense system against infection. In mammals, neutrophils and macrophages are the two main phagocytic cell types.

12.2.2.1 Neutrophils

Polymorphonuclear neutrophils (PMNs) are among the first cells of the innate immune system to accumulate at sites of tissue injury and microbial entry. Morphologically they are characterized by their multilobed nuclei. Upon infection (or injury), PMNs migrate from the blood to the affected tissue within less than an hour after the initial insult, a response significantly more rapid than the recruitment of blood monocyte-derived macrophages (>8 h). A peculiarity of PMNs is their unusually short lifespan compared with rather long-lived macrophages or DCs. In humans, PMNs circulate only for a few days before becoming apoptotic and being eliminated by macrophages.

Although primarily associated with controlling infections by extracellular bacteria such as streptococci and staphylococci, PMNs are also significantly involved in infections by intracellular pathogens such as *Listeria monocytogenes*, *Salmonella typhimurium* and *Burkholderia pseudomallei*. PMNs were first described as potent phagocytes that engulf and kill microbes using a vast arsenal of antimicrobial substances stored in intracellular granules (reviewed in [14]). Killing occurs when PMNs release the content of their granules into the phagosome lumen. Fusion of the cytoplasmic granules with phagosomes and activation of their content relies on NADPH oxidase (NOX)-dependent production of reactive oxygen intermediates (ROIs) such as H_2O_2, O_2^- and others, and neutrophil myeloperoxidase (MPO) activity generating hypochloride (HOCl). Several types of cytoplasmic granules have been described, including azurophilic and gelatinase-containing granules. They contain a plethora of effector molecules that contribute to their antimicrobial activity: antimicrobial peptides such as the α-defensins, human neutrophil peptides (HNP) 1, 2 and 3, gelatinase, lysozyme and neutral proteases (elastase, cathepsin G, proteinase 3), and the vitamer-binding protein transcobalamin II. The iron-binding protein lactoferrin and the neutrophil gelatinase-associated lipocalin (NGAL), which can bind bacterial siderophores, can sequester iron, thereby preventing pathogens from using it.

In addition to their direct antimicrobial activity, PMNs have other functions (reviewed in [15]). As potent producers of chemokines and cytokines they are proficient initiators and regulators of immunity. Upon activation by microbial stimuli they produce numerous chemokines, including MCP-1, MIP-1α/β and IP-10, to attract monocytes, DCs, NK and T cells. PMNs can also produce cytokines such as IFNγ, TNFα and IL-4. These cytokines induce activation, maturation and proliferation of lymphocytes, NK cells, DCs and monocytes/macrophages. Recently it was reported that PMNs can also act as antigen-presenting cells (APCs). Furthermore, in the case of infection with *Mycobacterium bovis* BCG and *Salmonella*, PMNs can carry the microbes to the draining lymph nodes and promote efficient antigen presentation and T cell activation. It is not clear, however, whether this is due to direct presentation by PMNs or through cross-presentation involving DCs as ultimate APCs.

PMNs can serve as host cells to various intracellular microbes, albeit only temporarily due to their short lifespan. *Anaplasma phagocytophilum*, *Leishmania major*, *Chlamydia pneumoniae* and the agent of human granulocytic ehrlichiosis, *Ehrlichia granulosum*, have all been reported to delay spontaneous PMN apoptosis and to reside and replicate in PMN phagosomes. Interestingly, some reports suggest that pathogens such as *L. major* can exploit PMNs as "Trojan horses" in order to gain entry into macrophages, thereby promoting infection [16]. The induction of transforming growth factor β (TGFβ) secretion upon uptake of apoptotic PMNs has been suggested to allow a "silent" entry by deactivating the macrophages. A recent intravital imaging study revealed that *L. major* parasites are taken up *in vivo* by PMNs soon after transmission. However, PMNs promote the establishment of a productive infection rather than eliminate the parasites [17].

12.2.2.2 Macrophages

Macrophages represent the major class of tissue-resident phagocytes. Examples include alveolar macrophages in the lungs, Kupffer cells in the liver and microglia cells in the central nervous system. It is generally believed that bone marrow-derived monocytes in peripheral blood migrate to tissue sites along a gradient of chemotactic molecules and then differentiate into macrophages. Macrophages are central to tissue reconstruction and repair by homeostatic clearance of cells (including senescent erythrocytes and apoptotic PMNs), and to phagocytosis and elimination of microbes. Upon infection, proinflammatory chemokines attract peripheral blood monocytes to the affected tissue, where they mature to effector macrophages.

During their phagocytic uptake by macrophages, microbes are ingested into phagosomes that are then gradually acidified, enriched with acid hydrolases and eventually fused with lysosomes, a sequence of events very similar to that observed in PMNs. In macrophages, fusion of phagosomes with lysosomes is a slower process than in PMNs, where phagosome–lysosome fusion often happens before closure of the phagosomal cup. As a result of phagocytosis, the majority of microbes, in particular nonpathogenic ones and those with an extracellular lifestyle, are rapidly killed. As detailed in this book, many pathogenic intracellular microbes, however, have developed ways to circumvent these effector mechanisms and instead exploit macrophages as host cells. Employing distinct surface receptors, some microbes direct uptake without activating macrophages to their full microbicidal potential. Fixing of C3 by *M. tuberculosis* and binding to the complement receptors 3 (CR3) promotes internalization without production of the proinflammatory cytokine IL-12 and inducing ROIs. Engagement of DC-SIGN and/or MMR grants mycobacteria entry into host cells and, at the same time, induces production of anti-inflammatory IL-10 rather than proinflammatory cytokines. Many pathogens express molecules that inactivate ROIs, such as bacterial catalases and superoxide dismutases. Mycobacterial cell wall glycolipids such as trehalose dimycolate (TDM) and LAM have been suggested to capture ROIs and reactive nitrogen intermediates (RNIs). Some intracellular pathogens (e.g., *Leishmania* parasites and mycobacteria) manipulate the macrophage vesicular system and exploit phagosomes as their intracellular niche by arresting or deviating phagosome maturation while others (e.g., *Listeria monocytogenes, Shigella* and *Trypanosoma cruzi*) escape from phagosomes into the cytoplasm using pore-forming toxins such as listeriolysin O (LLO). Since macrophages can also act as antigen-presenting cells (APCs), the nature of the intracellular niche chosen by the pathogen determines which pathway is employed to process and present pathogen antigens and, ultimately, what type of T cell response is induced predominantly.

In order to avoid humoral immune effectors and to exploit them as a nutritional niche, intracellular microbes require macrophages to remain in a resting state. Macrophages activated by IFNγ and/or TNFα, on the other hand, become fully competent to kill and eliminate the microbes. Activated macrophages produce more ROIs and defensins than resting macrophages, and they increase expression of the RNI-generating nitric oxide synthase (NOS-2) and of small IFN-induced GTPases such as the autophagy-promoting enzyme Irgm1 (Lrg47). Downregulation of surface transferrin receptor and expression of indoleamine 2,3-dioxygenase (IDO) deprives activated macrophages of nutrients (iron and tryptophan, respectively), which otherwise would be accessible to microbes. The capacity of IFNγ to activate macrophages therefore makes this cytokine central to protective immunity against intracellular microbes, which with no exception fail to grow in activated macrophages. IFNγ is predominantly produced by NK cells and T cells and thus these immune cells are instrumental in the efficient control of intracellular pathogens.

12.2.3 Other Cells of the Innate Immune System

12.2.3.1 Dendritic Cells

Dendritic cells are the most important class of professional APCs. They are not directly involved in the elimination of pathogens but, by priming and activating T lymphocytes, DCs are pivotal in the initiation of adaptive immunity (reviewed in [18]). DCs are found in lymphoid organs, in peripheral blood and in nonlymphoid tissues. Tissue-specific DCs include Langerhans cells in the epidermis, dermal interstitial DCs and resident DCs in other organs. Based on their origin and function, they are classified as monocyte-derived myeloid DCs and plasmacytoid DCs of lymphoid origin [18].

Immature DCs engulf antigen by macropinocytosis or through receptor-mediated phagocytosis and endocytosis. Concomitant activation of DCs through PRR ligation and/or through cytokines such as TNFα or IL-1β initiates maturation of tissue DCs, thereby reducing antigen uptake and upregulating the chemokine receptor CCR7. Subsequently, chemokine-guided migration directs activated DCs to draining lymph nodes where they interact with T cells. Mature DCs increase surface expression of molecules involved in antigen presentation (including major histocompatibility complex (MHC) and the costimulatory molecules CD40, CD80 and CD86) and produce chemokines and cytokines such as IFNα/β, IL-12, and IL-18.

The role of DCs in immunity to infection is twofold. Upon migration to draining lymph nodes DCs present antigen to and stimulate differentiation of naïve T cells into "armed" effector cells of adaptive immunity – an ability unique to DCs. On the other hand, DCs are also engaged in cross-talk with innate immune cells, in particular with NK cells. Interaction between DCs and NK cells can increase NK cell activity (i.e., cytotoxicity against infected cells and secretion of cytokines) but also enhances DC activity such as antigen presentation and cytokine secretion. Finally,

DCs can also be exploited as host cells by intracellular pathogens. However, it appears that they are less favorable to supporting growth of microbes such as *M. tuberculosis* when compared to resting macrophages for not yet identified reasons.

12.2.3.2 **Natural Killer Cells**

In addition to their role in allograft rejection and killing of tumor cells, NK cells play an important role in immunity to viral, bacterial and protozoal invaders cells. NK cells are granular lymphocytes which, in humans, are classically defined by surface expression of the N-CAM molecule CD56 (CD49b/NK1.1/ASGM1 in mice) and absence of T cell receptor and CD3 expression [19–21].

NK-mediated lysis of virally infected host cells is well documented but NK cells can also kill cells infected with nonviral microorganisms. The cytotoxic mechanisms employed by NK cells are: (i) engagement of death receptors on the infected cell and initiation of apoptosis; (ii) release of the pore-forming molecule perforin, which facilitates entry of granzymes and proteases into the target cell, resulting in the degradation of histones and the induction of apoptosis; and (iii) perforin-mediated delivery of the antimicrobial peptide granulysin and subsequent killing of intracellular microbes (e.g., *M. tuberculosis, M. leprae, S. typhimurium* and *L. major* [22]). The physiological significance *in vivo* of NK-mediated cytotoxicity in bacterial and protozoal infections, however, is still a matter of debate. While direct activation of NK cells by interactions between NK cell receptors and microbial ligands have been described *in vitro*, accessory cells such as DCs and macrophages are required to fully activate NK cells in the majority of bacterial and parasitic infections [21].

The second major effector function of NK cells is the secretion of inflammatory chemokines and cytokines including IFNγ, TNFα and β, lymphotoxin α (LTα), granulocyte-monocyte colony-stimulating factor (GM-CSF) and the macrophage inflammatory protein (MIP-1α/CCL3). Thereby, NK cells contribute to the chemoattraction and activation of macrophages and T lymphocytes. In many infections, NK cells represent the initial source of IFNγ in the early phase of an immune response.

While stimulation by proinflammatory cytokines (e.g., IL-12, IL-18, IFNα) contributes significantly to their activation, NK cells also have the ability to directly sense transformed, foreign or infected cells and immediately respond to this stimulus. Engagement of an elaborate repertoire of receptors by an equally complex range of ligands enables NK cells to discriminate between "normal" host cells and abnormal, potentially dangerous cells without prior sensitization. A delicate balance of inhibitory and stimulatory signals mediated by NK receptors is maintained to fine-tune the activation state of NK cells. Transformed and infected cells often downregulate MHC surface expression. MHC class I-binding NK receptors (e.g., CD94-NKG2 receptors, killer immunoglobulin-like receptors in humans, Ly49 receptors in mice) can detect these changes and instruct NK cells to kill the target cells. The activating receptor NKG2D binds to retinoic acid early transcript 1 (RAE1) ligands and UL16-binding proteins (ULPB) on altered cells (including macrophages infected with *L. monocytogenes* and *M. tuberculosis*), thereby mediating their lysis. NK cell cytotoxicity has also been demonstrated against cells infected with *M. bovis, L. pneumophila, Chlamydiae, T. gondii* and *T. cruzi*. However, in many infections with intracellular bacteria (listeriosis, tuberculosis) and in particular in parasitic infections (leishmaniasis, Chagas disease and malaria), secretion of proinflammatory cytokines and IFNγ is the more relevant NK cell function for the efficient control of pathogen replication when compared with cytotoxicity (reviewed in [23]).

12.2.3.3 Basophils and Mast Cells

Another class of innate immune cells which patrol tissues are blood basophils and mast cells (MCs) of the gut, the lungs, the skin and other tissues that are exposed to the environment. Their main function was thought to be the rapid acceleration of acute inflammatory responses, particularly the propagation of allergic reactions. Upon activation, they release the content of their granules (including proteoglycans such as heparin and chondroitin sulfates; leukotrienes; prostaglandins and histamine; carboxypeptidase A and other neutral proteases) which in turn increase blood flow, recruit other immune cells and promote inflammation. It has recently emerged that, in addition to their role in allergy, both cell types are involved in immune responses to infection [24–27]. MCs engulf enterobacteria and are able to present their antigens to T cells [27]. Both MCs and basophils have also been identified as potent producers of proinflammatory and regulatory cytokines [24, 25].

The role of MCs in infection is reflected in their tissue distribution, with markedly more MCs found in more superficial tissue layers that come into direct contact with microbes and fewer in deeper tissue layers where they are less likely to encounter invading pathogens. MCs can be activated by microbial stimuli, such as streptococcal superantigens, various bacterial toxins and *L. major* promastigotes through PRRs, complement receptor (CR) and other receptors [28].

12.2.4 Innate Effector Mechanisms

12.2.4.1 Defensins and Other Antimicrobial Molecules

Defensins are small, cationic peptides with antimicrobial potential. In mammals, three families of defensins can be distinguished, α-, β- and θ-defensins (reviewed in [29]). The latter family is absent in humans. Defensins are expressed in a number of epithelial tissues and leukocytes and are induced after PRR ligation by a variety of MAMPs. *M. bovis* BCG triggers expression of human β-defensin 2 in lung epithelial cells, and resistance to *S. typhimurium* in the gut is conferred by the α-defensin HD-5. Granulysin, another small antimicrobial peptide, is found in human cytotoxic T lymphocyte (CTL) and NK cell granules and is effective against intracellular mycobacteria, *Leishmania* and *Salmonella* (reviewed in [22]). Other antimicrobial peptides include members of the cathelicidin family, such as LL37 which is microbicidal against *M. tuberculosis* and whose expression is induced by mycobacterial lipoproteins, albeit not directly. TLR1/2 signaling in human monocytes initiates expression of the vitamin D receptor (VDR) and the vitamin D-converting enzyme (CYP 27B1), which converts pro-vitamin 25-D3 into its active form, 1,25-D3. VDR in complex with 1,25-D3 binds to the promoter of the cathelicidin gene and thereby mediates its expression [30, 31]. In the skin, 1,25-D3 can be generated by sunlight. Lower concentrations of serum vitamin D have been observed in African Americans compared with Caucasians and may explain their higher susceptibility to tuberculosis [30]. This phenomenon is thought to be a result of the enhanced protection from sunlight afforded by the higher melanin content of darker skin. Interestingly, mutations in the VDR gene have also been linked to higher susceptibility to tuberculosis in certain African populations. In the early twentieth century, before the introduction of antibiotic treatment, exposure to intense sunlight in high mountainous areas such as Davos, Switzerland and Saranac Lake, USA was a routine treatment for tuberculosis patients. The link between exposure to sunlight, serum levels of vitamin D and better control of tuberculosis by LL37 could potentially explain the benefit of this treatment.

12.2.4.2 The Complement System and Acute Phase Proteins

The complement response is a cardinal feature of humoral innate immunity. It is a cascade of more than 20 small circulating proteins which sequentially activate each other by proteolytic cleavage in direct response to microbial stimuli such as surface carbohydrates, to collectin (C1q, soluble mannose receptor) or to antibody-coated microbes. The central event of the cascade is the conversion of the complement protein C3 into two fragments, C3a and C3b, which ultimately induce inflammation and increased phagocytic activity, and/or direct lysis of microbes, respectively (reviewed in [32]). Although mainly protected from complement by their chosen niche within the cell, intracellular pathogens such as *Leishmania* parasites and mycobacteria can exploit C3b in order to gain entry into phagocytes by CR3-mediated uptake without triggering ROI induction and release of proinflammatory cytokines.

Other components of the innate humoral response are the acute phase proteins (e.g., C-reactive protein and other pentraxins) which have direct antimicrobial effects. The acute phase response is triggered by proinflammatory cytokines, in particular IL-1, IL-6 and TNFα. While components of the complement system and the acute phase response are effective against a plethora of extracellular pathogens including *Escherichia coli, Enterococcus faecalis* and *Streptococcus pyogenes*, their relevance to the control of infection with intracellular microbes seems to be limited.

12.2.4.3 Immunity-Related GTPases

Immunity-related GTPases (IRGs) are small GTP-binding proteins of 47–48 kDa (p47 GTPases) which can associate with intracellular membranes (reviewed in [33]). In mice, over 20 separate IRG genes have been described [34], the most extensively studied of which are *Irgm3* (*Igtp*), *Irgm1* (*Lrg47*), *Irgd* (*Irg47*), *Irgb6* (*Tgtp*), *Irga6* (*Iigp*), *Irgm2* (*Gtpi*) and *Irgb10*.

IRGs are induced within minutes or hours of infection by IFNγ, and to a lesser extent by type I IFN in a STAT1-dependent manner. They are involved in the expression of IDO, natural resistance-associated macrophage protein 1 (NRAMP1), NOS2, NADPH oxidase and other molecules with microbicidal activity. Recent reports ascribe a role to IRGs in promoting phagosome maturation and phagosome–lysosome fusion by the means of macroautophagy (as discussed below) and intraluminal vesicle formation within endosomes/phagosomes [35].

There are striking differences between the various IRGs in terms of intracellular localization and their relative importance in different infections. *Irgm1* (*Lrg47*), for example, is localized to phagosomes containing *M. tuberculosis* [36] whereas *Irga6*, *Irgb6*, *Irgd*, *Irgm2* and *Irgm3* colocalize with *T. gondii*-containing vacuoles [37, 38]. Studies using IRG-deficient mice suggest that *Irgm1* and *Irgm3* are important in host defense against acute *T. gondii* infection, even though *Irgm1* does not appear to be localized to the parasitophorous vacuole of *T. gondii*. Lack of *Irgm1* and *Irgm3* renders mice more susceptible to *L. major* and *C. trachomatis* but *Irgm3* does not compromise host responses against *Listeria* and *Salmonella*. *Irgm1* is not only relevant for protective responses to *T. gondii* and *L. major* but also to *Listeria, Salmonella, M. tuberculosis* and *M. avium*. It has been suggested that *Irgm1* is involved in the IFNγ-induced maturation of phagosomes, probably by promoting autophagosome formation as demonstrated for *M. tuberculosis* [35, 39]. Activated macrophages from *Irgm1*-deficient animals cannot promote maturation of mycobacterial phagosomes [40, 41].

12.2.4.4 **Macroautophagy**

The vesicular transport of cytoplasmic content to lysosomes and its subsequent degradation and recycling is termed autophagy. Autophagy was first described as a physiological response, on the cellular level, to nutrient starvation. Later it was demonstrated that autophagy is a tightly regulated process that contributes to cell homeostasis by degrading senescent organelles and proteins (reviewed in [35]). There are two types of autophagy: (i) Macroautophagy describes the formation of double-membrane vesicles – autophagosomes – from a cup-shaped membrane structure (the phagophore) probably derived from the endoplasmic reticulum (ER). Similar to phagosomes and endosomes, autophagosomes subsequently acidify and mature, ultimately leading to autophagolysosome formation and degradation of the cytoplasmic cargo. (ii) Microautophagy is a process where lysosomes invaginate to generate cytosol-containing vesicles, which bud into the lumen and are degraded.

Intracellular microbes, including *L. monocytogenes* and *Shigella flexneri*, avoid lysosomal destruction by escaping into the cytoplasm but can be targeted by macroautophagy [42–45]. Rather more surprisingly, phagosome-resident pathogens, such as *M. tuberculosis, S. enterica, T. gondii* and *L. pneumophila*, can also become targets of the autophagic process [46–48]. Whereas autophagosomes may represent niches for survival for *L. pneumophila*, autophagy of phagosomes containing *M. tuberculosis* and *T. gondii* requires the IFNγ-dependent activation of the GTPases Irgm1 and Irgm3, respectively [35, 39, 46, 49]. Autophagy-mediated killing of mycobacteria is – at least in part – executed by ubiquitin-derived antibacterial peptides [50]. While IFNγ has been shown to promote autophagy, it is unclear whether IFNγ-mediated autophagy directly contributes to killing, for example, of mycobacteria and promotion of mycobacterial phagosome maturation, or whether it is an epiphenomenon concomitant to homotypic fusion between maturing phagosomes and autophagosomes. It should be noted that the amastigote stage of another intracellular parasite, *L. mexicana*, lives and proliferates in autophagolysosomes, probably to access purines which these parasites are unable to synthesize [51].

12.3
Adaptive Immunity

Innate immune mechanisms to defend the host against invading microbes exist even in evolutionary primitive invertebrates. The advent of the gnathostomata (jawed vertebrates) in evolution was accompanied by the introduction of an entirely novel concept of immune defense: adaptive immunity. Adaptive (or acquired) immune responses are characterized by specific antigen reactivity and immunological memory. The diversity of specific antibody responses, the highly refined mechanism of self/non-self discrimination by polymorphic major histocompatibility (MHC) molecules and diverse T cell receptors, and the elaborate organization of lymphoid tissues found in mammals are examples of this increase in complexity.

The chief mediators of adaptive immunity are B and T lymphocytes. They fall into several subpopulations with effector, memory and regulatory function, the most important of which are discussed in the following sections with special focus on their function in immunity to intracellular pathogens (Figure 12.2).

12.3.1
Cells of the Adaptive Immune System

12.3.1.1 Conventional αβT Lymphocytes

T lymphocytes (or T cells) originate from bone marrow precursor cells and mature in the thymus where they are positively and negatively selected on the basis of their affinity to MHC molecules and self-antigens, respectively. T cells are characterized by surface expression of the T cell receptor (TCR). The TCR is central to the specific recognition of an antigen presented to T cells, in the case of αβ T cells in the context of MHC class I and class II molecules. Antigens are usually presented to T cells as peptide fragments, generated from proteins by proteolytic processing, and bound to MHC molecules on antigen-presenting cells (APCs). A complex sequence of genetic recombination steps during thymic maturation of T cells generates a remarkable repertoire of TCR molecules of different specificity. The vast majority of mature T cells express a TCR heterodimer composed of α and β chains, while a minority make use of the alternative γ and δ chains. It is estimated that, in humans, the number of different αβ TCR specificities exceeds 10 000 000. In the periphery, they usually exist as either CD4- or CD8-single-positive T cells which concomitantly express the αβ TCR. CD4-expressing T cells are known as T helper (T_H) cells whereas $CD8^+$ T cells have cytolytic functions and are therefore called cytotoxic T lymphocytes (CTLs).

12.3.1.2 $CD4^+$ T_H1 and T_H2 Lymphocytes

The main function of T helper cells is to secrete cytokines, thereby instructing other immune cells. The IFNγ- and TNFα-producing T_H1 subset of $CD4^+$ T cells enhances activation of macrophages, induction of cytotoxic T cells and production of opsonizing antibodies. The T_H2 $CD4^+$ T cell subset secretes IL-4, IL-5, IL-6 and other T_H2 signature cytokines which promote differentiation of B cells into immunoglobulin

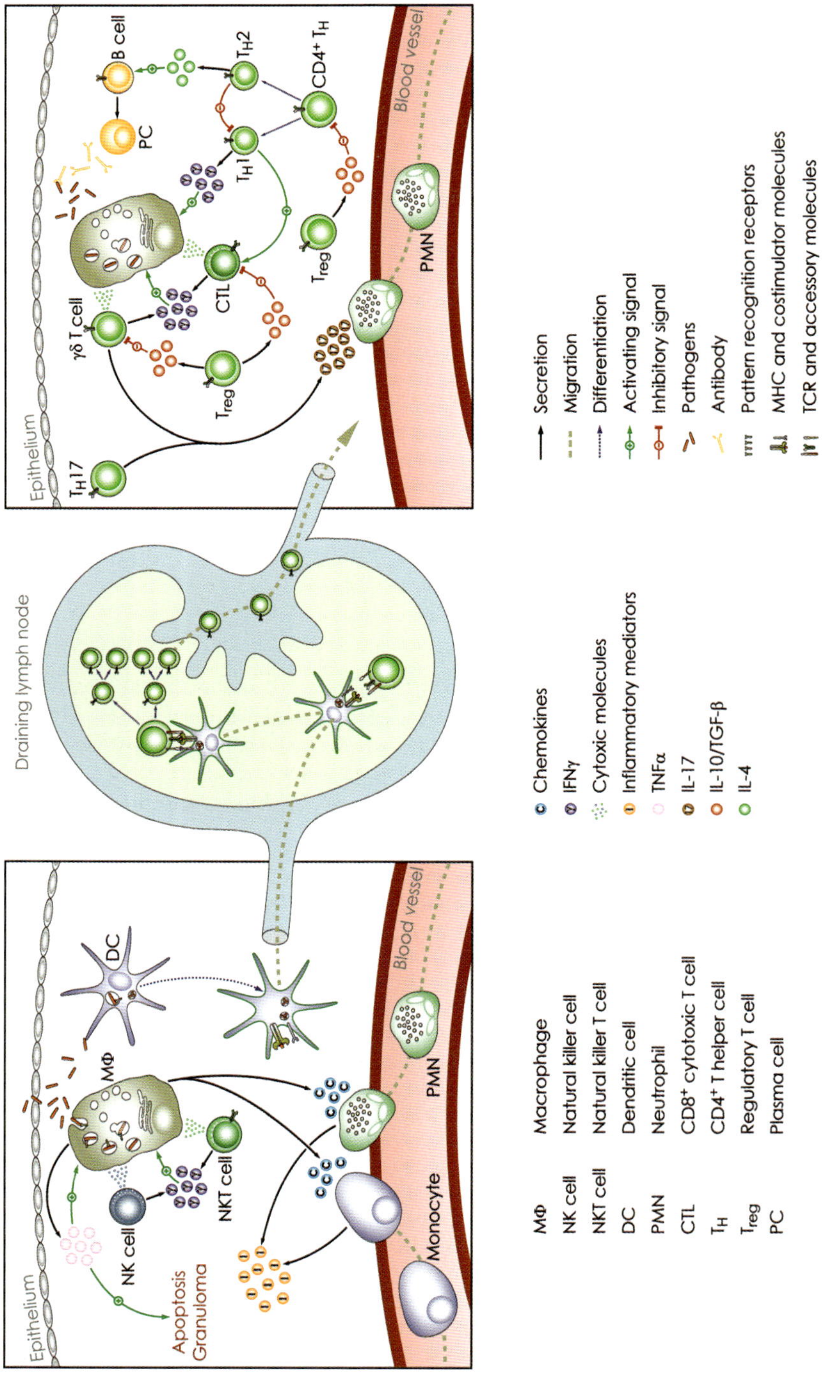
Epithelium
NK cell
Apoptosis
Granuloma
MΦ
DC
NKT cell
Monocyte
PMN
Blood vessel
Draining lymph node
Epithelium
T_H17
T_{reg}
γδ T cell
CTL
T_{reg}
T_H1
PC
B cell
T_H2
$CD4^+ T_H$
PMN
Blood vessel
MΦ Macrophage
NK cell Natural killer cell
NKT cell Natural killer T cell
DC Dendritic cell
PMN Neutrophil
CTL $CD8^+$ cytotoxic T cell
T_H $CD4^+$ T helper cell
T_{reg} Regulatory T cell
PC Plasma cell
Chemokines
IFNγ
Cytoxic molecules
Inflammatory mediators
TNFα
IL-17
IL-10/TGF-β
IL-4
Secretion
Migration
Differentiation
Activating signal
Inhibitory signal
Pathogens
Antibody
Pattern recognition receptors
MHC and costimulator molecules
TCR and accessory molecules

(Ig)-producing plasma cells, Ig isotype switching and affinity maturation and alternative macrophages activation (see below). Both T helper subsets are MHC class II-restricted, that is, their TCRs recognize antigen only when presented by MHC II molecules. Consequently, infection of MHC class II-negative cells such as hepatocytes, Schwann cells, endothelial and epithelial cells with intracellular pathogens such as *L. monocytogenes, M. leprae, Chlamydiae,* or *Salmonellae,* respectively, cannot be directly detected by $CD4^+$ T cells. For this reason, T helper cells specialize in responses to microbes residing in and antigen presented by professional APCs, that is, DCs, activated macrophages and B cells. In the case of pathogens dwelling in nonprofessional APCs, cross-presentation becomes important in the induction of a specific T helper response (see below).

T helper cells are instrumental in immunity against extracellular pathogens but also play a crucial part in the fight against intracellular infection. This concept is supported by a wealth of data from models of intracellular infection in $CD4^+$ T cell- or MHC class II-deficient mice – or animals lacking $CD4^+$ T cell-derived cytokines such as IFNγ – and the dramatically increased susceptibility to intracellular pathogens of patients with HIV/AIDS who have substantially reduced numbers of T helper cells. Since $CD4^+$ T cells respond to antigen derived from endocytic compartments, they are ideally placed to recognize intracellular pathogens that reside in phagosomes of MHC class II-expressing APCs. It is therefore not surprising that people with AIDS are particularly prone to developing severe disease caused by intracellular pathogens such as *M. tuberculosis, M. avium-intracellulare* or *Leishmania* spp., and there is increasing evidence in the literature that this may also be true for HIV-positive patients co-infected with *M. leprae* or *S. enterica.* Mice selectively depleted of $CD4^+$ T cells develop nonhealing cutaneous leishmaniasis caused by *L. major* [52], and severely immunocompromised SCID (severe combined immunodeficiency) animals that lack T cells altogether are able to, at least partially, limit parasite burden in *L. donovani* infection when repleted with $CD4^+$ T cells [53]. Control of tuberculosis strictly depends on T helper cells accumulating in granulomas. These cells are

Figure 12.2 Orchestration of the immune response against intracellular pathogens. Upon the encounter of a pathogen by a tissue macrophage the innate immune response is launched to rapidly tackle the invader. Proinflammatory cytokines and chemokines are released upon activation through pattern recognition receptors (PRRs). Monocytes and polymorphonucelar neutrophils (PMNs) are attracted along chemokine gradients to the site of infection, further perpetuating the inflammatory reaction. NK, NKT and γδ T cells produce cytokines including interferon γ, (IFNγ) and kill infected target cells. Dendritic cells (DCs) engulf the microbes, become activated through PRRs, differentiate and migrate to the draining lymph node to present antigens to T cells, priming them to become effector cells. T helper 1 (T_H1) cells produce IFNγ. IFNγ activates macrophages to express the full array of microbicidal effectors. Activated macrophages also produce TNFα, which can induce apoptosis and mediates granuloma formation. $CD8^+$ T cells (CTLs) also produce IFNγ, but also recognize and lyse infected cells. T_H2 cells promote B cell maturation and generation of high affinity antibodies by secretion of interleukin 4 (IL-4), and they control T_H1 cells. T_H17 and γδ T cells produce the PMN-attracting cytokine, IL-17. Finally, regulatory T cells (T_{reg}) limit exacerbation of the inflammatory response through the deactivating cytokines IL-10 and TGFβ.

essential for protective immunity to *M. tuberculosis* since animals lacking either $CD4^+$ T cells or MHC II expression quickly succumb to infection.

While T helper cell responses contribute to immunity against other important intracellular pathogens such as *S. enterica* and *Francisella tularensis* [54, 55], only limited data are available as to how crucial T helper cell responses are in the control of legionnaires' disease, which is partially due to the lack of an appropriate animal model [56]. Experimental evidence indicates that $CD4^+$ T helper cells also play a crucial role in promoting immunity against the cytoplasmic bacterium *L. monocytogenes*, although $CD8^+$ T cells are key mediators of the protective response in this infection [57].

A wealth of evidence from experimental models of infection has shown that $CD4^+$ T cells are also essential for the efficient control of blood-stage malaria. For example, SCID mice that lack T and B cells survived otherwise lethal doses of *Plasmodium chabaudi* when they received adoptively transferred $CD4^+$ T cells, and C57BL/6 animals which are genetically-resistant to *P. chabaudi* succumbed to infection when they were depleted of $CD4^+$ T cells. Malaria antigen-primed $CD4^+$ T cells provide help to B cells to mount an antimalarial antibody response, and protection against liver-stage malaria depends on $CD4^+$ T cell-derived IFNγ production as well as T helper-mediated $CD8^+$ T cell effector function.

Overwhelming evidence both from experimental models of infection and from clinical studies demonstrates that T helper cell-mediated protection from intracellular infection is largely associated with a T_H1-biased response and that overproduction of T_H2 cytokines and activation of B cells can correlate with exacerbated pathology. A very prominent example of this T_H1–T_H2 paradigm is the clinical presentation of leprosy, with a shift from a protective, T_H1 cytokine-mediated immune response that results in efficient granuloma formation and control of bacilli growth (tuberculoid leprosy) to the severe form (lepromatous leprosy) that is characterized by a strong T_H2 response, immunopathology and bacterial propagation and dissemination (reviewed in [58]). A similar dichotomy has been proposed for human tuberculosis. Equally, localized parasite containment and control of cutaneous *L. major* infection in mouse and human appear to be linked to vigorous T_H1 responses [52]. T_H2 cytokines, on the other hand, are predominantly found in patients with disseminated infection, and animals with a genetically hardwired bias towards T_H2 responses develop uncontrolled leishmaniasis. It should be noted, however, that the clear T_H1–T_H2 dichotomy described for *L. major* infection in mice does not hold true for other *Leishmania* species and even other *L. major* strains, demonstrating that immunity to intracellular pathogens is multifaceted, involving various different immune cell populations and effectors.

12.3.1.3 IL-17-Producing T_H17 Lymphocytes

T_H17 lymphocytes are the most recent addition to the family of T lymphocyte subsets. The cytokine repertoire that defines T_H17 cells is distinct from the ones that characterize T_H1 and T_H2 cells. It includes the eponymous inflammatory cytokine IL-17 and, in addition, TNFα, IL-6 and IL-22. IL-17 mainly targets PMNs, inducing their migration, maturation and activation. IL-12, IL-23 and IL-27 – mainly derived from DCs and macrophages – synergistically trigger production of IL-17 by T_H17 and $\gamma\delta$ T cells (see below). In tuberculosis, the latter T cells were identified as the main source of IL-17 [59]. Protection against tuberculosis in mice, as elicited by the vaccine strain *M. bovis* BCG, depends on a balanced response by IL-17-producing cells and T_H1 cells, which is promoted synergistically by IL-12p70, IL-23 and IL-27 [60]. An IL-27-dominated response, however, leads to elimination of the bacilli and premature death due to strong PMN-driven immunopathology.

Although these recent studies of experimental tuberculosis and also others on toxoplasmosis have implicated T_H17-derived IL-17 in both protection and pathology, our general understanding of the impact of T_H17 cell activation and IL-17 production in the context of intracellular infection is still poor.

12.3.1.4 Cytotoxic $CD8^+$ T Lymphocytes

The term "cytotoxic T lymphocyte" is used to primarily refer to CD8-expressing $\alpha\beta$ T cells, as their main function is destruction of infected or otherwise aberrant cells by cytolysis. However, other T cells (including $CD4^+$ ones) can also express cytolytic function. The recognition of MHC class I-restricted foreign antigen by their TCR enables CTLs to detect and destroy infected cells. CTLs are equipped with a vast arsenal of cytotoxic effector molecules very similar to the ones employed by NK cells [61] and described in more detail in the previous section.

There are two ways to kill a target cell. Upon TCR-mediated antigen recognition, CTLs deliver the content of their cytotoxic granules (granzymes, perforin, granulysin) to the cleft between CTL and target cell and directly lyse it. This cleft is at the immunological synapse formed by binding of MHC class I to TCR/CD8. Alternatively, Fas ligand expressed on the surface of the CTL can bind Fas on the target cell plasma membrane, which kills the target cell by apoptosis.

It is now widely recognized, that in many infections CTLs are not only cytotoxic but also produce proinflammatory cytokines, in particular IFNγ and TNFα. Because of their restriction to MHC class I-mediated antigen presentation, $CD8^+$ T cells can, similar to NK cells, monitor almost all cells of the body for infection (apart from MHC class I-negative erythrocytes and neurons). MHC class I molecules primarily present endogenous cytosolic antigens such as those produced during viral replication. Similarly, those intracellular pathogens which escape into the host cell cytoplasm (e.g., *L. monocytogenes*, *Rickettsia* div. spec. or *T. cruzi*) induce CTL responses, and protection against these infections depends – at least partially – on $CD8^+$ T cells [62, 63]. Nevertheless, $CD8^+$ T cells are also involved in adaptive immunity to and control of many intracellular pathogens dwelling within phagosomes (including *Brucella*, *M. tuberculosis*, *M. leprae*, *Plasmodium*, *Toxoplasma* and *Histoplasma* [64], reviewed in [65, 66]). $CD8^+$ T cells are primarily involved in visceral leishmaniasis, *L. donovani*, and low- but not high-dose challenge by *L. major* (reviewed in [66, 67]). In the murine brucellosis model $CD8^+$ T cells are critical to control infection, whereas $CD4^+$ T cells are dispensable [68]. Human CTLs can kill infected macrophages as well as intracellular tubercle bacilli, *Salmonella* bacteria and *Leishmania* parasites [69]. Similarly, mice lacking $CD8^+$ T cells or MHC class I molecules are more susceptible to tuberculosis, although the effect of $CD8^+$ T cell deficiency was much less drastic than that observed in animals lacking $CD4^+$ T cells [70]. Taken together, although predominantly involved in the elimination of virus-infected cells, $CD8^+$ T cells play a pivotal role – both as CTLs and as cytokine producers – in immunity to intracellular pathogens but their exact contribution needs further investigation.

12.3.1.5 γδ TCRT Lymphocytes

γδ TCR T cells are an unconventional subset within the T lymphocyte pool (reviewed in [71]). They express T cell receptors composed of γ and δ chains (rather than α and β chains) whose repertoire is far less diverse than that of αβ TCRs. γδ T cells exert both CTL and cytokine-secreting function and are generally believed to link innate and adaptive immune responses [72, 73]. Human γδ cells primarily recognize structurally similar nonprotein phosphate-containing antigens derived from metabolic pathways of bacteria and parasites [74]. These antigens include schizont stage-associated molecules of *P. falciparum* and an isopentenyl pyrophosphate (IPP) derived from *M. tuberculosis*. While recognition of such antigens seems to be independent of antigen-presenting molecules and probably even of APCs, a human γδ T cell subset was found to respond to bacterial isoprenoids in the context of CD1c. In contrast, murine γδ T cells recognize heat shock protein-derived peptides in the context of nonpolymorphic MHC molecules.

The role of γδ T cells in infection with intracellular pathogens is still under investigation. However, γδ T cells are expanded in response to infection as suggested by the increase of the phosphoantigen-reactive Vγ2/Vδ2 subset in humans in their first 20 years of life to frequencies of up to 50% of the peripheral T cell pool [74].

There is evidence that γδ T cells can act as early effector T cells in toxoplasmosis, tuberculosis and listeriosis by directly killing infected target cells and by producing T_H1-type cytokines such as IFNγ and TNFα. In experimental infections with *M. tuberculosis* and *M. bovis* BCG, γδ T cells appear to contribute to the formation of protective granulomas by producing IL-17 in an IL-23-dependent manner. Other data indicate that some subsets of γδ T cells are less able to protect from infection and may even contribute to pathology (e.g., in tuberculosis and malaria; [75]).

12.3.1.6 Natural Killer T Lymphocytes

Natural killer T (NKT) cells comprise a subset of lymphocytes that share characteristics with classic αβ T cell receptor (TCR) T lymphocytes and NK cells (reviewed in [76]). It has long been assumed that all NKT cells are restricted by CD1d, a nonclassical MHC class I-like membrane protein. However, it is becoming apparent that they are a heterogeneous cell population and, in some cases, can recognize alternative MHC molecules (reviewed in [77]). Classical NKT cells indiscriminately respond to the glycolipid α-galactosylceramide (α-GalCer) derived from a marine sponge. While its physiological relevance is uncertain, α-GalCer may have some potential as a therapeutic agent or as a vaccine adjuvant. NKT cells concomitantly secrete IFNγ and IL-4 in response to α-GalCer. It was recently demonstrated that human and murine NKT cells can respond to bacterial products in a CD1d-dependent manner, recognizing sphingolipids derived from LPS-free Gram-negative bacteria (such as *Ehrlichia* and *Sphingomonas* species) and mycobacterial phosphatidylinositol mannosides (PIMs) [78].

It has also been proposed that TLR-mediated signals such as those induced by LPS render CD1d-expressing APCs competent to activate NKT cells. This could probably be due to the presentation of endogenous glycolipid antigens [79]. Some NKT subpopulations can contribute to the control of protozoal and bacterial infections (e.g., leishmaniasis, toxoplasmosis and listeriosis). In other infections, NKT cells may not be essential for resistance. In experimental tuberculosis, for example, mice lacking CD1d or the invariant TCR Jα18 were not found to be more susceptible to infection than wild-type animals. NKT cells may even be responsible for infection-related immunopathology, as suggested by the observation that NKT cells contribute to splenomegaly and liver injury in some models of murine malaria. Upon microbial challenge, NKT cells are rapidly activated to secrete cytokines, placing them into the innate rather than the acquired immune armada. However, the general role of NKT cells in responses to infections with intracellular pathogens remains unclear.

12.3.1.7 **Regulatory T Cells**

Immune responses against invading microbes need to be vigorous and sustained to efficiently control infection. Many of the symptoms associated with infection, however, are not caused by microbial growth or toxin production but are immune-mediated. The immune system has regulatory mechanisms in place to regulate immune cells and dampen exacerbated inflammation and immunopathology. There is also increasing evidence from infection models that a tight regulation of immune responses can be necessary not only to prevent overreaction to infection but also for the development of an efficient memory T cell pool. In the case of experimental infections with *Leishmania* spp., this is achieved by reducing the parasite burden to a subsymptomatic level as complete elimination would prevent antigen exposure [80]. It is also possible that regulatory responses account for the observation that only a small proportion of those individuals latently infected with *M. tuberculosis* develop active tuberculosis by maintaining equilibrium between inflammatory effectors against mycobacteria and exacerbated pathology. Active disease and hence transmission of tubercle bacilli to the next host are most probably caused by immunopathology destroying the lower airways [81].

Immune regulation appears to be mediated mainly by anti-inflammatory cytokines such as transforming growth factor β (TGFβ), IL-10 and IL-35, and by the interaction of cytotoxic T lymphocyte-associated antigen 4 (CTLA-4) with its ligands. Relatively recently, several subsets of T lymphocytes were identified as the main source of TGFβ and IL-10 (reviewed in [82, 83]). Collectively known as regulatory T cells, these comprise mainly naturally occurring T_{reg} and inducible (or adaptive) regulatory T cells. Natural T_{reg} develop in the thymus and express CD4, the IL-2 receptor subunit CD25 and Forkhead box transcription factor P3 (FoxP3). The importance of natural T_{reg} is highlighted by the observation that mice deficient in $CD4^+$ $FoxP3^+$ cells are not viable and that humans develop a fatal multisystem autoimmune disease in the absence of a functional form of FoxP3. Similar to the role of T_{reg} cells in experimental leishmaniasis, $CD4^+$ $CD25^+$ $FoxP3^+$ T_{reg} cells impair protective T_H1-driven immunity in the murine tuberculosis model [84]. Infection with the murine malaria parasite *Plasmodium yoelii*, on the other hand, has shown that natural T_{reg} can also contribute to pathology. Their role in infections with other intracellular parasites remains to be established.

Other regulatory T cells are inducible ones and develop in the periphery. They are a heterogeneous population of antigen-specific T cells that include TGFβ-producing T helper 3 (T_H3) cells, IL-10-secreting T regulatory 1 (T_R1) cells, regulatory $CD8^+$ T cells and naïve $CD4^+$ T cells that were converted into $FoxP3^+$ regulatory cells. Inducible regulatory T cells, which are an IL-10 producing T_H1 subpopulation, were required for the control of *T. gondii* infection in mice. Inducible $CD4^+$ $CD25^-FoxP3^-CD127^+$ T_R1 cells confer protection against murine *P. yoelii* malaria in an IL-10-dependent manner. Evidence is accumulating that T cell-derived IL-10 is also important in tuberculosis and leishmaniasis. Although it has been known for decades that TGFβ is a crucial element of the immune response to many infections (e.g., in malaria), the contribution of TGFβ-secreting T_H3 cells to control immunopatholgy of infection with intracellular pathogens is still ill-defined.

12.3.1.8 **B Cells, Antibodies and Immunity Against Intracellular Pathogens**

In addition to being professional APCs, B lymphocytes are also the central element of the acquired humoral immune response. They develop in the bone marrow and are responsible for the production and secretion of antibodies in response to specific antigen recognition via the B cell receptor (BCR). Similar to the TCR locus, the genetic locus encoding the BCR undergoes complex recombination which generates a vast repertoire of B cells and antibodies of diverse antigen specificities.

Binding of their cognate antigen to and signaling through the BCR induces B cell activation. They are further stimulated by direct interaction with T_H2 cells and in response to T_H2-derived cytokines (e.g., IL-4). Fully activated B cells differentiate and secrete antibodies of the same antigen specificity as their respective BCR. Antibodies are molecules of the immunoglobulin family that can bind antigen with high affinity. They have the ability to neutralize microbial toxins (e.g., staphylococcal toxins), block invasion of host cells (e.g., by binding the *Plasmodium* merozoite surface protein 1) and target pathogens for destruction (i.e., by complement activation and Fc-receptor-mediated phagocytosis).

Unlike extracellular pathogens, pathogens that dwell within host cells are often not directly accessible to antibodies. Therefore, the protective role of B cells in the immune response to intracellular pathogens is generally limited to those periods when the respective microbe is extracellular as part of its lifecycle, for example, during the initial invasion of the host. There are, however, exceptions to this rule. Recent data have demonstrated, for example, that intracellular antibodies can limit growth of *L. monocytogenes* [85]. It should be noted, however, that during active infection a number of intracellular pathogens – including *Listeria*, *Salmonella* and *Mycobacterium* – live and proliferate extracellularly. Parasites with extracellular life stages such as *Plasmodium* are also vulnerable to antibodies. Hence, there is strong evidence that B cells contribute to protection from malaria, listeriosis and salmonellosis.

In tuberculoid leprosy, granuloma formation and control of mycobacterial growth are a result of a strong T_H1 response, involving the release of IL-1, IL-2, IL-6, IFNγ and TNFα. In patients with the more severe lepromatous leprosy, on the other hand, overproduction of the T_H2 cytokines IL-4, IL-5 and IL-10 and high antibody titers are associated with disseminated infection [58]. This suggests that B cell responses can also be part of the immunopathological outcome of infection.

12.3.2
Antigen Processing and Presentation to T Cells

Originally, the term "antigen" exclusively referred to molecules with the ability to induce the production of antibodies ("*anti*body *gen*erating"). Antigens *sensu strictu* are those molecules that can be bound by T cell receptors, B cell receptors or antibodies. Antigen classes range from peptide fragments and *N*-formylmethionine-containing peptides of *L. monocytogenes* (pentapeptides) to lipopeptides such as the *M. tuberculosis* mycobactin precursor and *M. leprae* LpK, carbohydrates and glycolipids including parasite-derived lipophosphoglycans (LPGs) and mycobacterial manLAM. Antigens also include small, nonpeptide phosphoantigens such as the microbial metabolites isopentenyl diphosphate (IPP) and dimethylallyl diphosphate (DMAPP).

While B cells often recognize free antigens in their native, three-dimensional conformation, recognition of antigen by T lymphocytes relies on the presence of antigen-presenting molecules. In order for a specific $\alpha\beta$ T cell-dependent immune response to occur, peptide antigen needs to be bound to MHC molecules and presented on the APC surface to the specific TCR. Classical MHC molecules can be either class I or class II. While MHC I is expressed by almost all somatic nucleated cells, expression of MHC II molecules is largely restricted to professional APCs such as DCs, macrophages and B cells. These cells endocytose and phagocytose antigen for intracellular processing by enzymatic cleavage and, upon binding to MHC class II molecules, presentation to T_H cells. MHC I-restricted antigen presentation allows "sampling" of the otherwise difficult to access intracellular environment by CTLs, thereby enabling them to detect infected cells. This is especially relevant in the case of virus-infected cells where pathogen replication and biosynthesis occur within the host cell cytoplasm. In the classical MHC I pathway, cytosolic proteins are degraded by the proteasome complex and thereby generated peptides are translocated into the lumen of the ER via the transporters associated with antigen processing (TAP-1 and 2) pores. Peptides of 8 to 10 amino acids are preferentially bound by MHC I molecules. MHC molecules are formed of three α chains noncovalently linked to β_2 microglobulin, with the binding groove for antigenic peptides between the α_2 and α_3 chain. In the ER, MHC I molecules are assembled with the help of chaperones (e.g., calreticulin, Erp56 and calnexin). Tapasin mediates the interaction of MHC I with the antigenic peptide transporters, TAP, and facilitates peptide loading. After loading onto newly synthesized MHC I molecules, further processing of the peptides can occur within the ER and Golgi apparatus. Finally, MHC I molecules are exposed on the plasma membrane for sampling by $CD8^+$ T cells.

MHC II molecules, on the other hand, bind antigenic peptides within the late endosomal/lysosomal system. MHC II molecules are transported to late endosomes with the help of the invariant chain. The invariant chain contains a peptide (CLIP) which serves as a spacer to block the antigen binding site between the MHC II α_1 and β_1 chain, while foreign antigen is not yet available. Once the MHC molecule is targeted to endosomes/lysosomes, the invariant chain is cleaved by cathepsins S and L. Finally, HLA-DM mediates exchange of CLIP for antigenic peptides of exogenous origin, and the MHC II–peptide complex is transported to the plasma membrane. MHC II-bound peptides are generated by acidic hydrolases such as cathepsins B, D and L within late endosomal/lysosomal compartments. MHC class II-presented peptides are longer, that is, 20–24 amino acids, than those bound to MHC I, and they often reach beyond the antigen-binding groove. Intriguingly, there is some evidence that association with MHC II molecules can protect antigen from further degradation by lysosomal proteases.

MHC II-bound antigens are derived from phagocytosed pathogens sampled from the extracellular environment for presentation to $CD4^+$ T cells. In most cases, this process is sufficient to induce protective immunity to extracellular pathogens. However, MHC II-restricted $CD4^+$ T cell responses are also vital for the fight against many intracellular pathogens since they often have extracellular stages during which they can be targeted by antibodies. Antigen derived from intracytoplasmic pathogens such as *L. monocytogenes*, *Rickettsia* spp. and *T. cruzi* are secreted during intracellular growth and can therefore be presented by MHC I and recognized by $CD8^+$ T cells. Nevertheless, immune control of these pathogens also depends on T helper cells, activated macrophages and even antibodies in addition to $CD8^+$ T cells. Similarly, viruses – which entirely depend on the transcriptional and translational machinery of the host cell – provide antigens for MHC I presentation to $CD8^+$ T cells. However, while virus-infected cells can be eliminated by $CD8^+$ T cells, antibodies and T helper cell-mediated macrophage activation are often at least as important to fully control a virus infection.

12.3.2.1 **CD1-Mediated Antigen Presentation**

MHC class I-like glycoproteins of the CD1 family are responsible for the MHC-independent presentation of microbial and endogenous lipid antigens (reviewed in [77]). CD1 molecules are primarily found on APCs such as DCs and some B cells. Human APCs express five CD1 isoforms, CD1a, b, c, d and e. CD1d is the only member of the CD1 family that is also found in mouse and rat. CD1d presents endogenous as well as exogenous lipids to NKT cells. While CD1a, CD1b and CD1c can present antigens to clonally diverse and either $CD4^+$, $CD8^+$ or $CD4^-/CD8^-$ double-negative $\alpha\beta$ T cells. CD1c is also a restriction element for $\gamma\delta$ T cells.

Endogenous lipid antigens presented by CD1 molecules are phosphoglycerides such as phosphatidylethanolamine and phosphatidylinositol, and glycosphingolipids (GM1, GD3). Most microbial CD1 ligands identified so far, such as mannosylated lipoarabinomannan (LAM; CD1b), PIM (CD1b, CD1d), arabinogalactan (AGL;

CD1b), glucose monomycolate (GMM; CD1b) and mycolic acids (CD1b), originate from pathogenic mycobacteria. Other CD1d ligands include a sphingolipid of *Ehrlichia* spp., *Borrelia burgdorferi*-derived α-galactosyldiacylglycerol and lipophosphoglycan (LPG) of *Leishmania donovani*. Bacterial isoprenoids, sulfoglycolipids or lipopeptides are ligands for CD1c and CD1a, respectively.

CD1 molecules have varying degrees of antigen specificity, with CD1c being restricted to lipids with one fatty acid, CD1b and CD1d binding lipids with two fatty acid chains, and CD1a being more promiscuous. CD1a shuttles between plasma membrane and early endosomal compartments and therefore binds lipids under non-acidic conditions. In contrast, lipids from late endosomes/lysosomes are sampled by CD1b, CD1c or CD1d, which all shuttle through those compartments using their C-terminal endosomal targeting sequence YXXZ. Loading of CD1 molecules with lipids involves lysosomal lipid transfer proteins. Saposin A is essential for loading of mycobacterial lipids onto CD1b, whereas loading of endogenous lipids as well as of α-GalCer onto CD1d is facilitated by ganglioside activator protein and saposins [86–88]. Saposins (A–D) are lipid transfer proteins in the lysosome which are involved in sphingolipid metabolisms [89]. Serum apoprotein 1 is also involved in the transfer of lipid antigens to CD1 molecules [90]. In contrast to other CD1 molecules, functional CD1e is not found on the cell surface but is primarily found in late endosomes and lysosomes of mature DCs. It is thought that CD1e is specifically involved in the antigen processing of exogenous lipids such as mycobacterial PIM, probably acting as a helper molecule for enzymatic processing [91].

The CD1 system of lipid antigen recognition is thought to have evolved specifically to recognize bacterial – if not primarily mycobacterial – infections. Enhanced numbers of CD1b- and CD1c-restricted T cells have been found in patients with skin tests positive for *M. tuberculosis* antigens, indicative of a recent exposure to mycobacteria. This suggests a role of CD1-restricted T cells in tuberculosis.

12.3.2.2 **Cross-presentation**

The mode of microbe-derived antigen processing and presentation is largely determined by the intracellular niche chosen by the particular microbe. The classical MHC I pathway confers antigen presentation to and activation of $CD8^+$ T cells in infections with pathogens that escape into the cytosol after phagocytosis, that is, *T. cruzi*, *Rickettsia* spp. and *L. monocytogenes*. Intracellular pathogens residing in noncytoplasmic compartments such as phagosomes or parasitophorous vacuoles cannot be accessed directly by the MHC I pathway. Therefore it seems intuitive to conclude that cellular immunity to these pathogens largely relies on the specific activation of $CD4^+$ T cell via the MHC II pathway. However, adaptive immune responses against many intracellular pathogens that do not reside in the cytosol also activate $CD8^+$ T cells through the MHC I pathway, a surprising phenomenon called cross-presentation.

A number of recent reports on $CD8^+$ T cell activation in response to phagosomal pathogens gave rise to three models of antigen cross-presentation by MHC I. One is that upon phagocytosis of microbes, ER and late endosome membranes fuse with the plasma membrane in the vicinity of the phagocytic cup. This allows membrane-spanning molecules of the proteasomal and MHC I antigen-processing/presentation machinery (e.g., TAP and Sec61) to access the phagosome and channel microbial antigen to the classical MHC I pathway. Recent studies, however, have strongly argued against a role of ER membranes in phagocytosis and their possible function in cross-presentation. An alternative model is based on the observation that antigenic peptides can be transferred from cell to cell through gap junctions formed by connexins. Yet a third model places apoptosis of the infected host cell in the center of the action (see Figure 19.4). Many intracellular pathogens promote apoptosis of their host cell, if not specifically through activation of death pathways, then by nutrient deprivation and overgrowth. Host cell apoptosis has been observed in infection with *M. tuberculosis*, *S. typhimurium* and *Leishmania* spp.

Induction of cell death promotes membrane blebbing, that is, formation and release of apoptotic vesicles containing microbial antigens and TLR ligands. These vesicles are engulfed by APCs, which are activated by he TLR ligands, and channel their antigenic cargo into the classical MHC I pathway to prime pathogen-specific T cells [92, 93]. Although important in $CD8^+$ T cell activation by phagosomal pathogens, cross-presentation/-priming is also involved in the activation of $CD4^+$ T cells and those restricted by CD1 molecules. Many intracellular pathogens such as *Chlamydia*, *Rickettsia* or *Toxoplasma* dwell inside nonprofessional APCs (e.g., epithelial and endothelial cells) incapable of expressing MHC II or costimulatory molecules. Other microbes actively interfere with antigen presentation (e.g., *M. tuberculosis*). Therefore, shuttling antigens to non-infected professional APCs through apoptotic blebs represents a vital pathway for cross-priming of the full array of different T cell subsets against intracellular pathogens.

12.3.2.3 Granuloma Formation

A granulomatous tissue reaction is formed during chronic infections with many intracellular bacteria. Granulomas are well-organized sites of inflammation and immune cell interaction, and they are an essential requirement for effective control of infection. The function of granulomas is containment of microbes and restriction of their replication. This may ultimately contribute to pathogen eradication. However, in many infections (e.g., tuberculosis and leprosy), the granuloma can also be the site of latent infection, which can eventually be reactivated.

Inflammation caused by microbial stimuli recruits phagocytes to the site of infection such as the skin in leprosy or the lower respiratory tract interstitium in tuberculosis. As more and more phagocytes – and subsequently lymphocytes – are drawn to the lesion by locally produced chemokines, granulomatous structures form around the infectious foci. Phagocytes – in later stages of granuloma formation primarily macrophages recruited from blood monocytes – form the center of the lesion, eventually becoming multinucleated epitheloid cells and foamy macrophages, while lymphocytes are mainly found in the periphery of the granuloma. It has been suggested that inside the granuloma intensive cross-talk between T cells and APCs – in particular via IL-2, IL-12, IFNγ and TNFα – leads to inhibition of microbial growth. In later stages of granuloma formation, the lesion is enclosed by a fibrous wall and eventually becomes necrotic. Thereby an anaerobic environment is created and the contained microbes are cut off from nutrient supply to further limit their growth. These conditions may however also compromise intragranuloma immunity.

Granulomas can cause broader tissue damage upon unrestrained spread and thereby contribute to pathology, as frequently observed in mycobacterial infections. In addition, overactivation of immune cells can lead to rupture of the lesion and dissemination of the pathogens. Tight, regulatory T cell-mediated control of the immunological events encompassing granuloma formation and maintenance is therefore of vital importance to keep the balance between antimicrobial defense and immunopathology.

12.3.2.4 Immunological Memory and Vaccines

Memory is one of the hallmarks of adaptive immunity. It is defined as the capacity of immune cells to respond more rapidly and more vigorously to re-exposure to previously encountered antigen compared to the primary response of naïve lymphocytes. Both B and T lymphocytes have the ability to differentiate into memory cells upon primary stimulation by antigen. These populations of memory cells can remain in a dormant state in lymphoid tissues (central memory cells) or in peripheral, extralymphoid tissues (effector memory cells). Central memory T cells have the capacity to replicate efficiently and to develop into effector cells. This differentiation process happens significantly more rapidly than in the case of the initial priming of naïve T cells. Recently activated central memory-derived cells migrate to the periphery as effector cells and subsequently undergo activation-induced cell death by apoptosis to avoid exacerbated immunopathology. Effector memory T cells, on the other hand, are already situated in the periphery. Following activation, they can either carry out effector functions and then die of activation-induced apoptosis or migrate into lymphoid tissues to become central memory cells.

Successful vaccination against infection requires induction of long-lasting immunological memory. Vaccines against viruses (such as measles and polio), extracellular bacteria (such as *Haemophilus influenzae* and *S. pneumoniae*) and bacterial toxins (such as diphtheria) have had a global impact on disease burden and thus provide evidence that immunization can be an effective tool in the fight against pathogens. The development of vaccines against intracellular pathogens, however, has yielded only limited success. The most prominent example is the attenuated vaccine against tuberculosis, BCG (bacillus Calmette-Guérin) derived from the causative agent of cattle tuberculosis, *Mycobacterium bovis*. Without a doubt, BCG has – together with improved housing, nutrition and medical infrastructure – contributed to the control of tuberculosis in industrialized countries and was proven to protect against childhood tuberculosis meningitis. However, protection against lung tuberculosis in adults – the typical clinical outcome of tuberculosis – is rather limited and varies tremendously from one geographical region to another. More importantly, after successful treatment, patients can become re-infected, suggesting that the development of immunological memory in response to primary infection is inadequate.

Current strategies to develop improved antituberculosis vaccines include the enhancement of the memory T cell pool against an array of *M. tuberculosis*-specific antigens by means of novel prime-boost strategies building on pre-existing BCG vaccination worldwide using recombinant bacterial or viral carriers [94]. Rational vaccine design also takes advantage of the knowledge of the intracellular behavior of the tubercle bacillus. Genetically modified mycobacterial strains with enhanced capacity to induce apoptosis have been developed to promote cross-presentation. Examples of such strains are recombinant BCG-expressing listeriolysin and an attenuated *M. tuberculosis* strain that is unable to secrete superoxide dismutase [95, 96].

Live attenuated strains are also employed as vaccines against typhoid fever (Ty2) or leishmaniasis but only induce a short-lasting protective effect. It can be hypothesized that the presence of the infectious agent such as dormant mycobacteria or *Leishmania* parasites may be required to continuously boost the immune system and thereby keep the infection at bay. This is corroborated by experiments in the *Leishmania* model showing that T_{reg} cells promote the development of a *Leishmania*-specific memory T cell pool by avoiding total elimination of the parasites [97, 98].

12.4 Conclusion

The immune system protects the integrity of multicellular organisms by recognizing pathogenic invaders and by immediately launching a broad (innate) counterattack. In higher vertebrates (gnathostomata), this process is followed by pathogen-specific, adaptive immune responses and the induction of immunological memory, which enables the immune system to mount a faster and more specific response upon reencounter of the same antigen. Vaccination against infectious agents seeks to induce long-lasting immunological memory without causing pathology. An adequate vaccine against intracellular pathogens, however, has yet to be developed. In light of the increasing socio-economic and public health burden imposed on humanity by diseases such as malaria, tuberculosis, leishmaniasis and Chagas disease, there is an urgent need for novel strategies to control intracellular infections. Protective immunity to such pathogens is determined by their localization within the host cell. Better knowledge of the intracellular niches, survival strategies and virulence factors of these pathogens – as outlined in this book – will improve our understanding of the immune mechanisms that may be effective against them, and will thus assist us in the rational design of novel therapeutic and prophylactic immune-based approaches.

12.5 Excursus – Immunity in Plants

Adaptive immunity is an evolutionary acquisition exclusive to modern vertebrates. Thus, host defense in invertebrates solely relies on innate mechanisms. It is becoming more and more apparent that similar innate mechanisms to sense and counteract microbial invaders exist outside the animal kingdom, that is, in plants. Some of these mechanisms appear to be evolutionarily very ancient (reviewed in [99]).

The genomes of many well-studied plants encode numerous PRRs that may have the potential to recognize MAMPs and virulence effectors. Prominent examples are receptor-like kinases on the cell surface and the large family of intracellular nucleotide-binding site leucine-rich repeat (NBS-LRR) pathogen resistance proteins which are homologous to NODs in animals. Of those PRRs which specifically respond to pathogen-associated virulence factors, the genome of *Arabidopsis* contains 140 potential candidate genes and that of *Oryza* (domesticated rice) around 500 different ones.

One of the best-studied examples of a membrane-spanning plant PRR is FLS2, the receptor for bacterial flagellin. In some of its functional domains, FLS2 bears striking similarity on the structural level to its mammalian counterpart, TLR5. As is the case for other plant PRRs, its degree of global sequence similarity to animal PRRs, however, is very limited.

In response to MAMP stimulation, membrane-bound PRRs use mitogen-associated protein kinase (MAPK) domains to trigger downstream signaling. Other PRRs – notably the intracellular pathogen resistance proteins – are believed to employ either N-terminal coiled-coil regions or Toll/IL-1 receptor (TIR) domains to signal. Interestingly, the signaling pathways downstream of plant TIR domains do not bear any resemblance to the classical TLR/TIR pathway in animals. Most importantly, instead of relying on the recruitment of NFκB-like molecules, plants utilize a complex network of WRKY transcription factors (reviewed in [100]). Ultimately, activation of the innate response system triggers calcium influx, effector gene expression, apoptosis, cell wall reinforcement, RNI and ROI.

The plant defense system may also bear dangers for symbionts such as intracellular rhizobia. It is beginning to emerge, however, that legume species can discriminate between beneficial symbionts and other bacteria, and rhizobial poly- and oligosaccharides seem to antagonize triggers of the host defense [101, 102].

With respect to the topic of this book, classical intracellular pathogens of plants have not been described to date. However, it has recently been demonstrated that *Salmonella* can readily invade cells derived from various tissues of *A. thaliana*. The implications of this finding may be far-reaching. For example, contaminated vegetables may still pose a risk to human health, even after thorough washing or surface sterilization [103].

Research on immunity in plants is a new but vastly developing field of great importance not only from an agricultural but also from a public health point of view. Also, future findings will certainly surprise us with more parallels between animals and our green and rooted cohabitants of the earth.

Acknowledgments

US thanks the present and past members of his lab for their great enthusiasm and scientific interest, and acknowledges the financial support from the Medical Research Council UK, the Wellcome Trust UK, the Bill and Melinda Gates Foundation and the Deutsche Forschungsgemeinschaft.

References

1 Kanzok, S.M., Hoa, N.T., Bonizzoni, M., Luna, C. *et al.* (2004) *J Mol Evol*, **58**, 442–448.

2 Medzhitov, R., Preston-Hurlburt, P. and Janeway, C.A. Jr (1997) *Nature*, **388**, 394–397.

3 Medzhitov, R. and Janeway, C.A. Jr (1998) *Curr Opin Immunol*, **10**, 12–15.

4 Takeda, K., Kaisho, T. and Akira, S. (2003) *Annu Rev Immunol*, **21**, 335–376.

5 Delbridge, L.M. and O'Riordan, M.X. (2007) *Curr Opin Immunol*, **19**, 10–16.

6 Gazzinelli, R.T. and Denkers, E.Y. (2006) *Nat Rev Immunol*, **6**, 895–906.

7 Noppert, S.J., Fitzgerald, K.A. and Hertzog, P.J. (2007) *Immunol Cell Biol*, **85**, 446–457.

8 Trinchieri, G. and Sher, A. (2007) *Nat Rev Immunol*, **7**, 179–190.

9 Sirard, J.C., Vignal, C., Dessein, R. and Chamaillard, M. (2007) *PLoS Pathog*, **3**, e152.

10 Lee, M.S. and Kim, Y.J. (2007) *Mol Cells*, **23**, 1–10.

11 Mariathasan, S. and Monack, D.M. (2007) *Nat Rev Immunol*, **7**, 31–40.

12 Strober, W., Murray, P.J., Kitani, A. and Watanabe, T. (2006) *Nat Rev Immunol*, **6**, 9–20.

13 Underhill, D.M. (2007) *Immunol Rev*, **219**, 75–87.

14 Segal, A.W. (2005) *Annu Rev Immunol*, **23**, 197–223.

15 Appelberg, R. (2007) *Trends Microbiol*, **15**, 87–92.

16 Laskay, T., van Zandbergen, G. and Solbach, W. (2008) *Immunobiology*, **213**, 183–191.

17 Peters, N.C., Egen, J.G., Secundino, N., Debrabant, A. *et al.* (2008) *Science*, **321**, 970–974.

18 Ueno, H., Klechevsky, E., Morita, R., Aspord, C. *et al.* (2007) *Immunol Rev*, **219**, 118–142.

19 Lodoen, M.B. and Lanier, L.L. (2006) *Curr Opin Immunol*, **18**, 391–398.

20 Moretta, A., Marcenaro, E., Parolini, S., Ferlazzo, G. and Moretta, L. (2008) *Cell Death Differ*, **15**, 226–233.

21 Newman, K.C. and Riley, E.M. (2007) *Nat Rev Immunol*, **7**, 279–291.

22 Clayberger, C. and Krensky, A.M. (2003) *Curr Opin Immunol*, **15**, 560–565.

23 Korbel, D.S., Finney, O.C. and Riley, E.M. (2004) *Int J Parasitol*, **34**, 1517–1528.

24 Min, B., Le Gros, G. and Paul, W.E. (2006) *Allergol Int*, **55**, 99–104.

25 Metz, M. and Maurer, M. (2007) *Trends Immunol*, **28**, 234–241.

26 Anthony, R.M., Rutitzky, L.I., Urban, J.F. Jr., Stadecker, M.J. and Gause, W.C. (2007) *Nat Rev Immunol*, **7**, 975–987.

27 Malaviya, R. and Abraham, S.N. (2001) *Immunol Rev*, **179**, 16–24.

28 Galli, S.J., Kalesnikoff, J., Grimbaldeston, M.A., Piliponsky, A.M. *et al.* (2005) *Annu Rev Immunol*, **23**, 749–786.

29 Lehrer, R.I. (2004) *Nat Rev Microbiol*, **2**, 727–738.

30 Liu, P.T., Stenger, S., Li, H., Wenzel, L. *et al.* (2006) *Science*, **311**, 1770–1773.

31 Liu, P.T., Stenger, S., Tang, D.H. and Modlin, R.L. (2007) *J Immunol*, **179**, 2060–2063.

32 Lambris, J.D., Ricklin, D. and Geisbrecht, B.V. (2008) *Nat Rev Microbiol*, **6**, 132–142.

33 Taylor, G.A. (2007) *Cell Microbiol*, **9**, 1099–1107.

34 Bekpen, C., Hunn, J.P., Rohde, C., Parvanova, I. *et al.* (2005) *Genome Biol*, **6**, R92.

35 Levine, B. and Deretic, V. (2007) *Nat Rev Immunol*, **7**, 767–777.

36 MacMicking, J.D., Taylor, G.A. and McKinney, J.D. (2003) *Science*, **302**, 654–659.

37 Butcher, B.A., Greene, R.I., Henry, S.C., Annecharico, K.L. *et al.* (2005) *Infect Immun*, **73**, 3278–3286.

38 Ling, Y.M., Shaw, M.H., Ayala, C., Coppens, I. *et al.* (2006) *J Exp Med*, **203**, 2063–2071.

39 Singh, S.B., Davis, A.S., Taylor, G.A. and Deretic, V. (2006) *Science*, **313**, 1438–1441.

40 Gutierrez, M.G., Master, S.S., Singh, S.B., Taylor, G.A. *et al.* (2004) *Cell*, **119**, 753–766.

41 Deretic, V. (2006) *Curr Opin Immunol*, **18**, 375–382.

42 Ogawa, M., Yoshimori, T., Suzuki, T., Sagara, H. *et al.* (2005) *Science*, **307**, 727–731.

43 Ogawa, M. and Sasakawa, C. (2006) *Autophagy*, **2**, 171–174.

44 Birmingham, C.L., Canadien, V., Gouin, E., Troy, E.B. *et al.* (2007) *Autophagy*, **3**, 442–451.

45 Py, B.F., Lipinski, M.M. and Yuan, J. (2007) *Autophagy*, **3**, 117–125.

46 Vergne, I., Singh, S., Roberts, E., Kyei, G. *et al.* (2006) *Autophagy*, **2**, 175–178.

47 Birmingham, C.L., Smith, A.C., Bakowski, M.A., Yoshimori, T. and Brumell, J.H. (2006) *J Biol Chem*, **281**, 11374–11383.

48 Andrade, R.M., Wessendarp, M., Gubbels, M.J., Striepen, B. and Subauste, C.S. (2006) *J Clin Invest*, **116**, 2366–2377.

49 Amer, A.O. and Swanson, M.S. (2005) *Cell Microbiol*, **7**, 765–778.

50 Alonso, S., Pethe, K., Russell, D.G. and Purdy, G.E. (2007) *Proc Natl Acad Sci USA*, **104**, 6031–6036.

51 Schaible, U.E., Schlesinger, P.H., Steinberg, T.H., Mangel, W.F. *et al.* (1999) *J Cell Sci*, **112** (Pt 5), 681–693.

52 Von Stebut, E. (2007) *Eur J Dermatol*, **17**, 115–122.

53 Stanley, A.C. and Engwerda, C.R. (2007) *Immunol Cell Biol*, **85**, 138–147.

54 Mastroeni, P. and Menager, N. (2003) *J Med Microbiol*, **52**, 453–459.

55 Elkins, K.L., Cowley, S.C. and Bosio, C.M. (2003) *Microbes Infect*, **5**, 135–142.

56 Neild, A.L. and Roy, C.R. (2004) *Cell Microbiol*, **6**, 1011–1018.

57 Zenewicz, L.A. and Shen, H. (2007) *Microbes Infect*, **9**, 1208–1215.

58 Modlin, R.L. (1994) *J Invest Dermatol*, **102**, 828–832.

59 Lockhart, E., Green, A.M. and Flynn, J.L. (2006) *J Immunol*, **177**, 4662–4669.

60 Khader, S.A. and Cooper, A.M. (2008) *Cytokine*, **41**, 79–83.

61 Smyth, M.J., Kelly, J.M., Sutton, V.R., Davis, J.E. *et al.* (2001) *J Leukoc Biol*, **70**, 18–29.

62 Pamer, E.G. (2004) *Nat Rev Immunol*, **4**, 812–823.

63 Bustamante, J.M., Bixby, L.M. and Tarleton, R.L. (2008) *Nat Med*, **14**, 542–550.

64 Lang, C., Gross, U. and Luder, C.G. (2007) *Parasitol Res*, **100**, 191–203.

65 Schaible, U.E., Kaufmann, S.H. *et al.* (2008) in *Clinical Immunology – Principles and Practice* (eds R.R. Rich, T.A. Fleischer, W.T. Shearer and H.W. Schroeder), Mosby Elsevier, Philadelphia, pp. 389–410.

66 Melby, P.C., Anstaed, G.M. *et al.* (2008) in *Clinical Immunology – Principles and Practice* (eds R.R. Rich, T.A. Fleischer, W.T. Shearer and H.W. Schroeder), Mosby Elsevier, Philadelphia, pp. 433–446.

67 Belkaid, Y., Von Stebut, E., Mendez, S., Lira, R. *et al.* (2002) *J Immunol*, **168**, 3992–4000.

68 Golding, B., Scott, D.E., Scharf, O., Huang, L.Y. *et al.* (2001) *Microbes Infect*, **3**, 43–48.

69 Stenger, S., Rosat, J.P., Bloom, B.R., Krensky, A.M. and Modlin, R.L. (1999) *Immunol Today*, **20**, 390–394.

70 Stenger, S. (2001) *Scand J Infect Dis*, **33**, 483–487.

71 Hayday, A.C. (2000) *Annu Rev Immunol*, **18**, 975–1026.

72 Moser, B. and Eberl, M. (2007) *Immunol Rev*, **215**, 89–102.

73 Konigshofer, Y. and Chien, Y.H. (2006) *Curr Opin Immunol*, **18**, 527–533.

74 Morita, C.T., Jin, C., Sarikonda, G. and Wang, H. (2007) *Immunol Rev*, **215**, 59–76.

75 Salerno, A., Sireci, G., Milano, S., Di Sano, C. *et al.* (2001) *Chem Immunol*, **79**, 87–98.

76 Kronenberg, M. (2005) *Annu Rev Immunol*, **23**, 877–900.

77 Barral, D.C. and Brenner, M.B. (2007) *Nat Rev Immunol*, **7**, 929–941.

78 Brigl, M. and Brenner, M.B. (2004) *Annu Rev Immunol*, **22**, 817–890.

79 Brigl, M., Bry, L., Kent, S.C., Gumperz, J.E. and Brenner, M.B. (2003) *Nat Immunol*, **4**, 1230–1237.

80 Belkaid, Y., Piccirillo, C.A., Mendez, S., Shevach, E.M. and Sacks, D.L. (2002) *Nature*, **420**, 502–507.

81 Hunter, R.L., Jagannath, C. and Actor, J.K. (2007) *Tuberculosis (Edinb)*, **87**, 267–278.

82 Tang, Q. and Bluestone, J.A. (2008) *Nat Immunol*, **9**, 239–244.

83 Rouse, B.T. and Suvas, S. (2007) *Expert Opin Biol Ther*, **7**, 1301–1309.

84 Kursar, M., Koch, M., Mittrucker, H.W., Nouailles, G. *et al.* (2007) *J Immunol*, **178**, 2661–2665.

85 Mittrucker, H.W., Raupach, B., Kohler, A. and Kaufmann, S.H. (2000) *J Immunol*, **164**, 1648–1652.

86 Winau, F., Schwierzeck, V., Hurwitz, R., Remmel, N. *et al.* (2004) *Nat Immunol*, **5**, 169–174.

87 Kang, S.J. and Cresswell, P. (2004) *Nat Immunol*, **5**, 175–181.

88 Zhou, D., Cantu, C. 3rd, Sagiv, Y., Schrantz, N. *et al.* (2004) *Science*, **303**, 523–527.

89 Kolter, T., Winau, F., Schaible, U.E., Leippe, M. and Sandhoff, K. (2005) *J Biol Chem*, **280**, 41125–41128.

90 van den Elzen, P., Garg, S., Leon, L., Brigl, M. *et al.* (2005) *Nature*, **437**, 906–910.

91 de la Salle, H., Mariotti, S., Angenieux, C., Gilleron, M. *et al.* (2005) *Science*, **310**, 1321–1324.

92 Winau, F., Weber, S., Sad, S., de Diego, J. *et al.* (2006) *Immunity*, **24**, 105–117.

93 Schaible, U.E., Winau, F., Sieling, P.A., Fischer, K. *et al.* (2003) *Nat Med*, **9**, 1039–1046.

94 Baumann, S., Nasser Eddine, A. and Kaufmann, S.H. (2006) *Curr Opin Immunol*, **18**, 438–448.

95 Grode, L., Seiler, P., Baumann, S., Hess, J. *et al.* (2005) *J Clin Invest*, **115**, 2472–2479.

96 Hoft, D.F. (2008) *Lancet*, **372**, 164–175.

97 Huster, K.M., Stemberger, C. and Busch, D.H. (2006) *Curr Opin Immunol*, **18**, 458–464.

98 Kaufmann, S.H. (2007) *Nat Rev Microbiol*, **5**, 491–504.

99 Ausubel, F.M. (2005) *Nat Immunol*, **6**, 973–979.

100 Eulgem, T. and Somssich, I.E. (2007) *Curr Opin Plant Biol*, **10**, 366–371.

101 Leitner, M., Kaiser, R., Rasmussen, M.O., Driguez, H. *et al.* (2008) *Phytochemistry*, **69**, 2029–2040.

102 Mithofer, A. (2002) *Trends Plant Sci*, **7**, 440–444.

103 Schikora, A., Carreri, A., Charpentier, E. and Hirt, H. (2008) *PLoS ONE*, **3**, e2279.

Part II
Selected Pathogens

IIA Vacuolar Bacteria

Intracellular Niches of Microbes. Edited by Ulrich E. Schaible and Albert Haas

ISBN: 978-3-527-32207-7

13
Afipia felis

Bianca E. Schneider and Albert Haas

13.1
Introduction of *Afipia felis* and its Role in Cat Scratch Disease

In 1991, *Afipia felis* was isolated as *the* cat scratch disease bacterium [1], yet its role in the pathogenesis of the disease has been questioned repeatedly since then. To discuss this, we will introduce the bacterium, give a brief historic overview of current *Afipia* research and place *A. felis* in the current picture of cat scratch disease.

13.1.1
Systematics

Afipia spp. are Gram-negative, oxidase-positive and nonfermentative rods, often with a single polar flagellum (for taxonomic details, see [1]). The genus belongs to the α2-subgroup of proteobacteria and is phylogenetically close to the genera *Bosea, Bradyrhizobium, Methylobacterium, Nitrobacter* and *Rhodopseudomonas*. There are now five named species (*A. birgiae, A. broomeae, A. clevelandensis, A. felis, A. massiliensis*), a candidate species (*A. lausanniensis*), *A. felis* genospecies A, plus three unnamed *Afipia* genospecies (1–3; Table 13.1). Although most *Afipia* isolates are from patients with diverse symptoms, only *A. felis* has been implicated in cat scratch disease. *A. felis* has never been described in samples from a human not having cat scratch disease symptoms. Recent research has led, however, to the isolation of *A. felis* (or species very closely related to it) from the environment, including several locations on Signy Island in the Antarctic region [4] where the bacteria are possibly present in amoebae rather than free-living [3]. *Afipia* strains have also been isolated from freshwater lakes, from river sediment, from freshwater biofilms, and from tap water (reviewed in [4]).

13.1.2
Cat Scratch Disease

Cat scratch disease is a self-limiting lymphadenopathy most commonly found in children and usually results from a scratch or bite either by an infected kitten or from

Intracellular Niches of Microbes. Edited by Ulrich E. Schaible and Albert Haas

ISBN: 978-3-527-32207-7

Table 13.1 The *Afipia* species.

Species	Source of type strain	Reference
Afipia birgiae	Drinking water supplies, likely within amoebae	[2]
Afipia broomeae	Human bone marrow, sputum	[1]
Afipia clevelandensis	Human tibial biopsy	[1]
Afipia felis	Cat scratch disease lymph node aspirates	[1]
Afipia felis genospecies A	Drinking water supplies, likely within amoebae	[2]
Afipia genospecies 1	Human pleural fluid, fatal	[1]
Afipia genospecies 2	Human bronchial lavage	[1]
Afipia genospecies 3	Environmental water	[1]
Candidatus *Afipia lausannensis*	Co-culture hospital water sample with amoeba	[3]
Afipia massiliensis	Drinking water supplies, likely within amoebae	[2]

contaminated objects such as barbed wire. The hallmarks of cat scratch disease are enlarged, tender lymph nodes found under toughened, warm, red skin close to the area of the inoculation lesion, and occasionally the lymph nodes drain pus. In most immunocompetent patients, long-lasting swollen lymph nodes are the main symptom of the disease and the illness is mild. Most cases of cat scratch disease are self-resolving and do not require antibiotic treatment. About a third of patients with cat scratch disease have other general symptoms including fever, fatigue, loss of appetite, headache, rash, sore throat and an overall ill feeling. Immunocompromised individuals, however, may experience a dramatic and potentially life-threatening course of disease [5].

The association between cats and the disease was first made by the French physician Debré and colleagues in the early 1930s [6]. Accordingly, Debré named the disease "La maladie des griffes de chat" (cat scratch disease). However, the causative bacterial agent remained obscure.

In 1983, Wear and colleagues visualized pleomorphic microorganisms in lymph node biopsy specimens stained with the Warthin–Starry silver stain [7]. Five years later, English and colleagues (1988) cultivated a bacterium in a long-lasting procedure using lymph node material from patients with cat scratch disease, succeeding with 10 out of 19 samples. This motile, Gram-negative rod was subsequently named *Afipia felis* in honour of the Armed Forces Institute of Pathology (AFIP), where the initial isolation succeeded. In 1991, Brenner *et al.* [1] introduced six *Afipia* species from a broad range of clinical (and one environmental) specimens, including lymph node tissue, lymph node aspirate, bone, sputum, pleural fluid and bone marrow. The next year, *A. felis* was isolated from 10 clinical specimens with the use of a tissue-culture technique (listed in [8]), later a completely independent *A. felis* strain was isolated from a cat scratch disease case in Israel [9].

There was little doubt that this new bacterium, *A. felis*, was the causative agent of cat scratch disease, as the experiments fulfilled all of Koch's postulates, including the development of cat scratch disease-like symptoms after intradermal application of *Afipia felis* [5]. Unfortunately, the susceptible animal was a nine-banded armadillo, which is not a first choice experimental animal, but which probably was a good target for *Afipia*, as armadillos have an average body temperature in the low 30s centigrade. *Afipia* spp. multiply best at 25–30 °C and only extremely slowly at 37 °C. Nevertheless, although *A. felis* would not multiply in rodents, it can persist there for a very long time [5].

Soon after the initial isolation of *A. felis*, conflicting information emerged, suggesting that *Bartonella henselae* (then *Rochalimaea henselae*) was a more common cause of cat scratch disease [10, 11]. *B. henselae* is also an α-2 protobacterium with similar morphology to *A. felis* when viewed in Warthin–Starry stain samples [12]. Patients with cat scratch disease often have high serum titers of *B. henselae* [13].

A. felis as a cause of cat scratch disease may be missed in practice because most skin-testing preparations contain *Bartonella* antigens only [14], and even when animals are experimentally infected, antibody titers to *A. felis* are low [5], making serology a difficult tool for identification. Furthermore, polymerase chain reaction (PCR) assays for *A. felis* seem much less sensitive than those for *B. henselae* [15] although there are documented cases of PCR detection of *A. felis* [16, 17]. Nevertheless, it has now been established beyond doubt that *Bartonella* is the main cause of cat scratch disease.

13.1.3 *Afipia felis* as an Intracellular Bacterium in its Own Right

The first paper giving a more detailed analysis of the interaction between *A. felis* and mammalian cells was by Birkness and colleagues [8]. They demonstrated that *A. felis* can enter nonprofessional phagocytic cells and that it is, in principle, able to multiply in HeLa cells and possibly also in other epithelial and endothelial cell lines [8]. A further study demonstrated that *A. felis* localizes to a late endocytic compartment in some endothelial and epithelial cells but would not confirm multiplication in these cells at 37 °C [18], although the bacteria may exist in a "dormant" state. As for the phagosome identity in macrophages, see below.

Interestingly, *A. felis* and some of its very close relatives also persist and may multiply in aquatic amoebae [19] which may help *Afipia* to survive in drinking water [20, 21], sea water [4], and river sediment [4] since they are sheltered from bactericidal environmental conditions. As stated above, *A. felis* grows best at temperatures below 37 °C with an optimum at 25–30 °C, a temperature at which most amoebae are very comfortable, whereas many amoeba species do not survive temperatures above 30 °C. Infections of amoebae may actually explain the evolution of the interference of *A. felis* with macrophage phagolysosome formation, reminiscent of the situation with *Legionella pneumophila*, the causative agent of legionnaires' disease. For *Legionella*, amoebae were likely the "training ground" for the exposition to human alveolar macrophages, whose infection can be seen more like an accident [20] (Chapters 4 and 18). Amoeba-associated bacteria are generally implicated in community- or hospital-acquired pneumonia [22]. They colonize hospital water supplies and have previously been shown to be causally associated with nosocomial infections. *A. felis* has also been isolated from hospital water supplies in intensive care units [23] and hence is a potential cause of nosocomial infections.

13.1.4
Culturable or Nonculturable

Biopsies taken from lymph nodes of cat scratch disease patients can be laden with bacteria [5]. Still, the odds are small that bacteria can be readily cultured from such material [8] and this is true not only for cat scratch disease accompanied by *A. felis* but also that accompanied by *B. henselae* [24], suggesting that these bacteria might be "noncultivatable" once they have adapted to the intracellular environment. Birkness *et al.* [8] found that the number of *A. felis* in experimentally infected epithelial HeLa cells steadily increased when quantified by a histological stain. Surprisingly, though, when these preparations were plated on the usual buffered charcoal yeast extract (BCYE) agar on which the bacteria were grown before infection, most of the bacteria were not able to grow colonies. It would take the bacteria only a few hours to transform into a nonculturable state within the HeLa cells. This is different from the situation in macrophages: *A. felis* yields static live cell counts over the time course of 4 days when harvested from infected murine macrophage-like cells (infection at 37 °C) and plated onto nutrient agar [25, 26].

13.2
Uptake and Intracellular Compartmentation of *Afipia felis* in Murine Macrophages

Intracellular bacteria have been observed microscopically within endothelial cells and macrophages in clinical specimens from patients with cat scratch disease [5]. *A. felis* not only enters and survives in nonprofessional phagocytic cells, but also multiplies in macrophages, where it inhibits phagosome–lysosome fusion [27]. Recent research has helped to characterize the *Afipia*-containing phagosome in detail and has shed light on the cellular events involved in the biogenesis of this intramacrophagic vacuole. Importantly, the intracellular compartmentation of *Afipia* is already (co-) determined during uptake [28]. Changing the port of entry by for example, IgG-opsonization of *A. felis*, thus forcing the bacteria to enter by immunoglobulin receptor-mediated phagocytosis, renders the *Afipia*-containing phagosome accessible to the endocytic degradative pathway and hence to the bactericidal machinery of the macrophage, resulting in phagolysosome formation [26, 28]. In clear contrast, the majority of unopsonized *Afipia* enter macrophages into a phagosome that is disconnected from the canonical endosome–lysosome continuum, indicating that the port of entry is pivotal for the intracellular compartmentation of *Afipia* in macrophages. It should be emphasized that "canonical" refers to the classical endocytic degradation and recycling pathways. It is understood that recent studies have characterized clathrin-independent uptake pathways to some extent [29], but their relationship to the *Afipia* phagosome remains to be solved.

The following paragraphs will introduce the unusual compartmentation of *A. felis* inside macrophages, illuminate the uptake process and discuss how this determines compartmentation and the survival of *A. felis*. Finally, we will indicate how *Afipia*-containing vacuoles (ACVs) in macrophages differ from those in nonprofessional phagocytes.

13.2.1
The *Afipia felis*-Containing Phagosome – Out of Touch with the Endocytic System

In 1993, Brouqui and Raoult reported that the majority of ACVs did not fuse with cationized ferritin-labeled lysosomes in either human macrophages or the murine macrophage cell line P388D1. However, after this first attempt to shed light on the intracellular life of *A. felis*, several years passed until a more detailed report on the molecular nature and composition of the ACV was published [28]. This comprehensive analysis presented evidence that the vast majority of ACVs did not have any endosomal or lysosomal characteristics, thus being disconnected from the endosomal–lysosomal continuum. Confocal immunofluorescence microscopy of *Afipia*-infected macrophages revealed that established ACVs (2 h post infection) acquired neither of the prominent early endosomal maker proteins early endosomal antigen-1 (EEA1), Rab5 or transferrin receptor (TfR), nor any of the late endosomal/lysosomal marker proteins Rab7, mannose-6-phosphate receptor (M6PR) or lysosome-associated membrane protein 1 (LAMP1) (and others), demonstrating a lack of all characteristics of typical early and late endocytic compartments. This was further supported by the fact that ACVs were not accessible to fluid-phase markers such as ovalbumin Texas Red loaded into macrophage endo- or lysosomes before or after establishment of *Afipia* infection, suggesting that ACVs not only avoid contact with pre-existing endocytic organelles but also with endosomes formed after establishment of an infection.

Finally, analysis of purified ACVs for their content in various subcellular marker proteins showed that lysosomal hydrolases like cathepsin D were excluded from ACVs as well as the proton-pumping vacuolar ATPase which normally acidifies the phagosome lumen. Furthermore, isolated ACVs contained less lysosomal enzyme activity when compared to *Listeria innocua*-containing phagosomes which undergo normal phagosome maturation and served as a reference. Selective enrichment of endoplasmic reticulum (ER) or Golgi proteins on the ACVs could be ruled out by this approach as well. Taken together, this study not only revealed that ACVs prevent phagolysosome formation but that they are not early endosomal compartments either, suggesting that *Afipia* does not simply delay phagosomal maturation but rather establishes a compartment that does not interact with the degradative pathway. In line with this, it was shown that during uptake *Afipia* avoids any contact with the endocytic system by directly entering into an EEA1-negative and, hence, unusual compartment. A later study [26] underlined this previous notion that biogenesis of ACVs is typical from the moment of ingestion with the finding that the transferrin receptor (TfR), shuttling between plasma membrane and early endosomes, is absent from the newly formed ACV. The fact that *Afipia* enters macrophages via TfR-negative membrane domains into a TfR- and EEA1-negative early phagosome gave reason to elucidate the mode of uptake, which turned out to be a prerequisite for *Afipia* to bypass the degradative phagocytic pathway, thereby establishing its atypical intracellular niche inside macrophages (see below).

Like *A. felis*, *Bartonella henselae*, the main cause of cat scratch disease, establishes a non-endocytic compartment in macrophages 2 h after infection, with the *Bartonella*-containing vacuole (BCV) being negative for EEA1, TfR and LAMP1 [30].

BCVs are neither accessible to the fluid-phase marker dextran Texas Red loaded into the endosomal system nor to the lysosomotropic marker LysoTracker, which is specific for acidified compartments, further demonstrating the failure of BCVs to interact with compartments of the endosomal system 2 h after infection. However, unlike ACVs, lysosomal fusion is not completely avoided by BCVs since after 24 h, the vast majority of *Bartonella* is found in LAMP1-positive phagolysosomes [30]. In line with this phagolysosomal localization, most of the internalized *Bartonella* were dead by 24 h, as judged by electron microscopy and the quantification of viable bacteria from infected macrophages.

Similar to what was observed for *A. felis*, opsonization of *B. henselae* with anti-*B. henselae* antibodies before infection resulted in increased maturation of BCVs into phagolysosomes [30]. Thus, the port of entry influences intracellular location of *Bartonella*, further underlining the need to understand how entry determines the fate of intracellular pathogens in more detail.

13.3 The Port of Entry Paves the Way for Non-Endocytic Compartmentation

13.3.1 *A. felis* Enters via a Low-Efficiency Uptake Pathway

Understanding how *A. felis* manages to interfere with phagolysosome formation has also helped to understand its mode of entry. The first step leading to phagocytosis of a certain particle or microorganism is making contact with the plasma membrane of the phagocytic cell. Phagocytes are endowed with a variety of receptors such as mannose-, Fc-, Toll-like or complement receptors and, depending on the particle to be taken up, different receptor subsets are involved in binding and/or signaling activities that finally lead to internalization and certain intracellular trafficking events. The finding that *A. felis* establishes its non-endocytic vacuole during entry (see above) raised questions about whether binding to a particular plasma membrane domain or set of surface receptors might direct uptake into this compartment.

A prominent feature of *A. felis* uptake by macrophages is its relatively slow progression when compared to that of other microorganisms or to *Afipia* that are opsonized by specific monoclonal antibodies [26]. Immunoglobulin receptor-mediated uptake is a very fast and highly efficient process, leading to rapid internalization and the formation of tightly fitting phagosomes through a zipper-like mechanism. These tightly fitting *Afipia*-containing phagosomes undergo gradual maturation into phagolysosomes, which was shown by their colocalization with TfR during uptake and shortly thereafter, their accessibility to endo- and lysosomal dextran or ovalbumin, respectively [26, 28] and the acquisition of late endosomal/lysosomal LAMP1 [26]. In contrast, *Afipia* that were left non-opsonized were found loosely attached to the macrophage plasma membrane, entangled in very long cellular projections or membrane ruffles which eventually seemed to collapse back and fuse with the plasma membrane, thereby forming spacious *Afipia*-containing vacuoles. This process was very slow and relatively inefficient in terms of numbers of internalized bacteria. When compared with antibody-opsonized *Afipia*, with some 90% being internalized within 5 min of infection, only some 40% of non-opsonized *Afipia* were found intracellularly after 1 h. Uptake of antibody-opsonized *A. felis* was accompanied by strong actin polymerization, whereas production of filamentous actin beneath attached non-opsonized *Afipia* was rare and less pronounced. However, phagocytosis of either opsonized or non-opsonized *Afipia* strictly depends on the actin cytoskeleton, since macrophage treatment with cytochalasin D inhibited uptake almost completely. These observations suggested an uptake mechanism that was most likely not triggered by *Afipia* itself which probably hitchhikes an existing low-efficiency pathway.

13.3.2 The Uptake Process is Reminiscent of Macropinocytosis

As already mentioned, *A. felis* is taken up into a spacious vacuole (Figure 13.1), and this vacuole shrinks within minutes after formation, leading to a tightly fitting phagosome [26]. Such behavior is reminiscent of macropinocytosis, which is known to require actin polymerization and the formation of membrane ruffles which, when they collapse and fuse with the plasma membrane, form spacious vacuoles called macropinosomes. These can be specifically labeled with fluorescent high molecular weight dextran which shrinks within minutes, just as is seen with the *Afipia* phagosome. While macropinocytosis is a constitutive process in macrophages and even more so in dendritic cells, it can also occur in endothelial or epithelial cells when triggered by growth factors or other stimuli [31]. Some intracellular pathogens, for example, *Salmonella* and *Shigella*, trigger macropinocytosis by injecting effector proteins via a secretion system into the phagocytes [32–34]. Macropinosomes can be specifically labeled with fluorescent high molecular weight dextran which is selectively taken up via macropinocytosis due to its very large size. When added to the macrophage environment during infection, this probe was taken up into the same large phagosomes as *Afipia* and again, these large organelles shrank quickly (Figure 13.2) [26].

To further test whether *Afipia* is taken up by macropinocytosis, amiloride, a commonly used drug which inhibits macropinocytosis without affecting other endocytic mechanisms, was added. As predicted, amiloride abrogated endocytosis of the 70 000 MW dextran macropinocytic probe, which was taken up into the same compartment as *Afipia* in the absence of the drug. Again, these large organelles shrank quickly [26]. Importantly, amiloride-treated macrophages also failed to ingest *A. felis* without affecting the attachment of the bacteria to the macrophage plasma membrane [26]. On an ultrastructural level, it was shown that amiloride treatment of macrophages led to an accumulation of cell protrusions such as membrane ruffles, which were apparently inhibited in fusing with the plasma membrane to form macropinosomes. Hence, amiloride did not block the generation of membrane ruffles but rather their fusion with the plasma membrane and thus the closure of the macropinosomes. According to this, only very few afipiae were seen inside the

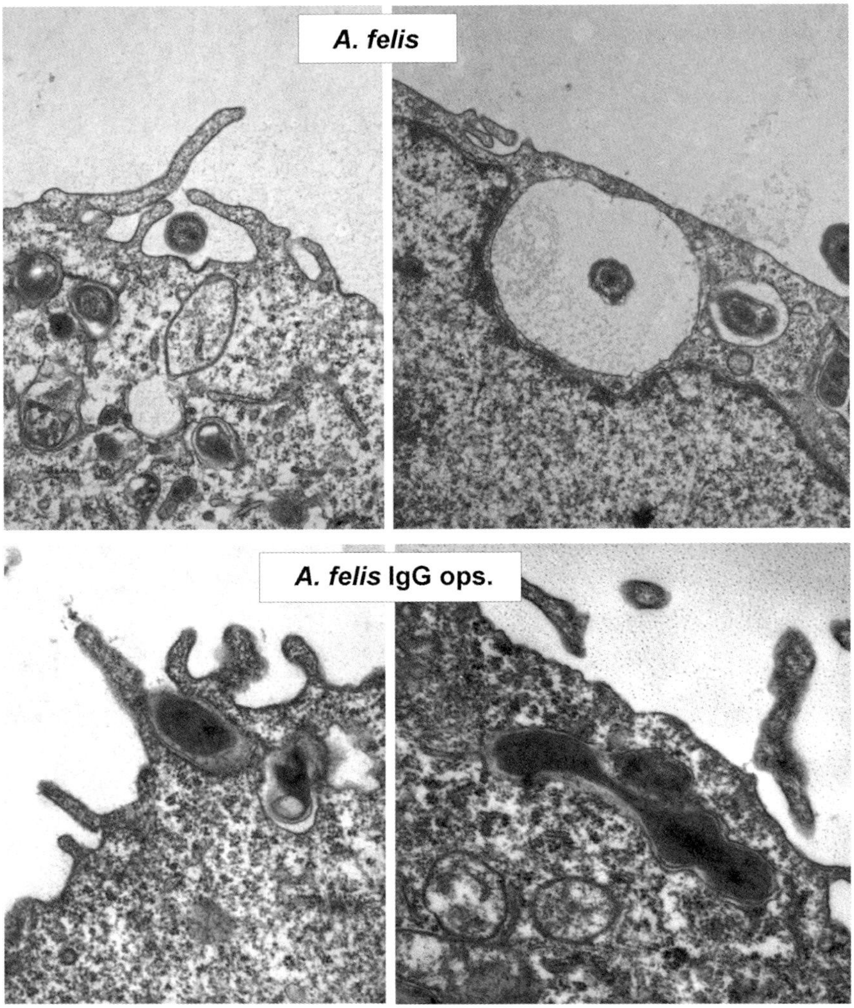

Figure 13.1 Transmission electronic micrograph of J774E macrophages infected with *Afipia felis*. While untreated *A. felis* were taken up by macropinocytosis into large, "loose" vacuoles (top), antibody-opsonized A. felis were taken up by "zipper-mode" phagocytosis into phagosomes with tightly apposed membranes (bottom). From [26] with permission.

treated macrophages but many became entangled between the numerous cellular projections. In contrast, uptake of antibody-opsonized *Afipia* progressed normally after amiloride treatment. Macropinocytosis is known to be dependent on phosphatidylinositol 3-kinases (PI3Ks), and so is phagocytosis. Consequently, both uptake of immunoglobulin-opsonized and non-opsonized afipiae were blocked when macrophages were treated with wortmannin [26], a classical inhibitor of type I PI3Ks [35]. On the other hand, dynamin-2 is known to play a crucial role in Fc receptor-mediated phagocytosis but is dispensable for macropinocytosis [36]. Consistently, dynamin-2

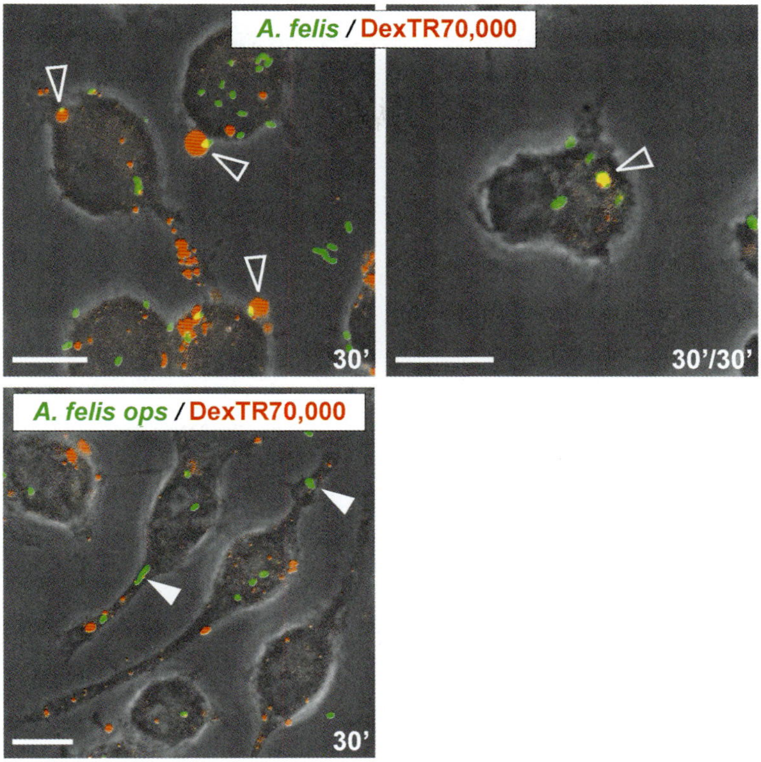

Figure 13.2 Confocal immunofluorescence analysis of J774E macrophages, infected with *Afipia felis* in the presence of red fluorescent 70 000 molecular weight dextran. Non-opsonized or immunoglobulin G-opsonized *A. felis* were added together with 70 000 molecular weight dextran for 30 min (30′) or for 30 min with a 30 min incubation in the absence of dextran (30′/30′) coupled to Texas Red. The samples were rinsed, fixed, bacteria were labeled in green, and the samples were analyzed by confocal microscopy. Untreated *A. felis* often colocalized to large surface-located dextran-filled bubbles, macropinosomes (open arrows), whereas antibody-covered bacteria did not (closed arrows). Bars: 10 μm. From [26] with permission.

was recruited to early phagosomes containing antibody-opsonized *Afipia* but never to those containing non-opsonized *Afipia* [26].

13.3.3
The Role of Lipid Microdomains in *Afipia* Uptake

A prerequisite for macropinocytosis is membrane ruffling [31, 37], and membrane ruffling in turn requires cholesterol [38]. Hence, membrane ruffling is more likely to occur in cholesterol-rich membrane domains and according to this, membrane ruffles contain many lipid microdomains [39]. Lipid microdomains are rich in cholesterol and also known as detergent resistant membranes (DRMs) due to their resistance to treatment with anionic detergents such as Triton X-100 at 4 °C. Apart from cholesterol, lipid microdomains contain high amounts of (glyco-)sphingolipids with mostly saturated fatty acid moieties, leading to tightly packed membrane domains that float in the surrounding membrane like rafts (hence the term "lipid raft"). Certain transmembrane and peripheral membrane proteins are permanently or transiently associated with these microdomains while others, such as TfR, are excluded. Uptake of *A. felis* occurs via TfR-free plasma membrane domains [26]. As TfR is a typical resident of membrane domains outside lipid rafts, this suggested that lipid microdomains were involved in the process. Since cholesterol is essential for the integrity of lipid rafts, the role of such domains in a certain cellular process can be assessed through selective membrane cholesterol depletion by drugs such as methyl-β-cyclodextrin (MCD).

Depleting the macrophage plasma membrane of cholesterol and subsequent *Afipia* infection resulted in significantly reduced internalization which could be restored by replenishing depleted host cell plasma membrane cholesterol before infection [26]. Similar results were obtained when some plasma membrane cholesterol was oxidized to cholestenone by the enzyme cholesterol oxidase. Such treatment can abolish the hydrogen bonds between cholesterol hydroxyl groups and neighboring sphingolipid amino groups and thus impair microdomain integrity. According to this, functional lipid microdomains are essential for internalization of *A. felis*. In addition, early ACVs colocalized with cholera toxin B (CTB), which, due to its pentameric structure, specifically binds to the ganglioside GM1 that is clustered in lipid rafts. However, when macrophages were treated with CTB prior to infection with *Afipia*, bacterial uptake was dramatically decreased, most likely because binding of CTB to lipid raft resident GM1 leads to endocytosis of these microdomains [40], thereby depleting the plasma membrane of domains required for *Afipia* to bind to the host cell surface.

Importantly, cholesterol depletion of macrophage plasma membrane not only inhibits *Afipia* entry but almost completely abrogates macropinocytic uptake of 70 000 MW dextran [26]. Schneider and colleagues were thus able to demonstrate that functional lipid microdomains are a prerequisite for macropinocytosis in macrophages, thereby bringing in line their findings that both blocking macropinocytosis and disrupting lipid microdomains significantly decreased *Afipia* uptake by macrophages. Hence, cholesterol depletion and amiloride treatment inhibit at least to some extent the same uptake process, which can be described as lipid microdomain-dependent macropinocytosis.

13.3.4 Uptake via Lipid Microdomain-Dependent Macropinocytosis Determines the Intracellular Fate of *A. felis* in Macrophages

Although uptake of *Afipia* is significantly reduced by cholesterol depletion, CTB binding or amiloride treatment, some 20% of afipiae are still taken up by macrophages, suggesting that they are internalized via a different route, independent of lipid microdomain-mediated macropinocytosis. Interestingly, these few bacteria are predominantly delivered to a lysosomal (LAMP1-positive) compartment, indicating that they are not able to interfere with phagosome maturation [26]. Thus, blocking the preferred way of entry or redirecting it by antibody opsonization as stated above results in delivery of internalized *A. felis* to phagolysosomes, indicating that intracellular compartmentation is intimately connected with the method of uptake.

Although *A. felis* does not necessarily replicate in murine macrophages, it can survive for at least 4 days [25, 26]. It is not clear whether localization to a non-endocytic compartment is a prerequisite for survival of *Afipia*. However, in amiloride-treated macrophages afipiae that are still taken up were almost completely eliminated within 24 h [26]. This finding strongly indicates that macropinocytosis not only influences compartmentation but also favors bacterial survival inside macrophages. Taken together, uptake via lipid microdomain-dependent macropinocytosis is essential for *A. felis* to enter a non-endocytic compartment and to survive inside the macrophage host cell.

As mentioned above, *A. felis* likely does not trigger macropinocytosis, but rather is a passive hitchhiker, as *Afipia*-infected macrophages did not seem to have more ruffle-like structures than uninfected macrophages. We propose that *A. felis* binds to certain lipid microdomains, which facilitates its uptake by exploiting macropinocytosis that constitutively occurs at a low level in macrophages. The finding that depletion of macrophage plasma membrane cholesterol inhibits not only uptake of *A. felis* but also its binding to the macrophage cell surface emphasizes the role of lipid microdomains as binding sites for successful *A. felis* [26]. In agreement with *Afipia* being a passive load neither actin polymerization nor ruffle formation were induced when macrophages were infected with *A. felis*.

Watarai and co-workers [41] reported a comparatively time-consuming macropinocytic uptake of *Brucella abortus* designated as swimming internalization. *Brucella* moves along the plasma membrane before being taken up into a large vacuole, which, like the ACV, shrinks over time and evades phagolysosome formation. However, in the case of *Brucella*, the macropinocytic uptake fully depends on a functional type IV secretion system. The authors suggest that in the course of swimming internalization, *Brucella* actively remodels the host cell plasma membrane and recruits lipid microdomains into the forming *Brucella*-containing phagosome. To date, it is not known whether *Afipia* expresses any type of secretion system but it has been shown that *A. felis* has to be alive to establish its unusual intracellular compartment [28]. Heat-killing of *A. felis* leads to delivery of *Afipia* to phagolysosomes. It is therefore possible that *A. felis* expresses some sort of effector or surface molecules that influences the method of uptake at least by facilitating binding to certain lipid microdomains on the host cell surface, thus allowing uptake via macropinocytosis but without triggering it.

It has been reported that in macrophages, macropinosomes are part of the endolysosomal system and are not disconnected as the *Afipia*-containing macropinosome is. However, J774E cells (which were used for the described studies) are also able to form macropinosomes that do not seem to mature into late endocytic organelles [42] and these may be identical to the *A. felis*-ingesting ones. On the other hand, *A. felis* may further influence its fate after the uptake process, for example, by secretion of effector proteins. In line with such hypotheses, heat-killed afipiae were

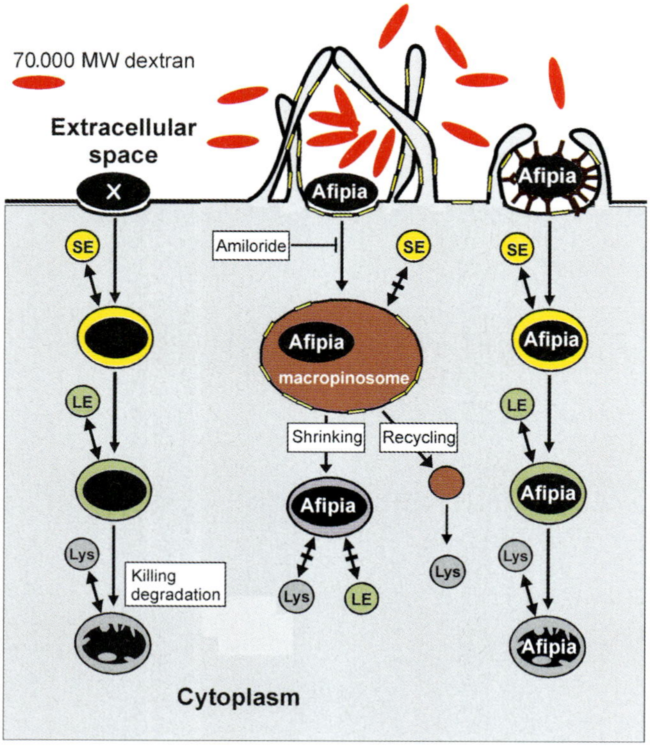

Figure 13.3 Model of uptake and intracellular compartmentation of *Afipia felis* in macrophages. After binding to lipid microdomains (yellow boxes) on the macrophage plasma membrane, *A. felis* is entangled in long cellular protrusions and taken up into a large vacuole via lipid microdomain-dependent macropinocytosis. This process is inhibitable by amiloride. The resulting spacious *Afipia*-containing macropinosomes shrink very rapidly by liquid loss and membrane recycling, and *A. felis* resides eventually in tightly fitting phagosomes which interact with neither early nor late endosomal/lysosomal compartments. The ingested liquid will eventually appear in the lysosomes. In contrast, anti-*Afipia* IgG opsonized bacteria (right) or other (harmless) particles ("X") are ingested by Fc receptor-mediated phagocytosis. Subsequent maturation of the *Afipia*-containing phagosome progresses along the degradative endocytic pathway. SE, sorting endosome; LE, late endosome; Lys, lysosome.

also taken up by macrophages into a TfR-negative compartment and internalization was inhibited by amiloride as well, however they were delivered to phagolysosomes straight away, suggesting an active role of *A. felis* metabolism in subversion of phagolysosome formation that remains after uptake. Figure 13.3 summarizes the data on ACV trafficking.

13.3.5
A. felis is Delivered to Lysosomes in Nonprofessional Phagocytes

A. felis is able to enter and maybe even multiply inside nonprofessional phagocytic cells [8]. The ability to invade such cells is not a common event among bacteria and likely contributes to pathogenicity. However, the intracellular compartmentation of *A. felis* in nonprofessional phagocytes was unknown, but became particularly interesting once it was known that *A. felis* resides in an unusual compartment inside macrophages. One would assume that if a pathogen can enter a nonprofessional phagocyte and, at the same time, has managed to reprogramme a professional phagocyte, then it should be very successful in establishing a suitable intracellular niche in the more harmless host cells.

However, *A. felis* is delivered to phagolysosomes in nonprofessional phagocytes [18]. Invasion of several endothelial and epithelial cells by *A. felis* was confirmed and the interaction of the ACVs and the endosomal pathway of the respective host cells was observed: ACVs in these nonprofessional phagocytes were transiently positive for EEA1 and then became and remained positive for LAMP1 and the proton pumping "vacuolar" ATPase, suggesting undisturbed, albeit delayed, phagolysosome biogenesis (Figure 13.4). Furthermore, judged by light microscopy, most of the intracellular *A. felis* were degraded within 72 h of infection whereas they survived an even longer period of time inside macrophages (see above). These data were in contrast to an earlier study reporting survival and multiplication of *Afipia* in HeLa cells and other epithelial and endothelial cell lines [8].

Why do *A. felis*-containing phagosomes mature differently in macrophages than in most other mammalian cells? The most obvious explanation would be the port of entry. *Afipia* most likely does not trigger macropinocytosis in macrophages, but it needs to be taken up via this route, otherwise its phagosome will mature along the degradative pathway. Macropinocytosis is not constitutive in endothelial or epithelial cells but only occurs after addition of growth factor or alternative stimulation. Assuming that *A. felis* is not able to trigger such events, it is likely that the way it enters an endothelial or epithelial cell differs significantly from the way it enters macrophages. In line with this, Schueller and colleagues [18] observed that *Afipia* enters human microvascular epithelial cells (HMEC-1) and human umbilical vein endothelial cells (HUVEC) in a rather zipper-like mode, supporting the idea of a non-macropinocytic entry.

Whereas endothelial cells do not seem to be ideal host cells for *A. felis*, they serve as a primary target cell for *Bartonella*. While BCVs show delayed phagolysosome formation in macrophages, *B. henselae* prevents maturation of its vacuole even 24 h after infection of HUVECs [30]. Furthermore, *Bartonella* is able to replicate inside endothelial cells [30, 43] whereas phagosome maturation is accompanied by bacterial killing in macrophages [30].

It is not clear why *Bartonella* can block phagosome maturation and hence survive in endothelial cells and not in macrophages. Whether the differences between the cell types are due to different entry pathways of *B. henselae* has not been analyzed so far. Likely in contrast to *A. felis*, *B. henselae* expresses a type IV secretion system, and it is quite likely that *Bartonella* is able to actively modulate the host cell by injecting effector proteins into the host cell cytosol. Thereby, *Bartonella* could either influence its way of uptake or control phagosome maturation directly. Interestingly, *B. henselae* infection stimulates macrophages to produce vascular endothelial growth factor (VEGF), which stimulates proliferation of endothelial cells. Given

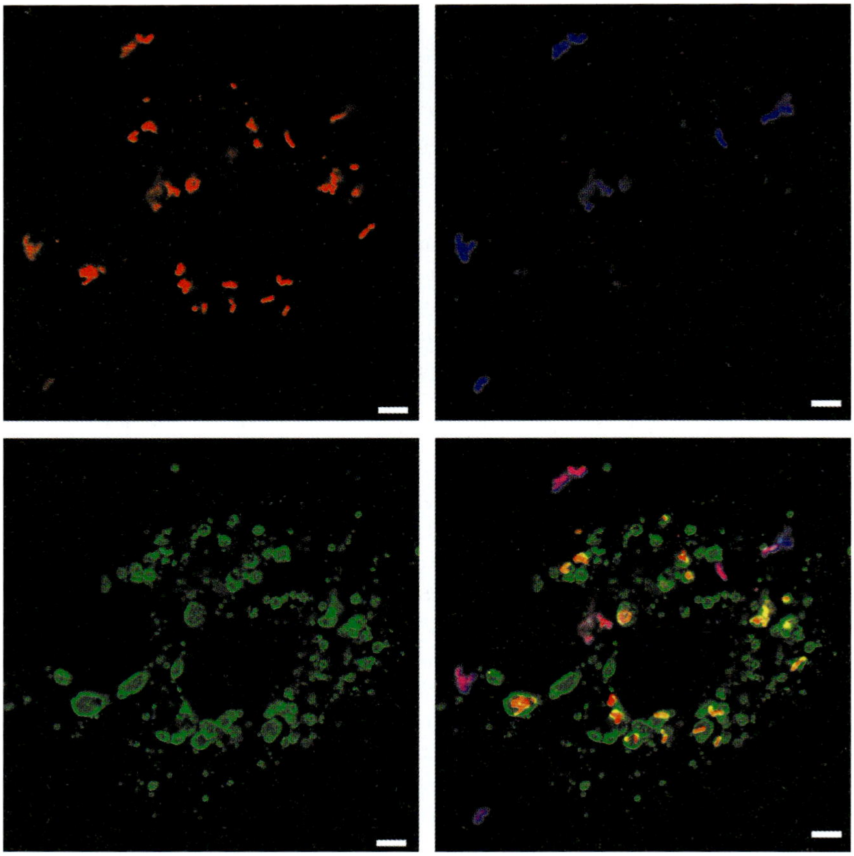

Figure 13.4 *Afipia*-containing vacuoles in endothelial (HMEC-1) cells are late endocytic (positive for LAMP1). Human endothelial (HMEC-1) cells were infected with *A. felis* for 16 h at 200 bacteria per endothelial cell. A differential stain was performed for bacteria that had not entered yet and others that were intracellular [26] and confocal microscopy was used to test for colocalization. Extracellular bacteria are visualized in blue (top left), intracellular bacteria in red (top right), LAMP1 in green (bottom left) and an overlay of all three channels is also shown (bottom right). Bar: 2.5 μm. Courtesy of Christian Schueller from the Haas laboratory.

that bartonellae survive and replicate inside endothelial cells, their ability to promote endothelial cell proliferation by triggering VEGF production by macrophages might represent a dedicated pathogenic strategy for expanding the bacterial host cell habitat.

13.4
Host and Pathogen Determinants Involved in Diverted Phagosome Biogenesis

Many details of *Afipia*–host cell interplay remain unsolved, such as the contribution of *Afipia* to intracellular compartmentation and survival but also the contribution of most host cell factors. In order to identify host cell factors, we performed a proteomics approach using one-dimensional gel electrophoresis and mass spectrometry of purified ACVs (unpublished data). Comparing the protein composition of the ACV with a control phagosome containing a microorganism that does not interfere with phagolysosome biogenesis will uncover proteins that are specifically acquired by the ACV, thus giving an idea of the nature of the compartment and why it is not interacting with other endocytic organelles. Based on this, functional analysis of the respective host cell factors can be performed, with gene silencing or knockout studies being potentially most important to gain deeper insight into the mechanisms that lead to the creation of this nonfusogenic organelle. It should be noted, however, that the ACV seems to be a compartment that is relatively void of specific proteins, possibly due to its natural role as a "vanishing macropinosome" which makes it, however, an appealing model for the progression of macropinosome biogenesis.

Genetic analysis of *A. felis* has been hindered mainly due to the lack of its genome sequence. Our lab, however, has now successfully applied transposon mutagenesis to generate random *Afipia* transposon mutants and to identify specific mutants, in this particular case flagellar mutants (C. Schueller and A. Haas, unpublished). Furthermore, the success of the establishment of recombinant *A. felis* harboring a broad host-range green fluorescent protein plasmid shows that genetics in *A. felis* are possible. It is now possible to screen the mutant pool for a desired bacterial phenotype, for example, for mutants that will no longer interfere with phagolysosome formation. The results from such studies will add to our general knowledge about how intracellular pathogens manipulate and reprogram intracellular trafficking events and will present yet another facet to our picture of how pathogens succeed.

13.5
Immunology of the *Afipia*-Containing Phagosome

Little is known about immunology to *A. felis* or how it is initiated. Given that *A. felis* resides in a compartment that is disconnected from the canonical endocytic pathway, antigen presentation might be impaired because antigens would not be accessible to the antigen-presenting machinery. However, although the vast majority of afipiae are taken up into a non-endocytic phagosome, some 20% are delivered to late endosomal compartments. Hence, killing and subsequent degradation and processing of *Afipia*-derived antigens for antigen presentation via major histocompatibility class II (MHC class II) molecules can occur in such compartments. Furthermore, it remains a possibility that antigens like secretory or peripheral cell surface proteins of *A. felis* end up in vesicles budding off the ACV, thereby delivering *Afipia* antigens to the endocytic system. This is likely since *Afipia*-containing macropinosomes shrink within minutes after formation and the bacteria become separated from the bulk macropinosomal content such as 70 000 molecular weight dextran which is delivered to a LAMP1-positive compartment whilst *A. felis* remains in the now tightly fitting but still non-endocytic phagosome [26]. By these means, antigen could be processed and presented via MHC class II to T cells to generate a T cell-mediated immune response, although such an event remains enigmatic in the case of *Afipia*.

To effectively control intracellular pathogens, a T_H1 T cell response is required. Interferon γ (IFNγ) is the hallmark T_H1 cytokine due to its ability to maximize the killing efficacy of macrophages and the proliferation of cytotoxic T cells. Hence, the T_H1 T cell response facilitates the clearance of the intracellular microbe by either directly killing infected nonprofessional phagocytes that themselves are unable to cope with the intracellular infection or by helping the macrophage to kill an ingested pathogen due to stimulating its bactericidal activities. In this regard we observed that *in vitro*, colocalization of lysosomal tracer with live *A. felis* was more than twice as frequent in murine macrophages preactivated with IFNγ and lipopolysaccharide than in resting macrophages (unpublished data). Thus, preactivation with IFNγ allows the macrophage to overcome some of the inhibition in phagosome formation, suggesting that *in vivo* a T_H1 immune response may control *A. felis* infection.

A T_H2 response will tend to stimulate the humoral immunity with the B cells being the main effector cells, producing antibodies to neutralize toxins or to opsonize extracellular pathogens to facilitate phagocytosis and subsequent killing by macrophages. *A. felis* does not seem to induce a pronounced antibody response since the antibody titers to *A. felis* are rather low in both cat scratch disease patients and experimentally infected animals [5]. Hence, humoral immunity may not play a major role during *A. felis* infection.

13.6
Summary and Outlook

There is no doubt that *A. felis* is a rare cause of human cat scratch disease, even though *Bartonella henselae* turned out to be responsible for the majority of cases. Most likely because *Afipia* favors temperatures between 25 and 30 °C for optimal growth it is much less successful as a human pathogen than *B. henselae*, with an optimal growth at 37 °C.

In patients' specimens, however, *A. felis* was found in endothelial cells and macrophages [5], making the investigation of its intracellular fate in these host cells particularly interesting. A prerequisite to be a successful intracellular pathogen is to escape killing in a phagolysosome by either subverting phagosomal trafficking events or by adapting to the harsh environment in such a compartment. *A. felis* avoids any contact with the degradative pathway inside macrophages from the time of uptake while it is delivered to a lysosomal compartment in nonprofessional phagocytes. The difference in intracellular compartmentation is most likely due to a different entry mode of *A. felis*. In macrophages, *A. felis* binds to lipid microdomains on the cell surface, thereby facilitating its uptake via macropinocytosis, which is a constitutive process in macrophages and mainly occurs at cholesterol-rich plasma membrane domains. Since *A. felis* most likely is not able to trigger such events but rather exploits constitutive macropinocytosis in macrophages it enters host cells other than professional phagocytes via a different route because macropinocytosis has to be triggered by growth factors or other stimuli in such cell types. Macropinocytic uptake is, however, the prerequisite for intracellular compartmentation and survival, since both blocking macropinocytosis and forcing *Afipia* uptake via Fc receptors by antibody opsonization lead to delivery of afipiae to a lysosomal compartment.

Preactivation of macrophages with IFNγ and lipopolysaccharide and subsequent infection with *A. felis* leads to enhanced phagolysosome formation compared with untreated macrophages. It would be interesting to know if IFNγ activation forces *Afipia* to enter the macrophage via a route other than macropinocytosis or if the activated macrophage is able to overcome fusion inhibition between *Afipia*-containing phagosomes and lysosomes independently of the uptake mechanism. IFNγ activation of macrophages is known to downregulate certain types of receptors such as the macrophage mannose receptor or transferrin receptor. This way, IFNγ may lead to altered binding receptor expression or even decreased macropinocytosis, thus modulating the *Afipia* binding and uptake and subsequent intracellular trafficking of *Afipia*-containing phagosomes. In this context, it will be interesting to determine the precise macrophage surface receptor(s) for *Afipia* and also the respective bacterial ligand. Furthermore, these host cell receptors, although only very rare, might be already clustered and hence concentrated in certain lipid microdomains, which could be a reason for *A. felis* to preferentially bind to such domains on the macrophage surface. After disintegrating these domains, for example by cholesterol depletion, such receptors would no longer be clustered but evenly distributed in the plasma membrane, making it difficult for *A. felis* to engage an appropriate amount of receptors for proper binding and uptake.

Taken together, *A. felis* can be used as a tool to study a particular, and likely constitutively occurring, macropinocytic uptake pathway and further studies of *Afipia*–host cell interaction may help to elucidate candidate factors on both the host side and the pathogen side that are involved in the biogenesis and the maintenance of the *Afipia*-containing phagosome.

References

1 Brenner, D.J., Hollis, D.G., Moss, C.W., English, C.K., Hall, G.S., Vincent, J., Radosevic, J., Birkness, K.A., Bibb, W.F., Quinn, F.D. *et al.* (1991) *J Clin Microbiol*, **29**, 2450–2460.

2 La Scola, B., Mallet, M.N., Grimont, P.A. and Raoult, D. (2002) *Int J Syst Evol Microbiol*, **52**, 1773–1782.

3 Thomas, V., Casson, N. and Greub, G. (2007) *Syst Appl Microbiol*, **30**, 572–579.

4 Moosvi, S.A., Pacheco, C.C., McDonald, I.R., De Marco, P., Pearce, D.A., Kelly, D.P. and Wood, A.P. (2005) *Environ Microbiol*, **7**, 22–33.

5 English, C.K., Wear, D.J., Margileth, A.M., Lissner, C.R. and Walsh, G.P. (1988) *JAMA*, **259**, 1347–1352.

6 Jerris, R.C. and Regnery, R.L. (1996) *Annu Rev Microbiol*, **50**, 707–725.

7 Wear, D.J., Margileth, A.M., Hadfield, T.L., Fischer, G.W., Schlagel, C.J. and King, F.M. (1983) *Science*, **221**, 1403–1405.

8 Birkness, K.A., George, V.G., White, E.H., Stephens, D.S. and Quinn, F.D. (1992) *Infect Immun*, **60**, 2281–2287.

9 Giladi, M., Avidor, B., Kletter, Y., Abulafia, S., Slater, L.N., Welch, D.F., Brenner, D.J., Steigerwalt, A.G., Whitney, A.M. and Ephros, M. (1998) *J Clin Microbiol*, **36**, 2499–2502.
10 Dolan, M.J., Wong, M.T., Regnery, R.L., Jorgensen, J.H., Garcia, M., Peters, J. and Drehner, D. (1993) *Ann Intern Med*, **118**, 331–336.
11 Regnery, R.L., Olson, J.G., Perkins, B.A. and Bibb, W. (1992) *Lancet*, **339**, 1443–1445.
12 Tompkins, D.C. and Steigbigel, R.T. (1993) *Ann Intern Med*, **118**, 388–390.
13 Dalton, M.J., Robinson, L.E., Cooper, J., Regnery, R.L., Olson, J.G. and Childs, J.E. (1995) *Arch Intern Med*, **155**, 1670–1676.
14 Perkins, B.A., Swaminathan, B., Jackson, L.A., Brenner, D.J., Wenger, J.D., Regnery, R.L. and Wear, D.J. (1992) *N Engl J Med*, **327**, 1599–1601.
15 Bergmans, A.M., Groothedde, J.W., Schellekens, J.F., van Embden, J.D., Ossewaarde, J.M. and Schouls, L.M. (1995) *J Infect Dis*, **171**, 916–923.
16 Del Prete, R., Fumarola, D., Fumarola, L. and Miragliotta, G. (2000) *Eur J Clin Microbiol Infect Dis*, **19**, 964–967.
17 N.N. (1992) *N Engl J Med*, **326**, 1480–1489.
18 Schueller, C., Schneider, B., Kempf, V.A. and Haas, A. (2007) *Microbes Infect*, **9**, 355–363.
19 La Scola, B. and Raoult, D. (1999) *Lancet*, **353**, 1330.
20 Greub, G. and Raoult, D. (2004) *Clin Microbiol Rev*, **17**, 413–433.
21 La Scola, B., Mezi, L., Auffray, J.P., Berland, Y. and Raoult, D. (2002) *Infect Control Hosp Epidemiol*, **23**, 462–465.
22 Marrie, T.J., Raoult, D., La Scola, B., Birtles, R.J. and de Carolis, E. (2001) *Emerg Infect Dis*, **7**, 1026–1029.
23 La Scola, B., Barrassi, L. and Raoult, D. (2000) *FEMS Microbiol Ecol*, **34**, 129–137.
24 La Scola, B. and Raoult, D. (1999) *J Clin Microbiol*, **37**, 1899–1905.
25 Le Pocher, H., Brouqui, P. and Raoult, D. (1998) *J Antimicrob Chemother*, **42**, 825–829.
26 Schneider, B., Schueller, C., Utermoehlen, O. and Haas, A. (2007) *Traffic*, **8**, 226–240.
27 Brouqui, P. and Raoult, D. (1993) *Microb Pathog*, **15**, 187–195.
28 Lührmann, A., Streker, K., Schüttfort, A., Daniels, J.J. and Haas, A. (2001) *Proc Natl Acad Sci USA*, **98**, 7271–7276.
29 Mayor, S. and Pagano, R.E. (2007) *Nat Rev Mol Cell Biol*, **8**, 603–612.
30 Kyme, P.A., Haas, A., Schaller, M., Peschel, A., Iredell, J. and Kempf, V.A. (2005) *Cell Microbiol*, **7**, 1019–1034.
31 Swanson, J.A. and Watts, C. (1995) *Trends Cell Biol*, **5**, 424–428.
32 Alpuche-Aranda, C.M., Racoosin, E.L., Swanson, J.A. and Miller, S.I. (1994) *J Exp Med*, **179**, 601–608.
33 Garcia-del Portillo, F. and Finlay, B.B. (1994) *Infect Immun*, **62**, 4641–4645.
34 Nhieu, G.T. and Sansonetti, P.J. (1999) *Curr Opin Microbiol*, **2**, 51–55.
35 Araki, N., Johnson, M.T. and Swanson, J.A. (1996) *J Cell Biol*, **135**, 1249–1260.
36 Kirkham, M. and Parton, R.G. (2005) *Biochim Biophys Acta*, **1746**, 349–363.
37 Conner, S.D. and Schmid, S.L. (2003) *Nature*, **422**, 37–44.
38 Grimmer, S., van Deurs, B. and Sandvig, K. (2002) *J Cell Sci*, **115**, 2953–2962.
39 Manes, S., Mira, E., Gomez-Mouton, C., Lacalle, R.A., Keller, P., Labrador, J.P. and Martinez, A.C. (1999) *EMBO J*, **18**, 6211–6220.
40 Shin, J.S. and Abraham, S.N. (2001) *Microbes Infect*, **3**, 755–761.
41 Watarai, M., Makino, S., Fujii, Y., Okamoto, K. and Shirahata, T. (2002) *Cell Microbiol*, **4**, 341–355.
42 Albrecht, I., Gatfield, J., Mini, T., Jeno, P. and Pieters, J. (2006) *Int Immunol*, **18**, 755–765.
43 Kempf, V.A., Schaller, M., Behrendt, S., Volkmann, B., Aepfelbacher, M., Cakman, I. and Autenrieth, I.B. (2000) *Cell Microbiol*, **2**, 431–441.

14
Brucella

Monika Kalde, Edgardo Moreno, and Jean-Pierre Gorvel

14.1
Introduction

14.1.1
Brucellosis – An Old Disease with Many Names

Also known as Malta fever, Mediterranean fever, undulant fever and under many more synonyms, brucellosis is named after Lieutenant Sir David Bruce, a physician and microbiologist, who first identified the causative organisms as small α-proteobacteria. He originally named the isolated bacterium "*Micrococcus melitensis*" with "melitensis" being derived from the latin name for Malta, the place of the discovery, but it was soon renamed *Brucella melitensis* honoring Bruce's discovery. Nine species are currently recognized within the genus *Brucella*, classified according to their host preference and their pathogenicity. *B. melitensis* predominantly infects goats and sheep, but also camels and dogs. *B. abortus* causes bovine brucellosis and *B. suis* causes swine brucellosis. All three strains exhibit pathogenicity in humans with *B. melitensis* being the least host-specific and also the most infectious for humans. Other *Brucella* species display a host preference for dogs (*B. canis*), wood rats (*B. neotoma*), sheep (*B. ovis*), seals (*B. pinnipedialis*) cetaceans (*B. ceti*) and rodents (*B. microti*).

Brucella bacteria enter the host body via the oral and respiratory tracts and can penetrate the mucous membranes covering the conjunctiva and vagina. The most common source of human infection is through ingestion of contaminated dairy products and probably happens through local lymph nodes and the tonsils. Experimentally, *Brucella* has been shown to be internalized by M cells in the intestine, which subsequently release the bacteria into the gastrointestinal-associated basal space. Here, they are phagocytosed by macrophages, neutrophils and dendritic cells (DCs). The bacteria are able to avoid phagosomal killing and undergo efficient replication in both macrophages and DCs. The migratory abilities of the latter might achieve dissemination of the bacteria to specific locations of the body (spleen, heart, bones, brain, reproductive organs). A colonization of trophoblasts is often concomitant

Intracellular Niches of Microbes. Edited by Ulrich E. Schaible and Albert Haas

ISBN: 978-3-527-32207-7

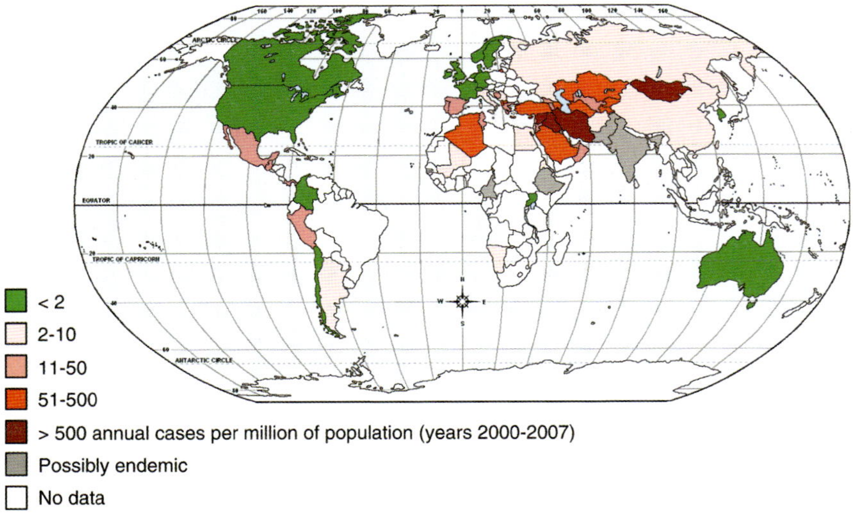

Figure 14.1 Global incidence of human brucellosis. Adapted from [2].

with abortions in pregnant animals. Such an abortion expels tissue and fluid containing high loads of *Brucella* spp. into the surrounding area and poses a serious health risk for flocks and their handlers, especially in crowded facilities. The infection may also reduce fertility, cause weight loss, and lower the milk yield of animals. The economic losses of *Brucella*-infected herds can thus be devastating, especially in low-income countries where brucellosis remains a serious problem, mainly due to exiguous control programs and deficient animal husbandry.

Brucellosis is classified as a zoonotic disease and not sustainable in humans. Transmission from person to person is uncommon. Nevertheless, together with infected livestock, the disease has accompanied human civilization for more than 2000 years [1]. Regions where brucellosis is endemic, such as the Mediterranean basin, the Middle East, Mongolia, Russia and Latin America [2] (see Figure 14.1), require a reservoir in their primary hosts. While *Brucella*-related abortions remain the exception in humans, classical symptoms include bouts of high fever, joint pain, spondilytis, physical weakness, and endocarditis and meningitis with fatal outcomes if untreated. Although diagnosis using immunological methods is well standardized, brucellosis is often misdiagnosed as drug-resistant malaria in tropical regions and, as a consequence, is not adequately treated with a combination therapy of antibiotics. The burden of the human disease adds to the economic impact of brucellosis.

14.1.2
Current Vaccination Strategies

The prevention of human brucellosis predominantly depends on the eradication of the disease in their primary hosts. Fortunately, and in contrast to many other bacterial diseases, there are efficient live vaccines that protect bovines, caprines and ovines against brucellosis. Although these live vaccines are less virulent than the field strains, they may cause abortions if administered to pregnant animals, and are potentially infectious for persons handling and administering the vaccine.

Strain 19 (S19) is an attenuated *B. abortus* strain lacking a gene required for erythritol metabolism [3] that was developed for use in cattle in the 1920s. For many years it was thought that this genetic defect was the key virulence factor; however, currently there are conflicting results regarding its role in virulence [4, 5]. Despite the fact that S19 is eliminated more quickly than virulent strains from vaccinated cattle than virulent strains, it causes protective immunity. Other live vaccines are Rev1 and the *B. suis* strain 2 (S2) vaccine, attenuated strains of *B. melitensis* and *B. suis*, respectively. Rev1 is the most effective vaccine against ovine and caprine brucellosis, while S2 was used as an oral vaccine in China to protect cattle, goats, sheep and pigs. All three attenuated strains possess a smooth phenotype, and vaccinated animals produce antibodies against the O-chain and native hapten (NH), two highly immunogenic components of the cell envelope of smooth *Brucella* spp.

Because the general surveillance to identify infected animals routinely applies immunoassays based on these very same antibodies, the utilization of smooth vaccine strains makes the distinction between vaccinated and infected animals impossible. To avoid the interference of the O-chain with disease control surveillance, efforts have therefore been undertaken to generate rough vaccine strains. A rough candidate strain was the strain 45/20, but this proved to be very inconsistent in expressing the O-chain or not, often causing the generation of antibodies against lipopolysaccharide (LPS) determinants in vaccinated animals. Therefore its application was not pursued [6]. The vaccine strain RB51 is a spontaneous rough mutant of the common laboratory strain *B. abortus* 2308. RB51 also produces low quantities of the O-chain [7]. Its protective potential in cattle and sheep remains highly controversial [8]. Overall, the development of effective vaccines that successfully protect a wide variety of hosts, including humans, without interfering with diagnostic procedures or causing harm, remains a challenge.

Human vaccines are employed in endemic areas where the vaccination of large livestocks is not feasible. Attenuated strains similar to S19 were formerly widely used in Asia and the former USSR, but the immunizing effect was short-lived and strong hypersensitivity reactions against the vaccine were frequently observed when the immunization was repeated [6]. Other approaches have their focus on subcellular fractionations of the bacteria and include sodium dodecylsulfate (SDS)-insoluble peptidoglycan fractions of *B. melitensis*, acetic acid-extracted polysaccharide–protein fractions, lipopolysaccharide–protein conjugates, purified protein antigens such as L7/L12, Cu-Zn superoxide dismutase (SOD) or glyceraldehyde dehydrogenase. The testing of candidate vaccines is a challenge because the current animal models of infection cannot be readily compared to the human situation, and challenging human subjects with *Brucella* is unethical. High costs and little reward for commercialism are likely to restrict the development of human vaccines to national defense agencies [6].

14.2
Characterization of Intra-Host Cell Compartmentation of *Brucella* and its Relationship to Virulence

14.2.1
Breaking and Entering

Brucella bacteria are able to enter and replicate efficiently in a variety of cells, including epithelial cells, placental trophoblasts, DCs and macrophages. The percentage of *Brucella* present in the culture medium that will successfully invade the host cell *in vitro* is low when compared to other intracellular bacteria [9]. The limiting step of invasion in epithelial cells seems to be the attachment to the host cell, as almost all of the *Brucella* bacteria that successfully attach will eventually be internalized [10]. In spite of such an obvious importance of the initializing step for virulence, the molecular determinants and mechanisms of the internalization process remain poorly understood. As is the case in other host–pathogen interactions, adhesion is likely to occur via surface-exposed proteins such as lectins and adhesins on the bacterium, and matrix proteins (e.g., fibronectin and vitronectin) and/or cellular receptors containing sialic acid on the host cell surface [11–13]. The internalization presumably requires specific receptors. Professional phagocytes are invaded by *Brucella* more easily than nonprofessional phagocytes, especially if the bacteria are opsonized and the phagocytes are activated [14]. This indicates a role of complement and Fc receptors for adhesion and entry into these cells. Non-opsonized smooth *Brucella* bacteria instead seem to employ a combination of functional lipid rafts, microdomains rich in cholesterol and sphingolipids, and yet unidentified receptors to enter macrophages [15].

Although the mechanisms of entry and the implication of lipid rafts in this process need to be further analyzed, their presence seems to be of major importance for intracellular survival (and therefore bacterial virulence) because it apparently promotes the bacterial ability to redirect cellular trafficking in their favor. Rough mutant strains, in contrast, seem to enter the host cells in a lipid raft-independent manner and, as a consequence, suffer the fate of many pathogenic bacteria: destruction in lysosomes. The same applies to smooth *Brucella* spp. internalized by phagocytes artificially depleted of lipid rafts [16]. The scavenger receptor A (SR-A), a pattern recognition receptor, has been suggested to promote bacterial binding and internalization of *Brucella* bacteria into macrophages. SR-A is not present in lipid rafts in naïve cells, but moves there as a consequence of the infection process. Although there are reports claiming the involvement of type IV secretion system (T4SS) in *Brucella* internalization [17], there are other studies that disregard this, pointing out that this system is only required at later stages of the infection and not for invasion [16].

In epithelial cells, rough *Brucella* bind and penetrate through the recruitment of coated pits and replicate extensively [18]. The involvement of specific receptors for invasion by *Brucella* bacteria is suggested [19], but none of these receptors have been identified to date. It is likely that the bacteria induce their own uptake via receptor-mediated endocytosis, and in some exceptions via the zipper-like mechanism [20, 21]. Uptake is accompanied by moderate rearrangements of the cytoskeleton and general membrane ruffling has been observed [22]. In accordance with a suspected involvement of the cytoskeleton, drug-inhibition studies demonstrated the requirement of both actin and microtubule filaments. Furthermore, *Brucella* invasion requires the activation of GTPases of the Rho subfamily, such as Rho, Rac and Cdc42, regulators of the cytoskeleton. A direct activation of Cdc42 by *Brucella* was shown to be influenced by the two-component sensor/regulator system BvrS/BvrR, which has major effects on the composition of the outer membrane of the bacteria [22] and is indispensable for efficient invasion [23].

14.2.2
The Manipulative Powers of *Brucella*

At first glance, macrophages seem to be very efficient at eliminating *Brucella* bacteria they have encountered and phagocytosed. The majority of *Brucella*-containing vacuoles (BCVs) fuse with lysosomes, and the bacteria are killed and degraded within a few hours after their internalization. About 10–30% of the internalized bacteria are, however, able to redirect the intracellular traffic of the host cell. By doing this, they not only avoid immediate degradation, but also reach an intracellular niche where they are undisturbed by lysosomal activity over a prolonged period of time and can replicate efficiently. The replication of this small fraction of surviving bacteria is so efficient that cell culture experiments usually display a net increase of bacteria within 30–40 h after infection.

The BCV undergoes complex membrane modifications, resulting in fast and transitory changes of its characteristics, especially during the first hours of infection (compare Figure 14.2). A complex mosaic of *Brucella*-containing compartments, representing different stages of maturation, can thus be found in the early infected cell. In combination with the high percentage of bacteria that do not escape lysosomes in professional phagocytes, the medley of maturing vacuoles versus late endosomal compartments makes it difficult to separate the pathway that allows the bacteria to reach their replication niche from those that target the bacteria for destruction. In the epithelial cell lines HeLa and Vero most of the BCVs will develop into a replicative niche, pruning the degradation pathway out of this mosaic. Because characteristics of the BCVs that are found in murine macrophages have also been identified in the interaction of *B. abortus* with these epithelial cell lines, the general consensus today is that trafficking is essentially the same in professional and nonprofessional phagocytes.

Within minutes after uptake of *Brucella* bacteria the BCV obtains the characteristics of the early endosomal network, like the presence of the Rab5 receptor, the transferrin receptor and early endosomal antigen 1 (EEA1) on its membrane. These early endosomal markers, however, gradually disappear and are no longer detectable 30 min after invasion. Instead, the BCV acquires the characteristics of the endosomal/lysosomal compartment. While the presence of LAMP1 in the early BCV is well documented, previous studies have reported the absence of additional lysosomal

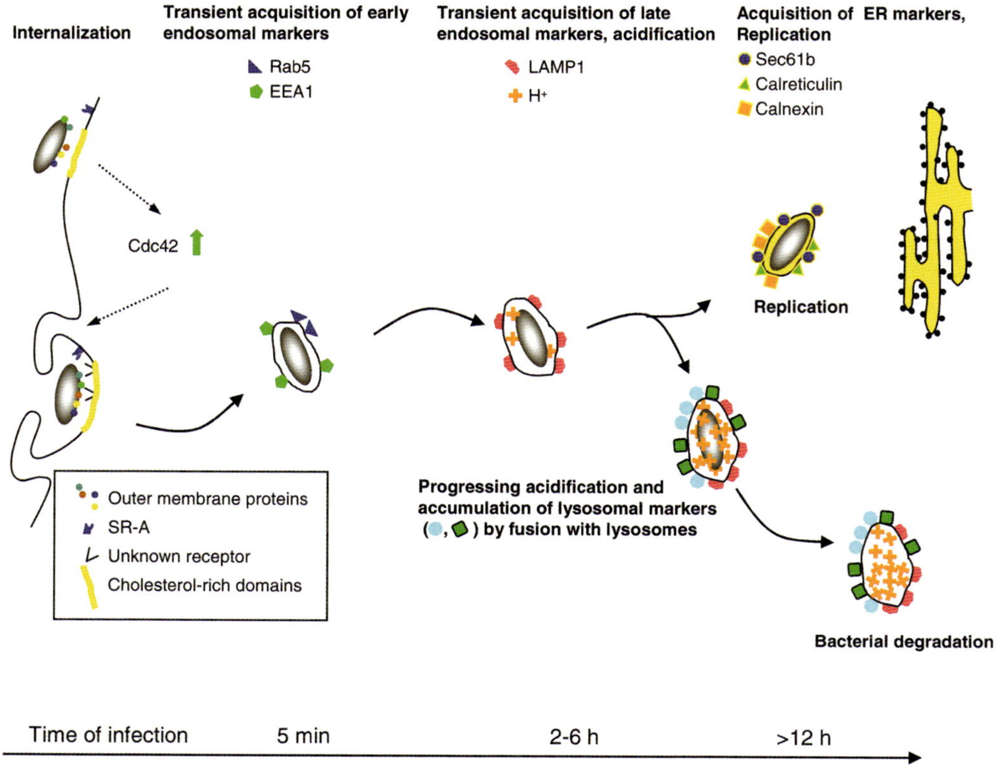

Figure 14.2 Schematic view of the maturation of the *Brucella*-containing vacuole in professional phagocytes.

markers such as Rab7 on the BCV [24, 25]. Recently, applying high-resolution live imaging and protein overexpression, additional markers for the lysosomal compartment, including Rab7, have been reported to colocalize with the maturing BCV [26]. The overexpression of Rab proteins, however, is able to modifiy intracellular membrane trafficking, therefore one should be cautious in the interpretation of data derived from such experiments. This recent study nevertheless shows that the BCV lumen experiences progressing acidification and an increase in proteolytic activity [26], typical characteristics of the lysosome. The hypothesized limited fusion of the maturing BCV with late endosomal compartments may provide an answer to the long-lasting question regarding the origin of LAMP1 on BCVs. The acidification and proteolysis possibly provides the environmental cues to activate expression and/or function of bacterial virulence genes which begin to modulate the intracellular trafficking of the host at this point. Indeed, expression of the T4SS encoded by the *Brucella virB* operon, for example, is induced under acidic conditions and plays a role in the ability of the BCVs to fuse with the ER. As a consequence of redirected cell trafficking, the lysosomal markers on the BCV disappear progressively.

At approximately 4–8 h after infection, markers for the ER such as calreticulin or Sec61b appear on the maturing BCV, indicating that the developing compartment, the replicative niche, becomes related to the ER. In macrophages, the bacteria achieve this by intercepting the early secretory pathway that is initiated at ER exit sites. The BCV interacts with the Sar1/COPII complex at those sites. In cells that express a variant of the small GTPase SAR1 in which the molecule stays locked in the GDP-bound (inactive) form, *B. abortus* cannot establish its replicative niche [27].

In epithelial cells, but never in professional phagocytes, a special variation of the BCV membrane has been observed. In this case, the BCV is at least partly turning into a multimembranous compartment with characteristics of an autophagosome [28]. Drugs that modulate autophagosomes, for example wortmannin, affect the intercellular fate of the internalized *Brucella* bacteria in these cells. It is tempting to speculate that epithelial cells use autophagy of the BCV as a line of defense and that the autophagosomal vacuoles might be formed by an invagination of the ER around the BCV. However, the way in which the membrane rearrangements are performed remains elusive. Because this autophagosomal compartment remains negative for endoplasmic markers such as BiP or ribophorin, it seems that only specialized regions of the ER, if any, are in involved in its formation. With progressing maturation the multimembranicity is resolved, while the ER markers that are already present are retained. It is not known how the bacteria survive in this compartment without being negatively affected, and how they succeed to transform it into the ER-derived replicative niche.

14.2.3
To be Fruitful, and Multiply

Besides hosting characteristic ER markers, the replicative niche shares other features of the ER; it is decorated with ribosomes, the drug proaerolysin which vacuolizes the ER has the same effect on the BCV, and treatment of infected cells with brefeldin A, which inhibits anterograde protein transport from the ER to the Golgi, results in the accumulation of Golgi markers around the BCV as well as in the ER membrane.

The association of intracellular pathogens with the ER in order to avoid the lysosomal pathway, as is also the case for *Legionella pneumophila* and *Toxoplasma gondii*, might give these pathogens access to components and substrates they require for their metabolism. Besides this potential nutritional aspect, the ER is an important source of membranous material for the replicating bacteria, each surrounded by what seems an individual vacuole. Once the bacteria start replicating, they do so until the BCV finally fills nearly the whole cytoplasm. The heavily parasitized host cells show no signs of degeneration, necrosis or apoptosis and are even able to divide. Nevertheless, as a consequence of *Brucella* overgrowth, a rupture of the plasma membrane may occur, expelling the bacteria into the extracellular space. If such cell disruption in fact occurs *in vivo*, it has the potential to be an important process for dissemination throughout the organism and is important for a progressing infection.

Depending on the cell type, alterations of this general process of cellular infection are possible. Neutrophils (or PMNs, polymorphonuclear cells), for example, do not allow replication of *Brucella*. However, *Brucella* hampers degranulation, avoids strong activation and withstands phagosomal killing for as long as possible in this short-lived cell type. Some bacteria outlive their host cell and spread within the organism [29]. Interestingly, depletion of PMNs does not alter the course of brucellosis, limiting the role of these cells in *Brucella* infections, at least in the murine model [27].

14.3 Factors Involved in Diverted Phagosome Maturation and Compartment Maintenance

Classical virulence factors found in other bacteria, such as capsules, fimbria, lysogenic phages, endotoxic lipopolysaccharides or proteins encoded by pathogencity islands are absent in *Brucella* spp. The sophisticated transit of *Brucella* from the outside of the cell to their replicative niche is, nevertheless, dependent on a number of bacterial factors and strongly linked to molecules that form building blocks of the cellular envelope. These provide *Brucella* bacteria with resistance against extracellular killing, allow them to be internalized, and guide the bacteria to their replicative niche. The absence of any of the *Brucella* virulence factors described below generally interferes with the ability of the bacteria to avoid lysosomal degradation. The mechanisms utilized, however, remain enigmatic.

14.3.1 *Brucella* LPS: The Key to Invisibility

Each LPS molecule contains a lipid A portion that is inserted into the outer membrane and is linked to an extracellular core polysaccharide that can be either associated with (smooth LPS) or not associated with (rough LPS) an O-polysaccharide chain (O-chain), a homopolymer composed of *N*-formylperosamine. The noncanonical LPS of *Brucella* differs from the canonical LPSs of other Gram-negative bacteria in several structural features. Its lipid A contains a diaminoglucose backbone rather than glucosamine and is linked to acyl groups that are longer (C18, C19 or C28) than in the case of classical LPS (C12 and C14). In *Brucella* LPS, the linkage of lipid A to the core consists of amide bonds only, whereas in the case of classical LPS both amide and ester bonds are possible.

The noncanonical structure of *Brucella* LPS has direct effects on intracellular bacterial survival as it lacks many of the pathogen-associated molecular patterns (PAMPs) that are typical of other proteobacteria. Usually, a number of pattern-recognition receptors on macrophages and DCs, such as Toll-like receptors (TLRs), detect potential pathogens and mount an immune response specific to the detected patterns. As a consequence of its structure, *Brucella* LPS it is a poor inducer of TLR4-mediated signaling [30, 31], and therefore does not stimulate infected cells to produce bactericidal nitrogen intermediates, to undergo a respiratory burst, or to release proinflammatory cytokines [31]. Compared with LPS derived from *Salmonella* spp. or *E. coli*, the endotoxicity induced by *Brucella* LPS is several hundred times lower and allows an undisturbed bacterial replication within the cell.

A number of genes have been identified that are involved in the synthesis, formylation, transport and polymerization of perosamine, in the formation of the core, and in the generation of the lipid A moiety [32]. Mutations that alter the core structure often affect the linkage to the O-chain and cause rough phenotypes [33]. This hampers the dissection of the specific roles of the LPS subunits on bacterial trafficking and survival, especially since the O-chain is of exceptionally high value for bacterial survival: smooth strains are generally more virulent than their rough variants.

Brucella LPS contains a relatively low number of negatively charged groups, such as phosphate, or acidic sugars. This low charge allows the formation of tight LPS aggregates via hydrophobic interactions of the lipid A. The O-chains of these aggregates are intertwined with native haptens (NHs) and together form a dense "shield" over the outer membrane, minimizing its contact with foreign molecules. The low charge also provides a weak target for cationic bactericidal peptides such as defensins, lactoferrin, lysozyme and the defensin-like antibiotic polymyxin B. A combination of both effects is able to explain the resistance of *Brucella* against these bactericidal substances. The formation of an LPS shield might also partially protect PAMPs present in the outer membrane from detection by their respective TLRs. Such a shield would furthermore block the access of C1q complement [34] to the bacterial cell surface. If the formation of LPS aggregates is abrogated, by for example the *bacA* or the BvrS/BvrR mutations which change the acylation pattern on lipid A, smooth bacteria become more susceptible to cationic peptides and complement-mediated attack by non-immune serum [35, 36].

The requirement of the O-chain for the redirection of cellular trafficking has already been mentioned. In short, rough mutants adhere more efficiently to host cells than smooth variants, but they enter the cells in a manner that is independent of lipid rafts [15, 37], and are not able to avoid lysosomal degradation in professional phagocytes [16]. All these results indicate a very important role of *Brucella* LPS for the infection process, the intracellular survival and for the maintenance and maturation of their intracellular habitat.

14.3.2
The BvrR-BvrS System: A Coordinator of the Cell Envelope

The *Brucella* BvrS/BvrR system codes for a two-component sensor and regulator system, in which the sensor is located in the inner membrane, while the regulatory protein is cytoplasmic. It is able to influence the gene expression of several cell envelope components such as periplasmic and outer membrane proteins (Omps) in response to different environmentally stimuli [38, 39]. It is also involved in the control of the acylation pattern of the lipid A moiety in the bacterial LPS [35]. In general, it seems that the BvrS/BvrR system is devoted to the homeostasis of the cell envelope, mainly the outer membrane (Figure 14.3).

Mutants affected in the BvrS/BvrR system were first identified by their increased sensitivity towards detergents, complement attack and bactericidal peptides,

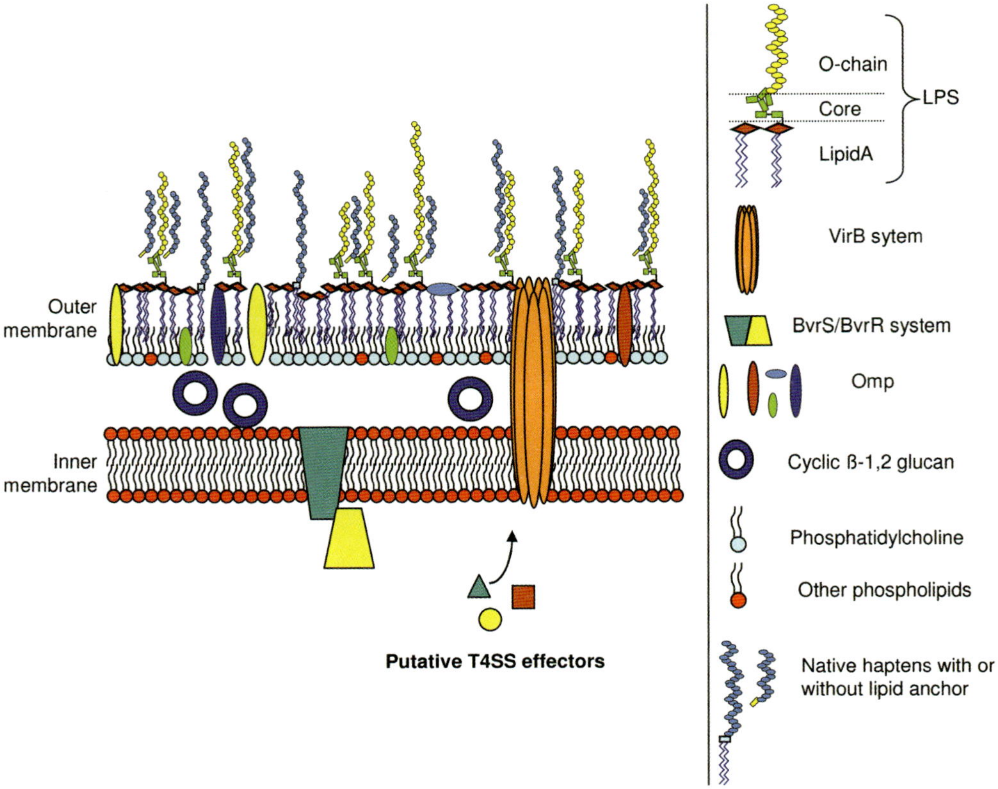

Figure 14.3 Schematic representation of the *Brucella* virulence factors connected to the bacterial cell envelope mentioned in the text.

probably a consequence of their altered LPS structure. Mutants are avirulent in mice and show decreased invasiveness into macrophages and epithelial cells. The penetration defect is, at least in epithelial cells, connected with their inability to activate the small GTPase Cdc42 [22]. Once internalized, *bvrR-bvrS* mutants are not able to avoid lysosomal degradation or to establish their replicative niche. The virulence conferred by this regulatory system is probably a phenomenon based on the coordinated participation of a number of discrete components, since mutations in distinct Omps regulated by BvrS/BvrR do not result in a loss of virulence in macrophages [40].

14.3.3
Phosphatidylcholine: Bacterial Camouflage

Phosphatidylcholine (PC) is typically absent from most proteobacteria, whose membranes are typically rich in phosphatidylethanolamine (PE). Nevertheless, in the case of *Brucella* and other α-proteobacteria, PC is the major phospholipid in the outer membrane. The *pmtA/pcs* mutants, which are unable to produce PC by either of the two pathways present in *Brucella*, display a reduced ability to avoid the fusion with lysosomes. In mice, this mutant is severely attenuated [41]. Like the nonclassical LPSs, PC replaces a potential PAMP, in this case PE, from the bacterial surface. In addition to limiting the detection by innate immune receptors, PC might be required to confer physical features to the membrane integrity that are indispensable for virulence [41].

14.3.4
Cyclic β-1,2 Glucan: Keeping an Eye on the BCV Membrane

Cyclic glucan is a constituent of the periplasmic space of *Brucella* spp., with osmoregulatory activity and an ability to sequester cholesterol. In nonpathogenic bacteria its function is mainly seen as a stabilizing agent of the membranous cell envelope. *Brucella* mutants in the gene encoding for its synthase, *cgs*, are attenuated in mice and not able to circumvent lysosomal destruction in macrophages unless cyclic glucan is added externally to the bacteria prior to infection [42]. Interestingly, different groups of professional phagocytes show a distinct requirement of cyclic glucan. It plays a role in virulence in macrophages but not in DCs [43], which possess lower intrinsic proteolytic and bactericidal activity than macrophages. The effects of the *cgs* mutation are not simply the results of a disturbed bacterial cell envelope function. Instead, cyclic glucan confers the ability to extract membrane cholesterol, as it is for example present in phagosomal lipid rafts, in a controlled manner. The perturbation of the lipid microdomains may not only alter the fusogenic properties of the BCVs and inhibit their fusion with the lysosome, but also affect general raft-induced signaling pathways. While the requirement of cyclic glucan for virulence is undoubted, it still remains unknown how such a high molecular weight component is able to access the BCV from the periplasmic space, and how the disruption of lipid rafts on the BCV prevents the fusion with lysosomes.

14.3.5
The VirB System: A Door for the Accomplices

The *virB* operon is composed of 12 genes that form a type IV secretion system (T4SS). Its expression seems to be regulated and induced in the early exponential phase and upon exposure to acidic conditions, such as those present in the early BCV (see Section 14.2). The VirB apparatus is an essential factor for virulence of *Brucella* spp. in mice and for intracellular replication *in vitro*. The invasion process into the cell is not affected in *virB* mutants, nor is their early survival lipid raft dependent [16], making a role of VirB for lipid raft-dependent invasion rather unlikely (compare Section 14.2). Instead, this T4SS is indispensable for the maturation of the BCVs. It is hypothesized that effector molecules that are secreted by the VirB system into the host cell actively redirect the cellular trafficking and target the bacteria to their replicative niche. Support for its role in late BCV maturation and replication comes from experiments in the mouse model of infection, where *virB* mutants are initially able to spread within the organism, but then fail to replicate and persist in the host [44]. T4SS from other pathogenic bacteria have been shown to secrete a wide range of molecules, including DNA, proteins and peptidoglycans [45]. Despite the dramatic effects caused by a disrupted *virB* operon in *Brucella*, none of its effectors have so far been identified. Their identification will finally shed more light on the mechanisms that guide the BCV to become an ER-associated compartment.

14.4
Immunology of Phagosomes Containing *Brucella*

Brucella is considered to be an extremely evasive pathogen. The bacteria are able to infect and replicate in cells of the immune system like macrophages and dendritic cells without causing their degradation or activation. As a consequence of this stealth on the one side and active strategies to keep immune responses at low level on the other, they can persist in their hosts for years without being cleared. This section unravels their currently known mechanisms of escape from the immune system.

14.4.1
Stealth

Brucella does not induce pronounced inflammatory responses in infected cells, mice do not undergo sepsis, and no recruitment of neutrophils to the site of infection can be observed. The reason for these missing hallmarks of bacterial infection is at least partly due to the exceptionally low activation profile of the *Brucella* surface [30]. As discussed above, the nonclassical structure of *Brucella* LPS is generally categorized as a poor ligand of TLRs, receptors that induce innate immune responses. Macrophages incubated with purified molecules of *Brucella* LPS produce and secrete, compared with LPS from *E. coli*, only moderate amounts of the proinflammatory cytokines TNFα and IL-12, and of the small and bactericidal p47 GTPases [31]. The weak secretion of cytokines is, however, dependent on the same TLR4- and MyD88-dependent pathway as in the case of classical LPS. Other potential PAMPs of *Brucella* are equally weak in inducing TLR responses [30]. The observations that heat-killed *Brucella* bacteria are able to mount a modest TLR9 response, and that a bacterial lumazine synthase causes activation of DCs via TLR4 [46], show that *Brucella* is not generally low in PAMPs, but that some of them, including DNA and peptidoglycan, are normally not accessible for recognition by the host. The secretive setup allows *Brucella* to replicate in macrophages and DCs, cell types that are usually highly potent for bacterial elimination. Even heavily infected cells are still able to undergo division, which strikingly shows that the replication of *Brucella* remains undetected and does not cause toxic side-effects to the cells. Once inside their replicative niche, *Brucella* bacteria are furthermore insensitive to fusion with lysosomes, even if the hosting macrophage becomes activated [30], making it possible for the bacteria to establish long-lasting infections.

14.4.2 Inhibition of DC activation

Dendritic cells are critical mediators of immune responses. They are strategically located at potential sites of infection, phagocytose pathogens and recognize them via a multitude of pattern recognition receptors. Upon detection of a pathogen, DCs undergo a maturation process during which the cell experiences extensive morphological changes and secretes cytokines. The cytokines activate other cells, mainly macrophages, causing them to become more efficient in pathogen uptake and elimination, and also signal to T cells to travel to the site of infection. Maturing DCs accumulate MHC and a number of costimulatory molecules on their cell surface. They present pathogen-derived antigens bound to the MHC II molecules to T lymphocytes, which will then activate the specific immune response.

The stealth strategy described above is an important factor in avoiding DC maturation and the onset of adaptive immunity, and allows *Brucella* replication in this cell type. However low the activation potential of living *Brucella* may be, in macrophages, and to a certain degree also in DCs, a high proportion of the internalized bacteria undergo lysosomal destruction. This makes potential PAMPs accessible at early stages of the infection and, despite their generally low recognition profile by TLR, potentially sabotages the stealth strategy. Consequently, mice lacking MyD88 show an enhanced susceptibility [47, 48], demonstrating that the innate immune system is, at least at a low level, induced and active in bacterial elimination. DCs infected with *Brucella* nevertheless fail to become fully activated, an effect linked with the low induction of TNFα during *Brucella* infection [40]. This indicates that *Brucella* might have developed, in addition to stealth, active strategies to interfere with innate immunity and specifically with the maturation of DCs and activation of macrophages [27]. In the case of DCs prepared from human monocytes and infected with *B. suis*, IL-12 and TNFα secretion and the capacity to present antigens is strongly induced when Omp25, expressed under the control of the BvrS/BvrR operon, is missing [49]. This result might be a pleiotropic effect caused by a potentially significant change in the outer membrane qualities, but it can be nevertheless taken as a hint that *Brucella* is able to control immune responses. [50].

Recently, another *Brucella* protein, Btp1 (*Brucella* Tir-protein 1) was found to directly limit maturation of DCs [43]. Bacterial proteins carrying TIR domains were only recently suggested to actively inhibit TLR responses [51]. Additional indications for the potential to actively repress innate immunity comes from the analysis of macrophages, where it has been proposed that *Brucella melitensis* interferes with apoptosis by blocking cytochrome *c* release in mitochondria, and thus the production of reactive oxygen [50].

14.4.3 General Role of Proinflammatory Cytokines for Immunity Against *Brucella*

Brucella is effective in avoiding the production of proinflammatory cytokines that have negative effects on bacterial survival. IL-12, [47, 48] secreted in a TLR- and

MyD88-dependent manner by DCs and macrophages, has a central role in controlling *Brucella* infection. It stimulates NK and CD4$^+$ T cells to produce IFNγ, which is a strong activator of macrophage killing and thus has adverse effects on intracellular survival of *B. abortus* [52]. However, depletion of NK cells does not alter the course of brucellosis [53]. Interesting insights about the way that adaptive immunity is geared with *Brucella* survival come from the analysis of BALB/c and C57BL/6 mice. BALB/c mice are highly susceptible to *Brucella* infection and die of the infection within a week, while the C57BL/6 strain has a resistant phenotype which, although it allows bacterial colonization, limits the infection to bacterial numbers at levels 10-fold lower than in BALB/c mice. One of the differences between the two mouse genotypes is that BALB/c mice do not produce IFNγ during the plateau phase of infection. Macrophages isolated from the two strains do not show different abilities to clear the bacteria, but if they are activated by INFγ the intracellular survival rate of the bacteria is reduced. By treating BALB/c mice throughout the infection with IL-12, a continued stimulation of IFNγ production at those late stages of infection is achieved, and BALB/c mice become as resistant against *B. abortus* as C57BL/6 mice [52]. Equally, IFNγ-deficient C57BL/6 mice are more susceptible towards infection with *Brucella* than wild-type mice [36].

An interesting twist in the role of IFNγ activation during *Brucella* infection comes from experiments using *B. abortus bacA* mutants. The length of the fatty acid chains is altered in these mutants, and consequentially, their LPS seems to possess a higher capacity to interact with host TLR4 and to induce proinflammatory cytokines. In correspondence with the negative effects of cytokines on bacterial replication, the *bacA* mutant is attenuated in the susceptible BALB/c mouse genotype, but unexpectedly displays the same degree of attenuation in C57BL/6 mice [36]. Moreover, the *bacA* mutant was more pathogenic in C57BL/6 IFNγ-deficient mice than 2308, causing abscesses and wasting even though the splenic loads of mutant were significantly lower. This result not only supports the importance of IFNγ to control *Brucella* infection, but also demonstrates that some attenuated bacterial strains are potentially inappropriate for use as vaccines in immunocompromised animals or humans [36].

14.4.4 A Potential Role of *virB* in Inflammation and Immunity

A number of inflammation- and immunity-specific genes are induced during early stages of infection of murine splenocytes with *B. abortus* and *B. melitensis*, but not with their respective *virB* mutants [54]. This suggests an active role of the *Brucella* T4SS in inflammation and immune responses at very early time points of infection, additional to its function in the maturation processes of the BCV. Nevertheless, it remains open whether such a transcriptional regulation leads to a final activation or repression of immune responses, which may effect the establishment of the chronic disease.

14.4.5 Abrogating the Antigen Presenting Efficiency

Avoiding antigen presentation renders the infected macrophages less visible to the immune system, reduces Tcell-mediated responses, and is generally compatible with the immunosuppressive stages that have been observed in mouse and human brucellosis. In addition to the indirect route in which *Brucella* interferes with antigen recognition by avoiding lysosomal fusion, *Brucella* is able to affect MHC II-dependent antigen presentation in a direct way as well, using structural components of the outer membrane. *Brucella* Omp19 seems to inhibit MHC II expression and antigen processing and presentation [55]. The nonclassical LPS is also an important inhibitor of antigen presentation. The LPS of *Brucella* bacteria degraded in the lysosome, or LPS derived from dead bacteria ingested by macrophages, first accumulates in the lysosome and becomes associated with MHC II. Following the MHC II pathway, it is then directed to the cell surface, where the LPS causes the associated MHC II to form large macrodomain clusters [56]. These highly stable clusters are also associated with GM1 and GPI-anchored proteins, characteristic proteins for lipid raft-like microdomains. It is hypothesized that the sequestration of MHC II molecules into these rigid, nonfunctional lipid raft-like structures impairs their function and inhibits the presentation of peptides to specific T cells [57].

14.5 Conclusions

Brucellosis is one of the world's most widespread zoonotic diseases, causing substantial losses in livestock and a febrile disease in humans. Broad knowledge of the infection process has been gathered since the discovery of *Brucella* spp. as the causative bacterium. In particular, the establishment of cell culture and murine virulence models have accelerated the progress of the identification of *Brucella* virulence mechanism. Virulence is strongly connected to the ability of *Brucella* to avoid lysosomal fusion by guiding the BCVs to an ER-related compartment. Many of the virulence factors discovered so far are related to the outer cell membrane, a not unexpected finding since this part of the bacterium establishes the first contact with the cell. Indeed, the mode of invasion is of great importance for intracellular survival and replication of the bacteria. The outer membrane of *Brucella* also carries distinct structural features that render *Brucella* poorly detectable by the innate immune system. The list of *Brucella* factors related to virulence is extensive, and many of the observed attenuations are probably due to pleiotropic effects. However, our basic understanding of the structure and function of the virulence factors, together with advances in life imaging in infection conditions may bring light to the mechanisms by which *Brucella* causes disease.

The trafficking of *Brucella* seems to be comparable in the majority of the cell types analyzed to date, although some differences were observed in some cell types using different mutants. These differences concern the invasion process that utilizes distinct mechanisms in phagocytic and nonphagocytic cells, the assumed formation of autophagosome-like structures in nonprofessional phagocytes, and the expression of virulence factors needed in macrophages but not in DCs. A slowly emerging picture hints that a distinction just between professional and nonprofessional phagocytes is an oversimplification of the actual situation. The presence of cyclic glucan is essential for bacterial survival in macrophages, but not in DCs. Infection of PMNs is different from that of both macrophages and DCs, since PMNs do not support bacterial replication. As long as the reasons for those differences in bacterial colonization between the various cell types remain unclear, we are far from constructing a complete picture of the *Brucella* cell cycle.

Some of the molecular determinants identified also enable the bacteria to resist host immunity. In recent years, the abilities of *Brucella* to be a low activator of innate and cell-mediated immune responses, or even to regulate those negatively, have come increasingly into focus and the knowledge that has been gathered on this subject is currently expanding rapidly. To fully understand the effect of *Brucella* on cell-mediated immunity, however, it will be important to integrate the data derived from cellular models into the *in vivo* situation, with its complex arsenal of highly regulated cellular factors. Knowledge gained about the mechanisms used by *Brucella* to survive within the host will eventually aid the identification of potential targets for developing more broadly active and effective vaccines.

References

1 Capasso, L. (2002) *J Infect*, **45**, 122–127.

2 Pappas, G., Papadimitriou, P., Akritidis, N., Christou, L. and Tsianos, E.V. (2006) *Lancet Infect Dis*, **6**, 91–99.

3 Nicoletti, P. (1990) *Adv Biotechnol Processes*, **13**, 147–168.

4 Burkhardt, S., Jimenez de Bagues, M.P., Liautard, J.P. and Kohler, S. (2005) *Infect Immun*, **73**, 6782–6790.

5 Sangari, F.J., Grillo, M.J., Jimenez De Bagues, M.P., Gonzalez-Carrero, M.I. *et al.* (1998) *Vaccine*, **16**, 1640–1645.

6 Schurig, G.G., Sriranganathan, N. and Corbel, M.J. (2002) *Vet Microbiol*, **90**, 479–496.

7 Cloeckaert, A., Zygmunt, M.S. and Guilloteau, L.A. (2002) *Vaccine*, **20**, 1820–1822.

8 Moriyon, I., Grillo, M.J., Monreal, D., Gonzalez, D. *et al.* (2004) *Vet Res*, **35**, 1–38.

9 Sola-Landa, A., Pizarro-Cerda, J., Grillo, M.J., Moreno, E. *et al.* (1998) *Mol Microbiol*, **29**, 125–138.

10 Moreno, E. and Moriyon, I. (2002) *Proc Natl Acad Sci USA*, **99**, 1–3.

11 Castaneda-Roldan, E.I., Avelino-Flores, F., Dall'Agnol, M., Freer, E. *et al.* (2004) *Cell Microbiol*, **6**, 435–445.

12 Castaneda-Roldan, E.I., Ouahrani-Bettache, S., Saldana, Z., Avelino, F. *et al.* (2006) *Cell Microbiol*, **8**, 1877–1887.

13 Rocha-Gracia, R.D., Castaneda-Roldan, E.I., Giono-Cerezo, S. and Giron, J.A. (2002) *FEMS Microbiol Lett*, **213**, 219–224.

14 Eze, M.O., Yuan, L., Crawford, R.M., Paranavitana, C.M. *et al.* (2000) *Infect Immun*, **68**, 257–263.

15 Naroeni, A. and Porte, F. (2002) *Infect Immun*, **70**, 1640–1644.

16 Porte, F., Naroeni, A., Ouahrani-Bettache, S. and Liautard, J.P. (2003) *Infect Immun*, **71**, 1481–1490.

17 Kim, S., Watarai, M., Suzuki, H., Makino, S. *et al.* (2004) *Microb Pathog*, **37**, 11–19.

18 Detilleux, P.G., Deyoe, B.L. and Cheville, N.F. (1990) *Infect Immun*, **58**, 2320–2328.

19 Gorvel, J.P. and Moreno, E. (2002) *Vet Microbiol*, **90**, 281–297.

20 Ackermann, M.R., Cheville, N.F. and Deyoe, B.L. (1988) *Vet Pathol*, **25**, 28–35.

21 Kusumawati, A., Cazevieille, C., Porte, F., Bettache, S. *et al.* (2000) *Microb Pathog*, **28**, 343–352.

22 Guzman-Verri, C., Chaves-Olarte, E., von Eichel-Streiber, C., Lopez-Goni, I. *et al.* (2001) *J Biol Chem*, **276**, 44435–44443.

23 Lopez-Goni, I., Guzman-Verri, C., Manterola, L., Sola-Landa, A. *et al.* (2002) *Vet Microbiol*, **90**, 329–339.

24 Celli, J., de Chastellier, C., Franchini, D.M., Pizarro-Cerda, J. *et al.* (2003) *J Exp Med*, **198**, 545–556.

25 Chaves-Olarte, E., Guzman-Verri, C., Meresse, S., Desjardins, M. *et al.* (2002) *Cell Microbiol*, **4**, 663–676.

26 Starr, T., Ng, T.W., Wehrly, T.D., Knodler, L.A. and Celli, J. (2008) *Traffic*, **9**, 678–694.

27 Celli, J., Salcedo, S.P. and Gorvel, J.P. (2005) *Proc Natl Acad Sci USA*, **102**, 1673–1678.

28 Pizarro-Cerda, J., Meresse, S., Parton, R.G., van der Goot, G. *et al.* (1998) *Infect Immun*, **66**, 5711–5724.

29 Riley, L.K. and Robertson, D.C. (1984) *Infect Immun*, **46**, 231–236.

30 Barquero-Calvo, E., Chaves-Olarte, E., Weiss, D.S., Guzman-Verri, C. *et al.* (2007) *PLoS ONE*, **2**, e631.

31 Lapaque, N., Takeuchi, O., Corrales, F., Akira, S. *et al.* (2006) *Cell Microbiol*, **8**, 401–413.

32 Moreno, E. and Moriyon, I. (2006) *Prokaryotes*, **5**, 315–456.

33 Monreal, D., Grillo, M.J., Gonzalez, D., Marin, C.M. *et al.* (2003) *Infect Immun*, **71**, 3261–3271.

34 Fernandez-Prada, C.M., Nikolich, M., Vemulapalli, R., Sriranganathan, N. *et al.* (2001) *Infect Immun*, **69**, 4407–4416.

35 Manterola, L., Moriyon, I., Moreno, E., Sola-Landa, A. *et al.* (2005) *J Bacteriol*, **187**, 5631–5639.

36 Parent, M.A., Goenka, R., Murphy, E., Levier, K. *et al.* (2007) *Microbes Infect*, **9**, 55–62.

37 Watarai, M., Makino, S. and Shirahata, T. (2002) *Microbiology*, **148**, 1439–1446.

38 Guzman-Verri, C., Manterola, L., Sola-Landa, A., Parra, A. *et al.* (2002) *Proc Natl Acad Sci USA*, **99**, 12375–12380.

39 Lamontagne, J., Butler, H., Chaves-Olarte, E., Hunter, J. *et al.* (2007) *J Proteome Res*, **6**, 1519–1529.

40 Manterola, L., Guzman-Verri, C., Chaves-Olarte, E., Barquero-Calvo, E. *et al.* (2007) *Infect Immun*, **75**, 4867–4874.

41 Conde-Alvarez, R., Grillo, M.J., Salcedo, S.P., de Miguel, M.J. *et al.* (2006) *Cell Microbiol*, **8**, 1322–1335.

42 Arellano-Reynoso, B., Lapaque, N., Salcedo, S., Briones, G. *et al.* (2005) *Nat Immunol*, **6**, 618–625.

43 Salcedo, S.P., Marchesini, M.I., Lelouard, H., Fugier, E. *et al.* (2008) *PLoS Pathog*, **4**, e21.

44 Rolan, H.G. and Tsolis, R.M. (2007) *Infect Immun*, **75**, 2965–2973.

45 Backert, S. and Meyer, T.F. (2006) *Curr Opin Microbiol*, **9**, 207–217.

46 Berguer, P.M., Mundinano, J., Piazzon, I. and Goldbaum, F.A. (2006) *J Immunol*, **176**, 2366–2372.

47 Weiss, D.S., Takeda, K., Akira, S., Zychlinsky, A. and Moreno, E. (2005) *Infect Immun*, **73**, 5137–5143.

48 Macedo, G.C., Magnani, D.M., Carvalho, N.B., Bruna-Romero, O. *et al.* (2008) *J Immunol*, **180**, 1080–1087.

49 Billard, E., Dornand, J. and Gross, A. (2007) *Infect Immun*, **75**, 4980–4989.

50 He, Y., Reichow, S., Ramamoorthy, S., Ding, X. *et al.* (2006) *Infect Immun*, **74**, 5035–5046.

51 Cirl, C., Wieser, A., Yadav, M., Duerr, S. *et al.* (2008) *Nat Med*, **14**, 399–406.

52 Sathiyaseelan, J., Goenka, R., Parent, M., Benson, R.M. *et al.* (2006) *Cell Immunol*, **243**, 1–9.

53 Baldwin, C.L. and Goenka, R. (2004) in *Brucella: Molecular and Cellular Biology* (eds I. Lopez-Goñi and I. Moriyon), Horizon Bioscience, New York, p. 341.

54 Roux, C.M., Rolan, H.G., Santos, R.L., Beremand, P.D. *et al.* (2007) *Cell Microbiol*, **9**, 1851–1869.

55 Barrionuevo, P., Cassataro, J., Delpino, M.V., Zwerdling, A. *et al.* (2008) *Infect Immun*, **76**, 250–262.

56 Forestier, C., Moreno, E., Meresse, S., Phalipon, A. *et al.* (1999) *Infect Immun*, **67**, 4048–4054.

57 Lapaque, N., Forquet, F., de Chastellier, C., Mishal, Z. *et al.* (2006) *Cell Microbiol*, **8**, 197–206.

15
Chlamydiae

Ted Hackstadt

15.1
Diseases

Chlamydiae are bacterial, obligate intracellular pathogens that are the causative agents of many human diseases. There are over 15 serologically defined variants, or serovars, of *Chlamydia trachomatis*. Distinct serovars are associated with different disease states. Serovars A, B, Ba and C are the etiologic agents of endemic, blinding trachoma, the leading cause of infectious blindness worldwide. Serovars D–K are most commonly associated with sexually transmitted disease and are occasionally associated with conjunctivitis. The ocular and urogenital strains of *C. trachomatis* are restricted to mucosal surfaces. The more serious sequelae of these infections are due to untreated or repeated infections that lead to an immunopathological response resulting in blinding trachoma or pelvic inflammatory disease, respectively. The lymphogranuloma venereum (LGV) strains L1, L2 and L3 of *C. trachomatis* are also sexually transmitted but cross the epithelial cell barrier to infect macrophages and ultimately localize to regional lymph nodes. *C. pneumoniae* is a more recently recognized species that is a common cause of community-acquired pneumonia. *C. psittaci* is a zoonotic disease; the infection is common in birds and mammals and occasionally transmitted to humans to cause a severe pneumonia (psittacosis) or, rarely, abortion [1].

15.2
Characterization of the Chlamydial Inclusion and its Relationship to Virulence

Despite differences in disease manifestations and cell tropisms, all chlamydiae share several basic pathogenic mechanisms, including a complex developmental cycle and a unique intracellular niche within the host cell [2]. The chlamydial developmental cycle consists of cell types specialized for extracellular survival and intracellular multiplication [3, 4]. Elementary bodies (EBs) are the infectious, metabolically dormant extracellular forms. They are characterized by their small size, about 0.3 μm in diameter, and a central core of condensed chromatin maintained by a pair of

Intracellular Niches of Microbes. Edited by Ulrich E. Schaible and Albert Haas

ISBN: 978-3-527-32207-7

histone H1-like proteins [5–9]. EBs are relatively resistant to harsh environmental conditions and exhibit no demonstrable transcriptional or metabolic activity. Once internalized, they differentiate over several hours to the larger replicative forms or reticulate bodies (RBs). By 2 h post infection, the nucleoid of EBs has become dispersed and the early RBs have initiated a regulated sequence of temporal gene expression [10, 11]. By 10 h post infection, the RBs begin to multiply by binary fission; after about 18 h, increasing proportions of the RBs begin to differentiate back to EBs, which then accumulate while the remainder of the RBs continue to multiply. Upon completion of their intracellular development, up to 500–1000 EBs are released from each infected cell by mechanisms that can involve either lysis or exocytosis [12].

The entire chlamydial developmental cycle takes place within a membrane-bound parasitophorous vacuole termed an inclusion (Figure 15.1). Intracellular parasites have evolved many different mechanisms to avoid antimicrobial responses of the host cell. Typically, pathogenic microbes that replicate within a vacuolar compartment block maturation of the endosome in which they are engulfed, such that it does not mature into a phagosome with its characteristic harsh antimicrobial environment [13–20]. The chlamydial inclusion is unique among parasitophorous vacuoles in that it does not interact with the endocytic pathway to any appreciable degree but instead intercepts a subset of exocytic vesicles to acquire sphingomyelin and

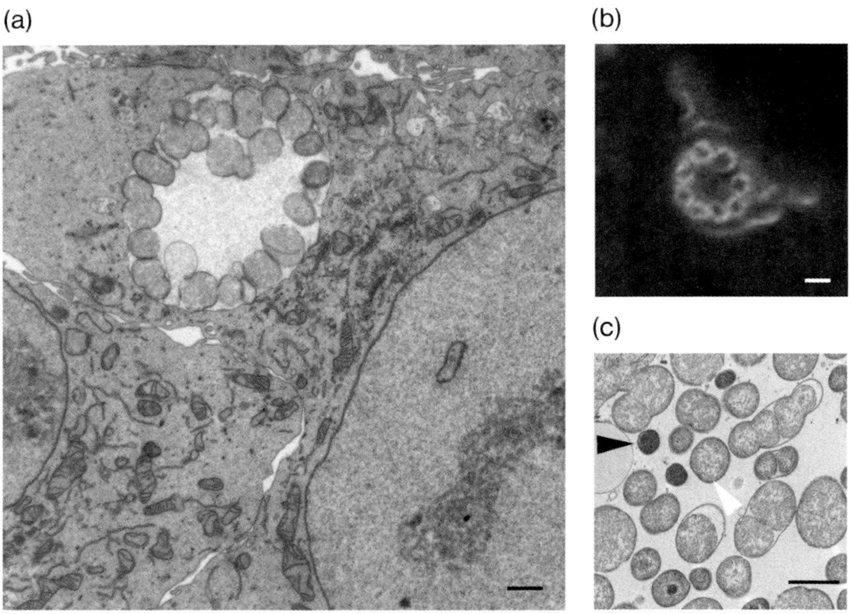

Figure 15.1 (a) Typical *Chlamydia trachomatis* inclusion at 18 h post infection showing characteristic association of reticulate bodies (RBs) with the inclusion membrane. (b) *C. trachomatis* inclusion at 18 h post infection after labeling with C6-NBD-ceramide showing labeling of the RBs within the inclusion and Golgi labeling around the inclusion. (c) *C. trachomatis* inclusion at 24 h post infection showing characteristic elementary bodies (EBs) (black arrow) and RBs (white arrow). Scale bars: 1 μm.

cholesterol [21–29]. Sphingomyelin and cholesterol are not typical prokaryotic lipids and chlamydiae lack the enzymatic machinery to synthesize them. This interaction is readily visualized using a fluorescent analog of ceramide called 6-[*N*-(7-nitrobenzo-2-oxa-1,3-diazol-4-yl)aminocaproyl sphingosine] (C_6-NBD-ceramide). C_6-NBD-ceramide was originally used as a vital stain for the Golgi apparatus, where it is the immediate biosynthetic precursor of sphingomyelin and glucosylceramide [30]. After synthesis in the *cis*-Golgi, sphingomyelin and glucosylceramide are transported to the plasma membrane by a vesicular process in which they are believed to be localized to the luminal side of the vesicle such that they are exposed to the outer leaflet of the plasma membrane upon fusion with the plasma membrane [31]. The topology of the fusion with the chlamydial inclusion is similar, such that the sphingomyelin is exposed on the inner surface of the inclusion membrane. The transport of sphingomyelin to the inclusion is temperature- and energy-dependent with all the properties of Golgi-to-plasma membrane transport [24]. Once exposed on the inner surface of the chlamydial inclusion, however, acquisition of sphingomyelin appears to be by passive transfer of the lipid to apposing membranes. Sphingomyelin is incorporated into the chlamydial cell wall without additional metabolism.

Previous studies of the lipid composition of chlamydial EBs had identified sphingomyelin as one of the lipid components of the cell wall [32]. Interaction with a secretory pathway has been proposed as a pathogenic mechanism by which chlamydiae avoid lysosomal destruction by occupying a pathway dissociated from the endosomal/lysosomal pathway [23]. This interaction is not known to occur with any other intracellular pathogen but is common to all species of chlamydiae [25, 33, 34].

Interestingly, chlamydiae appear to have a great need for lipids and may acquire them from multiple pathways in addition to a direct route from the Golgi. Host phospholipids are also trafficked to the inclusion membrane in a Golgi-independent manner, where they are deacylated and reacylated with a chlamydia-derived branched-chain fatty acid [35, 36]. It has also been proposed that chlamydiae may obtain sphingolipids from CD63-positive multivesicular bodies [37]. In addition, chlamydiae have recently been shown to secrete proteins incorporated into lipid droplets and to recruit lipid droplets as a potential source of neutral lipid [38, 39].

15.3
Genes Involved in Chlamydial Inclusion Maturation

Despite the inability to apply recombinant DNA technologies to questions of chlamydial pathogenesis and cellular interactions, there are several cell biology and proteomic approaches that may be applied to address questions related to chlamydia–host cell interactions. Virtually all stages of chlamydial interaction with the host cell are controlled to some degree by chlamydial gene products. The lack of genetic systems generally precludes definitive identification of function since the candidate genes for specific function cannot be specifically mutated or knocked out, yet there is enough evidence, sometimes circumstantial, to implicate chlamydial genes in regulation of cellular interactions. It may be best to systematically address each of the steps in chlamydial interaction regarding the role of chlamydial gene products in manipulating those interactions.

15.3.1 Attachment

Chlamydial EBs attach to and enter nonphagocytic host cells so efficiently that the term "parasite-specified phagocytosis" has been coined to describe this process [40]. Despite the necessity of these steps in the natural history of an obligate intracellular parasite, these aspects of chlamydial biology are among the most controversial in the field [41]. With a dearth of genetic tools and their obligate intracellular lifestyle, biological screens for chlamydial ligands functioning in attachment are limited. In an effort to identify host factors required for chlamydial uptake, CHO K1 cells were chemically mutagenized with ethylmethanesulfate (EMS) and cells resistant to chlamydial infection selected based upon resistance to the lethal effects of *C. trachomatis* infection. Different resistance phenotypes were selected and characterized. Some were resistant to only the LGV biovar that was used in the selection for resistance while others displayed resistance to multiple *C. trachomatis* serovars but not *C. caviae*, and yet others displayed resistance to both *C. trachomatis* biovars and *C. caviae* [42, 43].

In more detailed analysis of the CHO D4.1–3 cell line resistant to LGV serovar only, it was found that L2 EBs attached to the cell surface normally at 4 °C but were not internalized upon shifting the temperature to 37 °C as shown by their release following treatment with excess heparan sulfate. Typically, EBs are rapidly internalized at 37 °C. By using paraformaldehyde-fixed mutant D4.1–3 cells as the target monolayer, a temperature-dependent irreversible step in EB attachment to the cell surface was identified. Unfortunately, neither the bacterial ligand nor the host receptor were identified [42]. An independently generated mutant cell line, CHO6, resistant to both *C. trachomatis* biovars, *C. pneumoniae* and *C. psittaci* was found to be defective in processing protein disulfide isomerase [44].

15.3.2 Entry

As with the ligands and receptors potentially involved in chlamydial attachment, the receptors, signaling pathways and mechanisms of endocytosis are varied and incompletely understood. Again, experimental parameters may influence the conclusions but the implications are that multiple modes of entry may be utilized. For example, centrifugation of *C. psittaci* onto the monolayer versus static incubation can affect the mechanism of entry, that is, microfilament-dependent or microfilament-independent, respectively [45]. In these same studies, post-entry events, such as rates of fusion with lysosomes, were also found to be affected by the mode of inoculation, suggesting that mechanisms of entry can determine subsequent intracellular interactions and chlamydial survival. Several recent reviews discuss the entry processes of chlamydiae [41, 46–48].

Endocytosis of EBs is accompanied by a localized rearrangement of the actin cytoskeleton believed to be required for entry [49]. Ultrastucturally, this localized recruitment of actin leads to the formation of a pedestal-like structure with the EB at the apex. By largely unknown mechanisms, this structure recedes and culminates with the endocytosis of the EB enveloped within a plasma membrane-derived vesicle. The cytoskeletal remodeling initiated while the EBs are still extracellular implies some signal transduction across the plasma membrane to influence cytoskeletal dynamics. Among the earliest known cell-signaling processes in chlamydial infection is the tyrosine phosphorylation of proteins at the site of EB attachment. Proteins of 64–68, 97 and 140 kDa [50, 51] or 75–85 and 100 kDa [52] were observed by immunoblotting of *C. trachomatis*-infected cells with an anti-phosphotyrosine antibody and localized immunofluorescence observed at the site of attachment. These proteins were not identified and assumed to be of host origin. Recently, a chlamydial type III-secreted protein has been identified as tyrosine phosphorylated and may represent one of the phosphorylated protein species previously observed [53]. The protein, termed Tarp (for translocated actin-recruiting protein) was initially recognized in *C. trachomatis*-infected cells and implicated in the entry process. Tarp is preexisting in EBs in an unphosphorylated state and is a substrate for a heterologous, *Yersinia*-based type III secretion assay, thus is believed to be secreted via the chlamydial type III secretion system upon contact with host cells. Within minutes of contact of EBs with the cell surface, Tarp is translocated such that it is exposed to the cytosol of the host cell and rapidly phosphorylated. Ectopically expressed as a recombinant protein in HeLa cells, Tarp associates with actin and is tyrosine phosphorylated. Tarp is thus spatially and temporally associated with chlamydial internalization and thought to play an important role in the entry process but possibly also in subsequent events.

Although Tarp was initially identified in *C. trachomatis*-infected cells based upon tyrosine phosphorylation, an analysis of other chlamydial species revealed that all possess a Tarp ortholog, but only *C. trachomatis* Tarp is phosphorylated [54]. This was substantiated by demonstration that the tyrosine phosphorylated domain of *C. trachomatis* Tarp, which consists of up to six tandem tyrosine-rich repeat units of approximately 50 amino acids each, is absent from chlamydial species other than *C. trachomatis*. Even within *C. trachomatis*, the number of tyrosine-rich repeat units varies [53, 55, 56]. Transfection studies with unique Tarp domains coupled with biochemical analyses showed that the N-terminal tyrosine phosphorylated domain and the C-terminal actin-binding domain functioned independently of each other [54].

Many intracellular pathogens initiate remodeling of the actin cytoskeleton to mediate entry but generally through subversion of the host actin-nucleating machinery [57]. Chlamydial Tarp is unusual in that purified Tarp, or the C-terminal domain, independently nucleates long, linear actin filament formation in *in vitro* actin polymerization assays in the absence of additional host factors [58].

Despite the independent actin-nucleating activity of Tarp, host factors are also required to mediate entry. *C. trachomatis* has a requirement for the Rho family GTPase Rac [59], which activates Abi1, WAVE2 and the Arp2/3 complex to promote internalization [60]. The Arp2/3 complex is a cellular actin nucleator that is co-opted by many intracellular pathogens [57]. Thus there appears to be a requirement for both a bacterial and a host actin-nucleating machinery. A model was proposed taking these requirements into account that suggests a role for Tarp in nucleating actin filament formation at the site of attachment but requires Arp2/3, which nucleates branching filaments from an existing actin filament, to complete the changes in cytoskeletal structure promoting uptake [58]. The signals mediating the activation of Rac and subsequent downstream events are unclear. The absence of tyrosine phosphorylation of Tarp in *C. caviae, C. muridarum* and *C. pneumoniae* suggests that any signals mediated by Tarp phosphotyrosine are not necessary for entry. Of course these requirements could be supplanted by other, unknown host factors in species other than *C. trachomatis*. Indeed, there is evidence that the requirements for cellular factors differ between species. Whereas *C. trachomatis* activates and requires only the Rac-GTPase, *C. caviae* activates and requires both Rac and Cdc42 [61]. Interestingly, tyrosine phosphorylated proteins are transiently recruited near *C. caviae* EBs during uptake although the identify of these proteins are not known [61]. Furthermore, uptake of *C. caviae* or *C. pneumoniae* is blocked by inhibitors of phosphatidylinositol 3-kinase (PI3K) such as wortmannin [61, 62] while uptake of *C. trachomatis* is not [49].

Collectively, the role of tyrosine phosphorylation of Tarp in chlamydial uptake remains in question. Recent data suggest that chemical inhibition of Tarp phosphorylation has no demonstrable effect on internalization of *C. trachomatis* EBs [63]. Because Tarp is exposed and remains for some time on the cytosolic face of the nascent inclusion membrane, potential roles for post-entry events, such as inhibition of lysosomal fusion, have been suggested [53]. Demonstration of these roles awaits experimental confirmation.

15.3.3
Chlamydial Control of Inclusion Membrane Interactions with Vesicular Trafficking Pathways

Although the molecular mechanisms are largely unknown, it is clear that chlamydiae establish control of the cellular interactions of the inclusion membrane. Once EBs are endocytosed into a plasma membrane-derived vesicle, there are no known interactions with any host pathways or organelles [64]. Only after the onset of chlamydial protein synthesis are any interactions with the host cell observed. Notable among these are the readily visualized uptake of fluorescent sphingomyelin analogs and microtubule-mediated trafficking of the EBs to the microtubule-organizing center and peri-Golgi region of the cell [65–67]. If chlamydiae are permitted to interact with cells in the continuous presence of chloramphenicol to inhibit bacterial protein synthesis, they attach and enter normally; however, they fail to become fusogenic with vesicles containing sphingomyelin and remain dispersed in individual vesicles throughout the cytoplasm [67]. Unlike viable chlamydial inclusions, vesicles containing protein synthesis-inhibited EBs ultimately fuse with lysosomes although the rates of lysosomal fusion are very slow compared to control particles. Vesicles containing EBs inhibited with chloramphenicol acquire the lysosomal marker LAMP1 with a half-time of approximately 12 h; this can be compared to a *Yersinia pseudotuberculosis* control, which shows over 80% of the bacteria in LAMP1-positive vesicles by only an hour post infection [67]. These results imply that some component intrinsic to EBs is responsible for the delayed interactions with lysosomes.

Intriguingly, Tarp translocations and tyrosine phosphorylation are not interrupted by inhibition of chlamydial protein synthesis [53]. As described above, at least one chlamydial protein, Tarp, is exposed on nascent inclusions although a role for that presence is unknown.

Chlamydial EBs display no measured metabolic or transcriptional activity but upon entry into host cells they initiate a progression through a pattern of at least three temporal classes of gene expression [10, 11]. The early class of proteins is weighted toward those modifying the inclusion membrane as well as those involved in macromolecular synthesis. These are expressed as early as 2 h post infection. Expression of mid-cycle genes is not initiated until 10–12 h post infection and include many structural genes and those involved in metabolism. Late genes are considered those expressed during the terminal differentiation of RBs back to EBs which begins in individual RBs as early as 16 h post infection and continues asynchronously until the completion of development at 48–72 h post infection. Because chlamydiae begin to receive sphingomyelin within 2 h post infection [24], it is apparent that early gene products define the properties of the chlamydial inclusion. Inhibition of lysosomal fusion and acquisition of sphingomyelin are properties unique to the chlamydial inclusion and are not due to generalized disruption of vesicular trafficking in the host cell [26, 27, 68]. Indeed, co-infections of *C. trachomatis* with the intracellular pathogen *Coxiella burnetii*, the agent of Q fever, demonstrates typical delivery of sphingomyelin to the chlamydiae but not *Coxiella*, and the *Coxiella* parasitophorous vacuole acquires fluid phase markers normally as is typical of *C. burnetii* interaction with endosomal/lysosomal compartments [26]. Collectively, these observations suggest that cellular interactions of the chlamydial inclusion are specific to that compartment and the most probable mechanism for that specificity would be chlamydial modification of the inclusion membrane.

Chlamydiae express a large number of genes encoding proteins localized to the inclusion membrane [11, 69–73]. These inclusion membrane proteins, or Incs, are characterized by a long hydrophobic domain of 40 or more amino acid residues; theoretically sufficient to span a membrane bilayer twice. The *C. trachomatis* genome encodes over 40 inclusion membrane proteins bearing this motif [56]. Of these, over half have been confirmed as present in the inclusion membrane with specific antibodies and over 70% are expressed early in chlamydial development [11]. Microinjection of specific antibodies to a limited number of Incs has confirmed their exposure on the cytosolic face of the inclusion membrane [74–76]. Incs are also substrates for the type III secretion systems [77, 78]. Unlike Tarp, however, the Incs are not detected in EBs and are synthesized *de novo* soon after entry and then secreted out across the inclusion membrane [73]. Because of their location at the interface of the chlamydial inclusion and the host cytosol, the Incs are likely candidates for chlamydial factors regulating vesicular interaction of the inclusion membrane.

If Incs do indeed control the interactions of the inclusion, the mechanisms by which they do so are not clear. Database searches show little similarity to known proteins and few clues as to function. Surprisingly, there is little similarity in Inc repertoires between chlamydial species [56, 79–81]. Because the interactions of the chlamydial inclusion are conserved between chlamydial species and unique to chlamydiae, one would predict that proteins expected to influence interactions with the host cell would also be conserved. This seems not to be the case, however. Many of the Incs identified in *C. trachomatis* are absent from *C. pneumoniae* and those that are detected often show low levels of similarity or may lack the hydrophobic domain. Conversely, *C. pneumoniae* appears to encode many Incs not found in *C. trachomatis*. The diversity in Incs has yet to be explained. It has been hypothesized that a large number of Incs may keep the copy number of individual Incs low [46]. As the Incs are exposed on the cytoplasmic face of the inclusion membrane, they are susceptible to cleavage by proteases and processing through antigen processing pathways. Indeed, MHC class I restricted presentation of at least one Inc, CrpA, targeted by $CD8^+$ T cells from infected mice, implies processing from a cytosolic compartment [82].

Despite the large number of inclusion membrane proteins and their likelihood of controlling cellular interactions, relatively little is known regarding their control of vesicular interactions common to all chlamydiae. Part of this may be due to the diversity in Inc proteins. For example, IncA from *C. trachomatis* shares only about 21% identity with *C. psittaci* IncA and is considerably smaller; 30.3 kDa vs. 38.8 kDa, respectively. IncA is not represented in *C. pneumoniae* at all. IncA is one of the few inclusion membrane proteins with an ascribed function although it does not appear to be a required function. Microinjection of anti-*C. trachomatis* IncA antibodies into the cytosol of *C. trachomatis*-infected HeLa cells results in the formation of aberrant, nonfusogenic inclusions [74]. In addition, *C. trachomatis* IncA interacted with itself in a yeast two-hybrid system but not with *C. caviae* IncA, and *C. caviae* IncA did not interact with itself in the same system. Typically, in cells multiply infected with *C. trachomatis*, the inclusions fuse into a single large inclusion beginning at about 10–12 h post infection; consistent with the synthesis of IncA [74].

Collectively, the data support a model in which IncA on opposing inclusion membrane faces promotes fusion of *C. trachomatis* inclusions which is characteristic of *C. trachomatis* in multiply infected cells [83–85]. The detection of IncA oligomers coupled with molecular modeling has led to a proposal that IncA may form tetramers similar to the SNARE complexes mediating membrane fusion in eukaryotic cells [86]. Laboratory findings were corroborated in analyses of clinical isolates of *C. trachomatis* which displayed aberrant inclusion morphologies. In these isolates, IncA was found to be absent or mutated such that IncA was not detected on the inclusion membrane [87]. A surprisingly high number, about 1.5–2%, of *C. trachomatis* clinical isolates exhibited aberrant inclusion formation and were mutated in IncA, suggesting that IncA is non-essential.

The function(s) of the remainder of the Inc proteins is unclear although there are multiple examples of specific interactions with host molecules involved in membrane trafficking, cell signaling or vesicle fusion. IncG specifically bind cellular 14-3-3β [88], one of a family of highly conserved and broadly expressed proteins involved in diverse cellular functions such as cell cycle regulation, differentiation and apoptosis [89–91]. A role for 14-3-3 in regulation of vesicular transport has also been proposed [92]. 14-3-3 is recruited to the inclusion membrane via binding to conserved 14-3-3 phosphoserine-containing consensus binding domains on IncG. All *C. trachomatis* strains recruit IncG to the inclusion membrane. Surprisingly, IncG is not represented in other chlamydial species and these species also fail to recruit IncG to the inclusion membrane [88]. If 14-3-3 activity is essential for chlamydial replication, it is unclear how the other species supplant that function or whether it represents a process unique to *C. trachomatis.*

The chlamydial inclusion also specifically recruits several of the Rab-GTPases although there are some distinctions in the Rabs recruited between chlamydial strains and species [93]. Rabs 1, 4 and 11 are recruited to *C. trachomatis* serovars L2 and D, as well as *C. muridarum* and *C. pneumoniae.* In contrast, Rab6 is recruited to only *C. trachomatis* and Rab10 is recruited to only *C. muridarum* and *C. pneumoniae.* Notably, Rab5, Rab7 and Rab9, which are associated with the endocytic pathway, were not recruited, thus demonstrating the specificity of Rab recruitment and confirming the chlamydial dissociation from the endocytic pathway [93]. Rab4 is bound by *C. trachomatis* Inc CT229 [94], an inclusion membrane protein not represented in *C. pneumoniae.* In contrast, the *C. pneumoniae* inclusion membrane protein Cpn0585 binds Rab1, Rab10 and Rab11, but not Rab4 or Rab6 [76]. Chlamydial Inc proteins can thus bind individual or multiple Rab proteins.

Despite Rab6 being recruited to all *C. trachomatis* serovars examined [93], the Rab6 effector Bicaudal D1 is recruited only to an LGV serovar [95]. Like the unique Rab proteins associated with different species, the biovar specificity of Bicaudal D1 recruitment to LGV is surprising and suggests that any necessary function may be provided by other, unknown effectors in the ocular and urogenital biovars. Rab-GTPases, in their active GTP-bound state, act as positive or negative regulators of SNARE complex assembly [96, 97], thus Rab recruitment by the chlamydial inclusion likely indicates a role for these proteins in controlling the vesicular interactions of the inclusion. How the complex and varied assortment of Rab proteins recruited to the inclusion confers specificity in these interactions remains to be elucidated.

15.4
Immunology of the Chlamydial Inclusion

Due to the socio-economic burden of human chlamydial diseases, there is great interest in the development of effective vaccines and consequently the mechanisms of protective immunity. Unfortunately, development of a chlamydial vaccine has not been a trivial undertaking. Whole-cell vaccines have induced short-lived and strain-specific immunity in both humans and primate models but can lead to a deleterious hypersensitivity believed to contribute to the pathogenesis of the disease [98–101]. There are several factors that conspire to make development of a chlamydial vaccine challenging. *C. trachomatis* is strictly a human pathogen thus well-adapted to its host and the most widespread diseases are restricted to ocular or genital mucosal surfaces. Furthermore, there are over 15 serovars of *C. trachomatis*, and immunity, when it occurs, is typically serovar-specific. Nevertheless, development of an efficacious vaccine remains a significant goal in efforts to reduce chlamydial disease worldwide. Innate immune mechanisms as well as mucosal and serum antibodies, and $CD4^+$ and $CD8^+$ T cells have all been implicated in resolution of chlamydial infection. The reader is referred to several recent reviews on the status of vaccine development and mechanisms of immunity [102–106].

By remaining sequestered within a vacuole outside of the endocytic pathway, chlamydiae generally avoid cellular antigen processing and presentation pathways. From within the inclusion, however, chlamydiae secrete a number of proteins that are exposed to the cytosol and are thus susceptible to proteosomal processing and presentation via the MHC class I pathway to activate $CD8^+$ T cells. Indeed, at least one inclusion membrane protein has been shown to be recognized by $CD8^+$ T cells [82] and a random screen against a genomic expression library was used to identify an additional $CD8^+$ T cell antigen, Cap1. Cap1 is also associated with the inclusion membrane but does not display the characteristic bilobed hydrophobic domain typical of Incs [107]. In addition to actively secreted or exposed chlamydial protein antigens, antigen may also be released in blebs from the inclusion; a process accelerated by antibiotic treatment of infected cells [108].

In a sophisticated mechanism to avoid immune surveillance, chlamydiae secrete into the cytosol a chlamydial protease activity factor (CPAF) [109] that degrades the transcription factors RFX5 and USF-1 to downregulate MHC class I and II expression in response to interferon γ (IFNγ) [110, 111]. Chlamydiae modulate responses to IFNγ in other ways also [112]. IFNγ induces the synthesis of indoleamine 2,3-dioxygenase (IDO), which degrades tryptophan to effectively deplete intracellular tryptophan pools. IFNγ causes *C. trachomatis* to enter a persistent, noncultivable state with aberrant RBs that fail to undergo secondary differentiation to EBs [113, 114]. Sexually transmitted *C. trachomatis* strains, but not ocular strains, upregulate tryptophan synthase (trpA and trpB) in response to IFNγ and tryptophan depletion. Although all chlamydiae lack the full tryptophan biosynthetic operon, the sexually transmitted strains of *C. trachomatis* can overcome the activity of IDO by the synthesis of tryptophan from indole [115, 116]. All strains can recover from low-level IFNγ induction of the aberrant persistent state upon removal of IFNγ [117].

As obligate intracellular parasites with an intracellular developmental stage that is non-infectious, it is imperative that chlamydiae survive within infected cells for sufficient time to initiate secondary differentiation back to the infectious EB form. Accordingly, they have evolved several mechanisms to avoid immune surveillance and evade cellular defense mechanisms. Chlamydiae promote their intracellular survival by actively inhibiting caspase-3 activation and apoptosis [118–120]. The mechanism does not involve activation of NFκB but may involve inhibition of cytochrome *c* release from mitochondria [118–120]. Resistance to apoptosis appears to involve the proteosomal degradation of the proapoptotic BH3-only proteins Bim, Puma and Bad during chlamydial infection [121, 122].

15.5
Summary and Outlook

With their obligate intracellular lifestyle chlamydiae provide a number of challenges to research. Despite these limitations, the past decade has seen a tremendous increase in our knowledge of the cellular interactions of chlamydiae. Chlamydiae have clearly adapted to an unusual intracellular niche that provides them with a safe refuge from many cellular defenses. Understanding the mechanisms of chlamydial subversion of host cell activities may also yield new insights into cellular functions by providing a novel target organelle with which to study these interactions. The close and unique interactions of chlamydiae with the host cell offer several novel sites for potential intervention.

Acknowledgments

Work from the authors' laboratory was supported by the Intramural Research Program of the National Institute of Allergies and Infectious Diseases, National Institutes of Health.

References

1 Schachter, J. (1999) in *Chlamydia; Intracellular Biology, Pathogenesis, and Immunity* (ed. R.S. Stephens), American Society for Microbiology Press, Washington, D.C., pp. 139–169.

2 Moulder, J.W. (1991) *Microbiol Rev*, **55**, 143–190.

3 Friis, R.R. (1972) *J Bacteriol*, **110**, 706–721.

4 Hatch, T.P. (1999) in *Chlamydia: Intracellular Biology, Pathogenesis, and Immunity* (ed. R.S. Stephens), American Society for Microbiology Press, Washington D.C., pp. 29–67.

5 Barry, C.E.D., Hayes, S.F. and Hackstadt, T. (1992) *Science*, **256**, 377–379.

6 Brickman, T.J., Barry, C.E. 3rd and Hackstadt, T. (1993) *J Bacteriol*, **175**, 4274–4281.

7 Hackstadt, T., Baehr, W. and Ying, Y. (1991) *Proc Natl Acad Sci USA*, **88**, 3937–3941.

8 Perara, E., Ganem, D. and Engel, J.N. (1992) *Proc Natl Acad Sci USA*, **89**, 2125–2129.

9 Tao, S., Kaul, R. and Wenman, W.M. (1991) *J Bacteriol*, **173** (9), 2818–2822.

10 Belland, R.J., Zhong, G., Crane, D.D., Hogan, D. *et al.* (2003) *Proc Natl Acad Sci USA*, **100**, 8478–8483.

11 Shaw, E.I., Dooley, C.A., Fischer, E.R., Scidmore, M.A. *et al.* (2000) *Mol Microbiol*, **37**, 913–925.

12 Hybiske, K. and Stephens, R.S. (2007) *Proc Natl Acad Sci USA*, **104**, 11430–11435.

13 Deretic, V. and Fratti, R.A. (1999) *Mol Microbiol*, **31**, 1603–1609.

14 Duclos, S. and Desjardins, M. (2000) *Cell Microbiol*, **2**, 365–377.

15 Garcia-del Portillo, F. and Finlay, B.B. (1995) *Trends Microbiol*, **3**, 373–380.

16 Hackstadt, T. (2000) *Traffic (Review)*, **1**, 93–99.

17 Haas, A. (2007) *Traffic*, **8**, 311–330.

18 Meresse, S., Steele-Mortimer, O., Moreno, E., Desjardins, M. *et al.* (1999) *Nat Cell Biol*, **1**, E183–E188.

19 Russell, D.G. (1996) *Trends Cell Biol*, **5**, 125–128.

20 Sinai, A.P. and Joiner, K.A. (1997) *Annu Rev Microbiol*, **51**, 415–462.

21 Al-Younes, H.M., Rudel, T. and Meyer, T.F. (1999) *Cell Microbiol*, **1**, 237–247.

22 Carabeo, R.A., Mead, D.J. and Hackstadt, T. (2003) *Proc Natl Acad Sci USA*, **100**, 6771–6776.

23 Hackstadt, T., Fischer, E.R., Scidmore, M.A., Rockey, D.D. *et al.* (1997) *Trends Microbiol*, **5**, 288–293.

24 Hackstadt, T., Rockey, D.D., Heinzen, R.A. and Scidmore, M.A. (1996) *EMBO J*, **15**, 964–977.

25 Hackstadt, T., Scidmore, M.A. and Rockey, D.D. (1995) *Proc Natl Acad Sci USA*, **92**, 4877–4881.

26 Heinzen, R.A., Scidmore, M.A., Rockey, D.D. and Hackstadt, T. (1996) *Infect Immun*, **64**, 796–809.

27 Scidmore, M.A., Fischer, E.R. and Hackstadt, T. (1996) *J Cell Biol*, **134**, 363–374.

28 Taraska, T., Ward, D.M., Ajioka, R.S., Wyrick, P.B. *et al.* (1996) *Infect Immun*, **64**, 3713–3727.

29 van Ooij, C., Apodaca, G. and Engel, J. (1997) *Infect Immun*, **65**, 758–766.

30 Lipsky, N.G. and Pagano, R.E. (1985) *Science*, **228**, 745–747.

31 Lipsky, N.G. and Pagano, R.E. (1985) *J Cell Biol*, **100**, 27–34.

32 Newhall, W.J. (1988) in *Microbiology of Chlamydia* (ed. A.L. Barron), CRC Press, Inc., Boca Raton, Florida, pp. 47–70.

33 Rockey, D.D., Fischer, E.R. and Hackstadt, T. (1996) *Infect Immun*, **64**, 4269–4278.

34 Wolf, K. and Hackstadt, T. (2001) *Cell Microbiol*, **3**, 145–152.

35 Wylie, J.L., Hatch, G.M. and McClarty, G. (1997) *J Bacteriol*, **179**, 7233–7242.

36 Su, H., McClarty, G., Dong, F., Hatch, G.M. *et al.* (2004) *J Biol Chem*, **279**, 9409–9416.

37 Beatty, W.L. (2006) *J Cell Sci*, **119**, 350–359.

38 Cocchiaro, J., Kumar, Y., Fischer, E.R., Hackstadt, T. *et al.* (2008) *Proc Natl Acad Sci USA*, **105**, 9379–9384.

39 Kumar, Y., Cocchiaro, J. and Valdivia, R.H. (2006) *Curr Biol*, **16**, 1646–1651.

40 Byrne, G.I. and Moulder, J.W. (1978) *Infect Immun*, **19**, 598–606.

41 Hackstadt, T. (1999) in *Chlamydia: Intracellular Biology, Pathogenesis, and Immunity* (ed. R.S. Stephens), American Society for Microbiology Press, Washington, D.C., pp. 101–138.

42 Carabeo, R.A. and Hackstadt, T. (2001) *Infect Immun*, **69**, 5899–5904.

43 Fudyk, T., Olinger, L. and Stephens, R.S. (2002) *Infect Immun*, **70**, 6444–6447.

44 Conant, C.G. and Stephens, R.S. (2007) *Cell Microbiol*, **9**, 222–232.

45 Prain, C.J. and Pearce, J.H. (1989) *J Gen Microbiol*, **135**, 2107–2123.

46 Fields, K.A. and Hackstadt, T. (2002) *Annu Rev Cell Dev Biol*, **18**, 221–245.

47 Dautry-Varsat, A., Balana, M.E. and Wyplosz, B. (2004) *Traffic*, **5**, 561–570.

48 Dautry-Varsat, A., Subtil, A. and Hackstadt, T. (2005) *Cell Microbiol*, **7**, 1714–1722.

49 Carabeo, R.A., Grieshaber, S.S., Fischer, E. and Hackstadt, T. (2002) *Infect Immun*, **70**, 3793–3803.

50 Birkelund, S., Johnsen, H. and Christiansen, G. (1994) *Infect Immun*, **62**, 4900–4908.

51 Birkelund, S., Bini, L., Pallini, V., Sanchez-Campillo, M. *et al.* (1997) *Electrophoresis*, **18**, 563–567.

52 Fawaz, F.S., van Ooij, C., Homola, E., Mutka, S.C. *et al.* (1997) *Infect Immun*, **65**, 5301–5308.

53 Clifton, D.R., Fields, K.A., Grieshaber, S., Dooley, C.A. *et al.* (2004) *Proc Natl Acad Sci USA*, **101**, 10166–10171.

54 Clifton, D.R., Dooley, C.A., Grieshaber, S.S., Carabeo, R.A. *et al.* (2005) *Infect Immun*, **73**, 3860–3868.

55 Carlson, J.H., Porcella, S.F., McClarty, G. and Caldwell, H.D. (2005) *Infect Immun*, **73**, 6407–6418.

56 Stephens, R.S., Kalman, S., Lammel, C., Fan, J. *et al.* (1998) *Science*, **282**, 754–759.

57 Cossart, P. (2000) *Cell Microbiol*, **2**, 195–205.

58 Jewett, T.J., Fischer, E.R., Mead, D.J. and Hackstadt, T. (2006) *Proc Natl Acad Sci USA*, **103**, 15599–15604.

59 Carabeo, R.A., Grieshaber, S., Hasenkrug, A., Dooley, C.A. *et al.* (2004) *Traffic*, **5**, 418–425.

60 Carabeo, R.A., Dooley, C.A., Grieshaber, S.S. and Hackstadt, T. (2007) *Cell Microbiol*, **9**, 2278–2288.

61 Subtil, A., Wyplosz, B., Balana, M.E. and Dautry-Varsat, A. (2004) *J Cell Sci*, **117**, 3923–3933.

62 Coombes, B.K. and Mahony, J.B. (2002) *Cell Microbiol*, **4**, 447–460.

63 Jewett, T.J., Dooley, C.A., Mead, D.J. and Hackstadt, T. (2008) *Biochem Biophys Res Commun*, **371**, 339–344.

64 Scidmore, M.A., Fischer, E. and Hackstadt, T. (2003) *Infect Immun*, **71**, 973–984.

65 Clausen, J.D., Christiansen, G., Holst, H.U. and Birkelund, S. (1997) *Mol Microbiol*, **25**, 441–449.

66 Grieshaber, S., Grieshaber, N. and Hackstadt, T. (2003) *J Cell Biol*, **116**, 3793–3802.

67 Scidmore, M.A., Rockey, D.D., Fischer, E.R., Heinzen, R.A. *et al.* (1996) *Infect Immun*, **64**, 5366–5372.

68 Eissenberg, L.G. and Wyrick, P.B. (1981) *Infect Immun*, **32**, 889–896.

69 Bannantine, J.P., Griffiths, R.S., Viratyosin, W., Brown, W.J. *et al.* (2000) *Cell Microbiol*, **2**, 35–47.

70 Bannantine, J.P., Rockey, D.D. and Hackstadt, T. (1998) *Mol Microbiol*, **28**, 1017–1026.

71 Rockey, D.D., Heinzen, R.A. and Hackstadt, T. (1995) *Mol Microbiol*, **15**, 617–626.

72 Rockey, D.D., Scidmore, M.A., Bannantine, J.P. and Brown, W.J. (2002) *Microbes Infect*, **4**, 333–340.

73 Scidmore-Carlson, M.A., Shaw, E.I., Dooley, C.A., Fischer, E.R. *et al.* (1999) *Mol Microbiol*, **33**, 753–765.

74 Hackstadt, T., Scidmore-Carlson, M.A., Shaw, E.I. and Fischer, E.R. (1999) *Cell Microbiol*, **1**, 119–130.

75 Rockey, D.D., Grosenbach, D., Hruby, D.E., Peacock, M.G. *et al.* (1997) *Mol Microbiol*, **24**, 217–228.

76 Cortes, C., Rzomp, K.A., Tvinnereim, A., Scidmore, M.A. *et al.* (2007) *Infect Immun*, **75**, 5586–5596.

77 Fields, K.A., Mead, D., Dooley, C.A. and Hackstadt, T. (2003) *Mol Microbiol*, **48**, 671–683.

78 Subtil, A., Parsot, C. and Dautry-Varsat, A. (2001) *Mol Microbiol*, **39**, 792–800.

79 Kalman, S., Mitchell, W., Marathe, R., Lammel, C. *et al.* (1999) *Nat Genet*, **21**, 385–389.

80 Read, T.D., Brunham, R.C., Shen, C., Gill, S.R. *et al.* (2000) *Nucleic Acids Res*, **28**, 1397–1406.

81 Read, T.D., Myers, G.S., Brunham, R.C., Nelson, W.C. *et al.* (2003) *Nucleic Acids Res*, **31**, 2134–2147.

82 Starnbach, M.N., Loomis, W.P., Ovendale, P., Regan, D. *et al.* (2003) *J Immunol*, **171**, 4742–4749.

83 Blyth, W.A. and Taverne, J. (1972) *J Hyg (Lond)*, **70**, 33–37.

84 Ridderhof, J.C. and Barnes, R.C. (1989) *Infect Immun*, **57**, 3189–3193.

85 Van Ooij, C., Homola, E., Kincaid, E. and Engel, J. (1998) *Infect Immun*, **66**, 5364–5371.

86 Delevoye, C., Nilges, M., Dautry-Varsat, A. and Subtil, A. (2004) *Biol Chem*, **279**, 46896–46906.

87 Suchland, R.J., Rockey, D.D., Bannantine, J.P. and Stamm, W.E. (2000) *Infect Immun*, **68**, 360–367.

88 Scidmore, M.A. and Hackstadt, T. (2001) *Mol Microbiol*, **39**, 1638–1650.

89 Fu, H., Subramanian, R.R. and Masters, S.C. (2000) *Annu Rev Pharmacol Toxicol*, **40**, 617–647.

90 Pozuelo Rubio, M., Geraghty, K.M., Wong, B.H., Wood, N.T. *et al.* (2004) *Biochemistry*, **379**, 395–408.

91 van Hemert, M.J., Steensma, H.Y. and van Heusden, G.P. (2001) *Bioessays*, **23**, 936–946.

92 Gelperin, D., Weigle, J., Nelson, K., Roseboom, P. *et al.* (1995) *Proc Natl Acad Sci USA*, **92**, 11539–11543.

93 Rzomp, K.A., Scholtes, L.D., Briggs, B.J., Whittaker, G.R. *et al.* (2003) *Infect Immun*, **71**, 5855–5870.

94 Rzomp, K.A., Moorhead, A.R. and Scidmore, M.A. (2006) *Infect Immun*, **74**, 5362–5373.

95 Moorhead, A.R., Rzomp, K.A. and Scidmore, M.A. (2007) *Infect Immun*, **75**, 781–791.

96 Novick, P. and Zerial, M. (1997) *Curr Opin Cell Biol*, **9**, 496–504.

97 Schimmoller, F., Simon, I. and Pfeffer, S.R. (1998) *J Biol Chem*, **273**, 22161–22164.

98 Ward, M.E. (1999) in *Chlamydia: Intracellular Biology, Pathogenesis, and Immunity* (ed. R.S. Stephens), American Society for Microbiology Press, Washington, D.C., pp. 171–210.

99 Sowa, S., Sowa, J., Collier, L.H. and Blyth, W.A. (1969) *J Hyg*, **67**, 699–717.

100 Grayston, J.T., Wang, S.P., Yeh, L.J. and Kuo, C.C. (1985) *Rev Infect Dis*, **7**, 717–725.

101 Grayston, J.T., Woolridge, R.L. and Wang, S.-P. (1962) *Ann NY Acad Sci*, **98**, 352–367.

102 Brunham, R.C. and Rey-Ladino, J. (2005) *Nat Rev Immunol*, **5**, 149–161.

103 Ifere, G.O., He, Q., Igietseme, J.U., Ananaba, G.A. *et al.* (2007) *Microbiol Immunol Infect*, **40**, 188–200.

104 Morrison, R.P. and Caldwell, H.D. (2002) *Infect Immun*, **70**, 2741–2751.

105 Roan, N.R. and Starnbach, M.N. (2008) *Cell Microbiol*, **10**, 9–19.

106 Brunham, R.C., Zhang, D.J., Yang, X. and McClarty, G.M. (2000) *J Infect Dis*, **181**, S538–S543.

107 Fling, S.P., Sutherland, R.A., Steele, L.N., Hess, B. *et al.* (2001) *Proc Natl Acad Sci USA*, **98**, 1160–1165.

108 Giles, D.K., Whittimore, J.D., LaRue, R.W., Raulston, J.E. *et al.* (2006) *Microbes Infect*, **8**, 1579–1591.

109 Zhong, G., Fan, P., Ji, H., Dong, F. *et al.* (2001) *J Exp Med*, **193**, 935–942.

110 Zhong, G., Fan, T. and Liu, L. (1999) *J Exp Med*, **189**, 1931–1938.

111 Zhong, G., Liu, L., Fan, T., Fan, P. *et al.* (2000) *J Exp Med*, **191**, 1525–1534.

112 Beatty, W.L., Morrison, R.P. and Byrne, G.I. (1994) *Microbiol Rev*, **58**, 686–699.

113 Beatty, W.L., Belanger, T.A., Desai, A.A., Morrison, R.P. *et al.* (1994) *Infect Immun*, **62**, 3705–3711.

114 Beatty, W.L., Byrne, G.I. and Morrison, R.P. (1993) *Proc Natl Acad Sci USA*, **90**, 3998–4002.

115 Fehlner-Gardiner, C., Roshick, C., Carlson, J.H., Hughes, S. *et al.* (2002) *J Biol Chem*, **277**, 26893–26903.

116 Caldwell, H.D., Wood, H., Crane, D., Bailey, R. *et al.* (2003) *J Clin Invest*, **111**, 1757–1769.

117 Beatty, W.L., Morrison, R.P. and Byrne, G.I. (1995) *Infect Immun*, **63**, 199–205.

118 Fan, T., Lu, H., Hu, H., Shi, L. *et al.* (1998) *J Exp Med*, **187**, 487–496.

119 Fischer, S.F., Schwarz, C., Vier, J. and Hacker, G. (2001) *Infect Immun*, **69**, 7121–7129.

120 Geng, Y., Shane, R.B., Berencsi, K., Gonczol, E. *et al.* (2000) *J Immunol*, **164**, 5522–5529.

121 Fischer, S.F., Vier, J., Kirschnek, S., Klos, A. *et al.* (2004) *J Exp Med*, **200**, 905–916.

122 Ying, S., Seiffert, B.M., Hacker, G. and Fischer, S.F. (2005) *Infect Immun*, **73**, 1399–1403.

16
Coxiella burnetii

Stacey D. Gilk, Daniel E. Voth, and Robert A. Heinzen

16.1
Coxiella burnetii and Q Fever

Coxiella burnetii is an obligate intracellular bacterium and the causative agent of the zoonosis human Q fever. The presentation of acute disease is nondescript and includes flu-like symptoms, debilitating headaches and atypical pneumonia. Chronic Q fever, although much less frequent, is more severe as endocarditis is the usual outcome [1]. *Coxiella* is present worldwide with a broad host range that includes mammals, arthropods and birds [2]. Primarily transmitted by aerosols, *Coxiella* displays tropism for mononuclear phagocytes, in particular alveolar macrophages and Kupffer cells [3]. A low infectious dose (<10 organisms) [4], pronounced environmental stability [5] and aerosol transmission make *Coxiella* a potential biowarfare agent and a United States Centers for Disease Control and Prevention category B organism.

Coxiella was originally assigned to the α-proteobacterial order *Rickettsiales* [6], but it is now clear that the genus belongs with the γ-proteobacteria order *Legionellales*, which includes the facultative intracellular pathogen *Legionella pneumophila* [7]. Genetically diverse *Coxiella* isolates have been grouped according to various genotyping schemes [8, 9]. These studies revealed associations between genetic groups and the type of human disease caused (acute vs. chronic), suggesting that isolates may have unique pathogenic potential. Pre-existing host conditions, such as immune suppression, are also clearly involved in the development of chronic Q fever [10].

Lipopolysaccharide (LPS) is currently the only characterized *Coxiella* virulence factor [4]. Isolates from natural infections produce a full-length "phase I" LPS and are virulent in animal models of Q fever. Following serial *in vitro* passage of phase I *Coxiella* in tissue culture cells or embryonated eggs, the organism's LPS becomes increasingly truncated, culminating in the severely truncated LPS of "phase II" *Coxiella* [11]. Cloned phase II organisms are considered avirulent and recent cross-genome comparisons suggest that genetic defects in LPS biosynthesis are the sole reason for this attenuation [9]. Full-length LPS is nonstimulatory and promotes *Coxiella* virulence by masking the organism's outer surface. Specifically, the tetra-acyl

Intracellular Niches of Microbes. Edited by Ulrich E. Schaible and Albert Haas

ISBN: 978-3-527-32207-7

lipid A structure of *Coxiella* LPS is nonpyrogenic and does not bind Toll-like receptor 4 [12]. Moreover, phase I LPS masks other *Coxiella* surface components from innate recognition by dendritic cells [13], inhibits deposition of complement [14] and sterically hinders antibody binding to surface proteins [15].

Intracellularly, *Coxiella* undergoes a biphasic developmental cycle that gives rise to two morphologically distinct cell variants. The small cell variant (SCV) is rod-shaped, 0.2–0.5 μM in length, and has an electron-dense, compacted nucleoid [16]. The large cell variant (LCV) is pleomorphic, >1 μM in size and contains a dispersed nucleoid [16]. *In vitro*, the LCV is disrupted by procedures that the SCV resists, such as osmotic shock and sonication [17]. Following infection of Vero cells with purified SCVs, a 2-day lag phase ensues during which SCVs morphologically differentiate into LCVs [18]. LCVs then replicate slowly with a doubling time of 10–12 h during an exponential phase lasting approximately 4 days. During the early stationary phase (approximately 6 days post infection), a significant portion of LCVs have transitioned to SCVs, and when cell cultures are incubated for an additional two weeks, nearly all *Coxiella* are in the SCV form. Thus, the LCV appears specifically adapted for intracellular replication while the more resistant and metabolically inactive SCV is adapted for extracellular survival. Consequently, the SCV may be responsible for most environmentally acquired cases of Q fever.

16.2
Characterization of the Intrahost Compartment of *Coxiella* and its Relationship to Virulence

During classical phagocytosis, engulfed extracellular material is sequestered within a phagosome that traffics through the endocytic pathway, sequentially acquiring markers of early and late endosomes, and eventually lysosomes [19]. The end result is the phagolysosome, an acidic (pH ~4.5), hydrolytic, and in some cases oxidative compartment that degrades foreign material. This harsh environment is generally not suitable for microbial survival; thus, most successful intracellular pathogens avoid this fate by either actively preventing endocytic maturation of their replication vacuole (e.g., *Mycobacterium tuberculosis*) [20], escaping the nascent phagosome (e.g., *Shigella flexneri* and *Francisella tularensis*) [21, 22], or redirecting their vacuole to interact with other vesicular trafficking pathways (e.g., *Chlamydia* spp.) [23]. *Coxiella* is unique among bacterial pathogens in replicating within a vacuole that, to a first approximation, is indistinguishable from a phagolysosome. However, as discussed in more detail below, the *Coxiella* parasitophorous vacuole (PV) represents a specialized lysosome-like compartment that is actively modified by the pathogen.

16.2.1
Entry into the Host Cell

While *Coxiella* targets macrophages during natural infection [24, 25], *in vitro* the organism promiscuously infects a wide range of fibroblast, epithelial, and macrophage-like cell lines [3]. The results of studies using these different cell model systems has allowed development of a working model of *Coxiella* PV biogenesis.

Coxiella contact with fibroblasts and macrophages results in receptor–ligand interactions that trigger classical actin-dependent phagocytosis [26–28]. In mouse fibroblasts and macrophages, both live and dead bacteria are internalized at equal rates, indicating that *Coxiella* plays a passive role in cellular uptake [26, 29]. Rearrangement of host cell filamentous actin is required for pathogen uptake as internalization is inhibited in cells treated with actin-depolymerizing agents such as cytochalasin D [26]. The bacterial ligand(s) mediating attachment are unknown; however, it is likely proteinaceous as treatment of *Coxiella* with proteases hinders uptake [26]. LPS does not appear to act as a ligand as phase II bacteria are internalized 10–100 times more efficiently by cultured cells than phase I organisms [4]. The lower carbohydrate content of phase II LPS is proposed to make the organism hydrophobic, thereby facilitating nonspecific host plasma membrane interactions [30]. The more accessible surface protein ligands of phase II *Coxiella* may further enhance uptake [15].

During infection of human monocytes, phase II bacteria interact with both complement receptor 3 (CR3) and $\alpha_v\beta_3$ integrin monocyte receptors, while phase I bacteria engage only $\alpha_v\beta_3$ [31]. *Coxiella* encodes a number of proteins containing RGD motifs that could potentially interact with cellular integrin receptors [32]. The lack of phase I CR3 engagement is presumably due to masking of the *Coxiella* CR3 ligand by LPS [13, 15] and/or restricted engagement of the receptor by phase I-induced membrane projections [33]. Indeed, differential engagement of cellular receptors by *Coxiella* phase variants results in dramatically different host cell responses. For example, adherence of phase I, but not phase II, *Coxiella* induces pronounced host cell actin reorganization that manifests as membrane ruffling and protrusions [27, 34]. Moreover, phase I binding activates host cell protein tyrosine kinases that phosphorylate proteins associated with protrusions [34]. While actin rearrangements and protrusion formation induced by other intracellular bacteria, such as *Salmonella typhimurium*, facilitates uptake [35], this process actually appears to impede internalization of phase I *Coxiella* [31].

16.2.2
Maturation of the *Coxiella*-Containing Phagosome

Following internalization, the *Coxiella*-containing phagosome fully matures through the default endocytic pathway, albeit with delayed kinetics [36, 37]. Phagosome interactions with early and late endosomes are observed through the sequential recruitment of Rab5 and Rab7, respectively, two small GTPases involved in vesicle fusion [38]. The early endosomal marker Rab5 is found on the *Coxiella* phagosome as early as 5 min after bacterial uptake, with maximum association at 20 min [38]. Rab7, a late endosomal/lysosomal marker, has maximal phagosome association between 40 and 60 min post infection, when Rab5 levels are decreasing [38]. Phagosome recruitment of lysosomal hydrolases is a defining marker of lysosome fusion; however, trafficking of the lysosomal enzymes acid phosphatase and cathepsin D to the *Coxiella* phagosome is significantly delayed, taking approximately 2 h to reach high levels [37, 38]. These maturation kinetics are substantially longer than those of phagosomes containing inert particles such as latex beads, which acquire lysosomal markers by 15 min after uptake [39].

Interactions between the autophagic pathway and the *Coxiella* phagosome are proposed to delay phagosome maturation [38, 40]. The *Coxiella* phagosome acquires the autophagosomal marker LC3 by 5 min after pathogen uptake [38], and the marker remains associated through at least 2 days post infection [41]. An increasing body of evidence indicates the pathway is an important mediator of innate immunity that acts by delivering both intravacuolar and cytoplasmic bacterial pathogens to lysosomes for degradation [42]. However, *Coxiella* clearly benefits from interactions with the autophagic pathway as its induction, in addition to delaying *Coxiella* phagosome–-lysosome fusion [38, 40], increases the number of infected cells, the size of the PV, and enhances bacterial replication [40].

Coxiella actively participates in maturation of its nascent phagosome by producing protein(s) that mediate autophagosomal interactions and delayed lysosomal fusion. Phagosome acquisition of lysosomal acid phosphatase and cathepsin D are substantially accelerated in cells harboring formalin-fixed [37] or chloramphenicol-inactivated organisms [38], respectively. Moreover, phagosome acquisition of LC3, but not Rab5 or Rab7, is significantly inhibited in chloramphenicol-treated cells [38].

How might stalled phagosome maturation and associated autophagosome interactions augment *Coxiella* infection? By actively pausing PV maturation before lysosomal fusion, *Coxiella* may gain extra time to become metabolically active and express proteins required for survival within the phagolysosomal environment. A modified vacuole may be necessary for division of the replicative LCV developmental form, and LCV metabolic activation may be contingent upon autophagosome delivery of a suitable nutrient pool.

16.2.3
The Mature *Coxiella* Parasitophorous Vacuole

Coincident with *Coxiella* entry into the log phase of its growth cycle (~2 days post infection), the mature PV undergoes a dramatic expansion, resulting in a large, spacious vacuole that is easily visible by light microscopy [18]. At this time, the vacuole can occupy most of the host cell cytoplasm while harboring just a few organisms. Like early phagosome stalling, maintenance of the large, spacious PV and continued fusion with lysosomes relies on *Coxiella* protein synthesis [43]. This early generation of a seemingly unnecessarily large PV may be a mechanism to dilute toxic lysosomal constituents [44], an effect that may be particularly beneficial to the LCV developmental form present in the vacuole at this point [18]. Around the beginning of the stationary phase of *Coxiella's* growth cycle (5–6 days post infection), the vacuolar space is nearly filled with organisms [18]. The large *Coxiella* PV does not appear detrimental to the host cell, as cells continue to divide without alteration of the cell cycle, with the PV segregating to one of two daughter cells [45, 46]. Unlike vacuoles harboring *Chlamydia trachomatis*, segregation of the *Coxiella* PV does not apparently lead to defects in host cell cytokinesis [47]. (Electron micrographs of early and late *Coxiella* PV are shown in Figure 16.1a.)

The mature PV has numerous characteristics of a mature autophagolysosome including acidity (pH ~4.5), lysosomal hydrolases acid phosphatase, cathepsin D and 5′ nucleotidase, and a membrane containing the late endosomal/lysosomal and autophagosomal markers vacuolar H^+-ATPase, LAMP1, 2 and 3, Rab7, Rab24 and LC3 (Figures 16.1b and 16.2) [3]. Lipid for the expanding PV membrane is presumably delivered by promiscuous fusion with endolysosomal and autophagic vesicles. The PV does not appear to intercept Golgi-derived vesicles as ceramide and ceramide derivatives are not trafficked to the PV [36]. Moreover, unlike *L. pneumophila*, the endoplasmic reticulum is not in close association with the *Coxiella* PV [41].

The *Coxiella* PV membrane is structurally sturdy and impermeable to compounds as small as 500 Da [48]. The high cholesterol content of the PV membrane likely contributes to these characteristics [49]. *Coxiella* subversion of the host's cholesterol distribution machinery is critical for a productive infection. For example, pharmacological reagents that block host cell cholesterol biosynthesis or uptake dramatically impact PV development [49]. Indeed, complete inhibition of *Coxiella* PV development and replication is observed when infected cells are treated with imipramine or ketoconazole, inhibitors of cholesterol uptake and biosynthesis, respectively [49]. In addition to providing mechanical stability, cholesterol, along with sphingolipids, may form lipid raft microdomains in the PV membrane that partition signaling proteins important in PV biogenesis. This hypothesis is supported by decoration of the PV with the lipid raft proteins flotillin-1 and flotillin-2 [49].

16.2.4 Resistance to the Parasitophorous Vacuole Environment

The phagolysosome within a resting macrophage is considered toxic for the vast majority of bacterial pathogens. Bacteriostatic/cidal elements of this compartment include acid pH, hydrolytic enzymes, cationic peptides and potentially reactive oxygen and nitrogen species. *Coxiella* possesses both passive and active mechanisms to combat these threats. The pathogen produces a biochemically unusual peptidoglycan with covalently bound protease-resistant proteins that may confer intrinsic protection against lysosomal digestion [50]. *Coxiella* also maintains an intracellular pH near neutrality within the acidic milieu of the PV [51], a process that is likely aided by the activities of four predicted sodium ion/proton exchangers [32]. An abundance of basic proteins (approximately 25% of *Coxiella* proteins have a predicted isoelectric point of >10) is further proposed to help maintain pH homeostasis by acting as an intracellular buffer [32]. Osmotic stress elicited by the action of cationic peptides may be combated by the action of mechanosensitive ion channels and osmoprotectant transporters [32].

Infection of resting macrophages by opsonized or non-opsonized *Coxiella* does not elicit a respiratory burst or superoxide production [52], a behavior that may be mediated by a secreted protein tyrosine acid phosphatase that inhibits NADPH oxidase activity [53]. *Coxiella* possesses other enzymes capable of detoxifying deleterious

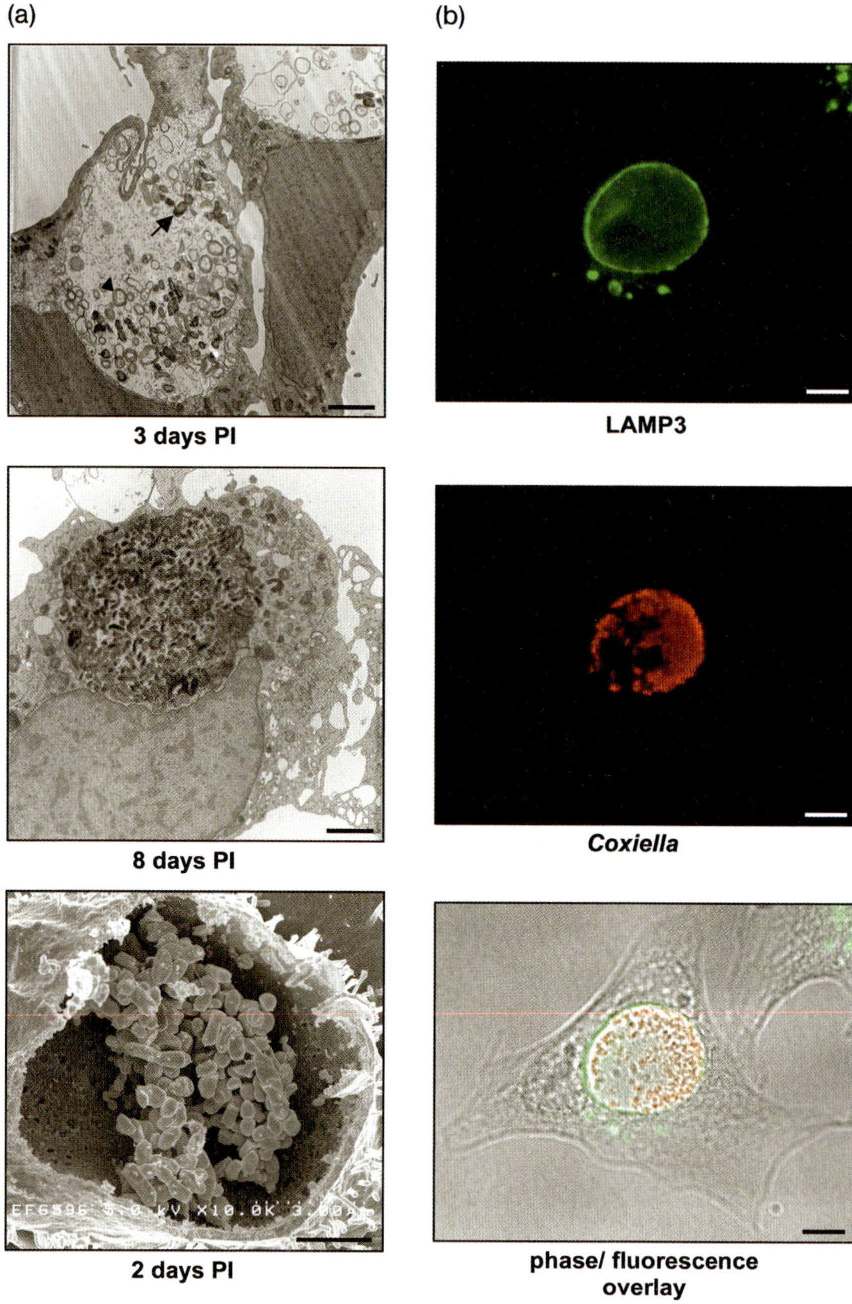

Figure 16.1 Fluorescence, phase contrast, and electron micrographs of host cells infected with the *Coxiella* Nine Mile phase II isolate. (a) Electron micrographs of *Coxiella*-infected Vero cells. Cells were infected for 3 days (top panel) or 8 days (center panel) and examined via transmission electron microscopy. At 3 days post infection (PI), cells harbor a large, spacious

reactive oxygen intermediates (ROIs) that may be present in the PV, including iron/manganese- and copper/zinc-containing superoxide dismutases (SODs), thioredoxin peroxidases, and two peroxide-scavenging alkyl hydroperoxide reductases (AphC and AphD) [32]. In macrophages activated with interferon γ, *Coxiella's* impressive array of antioxidant defenses appears overwhelmed as growth is severely inhibited by both ROI and reactive nitrogen species elicited by this cytokine [54].

16.2.5
Nutrient Acquisition and Metabolism

Coxiella is metabolically active under the moderately acidic (pH ~4.5) conditions of the PV while being metabolic quiescent at neutral pH [55]. This behavior is considered a "biochemical strategem" [55] that restricts the organism's metabolism in the extracellular environment, thereby conferring prolonged viability without replication. The mechanism of acid activation of *Coxiella* metabolism is still not understood. The proton gradient established by the PV's low luminal pH may provide a proton motive force that supports transport of metabolites [56]. Other possibilities include pH-dependent activation of transporters or sensory systems that affect gene regulation.

Relative to most obligate intracellular bacteria, *Coxiella* is metabolically complex, with pathways of central carbon metabolism and bioenergetics largely intact [32]. Data strongly suggest that amino acids are the primary carbon sources of *Coxiella*. First, amino acids (e.g., glutamate) are more efficiently transported and metabolized by *Coxiella* in a low pH buffer than sugars (e.g., glucose) [56]. Second, *Coxiella* encodes a large number of transporters associated with amino acid acquisition. For example, the pathogen has a large family of major facilitator superfamily (MSF) transporters that transport a variety of molecules including amino acids [32]. *L. pneumophila* mutants in *phtA*, an MSF transporter of threonine, have severe growth defects in macrophages [57] and it is logical to invoke a similar critical role for MSF transporters in *Coxiella* amino acid acquisition. In addition, *Coxiella* likely acquires amino acids via the activity of di-, tri- and oligo-peptide transporters. Finally, the hydrolytically active *Coxiella* PV is promiscuously fusogenic with autophagosomes and late endosomes/lysosomes that carry cargo expected to be rich in peptides [36, 40]. A similar mechanism for nutrient delivery is proposed for *Leishmania mexicana*, a protozoan pathogen that resides in a PV phenotypically similar to that of *Coxiella* [58].

parasitophorous vacuole (PV) that contains a few organisms (arrow) primarily in the large cell variant developmental form. Autophagosome fusion with the PV may result in the multilamellar bodies also observed in the vacuole (arrowhead). At 8 days PI, the PV has enlarged to occupy a large portion of the host cell and is filled with *Coxiella* displaying a continuum of developmental morphologies. The bottom panel shows a scanning electron micrograph of a Vero cell infected for 2 days with *Coxiella*. The lumen of the PV was exposed using a dry-cleave method. Scale bar: 2 μm. (b) Immunofluorescence and phase-contrast images of HeLa cells infected for 6 days and stained with antibodies directed against *Coxiella* (red) and the late endosomal/lysosomal marker LAMP3 (green). The *Coxiella* PV membrane is heavily labeled with LAMP3. Scale bar: 5 μm.

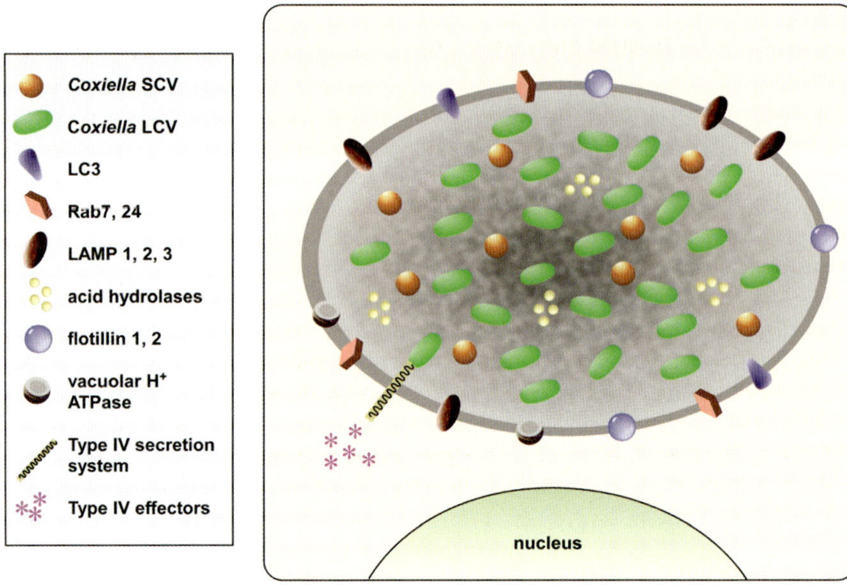

Figure 16.2 Model of the mature *Coxiella* parasitophorous vacuole (PV). The mature *Coxiella* PV is acidic (pH ~4.5) and contains various acid hydrolases. The PV membrane is decorated with markers of late endosomes/lysosomes (LAMP1, 2, 3; Rab7) and autophagosomes (Rab24, LC3), demonstrating interactions with these pathways. The lipid raft proteins flotillin-1 and -2 are also found within the cholesterol-rich PV membrane. Effector molecules, possibly secreted by the bacteria's type IV secretion apparatus, are required to maintain the PV architecture and manipulate host cell functions such as apoptosis.

16.2.6
Inhibition of Host Cell Apoptosis

Like other intracellular pathogens [59], *Coxiella* infection has a potent antiapoptotic effect on host cells. In both macrophages [60] and epithelial cells [61], *Coxiella* infection blocks caspase processing but does not cause degradation of proapoptotic Bcl-2-like proteins as has been observed during infection by *Chlamydia* spp. [62]. Mitochondrial cytochrome *c* is not released in *Coxiella*-infected cells, suggesting the pathogen directly blocks this proapoptotic event or manipulates upstream processes regulating release [61]. Upregulation of both pro- and antiapoptotic gene expression is observed during *Coxiella* infection [60]. The proapoptotic response may result from interaction between a *Coxiella* product and a component of the host's pathogen recognition system. The antiapoptotic gene response involves downregulation of proapoptotic *bax*, *bim*, and *bik* and upregulation of antiapoptotic *A1/bfl-1* and *c-iap2*. Moreover, A1/Bfl-1 and c-IAP2 protein levels are dramatically increased in *Coxiella*-infected cells [60]. In total, the apoptotic gene expression program in *Coxiella*-infected cells confers an overall potent antiapoptotic condition.

16.3
Genes Involved in Diverted Phagosome Maturation/Compartment Maintenance

16.3.1
Components of the Dot/Icm Type IV Secretion System

Multiple Gram-negative bacterial pathogens employ a type IV secretion system (T4SS) for translocation of bacterial proteins to the host cytosol where they modulate specific processes that benefit the pathogen [63]. *Coxiella* encodes homologs of 23 of the 26 Dot/Icm T4SS proteins of *L. pneumophila* [64]. The exceptions are IcmR, DotJ and DotV, which encode a chaperone and two predicted inner membrane proteins, respectively [65, 66]. *Coxiella dotB, icmS, icmW* and *icmT* complement corresponding mutants in *L. pneumophila*, indicating functional conservation of the Dot/Icm T4SS between these two intracellular pathogens [67, 68]. Specifically, conservation of the effector chaperones IcmS and IcmW indicate a similar mechanism of substrate recognition by these two pathogens. Indeed, *L. pneumophila* has recently been used as a surrogate host to identify *Coxiella* proteins that are secreted in a Dot/Icm-dependent fashion [69–71]. *Coxiella icmX, icmQ, dotM, dotL, dotN* and *dotO* do not rescue the corresponding *L. pneumophila* mutants, suggesting there are *Coxiella*-specific interactions between functional Dot/Icm components that are not conserved in *L. pneumophila* [67, 68]. The lack of *Coxiella icmQ* complementation is attributed to failure of the encoded protein to interact with *L. pneumophila* IcmR, an IcmQ chaperone that is absent in *Coxiella* [67]. However, a recent study showed that both *Coxiella* and *Legionella* spp. produce nonhomologous proteins that function similar to IcmR, suggesting that *Coxiella* IcmQ/IcmR-like chaperone function is conserved between these species [72].

16.3.2
Predicted Temporal Utilization of Dot/Icm Effectors During the *Coxiella* Infectious Cycle

As mentioned above, *Coxiella* protein synthesis is required for autophagosomal interactions and stalled phagosome maturation. Similarly, *Coxiella* actively modulates homotypic fusion of pathogen-containing phagosomes, maintenance of the mature PV and antiapoptotic signaling [3, 60]. It is logical to suspect that *Coxiella* protein effectors that regulate these and other host–pathogen interactions are delivered in a temporal fashion to the host cytosol via the pathogen's T4SS. In *L. pneumophila, dot/icm* expression is necessary during uptake by macrophages to result in a productive infection [73]. *In vivo, Coxiella* is metabolically inactive outside of the PV environment [55]; thus, it is unclear if energy-dependent Dot/Icm secretion modulates entry events. One possibility involves contact-dependent release of an effector(s) preloaded into the secretion apparatus. This behavior is demonstrated by the type III secretion system of *Chlamydia trachomatis* in which metabolically quiescent elementary bodies (EBs) secrete Tarp, an actin nucleating protein that induces EB uptake [74].

Early following internalization (1–2 h post infection), *Coxiella* likely employs Dot/Icm T4SS to mediate autophagic interactions and initially delay phagosome maturation. A secreted protein that alters autophagy function is not unprecedented, as *Shigella flexneri* secretes IcsV via its type III secretion system to directly inhibit autophagic sequestration of the pathogen [75]. Alternatively, *Coxiella's* T4SS may allow the indiscriminate release of a molecule containing a pathogen-associated molecular pattern that is detected by a nucleotide oligomerization domain (NOD) leucine-rich repeat (LRR) protein involved in cytosolic innate immune recognition. Type IV secretion systems of *Helocobacter pylori* [76] and *L. pneumophila* [77] are known to nonspecifically leak peptidoglycan and flagellin, respectively, into the host cytosol and NOD-LRR recognition of *L. pneumophila* is postulated to induce autophagy [77]. As the nascent phagosome develops into the spacious lysosome-like PV, Dot/Icm function is presumably necessary for continued recruitment of autophagosomal and endolysosomal vesicles that provide membrane for the enlarging PV and nutrients for *Coxiella* metabolism. *Coxiella* appears to be passively released following host cell demise. Therefore, the pathogen likely does not employ Dot/Icm T4SS substrates that promote host cell egress, as observed in *L. pneumophila* [78]. Finally, as *Coxiella*-infected cells can remain viable for weeks, T4SS substrates that inhibit apoptosis are likely deployed throughout the course of infection. *L. pneumophila* Dot/Icm T4SS substrates are known to delay macrophage apoptosis, with the effector SdhA suppressing caspase activity [79] and SidF binding to BNIP3 and Bcl-rambo to antagonize the proapoptotic effects of these host proteins [80].

16.3.3
Coxiella Dot/Icm Substrates

Over 30 *L. pneumophila* Dot/Icm substrates have been identified to date [81]; however, *Coxiella* does not encode homologs of these proteins, which is consistent with its unique vacuolar niche [82]. *Coxiella* lacks genetic systems to directly identify Dot/Icm substrates. Therefore, indirect approaches using bioinformatic screens and *L. pneumophila* as a surrogate host have been employed to identify *Coxiella* effector molecules [69–71]. Candidate T4SS substrates are identified as those delivered to the cytosol by *L. pneumophila* when fused to calmodulin-activated adenylate cyclase [83].

An emerging theme among secreted bacterial effectors is the presence of eukaryotic-like motifs/domains that functionally mimic host cell protein activity [84]. Eukaryotic domains exploited in this manner include coiled-coil domains (CCDs), tetratricopeptide repeats (TPRs), leucine-rich repeats, F-box domains and ankyrin repeats (Anks). Unifying characteristics of these motifs/domains are their involvement in protein : protein interactions and host signaling pathways. Bioinformatic analyses of the sequenced *Coxiella* Nine Mile isolate genome revealed a large number of proteins containing eukaryotic-like motifs/domains. These include at least eight TPR proteins, twelve Anks and three F-box proteins [3, 32, 85]. Preliminary results from our laboratory and the laboratory of C.R. Roy indicate that most *Coxiella* Anks are Dot/Icm substrates that, when ectopically expressed, traffic to a variety of subcellular locations including the PV membrane [69–71]. Identification of host cell targets of both Ank and non-Ank effectors, and the cellular functions they manipulate, will provide valuable insight into mechanisms of *Coxiella* PV establishment and successful parasitism.

16.4 Immunology of the *Coxiella* Parasitophorous Vacuole

The immune function of the *Coxiella* PV is poorly defined. In HeLa cells, the vacuole functions as the loading compartment of the class II antigen presentation pathway [86]. As the PV represents the primary lysosomal compartment in infected cells, class II MHC molecules such as HLA-DM and HLA-DR accumulate in the vacuole. This results in elevated levels of HLA-DM/HLA-DR complexes that are thought to alter the dynamics and repertoire of peptide loading, which consequently may lead to downstream defects in the immune recognition of *Coxiella*. The MHC class I antigen presentation pathway may also recognize and process *Coxiella* protein antigens that are delivered to the cytosol by cross-presentation [87] or by the T4SS.

16.5 Summary and Outlook

Coxiella is the only intracellular bacterium that thrives in a phagolysosome-like PV. *Coxiella* actively remodels its vacuolar compartment to create a highly fusogenic, specialized organelle that supports the pathogen's metabolism and subsequent differentiation into a replicative form. In addition to vesicular trafficking processes, *Coxiella* must also subvert cellular functions, such as apoptosis, to maintain the host cell for the duration of its lengthy infectious cycle. *Coxiella's* obligate intracellular nature and lack of genetic systems present significant experimental obstacles towards unraveling the molecular details of PV modification and host cell manipulation. However, *Coxiella* T4SS effectors and other secreted virulence factors are currently being identified using heterologous hosts (e.g., *L. pneumophila*) and other indirect means, and their functional characterization will dramatically aid our ability to model *Coxiella* cellular microbiology. This will ultimately lead to a better understanding of mechanisms used by the pathogen to cause animal and human disease.

Acknowledgments

We thank Dale Howe and Shelly Robertson for critical reading of the manuscript, Anita Mora for graphic illustrations, and Elizabeth Fischer for electron microscopy.

This research was supported by the Intramural Research Program of the National Institutes of Health, National Institute of Allergy and Infectious Diseases.

References

1 Maurin, M. and Raoult, D. (1999) *Clin Microbiol Rev*, **12**, 518–553.
2 Babudieri, C. (1959) *Adv Vet Sci*, **5**, 81–84.
3 Voth, D.E. and Heinzen, R.A. (2007) *Cell Microbiol*, **9**, 829–840.
4 Moos, A. and Hackstadt, T. (1987) *Infect Immun*, **55**, 1144–1150.
5 Williams, J.C. (1991) in *Q Fever: The Biology of Coxiella Burnetii (eds.* J.C. Williams and H.A. Thompson), CRC Press, Boca Raton, FL, pp. 21–71.
6 Drancourt, M. and Raoult, D. (1994) *FEMS Microbiol Rev*, **13**, 13–24.
7 Weisburg, W.G., Dobson, M.E., Samuel, J.E., Dasch, G.A. *et al.* (1989) *J Bacteriol*, **171**, 4202–4206.
8 Glazunova, O., Roux, V., Freylikman, O., Sekeyova, Z. *et al.* (2005) *Emerg Infect Dis*, **11**, 1211–1217.
9 Beare, P.A., Samuel, J.E., Howe, D., Virtaneva, K. *et al.* (2006) *J Bacteriol*, **188**, 2309–2324.
10 Raoult, D., Marrie, T. and Mege, J. (2005) *Lancet Infect Dis*, **5**, 219–226.
11 Hackstadt, T., Peacock, M.G., Hitchcock, P.J. and Cole, R.L. (1985) *Infect Immun*, **48**, 359–365.
12 Zamboni, D.S., Campos, M.A., Torrecilhas, A.C., Kiss, K. *et al.* (2004) *J Biol Chem*, **279**, 54405–54415.
13 Shannon, J.G., Howe, D. and Heinzen, R.A. (2005) *Proc Natl Acad Sci USA*, **102**, 8722–8727.
14 Vishwanath, S. and Hackstadt, T. (1988) *Infect Immun*, **56**, 40–44.
15 Hackstadt, T. (1988) *Infect Immun*, **56**, 802–807.
16 McCaul, T.F. and Williams, J.C. (1981) *J Bacteriol*, **147**, 1063–1076.
17 McCaul, T.F., Hackstadt, T. and Williams, J.C. (1981) in *Rickettsiae and Rickettsial Diseases* (eds W. Burgdorfer and R.L. Anacker), Academic Press, New York, pp. 267–280.
18 Coleman, S.A., Fischer, E.R., Howe, D., Mead, D.J. and Heinzen, R.A. (2004) *J Bacteriol*, **186**, 7344–7352.
19 Haas, A. (2007) *Traffic*, **8**, 311–330.
20 Via, L.E., Deretic, D., Ulmer, R.J., Hibler, N.S. *et al.* (1997) *J Biol Chem*, **272**, 13326–13331.
21 Sansonetti, P.J., Ryter, A., Clerc, P., Maurelli, A.T. and Mounier, J. (1986) *Infect Immun*, **51**, 461–469.
22 Checroun, C., Wehrly, T.D., Fischer, E.R., Hayes, S.F. and Celli, J. (2006) *Proc Natl Acad Sci USA*, **103**, 14578–14583.
23 Hackstadt, T., Rockey, D.D., Heinzen, R.A. and Scidmore, M.A. (1996) *EMBO J*, **15**, 964–977.
24 Stein, A., Louveau, C., Lepidi, H., Ricci, F. *et al.* (2005) *Infect Immun*, **73**, 2469–2477.
25 Khavkin, T. and Tabibzadeh, S.S. (1988) *Infect Immun*, **56**, 1792–1799.
26 Baca, O.G., Klassen, D.A. and Aragon, A.S. (1993) *Acta Virol*, **37**, 143–155.
27 Meconi, S., Jacomo, V., Boquet, P., Raoult, D. *et al.* (1998) *Infect Immun*, **66**, 5527–5533.
28 Capo, C., Moynault, A., Collette, Y., Olive, D. *et al.* (2003) *J Immunol*, **170**, 4217–4225.
29 Tujulin, E., Macellaro, A., Lilliehook, B. and Norlander, L. (1998) *Acta Virol*, **42**, 125–131.
30 Williams, J.C., Peacock, M.G. and McCaul, T.F. (1981) *Infect Immun*, **32**, 840–851.
31 Capo, C., Lindberg, F.P., Meconi, S., Zaffran, Y. *et al.* (1999) *J Immunol*, **163**, 6078–6085.
32 Seshadri, R., Paulsen, I.T., Eisen, J.A., Read, T.D. *et al.* (2003) *Proc Natl Acad Sci USA*, **100**, 5455–5460.
33 Capo, C., Amirayan, N., Ghigo, E., Raoult, D. and Mege, J. (1999) *Clin Exp Immunol*, **115**, 120–123.
34 Meconi, S., Capo, C., Remacle-Bonnet, M., Pommier, G. *et al.* (2001) *Infect Immun*, **69**, 2520–2526.
35 Steele-Mortimer, O., Knodler, L.A. and Finlay, B.B. (2000) *Traffic*, **1**, 107–118.

36 Heinzen, R.A., Scidmore, M.A., Rockey, D.D. and Hackstadt, T. (1996) *Infect Immun*, **64**, 796–809.

37 Howe, D. and Mallavia, L.P. (2000) *Infect Immun*, **68**, 3815–3821.

38 Romano, P.S., Gutierrez, M.G., Beron, W., Rabinovitch, M. and Colombo, M.I. (2007) *Cell Microbiol*, **9**, 891–909.

39 Oh, Y.K. and Swanson, J.A. (1996) *J Cell Biol*, **132**, 585–593.

40 Gutierrez, M.G., Vazquez, C.L., Munafo, D.B., Zoppino, F.C. *et al.* (2005) *Cell Microbiol*, **7**, 981–993.

41 Beron, W., Gutierrez, M.G., Rabinovitch, M. and Colombo, M.I. (2002) *Infect Immun*, **70**, 5816–5821.

42 Swanson, M.S. (2006) *J Immunol*, **177**, 4945–4951.

43 Howe, D., Melnicakova, J., Barak, I. and Heinzen, R.A. (2003) *Cell Microbiol*, **5**, 469–480.

44 Alpuche-Aranda, C.M., Racoosin, E.L., Swanson, J.A. and Miller, S.I. (1994) *J Exp Med*, **179**, 601–608.

45 Roman, M.J., Coriz, P.D. and Baca, O.G. (1986) *J Gen Microbiol*, **132**, 1415–1422.

46 Baca, O.G., Scott, T.O., Akporiaye, E.T., DeBlassie, R. and Crissman, H.A. (1985) *Infect Immun*, **47**, 366–369.

47 Grieshaber, S.S., Grieshaber, N.A., Miller, N. and Hackstadt, T. (2006) *Traffic*, **7**, 940–949.

48 Heinzen, R.A. and Hackstadt, T. (1997) *Infect Immun*, **65**, 1088–1094.

49 Howe, D. and Heinzen, R.A. (2006) *Cell Microbiol*, **8**, 496–507.

50 Amano, K. and Williams, J.C. (1984) *J Bacteriol*, **160**, 989–993.

51 Hackstadt, T. (1983) *J Bacteriol*, **154**, 591–597.

52 Baca, O.G., Li, Y.P. and Kumar, H. (1994) *Trends Microbiol*, **2**, 476–480.

53 Li, Y.P., Curley, G., Lopez, M., Chavez, M. *et al.* (1996) *Acta Virol*, **40**, 263–272.

54 Brennan, R.E., Russell, K., Zhang, G. and Samuel, J.E. (2004) *Infect Immun*, **72**, 6666–6675.

55 Hackstadt, T. and Williams, J.C. (1981) *Proc Natl Acad Sci USA*, **78**, 3240–3244.

56 Hackstadt, T. and Williams, J.C. (1983) *J Bacteriol*, **154**, 598–603.

57 Sauer, J.D., Bachman, M.A. and Swanson, M.S. (2005) *Proc Natl Acad Sci USA*, **102**, 9924–9929.

58 Schaible, U.E., Schlesinger, P.H., Steinberg, T.H., Mangel, W.F. *et al.* (1999) *J Cell Sci*, **112**, 681–693.

59 DeLeo, F.R. (2004) *Apoptosis*, **9**, 399–413.

60 Voth, D.E., Howe, D. and Heinzen, R.A. (2007) *Infect Immun*, **75**, 4263–4271.

61 Luhrmann, A. and Roy, C.R. (2007) *Infect Immun*, **75**, 5282–5289.

62 Fischer, S.F., Vier, J., Kirschnek, S., Klos, A. *et al.* (2004) *J Exp Med*, **200**, 905–916.

63 Backert, S. and Meyer, T.F. (2006) *Curr Opin Microbiol*, **9**, 207–217.

64 Sexton, J.A. and Vogel, J.P. (2002) *Traffic*, **3**, 178–185.

65 Vincent, C.D., Friedman, J.R., Jeong, K.C., Buford, E.C. *et al.* (2006) *Mol Microbiol*, **62**, 1278–1291.

66 Vogel, J.P. (2004) *Trends Microbiol*, **12**, 103–105.

67 Zamboni, D.S., McGrath, S., Rabinovitch, M. and Roy, C.R. (2003) *Mol Microbiol*, **49**, 965–976.

68 Zusman, T., Yerushalmi, G. and Segal, G. (2003) *Infect Immun*, **71**, 3714–3723.

69 Voth, D.E., Howe, D., Vogel, J.P., Unsworth, N. *et al.* (2008) American Society for Microbiology General Meeting, Abstract B-420.

70 Carey, K.L. and Roy, C.R. (2007) American Society for Microbiology General Meeting, Abstract B-332.

71 Luhrmann, A. and Roy, C.R. (2007) American Society for Microbiology General Meeting, Abstract B-344.

72 Feldman, M., Zusman, T., Hagag, S. and Segal, G. (2005) *Proc Natl Acad Sci USA*, **102**, 12206–12211.

73 Roy, C.R., Berger, K.H. and Isberg, R.R. (1998) *Mol Microbiol*, **28**, 663–674.

74 Jewett, T.J., Fischer, E.R., Mead, D.J. and Hackstadt, T. (2006) *Proc Natl Acad Sci USA*, **103**, 15599–15604.

75 Ogawa, M., Yoshimori, T., Suzuki, T., Sagara, H. *et al.* (2005) *Science*, **307**, 727–731.

76 Viala, J., Chaput, C., Boneca, I.G., Cardona, A. *et al.* (2004) *Nat Immunol*, **5**, 1166–1174.

77 Molofsky, A.B., Byrne, B.G., Whitfield, N.N., Madigan, C.A. *et al.* (2006) *J Exp Med*, **203**, 1093–1104.

78 Chen, J., de Felipe, K.S., Clarke, M., Lu, H. *et al.* (2004) *Science*, **303**, 1358–1361.

79 Laguna, R.K., Creasey, E.A., Li, Z., Valtz, N. and Isberg, R.R. (2006) *Proc Natl Acad Sci USA*, **103**, 18745–18750.

80 Banga, S., Gao, P., Shen, X., Fiscus, V. *et al.* (2007) *Proc Natl Acad Sci USA*, **104**, 5121–5126.

81 Ninio, S. and Roy, C.R. (2007) *Trends Microbiol*, **15**, 372–380.

82 Sauer, J.D., Shannon, J.G., Howe, D., Hayes, S.F. *et al.* (2005) *Infect Immun*, **73**, 4494–4504.

83 Sory, M.P. and Cornelis, G.R. (1994) *Mol Microbiol*, **14**, 583–594.

84 Bruggemann, H., Cazalet, C. and Buchrieser, C. (2006) *Curr Opin Microbiol*, **9**, 86–94.

85 Angot, A., Vergunst, A., Genin, S. and Peeters, N. (2007) *PLoS Pathog*, **3**, e3.

86 Lem, L., Riethof, D.A., Scidmore-Carlson, M., Griffiths, G.M. *et al.* (1999) *J Immunol*, **162**, 523–532.

87 Houde, M., Bertholet, S., Gagnon, E., Brunet, S. *et al.* (2003) *Nature*, **425**, 402–406.

17
Ehrlichia and *Anaplasma*

Yasuko Rikihisa

17.1
Ehrlichia and *Anaplasma* and the Illnesses They Cause in Humans

Ehrlichia chaffeensis primarily infects monocytes and macrophages, and it was originally isolated from the blood of a patient with human monocytic ehrlichiosis (HME) in 1990 [1]. *Anaplasma phagocytophilum* (previously known as the human granulocytic ehrlichiosis agent, *Anaplasma phagocytophila*) primarily infects granulocytes (also recently shown to infect endothelial cells [2]) and was originally isolated in 1994 from the blood of a patient with human granulocytic anaplasmosis (HGA), as known as human granulocytic ehrlichiosis [3]. As the human isolates are genetically, antigenically and biologically related to the previously known veterinary pathogens *Ehrlichia phagocytophila* and *Ehrlichia equi*, these closely related bacteria were later classified into the single species *A. phagocytophilum* [4].

HME and HGA are severe influenza-like illnesses characterized by fever, headache, myalgia, anorexia and chills and are frequently accompanied by leukopenia with a leftward shift, thrombocytopenia, anemia and elevated levels of serum hepatic aminotransferases. No vaccines exist for these diseases and the diagnoses are based on retrospective seroconversions or PCR analysis. Doxycycline remains the drug of choice for treating HME and HGA patients. However, delayed initiation of therapy, the presence of underlying illness or immunosuppression often lead to severe complications or death [5, 6]. HME cases are reported primarily in the southeastern and south central United States, as well as in Europe, Mexico, Argentina, Mali and Israel. Cases of HGA have been confirmed primarily in the upper Midwest, the Northeast and the Pacific regions of the United States. HGA has been reported in Europe more frequently than HME.

These bacteria alternate infection between two evolutionary divergent hosts (mammals and ticks). Wild animals such as white-tailed deer and white-footed mice are reservoirs for these bacteria [7, 8]. *E. chaffeensis* has most commonly been identified in the Lone Star tick (*Amblyomma americanum*) [7, 9]. *A. phagocytophilum* has been found in the deer tick (*Ixodes scapularis*) and other *Ixodes* spp. [8, 10].

Intracellular Niches of Microbes. Edited by Ulrich E. Schaible and Albert Haas

ISBN: 978-3-527-32207-7

E. chaffeensis and *A. phagocytophilum* are fastidious obligatory intracellular, small pleomorphic cocci. The *E. chaffeensis* and *A. phagocytophilum* genomes are 1.18 and 1.47 Mb, respectively [11]. They are closely related to obligate intracellular rickettsial pathogens (e.g., *Rickettsia prowazekii,* the agent of epidemic typhus), sharing many orthologous genes and the dual-host (vertebrate and invertebrate) life cycle. *E. chaffeensis* and *A. phagocytophilum* share approximately 640 genes, of which 461 (72%) are also shared with *R. prowazekii* [11]. Approximately 180 genes shared by *E. chaffeensis* and *A. phagocytophilum* are not present in *R. prowazekii. E. chaffeensis* and *A. phagocytophilum* as well as *R. prowazekii* have a diminished ability to synthesize amino acids, and lost most genes for central intermediary metabolism compared to their closest free-living relatives. However, unlike *R. prowazekii, E. chaffeensis* and *A. phagocytophilum* are capable of making all major vitamins, cofactors and nucleotides, and lack ATP/ADP translocase [11]. Another important difference from *R. prowazekii* as well as most other Gram-negative bacteria is that *E. chaffeensis* and *A. phagocytophilum* have lost all genes for the biosynthesis of lipopolysaccharides and most genes for the biosynthesis of peptidoglycans [12]. Other pathogen-associated molecules, such as lipoproteins, however, are present; proteomic analysis has shown that all 15 predicted lipoproteins are expressed by *E. chaffeensis* when cultured in the human promyelocytic leukemia cell line HL-60 [13]. Notably, *E. chaffeensis* and *A. phagocytophilum* have also acquired the ability to take up host cholesterol into their membranes [12], and cholesterol has become indispensable for their survival and infection cycle [12].

Not shared between *E. chaffeensis* and *A. phagocytophilum* are approximately 470 and 580 genes, respectively. These genes are mostly of unknown function and are thought to be responsible for the phenotypic differences between these species as described below.

17.2
Characterization of Host Cell Compartmentalization of *E. chaffeensis* and *A. phagocytophilum*

For *E. chaffeensis* and *A. phagocytophilum* to replicate exclusively in primary immune defense phagocytes, upon their binding, internalization and intracellular proliferation these bacteria must subvert host cell immune responses including the generation of reactive oxygen species [14–16]. Furthermore, this group of bacteria does not kill the host cell but rather inhibits apoptosis to maximize replication [17–20]. *E. chaffeensis* and *A. phagocytophilum* enter and replicate within membrane-bound inclusions that do not become integrated with host cell lysosomal vesicular trafficking. Despite sharing the features described above, *E. chaffeensis* and *A. phagocytophilum* use different pathways to create replicative intracellular compartments for exponential growth. Current knowledge suggests that the replicative inclusions of *E. chaffeensis* resemble early endosomes, whereas those of *A. phagocytophilum* resemble early autophagosomes (as summarized in Table 17.1) [21–23].

Table 17.1 Host intracellular compartments in which *Ehrlichia chaffeensis* and *Anaplasma phagocytophilum* replicate.

Markers	Cellular compartments labeled	*E. chaffeensis*	*A. phagocytophilum*
LAMP1, CD63	Late endosomes, lysosomes	−	−
Transferrin receptor	Early and recycling endosomes	+	−
Rab5, EEA1	Early endosomes	+	−
V-ATPase, DAMP	Endosomes, lysosomes, Golgi	+/−	−
Mannose-6-phosphate receptor	Lysosomes, plasma membrane, late endosomes, Golgi	−	−
Class I and II antigen	Endosomes, membrane vesicles	+/−	+/−
PLCγ2	Cytosol and plasma membrane	+	+
Phosphotyrosine	Cytosol and plasma membrane	+	+
p22^{phox}	Specific granules, secretory granules, plasma membrane, phagosomes, phagolysosomes	−	−
LC3	Autophagosomes, cytosol	ND	+
Beclin 1	Early autophagosomes, cytosol	ND	+/−

LAMP1, Lysosome-associated membrane protein-1; EEA1, early endosomal antigen 1; DAMP, *N*-(3-((2,4-dinitrophenyl)amino)propyl)-*N*-(3-aminopropyl)methylamine, dihydrochloride; PLCγ2, phospholipase-γ2; LC3, microtubule-associated protein light chain 3; ND, not done.

17.2.1 Replicative Inclusions of Dual-Infected Cells

When HL-60 cells are infected with *A. phagocytophilum* and *E. chaffeensis* concurrently, they do not share replicative inclusions following entrance into the host cell (Figure 17.1) [21]. Thus, these two species apparently avoid interfering with each other by entering and occupying distinct cytoplasmic compartments. Thus, the signals required for inclusion biogenesis are spatially restricted to individual inclusions. In addition, the distinct host cell signals induced by each bacterial species do not impede the other's replication.

17.2.2 Caveolae/Lipid Raft and GPI-Anchored Proteins

Treatment of human THP-1 and HL-60 cells for 30 min with methyl-β-cyclodextrin (which depletes the plasma membrane cholesterol), nystatin (which binds cholesterol) or NBD-cholesterol (a cholesterol derivative that can be incorporated in place of cholesterol) before addition of bacteria can block internalization and infection by these pathogens [24]. Caveolin-1 colocalizes with intracellular compartments containing *E. chaffeensis* and *A. phagocytophilum* as early as 5 min after incubation at 37 °C and are continuously detected in replicative inclusions 2 days post infection [24]. Thus, internalization and subsequent infection of leukocytes with *E. chaffeensis* and *A. phagocytophilum* are dependent on caveolae or lipid rafts.

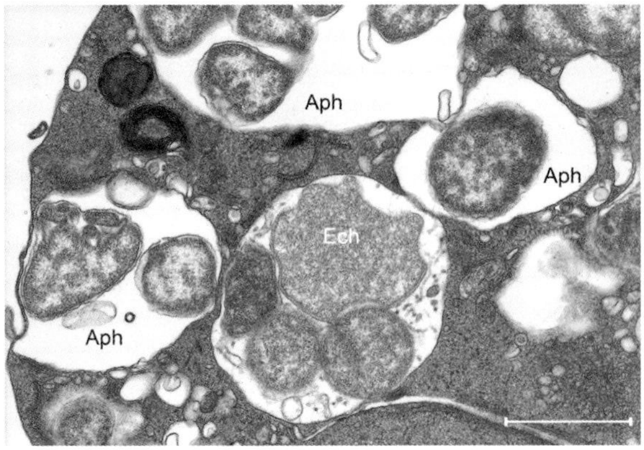

Figure 17.1 Electron microscopy analysis of dually *Anaplasma phagocytophilum*- and *Ehrlichia chaffeensis*-infected HL-60 cells. *A. phagocytophilum* and *E. chaffeensis* are shown in inclusions in an HL-60 cell. The tightly packed *E. chaffeensis* inclusion (Ech, center) is surrounded by three separate *A. phagocytophilum* inclusions (Aph) characterized by loose packing in the inclusions. Bar: 1 μm. Modified with permission from [21].

Glycosylphosphatidylinositol (GPI)-anchored proteins are located in lipid rafts or caveolae [25], and the removal of surface-exposed GPI-anchored proteins by phosphatidylinositol-specific phospholipase C (PI-PLC) pretreatment inhibits internalization and therefore proliferation of *E. chaffeensis* and *A. phagocytophilum* [24]. The replicative inclusions also contain monosialotetrahexosyl ganglioside. These findings indicate that the replicative inclusions of *E. chaffeensis* and *A. phagocytophilum* are maintained by the caveolar trafficking system. As lysosomes are not known to be directed to caveosomes [26], entry through a caveolar trafficking system allows *E. chaffeensis* and *A. phagocytophilum* inclusions to avoid fusion with lysosomes. The persistence of caveolin-1 in the bacterial inclusion membrane is similar to its presence during vesicular transcytosis [27]. These data suggest that *E. chaffeensis* and *A. phagocytophilum* can subvert and delay the transcytosis process in order to establish the environment and have sufficient time to complete their replicative cycle. The fact that inclusions become larger and more abundant in concert with bacterial growth suggests that lipid rafts are continuously infused into bacterial inclusions during intracellular growth.

17.2.3
Phospholipase Cγ2 and Other Tyrosine-Phosphorylated Proteins

Protein tyrosine phosphorylation and the activation of phospholipase Cγ2 (PLCγ2) are rapidly induced and are required for entry and infection of *E. chaffeensis* and *A. phagocytophilum* [28, 29]. Both tyrosine-phosphorylated proteins and PLCγ2 colocalize in the *E. chaffeensis* and *A. phagocytophilum* inclusions as early as 5 min after infection [24]. PLCγ2 is detected in the lipid raft fraction within 1 h of infection with *E. chaffeensis* or *A. phagocytophilum* [24]. As caveolin-1 colocalizes with these bacteria under the same conditions, caveolae are not only the sites of bacterial binding but are also the sites of signaling events for the internalization of *E. chaffeensis* and *A. phagocytophilum*. PLCγ2 is one of the rapidly tyrosine-phosphorylated proteins in *E. chaffeensis*-infected host cells [28]. As the substrate of PLCγ2, phosphatidylinositol 4,5-bisphosphate, is also enriched in caveolae [30], caveolae may facilitate PLCγ2 enzymatic action and a consequent increase in inositol 1,4,5-trisphosphate, leading to an increase in the intracellular Ca^{2+} level. Intracellular Ca^{2+} mobilization as well as calmodulin are essential for infection by these bacteria [28]. Another protein rapidly tyrosine-phosphorylated during *A. phagocytophilum* infection is the bacterially encoded ankyrin repeats-containing protein, AnkA [29, 31]. AnkA is translocated from bacteria cytoplasm to the host cell cytoplasm in a VirB/D-dependent manner, and numerous tyrosine residues are phosphorylated by host tyrosine kinases including Abl-1 [29, 31].

17.2.4 Lysosomal Membrane Markers

CD63 and human lysosome-associated membrane protein 1 (LAMP1) are unique membrane glycoproteins found in late endosomes and lysosomes [32]. Neither *A. phagocytophilum* nor *E. chaffeensis* inclusions colocalize with CD63 or LAMP1 [21, 22, 33] However, lysosome-like structures are strongly labeled throughout the cytoplasm of infected host cells, with a higher population around *A. phagocytophilum* inclusions Table 17.1.

17.2.5 Myeloperoxidase Localization

Myeloperoxidase is stored in azurophil granules inside leukocytes [34]. The localization of myeloperoxidase-positive granules differs between *A. phagocytophilum*- and *E. chaffeensis*-infected host cells. Electron microscopic cytochemical analyses in *A. phagocytophilum*-infected HL-60 cells revealed a large population of myeloperoxidase-positive granules surrounding the ehrlichial inclusions (Figure 17.2); however, myeloperoxidase activity is absent inside the inclusions, indicating the absence of granule fusion. In contrast, *E. chaffeensis* inclusions are not only negative for myeloperoxidase, but there is also no accumulation of myeloperoxidase-positive granules around the inclusions. Instead, more mitochondria are adjacent to the *E. chaffeensis* inclusions (Figure 17.2), as previously noted in infected DH82 cells [35].

17.2.6 Acidic Markers

Vacuolar-type H^+-ATPase is located in early and late endosomes, lysosomes and Golgi and is required for acidification of these compartments [36].

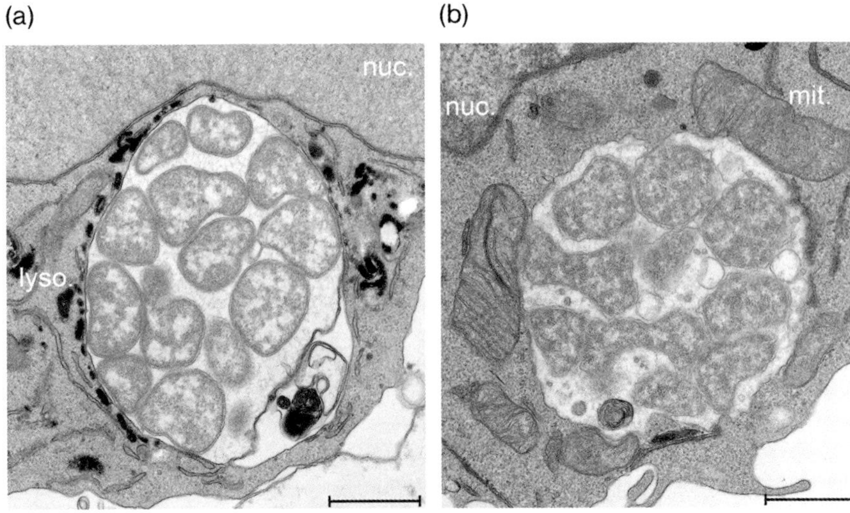

Figure 17.2 Localization of myeloperoxidase in *Anaplasma phagocytophilum*- or *Ehrlichia chaffeensis*-infected HL-60 cells. The electron microscopy images show the localization of myeloperoxidase in granules surrounding *A. phagocytophilum* inclusions (a), suggesting that lysosomes (lyso.) dock but do not fuse with *A. phagocytophilum* inclusions. By contrast, there is a lack of myeloperoxidase-positive granules around *E. chaffeensis* inclusions, with more mitochondria (mit.) adjacent to the inclusions (b). Nucleus (nuc.) is seen above. Bar: 1 μm. (Modified with permission from [21]).

A. phagocytophilum inclusions do not colocalize with vacuolar-type H^+-ATPase [21], and *E. chaffeensis* inclusions only weakly colocalize with this ATPase [22]. The weak base, 3-(2,4-dinitroanilino)-3′-amino-*N*-methyldipropylamine (DAMP), which accumulates in direct proportion to compartment acidity [37], accumulates in replicative inclusions of *E. chaffeensis,* but not those of *A. phagocytophilum* [21, 22]. Moreover, granules labeled with acridine orange dye that accumulate in the acidic compartment in late endosomes and lysosomes do not colocalize with *A. phagocytophilum* replicative inclusions, but are seen around the inclusions [21].

17.2.7 Early Endosomal Markers

The transferrin receptor (TfR), the cytoplasmic/membrane cycling small GTPase Rab5 and the hydrophilic peripheral membrane protein early endosomal antigen 1 (EEA1) are found in early endosomes [38]. The association of EEA1 with endosomal membranes requires Rab5-GTP [39]. EEA1 binds to phosphatidylinositol-3-phosphate [39] and is involved in endosome–endosome docking and fusion [40–42]. EEA1, Rab5 and TfR colocalize with *E. chaffeensis* inclusions [21, 22]. Of note, TfRs are not present during internalization of *E. chaffeensis,* but are progressively upregulated 6 h after infection and accumulate in ehrlichial inclusions [22, 24]. Iron is essential for ehrlichial survival; *E. chaffeensis* is sensitive to deferoxamine, a cell-permeable iron chelator and an actinomycete siderophore. This suggests that *E. chaffeensis* has access to the labile iron pool but does not produce siderophores with high iron-binding affinity [43]. TfR can deliver iron to the cytoplasm by shuttling transferrin between the endosomal compartments and the plasma membrane. In fact, *E. chaffeensis* replicative inclusions are part of the transferrin and TfR membrane recycling pathway, as indicated by the uptake of fluorescein isothiocyanate-conjugated iron-saturated transferrin by the inclusions [22]. Because the internalization and recycling of transferrin and TfR are inhibited by monodansyl cadaverine, a transglutaminase inhibitor [44], monodansyl cadaverine can block ehrlichial proliferation in host cells after entry [28].

E. chaffeensis is sensitive to interferon γ (IFNγ) treatment; in human monocytes, IFNγ reduces the intracellular labile iron pool by reducing the number of host cell surface TfRs [43, 45]. However, after infection, *E. chaffeensis* becomes refractory to IFNγ treatment [43, 45]. This is in part due to upregulation of TfR through activation of iron-responsive factor 1, which stabilizes TfR mRNA during *E. chaffeensis* infection [46]. Thus, *E. chaffeensis* infection alters host cell iron homeostasis. In contrast, *A. phagocytophilum* inclusions do not colocalize with TfR, Rab5 or EEA1 [21]. Neither the TfR mRNA level nor the iron responsive factor 1 activity are altered by *A. phagocytophilum* infection [46].

17.2.8 Rough Endoplasmic Reticulum–Golgi Pathway

Vacuoles containing the facultative intracellular bacteria, *Brucella* and *Legionella*, converge, respectively on the secretory and the retrieval membrane vesicle pathways between the rough endoplasmic reticulum (rER) and the Golgi complex. Vacuoles containing *Brucella* and *Legionella* interact with the Sar-1/CopII complex and the Arf/CopI complex on transport vesicles, respectively, thereby initiating fusion between the vacuoles and these vesicles [47]. Unlike *Brucella* and *Legionella*, *A. phagocytophilum* and *E. chaffeensis* inclusions do not colocalize with β-COP, a COPI complex protein [21, 48], or ribosomes. *Legionella* growth is sensitive to levels of brefeldin A, a fungal metabolite that inhibits anterograde (but not retrograde) vesicular transport from rER to Golgi by causing redistribution and loss of stacked cisternae in the Golgi [49, 50]. However, when *A. phagocytophilum*- or *E. chaffeensis*-infected HL-60 cells were treated with brefeldin A for 24 h despite a complete loss of the Golgi complex, *A. phagocytophilum*- or *E. chaffeensis*-containing inclusions are not affected [21]. Thus, the transitional rER in the rER–Golgi intermediate compartment is likely not involved in the biogenesis of *A. phagocytophilum* or *E. chaffeensis* inclusions.

17.2.9 Golgi–Plasma Membrane Pathway

C_6-NBD-ceramide, like endogenous ceramide, is processed to sphingomyelin or glucosylceramide within the Golgi apparatus prior to transport to the plasma membrane via a vesicle-mediated process [51]. C_6-NBD-ceramide labeling of

A. phagocytophilum-infected cells revealed accumulation in the Golgi and in a small population of vesicles concentrated around the Golgi. However, unlike *Chlamydia* inclusions [52, 53], C_6-NBD-ceramide or -spingomyelin does not accumulate in *A. phagocytophilum* inclusions [21]. Thus, Golgi-derived sphingomyelin-containing vesicles do not fuse with *A. phagocytophilum* inclusions.

17.2.10
Cation-independent Mannose-6-phosphate Receptor

Prelysosomal vesicles, late endosomes and *trans*-Golgi vesicles contain the cation-independent mannose-6-phosphate (CI-M-6-P) receptor, which delivers newly synthesized, soluble lysosomal enzymes to the prelysosomes from the *trans*-Golgi network [54]. The CI-M-6-P receptor releases its ligands to the acidic lumen of prelysosomes, and the receptor is rapidly cycled back to the *trans*-Golgi network [54]. CI-M-6-P receptor is not detected on the membranes of either *A. phagocytophilum* or *E. chaffeensis* replicative inclusions [21]. Taken together with results of aforementioned markers, Golgi and lysosome pathway does not seem to have a significant role in the biogenesis of *A. phagocytophilum* and *E. chaffeensis* inclusions. Webster *et al.* [33], however, reported that some *A. phagocytophilum* inclusions are positive for CI-M-6-P receptor. It is possible that CI-M-6-P is transiently present at certain developmental stages of *A.phagocytophilum* inclusions as CI-M-6-P is found in some autophagosomes [55, 56]. Alternatively, as these bacteria are sensitive to chemical and mechanical stresses and the host cell conditions, it is also possible that some inclusions containing inactivated *A. phagocytophilum* are delivered to late endosomes.

17.2.11
Autophagosome Markers

A. phagocytophilum replicative inclusions exhibit several hallmarks of early autophagosomes, including a double lipid bilayer membrane and colocalization with green fluorescent protein (GFP)-tagged LC3 and Beclin 1, the human homologs of *Saccharomyces cerevisiae* autophagy-related proteins Atg8 and Atg6, respectively [23]. While the levels of the membrane-associated form of LC3, LC3-II, increase during *A. phagocytophilum* infection, *A. phagocytophilum*-containing inclusions enveloped with punctate GFP-LC3 do not colocalize with a lysosomal marker. Stimulation of autophagy by rapamycin favors *A. phagocytophilum* infection. Inhibition of the autophagosomal pathway by 3-methyladenine does not inhibit *A. phagocytophilum* internalization, but reversibly arrests its growth [23]. Although autophagosomes are part of the innate immune system, *A. phagocytophilum* subverts this system to establish itself in an early autophagosome-like compartment segregated from lysosomes. Because autophagy engulfs cytosolic components [57], it is possible that the autophagosome provides *A. phagocytophilum* with direct access to host cytosolic nutrients without the need for transport across the inclusion membrane.

17.2.12 Vesicle-Associated Membrane Protein 2

Vesicle-associated membrane protein 2 (VAMP2), also known as synaptobrevin-2, is an integral membrane protein involved in vesicular transport and neurotransmission [58]. VAMP2 is a soluble *N*-ethylmaleimide-sensitive factor attachment protein receptor (SNARE) that localizes to both vesicle donor and acceptor membranes and is essential for membrane fusion. Several SNARE proteins including VAMP2 are found in lipid rafts [59]. In human blood neutrophils, VAMP2 concentration is high in secretory vesicles and tertiary granules but low in primary and secondary granules [58], and it has been implicated in the exocytosis of both of these vesicle/granule types [60]. Approximately 40% of *A. phagocytophilum* inclusions and 15–20% of *E. chaffeensis* inclusions are labeled with VAMP2 in a punctate pattern [21]. VAMP2 on ehrlichial replicative inclusions may act to acquire lipid membranes or essential nutrients for the growing ehrlichial organism through regulated vesicle trafficking. VAMP2 may also be involved in the fusion of bacterial inclusions with the plasma membrane to liberate the bacteria for transmission to the new host cells.

17.2.13 Clathrin

Clathrin heavy chain is attached to membrane proteins via the adapter protein α-adaptin [61], which is required for receptor-mediated endocytosis of some ligands, and vesicle transport to endosomes from the plasma membrane. *A. phagocytophilum* and *E. chaffeensis* replicative inclusions do not colocalize with either clathrin heavy chain or with α-adaptin, suggesting that clathrin-mediated endocytosis is not involved in biogenesis of inclusions [21, 22, 24].

17.2.14 Annexins

The annexins are a family of structurally related Ca^{2+}-regulated, phospholipid-binding proteins that may have roles in many cellular processes such as exocytosis, ion permeation, lipid metabolism and membrane fusion [62]. *E. chaffeensis* is sensitive to Ca^{2+}-calmodulin inhibitors [28]. However, during infection by *A. phagocytophilum* or *E. chaffeensis*, neither annexin I, II, IV nor VI are colocalized with the bacterial inclusions or areas surrounding the inclusions [21]. Therefore, it is unlikely that annexin-mediated vesicle fusion is involved in biogenesis of inclusions.

17.2.15 NADPH Oxidase Components

NADPH oxidase consists of cytochrome b_{558} (heterodimer of integral membrane proteins gp91^{phox} and p22^{phox}) and several cytosolic components (p67^{phox}, p47^{phox} and p40^{phox}) [63]. In resting neutrophils, the inactive oxidase components remain unassembled and segregated into membranes of secretory vesicles and specific granules, and a cytosolic complex [63]. Upon activation, secretory vesicles and specific granules containing gp91^{phox} and p22^{phox} rapidly fuse with the plasma membrane or the phagosomal membrane, and a complex of p47^{phox}, p67^{phox} and possibly p40^{phox} translocates from the cytoplasm to the plasma membrane or the phagosomal membrane, and associates with cytochrome b_{558} [63, 64]. A functional NADPH oxidase enzyme is assembled at the phagosomes and/or the plasma membrane, exerting lethal effects of O_2^- and related reactive oxygen intermediates on the extracellular space or on ingested bacteria in close proximity. The small GTP-binding protein Rac2, which is highly expressed in neutrophils, is required for oxidase activity through direct interaction with p67^{phox} and cytochrome b_{558} [65]. The overall intensity of the fluorescence labeling of p47^{phox}, p67^{phox} and Rac2 is similar between uninfected and infected HL-60 cells or neutrophils, whereas the intensity of p22^{phox} labeling is lower throughout infected cells compared with uninfected HL-60 cells and neutrophils.

Immunofluorescence labeling and western blotting results indicate that *A. phagocytophilum* reduces the levels of a protein identified by the p22^{phox} antibody in the host cell; the loss is not reversed by adding phorbol-12-myristate-13-acetate (PMA). p47^{phox}, p67^{phox}, p22^{phox} and Rac2 do not colocalize with *A. phagocytophilum* inclusions [66], and PMA treatment does not induce colocalization of these components. Cytoplasmic inclusions occupied by *A. phagocytophilum* lack gp91^{phox} and p22^{phox} [67]. *E. chaffeensis* also reduces the levels of the protein identified by the p22^{phox} antibody in human monocytes. *E. chaffeensis* inclusions do not colocalize with p22^{phox} [15]. Pretreatment of monocytes with lipopolysaccharaide (LPS) inhibits *E. chaffeensis* growth in human monocytes [68]. LPS treatment not only partially prevents the reduction in p22^{phox} labeling by *E. chaffeensis*, but also causes colocalization of p22^{phox} with *E. chaffeensis*, suggesting that NADPH oxidase assembles in *E. chaffeensis*-containing inclusions in LPS-treated monocytes [15].

17.3 Microbe Genes Involved in Subverting Phagosome Maturation/Compartment Maintenance

17.3.1 Enhanced Lysosomal Fusion by Closantel Treatment

The two-component system (TCS), composed of a histidine sensor kinase (HK) and a response regulator (RR), is a signal-sensing, transduction and response regulatory system that allows bacteria to sense a wide variety of environmental signals and respond rapidly to changes in the environment through specific gene activation or repression. Through analysis of amino acid sequences deduced from the genome sequences and protein domain analysis we identified three HKs and three RRs encoded in each of the genomes of *E. chaffeensis* and *A. phagocytophilum* [69, 70]. Biochemical analysis using wild-type and mutant proteins demonstrated that in each bacterium these six proteins comprise three functionally equivalent pairs of TCR, designated PleC-PleD, NtrY-NtrX and CckA-CtrA [69, 70]. Incubation of *E. chaffeensis*-infected THP-1 cells for 1 h with 50 μM closantel, an HK inhibitor, increased colocalization of *E. chaffeensis* inclusions with LAMP1 [69]. This result suggests that activation of at least one of three HKs, leading to activation of the cognate RR is required for inhibition of lysosomal fusion of *E. chaffeensis* inclusions.

17.3.2 Substrates of Type IV Secretion

The prokaryotic type IV secretion system (T4SS) regulates the biogenesis of inclusions containing facultative intracellular bacteria such as *Legionella* and *Brucella* [71–73]. Although a gene encoding a protein containing the Sec7 homology domain, similar to the *Legionella* T4S effector RalF, has been identified in *R. prowazekii*, no such gene has been identified in the *E. chaffeensis* or *A. phagocytophilum* genome. However, *E. chaffeensis* and *A. phagocytophilum* each have a T4SS [74–76]. The *A. phagocytophilum* eukaryote-like AnkA 160 kDa protein is rarely retained within *A. phagocytophilum* or its host cell inclusions, but rather localizes to the host cell cytoplasm [29]. Using a Cre recombinase reporter assay of *Agrobacterium tumefaciens*, we showed that AnkA can be secreted by a VirB/D4-dependent T4SS [29]. Yeast two-hybrid and co-immunoprecipitation analyses demonstrated that AnkA can bind to Abl-interactor 1 (Abi-1), an adapter protein that interacts with Abl-1 tyrosine kinase, thereby mediating AnkA phosphorylation in the host cell cytoplasm. AnkA and Abl-1 are critical for bacterial infection; infection is inhibited upon host cytoplasmic delivery of an antibody against AnkA, by Abl-1 knockdown via targeted siRNA, or by treatment with a specific pharmacological inhibitor of Abl-1 [29]. The elucidation of the inhibitory mechanism requires identification of host proteins that interact with tyrosine-phosphorylated AnkA.

Other studies have shown that *A. phagocytophilum* induces tyrosine phosphorylation of the 190 kDa AnkA [31] (AnkA varies between strains of *A. phagocytophilum*). AnkA also localizes within nuclei of infected HL-60 cells and binds to the inter-nucleosomal region of HL-60 cell DNA and of DNA fragments containing genes encoding ATPase, tyrosine phosphatase, NADH dehydrogenase-like proteins and nuclear proteins in a cell-free system [77, 78]. Additional potential substrates of the T4SS are being identified [11].

17.4 Immunology of Phagosomes Containing *E. chaffeensis* and *A. phagocytophilum*

17.4.1 Major Histocompatibility Complex Class I and II Molecules

Major histocompatibility complex (MHC) class I and II molecules are membrane glycoproteins that are constitutively present on the surface of monocytes and neutrophils and in the cytoplasmic membrane vesicle compartment, but are distributed differently in the endosomal–lysosomal pathway. MHC class I molecules are trafficked from the Golgi to the plasma membrane, and are excluded during endocytosis or are rapidly redirected back to the plasma membrane following endocytosis [79]. In contrast, MHC class II molecules are trafficked from the Golgi to specialized MHC class II-containing endosomes, which are subsequently directed to the plasma membrane [80]. Both class I and II molecules are located on a distinct class of endosomes separated from the TfR-containing endosomes during endocytosis. Approximately 20–25% of *A. phagocytophilum* inclusions contain the class I heavy chain associated with the β_2-microglobulin and a similar percentage of *A. phagocytophilum* inclusions contain class II molecules. A small percentage of *E. chaffeensis* inclusions in HL-60 cells colocalize with MHC class I and II molecules ($\sim$10% and 5%, respectively). Thus, *E. chaffeensis* and *A. phagocytophilum* inclusions may potentially harbor complexes between bacterial antigen peptides and MHC I or MHC II molecules in human monocytes and neutrophils or endothelial cells, and may be able to activate T cells that can recognize these bacterial peptides.

17.5 Summary and Outlook

Ehrlichia and *Anaplasma* spp. are host cholesterol-dependent bacteria that have evolved mechanisms to exploit caveolae or lipid raft systems throughout their replicative cycle. For *E. chaffeensis*, conversion on the early endosome recycling pathway provides safe and nutritionally rich inclusions for replication. For *A. phagocytophilum*, interaction with early autophagosomes provides a safe haven and nutrients. Detailed understanding of the mechanism(s) underlying the biogenesis of these inclusions will come from studies with bacterial TCS signal transduction, and bacterial T4SS effector molecules and high-throughput screens of host cell proteins. These studies will also provide information on regulation of membrane trafficking and functions of membrane organelles in neutrophils and monocytes. *E. chaffeensis* and *A. phagocytophilum* infections cause detrimental systemic illnesses, yet the therapeutic choice is limited. Further analysis of bacterial inclusion biogenesis is expected to identify potential new therapeutic targets for treatment of HME and HGA.

Acknowledgment

The author would like to thank M. Lin for help with figure preparation. Most of studies in the author's laboratory reported in this review were supported by grants R01AI30010 and R01AI47885 from the National Institutes of Health.

References

1 Dawson, J.E., Anderson, B.E., Fishbein, D.B., Sanchez, J.L. *et al.* (1991) *J Clin Microbiol*, **29**, 2741–2745.

2 Munderloh, U.G., Lynch, M.J., Herron, M.J., Palmer, A.T. *et al.* (2004) *Vet Microbiol*, **101**, 53–64.

3 Goodman, J.L., Nelson, C., Vitale, B., Madigan, J.E. *et al.* (1996) *N Engl J Med*, **334**, 209–215.

4 Dumler, J.S., Barbet, A.F., Bekker, C.P., Dasch, G.A. *et al.* (2001) *Int J Syst Evol Microbiol*, **51**, 2145–2165.

5 Paddock, C.D. and Childs, J.E. (2003) *Clin Microbiol Rev*, **16**, 37–64.

6 Bakken, J.S., Krueth, J., Wilson-Nordskog, C., Tilden, R.L. *et al.* (1996) *JAMA*, **275**, 199–205.

7 Ewing, S.A., Dawson, J.E., Kocan, A.A., Barker, R.W. *et al.* (1995) *J Med Entomol*, **32**, 368–374.

8 Telford, S.R. 3rd, Dawson, J.E., Katavolos, P., Warner, C.K. *et al.* (1996) *Proc Natl Acad Sci USA*, **93**, 6209–6214.

9 Anderson, B.E., Sims, K.G., Olson, J.G., Childs, J.E. *et al.* (1993) *Am J Trop Med Hyg*, **49**, 239–244.

10 Pancholi, P., Kolbert, C.P., Mitchell, P.D., Reed, K.D. Jr. *et al.* (1995) *J Infect Dis*, **172**, 1007–1012.

11 Dunning Hotopp, J.C., Lin, M., Madupu, R., Crabtree, J. *et al.* (2006) *PLoS Genet*, **2**, e21.

12 Lin, M. and Rikihisa, Y. (2003) *Infect Immun*, **71**, 5324–5331.

13 Huang, H., Lin, M., Wang, X., Kikuchi, T., Mottaz, H., Norbeck, A. and Rikihisa, Y. (2008) *Infect Immun*, **76**, 3405–3414.

14 Rikihisa, Y. (2006) *Curr Opin Microbiol*, **9**, 95–101.

15 Lin, M. and Rikihisa, Y. (2007) *Cell Microbiol*, **9**, 861–874.

16 Mott, J. and Rikihisa, Y. (2000) *Infect Immun*, **68**, 6697–6703.

17 Ge, Y. and Rikihisa, Y. (2006) *Cell Microbiol*, **8**, 1406–1416.

18 Ge, Y., Yoshiie, K., Kuribayashi, F., Lin, M. and Rikihisa, Y. (2005) *Cell Microbiol*, **7**, 29–38.

19 Yoshiie, K., Kim, H.Y., Mott, J. and Rikihisa, Y. (2000) *Infect Immun*, **68**, 1125–1133.

20 Xiong, Q., Bao, W., Ge, Y. and Rikihisa, Y. (2008) *J Infect Dis*, **197**, 1110–1118.

21 Mott, J., Barnewall, R.E. and Rikihisa, Y. (1999) *Infect Immun*, **67**, 1368–1378.

22 Barnewall, R.E., Rikihisa, Y. and Lee, E.H. (1997) *Infect Immun*, **65**, 1455–1461.

23 Niu, H., Yamaguchi, M. and Rikihisa, Y. (2008) *Cell Microbiol*, **10**, 593–605.

24 Lin, M. and Rikihisa, Y. (2003) *Cell Microbiol*, **5**, 809–820.

25 Anderson, R.G. (1998) *Annu Rev Biochem*, **67**, 199–225.

26 Duncan, M.J., Shin, J.S. and Abraham, S.N. (2002) *Cell Microbiol*, **4**, 783–791.

27 Minshall, R.D. and Malik, A.B. (2006) *Handb Exp Pharmacol* (176 Pt 1), 107–144.

28 Lin, M., Zhu, M.X. and Rikihisa, Y. (2002) *Infect Immun*, **70**, 889–898.

29 Lin, M., den Dulk-Ras, A., Hooykaas, P.J. and Rikihisa, Y. (2007) *Cell Microbiol*, **9**, 2644–2657.

30 Galbiati, F., Razani, B. and Lisanti, M.P. (2001) *Cell*, **106**, 403–411.

31 IJdo, J.W., Carlson, A.C. and Kennedy, E.L. (2007) *Cell Microbiol*, **9**, 1284–1296.

32 Chen, J.W., Chen, G.L., D'Souza, M.P., Murphy, T.L. and August, J.T. (1986) *Biochem Soc Symp*, **51**, 97–112.

33 Webster, P., IJdo, J.W., Chicoine, L.M. and Fikrig, E. (1998) *J Clin Invest*, **101**, 1932–1941.

34 Arnljots, K., Sorensen, O., Lollike, K. and Borregaard, N. (1998) *Leukemia*, **12**, 1789–1795.

35 Rikihisa, Y. (1996) *Ehrlichiae*, Slovak Academy of Sciences, Bratislava.

36 Lukacs, G.L., Rotstein, O.D. and Grinstein, S. (1990) *J Biol Chem*, **265**, 21099–21107.

37 Anderson, R.G., Falck, J.R., Goldstein, J.L. and Brown, M.S. (1984) *Proc Natl Acad Sci USA*, **81**, 4838–4842.

38 Mu, F.T., Callaghan, J.M., Steele-Mortimer, O., Stenmark, H. *et al.* (1995) *J Biol Chem*, **270**, 13503–13511.

39 Simonsen, A., Lippe, R., Christoforidis, S., Gaullier, J.M. *et al.* (1998) *Nature*, **394**, 494–498.

40 Wilson, J.M., de Hoop, M., Zorzi, N., Toh, B.H. *et al.* (2000) *Mol Biol Cell*, **11**, 2657–2671.

41 Simonsen, A., Gaullier, J.M., D'Arrigo, A. and Stenmark, H. (1999) *J Biol Chem*, **274**, 28857–28860.

42 Christoforidis, S., McBride, H.M., Burgoyne, R.D. and Zerial, M. (1999) *Nature*, **397**, 621–625.

43 Barnewall, R.E. and Rikihisa, Y. (1994) *Infect Immun*, **62**, 4804–4810.

44 Kohno, H. and Tokunaga, R. (1985) *J Biochem (Tokyo)*, **97**, 1181–1188.

45 Lee, E.H. and Rikihisa, Y. (1998) *Infect Immun*, **66**, 2514–2520.

46 Barnewall, R.E., Ohashi, N. and Rikihisa, Y. (1999) *Infect Immun*, **67**, 2258–2265.

47 Roy, C.R. (2005) *Proc Natl Acad Sci USA*, **102**, 1271–1272.

48 Duden, R., Allan, V. and Kreis, T. (1991) *Trends Cell Biol*, **1**, 14–19.

49 Kagan, J.C. and Roy, C.R. (2002) *Nat Cell Biol*, **4**, 945–954.

50 Fujiwara, T., Oda, K., Yokota, S., Takatsuki, A. and Ikehara, Y. (1988) *J Biol Chem*, **263**, 18545–18552.

51 Lipsky, N.G. and Pagano, R.E. (1983) *Proc Natl Acad Sci USA*, **80**, 2608–2612.

52 Hackstadt, T., Rockey, D.D., Heinzen, R.A. and Scidmore, M.A. (1996) *EMBO J*, **15**, 964–977.

53 Hackstadt, T., Scidmore, M.A. and Rockey, D.D. (1995) *Proc Natl Acad Sci USA*, **92**, 4877–4881.

54 Geuze, H.J., Stoorvogel, W., Strous, G.J., Slot, J.W. *et al.* (1988) *J Cell Biol*, **107**, 2491–2501.

55 Dorn, B.R., Dunn, W.A. Jr. and Progulske-Fox, A. (2002) *Cell Microbiol*, **4**, 1–10. 2

56 Dunn, W.A. Jr. (1990) *J Cell Biol*, **110**, 1935–1945.

57 Klionsky, D.J. and Emr, S.D. (2000) *Science*, **290**, 1717–1721.

58 Brumell, J.H., Volchuk, A., Sengelov, H., Borregaard, N. *et al.* (1995) *J Immunol*, **155**, 5750–5759.

59 Chamberlain, L.H., Burgoyne, R.D. and Gould, G.W. (2001) *Proc Natl Acad Sci USA*, **98**, 5619–5624.

60 Mollinedo, F., Calafat, J., Janssen, H., Martin-Martin, B. *et al.* (2006) *J Immunol*, **177**, 2831–2841.

61 Traub, L.M. (1997) *Trends Cell Biol*, **7**, 43–46.

62 Moss, S.E. (1997) *Trends Cell Biol*, **7**, 87–89.

63 Babior, B.M. (1999) *Blood*, **93**, 1464–1476.

64 DeLeo, F.R., Allen, L.A., Apicella, M. and Nauseef, W.M. (1999) *J Immunol*, **163**, 6732–6740.

65 Heyworth, P.G., Bohl, B.P., Bokoch, G.M. and Curnutte, J.T. (1994) *J Biol Chem*, **269**, 30749–30752.

66 Mott, J., Rikihisa, Y. and Tsunawaki, S. (2002) *Infect Immun*, **70**, 1359–1366.

67 IJdo, J. and Mueller, A.C. (2004) *Infect Immun*, **72**, 5392–5401.

68 Lin, M. and Rikihisa, Y. (2004) *Cell Microbiol*, **6**, 175–186.

69 Kumagai, Y., Cheng, Z., Lin, M. and Rikihisa, Y. (2006) *Infect Immun*, **74**, 5014–5022.

70 Cheng, Z., Kumagai, Y., Lin, M., Zhang, C. and Rikihisa, Y. (2006) *Cell Microbiol*, **8**, 1241–1252.

71 Celli, J., de Chastellier, C., Franchini, D.M., Pizarro-Cerda, J. *et al.* (2003) *J Exp Med*, **198**, 545–556.

72 Roy, C.R. (2002) *Int J Med Microbiol*, **291**, 463–467.

73 Ingmundson, A., Delprato, A., Lambright, D.G. and Roy, C.R. (2007) *Nature*, **450**, 365–369.

74 Cheng, Z., Wang, X. and Rikihisa, Y. (2008) *J Bacteriol*, **190**, 2096–2105.

75 Niu, H., Rikihisa, Y., Yamaguchi, M. and Ohashi, N. (2006) *Cell Microbiol*, **8**, 523–534.

76 Ohashi, N., Zhi, N., Lin, Q. and Rikihisa, Y. (2002) *Infect Immun*, **70**, 2128–2138.

77 Caturegli, P., Asanovich, K.M., Walls, J.J., Bakken, J.S. *et al.* (2000) *Infect Immun*, **68**, 5277–5283.

78 Park, J., Kim, K.J., Choi, K.S., Grab, D.J. and Dumler, J.S. (2004) *Cell Microbiol*, **6**, 743–751.

79 Neefjes, J.J., Stollorz, V., Peters, P.J., Geuze, H.J. and Ploegh, H.L. (1990) *Cell*, **61**, 171–183.

80 Amigorena, S., Drake, J.R., Webster, P. and Mellman, I. (1994) *Nature*, **369**, 113–120.

18
Legionella pneumophila

Alyssa Ingmundson and Craig R. Roy

18.1
Legionella pneumophila Biology

Legionella pneumophila are Gram-negative bacteria found ubiquitously in nature, where they survive and replicate within freshwater amoeba [1]. If these bacteria contaminate appliances such as cooling towers and water systems, they can become aerosolized and inhaled by humans. In elderly or immunocompromised individuals, this infection can result in the severe pneumonia known as Legionnaires' disease [2]. The disease occurs when *L. pneumophila* infect and replicate within alveolar macrophages; it affects between 8000 and 18 000 individuals in the United States annually [3]. *L. pneumophila* use the mechanisms evolved for parasitism of freshwater amoeba to evade the deleterious methods typically employed by macrophages to kill microorganisms.

18.2
The Intracellular Compartment of *L. pneumophila*

When either amoebae or mammalian macrophages phagocytose particles, the phagosome fuses with endocytic compartments and ultimately with degradative lysosomal compartments [4]. After phagocytosis, *L. pneumophila* is able to elude this canonical endocytic maturation to avoid the lethal environment of lysosomes. The bacteria reside in a compartment that avoids acidification and does not accumulate endocytic or lysosomal markers such as Rab5A, Rab7 and lysosome-associated membrane protein 1 (LAMP1) [5–8].

Meanwhile, endoplasmic reticulum (ER)-derived vesicles and mitochondria are recruited to *Legionella*-containing vacuoles (LCVs). These vesicles become closely associated with the *L. pneumophila*-containing membrane, and the LCV subsequently becomes decorated with ribosomes [9, 10]. The LCV actually fuses with and becomes morphologically indistinguishable from the rough ER [9–11]. This interaction of LCVs with the ER occurs in a biphasic manner. Shortly after infection, protein markers of vesicles that cycle from the ER appear on the LCV, while at later times, colocalization of resident ER markers is evident [12]. YFP-KDEL and GFP-p58,

Intracellular Niches of Microbes. Edited by Ulrich E. Schaible and Albert Haas

ISBN: 978-3-527-32207-7

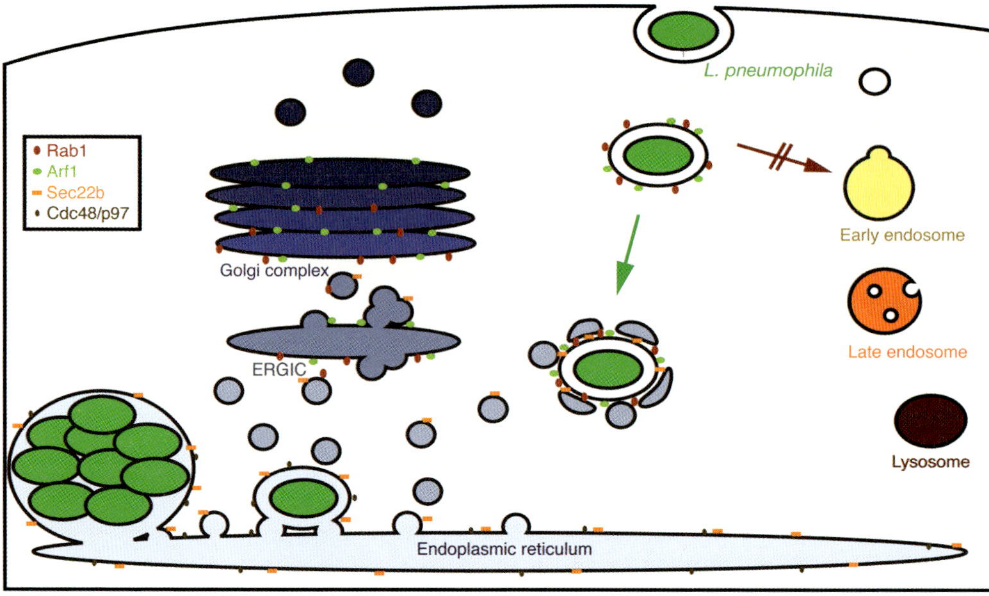

Figure 18.1 The intracellular lifecycle of *Legionella pneumophila*. When *L. pneumophila* are internalized, the *Legionella*-containing vacuoles (LCVs) immediately evade endocytic maturation. Vesicles transiting between the endoplasmic reticulutm (ER) and Golgi are recruited to the LCVs. These vesicles flatten and become closely associated with the LCV membrane. Subsequently, LCVs fuse with the ER where the bacteria are able to replicate. Several host proteins are required to direct the transport of the LCV. Rab1, Arf1, Sec22b and Cdc48/p97 are localized to LCVs during infection and are necessary to create a compartment conducive to *L. pneumophila* replication. ERGIC, ER–Golgi intermediate compartment.

proteins that cycle between the ER and Golgi, are enriched on the LCV 30 min after infection. In contrast, calnexin, a resident ER integral membrane protein that is excluded from ER-to-Golgi cycling vesicles, only colocalizes with the LCV 1–2 h after internalization [12]. At this time, luminal ER contents enter the *L. pneumophila*-containing compartment as seen by staining for the resident ER enzyme glucose-6-phosphatase [11]. Within the ER, *L. pneumophila* are protected from the defenses of the host cells and are able to replicate [9, 13]. Figure 18.1 provides a schematic of the intracellular life cycle of *L. pneumophila*.

18.3 Requirements for Intracellular Replication

18.3.1 Bacterial Virulence Factors

To determine the mechanisms by which *L. pneumophila* are able to direct LCV maturation, several methods to identify genes required for intracellular growth have been undertaken (reviewed in [14]). These screens identified a set of genes called *dot* (for defect in organelle trafficking) [15] and *icm* (for intracellular multiplication) [16]. The *dot* and *icm* genes were found to encode components of a type IVB secretion system termed the Dot/Icm system [17]. The Dot/Icm type IV system can translocate *L. pneumophila* proteins into the host cell cytosol, and this function is required for intracellular growth of *L. pneumophila*. The substrates of this secretion system are important for establishment of the intracellular niche of the bacteria and proper transport of the LCV within host cells (reviewed in [18]).

18.3.2
Host Factors Involved in Infection

Since the initial microscopy-based studies of *L. pneumophila* infection that described the escape from endocytic maturation and association with the ER, significant progress has been made in defining the host factors involved in establishment of LCVs. The use of dominant negative interfering proteins, small interfering RNA (siRNA) and biochemical inhibitors has identified host processes required for ER recruitment and fusion and, ultimately, *L. pneumophila* replication.

To discern the origin of the early secretory vesicles intercepted by *L. pneumophila*, LCV formation and bacterial replication were assessed in the presence of dominant interfering proteins against the GTPases that regulate early vesicle formation [12]. Sar1 and Arf1 control recruitment of various coat proteins to the early secretory compartments to mediate vesicle formation and budding (reviewed in [19, 20]). Arf1 controls membrane dynamics at the ERGIC and Golgi complex through the recruitment of coat proteins including COPI, adapter protein complexes and GGA proteins to mediate vesicle formation and budding. Arf1 can regulate both anterograde and retrograde transport among these compartments. In contrast, Sar1 is only known to regulate anterograde transport through the formation of COPII-coated vesicles at ER exit sites. Inhibition of either Arf1 or Sar1 function inhibits proper LCV transport. Expressing dominant negative interfering variants of Arf1 or Sar1 inhibited bacterial replication, as did depleting Arf1 or Sar1 with siRNA [12, 21].

Treating cells with brefeldin A (BFA), an inhibitor of Arf1 activation, reduced association of the transitional ER marker p58 with LCVs [12]. However, when infection was examined by electron microscopy, it became apparent that BFA treatment does not completely prevent the association of host vesicles with the LCV, but it does inhibit fusion of these vesicles with this compartment [11]. Therefore, while Arf1 function is not required for vesicle recruitment, Arf1 is required for fusion of the LCV with the ER and subsequent *L. pneumophila* replication. The importance of Arf1 during infection was further substantiated by the observation that Arf1 is actually present on LCVs [22]. *L. pneumophila* encodes and translocates a Dot/Icm substrate, RalF, that specifically activates and recruits Arf1 to LCVs [22]. Perhaps the direct recruitment of Arf1 aids in the latter step of the biphasic process of ER recruitment, the fusion of the LCV with the ER [11].

Interestingly, treatment of cells with nocodazol, which depolymerizes microtubules and therefore prevents vesicle transport to the ER–Golgi intermediate compartment (ERGIC) and Golgi, does not inhibit *L. pneumophila* replication [12]. In fact, nocodazol treatment causes accumulation of p58 at ER exit sites and results in increased association of p58 with LCVs [12, 23, 24]. This p58 recruitment in the presence of nocodazole is still sensitive to BFA treatment. Therefore, *L. pneumophila* intercepts vesicles budding from ER exit sites prior to their transport to downstream compartments [12]. These data demonstrate that vesicles derived from the ER are utilized by *L. pneumophila* but do not preclude the possibility that *L. pneumophila* obtains membrane from additional sources, such as the ERGIC.

The recruitment, tethering and fusion of vesicles to LCVs prompted researchers to investigate the involvement of Rab proteins in *L. pneumophila* infection. Rab proteins are small GTPases that regulate vesicle tethering and fusion with target membranes [25]. Rab1 was of particular interest because it regulates ER-derived vesicle interactions with the downstream compartments, the ERGIC and Golgi complex [26, 27]. Rab1 regulates these processes by recruiting and organizing effector molecules at specific membrane domains. These effectors include tethering factors that bring vesicles together with target membranes via interactions with each other and other Golgi matrix proteins [28]. These tethers subsequently facilitate the interaction of SNARE (soluble *N*-ethylmaleimide-sensitive factor attachment protein receptor) proteins that reside on the two opposing membranes, and this SNARE interaction drives membrane fusion [29].

When the localization of Rab1 was examined during infection, Rab1 was found to be present on LCVs shortly after uptake of the bacteria [30, 31]. At the LCV membrane, Rab1 could promote the recruitment and fusion of early secretory vesicles. Consistent with this prediction, perturbation of Rab1 function has deleterious effects on *L. pneumophila* intracellular replication [30, 31]. The localization of Rab1 to LCVs is directly controlled by *L. pneumophila* through the actions of the Dot/Icm substrate DrrA (also known as SidM) [32, 33]. DrrA recruits and activates Rab1 at the LCV during infection. In addition, *L. pneumophila* encodes another Dot/Icm substrate termed LepB that stimulates the deactivation of Rab1 [34]. This precise control of Rab1 by *L. pneumophila* further verifies of the importance of Rab1 to LCV maturation.

Interestingly, known effectors of Rab1 were not found on LCVs during infection [31]. Nevertheless, the SNARE Sec22b, a member of the SNAREpin complex that assembles downstream of Rab1 to drive membrane fusion, does localize to LCVs. Sec22b is the v-SNARE present on ER-derived vesicles and its function is important in LCV formation [21, 30]. Sec22b likely functions in LCV maturation by mediating fusion of the attached ER-derived vesicles with LCV membranes. However, the cognate SNAREs with which Sec22b typically forms fusion-competent SNAREpins are not found on LCVs [30, 31]. These t-SNAREs, membrin, syntaxin-5 and rBet1, are typically found on *cis*-Golgi membranes and do not appear to be required for *L. pneumophila* replication [30]. Therefore, the exact mechanism by which Sec22b influences LCV maturation is still unresolved.

Determining these mechanistic details of LCV biogenesis is likely complicated by the exploitation of multiple pathways by *L. pneumophila* to mediate vesicle tethering and fusion. This hypothesis that multiple pathways function in concert to provide membrane to LCVs was substantiated when RNA interference targeting various components of early secretory transport in *Drosophila* cells uncovered few individual factors absolutely required for *L. pneumophila* replication [21]. As mentioned, the small GTPases Arf1 and Sar1 were essential for LCV establishment; however, no other secretory factors, when knocked down individually, significantly reduced bacterial replication. For instance, all proteins tested that are directly involved in vesicle tethering and fusion were found to be dispensable for *L. pneumophila* replication. At first, this observation seems surprising given the importance of this function to LCV maturation. However, *L. pneumophila* may provide proteins that mediate these functions via Dot/Icm translocation, and multiple host pathways may be employed in parallel. In fact, when multiple eukaryotic secretory components were knocked down simultaneously, defects in bacterial replication were observed. For example, RNA interference targeting Sec22b only caused a defect in *L. pneumophila* replication when additional secretory components were knocked down in tandem.

Similarly, siRNA targeting members of the TRAPP complex negatively affected bacterial replication, but only in conjunction with siRNA against other secretory components [21]. TRAPP is an activator of Rab1 and is important in vesicle tethering to target membranes [35]. These results suggest that TRAPP activity is used in the establishment of LCVs, but functions within only one of presumably multiple pathways used to recruit early secretory vesicles [21].

One of the few host factors found to be crucial for *L. pneumophila* replication in this targeted RNA interference study was a component of the ER-associated protein degradation (ERAD) pathway [21]. ERAD is a cellular process responsible for removal and degradation of misfolded proteins from the ER [36]. Unfolded and misfolded proteins are recognized and delivered out of the ER into the cytoplasm where they are ubiquitinated and targeted to the proteosome for degradation. Cdc48/p97, a member of the AAA-ATPase family, acts on the cytosolic face of the ER and targets protein substrates to the proteosome for degradation [37]. Cdc48/p97 has an additional function in various membrane fusion processes [38–40]. Dorer and colleagues distinguished between these two functions by depleting cofactors of Cdc48/p97 that are specific to each process. Surprisingly, inhibiting the membrane fusion functions of Cdc48/p97 with siRNA against the binding partner p47 did not restrict bacterial replication [21, 41]. However, siRNA targeting Cdc48/p97 cofactors involved in the ERAD function, such as Npl4 and Ufd1, did inhibit *L. pneumophila* replication [21, 42, 43]. Cdc48/p97 actually localizes to the LCV and although it is not required for Dot/Icm substrate translocation, it does appear to be involved in removal of Dot/Icm substrates from the LCV membrane during infection [21]. Whether this role in bacterial factor regulation is the only function of Cdc48/p97 during infection remains to be determined.

In addition to proteins, lipids of the host cell are also expected to play an important role in LCV maturation, as they do in many cellular membrane transport processes [44]. Phosphoinositides are carefully regulated lipid species that contribute to the specificity of vesicle transport [45]. The role of phosphatidylinositol 3-kinase (PI3Ks) in LCV maturation was examined both with chemical inhibitors of PI3K and in PI3K mutants of *Dictyostelium discoideum* [44, 46]. Although inhibition of PI3K function had no effect on *L. pneumophila* phagocytosis, deletion of PI3K did have an effect on LCV biogenesis [44]. *L. pneumophila* were actually more successful at establishing replication-competent organelles in the absence of PI3Ks [44]. Perhaps this increased infectivity is attributable to the negative effect deleting PI3Ks has on the endocytic pathway. This disruption of endocytic transport may aid *L. pneumophila* in avoiding this degradative pathway [44, 47]. Alternatively, either the substrates or products of PI3K may be directly involved in LCV transport. Products of PI3Ks include phosphatidylinositol 3-phosphate (PI(3)P) and phosphatidylinositol 3,4,5-trisphosphate (PI(3,4,5)P_3) which promote phagosome maturation down the endocytic pathway [48]. In addition, one of the substrates of PI3K, phosphatidylinositol 4-phosphate (PI(4)P), is present on LCVs and is responsible for localization of at least one of the Dot/Icm substrates to this compartment. Either the decrease in PI(3)P and PI(3,4,5) P_3 or the increase in PI(4)P caused by the absence of PI3K could be the reason for the observed enhanced replication phenotype [44]. While the details of phosphoinositide function during infection have not been completely elucidated, these lipid species do seem to influence LCV intracellular transport.

Because LCV biogenesis involves recruitment of host membranes to a specific compartment, comparisons can be made between LCV maturation and the process of autophagy. Autophagy is used by cells to degrade cellular components such as proteins and organelles for metabolism and homeostasis and has been implicated in LCV maturation [49]. Autophagy involves the recruitment of vesicles to form a multimembrane-enclosed compartment [49] that is morphologically similar to LCVs [13]. When cells are in autophagy-inducing conditions, autophagic markers are present on LCVs [50]. In addition, treatment of cells with inhibitors of autophagy such as bafilomycin A and 3-methyladenine decrease the quantity of replication-competent LCVs [50]. However, the autophagy pathway does not seem to be required for LCV biogenesis. Studies using *D. discoideum* with mutations in several genes encoding proteins crucial for autophagy demonstrated that these genes are dispensable for *L. pneumophila* intracellular replication [51]. Therefore the role of autophagy in LCV biogenesis remains unclear.

18.4
The Immune Response to *L. pneumophila* Infection

When *L. pneumophila* gains access to the lung of healthy individuals, the immune system can successfully clear the infection. In the elderly or immunocompromised, however, the ability of *L. pneumophila* to infect and replicate within alveolar macrophages is essential to *L. pneumophila* pathogenesis. The avoidance of lysosomal destruction and the unique intracellular compartment formed by these bacteria may provide an unusual challenge to the immune system.

The innate immune system has developed several mechanisms to recognize and control bacterial infection. Toll-like receptors can detect extracellular or endosomal bacterial components, and signaling through these receptors is crucial for control of *L. pneumophila* replication *in vivo* [52, 53]. In addition, macrophages contain cytosolic surveillance systems, such as Nod-like receptors (NLRs), that can detect bacterial components that are translocated into the host cytosol [54, 55]. Detection of *L. pneumophila* type IV secretion by the NLRs Birc1e and Ipaf leads to the activation of caspase-1 and the restriction of intracellular replication [56]. Cytosolic detection of flagellin is also critical to Ipaf-mediated restriction of *L. pneumophila* replication [57, 58]. In addition to the cell-autonomous restriction of *L. pneumophila* replication, caspase-1 activation is also required for secretion of the proinflammatory cytokines interleukin 1 beta (IL-1β) and IL-18. IL-18 can stimulate interferon γ (IFNγ) production in T cells, B cells and NK cells [59–61].

The cytokine production in response to *L. pneumophila* infection has been characterized as a T_H1 response [62]. IFNγ, along with other cytokines such as tumor necrosis factor α (TNFα), IL-12 and IL-18, are central to the clearance of *L. pneumophila* infection. IFNγ-activated macrophages do not support replication of *L. pneumophila* [63]. This IFNγ-mediated restriction of intracellular replication is not caused by improper transport of the LCV, as *L. pneumophila* appears to traffic properly to the ER [64]. One mechanism that seems to contribute to the restriction of bacterial replication is the limited iron availability caused by the IFNγ-stimulated down-regulation of transferrin receptor in these cells [65]. IFNγ has also been shown to be required for the clearance of *L. pneumophila* infection *in vivo* [66]. In addition to IFNγ, type I IFNs can also result in the restriction of intracellular *L. pneumophila* replication [67–69]. Expression of type I IFNs is activated in response to *L. pneumophila* infection by a Dot/Icm-dependent mechanism [68, 70]. Interestingly, the cytokine response to *L. pneumophila* infection is enhanced in infection with virulent *L. pneumophila* that possess functional Dot/Icm systems relative to avirulent *L. pneumophila* [53, 71, 72].

Studies have revealed that adaptive immunity to *L. pneumophila* also develops during infection. Animals infected with *L. pneumophila* often produce a specific humoral response [73, 74]; however, the precise contribution of the antibody response to infection clearance is currently unclear. The response of T cells to *L. pneumophila* infection, however, does appear to influence infection outcome. Both CD4 and CD8 T cells appear to play critical roles in the control of *L. pneumophila* infection; when both of these cell types are depleted, mice suffer enhanced lethality in response to *L. pneumophila* infection [75].

Little is known about the way in which *L. pneumophila* antigens are presented to T cells. Classically, when phagocytosed bacteria traffic through endocytic compartments and are degraded in lysosomes, bacterial antigens are loaded onto major histocompatibility complex (MHC) class II molecules and are presented to CD4 T cells. The evasion of the endocytic pathway by *L. pneumophila* may undermine presentation of antigens by this canonical pathway. In fact, very low levels of MHC II are present on mature LCVs compared with MHC II present on the plasma membrane [76].

Nevertheless, *L. pneumophila* antigens are efficiently presented to CD4 T cells on MHC II by infected macrophages and dendritic cells (DCs), and *L. pneumophila*-specific CD4 T cells are generated in immunized mice [72, 77]. Interestingly, cells infected with wild-type *L. pneumophila* elicit greater T cell responses than cells infected with *dot/icm* mutant *L. pneumophila*, which would normally traffic to lysosomes, the compartment in which MHC II loading typically occurs. Replication of *L. pneumophila* is not required for this MHC II antigen presentation but *L. pneumophila* protein synthesis is necessary for optimal presentation [77]. Acidification of late endocytic compartments is also required for MHC II presentation of wild-type *L. pneumophila* antigens. This requirement for acidified endosomes indicates that the presented bacterial antigens are somehow exposed to this compartment during infection. Lysosomal markers have been found on *L. pneumophila* vacuoles at late timepoints after infection [78]. This phenomenon could occur either through direct fusion of LCVs with lysosomes, or through the phagocytosis and endocytic transport of dead cells in which *L. pneumophila* replication has occurred. Either of these possibilities could lead to loading of *L. pneumophila* antigens onto MHC II. Alternatively, vesicles containing *L. pneumophila* antigens may traffic between the LCV and endocytic compartments during infection, thus allowing MHC II presentation of these antigens [77].

Presentation of *L. pneumophila* antigens by MHC class I molecules to CD8 T cells is also likely to occur during *L. pneumophila* infection. *L. pneumophila* antigens could be processed and presented on MHC I through multiple mechanisms. Because MHC I normally present cytosolic antigens, substrates of the Dot/Icm system translocated into macrophages or DCs could be processed by this pathway. Loading of MHC I then typically occurs in the ER after antigens are transported from the cytosol into the ER via the TAP (transporter associated with antigen processing) pathway. Additionally, this TAP-dependent process can be bypassed, and exogenous *L. pneumophila* antigens may be cross-presented by MHC I. Presentation of *L. pneumophila* antigens on MHC I would lead to CD8 T cell activation and effector functions. Mice depleted of CD8 T cells exhibited increased morbidity after challenge with *L. pneumophila* [75], indicating that these cells do contribute to the immune response to *L. pneumophila*.

Because *L. pneumophila* evolved to replicate in protozoa, the ability of these bacteria to subvert mammalian immune processes is expected to be limited. This feature, along with the genetic amenability of *L. pneumophila*, makes *L. pneumophila* an excellent model for studying immunity against intracellular pathogens. Further investigations into the means by which hosts respond to *L. pneumophila* infection at both the cellular level and the organismal level will provide a greater understanding of how the immune system detects and controls these intracellular organisms.

18.5 Summary and Outlook

Legionella pneumophila possess a distinct ability to manipulate membrane transport. These bacteria are taken up by phagocytosis and are subsequently able to divert the phagosome in which they reside from the endocytic pathway. To maintain their distance from this degradative pathway, *L. pneumophila* usurp secretory transport to form a compartment that fuses with the ER where they are protected from host defenses. These bacteria use multiple mechanisms to create this intracellular compartment, and these mechanisms are mediated through the activities of the

Dot/Icm substrate proteins. None of these Dot/Icm substrates was identified in early screens for *L. pneumophila* factors required for intracellular growth; deletion of individual substrate proteins from the *L. pneumophila* genome rarely results in intracellular growth defects. This feature could indicate redundancy among Dot/Icm substrate function. Clearly, multiple routes of membrane recruitment are engaged by *L. pneumophila* during infection. Alternatively, each of these mechanisms of endocytic evasion and membrane recruitment is required to survive intracellularly in certain host species or during particular environmental stresses. Further examination of how the Dot/Icm substrates function in the context of infection will lead to a better understanding of their role in LCV biogenesis. Also, because *L. pneumophila* direct the fate of the vacuole in which they reside through direct interaction with host cell components, determining the molecular and cellular function of *L. pneumophila* virulence factors has the added benefit of providing novel discoveries on the function of eukaryotic factors involved in fundamental cellular processes.

References

1 Rowbotham, T.J. (1980) *J Clin Pathol*, **33**, 1179–1183.

2 Fraser, D.W., Tsai, T.R., Orenstein, W., Parkin, W.E. *et al.* (1977) *N Engl J Med*, **297**, 1189–1197.

3 Marston, B.J., Plouffe, J.F., File, T.M. Jr, Hackman, B.A. *et al.* (1997) *Arch Intern Med*, **157**, 1709–1718.

4 Kornfeld, S. and Mellman, I. (1989) *Annu Rev Cell Biol*, **5**, 483–525.

5 Horwitz, M.A. (1983) *J Exp Med*, **158**, 2108–2126.

6 Horwitz, M.A. and Maxfield, F.R. (1984) *J Cell Biol*, **99**, 1936–1943.

7 Clemens, D.L., Lee, B.Y. and Horwitz F M. A. (2000) *Infect Immun*, **68**, 5154–5166.

8 Roy, C.R., Berger, K.H. and Isberg, R.R. (1998) *Mol Microbiol*, **28**, 663–674.

9 Horwitz, M.A. (1983) *J Exp Med*, **158**, 1319–1331.

10 Tilney, L.G., Harb, O.S., Connelly, P.S., Robinson, C.G. *et al.* (2001) *J Cell Sci*, **114**, 4637–4650.

11 Robinson, C.G. and Roy, C.R. (2006) *Cell Microbiol*, **8**, 793–805.

12 Kagan, J.C. and Roy, C.R. (2002) *Nat Cell Biol*, **4**, 945–954.

13 Swanson, M.S. and Isberg, R.R. (1995) *Infect Immun*, **63**, 3609–3620.

14 Vogel, J.P. and Isberg, R.R. (1999) *Curr Opin Microbiol*, **2**, 30–34.

15 Berger, K.H. and Isberg, R.R. (1993) *Mol Microbiol*, **7**, 7–19.

16 Marra, A., Blander, S.J., Horwitz, M.A. and Shuman, H.A. (1992) *Proc Natl Acad Sci USA*, **89**, 9607–9611.

17 Vogel, J.P., Andrews, H.L., Wong, S.K. and Isberg, R.R. (1998) *Science*, **279**, 873–876.

18 Ninio, S. and Roy, C.R. (2007) *Trends Microbiol*, **15**, 372–380.

19 Sato, K. and Nakano, A. (2007) *FEBS Lett*, **581**, 2076–2082.

20 D'Souza-Schorey, C. and Chavrier, P. (2006) *Nat Rev Mol Cell Biol*, **7**, 347–358.

21 Dorer, M.S., Kirton, D., Bader, J.S. and Isberg, R.R. (2006) *PLoS Pathog*, **2**, e34.

22 Nagai, H., Kagan, J.C., Zhu, X., Kahn, R.A. *et al.* (2002) *Science*, **295**, 679–682.

23 Cole, N.B., Sciaky, N., Marotta, A., Song, J. *et al.* (1996) *Mol Biol Cell*, **7**, 631–650.

24 Hammond, A.T. and Glick, B.S. (2000) *Mol Biol Cell*, **11**, 3013–3030.

25 Zahraoui, A., Touchot, N., Chardin, P. and Tavitian, A. (1989) *J Biol Chem*, **264**, 12394–12401.

26 Moyer, B.D., Allan, B.B. and Balch, W.E. (2001) *Traffic*, **2**, 268–276.

27 Allan, B.B., Moyer, B.D. and Balch, W.E. (2000) *Science*, **289**, 444–448.

28 Grosshans, B.L., Ortiz, D. and Novick, P. (2006) *Proc Natl Acad Sci USA*, **103**, 11821–11827.

29 Pfeffer, S.R. (1999) *Nat Cell Biol*, **1**, E17–E22.

30 Kagan, J.C., Stein, M.P., Pypaert, M. and Roy, C.R. (2004) *J Exp Med*, **199**, 1201–1211.

31 Derre, I. and Isberg, R.R. (2004) *Infect Immun*, **72**, 3048–3053.

32 Murata, T., Delprato, A., Ingmundson, A., Toomre, D.K. *et al.* (2006) *Nat Cell Biol*, **8**, 971–977.

33 Machner, M.P. and Isberg, R.R. (2006) *Dev Cell*, **11**, 47–56.

34 Ingmundson, A., Delprato, A., Lambright, D.G. and Roy, C.R. (2007) *Nature*, **450**, 365–369.

35 Wang, W., Sacher, M. and Ferro-Novick, S. (2000) *J Cell Biol*, **151**, 289–296.

36 Meusser, B., Hirsch, C., Jarosch, E. and Sommer, T. (2005) *Nat Cell Biol*, **7**, 766–772.

37 Ye, Y., Meyer, H.H. and Rapoport, T.A. (2001) *Nature*, **414**, 652–656.

38 Rabouille, C., Levine, T.P., Peters, J.M. and Warren, G. (1995) *Cell*, **82**, 905–914.

39 Acharya, U., Jacobs, R., Peters, J.M., Watson, N. *et al.* (1995) *Cell*, **82**, 895–904.

40 Latterich, M., Frohlich, K.U. and Schekman, R. (1995) *Cell*, **82**, 885–893.

41 Kondo, H., Rabouille, C., Newman, R., Levine, T.P. *et al.* (1997) *Nature*, **388**, 75–78.

42 Meyer, H.H., Shorter, J.G., Seemann, J., Pappin, D. *et al.* (2000) *EMBO J*, **19**, 2181–2192.

43 Jarosch, E., Taxis, C., Volkwein, C., Bordallo, J. *et al.* (2002) *Nat Cell Biol*, **4**, 134–139.

44 Weber, S.S., Ragaz, C., Reus, K., Nyfeler, Y. *et al.* (2006) *PLoS Pathog*, **2**, e46.

45 De Matteis, M.A. and Godi, A. (2004) *Nat Cell Biol*, **6**, 487–492.

46 Khelef, N., Shuman, H.A. and Maxfield, F.R. (2001) *Infect Immun*, **69**, 5157–5161.

47 Simonsen, A., Wurmser, A.E., Emr, S.D. and Stenmark, H. (2001) *Curr Opin Cell Biol*, **13**, 485–492.

48 Gillooly, D.J., Simonsen, A. and Stenmark, H. (2001) *J Cell Biol*, **155**, 15–17.

49 Levine, B. and Klionsky, D.J. (2004) *Dev Cell*, **6**, 463–477.

50 Amer, A.O. and Swanson, M.S. (2005) *Cell Microbiol*, **7**, 765–778.

51 Otto, G.P., Wu, M.Y., Clarke, M., Lu, H. *et al.* (2004) *Mol Microbiol*, **51**, 63–72.

52 Archer, K.A. and Roy, C.R. (2006) *Infect Immun*, **74**, 3325–3333.

53 Sporri, R., Joller, N., Albers, U., Hilbi, H. *et al.* (2006) *J Immunol*, **176**, 6162–6171.

54 Kanneganti, T.D., Lamkanfi, M. and Nunez, G. (2007) *Immunity*, **27**, 549–559.

55 Ting, J.P. and Davis, B.K. (2005) *Annu Rev Immunol*, **23**, 387–414.

56 Zamboni, D.S., Kobayashi, K.S., Kohlsdorf, T., Ogura, Y. *et al.* (2006) *Nat Immunol*, **7**, 318–325.

57 Ren, T., Zamboni, D.S., Roy, C.R., Dietrich, W.F. *et al.* (2006) *PLoS Pathog*, **2**, e18.

58 Molofsky, A.B., Byrne, B.G., Whitfield, N.N., Madigan, C.A. *et al.* (2006) *J Exp Med*, **203**, 1093–1104.

59 Dinarello, C.A. (1998) *Ann N Y Acad Sci*, **856**, 1–11.

60 Nakanishi, K., Yoshimoto, T., Tsutsui, H. and Okamura, H. (2001) *Annu Rev Immunol*, **19**, 423–474.

61 Fantuzzi, G., Puren, A.J., Harding, M.W., Livingston, D.J. *et al.* (1998) *Blood*, **91**, 2118–2125.

62 Tateda, K., Matsumoto, T., Ishii, Y., Furuya, N. *et al.* (1998) *Clin Diagn Lab Immunol*, **5**, 401–403.

63 Horwitz, M.A. and Silverstein, S.C. (1981) *J Exp Med*, **154**, 1618–1635.

64 Bhardwaj, N., Nash, T.W. and Horwitz, M.A. (1986) *J Immunol*, **137**, 2662–2669.

65 Byrd, T.F. and Horwitz, M.A. (1991) *J Clin Invest*, **88**, 1103–1112.

66 Brieland, J., Freeman, P., Kunkel, R., Chrisp, C. *et al.* (1994) *Am J Pathol*, **145**, 1537–1546.

67 Schiavoni, G., Mauri, C., Carlei, D., Belardelli, F. *et al.* (2004) *J Immunol*, **173**, 1266–1275.

68 Opitz, B., Vinzing, M., van Laak, V., Schmeck, B. *et al.* (2006) *J Biol Chem*, **281**, 36173–36179.

69 Coers, J., Vance, R.E., Fontana, M.F. and Dietrich, W.F. (2007) *Cell Microbiol*, **9**, 2344–2357.

70 Stetson, D.B. and Medzhitov, R. (2006) *Immunity*, **24**, 93–103.

71 McHugh, S., Yamamoto, Y., Klein, T.W. and Friedman, H. (2000) *Can J Microbiol*, **46**, 885–891.

72 Neild, A.L. and Roy, C.R. (2003) *Immunity*, **18**, 813–823.

73 Brieland, J.K., Heath, L.A., Huffnagle, G.B., Remick, D.G. *et al.* (1996) *J Immunol*, **157**, 5002–5008.

74 Breiman, R.F. and Horwitz, M.A. (1987) *J Exp Med*, **165**, 799–811.

75 Susa, M., Ticac, B., Rukavina, T., Doric, M. *et al.* (1998) *J Immunol*, **160**, 316–321.

76 Clemens, D.L. and Horwitz, M.A. (1993) *Infect Immun*, **61**, 2803–2812.

77 Neild, A., Murata, T. and Roy, C.R. (2005) *Infect Immun*, **73**, 2336–2343.

78 Sturgill-Koszycki, S. and Swanson, M.S. (2000) *J Exp Med*, **192**, 1261–1272.

19
Mycobacterium tuberculosis and His Comrades

Ulrich E. Schaible

19.1
Pathogenic Mycobacteria

The genus *Mycobacterium* (*Mycobacteriaceae*) comprises mainly environmental, some opportunistic and few obligate pathogenic bacteria. Mycobacteria are small, non-motile, obligate aerobic rods. They can be distinguished from other bacteria by acid-fast staining (Ziehl–Neelsen) using phenolic Fuchsin or Auramin, which are linked to the hydrophobic wax-rich cell wall at high temperatures and cannot be destained by acidic/alcoholic solutions. This rigid cell wall makes mycobacteria also rather resistant to common disinfectants, UV light or desiccation. Members of this genus cause important human infections such as tuberculosis and leprosy.

19.1.1
The Species

Mycobacterium tuberculosis, the causative agent of tuberculosis, was first isolated 1882 by the German doctor Robert Koch (1843–1910). Along with its genetically close relatives, namely *Mycobacterium africanum*, *Mycobacterium canetti* and *Mycobacterium bovis*, which can also cause tuberculosis-like diseases in humans, it forms the *M. tuberculosis* complex (MTC). *M. bovis* is the causative agent of cattle tuberculosis. It is not restricted to ruminants and humans, but can also infect other mammals. This species has recently gained more public interest since it caused disastrous outbreaks of disease in African lion populations, which were put into danger by spread of the infection from infected wilderbeest and buffalos. Genetic studies have revealed that *M. bovis* became separated from *M. tuberculosis* during Neolithic times, when cattle were domesticated [1].

The whole genomes of both *M. tuberculosis* and *M. bovis* have been completely sequenced. The genomes within the *Mycobacterium* genus are highly conserved. The single chromosome of the tubercle bacillus contains between 3995 and 4249 open reading frames (ORFs) in the laboratory strain H37Rv or the clinical isolate CDC1551, respectively, whereas the *M. bovis* genome has 3951 ORFs [2–4]. It has

Intracellular Niches of Microbes. Edited by Ulrich E. Schaible and Albert Haas

ISBN: 978-3-527-32207-7

been noted that many genes encode for proteins involved in fatty acid metabolism and cell wall synthesis.

M. bovis is the parental species of the only currently available antituberculosis vaccine, the strain bacille Calmette-Guerin (BCG), which has been used extensively worldwide since 1921. The strain was attenuated by multiple passages for 13 years through broth cultures by the French bacteriologists Albert Calmette (1863–1933) and Camille Guérin (1872 1961). Attenuation of BCG is due to the deletion of genomic regions of difference (RD) including virulence-associated genes. Genomic analyses revealed that 11 RD regions, encompassing 91 ORFs, are deleted from the *M. bovis* genome but are otherwise present in wild-type *M. tuberculosis*. Five additional regions present in *M. bovis* have been additionally deleted from BCG (in total: RD1–16) [5, 6].

Mycobacterium leprae is the causative agent of leprosy and was the first bacterium to be associated with a human disease when in 1873 the Norwegian Gerhard Henrik Armauer Hansen (1841–1912) described the bacteria in skin biopsies from leprosy patients. He failed, however, to isolate the leprosy bacilli, which, unlike other pathogenic mycobacteria, can still not be cultured in artificial media. This inability to grow in broth culture is most probably due to the loss of genes essential for important metabolic pathways [7]. In comparison to the around 4000 ORFs in the *M. tuberculosis* genome, the genome of *M. leprae* has only 1600 ORFs [2, 4, 7]. Members of the MTC grow slowly, with replication times of 12 h for *M. tuberculosis* and *M. bovis*, and 12 days for *M. leprae*.

Nontuberculous, atypical mycobacteria are environmental microbes. Atypical mycobacteria comprise strains of the *Mycobacterium avium-intracellulare* complex (MAC), *Mycobacterium kansasii* and *Mycobacterium scrofulaceum*, which cause infections in immune-deficient patients. *Mycobacterium avium* ssp. *paratuberculosis* (MAP) is the causative agent of Johne's disease in cattle and other ruminants. The MAP genome is a single circular chromosome encoding 4350 predicted ORFs, of which more than 3000 genes are homologous to genes in *M. tuberculosis* plus 161 unique regions encoding 39 previously unknown MAP genes [8]. *M. avium* subspecies vary widely in both pathogenicity and host specificity. Genome comparisons between *M. avium* subspecies of different provenience have revealed a broader plasticity than found in the *M. tuberculosis* complex. Multiple large polymorphic regions have been found in the genomes of MAP isolates containing genes for biosynthesis of cell wall lipids involved in virulence [9, 10].

Other pathogenic mycobacteria include *Mycobacterium marinum*, a causative agent of tuberculosis-like lesions in cold-blooded amphibians and fish, which can also lead to granulomatous skin lesions in humans. Due to its intracellular lifestyle peculiarities, that is, cytoplasmic localization, this species is covered in a separate chapter (see Chapter 25). *Mycobacterium ulcerans* causes burulic ulcers endemic to tropical areas, but has an extracellular lifestyle.

It should be noted that most studies on mycobacteria, especially those analyzing their intracellular niche and virulence properties, have been performed using BCG or *M. avium*. Due to the category 3 safety measures required for work with *M. tuberculosis*, only recently have more studies been done using the *M. tuberculosis* lab strains H37Rv and Erdman. Future studies should focus on the analysis of clinical isolates of *M. tuberculosis*, which will most likely reveal tremendous differences between individual strains with respect to physiology, growth, survival in macrophages, intracellular niches, as well as induction of pathology, cell death and inflammation. This is indicated by the polymorphisms observed between the genomes of *M. tuberculosis* laboratory strains and clinical isolates [4]. Tuberculosis *in vivo* is studied primarily in the mouse and to some extent in the guinea-pig. Both systems have their advantages and disadvantages. Guinea-pigs develop active progressive tuberculosis with pathology similar to active tuberculosis in humans, whereas mice show a chronic infection but pathology dissimilar to the human disease. Neither model is sufficient to study the latent tuberculosis and reactivation that typically occurs in humans. Nevertheless, the murine tuberculosis model is the best one established, and benefits greatly from the availability of a vast number of mouse mutants and genetically engineered knockout (KO) and transgenic mouse strains and other tools such as monoclonal antibodies. Other animal models including rat, rabbit and non-primate monkeys are just now explored in more detail to study experimental tuberculosis.

19.1.2 The Diseases

19.1.2.1 Tuberculosis

Tuberculosis is usually a lung infection caused by inhalation of tubercle bacilli within microdroplets, but it can affect any organ. *M. bovis* infection in the past often occurred through the consumption of unpasteurized milk from infected cows, and therefore often led to infections in the gut. In the lung, tubercle bacilli are envisaged to be engulfed by cells patrolling the alveolar surface such as alveolar macrophages and dendritic cells (DCs). These cells transport the pathogens to the lung interstitium and draining lymph nodes. At both sites, primary lesions develop that rarely cause disease. Instead, the infection remains latent and is kept under the control of the immune system. The infectious foci are walled off by the formation of granulomas. These highly organized tissue reactions comprise macrophages, DCs, T and B lymphocytes, as well as macrophage-derived cells with peculiar morphologies such epitheloid and foamy multinucleated giant cells, and are often separated from the surrounding tissue by a calcified rim [11–13]. This situation can be maintained lifelong and usually remains clinically unrecognized. Tubercle bacilli persist in a dormant stage (see below). According to World Health Organization (WHO) estimates, one-third of the world population is considered to be latent carriers. Only 10% of those carriers will develop active tuberculosis during their lifetime [14, 15].

Tuberculosis is one of the most severe complications in patients with acquired immunodeficiency syndrome (AIDS), which changed the worldwide prevalence of tuberculosis drastically. Why tuberculosis infection switches from latent to active disease, called secondary tuberculosis, is still debated. It is probably not based on a common cause [11].

Observations from immunocompromised patients with AIDS or under treatment with immune suppressive therapeutics such as the antitumor necrosis factor antibody infliximab used to treat patients with autoimmune diseases (e.g., rheumatoid arthritis), suggest that once the immunity becomes debilitated, dormant bacteria are reactivated and pulmonary tuberculosis develops [16]. Upon reactivation, bacteria disseminate to other sites in the lung and occasionally to other organs, including bones, kidneys, liver and central nervous system. Rapid proliferation of mycobacteria and fast development of disease termed miliary tuberculosis is often observed in immunocompromised patients, especially those with AIDS, or in newborns. Although subtle immune compromising conditions such as malnutrition, micronutrient deficiencies, iron overload, drug- or alcohol abuse, stress, old age or co-infections with immune-modulating pathogens such as helminths can play a role in tuberculosis reactivation, these scenarios do not fully explain reactivation in otherwise immunocompetent patients [17]. A recent study in South Africa has demonstrated that following successful treatment, reinfection can occur later in life, suggesting preexisting predispositions in those patients [18].

It has been suggested that inflammatory immune responses can reactivate dormant mycobacteria [11]. The dormant state of tubercle bacilli is not fully understood, but a microaerophilic environment within the granuloma has been suggested to maintain dormancy. Induction of inflammation by mycobacteria or unrelated causes involving respiratory burst and influx of oxygen has been suggested to trigger reactivation. Iron overload is a risk factor for tuberculosis which could be due to free iron promoting the Fenton reaction, which generates reactive oxygen intermediates for subsequent reactivation of dormant bacteria [19, 20]. *In vitro* studies using different culture systems have demonstrated that dormant mycobacterie reduce protein synthesis, which can be rescued by oxygen (see below) [21]. Hypoxia has been demonstrated in tuberculosis lesions of rabbits, guinea-pigs and nonhuman primates, and it is suggested that lesions with central necrotic and caseous debris in humans also have low oxygen tension [13, 22]. In contrast, lesions in the lungs of infected mice fail to show signs of hypoxia, which is another disadvantage of this model and a putative reason why mice do not develop latent infection [23].

Tuberculosis represents a major health problem worldwide. According to a recent report by the World Health Organization, 3.9 million cases of active tuberculosis were notified worldwide in 2006, and 1.7 million people died of the disease. This number includes 0.7 million cases and 0.2 million deaths in people co-infected with the human immunodeficiency virus (HIV). For 2008, the WHO estimate there were more than 9.2 million new cases of tuberculosis globally [14]. The WHO estimate that up to 2 billion individuals worldwide are carriers of *M. tuberculosis*. This illustrates clearly the dissociation between infection and disease and reveals that the immune system can maintain infection in a dormant stage in an otherwise healthy individual.

The recent emergence of multi- and excessive drug-resistant isolates is hugely damaging for the global tuberculosis situation. Treatment of tuberculosis requires a patient to take two to three different drugs for at least six months, which necessitates a high degree of compliance. In poor countries with inadequate medical infrastructures and under-financed health systems, incomplete monitoring of medication and repeated dropping out of treatment promotes an increase in drug-resistant strains.

Currently, the antimycobacterial drugs employed are isoniazid, rifampicin, ethambutol, pyrazinamide and streptomycin. Only a few third-line therapeutics are available. Metronidazole is a novel drug currently in clinical trials; it is active under low oxygen pressures and it is hoped that it can kill mycobacteria within hypoxic lesions [22].

19.1.2.1.1 **BCG Vaccine** There is no doubt that improvements in housing conditions, health care, hygiene, education and nutrition as well as the introduction of chemotherapeutics, such as streptomycin and isoniazide, changed the tuberculosis situation in the industrialized countries in the first half of the twentieth century to a point where it was no longer recognized as an important health problem. In less privileged regions of the world, however, the situation is not much different from that in nineteenth-century Europe. The only available vaccine, BCG, shows only low protection against the classical outcome of infection, lung tuberculosis in adults, although it should be noted that its protective capacity is geographically highly variable, ranging from zero to 80%, with regions closer to the Equator showing the least satisfying effectiveness [24]. This phenomenon is probably due to early priming by environmental mycobacteria in those regions [25]. More importantly, the recent emergence of the *M. tuberculosis* strain Beijing, now responsible for the majority of current outbreaks, was blamed on worldwide BCG vaccination [26]. Nevertheless, BCG is protective against *M. tuberculosis* meningitis in small children and was probably responsible for decline of this complication in industrialized countries [24].

Currently, novel improved vaccine strains employing parental BCG to express recombinant *M. tuberculosis* antigens, proinflammatory cytokines or immune-stimulating virulence factors of other intracellular bacteria such as listeriolysin of *Listeria monocytogenes* are currently in phase 1 and 2 clinical trials [27]. It is hoped that these new vaccine strains in combination with novel therapeutics will finally help to control the global tuberculosis emergency [14].

19.1.2.2 **Leprosy**

Leprosy primarily affects skin, nasal epithelia and peripheral nerves and frequently leads to stigmatizing deformities. Its agent, *M. leprae*, is most likely transmitted by contact with patients who shed organisms in nasal secretions and lesion exudates. Infection of Schwann cells causes nerve damage and anesthesia, which leads to local injuries followed by secondary infection, inflammation and scarring, which significantly exaggerating the disease. The incidence of leprosy has declined dramatically over the past decade as a result of successful therapeutic regimens implemented by the WHO. In 1996, the WHO estimated a global case load of 1 million.

Leprosy consists of a spectrum of clinically, histologically, immunologically and bacteriologically distinct subtypes. The more benign tuberculoid pole is associated with strong T helper 1 (T_H1) cell responses, production of proinflammatory cytokines including interleukin 12 (IL-12), IL-18 and interferon γ and macrophage activation, which restrict mycobacterial growth in well-defined lesions containing few acid-fast bacilli. In contrast, at the lepromatous pole bacterial growth is unrestricted and lesions contain abundant acid-fast bacilli within macrophages obviously in the resting state. Regulatory suppressor T cells have been implicated in the lepromatous disease state.

19.1.2.3 Atypical Mycobacteria

Atypical mycobacteria are environmental microbes, which, because they are restricted by activated macrophages, rarely cause disease in immunocompetent individuals. Members of the MAC have gained clinical importance in the wake of the AIDS pandemic, being most prominent epidemiologically as the opportunistic cause of infections in immunocompromised patients. MAC infections of patients with AIDS primarily target the gastrointestinal tract. *M. kansasii* primarily causes infections in elderly men with pre-existing lung disease. *M. scrofulaceum* is an occasionally cause of granulomatous skin infections and lymphadenitis in children.

19.2 Intracellular Compartment and Virulence Properties of Mycobacteria

19.2.1 Host Cells

Pathogenic mycobacteria are facultative intracellular pathogens of cells capable of phagocytosing (Figures 19.1 and 19.2). During chronic stages of infection and disease, *M. tuberculosis, M. bovis* and MAC are predominantly found within macrophages. It is hypothesized that alveolar macrophages, which patrol the lung epithelium, engulf inhaled mycobacteria and carry them into the lung interstitium. Upon onset of infection, the prime host cells are thought to be newly invading monocyte-derived macrophages, which are attracted to the site of mycobacterial entry through proinflammatory chemokines. Mycobacteria can also readily be engulfed by other phagocytes such as DCs and polymorphonuclear neutrophils (PMNs), as well as by amoebae as shown for *M. avium* and *M. marinum* (see Chapter 4). In DCs and PMNs though, *M. tuberculosis* does not appear to grow very much and it has been suggested that these cells are able to restrict mycobacterial growth. PMNs are important producers of the cathelizidin LL37, which is a potent antimicrobial peptide (AMP, defensin), and iron or siderophore-scavenging molecules such as lactoferrin and lipocalin, respectively (see Chapter 12). All three effectors have been considered involved in antimycobacterial defense [28–30]. A residual function of PMNs in protection against *M. tuberculosis* has been shown in immunodeficient mice [31]. DCs are considered nonsupportive host cells for mycobacterial growth [32]. Nevertheless, *ex vivo* analysis of lung cells from *M. tuberculosis*-infected mice revealed that most infected cells are DCs [33]. DCs are also thought to be responsible for transport of mycobacteria to the draining lymph nodes, where priming of T lymphocytes and induction of acquired immunity occurs (see Chapter 12). Migration of DCs to the lymph nodes is triggered by Toll-like receptor signals upon interaction with microbes and microbe-associated molecular patterns (MAMPs). Upon expression

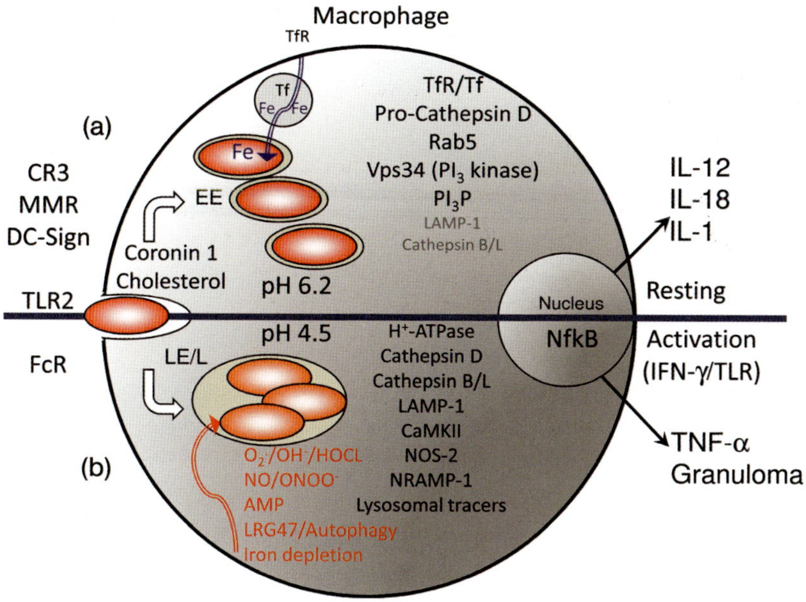

Figure 19.1 (a) The resting macrophage is the preferred host cell for mycobacteria. Entry by phagocytosis is facilitated by an array of phagocytosis receptors, which allow entry without activating the macrophage, including the macrophage mannose receptor (MMR), complement 3 receptor (CR3) and DC-SIGN. The latter is only present on alveolar macrophages and dendritic cells. Toll-like receptor 2 (TLR2) can sense mycobacterial lipoproteins and induce proinflammatory cytokines such as interleukin 1 (IL-1), IL-12 and IL-18. Upon phagocytosis, which involve coronin and cholesterol-rich domains, mycobacteria reside in tight phagosomes, which maintain early/recycling endosomal characteristics such as a pH of around 6.2, low amounts of the vesicular proton ATPase (v-H$^+$-ATPase) and lysosome-associated membrane protein 1 (LAMP1). It intersects with the transferrin (Tf)/transferrin receptor (TfR) pathway for iron (Fe) import. TfR-imported iron is taken up by mycobacteria. Cathepsin D and L activity is low and cathepsin D is only present in its inactive pro-form. The vacuolar membrane retains the GTPase Rab5 and phosphatidylinositol 3-kinase Vps34 generating phosphatidylinositol 3-phosphate (PI(3)P). (b) The interferon γ-activated macrophage is an effector cell against mycobacteria. The activated macrophage expresses nitric oxide synthase (NOS2) and NADPH oxidase, which generate reactive nitrogen and oxygen intermediates (red). The GTPase LRG47/IRGM1 promoting autophagosome formation to fuse with mycobacterial phagosomes. Autophagosomes deliver antimicrobial peptides (AMPs), either cathelizidins or peptides of ubiquitin origin, which are generated by cathepsins B and L, to the mycobacterial phagosome. Phagosomes fail to maintain inhibition of phagosome maturation and acquire late endosomal/phagolysosomal characteristics such as LAMP1, active cathepsins B, L and D. Enhanced presence of v-H$^+$-ATPase facilitates phagosome acidification. Increased calcium/calmodulin-dependent protein kinase II (CaMKII) activity is associated with phagolysosome fusion and killing of bacteria. Downregulation of TfR surface expression, segregation of mycobacteria from the iron import pathway and exposition to NRAMP-1, a putative iron exporter from late endosomes, deplete the phagosome of iron. Fc receptor-mediated uptake in the presence of specific antibodies has been suggested to promote phagolysosome fusion. Activation of macrophages promotes nuclear import of the transcription factor NFκB and subsequent enhanced expression of antibacterial effector mechanisms and proinflammatory mediators, including tumor necrosis factor α (TNFα) essential for granuloma formation.

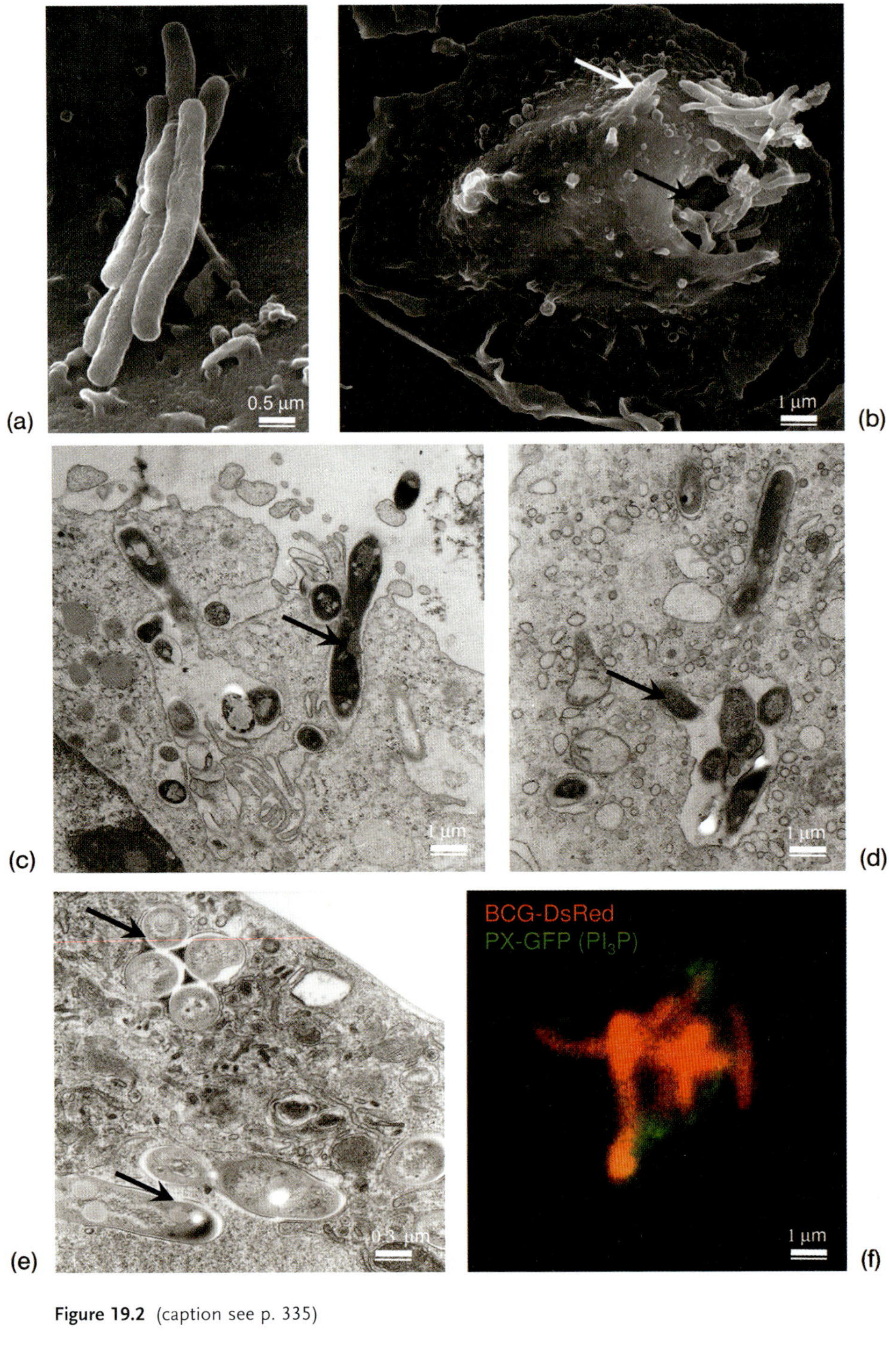

Figure 19.2 (caption see p. 335)

of the chemokine receptor CCR7, DCs migrate towards a gradient formed by the chemokines CCL19 and CCL21 to enter the drainig lymph nodes, from where mycobacteria can also be isolated.

Although there are reports of mycobacteria within nonphagocytic cells, *in vitro* infection of nonphagocytic cells requires enormously high and probably nonphysiological multiplicities of infection (MOI) per cell. Occasionally, tubercle bacilli have been observed in lung parenchymal cells. Both, tubercle bacilli and MAC have also been detected in adipose tissue, and *in vitro* studies revealed their ability to invade and persist in adipocytes [34]. *M. leprae* is observed in macrophages but also frequently in nonprofessional phagocytes, notably Schwann cells.

Although mycobacteria are able to survive and proliferate within phagocytes, it should be noted that during active pulmonary tuberculosis they are preferentially found in the extracellular space. The necrotic centers of infection sites during active disease contain liquefied, so-called caseating, debris. These lesions are rich in cellular detritus, membrane phospholipids and cholesterol, which serve as an important carbon source for mycobacteria. Transcriptome analysis of tuberculosis lesion material suggests a metabolic switch of mycobacteria towards the use of fatty acids as carbon source under these conditions [35, 36].

19.2.2
Recognition and Phagocytosis of Mycobacteria

Phagocytes possess a set of germline-encoded pattern recognition receptors (PRR), which recognize invariant MAMPs. Mycobacteria engage an array of surface receptors, which either induce phagocytosis or recognition/signaling of their presence, or both. Phagocytosis is a defense mechanism for elimination of pathogens, but also a prerequisite for mycobacteria to access their intracellular niche. The receptors that facilitate mycobacterial phagocytosis comprise C-type lectins such as the macrophage mannose receptor (MMR) and the DC-specific ICAM-3 grabbing non-integrin (DC-SIGN), which both bind mannose residues. Further receptors,

Figure 19.2 Phagocytosis and phagosomes of *Mycobacterium tuberculosis*. (a) Scanning electron micrograph (SEM) of a bundle of *M. tuberculosis*. (b) SEM showing a macrophage phagocytose bundles of *M. tuberculosis* (black arrow) and single bacilli (white arrow) revealing the differences in pseudopod morphology between both uptake processes. (c and d) Transmission electron micrographs (TEMs) of a macrophage in the process of phagocytosing *M. tuberculosis* cells, and already established (arrows). (e) TEMs of *M. tuberculosis* phagosomes in macrophages 24 h after uptake reveal that the mycobacteria reside in single-bacteria compartments with intact phagosomal membranes. (f) Confocal laser scanning micrograph of phagosomes containing *M. bovis* BCG expressing DsRed (kind gift of Dr. Nathalie Winter, Institute Pasteur, Paris) in a PX-GFP-expressing RAW macrophage 2 h after infection. PX is a domain of the NADPH oxidase, which binds PI(3)P [61]. PI(3)P is generated at early endosomal membranes by the PI3K Vsp34. This figure indicates that the BCG phagosome depicted is a PI(3)P-positive early endosomal compartment. Arrows indicate mycobacteria. Figures by Brinkmann and Schaible (Max-Planck-Institute of Infection Biology, Berlin and Korbel and Schaible, London School of Hygiene and Tropical Medicine).

which have also been associated with uptake of nontypical mycobacteria, are surfactant protein A, CD14-binding lipoarabinomannan (LAM), scavenger receptors (SR-A, SR-B) and CD36-binding microbial diacyl glycerides [37, 38].

Toll-like receptors (TLRs) are also engaged by mycobacterial MAMPs. Rather than leading to phagocytosis, they recognize the pathogens and provide the necessary intracellular signaling events. Upon ligation of TLR, phagocytes secrete proinflammatory cytokines such as IL-1, IL-12 and IL-18. TLR1/2 heterodimers recognize mycobacterial diacylic lipoproteins including the redundant lipoprotein p19. Mycobacterial cell wall lipids such as LAM, trehalose dimycolate (TDM) and phosphatidylinositol mannosides (PIMs) bind either TLR2 or TLR4. The endosomal TLR9 binds sugar backbones of low methylated bacterial DNA containing CpG motifs (see Chapter 12). The intracellular nucleotide-binding oligomerization domain (NOD)-like receptor NOD-2 can also recognize mycobacteria-derived ligands (probably muramyl dipeptides) and promote IL-1β secretion [39, 40]. It should be noted that intracellular signaling events can be triggered by ligand binding to MMR or DC-SIGN. However, these signals are anti-inflammatory by reducing IL-12 expression [41–44].

The relevance of TLR for the outcome of mycobacterial infections is, however, rather controversial. Triple knockout mice for TLR2/4/9 are only slightly more susceptibel to tuberculosis than wild type mice in one study, whereas others found higher susceptibility of mice lacking TLR2/9 and even TLR2 or TLR4 alone [45–48]. More importantly however, mice lacking the adapter molecule for TLR signaling, MyD88, have been found to be highly susceptible to tuberculosis by several authors [45, 47–49]. This suggests additional function of MyD88 in host responses such as in the IL-1/IL-18 receptor signaling cascade (see below).

19.2.3 The Mycobacterial Phagosome: The Resting Macrophage

A particle or non-pathogenic bacterium engulfed by a macrophage follows a programmed path through the cellular vesicular system controlled by fusion and fission events, as described in detail in Chapters 3 and 10 of this book.

Early on in tuberculosis research, it became obvious that *M. tuberculosis* interferes with phagosome maturation [50, 51]. These seminal studies brought cell biology and microbiology together. Follow-up studies revealed that phagosomes containing mycobacteria are characterized as early endosomal compartments due to their selection of endosomal markers (Figure 19.1) [52, 53]. Because of the paucity of the vesicular proton ATPase (v-H^+-ATPase) on the membrane of mycobacterial phagosomes, *M. avium* and *M. tuberculosis* phagosomes fail to acidify in the way that phagosomes with inert particles do (pH 4.5–5) and maintain a pH of around 6.3 [54]. Moreover, mycobacterial phagosomes carry only limited amounts of lysosome-associated membrane proteins such as LAMP1, and have limited hydrolytic enzyme activity such as cathepsin B/L. Cathepsin D, which requires proteolytic and (auto-) catalytic processing at low pH lysosomes to generate the enzymatically active form, is only present in its high molecular weight pro-form in mycobacterial phagosomes [55]. This indicates that mycobacterial phagosomes are restricted from fusion events with downstream late endosomal/lysosomal compartments. They are in dynamic exchange, however, with the host cell's plasma membrane. Both *M. tuberculosis* and *M. avium* phagosomes are readily accessible to GM1 gangliosides complexed to the surface by the cholera toxin B subunit [56]. Furthermore, mycobacterial phagosomes have access to the transferrin receptor pathway for iron import by transferrin as shown in murine and human macrophages [55, 57, 58].

Rab5, a small GTPase important in early endosomal fusion events, is also present on *M. tuberculosis* and *M. bovis* BCG phagosomes, whereas Rab7 usually associated with late endosomes is absent [59, 60]. Independent studies have demonstrated synthesis of the early endosomal lipid phosphatidylinositol 3-phosphate (PI(3)P) by the phosphatidylinositol 3-kinase (PI3K) Vsp34 at the cytoplasmic side of mycobacterial phagosome membranes, and its retention on mycobacterial phagosomes (Figure 19.2) [61–63]. PI3P interacts with Rab5 and its effector, early endosomal antigen 1 (EEA1), to promote endosome maturation. However, EEA1 has not been associated with mycobacterial phagosomes, suggesting that it is maintained at a Rab5-positive but EEA1-negative stage. Another small GTPase, Rab14, has been associated with homotypic early endosomal fusion and nonfusogenic mycobacterial phagosomes [64]. Overexpression of Rab14 in cell lines maintained mycobacterial phagosomes at an early endosomal stage, whereas silencing by RNA interference promoted phagosome maturation. Finally, the early endosomal v-SNARE cellubrevin is degraded on mycobacterial phagosome membranes, making it unavailable for fusion events and delivery of transport vesicles from the *trans*-Golgi [65].

The early endosomal characteristics of mycobacterial phagosomes prompted follow-up studies addressing how mycobacteria maintain their phagosomal niche, and which bacterial factors target host cell functions.

19.2.4
How to Inhibit Phagosome Maturation

Several explanations have been offered to explain how mycobacteria interfere with or block phagosome maturation. The current pictures suggest that mycobacteria employ different strategies, which act synergistically in order to secure their residency in an early endosomal niche and to avoid transfer to lysosomes.

Modes of uptake of mycobacteria have been associated with determining their phagosomal fate. The actin-binding protein coronin 1 (TACO) is associated with formation of the *M. bovis* BCG phagosome. It was suggested that its stable formation could prevent in a cholesterol-dependent manner further fusion events and maturation [66, 67]. Mycobacteria phagosomes labeled intensively with the cholesterol marker filipin, but this could, however, also be due to uptake through cholesterol-rich membrane moieties. Cells devoid of coronin 1 as well as those depleted of cholesterol promoted transfer of mycobacteria to phagolysosomes. A different scenario indicates that ligation of MMR by mannosylated glycolipids of the mycobacterial cell wall can deliver mycobacteria to early endosomal compartments [68]. Mannosylated ligands for MMR include phosphatidylinositol-mannosides (PIM) and mannose-capped lipoarabinomannans (manLAM). Finally, a hydrophobic particle surface similar to the mycobacterial cell wall can maintain close contact to the phagosomal membrane in a cholesterol-dependent manner. This direct bacteria–membrane interaction was suggested to interfere with phagosome maturation [69, 70]. Mycobacteria as well as beads coated with the mycobacterial glycolipid trehalose dimycolate (TDM) also maintain a close contact to the phagosomal membrane in resting macrophages [61, 69, 71] (Figure 19.2).

The intraphagosomal environment has also been put up on the table of explanations for mycobacteria's influence on phagosome maturation. Mycobacteria express a urease generating ammonia, which could buffer the phagosomal lumen to counteract proton influx [72]. Other scenarios suggest that mycobacterial products interfere with the vesicular fusion machinery on the cytoplasmic phase of the phagosomal membrane. These models require that the mycobacterial effectors are translocated into the cytoplasm. Mycobacteria have several systems for protein transport, including five ESX systems (1–5), which are responsible for secretion of the virulence-associated antigens ESAT-6 and CFP-10. Of those, ESX-1 and -5 are virulence associated, and may also facilitate transfer of effector proteins into the host cell cytosol [73]. The low molecular weight proteins ESAT-6/CFP-10 are virulence factors on their own right by binding to TLR2 and inhibiting TLR2 signaling [74]. A mycobacterial eukaryote-like lipid phosphatase (SapM) was proposed as an inhibitor of vesicle fusion through cleavage of PI(3)P in the phagosomal membrane [75]. The secreted tyrosine phosphatase PtpA of *M. tuberculosis* phosphorylates the human VPS33B, a regulator of membrane fusion, and thereby inhibits phagosome–lysosome fusion. A PtpA deletion mutant is attenuated and fails to inhibit phagosme maturation [76]. The *M. tuberculosis* genome also encodes 11 members of eukaryote-like serine/threonine kinases probably acquired by horizontal gene transfer which are involved in cell division, stress response and host cell modulation (see Chapter 2) [77]. Of those 11 genes, PknG is present in both the *M. tuberculosis* and *M. leprae* genome, and the protein is secreted and translocated into the host cell cytosol. A PknG deletion mutant failed to inhibit phagosome maturation, and a selective inhibitor for PknG abrogated inhibition of phagosome maturation by wild-type mycobacteria [78, 79].

Mycobacterial cell wall lipids have also been implicated in interfering with phagosome maturation, since delipidated mycobacteria lost their ability to arrest phagosome maturation [80]. The mycobacterial cell wall contains an array of genus- or even species-specific fatty acids, glycolipids and waxes such as PIM, LAM, TDM, mycolic acids, sulfatides, mycoside B, phthiocerol dimycocerosates (PDIMs) phthiocerol diphthiceranates (PDIPs) and phenolic glycolipids (PGLs; phthiocerol diesters) [81]. Analysing the function of mycobacterial cell wall lipids in phagosome maturation has usually been addressed by coating model particles such as beads with the respective lipids to follow the fate of the beads in macrophages. LAM- and PIM-coated beads have been shown to inhibit maturation of bead phagosomes. The phosphatidyl-anchored PIMs and LAMs may compete for EEA1 binding to the phagosomal membrane lipid PI3P or interfere with Vps34 activity [82–86]. The PGLs of *M. marinum* also inhibited bead phagosome maturation [87]. Intracellular *M. bovis* BCG enhance the production of TDM (cord factor) in the first 24 h after uptake into macrophages, indicating TDM's function in response to the intracellular environment [88]. Reconstitution with the TDM of delipidated mycobacteria, which failed to arrest phagosomal maturation, restored this ability [80]. An attenuated *M. tuberculosis* deletion mutant for antigen 85A (fbpA), a mycoloyl transferase involved in TDM synthesis, was unable to block phagosome maturation [89]. We found that TDM-coated beads decelerate bead phagosome maturation and keep the phagosomes at an early endosomal stage, that is, non-acidified, carrying transferrin/transferrin receptor, Rab5, PI3P, procathepsin D and low amounts of lysosome-associated membrane protein 1 (LAMP1), cathepsin B/L activity and v-H^+-ATPase [61].

In contrast to the scenarios involving interactions between phosphatidyl-anchored LAMs and PIMs with EEA1 or Vsp34, release of TDM from the particle/bacterium is not required to interfere with phagosome maturation. Of interest for the putative action of TDM are the observations that trehalose interacts with hydrophilic head groups of polar phospholipids, thereby slowing down membrane dynamics, which may affect fusion events required for phagosome maturation [90].

It has been shown that intracellular mycobacteria release a vast array of their cell wall lipids. These lipids travel inside the infected cell as far as lysosomes and further – via exosomes and apoptotic blebs – to neighboring cells [88, 91–95] (Figures 19.3 and 19.4). It can therefore be hypothesized that these bioactive molecules, which can interact with membrane moieties, membrane receptors or intercalate into phospholipid membranes impose a strong influence on host cell functions. Further analysis of the functions of mycobacterial cell wall lipids alone or in synergy will reveal their full function in maintaining the mycobacterial niche inside host cells.

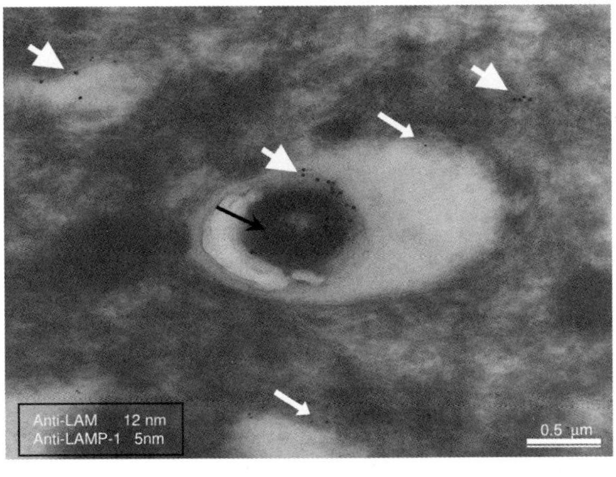

Figure 19.3 Intracellular mycobacteria release cell wall lipids. Transmission electron micrograph (TEM) of a *Mycobacterium tuberculosis* phagosome in a macrophage immunogold labeled using antibodies to the mycobacterial cell wall glycolipid lipoarabinomannan (LAM, antibody CS35, a kind gift of Delphi Chatterjee, Fort Collins, Colorado State University; 12 nm gold, large arrows) and LAMP1 (5 nm gold; small arrows). LAM is not only labeled within the mycobacterial cell wall but also in other compartments within the macrophage. Mycobacterial cell = black arrow. Figure by Fischer, Brinkmann and Schaible (Max-Planck-Institute of Infection Biology, Berlin).

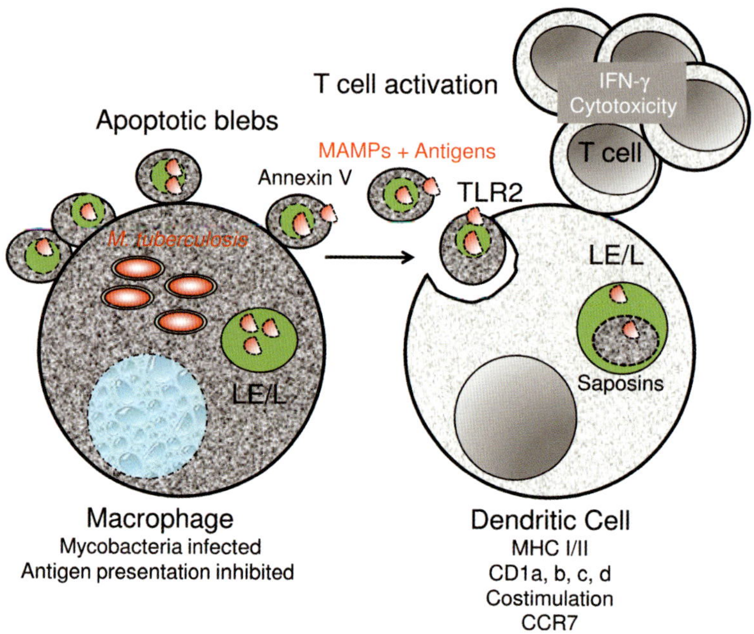

Figure 19.4 Cross-presentation by apoptotic blebs facilitates T cell activation in tuberculosis (schematic drawing). *Mycobacterium tuberculosis* (red)-infected macrophages are handicapped in presenting mycobacterial antigens to T cells to activate an immune response. Mycobacteria (red) residing in early endosomal phagosomes release antigens that traffic to lysosomes (green). Infected cells, however, succumb to apoptosis and annexin V-positive apoptotic blebs carry mycobacterial antigens and Toll-like receptor 2 (TLR2) ligands, that is, microbial-associated molecular patterns (MAMPs) to non-infected dendritic cells. Thereby, dendritic cells are prompted to migrate – guided by their chemokine receptor CCR7 – to draining lymph nodes where they cross-present mycobacterial (red) antigens to prime and activate CD4 and CD8 T cells through MHC class I, II and CD1 molecules and induce cytotoxicity and interferon γ (IFNγ) production [92, 162]. Presentation of mycobacterial antigens, which are delivered by apoptotic blebs to dendritic cells, require lysosomal processing with the help of lysosomal lipid transfer proteins, the saposins, to unwrap bleb membranes. The outcome is *M. tuberculosis*-specific and, in 90% of infected individuals, protective immunity able to keep the infection in a dormant stage.

19.2.4.1 **Intracellular Peculiarities**

Several authors have revealed – primarily by electron microscopy – that the phagosomal membranes remain intact during the intracellular life stage [52, 93, 96]. However, a recent study demonstrated that virulent *M. tuberculosis* and *M. leprae* but not attenuated *M. bovis* BCG can translocate 2 days after infection into the host cell cytoplasm [97]. A similar phenomenon has been observed for *M. marinum* (see Chapter 25). This feature is predominantly observed in DCs, which do not support mycobacterial growth [32]. Future studies are required to confirm whether the cytoplasmic localization of virulent mycobacteria is a common outcome of infection or associated with a certain virulence phenotypes or host cell types. It should be noted that most studies on mycobacterial phagosomes have been performed using either cell lines such as the RAW or J774 murine macrophage-like cells or murine bone marrow-derived macrophages, whereas the intracytoplasmic localization was observed in human monocyte-derived cells [97]. The authors also revealed that intracellular localization depends on the expression of ESAT-6 and CFP-10, low molecular weight secreted virulence factors specific for *M. tuberculosis* [97].

19.2.5
The Activated Macrophage

Mycobacteria are unable to maintain the early endosomal stage in IFNγ-activated macrophages. *M. avium* phagosomes lose the association with transferrin/transferrin receptor, gain mature cathepsin D, LAMP1 and v-H^+-ATPase, and consequently acidify to pH 5.5 [71]. In activated macrophages, *M. avium* does not grow and the number of viable bacteria is slowly reduced. *M. bovis* BCG is killed in activated macrophages, whereas *M. tuberculosis* remains in a nongrowing but obviously viable stage.

It is still a matter of debate whether promotion of phagosome maturation in activated macrophages precedes or follows killing of the mycobacteria. The activated macrophage is equipped with a number of microbicidal effector mechanisms, which may attack mycobacteria (Figure 19.1). Reactive oxygen intermediates generated by the early endosomal NADP oxidase can kill a number of microbes. Mice lacking this enzyme, however, are only slightly more, if at all, susceptible to *M. tuberculosis* infection in the initial phase of the infection [98]. IFNγ activation induces expression of many genes involved in host defense functions such as nitric oxide synthase (NOS2) and the small intracellular 47 kDa GTPases, of which IRGM1/LRG47 has been linked to antimycobacterial activity and phagolysosome fusion [99–103].

Mice deficient in NOS2 are more susceptible to tuberculosis than wild-type mice [99]. NOS2-derived reactive nitrogen intermediates (RNIs) can kill attenuated mycobacterial strains but this imposes a stress response in virulent *M. tuberculosis* [104]. TDM-mediated inhibition of phagosome maturation is abolished by nitric oxide in a NOS2-dependent manner [61]. Nitric oxide appears to target the hydroxyl groups of trehalose residues, which could abolish their direct interaction with the phagosomal phospholipid membrane [90].

IFNγ-activated LRG47 knockout macrophages failed to promote phagolysosome fusion and *M. tuberculosis* killing [103]. LRG47-mediated antimycobacterial activity was independent of NOS2, however. IRGM1/LRG47 has been linked to the induction of autophagy, which is induced by IFNγ activation and starvation [100–102]. Autophagosomes are formed from endoplasmic reticulum sheets engulfing regions of cytoplasm and organelles controlled by LC3/Atg8, a lipidated membrane protein. Autophagy is the waste and recycling system of eukaryotic cells for misfolded proteins, membranes and aged organelles. In activated macrophages, nascent autophagosomes fuse with mycobacterial phagosomes, thereby promoting their maturation and killing of the bacteria. As antibacterial effectors within autophagosomes, ubiquitin-derived peptides have been identified. These antibacterial peptides

(AMPs) are generated by the proteolytic activity of cathepsin D and L [105]. Alternative activation of macrophages by the T helper 2 cytokine IL-4 (see Chapter 12) masters a different program, which hinders antibacterial activity by autophagy and promotes mycobacterial survival [106].

TLR-mediated signals for activating macrophages have been indicated to promote maturation of phagosomes. Phagosomes containing Gram-positive or Gram-negative bacteria such as *Salmonella* and *Staphylococcus* matured less efficiently in macrophages lacking TLR2/4 when compared with wild-type cells, suggesting a local and autonomous signaling system to distinguish between pathogenic or harmless cargo [107]. However, using kinetic assays to assess phagosome maturation by measuring pH, hydrolytic activity and fluorescence resonance energy transfer (FRET), a follow-up study revealed similar kinetics of *Staphylococcus* phagosome maturation in TLR2 knockout and wild-type cells [108]. *M. tuberculosis* phagosome maturation was only partially influenced by the lack of TLR2 or TLR4 [45]. All three studies revealed, however, that in the absence of MyD88, an adapter molecule for the TLR signaling cascade, phagolysosome fusion was reduced even in the presence of IFNγ. This was also shown in *M. tuberculosis* phagosomes [45]. In human macrophages, TLR2 ligation induces expression of the AMP cathelizidin LL37 in a vitamin D/vitamin D receptor-dependent manner (see Chapter 12). LL37 is transferred to the mycobacterial phagosome, where it kills *M. tuberculosis* [29, 109]. Taken together, an array of effector mechanisms can interfere with mycobacterial growth and viability, and usually target the intracellular niche of the bacilli.

19.2.6
Expression Profiling of the Mycobacterial Responses to the Intracellular Environment

The knowledge of *M. tuberculosis* genomes spawned a number of follow-up studies focusing on the mycobacterial transcriptome and proteome under various living conditions and comparing mycobacterial species [2, 4]. Comparative transcriptome and proteome analyses between *M. tuberculosis* H37Rv and BCG revealed a number of genes/proteins involved in virulence as well as those putatively employable as vaccine candidates to replete the restricted BCG repertoire [110–113].

19.2.6.1 The Transcriptome/Proteome

Of special interest for this chapter, the transcriptome/proteome of intraphagosomal mycobacteria has been analyzed. These studies revealed upregulation of a number of signature response genes/proteins [114–117]. Inside macrophages, *M. tuberculosis* switches on an iron-scavenging, anaerobic respiration and dormancy gene expression program, including genes for mycobactin synthesis and β-oxidation of fatty acids as carbon and energy source [114, 115]. Upregulation of genes involved in mycolic acid synthesis and modification suggests enhanced production of cell wall lipids to conquer antibacterial effectors of macrophages and cell envelope damage [115]. This is in line with increased production of TDM by intracellular mycobacteria and the attenuation of deletion mutants for genes involved in cell wall synthesis [88, 89]. Genes enhanced in macrophages were also found to be upregulated in lungs from

M. tuberculosis-infected mice [115]. *In vitro* growth conditions mimicking phagosomal conditions such as low iron, low oxygen, acidification and exposure to reactive oxygen and nitrogen intermediates have further completed our knowledge of the *M. tuberculosis* stress response [118].

19.2.6.2 **Transcriptome Profiling**

A recent detailed transcriptome profiling study of the *M. tuberculosis*–macrophage interaction combined with pharmacological manipulation of the phagosomal environment and *in vitro* studies revealed interesting clues to the tubercle bacilli's response to intracellular environment [52, 119]. There was no overall change in gene expression patterns upon binding to macrophages without internalization, which is the opposite of what is seen in other intracellular bacteria. However, the subsequent transcriptional response monitored in short time intervals occurred only a few minutes after phagocytosis, indicating that the (small) change in phagosomal pH is an important trigger for differential gene expression [52, 119]. Intracellular bacilli upregulated genes linked to the β-oxidation pathway in the fatty acid metabolism including *fadD*, *fadE* and the isocitrate lyase (*icl*) genes. The glyoxylate shunt pathway allows bacteria to utilize fatty acids as prime carbon sources and to avoid loss of carbon through carbon dioxide during the Krebs cycle. Isocitrate lyase function in *M. tuberculosis* virulence was demonstrated by the inability of an *icl1/icl2* deletion mutant to survive in macrophages and mice [120]. Interestingly, the single *icl1* deletion mutant, though also attenuated for survival in mice, did grow normally in resting macrophages but was attenuated for growth in IFNγ-activated macrophages. This result suggests a more urgent need for this pathway under host response pressure, which may limit carbon source availability within the phagolysosome [36].

Analyses of both the transcriptome and the proteome of intracellular *M. tuberculosis* revealed upregulation of genes/proteins of the methyl citrate cycle [115, 116]. The methyl citrate cycle metabolizes propionyl-CoA generated by beta-oxidation of amino acids and odd-chain fatty acids. A double deletion mutant for the methyl citrate synthase (a prpC homolog) and methyl citrate dehydratase (a prpD homolog) in *M. tuberculosis* is attenuated in macrophages but not in mice [121, 122]. In our hands, a single methyl citrate dehydratase deletion mutant is attenuated in both macrophages and mice, and is unable to incorporate propionate (Hagens, Mattow, Kaufmann, Schaible; unpublished results). Interestingly, the lack of the methyl citrate lyase in *M. tuberculosis* is compensated by the redundant function of Icl1 [121]. Switching to alternative carbon/energy sources and distinct metabolic pathways is probably a common theme when *M. tuberculosis* enters the intracellular environment.

19.2.6.3 **The Intracellular Environment**

The intracellular environment and antibacterial effectors – especially in the activated macrophage – seem to impose a higher degree of stress to mycobacteria as shown by studies where the *M. tuberculosis* prcBA genes encoding the core of the proteasome were silenced *in vivo* using an conditional system. Proteasome depletion attenuated the mycobacteria for growth in both normal as well as INFγ knockout mice, suggesting proteasomal function beyond adaptive immunity. The mycobacterial proteasome is an essential response molecule probably to remove proteins damaged by host response effectors such as reactive oxygen/nitrogen intermediates [123]. Although *M. tuberculosis* is able to survive *in vitro* at pH 4.5, the shift to a low pH in phagolysosomes of IFNγ-activated macrophages puts the bacteria under stress. A number of mycobacterial genes for resistance to low pH conditions have been identified using transposon mutagenesis [124] [125]. *M. tuberculosis* deletion mutants for four of those genes, including the membrane protein Rv3671c, were sensitive to low pH and failed to maintain intrabacterial pH in acidic solutions and activated macrophages. The mutants were also attenuated in mice [124, 125]. Genes involved in resistance to host defense mechanisms and the intracellular environment are attractive targets for chemotherapy.

19.2.6.4 **Regulatory Genes**

Regulatory genes upregulated upon engulfment by macrophages included the transcriptional regulators of the WhiB (*whib3*, 6 and 7), PhoR and DosR families [52, 119].

- WhiB genes are iron–sulfur cluster proteins unique to *Actinomycetes* and have been characterized in *M. tuberculosis* as response genes to various stimuli including antibiotics, iron starvation and growth phase [126, 127]. Maximal expression of mycobacterial *whiB3* is observed 6 h after engulfment by macrophages and declines with increased mycobacterial numbers. Similarly, in the murine tuberculosis model, *whiB3* is maximally expressed in lungs within the first two weeks of infection but less at later time points of infection, suggesting its function in early regulation of adaptation to survival within host cells [126].
- Deletions within the PhoP two-component signal transduction system abolished expression of gene clusters encoding enzymes for biosynthesis of certain complex lipids involved in virulence such as sulfatides, diacyltrehalose and polyacyltrehalose, and attenuated respective deletion mutants [128, 129].
- Finally, the DosR regulon, another two-component signal transduction system, is involved in the response of *M. tuberculosis* to hypoxia and nitric oxide and has been suggested to control the mycobacterial dormancy program [130, 131]. *M. tuberculosis* can become dormant and persist for years within humans in a clinically latent state.

19.2.6.5 **Dormancy**

Dormancy has been linked to hypoxic conditions in tuberculosis lesions, as suggested to occur in humans and demonstrated in lesions of infected rabbits (but not in mice) [13, 22]. Pathogenic mycobacteria are aerobic. Consequently, microaerophilic conditions induce heat shock protein expression such as the 16 kDa α-crystalline, and finally shut down of protein synthesis and dormancy [21]. In mycobacteria grown under hypoxia, the cell wall became thicker and α-crystalline was associated with ribosomes [132, 133]. The Wayne culture model has been proven to be useful to mimick *in vitro* dormancy of mycobacteria [134, 135]. A number of dormancy-associated genes are regulated by the DosR (Rv3133) gene, a two-component response regulator and transcription factor [136]. DosR binds to two DNA-binding sites located upstream of the hypoxia-induced α-crystalline gene. Transcriptome analysis of bacilli within sputum from tuberculosis patients revealed their dormant stage, which is morphologically characterized by triacylglycerol storage inclusions [137]. This suggests that during transmission to the next person tubercle bacilli are in a nonproliferating stage but carry a potential carbon source with them. Another interesting finding indicates that BCG vaccination does not induce T cell response against dormancy-associated proteins [138]. This finding points out the feasibility that dormancy proteins could be used to improve current vaccination strategies, which could even target latent infections [139]. Although immunocompromising conditions are important determinants for reactivation of latent tuberculosis, future studies have to answer the enigmatic question how dormant tubercle bacilli are reactivated in the immunocompetent host.

19.2.7 Nutritional Benefit of Mycobacterium's Intracellular Niche

The intracellular niche of mycobacteria within resting macrophages allows access to the host's iron source. Transferrin transports iron into early endosomes but also into mycobacterial phagosomes where it is taken up by the bacilli [58]. Pathogenic mycobacteria require iron for vital biochemical processes, as does the host cell, causing competition for the intra- and extracellular iron pool. Active tuberculosis in humans as well as in mice is promoted by genetic and dietary iron overload, including nutritional iron supplementation [19, 20, 28, 140]. Host and pathogen possess specific iron-binding molecules for uptake and intracellular distribution of iron. Extracellular iron occurs usually in a form bound to transferrin. The host cell binds and internalizes transferrin-bound iron through the transferrin receptor into an early recycling compartment, where a mildly acidic pH permits uncoupling of iron from its receptor, and recycling of transferrin/transferrin receptor to the cell surface. Free iron is transported into the cytoplasm where it is bound to ferritin. This cytoplasmic iron pool can also be recruited by *M. tuberculosis*, though the pathway to do so has not been identified yet [141]. Mycobacteria produce a number of high-affinity iron-binding molecules, siderophores or mycobactins, production of which is upregulated in the intracellular environment. Released mycobactin captures free iron in the environment and is bound to membrane-bound siderophores for transport across the membrane [19, 142].

As countermeasures to segregate free iron from mycobacteria, activated macrophages downregulate surface expression of the TfR to limit intracellular iron [143]. In IFNγ-activated macrophages, *M. tuberculosis* becomes deprived of iron, which leads to upregulation of expression of those genes involved in iron scavenging and decrease of those for iron storage such as bacterioferritins [114, 115]. Lactoferrin is a high-affinity iron-binding protein in milk, tears and neutrophil granules which has been suggested to limit free iron in order to reduce availability to microbes [19].

In the iron-overloaded host, lactoferrin reduces free iron and inhibits mycobacterial load [28]. Moreover, neutrophil-derived lipochalins bind bacterial siderophores, including those from mycobacteria, to seclude iron from their usage [30]. Anemia of chronic inflammation is a symptom observed in a number of inflammatory diseases, including tuberculosis. It is promoted by a hormone-like peptide, hepcidin, released by hepatocytes and macrophages [144, 145]. Hepcidin is an AMP and binds to and reduces surface expression of ferroportin-1 (FPN1), the only cellular iron exporter, thereby limiting iron absorption in the gut and release from macrophages and enhancing intracellular iron stores. Whether these iron stores could be beneficial for intracellular mycobacteria remains to be elucidated.

FPN1 has been observed in mycobacterial phagosomes, which suggests there may be a pathway to target iron to phagosomal mycobacteria [146]. Hepcidin has also been found in the mycobacterial phagosome and direct antimycobacterial activity has been observed [147]. However, both these observations require further studies in order to establish their relevance for tuberculosis.

19.3
Host Response and Transmission

19.3.1
Immune Responses to *M. tuberculosis*

During mycobacterial infection, the full array of acquired immune cells are activated, including peptide-specific MHC class II-restricted CD4, MHC class I-restricted CD8 T cells as well as T cells responding to mycobacterial lipids in the context of CD1 molecules and T cells carrying a γδ instead of an αβ T cell receptor (TCR) (see Chapter 12). This effective immune activation is most probably the reason why in the majority of infected people, the infection is immune-controlled in a latent/dormant stage. Many T cell antigens of *M. tuberculosis* have been identified and associated with protective immunity, including the secreted proteins ESAT-6 and CFP-10 and the mycolyl-transferase Ag85 [148]. In contrast to polymorphic MHC molecules, the MHC I-like CD1a, CD1b, CD1c and CD1d molecules are nonpolymorphic and present lipid antigens [149, 150]. CD1d is a specific restriction element for NK/T cells. Glycolipids of mycobacterial cell wall origin recognized by CD1-restricted T cells include sulfatides and dideoxy-mycobactin by CD1a, diacylated sulfolipid, PIM, LAM, mycolic acids and glucose monomycolate by CD1b, mannose-isoprenoids by CD1c and PIMs by CD1d (see Chapter 12) [150, 151]. Mycobacteria also induce specific B cells – often recognizing various cell wall lipids – and B cells are abundant cells in the granulomas encaging the tuberculosis lesion [13]. In protection, B cells seem to play only a minor role if at all [152]. The innate and acquired immune response to *M. tuberculosis* and *M. leprae* is covered in many previous reviews and the interested reader is therefore referred to Chapter 12 [148, 153, 154]. In the context of the mycobacterial phagosome, however, the influence of the intracellular localization deserves further elaboration.

The early endosomal phagosome sequesters the mycobacteria from the classical MHC class I and class II-processing pathways, that is, within the cytoplasm or the late endosomes/lysosomes, respectively. Secreted proteins and released lipids, however, are delivered to late endosomes/lysosomes where antigen binding to MHC II, CD1b, CD1c and CD1d molecules occurs [93, 94, 155]. CD1a is shuttling into the mycobacterial phagosome and can directly sample lipid antigens there [148, 150, 155]. Therefore, access to mycobacterial antigens is still warranted in resting macrophages, although the phagosomes are prevented from reaching late endosomes/lysosomes. A bigger problem for inducing proper T cell responses is the fact that *M. tuberculosis* infection hampers the antigen-presenting function of macrophages and DCs. More importantly, mycobacteria interfere in a TLR2-mediated manner with IFNγ activation of macrophages needed to bring on the MHC II antigen processing and presentation machinery [92, 156–160]. Nevertheless, the full array of T cell populations is activated by tubercle bacilli, including CD8 T cells. This suggests that mycobacteria-specific T cells are induced by cross-presentation of antigens through cells others than the infected ones. We observed that *M. tuberculosis*-infected macrophages succumb to apoptosis and release apoptotic blebs [92, 161]. These blebs contain mycobacterial antigens – proteins and glycolipids – as well as ligands for TLR2 such as the p19 lipoprotein. Upon engulfment by non-infected DCs, these cells can present the mycobacterial antigens via MHC I, II and CD1b to T cells (Figure 19.4) [161, 162]. This detour pathway of cross-presentation for T cell activation by mycobacteria is functional in both the murine and human systems. Interaction with bleb-transported TLR ligands, activate DCs to secrete proinflammatory cytokines such as tumor necrosis factor α (TNFα) and IL-12, and to migrate in a CCR7-mediated manner to the draining lymph nodes [161]. We hypothesize that apoptosis is a prerequisite of T cell activation by *M. tuberculosis* – especially of CD8 T cells [163]. Indeed, proapoptotic vaccine strains of BCG or *M. tuberculosis* induce protective T cells including MHC I-restricted CD8 T cells [27, 164].

19.3.2
Cell Death and Transmission

At one point in the life of an intracellular tubercle bacillus, there comes the time to exit the cell and infect the next one. In the larger scale of the human infection, mycobacteria need to access a novel host organism. In the infectious stage bacilli are coughed up and aerosolized within microdroplets. Dissemination of the infection therefore requires leaking of bacilli out of caseating necrotic foci or caverns into the alveolar space, which is promoted by exacerbated inflammation and massive cell death. Intranasal lipopolysaccharide treatment simultaneous to *M. tuberculosis* infection in mice induced necrotic lesions, otherwise not seen in mice, indicating the importance of inflammatory stimuli for caseating necrosis [165].

Mycobacteria induce different types of cell death, apoptosis and necrosis. The capability of *M. tuberculosis* to induce either apoptosis or necrosis varies between different strains, and virulence is associated with inhibition of apoptosis and induction of necrosis rather than with proapoptotic properties [166–169]. Apoptosis of infected cells affects viability of intracellular mycobacteria [170]. Therefore it is more than consequent that virulent *M. tuberculosis* strains prevent apoptosis by inducing the antiapoptotic protein Mcl-1 (a Bcl-2 family member) or release of tumor necrosis factor receptor 2 (TNF-R2) [166, 169, 171]. In contrast, attenuated *M. tuberculosis* H37Ra, MAC or BCG induce classical apoptosis in a TLR2- and caspase-dependent manner [168, 172, 173]. In an *in vitro* cell culture system, apoptosis prevents spreading of MAC to macrophages newly added to the culture [174]. In cells infected with attenuated *M. tuberculosis* H37Ra, the apoptotic envelope is formed by exposure of phosphatidylserine on the cell surface and cross-linking with annexin 1. In cells infected with virulent *M. tuberculosis* H37Rv, the N-terminal domain of annexin 1 is removed by proteolysis, which abolishes apoptotic envelope formation and causes cell death by necrosis [167, 175]. In contrast, infection with an attenuated strain prevents proteolytic truncation of annexin 1 by the serine protease inhibitor PAI2 (plasminogen activator inhibitor type 2/SerpB2). Following high-dose infections, virulent tubercle bacilli seem just to overgrow infected cells and induce a caspase-independent type of cell death [176, 177]. Initial studies revealed that clinical isolates grow faster, induce necrosis rather than apoptosis, and may be more easily transmitted [178, 179]. Future studies have to reveal whether virulence-associated antiapoptotic properties in different clinical isolates of *M. tuberculosis* correlate with enhanced spreading within the community.

19.4 Outlook

Tuberculosis remains a major challenge worldwide for public health workers and researchers alike. Despite intensive research, novel therapeutics and vaccines will not be available immediately. Detailed knowledge of the host cell–pathogen interaction between tubercle bacilli and macrophages will contribute to improved vaccine design (see Chapter 12), more efficient drug delivery systems to target the intracellular niche (see Chapter 11) and compounds affecting dormant mycobacteria [27, 164, 180]. It could well be envisaged that transferrin is employed to target drug carriers into the early endosomal niche of the tubercle bacillus. It should be noted that the current picture we have of mycobacteria phagosome characteristics is based on *in vitro* work. Only one study showed that tubercle bacilli in alveolar macrophages from tuberculosis/HIV co-infected patients reside in non-mature phagosomes [181]. In the future, mass spectrometry of patient samples combined with high-end imaging technologies and metabolome analysis will reveal the microecology of tuberculosis lesions and localization of mycobacteria. Studies on mycobacterial virulence properties and intracellular trafficking should employ relevant clinical isolates. The influence of co-infections by pathogens such as malaria parasites, viruses and helminths, as well as exogenous factors such as micronutrients on the outcome of tuberculosis is a future challenge for translational research. It is hoped that orchestrated scientific effort will lead to better control measures of this important human infection.

Acknowledgments

The author would like to thank all present and past members of his lab for their great enthusiasm and scientific interest in all the projects, his mentors during his PhD and postdoctoral period, and his highly supportive family. The work of his group was supported over the last years by the following funders: the Deutsche Forschungsgemeinschaft, the Deutsches Bundesministerium für Forschung und Technologie, the Medical Research Council UK, the Wellcome Trust UK and the Bill and Melinda Gates Foundation.

References

1 Smith, N.H., Gordon, S.V., de la Rua-Domenech, R., Clifton-Hadley, R.S. and Hewinson, R.G. (2006) *Nat Rev Microbiol*, **4**, 670–681.

2 Cole, S.T., Brosch, R., Parkhill, J., Garnier, T. *et al.* (1998) *Nature*, **393**, 537–544.

3 Garnier, T., Eiglmeier, K., Camus, J.C., Medina, N. *et al.* (2003) *Proc Natl Acad Sci USA*, **100**, 7877–7882.

4 Fleischmann, R.D., Alland, D., Eisen, J.A., Carpenter, L. *et al.* (2002) *J Bacteriol*, **184**, 5479–5490.

5 Behr, M.A. and Small, P.M. (1999) *Vaccine*, **17**, 915–922.

6 Behr, M.A., Wilson, M.A., Gill, W.P., Salamon, H. *et al.* (1999) *Science*, **284**, 1520–1523.

7 Cole, S.T., Eiglmeier, K., Parkhill, J., James, K.D. *et al.* (2001) *Nature*, **409**, 1007–1011.

8 Li, L., Bannantine, J.P., Zhang, Q., Amonsin, A. *et al.* (2005) *Proc Natl Acad Sci USA*, **102**, 12344–12349.

9 Paustian, M.L., Kapur, V. and Bannantine, J.P. (2005) *J Bacteriol*, **187**, 2406–2415.

10 Paustian, M.L., Zhu, X., Sreevatsan, S., Robbe-Austerman, S. *et al.* (2008) *BMC Genomics*, **9**, 135.

11 Hunter, R.L., Jagannath, C. and Actor, J.K. (2007) *Tuberculosis (Edinb)*, **87**, 267–278.

12 Ulrichs, T. and Kaufmann, S.H. (2006) *J Pathol*, **208**, 261–269.

13 Tsai, M.C., Chakravarty, S., Zhu, G., Xu, J. *et al.* (2006) *Cell Microbiol*, **8**, 218–232.

14 World Health Organization (2008) *Global Tuberculosis Control: Surveillance, Planning, Financing, WHO Report 2008*. World Health Organization, Geneva.

15 Kaufmann, S.H. (2007) *Nat Rev Microbiol*, **5**, 491–504.

16 Keane, J., Gershon, S., Wise, R.P., Mirabile-Levens, E. *et al.* (2001) *N Engl J Med*, **345**, 1098–1104.

17 Schaible, U.E. and Kaufmann, S.H. (2007) *PLoS Med*, **4**, e115.

18 Fine, P.E. and Small, P.M. (1999) *N Engl J Med*, **341**, 1226–1227.

19 Schaible, U.E. and Kaufmann, S.H. (2004) *Nat Rev Microbiol*, **2**, 946–953.

20 Gangaidzo, I.T., Moyo, V.M., Mvundura, E., Aggrey, G. *et al.* (2001) *J Infect Dis*, **184**, 936–939.

21 Hu, Y.M., Butcher, P.D., Sole, K., Mitchison, D.A. and Coates, A.R. (1998) *FEMS Microbiol Lett*, **158**, 139–145.

22 Via, L.E., Lin, P.L., Ray, S.M., Carrillo, J. *et al.* (2008) *Infect Immun*, **76**, 2333–2340.

23 Aly, S., Wagner, K., Keller, C., Malm, S. *et al.* (2006) *J Pathol*, **210**, 298–305.

24 Fine, P.E. and Rodrigues, L.C. (1990) *Lancet*, **335**, 1016–1020.

25 Black, G.F., Weir, R.E., Floyd, S., Bliss, L. *et al.* (2002) *Lancet*, **359**, 1393–1401.

26 Abebe, F. and Bjune, G. (2006) *Clin Exp Immunol*, **145**, 389–397.

27 Grode, L., Seiler, P., Baumann, S., Hess, J. *et al.* (2005) *J Clin Invest*, **115**, 2472–2479.

28 Schaible, U.E., Collins, H.L., Priem, F. and Kaufmann, S.H. (2002) *J Exp Med*, **196**, 1507–1513.

29 Liu, P.T., Stenger, S., Li, H., Wenzel, L. *et al.* (2006) *Science*, **311**, 1770–1773.

30 Flo, T.H., Smith, K.D., Sato, S., Rodriguez, D.J. *et al.* (2004) *Nature*, **432**, 917–921.

31 Feng, C.G., Kaviratne, M., Rothfuchs, A.G., Cheever, A. *et al.* (2006) *J Immunol*, **177**, 7086–7093.

32 Tailleux, L., Neyrolles, O., Honore-Bouakline, S., Perret, E. *et al.* (2003) *J Immunol*, **170**, 1939–1948.

33 Wolf, A.J., Linas, B., Trevejo-Nunez, G.J., Kincaid, E. *et al.* (2007) *J Immunol*, **179**, 2509–2519.

34 Neyrolles, O., Hernandez-Pando, R., Pietri-Rouxel, F., Fornes, P. *et al.* (2006) *PLoS ONE*, **1**, e43.

35 Honer zu Bentrup, K. and Russell, D.G. (2001) *Trends Microbiol*, **9**, 597–605.

36 McKinney, J.D., Honer zu Bentrup, K., Munoz-Elias, E.J., Miczak, A. *et al.* (2000) *Nature*, **406**, 735–738.

37 Ernst, J.D. (1998) *Infect Immun*, **66**, 1277–1281.

38 Tailleux, L., Maeda, N., Nigou, J., Gicquel, B. and Neyrolles, O. (2003) *Trends Microbiol*, **11**, 259–263.

39 Ferwerda, G., Girardin, S.E., Kullberg, B.J., Le Bourhis, L. *et al.* (2005) *PLoS Pathog*, **1**, 279–285.

40 Ferwerda, G., Kramer, M., de Jong, D., Piccini, A. *et al.* (2008) *Eur J Immunol*, **38**, 184–191.

41 Quesniaux, V.J., Nicolle, D.M., Torres, D., Kremer, L. *et al.* (2004) *J Immunol*, **172**, 4425–4434.

42 Nigou, J., Zelle-Rieser, C., Gilleron, M., Thurnher, M. and Puzo, G. (2001) *J Immunol*, **166**, 7477–7485.

43 Geijtenbeek, T.B., Van Vliet, S.J., Koppel, E.A., Sanchez-Hernandez, M. *et al.* (2003) *J Exp Med*, **197**, 7–17.

44 Tailleux, L., Schwartz, O., Herrmann, J.L., Pivert, E. *et al.* (2003) *J Exp Med*, **197**, 121–127.

45 Holscher, C., Reiling, N., Schaible, U.E., Holscher, A. *et al.* (2008) *Eur J Immunol*, **38**, 680–694.

46 Bafica, A., Scanga, C.A., Feng, C.G., Leifer, C. *et al.* (2005) *J Exp Med*, **202**, 1715–1724.

47 Abel, B., Thieblemont, N., Quesniaux, V.J., Brown, N. *et al.* (2002) *J Immunol*, **169**, 3155–3162.

48 Drennan, M.B., Nicolle, D., Quesniaux, V.J., Jacobs, M. *et al.* (2004) *Am J Pathol*, **164**, 49–57.

49 Fremond, C.M., Yeremeev, V., Nicolle, D.M., Jacobs, M. *et al.* (2004) *J Clin Invest*, **114**, 1790–1799.

50 Armstrong, J.A. and Hart, P.D. (1975) *J Exp Med*, **142**, 1–16.

51 Armstrong, J.A. and Hart, P.D. (1971) *J Exp Med*, **134**, 713–740.

52 Rohde, K., Yates, R.M., Purdy, G.E. and Russell, D.G. (2007) *Immunol Rev*, **219**, 37–54.

53 Vergne, I., Chua, J., Singh, S.B. and Deretic, V. (2004) *Annu Rev Cell Dev Biol*, **20**, 367–394.

54 Sturgill-Koszycki, S., Schlesinger, P.H., Chakraborty, P., Haddix, P.L. *et al.* (1994) *Science*, **263**, 678–681.

55 Sturgill-Koszycki, S., Schaible, U.E. and Russell, D.G. (1996) *EMBO J*, **15**, 6960–6968.

56 Russell, D.G., Dant, J. and Sturgill-Koszycki, S. (1996) *J Immunol*, **156**, 4764–4773.

57 Clemens, D.L. and Horwitz, M.A. (1996) *J Exp Med*, **184**, 1349–1355.

58 Wagner, D., Maser, J., Lai, B., Cai, Z. *et al.* (2005) *J Immunol*, **174**, 1491–1500.

59 Via, L.E., Deretic, D., Ulmer, R.J., Hibler, N.S. *et al.* (1997) *J Biol Chem*, **272**, 13326–13331.

60 Clemens, D.L., Lee, B.Y. and Horwitz, M.A. (2000) *Infect Immun*, **68**, 2671–2684.

61 Axelrod, S., Oschkinat, H., Enders, J., Schlegel, B. *et al.* (2008) *Cell Microbiol*, **10**, 1530–1545.

62 Kelley, V.A. and Schorey, J.S. (2004) *Cell Microbiol*, **6**, 973–985.

63 Purdy, G.E., Owens, R.M., Bennett, L., Russell, D.G. and Butcher, B.A. (2005) *Cell Microbiol*, **7**, 1627–1634.

64 Kyei, G.B., Vergne, I., Chua, J., Roberts, E. *et al.* (2006) *EMBO J*, **25**, 5250–5259.

65 Fratti, R.A., Chua, J. and Deretic, V. (2002) *J Biol Chem*, **277**, 17320–17326.

66 Ferrari, G., Langen, H., Naito, M. and Pieters, J. (1999) *Cell*, **97**, 435–447.

67 Gatfield, J. and Pieters, J. (2003) *Adv Immunol*, **81**, 45–96.

68 Kang, P.B., Azad, A.K., Torrelles, J.B., Kaufman, T.M. *et al.* (2005) *J Exp Med*, **202**, 987–999.

69 de Chastellier, C. and Thilo, L. (2006) *Cell Microbiol*, **8**, 242–256.

70 de Chastellier, C. and Thilo, L. (1997) *Eur J Cell Biol*, **74**, 49–62.

71 Schaible, U.E., Sturgill-Koszycki, S., Schlesinger, P.H. and Russell, D.G. (1998) *J Immunol*, **160**, 1290–1296.

72 Gordon, A.H., Hart, P.D. and Young, M.R. (1980) *Nature*, **286**, 79–80.

73 Abdallah, A.M., Gey van Pittius, N.C., Champion, P.A., Cox, J. *et al.* (2007) *Nat Rev Microbiol*, **5**, 883–891.

74 Pathak, S.K., Basu, S., Basu, K.K., Banerjee, A. *et al.* (2007) *Nat Immunol*, **8**, 610–618.

75 Vergne, I., Chua, J., Lee, H.H., Lucas, M. *et al.* (2005) *Proc Natl Acad Sci USA*, **102**, 4033–4038.

76 Bach, H., Papavinasasundaram, K.G., Wong, D., Hmama, Z. and Av-Gay, Y. (2008) *Cell Host Microbe*, **3**, 316–322.

77 Pieters, J. (2008) *Cell Host Microbe*, **3**, 399–407.

78 Walburger, A., Koul, A., Ferrari, G., Nguyen, L. *et al.* (2004) *Science*, **304**, 1800–1804.

79 Scherr, N., Honnappa, S., Kunz, G., Mueller, P. *et al.* (2007) *Proc Natl Acad Sci USA*, **104**, 12151–12156.

80 Indrigo, J., Hunter, R.L. Jr and Actor, J.K. (2002) *Microbiology*, **148**, 1991–1998.

81 Brennan, P.J. (2003) *Tuberculosis (Edinb)*, **83**, 91–97.

82 Fratti, R.A., Chua, J. and Deretic, V. (2003) *J Biol Chem*, **278**, 46961–46967.

83 Vergne, I., Fratti, R.A., Hill, P.J., Chua, J. *et al.* (2004) *Mol Biol Cell*, **15**, 751–760.

84 Vergne, I., Chua, J. and Deretic, V. (2003) *J Exp Med*, **198**, 653–659.

85 Fratti, R.A., Chua, J., Vergne, I. and Deretic, V. (2003) *Proc Natl Acad Sci USA*, **100**, 5437–5442.

86 Vergne, I., Chua, J. and Deretic, V. (2003) *Traffic*, **4**, 600–606.

87 Robinson, N., Kolter, T., Wolke, M., Rybniker, J. *et al.* (2008) *Traffic*, **9**, 1936–1947.

88 Fischer, K., Chatterjee, D., Torrelles, J., Brennan, P.J. *et al.* (2001) *J Immunol*, **167**, 2187–2192.

89 Copenhaver, R.H., Sepulveda, E., Armitige, L.Y., Actor, J.K. *et al.* (2004) *Infect Immun*, **72**, 7084–7095.

90 Doxastakis, M., Sum, A.K. and de Pablo, J.J. (2005) *J Phys Chem B Condens Matter Mater Surf Interfaces Biophys*, **109**, 24173–24181.

91 Schaible, U.E. and Kaufmann, S.H. (2000) *Trends Microbiol*, **8**, 419–425.

92 Schaible, U.E., Winau, F., Sieling, P.A., Fischer, K. *et al.* (2003) *Nat Med*, **9**, 1039–1046.

93 Beatty, W.L., Rhoades, E.R., Ullrich, H.J., Chatterjee, D. *et al.* (2000) *Traffic*, **1**, 235–247.

94 Beatty, W.L. and Russell, D.G. (2000) *Infect Immun*, **68**, 6997–7002.

95 Beatty, W.L., Ullrich, H.J. and Russell, D.G. (2001) *Eur J Cell Biol*, **80**, 31–40.

96 Clemens, D.L., Lee, B.Y. and Horwitz, M.A. (2002) *Infect Immun*, **70**, 5800–5807.

97 van der Wel, N., Hava, D., Houben, D., Fluitsma, D. *et al.* (2007) *Cell*, **129**, 1287–1298.

98 Cooper, A.M., Segal, B.H., Frank, A.A., Holland, S.M. and Orme, I.M. (2000) *Infect Immun*, **68**, 1231–1234.

99 MacMicking, J.D., North, R.J., LaCourse, R., Mudgett, J.S. *et al.* (1997) *Proc Natl Acad Sci USA*, **94**, 5243–5248.

100 Singh, S.B., Davis, A.S., Taylor, G.A. and Deretic, V. (2006) *Science*, **313**, 1438–1441.

101 Deretic, V. (2006) *Curr Opin Immunol*, **18**, 375–382.

102 Gutierrez, M.G., Master, S.S., Singh, S.B., Taylor, G.A. *et al.* (2004) *Cell*, **119**, 753–766.

103 MacMicking, J.D., Taylor, G.A. and McKinney, J.D. (2003) *Science*, **302**, 654–659.

104 Schnappinger, D., Schoolnik, G.K. and Ehrt, S. (2006) *Microbes Infect*, **8**, 1132–1140.

105 Alonso, S., Pethe, K., Russell, D.G. and Purdy, G.E. (2007) *Proc Natl Acad Sci USA*, **104**, 6031–6036.

106 Harris, J., De Haro, S.A., Master, S.S., Keane, J. *et al.* (2007) *Immunity*, **27**, 505–517.

107 Blander, J.M. and Medzhitov, R. (2004) *Science*, **304**, 1014–1018.

108 Yates, R.M. and Russell, D.G. (2005) *Immunity*, **23**, 409–417.

109 Liu, P.T., Stenger, S., Tang, D.H. and Modlin, R.L. (2007) *J Immunol*, **179**, 2060–2063.

110 Jungblut, P.R., Schaible, U.E., Mollenkopf, H.J., Zimny-Arndt, U. *et al.* (1999) *Mol Microbiol*, **33**, 1103–1117.

111 Mattow, J., Jungblut, P.R., Schaible, U.E., Mollenkopf, H.J. *et al.* (2001) *Electrophoresis*, **22**, 2936–2946.

112 Mattow, J., Schaible, U.E., Schmidt, F., Hagens, K. *et al.* (2003) *Electrophoresis*, **24**, 3405–3420.

113 Mollenkopf, H.J., Grode, L., Mattow, J., Stein, M. *et al.* (2004) *Infect Immun*, **72**, 6471–6479.

114 Rachman, H., Strong, M., Schaible, U., Schuchhardt, J. *et al.* (2006) *Microbes Infect*, **8**, 747–757.

115 Schnappinger, D., Ehrt, S., Voskuil, M.I., Liu, Y. *et al.* (2003) *J Exp Med*, **198**, 693–704.

116 Mattow, J., Siejak, F., Hagens, K., Becher, D. *et al.* (2006) *Proteomics*, **6**, 2485–2494.

117 Sturgill-Koszycki, S., Haddix, P.L. and Russell, D.G. (1997) *Electrophoresis*, **18**, 2558–2565.

118 Garbe, T.R., Hibler, N.S. and Deretic, V. (1996) *Mol Med*, **2**, 134–142.

119 Rohde, K.H., Abramovitch, R.B. and Russell, D.G. (2007) *Cell Host Microbe*, **2**, 352–364.

120 Munoz-Elias, E.J. and McKinney, J.D. (2005) *Nat Med*, **11**, 638–644.

121 Gould, T.A., van de Langemheen, H., Munoz-Elias, E.J., McKinney, J.D. and Sacchettini, J.C. (2006) *Mol Microbiol*, **61**, 940–947.

122 Munoz-Elias, E.J., Upton, A.M., Cherian, J. and McKinney, J.D. (2006) *Mol Microbiol*, **60**, 1109–1122.

123 Gandotra, S., Schnappinger, D., Monteleone, M., Hillen, W. and Ehrt, S. (2007) *Nat Med*, **13**, 1515–1520.

124 Vandal, O.H., Pierini, L.M., Schnappinger, D., Nathan, C.F. and Ehrt, S. (2008) *Nat Med*, **14**, 849–854.

125 Vandal, O.H., Roberts, J.A., Odaira, T., Schnappinger, D. *et al.* (2008) *J Bacteriol.*, **191**, 625–631.

126 Banaiee, N., Jacobs, W.R. Jr. and Ernst, J.D. (2006) *Infect Immun*, **74**, 6449–6457.

127 Geiman, D.E., Raghunand, T.R., Agarwal, N. and Bishai, W.R. (2006) *Antimicrob Agents Chemother*, **50**, 2836–2841.

128 Perez, E., Samper, S., Bordas, Y., Guilhot, C. *et al.* (2001) *Mol Microbiol*, **41**, 179–187.

129 Walters, S.B., Dubnau, E., Kolesnikova, I., Laval, F. *et al.* (2006) *Mol Microbiol*, **60**, 312–330.

130 Voskuil, M.I., Schnappinger, D., Visconti, K.C., Harrell, M.I. *et al.* (2003) *J Exp Med*, **198**, 705–713.

131 Voskuil, M.I., Visconti, K.C. and Schoolnik, G.K. (2004) *Tuberculosis (Edinb)*, **84**, 218–227.

132 Cunningham, A.F. and Spreadbury, C.L. (1998) *J Bacteriol*, **180**, 801–808.

133 Tabira, Y., Ohara, N., Kitaura, H., Matsumoto, S. *et al.* (1998) *Res Microbiol*, **149**, 255–264.

134 Wayne, L.G. and Sohaskey, C.D. (2001) *Annu Rev Microbiol*, **55**, 139–163.

135 Wayne, L.G. and Hayes, L.G. (1996) *Infect Immun*, **64**, 2062–2069.

136 Park, H.D., Guinn, K.M., Harrell, M.I., Liao, R. *et al.* (2003) *Mol Microbiol*, **48**, 833–843.

137 Garton, N.J., Waddell, S.J., Sherratt, A.L., Lee, S.M. *et al.* (2008) *PLoS Med*, **5**, e75.

138 Lin, M.Y., Geluk, A., Smith, S.G., Stewart, A.L. *et al.* (2007) *Infect Immun*, **75**, 3523–3530.

139 Lin, M.Y. and Ottenhoff, T.H. (2008) *Biol Chem*, **389**, 497–511.

140 Murray, M.J., Murray, A.B., Murray, M.B. and Murray, C.J. (1978) *BMJ*, **2**, 1113–1115.

141 Olakanmi, O., Schlesinger, L.S., Ahmed, A. and Britigan, B.E. (2002) *J Biol Chem*, **277**, 49727–49734.

142 Ratledge, C. (2004) *Tuberculosis (Edinb)*, **84**, 110–130.

143 Byrd, T.F. and Horwitz, M.A. (1989) *J Clin Invest*, **83**, 1457–1465.

144 De Domenico, I., Ward, D.M., Langelier, C., Vaughn, M.B. *et al.* (2007) *Mol Biol Cell*, **18**, 2569–2578.

145 Nemeth, E. and Ganz, T. (2006) *Annu Rev Nutr*, **26**, 323–342.

146 Van Zandt, K.E., Sow, F.B., Florence, W.C., Zwilling, B.S. *et al.* (2008) *J Leukoc Biol*, **84**, 689–700.

147 Sow, F.B., Florence, W.C., Satoskar, A.R., Schlesinger, L.S. *et al.* (2007) *J Leukoc Biol*, **82**, 934–945.

148 Kaufmann, S.H. and Schaible, U.E. (2005) *Curr Opin Immunol*, **17**, 79–87.

149 Barral, D.C. and Brenner, M.B. (2007) *Nat Rev Immunol*, **7**, 929–941.

150 Brigl, M. and Brenner, M.B. (2004) *Annu Rev Immunol*, **22**, 817–890.

151 Gilleron, M., Stenger, S., Mazorra, Z., Wittke, F. *et al.* (2004) *J Exp Med*, **199**, 649–659.

152 Bosio, C.M., Gardner, D. and Elkins F K.L. (2000) *J Immunol*, **164**, 6417–6425.

153 North, R.J. and Jung, Y.J. (2004) *Annu Rev Immunol*, **22**, 599–623.

154 Korbel, D.S., Schneider, B.E. and Schaible, U.E. (2008) *Microbes Infect*, **10**, 995–1004.

155 Schaible, U.E., Hagens, K., Fischer, K., Collins, H.L. and Kaufmann, S.H. (2000) *J Immunol*, **164**, 4843–4852.

156 Ting, L.M., Kim, A.C., Cattamanchi, A. and Ernst, J.D. (1999) *J Immunol*, **163**, 3898–3906.

157 Fortune, S.M., Solache, A., Jaeger, A., Hill, P.J. *et al.* (2004) *J Immunol*, **172**, 6272–6280.

158 Gercken, J., Pryjma, J., Ernst, M. and Flad, H.D. (1994) *Infect Immun*, **62**, 3472–3478.

159 Banaiee, N., Kincaid, E.Z., Buchwald, U., Jacobs, W.R. Jr, and Ernst, J.D. (2006) *J Immunol*, **176**, 3019–3027.

160 Stenger, S., Niazi, K.R. and Modlin, R.L. (1998) *J Immunol*, **161**, 3582–3588.

161 Winau, F., Weber, S., Sad, S., de Diego, J. *et al.* (2006) *Immunity*, **24**, 105–117.

162 Winau, F., Kaufmann, S.H. and Schaible, U.E. (2004) *Cell Microbiol*, **6**, 599–607.

163 Winau, F., Hegasy, G., Kaufmann, S.H. and Schaible, U.E. (2005) *Apoptosis*, **10**, 707–715.

164 Hinchey, J., Lee, S., Jeon, B.Y., Basaraba, R.J. *et al.* (2007) *J Clin Invest*, **117**, 2279–2288.

165 Cardona, P.J., Llatjos, R., Gordillo, S., Diaz, J. *et al.* (2001) *Scand J Immunol*, **53**, 65–71.

166 Keane, J., Remold, H.G. and Kornfeld, H. (2000) *J Immunol*, **164**, 2016–2020.

167 Gan, H., Lee, J., Ren, F., Chen, M. *et al.* (2008) *Nat Immunol*, **9**, 1189–1197.

168 Keane, J., Balcewicz-Sablinska, M.K., Remold, H.G., Chupp, G.L. *et al.* (1997) *Infect Immun*, **65**, 298–304.

169 Balcewicz-Sablinska, M.K., Keane, J., Kornfeld, H. and Remold, H.G. (1998) *J Immunol*, **161**, 2636–2641.

170 Oddo, M., Renno, T., Attinger, A., Bakker, T. *et al.* (1998) *J Immunol*, **160**, 5448–5454.

171 Spira, A., Carroll, J.D., Liu, G., Aziz, Z. *et al.* (2003) *Am J Respir Cell Mol Biol*, **29**, 545–551.

172 Riendeau, C.J. and Kornfeld, H. (2003) *Infect Immun*, **71**, 254–259.

173 Fratazzi, C., Arbeit, R.D., Carini, C., Balcewicz-Sablinska, M.K. *et al.* (1999) *J Leukoc Biol*, **66**, 763–764.

174 Fratazzi, C., Arbeit, R.D., Carini, C. and Remold, H.G. (1997) *J Immunol*, **158**, 4320–4327.

175 Chen, M., Gan, H. and Remold, H.G. (2006) *J Immunol*, **176**, 3707–3716.

176 O'Sullivan, M.P., O'Leary, S., Kelly, D.M. and Keane, J. (2007) *Infect Immun*, **75**, 1984–1993.

177 Lee, J., Remold, H.G., Ieong, M.H. and Kornfeld, H. (2006) *J Immunol*, **176**, 4267–4274.

178 Park, J.S., Tamayo, M.H., Gonzalez-Juarrero, M., Orme, I.M. and Ordway, D.J. (2006) *J Leukoc Biol*, **79**, 80–86.

179 Theus, S.A., Cave, M.D., Eisenach, K., Walrath, J. *et al.* (2006) *Infect Immun*, **74**, 6865–6876.

180 Boshoff, H.I. and Barry, C.E. 3rd (2005) *Nat Rev Microbiol*, **3**, 70–80.

181 Mwandumba, H.C., Russell, D.G., Nyirenda, M.H., Anderson, J. *et al.* (2004) *J Immunol*, **172**, 4592–4598.

20
Rhodococcus equi and *Nocardia asteroides*

Kristine von Bargen and Albert Haas

While both *Rhodococcus equi* and *Nocardia asteroides* are important related pathogens, this review will largely concentrate on *R. equi*, because much more is known about the intracellular compartmentation of this pathogen.

20.1
Rhodococcus and *Nocardia* and Illnesses Caused by them

The genus *Rhodococcus* was first described by Zopf in 1891 and redefined in 1977 by Goodfellow and Alderson [1]. Rhodococci are ubiquitous, Gram-positive, aerobic, non-motile and metabolically diverse soil organisms [2, 3]. The genus *Rhodococcus* belongs to the group of nocardioform actinomycetes [4] and comprises more than 20 species [5–11]. It can be further classified as part of the mycolic acid-containing genera (mycolata) together with *Mycobacterium, Nocardia, Corynebacterium, Tsukamurella, Dietzia, Williamsia, Turicella, Gordonia* and *Skermania* [6]. Rhodococci have mycolic acids containing 30–54 carbons, which is in between the length of mycolic acids from corynebacteria (22–38 carbons) and that of mycobacteria (60–90 carbons) [12]. The genus comprises symbionts (*Rhodococcus rhodnii*) as well as pathogens of animals (e.g., *Rhodococcus equi*), plants (*Rhodococcus fascians*) and humans (e.g., *R. equi, Rhodococcus rhodochrous, Rhodococcus erythropolis* and several unidentified *Rhodococcus* species) [2, 13].

R. equi is the species in the genus with the highest pathogenic potential for animals and humans [3]. It was first isolated by Magnusson in 1923 as *Corynebacterium equi* from the lung of a foal with pyogranulomatous pneumonia [3] and reclassified in 1977 as *Rhodococcus equi* [1]. *R. equi* are frequently ingested by grazing animals and are excreted back into their natural soil habitat [14]. They grow well on simple organic acids such as proprionate or acetate [15], which are abundant in herbivore manure [16, 17].

R. equi are rather distantly related to other *Rhodococcus* species as assessed by 16S rRNA sequencing [5] yet analysis of the recently sequenced genome [18] identified the Rhodococcus strain RHA1 to be the closest relative [6]. *R. equi* strains possess a

Intracellular Niches of Microbes. Edited by Ulrich E. Schaible and Albert Haas

ISBN: 978-3-527-32207-7

circular chromosome of about 5 Mb (strain 103S [18]) and can harbor a variety of different plasmids, some of which are involved in virulence and will be discussed below. *R. equi* is genetically very variable [19–21].

R. equi are highly resistant to some extreme environmental conditions such as low pH and oxidative stress [22, 23]. They possess a complex hydrophobic cell wall composed of mycolic acid-containing lipids and lipoglycans, such as trehalose dimycolate and mannosylated lipoarabinomannan (*R. equi* ManLAM) [12, 24, 25]. The cell wall is surrounded by an antigenic variable lamellar polysaccharide capsule [3]. Twenty-seven different capsule serotypes have been identified [26, 27].

Nocardia asteroides is Gram-positive, chemoautotrophic, variably acid fast and strictly aerobic. Like *Rhodococcus*, it is a member of the mycolata and a physiologically heterogeneous genus [28]. *Nocardia*, more than *Rhodococcus* or *Mycobacterium*, forms filamentous, branched cells which fragment into coccoid smaller cells [28]. Nocardiae possess an extensive cell wall containing unusual (glyco)lipid compounds, in particular mycolic acids with a typical chain length of 44–66 carbon atoms. These make the cell wall impermeable to many hydrophilic substances and *Nocardia* resistant to lysozyme in growth media whereas other actinomycetes are sensitive [28].

20.1.1
R. equi Infection in Horses

Foals between one and four months are most commonly infected and are uniquely susceptible to *R. equi* infection [29, 30]. *R. equi* are the most common cause of lung abscesses in these animals [31]. Before the introduction of rifampin/erythromycin therapy, mortality of foals with *R. equi* pneumonia was as high as 80% but could be reduced to about 12% by the use of these antibiotics in treatment [32]. Infection is thought to occur by inhalation of aerolized dust contaminated with *R. equi* [3] while transmission of *R. equi* pneumonia from foal to foal has not been observed.

R. equi pneumonia is classically characterized by massive pulmonary abscessation with high fever, neutrophilia, mucopurulent respiratory discharge and cough [29]. When *R. equi* reach the foal lung by inhalation, they are internalized by alveolar macrophages but may not be killed [33, 34]. The early stage of disease is characterized by pulmonary lesions and influx of neutrophils, macrophages and giant cells into the alveoli [35]. In most of these cells, intracellular *R. equi* can be observed [35]. Later on, necrosis of the lung parenchyma occurs [30, 35]. Quite often the disease involves other organs after granulomatous foci in the lung have broken open and the bacteria spread through the body [3] causing, for example, an ulcerative enteritis with severe diarrhea [30, 36]. Cases of infected adult horses are rare and likely associated with immunological deficiencies [30, 37–39]. Adult horses – as well as the majority of foals – develop a protective immunity and quickly clear infection when exposed to virulent *R. equi* [40].

Nearly all foal isolates possess a plasmid encoding virulence-associated protein A (VapA) while barely any isolates from places that are not associated with horses contain virulence plasmids [41].

20.1.2
R. equi Infection in Species Other than Horse

Since the first case of human infection with *R. equi* was described in 1967 [42], a three figure number of cases have been reported, particularly since the start of the acquired immune deficiency syndrome (AIDS) pandemic. *R. equi* infect AIDS patients with a high mortality rate of some 60% [43–46]. Infections of immunocompromised individuals with *R. equi* mostly result in pneumonia with fever, cough and chest pain [46]. However, as in foals, infection can spread and cause lesions in other organs as well as bacteremia [47]. Immunocompetent humans are only rarely affected [48–55].

R. equi can frequently be found in pigs where they can cause a chronic cervical lymphadenitis with tuberculous lesions [56–58] but can also be isolated from submaxillary lymph nodes and tonsils of healthy animals [56, 59]. Isolates from pigs show intermediate virulence in the mouse infection model and often possess a virulence plasmid encoding a 20 kDa virulence-associated protein (VapB) [60]. *R. equi* has also been found to cause tuberculosis-like lesions in lymph nodes of cattle and in the livers of young goats [3].

20.1.3
Experimental Infection Models for *R. equi* Infection

The currently prevailing animal model for *R. equi* virulence is based on the persistence (but not the multiplication) of *R. equi* in the organs of mice [61–65]: The bacteria are injected intravenously and the animals are killed after a number of days. Mouse organs (particularly the liver) are homogenized and persistence of *R. equi* strains is evaluated by live cell determination. Whereas avirulent strains are cleared within 4 days, clinical isolates of *R. equi* persist for several days. This model has been used successfully to identify *R. equi* factors involved in virulence.

20.1.4
N. asteroides Infection

Although nocardiae are primarily soil microorganisms, they can cause localized and disseminated infections, often after trauma or inhalation of infested soil. In humans, these infections usually involve the lung, brain and cutaneous tissues. Many of these infections are caused by *N. asteroides* which is also a suspected cause of Parkinson's disease [28]. A variety of animals have been found to be infected, including goats, pigs, sheep, horses, dogs, cats and fish [28].

20.2
Characterization of Intra-Host Cell Compartmentation and its Relationship to Virulence

The compartments of *Rhodococcus equi* and *Nocardia asteroides* are illustrated in Figure 20.1.

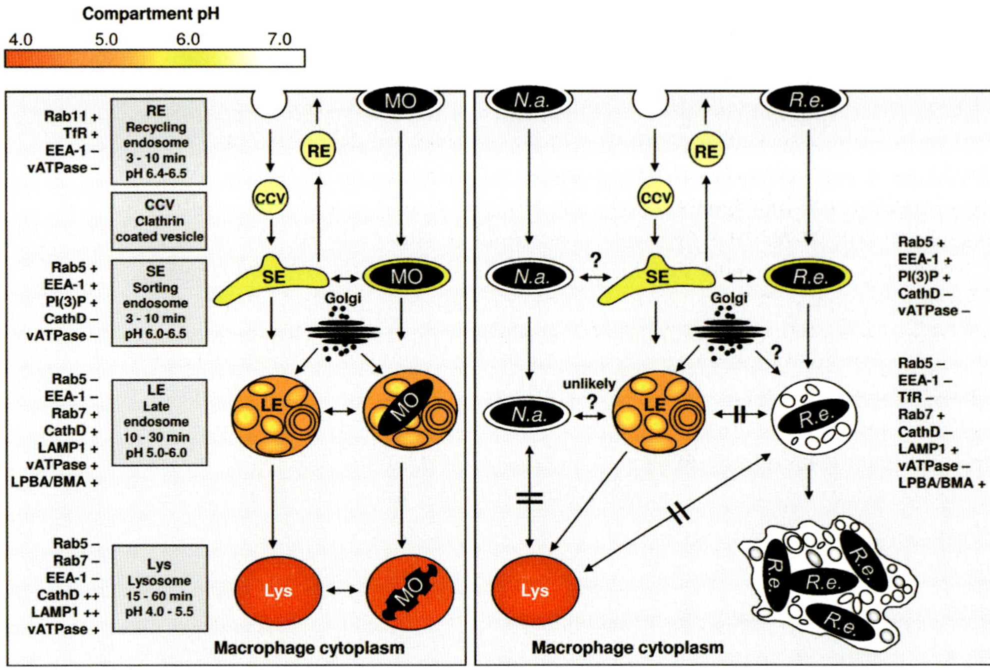

Figure 20.1 The compartments of *Rhodococcus equi* and *Nocardia asteroides*. Right section: Pathway of endosome maturation and the interactions of the endocytic pathway with normally maturing phagosomes of apathogenic microorganisms (MO). Left section: The compartments of *N. asteroides* (*N.a.*) and *R. equi* (*R.e.*) and their positioning in the endocytic system. Whereas little is known about the identity of the *N. asteroides* compartment, that of *R. equi* shows the initial hallmarks of normal maturation and acquires and loses early markers during its interaction with early endocytic vesicles. However, no fusion with lysosomes occurs and v-ATPase or lysosomal hydrolases such as cathepsin D are not acquired. *R. equi* multiplies in a compartment of neutral pH and the host cell is finally destroyed in a necrotic process.

20.2.1
The Compartment of *R. equi*

R. equi is a facultative intracellular bacterium and infectivity seems to be limited to mononuclear phagocytes [66]. Phagocytosis of *R. equi* can be enhanced by engaging the complement receptor CR3 (Mac-1) [66]. Binding of *R. equi* mannosylated lipoarabinomannan (ManLAM) to the mannose-binding protein may also lead to opsonization of *R. equi* by C3b and uptake via Mac-1 [25]. However, binding and internalization of *R. equi* does also occur in the absence of active complement without any changes in compartment identity [67], possibly because mannose receptors bind directly to *R. equi* mannose compounds [68]. Uptake of antibody-opsonized *R. equi* via Fcγ receptors has been reported to increase phagosome–lysosome fusion and killing of the bacteria by macrophages, suggesting that the mode of entry of *R. equi* into the host cell is crucial for the intracellular fate of the bacterium [34].

Once the bacteria have been internalized, their phagosomes initially mature normally. They acquire and lose the early endocytic markers early endosomal antigen 1 (EEA1), transferrin receptor (TfR), phophatidylinositol 3-phosphate (PI(3)P), coronin and Rab5 with normal kinetics and acquire some late marker proteins, such as lysosome-associated membrane protein 1 (LAMP1), LAMP2, Rab7 and, with a slight delay as compared to control phagosomes, lysobisphosphatidic acid (LBPA)/ bis(monoacylglycero)phosphate (BMP) [67]. However, most phagosomes containing virulent *R. equi* are negative for the proton-pumping vacuolar ATPase (v-ATPase) at 2 and 24 h post infection and they do not acidify (Figure 20.2(b)) [67, 69]. These observations agree with the finding that *R. equi*-containing vacuoles (RCVs) do not fuse with lysosomes in murine and equine macrophages [33, 34, 67], and therefore do not acquire ferritin or fluid-phase markers preloaded into lysosomes, lysosomal acid β-galactosidase or cathepsin D (CathD). Since fusion of the RCV with lysosomes occurs rarely, acquisition of LAMP proteins and Rab7 by most RCVs might be the result of trafficking of these proteins from the *trans*-Golgi network [67].

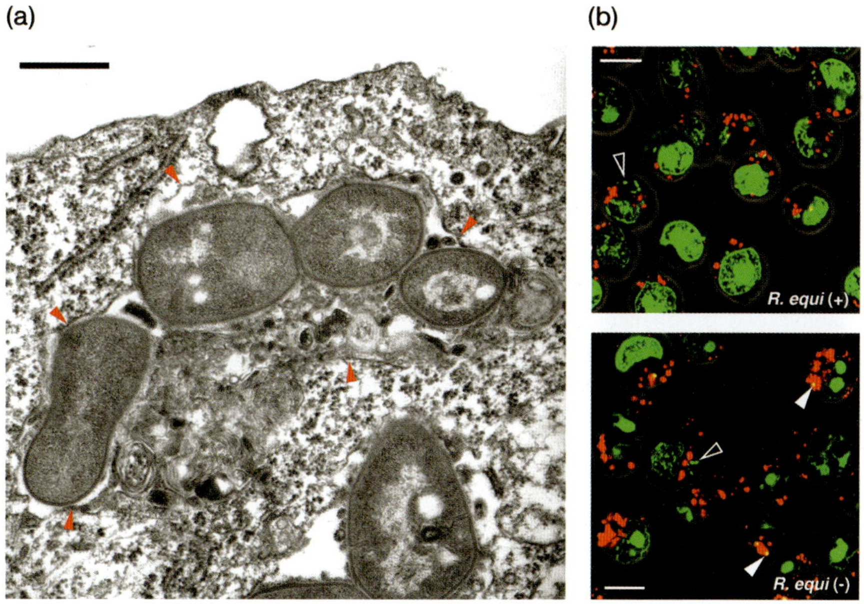

Figure 20.2 The *Rhodococcus equi*-containing vacuole. (a) Transmission electron micrograph of the *R. equi*-containing phagosome in J774E macrophages at 8 h post infection. The spacious vacuole is filled with membraneous material and vesicles. Red arrowheads point to the phagosome membrane. Bar: 500 nm. (b) Fluorescence micrographs of J774E macrophages infected with *R. equi* at 2 h post infection. Syto13 stain (green) visualizes the DNA of bacteria and macrophage nuclei and the lysosomotropic dye LysoTracker (red) visualizes acidic compartments. Virulent *R. equi* (*R. equi* 103+) reside in phagosomes that are non-acidic and do not stain with LysoTracker. Phagosomes containing plasmid-negative avirulent *R. equi* (*R. equi* 103−) are more frequently acidic, which is reflected in a higher percentage of colocalization of the bacteria with LysoTracker. Filled arrowheads point to bacteria colocalizing with the lysosomotropic dye; open ones to those in non-acidic compartments. Bars: 10 μm.

The unusual compartmentation of *R. equi* seems to be correlated to the possession of the *R. equi* virulence plasmid (discussed in Section 20.3.1.1) and bacterial viability. In early infection, phagosomes containing avirulent plasmid-negative strains fuse only slightly more frequently with lysosomes than phagosomes with virulent plasmid-positive strains. At 2 h post infection, absence or presence of v-ATPase on the phagosomal membrane is the only clear difference between virulent or avirulent *R. equi* [67], which becomes more pronounced at prolonged incubation times. RCVs containing avirulent plasmid-cured strains acidify progressively and fuse with lysosomes, suggesting a delayed phagosome maturation [67].

In murine macrophages, phagosomes containing heat- or formalin-killed *R. equi* acquire lysosomal fluid-phase markers as often as those containing plasmid-negative *R. equi* at 2 h post infection [67], but contain more acid β-galactosidase and CathD [67]. In equine alveolar macrophages, phagosome–lysosome formation is equally inhibited with formalin-killed or viable *R. equi* [33].

Virulent *R. equi* reside in "loose" phagosomes (equine macrophages) [33] or equally often in phagosomes with loose or tightly apposed membranes (murine macrophages) [67]. At 24 h post infection, RCVs are mostly communal multilobed vacuoles with loose membranes and much internal membrane material (Figure 20.2(a)) [67]. Bacteria do not escape into the host cell cytoplasm, and the phagosomal membrane stays morphologically intact [33, 34, 67].

After an initial lag-phase, virulent bacteria multiply intracellularly in peritoneal mouse, equine alveolar or J774E macrophages [33, 34, 67, 70]. Plasmid-negative strains do not multiply but survive intracellularly for about 24 h [62, 67, 70].

Infection with *R. equi* is cytotoxic for macrophages [33, 34, 71]. With large microbial loads, host cell degeneration becomes apparent at about 8 h post infection [33, 71]. Death is due to necrosis rather than apoptosis [70, 71], strictly depends on bacterial viability and is enhanced by the possession of a virulence plasmid [71]. *R. equi*-infected macrophages seem to contain fewer lysosomes than non-infected cells [33, 34, 72], either because of degranulation of lysosome material into the macrophage cytoplasm or because of enhanced lysosome secretion. Intracellular degranulation could contribute to macrophage degeneration in response to *R. equi* infection [33].

20.2.2
The Compartment of *N. asteroides*

Nocardia-containing phagosomes were heavily investigated in the late 1970s and early 1980s when most compartments containing other intracellular pathogens had not been analyzed at all. Major contributions came from Beaman and co-workers at UCSD (San Diego, USA) [28] who routinely used three unrelated isolates of *N. asteroides* with different virulence (GUH-2 being most virulent, strain ATCC 14795 less and strain ATCC 10905 least virulent) to study phagosome maturation and intracellular multiplication.

Just like *Mycobacterium tuberculosis* and *R. equi*, *N. asteroides* inhibits phagolysosome formation. While the most virulent GUH-2 strain does this most efficiently, the less virulent ATCC 14795 inhibited less and strain ATCC 10905 does not at all, as judged by transmission electron microscopic analysis of the acquisition of lysosomal acid phosphatase or lysosomal peroxidase tracer by *Nocardia* phagosomes [73]. Log-phase bacteria inhibit phagolysosome formation stronger than stationary cells [74] and viability is required for efficient inhibition [74]. Intramacrophage multiplication is most pronounced when the infecting bacteria are from late-logarithmic stage cultures compared with stationary bacteria. Opsonization of bacteria with hyperimmune serum increases their uptake by macrophages and slightly stimulates phagolysosome formation but does not much affect intracellular multiplication [28]. Opsonization does, however, promote phagolysosome formation with activated macrophages and is necessary for efficient killing of *N. asteroides* by macrophages *in vitro* [28].

In a study from 1986, the pH of *N. asteroides*-containing phagosomes was determined in murine peritoneal macrophages using cytophotometry. At 90 min of infection, phagosomes containing the GUH-2 had an average pH of 7.2, those containing ATCC 19247 of 6.5, and those containing heat-killed baker's yeast of 5.3 [75]. Whether the acidification-blocking bacterial effector would also inhibit phagolysosome formation remained unresolved, as acidification of phagosomes and their fusion with lysosomes are not necessarily coupled (see, for example, the case of *Salmonella* vacuoles).

So, the *N. asteroides* vacuole maintains its limiting membrane, does not acidify and does not fuse with lysosomes. Considering the close biology of *Nocardia, Rhodococcus* and *Mycobacterium*, it is not far-fetched to assume that the *Nocardia*-containing vacuole is part of the endocytic continuum and that phagosome maturation is arrested at a stage before fusion with lysosomes. *R. equi* is likely closest to virulent *Nocardia* strains in this respect.

20.3
Genes Involved in Diverted Phagosome Maturation or Compartmentation and its Maintenance

20.3.1
Genes of *R. equi* Involved in Diverted Phagosome Maturation

Some of the virulence determinants of *R. equi* have already been identified. However, although most of them were assessed for their relevance in intracellular multiplication, there are few data with respect to their relevance for the formation of the RCV.

20.3.1.1 The VapA-Encoding Plasmid

The virulence of *R. equi* depends on the possession of extrachromosomal virulence-associated plasmids [76–80]. Plasmids containing *vapA* have been most intensely studied due to their importance for virulence in foals. VapA-producing strains possess an 80–90 kb virulence plasmid harboring the *vapA* gene [77, 80, 81]. This plasmid confers the ability to replicate in macrophages [62, 70], enhances cytotoxicity [71] and is involved in establishment and maintenance of the RCV (described above).

Two VapA-encoding plasmids have been sequenced [82] and are nearly identical. They contain 73 predicted open reading frames (ORFs) including eight pseudogenes [83]. The plasmids can be divided into four major regions, based on amino acid homologies with known gene products. One part is responsible for replication, one for conjugation and one region has unknown functions [83]. A fourth region shows typical features of a pathogenicity island [82]. It is characterized by a lower G + C content than the rest of the plasmid and has probably been acquired from an unknown source [82, 83]. It contains 26 ORFs, many with unknown function [82, 83]. Nine of these ORFs code for members of the family of virulence-associated proteins (Vaps) [82, 84–86]. Three Vaps, that is, VapF, pseudo-VapE and VapI, are likely to be non-functional frameshift mutants [82, 85, 86]. The Vaps assumed to be functional – Vaps A, C, D, E, G and H – possess export signal sequences, indicating that they are secreted [82]. VapC, D and E can be found in the culture supernatant of virulent *R. equi* as three of about ten to twelve strongly immunogenic proteins [84, 87].

VapA is a highly antigenic protein that has long been connected to *R. equi* virulence [76, 78, 88] and is likely involved in establishment of the RCV (Polidori and co-workers, unpublished data). The *vapA* gene encodes a protein with a putative export signal peptide [81, 89]. In contrast to VapC, D and E, the VapA protein cannot be found in the culture broth but remains on the bacterial surface [78] and is probably covalently lipid modified, possibly even at multiple sites [89]. This modification might be important for anchoring VapA to the cell wall [89]. However, the protein lacks a "lipobox," the typical lipoprotein signal peptide cleavage site of lipoproteins of Gram-positive bacteria [90].

The function of the Vap proteins is not known. VapA is indispensable for *R. equi* virulence [63], yet expression of VapA in a plasmid-negative background does not confer virulence [62], suggesting that VapA is indispensable but not sufficient for pathogenicity. Expression of VapA is induced by temperatures above 32 °C, low pH, oxidative stress and a low level of divalent cations such as Fe^{2+}, Ca^{2+} and Mg^{2+} [23, 78, 91–94], conditions which might be encountered intraphagosomally [93]. Other factors regulating VapA expression are the plasmid genes *ORF4* or *virR* (a LysR type transcription regulator), *ORF8* (a putative response regulator of a two-component regulatory system) [82, 85] and probably *ideR* (a chromosomally encoded protein that might regulate gene expression depending on the availability of iron [93, 95]). Deletion mutants of *virR* or *ORF8* are fully attenuated in mice [64]. In general, expression of virulence-related genes seems to be controlled by a negatively regulating network [64].

20.3.1.2 **Chromosomally Encoded Virulence Factors**

A partial genomic analysis of *R. equi* [96] and the deciphering of the complete genome sequence in 2006 (unpublished and not annotated [18]) revealed the homology of many genes to known or suspected virulence genes of other pathogens, especially of *M. tuberculosis*. The role of selected chromosomally encoded factors for *R. equi* pathogenicity could already be confirmed. We speculate that some of the central chromosomal virulence genes might also be crucial for intracellular compartmentation, but this remains to be shown.

20.3.1.2.1 **Mycolic Acid-Containing Glycolipids** Structural components such as mycolic acid-containing glycolipids possibly contribute to the virulence of *Rhodococcus* species. Using unrelated *R. equi* isolates it was shown that their virulence paralleled the carbon chain length of their mycolic acids [97, 98]. In line with this, we have recently identified mutations in the biogenesis of mycolic acid compounds which would not affect the general fitness of *R. equi*, but which affect its inhibition of phagolysosome formation (T. Sydor and co-workers, unpublished data). Another structural component, *R. equi* ManLAM induces similar cytokines in equine macrophages as viable bacteria and therefore might play a role in pathogenesis [25].

The chromosomally encoded polysaccharide capsule of *R. equi* is unlikely to be a virulence factor [76, 99].

20.3.1.2.2 **Isocitrate Lyase** The enzyme isocitrate lyase is particularly important for intracellular long-term survival and growth [100]. Although the intraphagosomal nutrients of *R. equi* are not known, host cell lipids in the phagosome lumen are a likely source for carbon and other nutrients, especially since *R. equi* is known to grow well on lipid compounds in its natural environment (e.g., horse feces, see above). The key enzyme of lipid metabolism and carbon assimilation via the glyoxylate shunt is isocitrate lyase (gene: *aceA*). An *aceA* *R. equi* deletion mutant initially multiplies in mouse macrophages but the bacteria are killed after 48 h. The mutant is somewhat attenuated in mice and more so in foals, where mutant bacteria are present in the lung but no fever or visible lesions develop [100].

Secreted cholesterol oxidase (ChoE) was long suspected to damage host membranes in combination with bacterial phosphatases and in that way contribute to *R. equi* pathogenicity [101, 102]. However, virulent as well as avirulent strains secrete this enzyme [103] and ChoE can be excluded as a virulence factor by gene knockout analysis [104].

20.3.1.2.3 **Other Chromosomally Encoded Virulence Factors** Other chromosomally encoded virulence factors confirmed in the mouse model are nitrate reductase (NarG), peptidase D (PepD) and high-temperature-requirement A protein (HtrA) [65]. NarG plays a role in respiration in the absence of oxygen and might be involved in coping with microaerophilic conditions in the course of infection [65]. Expression of the *pepD* homolog in *M. tuberculosis* is controlled by a regulatory system (*mprA-mprB*) important for persistent infection as well as an alternative sigma factor involved in intracellular growth of the bacteria [105, 106]. An *R. equi* deletion mutant of *htrA* is fully attenuated in mice [65]. HtrA is a serine protease which might be important for resistance to oxidative stress. Homologous genes have been shown to be involved in virulence of many Gram-negative bacteria [65].

20.3.2 Genes of *N. asteroides* Involved in Diverted Phagosome Maturation

The contributions of *Nocardia* genes to phagosome maturation diversion are unknown. Comparison of the available related *Nocardia farcinica* genome with that of *M. tuberculosis* yields a number of candidate genes [107] that await further analysis.

It has long been suspected that glycolipids such as cord factor (trehalose dimycolate, TDM) could be important factors in interactions between phagocytes and nocardiae, largely because of their high toxicity when injected into mice [28]. Furthermore, TDM is found at the surface of mycolata and can be released during infection [108] which is assumed to be a prerequisite for an interacting factor. Spargo *et al.* [109] showed that purified TDM from GUH-2 inhibited Ca^{2+}-driven fusion between liposomes *in vitro*, likely through its membrane-active properties [110]. TDM seems to be abundant in virulent GUH-2 but not in ATCC 10905, which makes for a nice correlate, although it does not directly prove this role [111]. Surprising experiments showed that extraction of cord factor from GUH-2 cells with petroleum ether did not affect viability of the bacteria but abrogated their ability to inhibit phagolysosome formation [28]. The same is seen, although less pronouncedly, in *M. tuberculosis* [112].

There is little question that the unusual lipids of *Nocardia* contribute to inhibition of phagolysosome formation, as is the case with virulent *Mycobacterium* [113] and probably *R. equi*, but it is also likely that lipids alone cannot interfere with macrophage functions for prolonged periods of time and that proteins also play a role.

20.4 Immunology of Phagosomes Containing *R. equi* or *N. asteroides*

20.4.1 Host Genes Relevant for the Outcome of *R. equi* Infection

The specificity for foals and macrophages strongly suggests that specific host (cell) factors are decisive for the outcome of *R. equi* infection. Only a small number of foals on endemic farms develop pneumonia, suggesting a predisposition to *R. equi* infection of certain foal subpopulations. *R. equi* largely infect monocyte and macrophage cells [68], whereas they are efficiently killed by mature horse and foal neutrophils [114–118]. These cells probably play a central role for infection clearance in mice since antibodies against neutrophils result in more severe disease [119]. Macrophages of foals exposed to *R. equi* kill the bacteria better than those of non-exposed foals and opsonization of *R. equi* increases phagolysosome formation [34]. Host genes possibly involved in resistance to *R. equi* are:

- *Inducible nitric oxide synthase* (NOS2) and NADPH oxidase. *R. equi* multiplication in murine macrophages can be completely inhibited by activation using interferon γ (IFNγ) and lipopolysaccharide [120] in a process involving NOS2 and the NADPH oxidase [120] as shown using knockout mice. Infection control by activated macrophages has been ascribed to the bactericidal effect of peroxynitrite formed from NO radicals (produced by NOS2) and superoxide anions (produced by NADPH oxidase) [120].

- The *natural resistance-associated macrophage protein 1* (NRAMP1). Mutation of this divalent cation transporter is associated with increased susceptibility to different intracellular pathogens including mycobacteria [121]. NRAMP1 transcription is downregulated in equine macrophages infected with *R. equi* [68] and certain variants of the gene make horses more susceptible to *R. equi* infection [122]. However, deletion of the NRAMP1 gene does not increase susceptibility of mice to *R. equi* infection [123].
- The iron-binding transferrin protein which shows biochemical polymorphisms in various species including the horse [124]. A subset of specific transferrin alleles might render foals more susceptible to *R. equi* infection [124]. Similarly, a crucial role for iron availability in human tuberculosis has been demonstrated [125, 126].

20.4.2 Antigen Presentation from the RCV

Antigen presentation from *R. equi*-infected cells possibly occurs in a major histocompatibility complex (MHC)-I independent fashion [127]. Adult horses that had been challenged with virulent *R. equi* have more cytotoxic $CD8^+$ lymphocytes when restimulated with *R. equi* antigens [127]. $CD8^+$ cells seem to be crucial for the specific lysis of *R. equi*-infected target cells [127]. Interestingly, bacterial components present in virulent and avirulent *R. equi* are recognized by cytotoxic lymphocytes and mediate efficient lysis of infected cells [127]. Since lysis via $CD8^+$ lymphocytes occurs no matter whether the target cells are autologous or mismatched with respect to their MHC class I (horses: equine lymphocyte alloantigen A, ELA-A) haplotype, a novel way of antigen presentation or recognition was suspected [127].

In contrast to mycobacteria, nothing is known about the involvement of MHC class II in antigen presentation from the RCV. It has been shown that mycobacterial surface moieties leave the bacterial vacuole and can be found in endocytic compartments as well as extracellular vesicles [128, 129] which might be true for *R. equi*-infected cells as well.

20.4.3 Immunity to *R. equi*

Immunity to *R. equi* is likely mediated by both types of immune responses, the T_H1 (cell-mediated) and the T_H2 (humoral) response. A well-supported paradigm states that resistance to infection with intracellular pathogens such as *R. equi* basically relies on a T_H1 response and includes the action of cytotoxic T cells, inflammatory cytokines such as IFNγ and tumor necrosis factor α (TNFα), macrophages activated by these cytokines, and NK cells, whereas extracellular pathogens face an antibody-based T_H2 response [130]. However, it seems that effectors of both response types are involved in immunity to *R. equi* depending on the species (e.g., mouse vs. horse), age of the individual (e.g., young foals vs. mature horses), experimental condition (e.g., low vs.

high numbers of bacteria used for inoculation) or experimental system (e.g., whole organism vs. isolated cell populations).

The relevance of T_H2 responses in *R. equi* infection control has long been suspected since the time of onset of *R. equi* pneumonia coincides with and might in part be related to the decline of maternal antibodies in foals [131]. A passive immunization of foals using serum of immune adult horses is sometimes protective [132, 133], suggesting an important role of antibodies in protection. Likewise, specific antibody subtypes occurring in response to *R. equi* infection seem to depend on the age of the individuals and might be involved in mediating a functional (adult horse) or ineffective (foal) immune response [40, 134–136].

In addition to T_H2 effectors, many studies support the central role for T_H1 responses, and particularly for T cells in control of *R. equi* pneumonia. Whereas deficiencies in the complement system, in phagocytic or NK cells have no effect on clearance of virulent *R. equi*, mice with T cell deficiencies such as SCID (severe combined immune deficiency) mice (lacking B and T lymphocytes), nude mice (lacking T lymphocytes) or mice lacking either $CD8^+$ or $CD4^+$ cells show impaired or no clearance of the bacteria from their lungs [137–140]. In contrast, the transfusion of SCID mice with $CD4^+$ or $CD8^+$ cells from immunocompetent mice previously challenged with *R. equi* significantly enhances clearance [141]. Avirulent *R. equi*, however, are rapidly eliminated in either normal, athymic nude or SCID mice, suggesting a T cell-mediated clearance mechanism for virulent vs. mechanisms of the innate immune system for clearance of avirulent *R. equi* [139, 142].

The central role for T_H1 responses in infection with virulent *R. equi* is also reflected in the importance of corresponding cytokines: whereas those associated with T_H2 responses (such as IL-4) seem to be dispensable for infection control, those involved in T_H1 responses (particularly IFNγ or TNFα) are critical [143–146].

In immune adult horses, clearance of *R. equi* infection is associated with an antigen-specific proliferation of leukocytes, increased levels of IFNγ expression by these cells and an increase in the level of opsonizing IgG antibodies without a rise in IgM antibodies, suggesting a stimulation of a protective and locally restricted recall response [40, 147, 148]. In mice, immune response to *R. equi* infection seems to be dependent on plasmid possession. Infection with plasmid-negative or heat-killed rhodococci does not confer protective immunity, whereas a sublethal dose of virulent *R. equi* does [140, 149]. Immunocompetent mice clear plasmid-negative bacteria but not plasmid-positive *R. equi* without any IFNγ response and without induction of TNFα [145, 148]. A study using young foals observed a cytokine response to virulent as well as avirulent *R. equi* [150]. However, some differences in the response were observed, particularly with respect to IFNγ expression and a possible immunomodulatory role of virulent *R. equi* was assumed [150].

Interestingly, in spite of the different cytokine production profiles found in mice for plasmid-positive and plasmid-negative strains, experiments with isolated resting murine macrophages showed a similar cytokine induction after infection with such strains [151]. Likewise, infection of mouse macrophages with plasmid-positive as well as plasmid-negative *R. equi* leads to the translocation of NFκB into the macrophage nucleus [152].

20.4.4
Immune Responses to *Nocardia*

As for *R. equi*, cellular immunity is central for the control of *N. asteroides*: athymic nude mice can be protected by transfer of T lymphocytes from preimmunized heterozygous mice, probably by macrophage activation and direct cytotoxic activity [153]. *Nocardia*, *M. tuberculosis* and *R. equi* are also similar with respect to infection control after immunological activation. In resting macrophages, nocardial fragments extend from within the macrophage to the outside, whereas in activated macrophages, most bacteria are in the coccobacillary form, indicating that bacterial multiplication is inhibited [154]. Efficient killing *in vitro* requires the simultaneous presence of immune lymphocytes, immune serum and alveolar lining material [28].

20.5
Summary and Outlook

The close relationship between *Mycobacterium*, *Rhodococcus* and *Nocardia* offers an opportunity to exploit the obvious parallels in experimenting with these pathogens. However, in spite of a similar disease pattern of *M. tuberculosis* and *R. equi*, their compartmentation and virulence mechanisms seem to be different. Although both vacuole types do not fuse with lysosomes and lack the proton pump, the RCV is characterized by several more late endocytic markers (Rab7, LBPA/BMP, LAMP1) as well as a neutral pH at 2 h post infection while the *M. tuberculosis*-containing vacuole resembles more an early endocytic compartment (Rab5, TfR, LAMP1, pH 6.2). The *Nocardia*-containing compartment has a pH similar to that of the RCV, yet nothing is known about its position in the endocytic system. How these different, yet similar compartments are established and maintained and how similar factors of these closely related pathogens interact with the host cells deserves further attention. Some common principles are most likely. Further studies should also look into how nutrients are secured by the pathogens residing in these privileged compartments.

Although considerable progress has been made in the field of immune responses to *R. equi*, confusion still prevails. All those seemingly contradictory results with respect to similarities or differences in cytokine responses to avirulent or virulent strains, the specific (and maybe distinct) roles of T_H1 and T_H2 responses in different species of different age as well as in diverse infection models for the clearance of *R. equi* infection will have to be resolved to explain how immunity to *R. equi* is established and what might account for the particular susceptibility of a subpopulation of foals.

Some virulence factors on the *R. equi* virulence plasmid as well as on the bacterial chromosome have been identified, including VapA. Yet VapA is necessary but not sufficient for virulence and its mode of action is still obscure. More effort will have to be invested in the analysis of chromosomally encoded proteins and their interplay with *R. equi* virulence plasmid factors, for many of which no homologs can be found in the databases. Factors specific for phagosome establishment and factors specific for cytotoxicity (if they exist) will have to be identified. New techniques of random or targeted gene manipulation of *R. equi* established in the last years [63, 99, 155], the availability of sequence data of the *R. equi* chromosome and the virulence plasmid [18] as well as recently published data on the genome and proteome of *Nocardia* [107, 156] will certainly accelerate our understanding of how these pathogens establish and maintain their unusual compartments. Cross-boundary comparisons of new findings between all pathogenic mycolata and cooperative research could clearly contribute to a faster unfolding of our understanding of all their virulence mechanisms.

Acknowledgments

We thank the Deutsche Forschungsgemeinschaft (SFB670) and the German National Academic Foundation for supporting our studies on *R. equi* phagosome biology and immune activation.

References

1 Goodfellow, M. and Alderson, G. (1977) *J Gen Microbiol*, **100**, 99–122.
2 Finnerty, W.R. (1992) *Annu Rev Microbiol*, **46**, 193–218.
3 Prescott, J.F. (1991) *Clin Microbiol Rev*, **4**, 20–34.
4 Goodfellow, M. (1989) Genus Rhodococcus, in *Bergey's Manual of Systematic Bacteriology* (eds S.T. Williams, M.E. Sharpe and J.G. Holt), Williams and Wilkins, Baltimore, pp. 2362–2371.
5 Goodfellow, M., Alderson, G. and Chun, J. (1998) *Antonie Van Leeuwenhoek*, **74**, 3–20.
6 Vazquez Boland, J.A., Prescott, J.F., Meijer, W.G., Leadon, D.P. and Hines, S.A. (2009) *Equine Vet J*, **41**, 93–95.
7 Xu, J.L., He, J., Wang, Z.C., Wang, K. *et al.* (2007) *Int J Syst Evol Microbiol*, **57**, 2754–2757.
8 Li, B., Furihata, K., Ding, L.X. and Yokota, A. (2007) *Int J Syst Evol Microbiol*, **57**, 1956–1959.
9 Ghosh, A., Paul, D., Prakash, D., Mayilraj, S. and Jain, R.K. (2006) *Int J Syst Evol Microbiol*, **56**, 1965–1969.
10 Rehfuss, M. and Urban, J. (2005) *Syst Appl Microbiol*, **28**, 695–701.
11 Yassin, A.F. (2005) *Int J Syst Evol Microbiol*, **55**, 1575–1579.
12 Sutcliffe, I.C. (1998) *Antonie Van Leeuwenhoek*, **74**, 49–58.
13 Bell, K.S., Philp, J.C., Aw, D.W. and Christofi, N. (1998) *J Appl Microbiol*, **85**, 195–210.
14 Muscatello, G., Leadon, D.P., Klayt, M., Ocampo-Sosa, A. *et al.* (2007) *Equine Vet J*, **39**, 470–478.
15 Hughes, K.L. and Sulaiman, I. (1987) *Vet Microbiol*, **14**, 241–250.
16 Barton, M.D. and Hughes, K.L. (1984) *Vet Microbiol*, **9**, 65–76.
17 Prescott, J.F. (1987) *Vet Microbiol*, **14**, 211–214.
18 *Rhodococcus equi* Sequencing Group at the Sanger Institute (2006) ftp://ftp.sanger.ac.uk/pub/pathogens/re/. (accessed 28 February 2008).
19 Fuhrmann, C. and Lammler, C. (1997) *Berl Munch Tierarztl Wochenschr*, **110**, 54–59.
20 Morton, A.C., Begg, A.P., Anderson, G.A., Takai, S. *et al.* (2001) *Appl Environ Microbiol*, **67**, 2167–2175.
21 Cohen, N.D., Smith, K.E., Ficht, T.A., Takai, S. *et al.* (2003) *Am J Vet Res*, **64**, 153–161.

22 Benoit, S., Taouji, S., Benachour, A. and Hartke, A. (2000) *Int J Food Microbiol*, **55**, 295–298.

23 Benoit, S., Benachour, A., Taouji, S., Auffray, Y. and Hartke, A. (2002) *Infect Immun*, **70**, 3768–3776.

24 Barton, M.D., Goodfellow, M. and Minnikin, D.E. (1989) *Zentralbl Bakteriol*, **272**, 154–170.

25 Garton, N.J., Gilleron, M., Brando, T., Dan, H.H. *et al.* (2002) *J Biol Chem*, **277**, 31722–31733.

26 Prescott, J.F. (1981) *Can J Comp Med*, **45**, 130–134.

27 Nakazawa, M., Kubo, M., Sugimoto, C. and Isayama, Y. (1983) *Microbiol Immunol*, **27**, 837–846.

28 Beaman, B.L. and Beaman, L. (1994) *Clin Microbiol Rev*, **7**, 213–264.

29 Yager, J.A. (1987) *Vet Microbiol*, **14**, 225–232.

30 Zink, M.C., Yager, J.A. and Smart, N.L. (1986) *Can Vet J*, **27**, 213–217.

31 Lavoie, J.P., Fiset, L. and Laverty, S. (1994) *Equine Vet J*, **26**, 348–352.

32 Hillidge, C.J. (1987) *Vet Microbiol*, **14**, 337–342.

33 Zink, M.C., Yager, J.A., Prescott, J.F. and Fernando, M.A. (1987) *Vet Microbiol*, **14**, 295–305.

34 Hietala, S.K. and Ardans, A.A. (1987) *Vet Microbiol*, **14**, 307–320.

35 Johnson, J.A., Prescott, J.F. and Markham, R.J. (1983) *Vet Pathol*, **20**, 440–449.

36 Cimprich, R.E. and Rooney, J.R. (1977) *Vet Pathol*, **14**, 95–102.

37 Roberts, M.C., Hodgson, D.R. and Kelly, W.R. (1980) *Aust Vet J*, **56**, 96–97.

38 Freestone, J.F., Hietala, S., Moulton, J. and Vivrette, S. (1987) *J Am Vet Med Assoc*, **190**, 689–691.

39 Vengust, M., Staempfli, H. and Prescott, J.F. (2002) *Can Vet J*, **43**, 706–708.

40 Lopez, A.M., Hines, M.T., Palmer, G.H., Alperin, D.C. and Hines, S.A. (2002) *Clin Diagn Lab Immunol*, **9**, 1270–1276.

41 Takai, S., Fukunaga, N., Ochiai, S., Sakai, T. *et al.* (1996) *J Vet Med Sci*, **58**, 669–672.

42 Golub, B., Falk, G. and Spink, W.W. (1967) *Ann Intern Med*, **66**, 1174–1177.

43 Donisi, A., Suardi, M.G., Casari, S., Longo, M. *et al.* (1996) *AIDS*, **10**, 359–362.

44 Votava, M., Skalka, B., Hrstkova, H., Tejkalova, R. and Dvorska, L. (1997) *Cas Lek Cesk*, **136**, 51–53.

45 Emmons, W., Reichwein, B. and Winslow, D.L. (1991) *Rev Infect Dis*, **13**, 91–96.

46 Harvey, R.L. and Sunstrum, J.C. (1991) *Rev Infect Dis*, **13**, 139–145.

47 McNeil, M.M. and Brown, J.M. (1992) *Eur J Epidemiol*, **8**, 437–443.

48 Kedlaya, I., Ing, M.B. and Wong, S.S. (2001) *Clin Infect Dis*, **32**, E39–E46.

49 Gabriels, P., Joosen, H., Put, E., Verhaegen, J. *et al.* (2006) *Eur J Clin Microbiol Infect Dis*, **25**, 46–48.

50 Ulivieri, S. and Oliveri, G. (2006) *J Neurosurg Sci*, **50**, 127–129.

51 Walsh, R.D. and Cunha, B.A. (1994) *Heart Lung*, **23**, 519–520.

52 Tunger, A., Ozkan, F., Vardar, F., Burhanoglu, D. *et al.* (1997) *APMIS*, **105**, 705–707.

53 Nasser, A.A. and Bizri, A.R. (2001) *J Infect*, **42**, 67–68.

54 Napoleao, F., Damasco, P.V., Camello, T.C., do Vale, M.D. *et al.* (2005) *J Clin Microbiol*, **43**, 1002–1004.

55 Obon Azuara, B., Gutierrez Cia, I., Villanueva Anadon, B., Martin Villel, L. and Montoiro Allue, R. (2007) *An Med Interna*, **24**, 75–76.

56 Karlson, A.G. and Feldman, W.H. (1940) *J Bacteriol*, **39**, 461–472.

57 Pate, M., Pirs, T., Zdovc, I., Krt, B. and Ocepek, M. (2004) *J Vet Med B Infect Dis Vet Public Health*, **51**, 249–250.

58 Komijn, R.E., Wisselink, H.J., Rijsman, V.M., Stockhofe-Zurwieden, N. *et al.* (2007) *Vet Microbiol*, **120**, 352–357.

59 Takai, S., Takeuchi, T. and Tsubaki, S. (1986) *Nippon Juigaku Zasshi*, **48**, 445–448.

60 Takai, S., Fukunaga, N., Ochiai, S., Imai, Y. *et al.* (1996) *J Clin Microbiol*, **34**, 1034–1037.

61 Prescott, J.F., Patterson, M.C., Nicholson, V.M., Morein, B. and Yager, J.A. (1997) *Vet Microbiol*, **56**, 213–225.

62 Giguere, S., Hondalus, M.K., Yager, J.A., Darrah, P. *et al.* (1999) *Infect Immun*, **67**, 3548–3557.
63 Jain, S., Bloom, B.R. and Hondalus, M.K. (2003) *Mol Microbiol*, **50**, 115–128.
64 Ren, J. and Prescott, J.F. (2004) *Vet Microbiol*, **103**, 219–230.
65 Pei, Y., Parreira, V., Nicholson, V.M. and Prescott, J.F. (2007) *Can J Vet Res*, **71**, 1–7.
66 Hondalus, M.K., Diamond, M.S. Rosenthal, L.A., Springer, T.A. and Mosser, D.M. (1993) *Infect Immun*, **61**, 2919–2929.
67 Fernandez-Mora, E., Polidori, M., Luhrmann, A., Schaible, U.E. and Haas, A. (2005) *Traffic*, **6**, 635–653.
68 Meijer, W.G. and Prescott, J.F. (2004) *Vet Res*, **35**, 383–396.
69 Toyooka, K., Takai, S. and Kirikae, T. (2005) *J Med Microbiol*, **54**, 1007–1015.
70 Hondalus, M.K. and Mosser, D.M. (1994) *Infect Immun*, **62**, 4167–4175.
71 Luhrmann, A., Mauder, N., Sydor, T., Fernandez-Mora, E. *et al.* (2004) *Infect Immun*, **72**, 853–862.
72 Polidori, M. (2007) Biochemische und genetische Analyse der Wechselwirkung von Rhodococcus equi mit Makrophagen. PhD thesis. Bonn.
73 Davis-Scibienski, C. and Beaman, B.L. (1980) *Infect Immun*, **28**, 610–619.
74 Davis-Scibienski, C. and Beaman, B.L. (1980) *Infect Immun*, **29**, 24–29.
75 Black, C.M., Paliescheskey, M., Beaman, B.L., Donovan, R.M. and Goldstein, E. (1986) *J Infect Dis*, **154**, 952–958.
76 Takai, S., Koike, K., Ohbushi, S., Izumi, C. and Tsubaki, S. (1991) *J Clin Microbiol*, **29**, 439–443.
77 Takai, S., Sekizaki, T., Ozawa, T., Sugawara, T. *et al.* (1991) *Infect Immun*, **59**, 4056–4060.
78 Takai, S., Iie, M., Watanabe, Y., Tsubaki, S. and Sekizaki, T. (1992) *Infect Immun*, **60**, 2995–2997.
79 Takai, S., Watanabe, Y., Ikeda, T., Ozawa, T. *et al.* (1993) *J Clin Microbiol*, **31**, 1726–1729.
80 Tkachuk-Saad, O. and Prescott, J. (1991) *J Clin Microbiol*, **29**, 2696–2700.
81 Sekizaki, T., Takai, S., Egawa, Y., Ikeda, T. *et al.* (1995) *Gene*, **155**, 135–136.
82 Takai, S., Hines, S.A., Sekizaki, T., Nicholson, V.M. *et al.* (2000) *Infect Immun*, **68**, 6840–6847.
83 Letek, M., Ocampo-Sosa, A., Sanders, M., Fogarty, U. *et al.* (2008) *J Bacteriol*, **190**, 5797–5805.
84 Byrne, B.A., Prescott, J.F., Palmer, G.H., Takai, S. *et al.* (2001) *Infect Immun*, **69**, 650–656.
85 Russell, D.A., Byrne, G.A., O'Connell, E.P., Boland, C.A. and Meijer, W.G. (2004) *J Bacteriol*, **186**, 5576–5584.
86 Polidori, M. and Haas, A. (2006) *Antonie Van Leeuwenhoek*, **90**, 299–304.
87 Kohler, A.K., Stone, D.M., Hines, M.T., Byrne, B.A. *et al.* (2003) *Infect Immun*, **71**, 6329–6337.
88 Chirino-Trejo, J.M. and Prescott, J.F. (1987) *Can J Vet Res*, **51**, 297–300.
89 Tan, C., Prescott, J.F., Patterson, M.C. and Nicholson, V.M. (1995) *Can J Vet Res*, **59**, 51–59.
90 Sutcliffe, I.C. and Russell, R.R. (1995) *J Bacteriol*, **177**, 1123–1128.
91 Takai, S., Fukunaga, N., Kamisawa, K., Imai, Y. *et al.* (1996) *Microbiol Immunol*, **40**, 591–594.
92 Benoit, S., Benachour, A., Taouji, S., Auffray, Y. and Hartke, A. (2001) *Res Microbiol*, **152**, 439–449.
93 Ren, J. and Prescott, J.F. (2003) *Vet Microbiol*, **94**, 167–182.
94 Jordan, M.C., Harrington, J.R., Cohen, N.D., Tsolis, R.M. *et al.* (2003) *Am J Vet Res*, **64**, 1337–1346.
95 Boland, C.A. and Meijer, W.G. (2000) *FEMS Microbiol Lett*, **191**, 1–5.
96 Rahman, M.T., Herron, L.L., Kapur, V., Meijer, W.G. *et al.* (2003) *Vet Microbiol*, **94**, 143–158.
97 Gotoh, K., Mitsuyama, M., Imaizumi, S., Kawamura, I. and Yano, I. (1991) *Microbiol Immunol*, **35**, 175–185.
98 Ueda, S., Fujiwara, N., Naka, T., Sakaguchi, I. *et al.* (2001) *Microb Pathog*, **30**, 91–99.

99 Sydor, T., von Bargen, K., Becken, U., Spuerck, S. *et al.* (2007) *Vet Microbiol*, **128**, 327–341.

100 Wall, D.M., Duffy, P.S., Dupont, C., Prescott, J.F. and Meijer, W.G. (2005) *Infect Immun*, **73**, 6736–6741.

101 Hondalus, M.K. (1997) *Vet Microbiol*, **56**, 257–268.

102 Navas, J., Gonzalez-Zorn, B., Ladron, N., Garrido, P. and Vazquez-Boland, J.A. (2001) *J Bacteriol*, **183**, 4796–4805.

103 Prescott, J.F., Lastra, M. and Barksdale, L. (1982) *J Clin Microbiol*, **16**, 988–990.

104 Pei, Y., Dupont, C., Sydor, T., Haas, A. and Prescott, J.F. (2006) *Vet Microbiol*, **118**, 240–246.

105 He, H. and Zahrt, T.C. (2005) *J Bacteriol*, **187**, 202–212.

106 Manganelli, R., Voskuil, M.I., Schoolnik, G.K. and Smith, I. (2001) *Mol Microbiol*, **41**, 423–437.

107 Ishikawa, J., Yamashita, A., Mikami, Y., Hoshino, Y. *et al.* (2004) *Proc Natl Acad Sci USA*, **101**, 14925–14930.

108 Russell, D.G., Mwandumba, H.C. and Rhoades, E.E. (2002) *J Cell Biol*, **158**, 421–426.

109 Spargo, B.J., Crowe, L.M., Ioneda, T., Beaman, B.L. and Crowe, J.H. (1991) *Proc Natl Acad Sci USA*, **88**, 737–740.

110 Crowe, L.M., Spargo, B.J., Ioneda, T., Beaman, B.L. and Crowe, J.H. (1994) *Biochim Biophys Acta*, **1194**, 53–60.

111 Ioneda, T., Niigaki, N.T., Almeida, E.T. and Levy, C.E. (1989) *Rev Microbiol*, **20**, 345–348.

112 Indrigo, J., Hunter, R.L. Jr. and Actor, J.K. (2003) *Microbiology*, **149**, 2049–2059.

113 Axelrod, S., Oschkinat, H., Enders, J., Schlegel, B. *et al.* (2008) *Cell Microbiol.*, **10**, 1530–1545.

114 Yager, J.A., Foster, S.F., Zink, M.C., Prescott, J.F. and Lumsden, J.H. (1986) *Am J Vet Res*, **47**, 438–440.

115 Hietala, S.K. and Ardans, A.A. (1987) *Vet Immunol Immunopathol*, **14**, 279–294.

116 Yager, J.A., Duder, C.K., Prescott, J.F. and Zink, M.C. (1987) *Vet Microbiol*, **14**, 287–294.

117 Martens, R.J., Martens, J.G., Renshaw, H.W. and Hietala, S.K. (1987) *Vet Microbiol*, **14**, 277–286.

118 Martens, J.G., Martens, R.J. and Renshaw, H.W. (1988) *Am J Vet Res*, **49**, 295–299.

119 Martens, R.J., Cohen, N.D., Jones, S.L., Moore, T.A. and Edwards, J.F. (2005) *Infect Immun*, **73**, 7040–7042.

120 Darrah, P.A., Hondalus, M.K., Chen, Q., Ischiropoulos, H. and Mosser, D.M. (2000) *Infect Immun*, **68**, 3587–3593.

121 Nevo, Y. and Nelson, N. (2006) *Biochim Biophys Acta*, **1763**, 609–620.

122 Halbert, N.D., Cohen, N.D., Slovis, N.M., Faircloth, J. and Martens, R.J. (2006) *J Vet Intern Med*, **20**, 974–979.

123 Cohen, N.D., Harrington, J., Gros, P., Del Castro, L. *et al.* (2004) *Immunogenetics*, **56**, 65–67.

124 Mousel, M.R., Harrison, L., Donahue, J.M. and Bailey, E. (2003) *J Vet Diagn Invest*, **15**, 470–472.

125 Lounis, N., Truffot-Pernot, C., Grosset, J., Gordeuk, V.R. and Boelaert, J.R. (2001) *J Clin Virol*, **20**, 123–126.

126 Boelaert, J.R., Vandecasteele, S.J., Appelberg, R. and Gordeuk, V.R. (2007) *J Infect Dis*, **195**, 1745–1753.

127 Patton, K.M., McGuire, T.C., Fraser, D.G. and Hines, S.A. (2004) *Infect Immun*, **72**, 7073–7083.

128 Beatty, W.L., Rhoades, E.R., Ullrich, H.J., Chatterjee, D. *et al.* (2000) *Traffic*, **1**, 235–247.

129 Beatty, W.L. and Russell, D.G. (2000) *Infect Immun*, **68**, 6997–7002.

130 Igietseme, J.U., Eko, F.O., He, Q. and Black, C.M. (2004) *Expert Rev Vaccines*, **3**, 23–34.

131 Hietala, S.K., Ardans, A.A. and Sansome, A. (1985) *Am J Vet Res*, **46**, 13–15.

132 Prescott, J.F., Nicholson, V.M., Patterson, M.C., Zandona Meleiro, M.C. *et al.* (1997) *Am J Vet Res*, **58**, 356–359.

133 Madigan, J.E., Hietala, S. and Muller, N. (1991) *J Reprod Fertil Suppl*, **44**, 571–578.

134 Jacks, S., Giguere, S., Crawford, P.C. and Castleman, W.L. (2007) *Clin Vaccine Immunol*, **14**, 669–677.

135 Sheoran, A.S., Timoney, J.F., Holmes, M.A., Karzenski, S.S. and Crisman, M.V. (2000) *Am J Vet Res*, **61**, 1099–1105.

136 Hooper-McGrevy, K.E., Wilkie, B.N. and Prescott, J.F. (2003) *Clin Diagn Lab Immunol*, **10**, 345–351.

137 Yager, J.A., Prescott, C.A., Kramar, D.P., Hannah, H. *et al.* (1991) *Vet Microbiol*, **28**, 363–376.

138 Nordmann, P., Kerestedjian, J.J. and Ronco, E. (1992) *Antimicrob Agents Chemother*, **36**, 1244–1248.

139 Kanaly, S.T., Hines, S.A. and Palmer, G.H. (1993) *Infect Immun*, **61**, 4929–4932.

140 Nordmann, P., Ronco, E. and Nauciel, C. (1992) *Infect Immun*, **60**, 2748–2752.

141 Ross, T.L., Balson, G.A., Miners, J.S., Smith, G.D. *et al.* (1996) *Can J Vet Res*, **60**, 186–192.

142 Madarame, H., Takai, S., Matsumoto, C., Minamiyama, K. *et al.* (1997) *FEMS Immunol Med Microbiol*, **17**, 251–262.

143 Kanaly, S.T., Hines, S.A. and Palmer, G.H. (1995) *Infect Immun*, **63**, 3037–3041.

144 Nordmann, P., Ronco, E. and Guenounou, M. (1993) *J Infect Dis*, **167**, 1456–1459.

145 Kasuga-Aoki, H., Takai, S., Sasaki, Y., Tsubaki, S. *et al.* (1999) *Immunology*, **96**, 122–127.

146 Kanaly, S.T., Hines, S.A. and Palmer, G.H. (1996) *Infect Immun*, **64**, 1126–1132.

147 Hines, M.T., Paasch, K.M., Alperin, D.C., Palmer, G.H. *et al.* (2001) *Vet Immunol Immunopathol*, **79**, 101–114.

148 Hines, S.A., Stone, D.M., Hines, M.T., Alperin, D.C. *et al.* (2003) *Clin Diagn Lab Immunol*, **10**, 208–215.

149 Takai, S., Kobayashi, C., Murakami, K., Sasaki, Y. and Tsubaki, S. (1999) *FEMS Immunol Med Microbiol*, **24**, 1–9.

150 Giguere, S., Wilkie, B.N. and Prescott, J.F. (1999) *Infect Immun*, **67**, 5041–5047.

151 Giguere, S. and Prescott, J.F. (1998) *Infect Immun*, **66**, 1848–1854.

152 Darrah, P.A., Monaco, M.C., Jain, S., Hondalus, M.K. *et al.* (2004) *J Immunol*, **173**, 1914–1924.

153 Beaman, B.L. and Black, C. (1985) *Curr Top Microbiol Immunol*, **122**, 138–147.

154 Filice, G.A., Beaman, B.L. and Remington, J.S. (1980) *Infect Immun*, **27**, 643–649.

155 Ashour, J. and Hondalus, M.K. (2003) *J Bacteriol*, **185**, 2644–2652.

156 Koenig, S., Mehlich, A.M., Ackermann, D., Hamid, M.E. *et al.* (2008) *Front Biosci*, **13**, 842–855.

21
Salmonella

Olivia Steele-Mortimer

21.1
Introduction

The genus *Salmonella* is a large group of highly similar Gram-negative bacteria that are closely related to *Escherichia* spp., *Shigella* spp. and *Citrobacter* spp. Taxonomically, the genus *Salmonella* is divided into two species: *Salmonella bongori* and *Salmonella enterica*. *S. bongori* is the older ancestral species, believed to have evolved from *Escherichia coli* with the acquisition of virulence factors encoded on a pathogenicity island known as *Salmonella* pathogenicity island 1 (SPI1) [1]. Subsequently *S. enterica* evolved by acquiring additional virulence factors and ultimately diverging into over 2500 closely related serovars, many of which cause disease in humans and other animals. The ability of *Salmonella* to invade mammalian host cells and subsequently to establish an intracellular replicative niche is dependent on two type III secretion systems (T3SS). Bacterial effector proteins translocated directly into the host cell by these systems are essential for invasion of nonphagocytic cells and for correct maturation of the *Salmonella*-containing vacuole (SCV), a modified phagosome within which *Salmonella* survive and replicate. Effector proteins translocated by T3SS1 mediate invasion and initial SCV formation whereas the second T3SS, T3SS2, is not active until after invasion.

Intracellular replication occurs within 3–6 h post infection and is preceded by changes in the SCV: notably acquisition of lysosomal glycoproteins (LGP), juxtanuclear positioning and luminal acidification. Coincident with the initiation of replication, LGP-enriched membrane tubules, known as SIFs, extend from the surface of SCVs and radiate throughout the host cell. SCV maturation, and the roles of T3SS effectors in this process, has been studied intensively in the last 10 years although many critical questions remain unanswered. In this chapter our current knowledge of the SCV and its role in *Salmonella* pathogenesis is summarized.

21.1.1
Salmonella Infections in Humans

S. enterica serovars cause two main disease syndromes in humans (Table 21.1). Enteric fever (typhoid or paratyphoid) affects over 21 million people annually with

Intracellular Niches of Microbes. Edited by Ulrich E. Schaible and Albert Haas

ISBN: 978-3-527-32207-7

Table 21.1 *Salmonella enterica* serovars cause different diseases in different hosts.

Serovar	Host	Disease manifestation			Refs
		Localized	Systemic	Persistence	
Typhimurium (ubiquitous)	Human	Gastroenteritis	In immune deficient, for example advanced AIDS	Shedding for several weeks	[163–165]
	Mouse	Orally inoculated Strep-treated mice develop colitis but also progress to systemic disease (BALB/c[a])	In susceptible strains (Slc11a1$^{-/-}$, TLR4$^{-/-a}$). Increased susceptibility associated with pregnancy	In resistant strains (Slc11a1$^{+/+}$, TLR4$^{+/+a}$)	[13, 166–169]
	Bovine	Older cattle – localized infection asymptomatic	Calves – enteric and systemic, highly mortality	Transient carriage	[168, 170–172]
	Fowl	Asymptomatic caecal colonization in older chicks	Systemic in very young chicks (1 d), high mortality rate	Cecal colonization accompanied by persistent shedding	[11, 88, 172, 173]
Typhi (host restricted)	Human		Typhoid fever. Various extraintestinal complications	Asymptomatic carriers	[163, 174]
Dublin (host restricted)	Humans		Rare infection – associated with bacteremia		[168, 175]
	Bovine	Diarrhea in young calves	Gastroenteritis progresses to systemic. Abortion	Asymptomatic carriers. Chronic carriage	[168, 176, 177]
Gallinarum biovar *gallinarum* (host restricted)			Typhoid and diarrhea in poultry		[175]

[a]BALB/c mice carry the *Slc11a1*^{s} allele and are highly susceptible to *Salmonella* infection. Slc11a1 was formerly known as natural resistance-associated macrophage protein 1 (NRAMP1).
AIDS, acquired immune deficiency syndrome; TLR, Toll-like receptor.

about 200 000 deaths almost all in the developing world [2]. This systemic disease is caused by serovars Typhi or Paratyphi, which infect only humans or higher primates [3], and are usually transmitted through contaminated water or food. Many other serovars cause non-typhoidal salmonellosis (NTS), which most often presents as acute gastroenteritis [4]. In the United States there are an estimated 1.4 million cases of NTS per year although the vast majority of these go unreported [5]. In contrast to enteric fever, *Salmonella*-induced gastroenteritis in humans is a localized infection that does not spread beyond the gastrointestinal tract and is usually self-resolving in healthy adults. Other *Salmonella* infections are uncommon but can present in a variety of manifestations such as abscesses or endocarditis [6–8].

21.1.2
Salmonella Infections in Other Animals

The vast majority of salmonellosis cases in the United States are due to the consumption of contaminated foods, particularly poultry, beef, pork and eggs. *Salmonella* carriage in food animal species is probably the major source of this contamination, especially since the strains most often associated with human disease, including Typhimurium, Enteritidis, Newport, Agona, and Infantis, are commonly found in food animal species [9]. Swine and poultry in particular can be asymptomatic carriers of *Salmonella* serovars that cause significant human disease [9]. For example, in the United States the emergence of serovar Enteritidis as one of the predominant serovars in human infections in the 1980s closely followed the widespread appearance of Enteritidis in commercial poultry flocks [10].

A remarkable aspect of *Salmonella* serovars is that, despite the strong identity at the DNA level, they have a high degree of host specificity. Some serovars, in particular Typhi, are extremely host restricted whereas others, such as Typhimurium, infect a wide variety of hosts but often cause dramatically different diseases (see Table 21.1). For example, Typhimurium causes a typhoid-like systemic disease in susceptible strains of laboratory mice and similar severe systemic infections with high mortality in young calves and chicks. Conversely, in humans, adult cattle or chickens Typhimurium causes a self-limiting localized enteric infection [11, 12].

21.1.3
Development of Persistent Infection

Salmonella serovars often establish long-term systemic infections characterized by bacterial persistence and replication in the presence of an adaptive immune response [13–18]. How the carrier state is established and the bacterial and host factors involved in maintenance of asymptomatic infection are still not well understood and may vary significantly for different serovar/host combinations. In the carrier state of Typhoid fever, during which the individuals can intermittently shed bacteria for months or even years [19], the gallbladder often contains *Salmonella* and carriers with gallstones are typically resistant to antibiotic treatment [20]. Studies using a murine model for persistent infection in genetically resistant Slc11a1$^{+/+}$ (NRAMP1$^{+/+}$) mice, have shown that a non-fimbrial adhesin, ShdA, and at least six fimbrial operons are involved in persistent intestinal carriage [21–23] as is bacterial survival in hemophagocytic macrophages [24]. The role of Slc11a1 itself is complex; it may be determined by its effect on T_H1 differentiation [17] or by counteracting the ability of *Salmonella* to direct SCV biogenesis through depletion of iron and/or other cations [25]. Persistence is often accompanied by fecal shedding, which is a major cause of transmission and can be affected by a variety of factors such as stress or reproductive state [15].

21.2
Characterization of the SCV and its Role in Virulence

The mechanism of internalization (host cell-mediated phagocytosis vs. T3SS1-mediated invasion) as well as the host cell type (epithelial, macrophage, dendritic cell, fibroblast, etc.) can influence the establishment of the intracellular niche and replicative potential of the bacteria and is likely to have significant effects on SCV biogenesis [26–29]. Here we will primarily focus on SCV biogenesis in epithelial cells since T3SS1-mediated invasion of this cell type is believed to be a critical step in the establishment of infection and it is in these cells that SCV biogenesis has been most intensively studied. The initial invasion step, and consequent formation of the nascent SCV, is mediated by a cohort of T3SS1 effector proteins translocated across the plasma membrane in the host cell [30]. Localized actin rearrangements lead to the formation of membrane ruffles on the surface of the host cell and the bacteria is rapidly engulfed into a phagosome, the SCV (Figure 21.1). Immediately following formation, the SCV contains early endosomal markers, such as early endosomal antigen 1 (EEA1), Rab5 and the transferrin receptor, which are then rapidly replaced by late endosome/lysosome (LE/Lys) markers, including the vacuolar ATPase (v-ATPase) and LGPs such as lysosome-associated membrane protein 1 (LAMP1) [31]. Acidification of the SCV lumen also occurs during the initial 1–2 h following invasion [26]. According to these observations SCV maturation appears similar to that of phagosomes containing dead bacteria or inert particles, nevertheless, several studies have shown that in contrast to those "canonical" phagosomes other LE/Lys markers, such as lysosomal hydrolases/cathepsins, mannose 6-phosphate receptors and internalized fluorescent dextrans, are apparently excluded from the SCV [32–34]. Based on these findings the working model has been that *Salmonella* inhibit SCV–lysosome fusion and that this is a crucial step in the establishment of the intracellular niche. However, other studies using live cell imaging have shown that lysosomes can fuse directly with the SCV, indicating that SCV maturation is more complex than indicated by the current model [35–37].

Within 1–2 h following invasion the SCV relocates from the periphery of the cell to a juxtanuclear position adjacent to the microtubule organizing center (MTOC) [38, 39]. Replication is initiated 4–6 h post invasion and is accompanied by the appearance of extensive membrane tubules or SIFs (*Salmonella*-induced filaments), which emanate from the SCV and extend throughout the cell [40]. *Salmonella* mutants unable to form SIFs have decreased virulence in the mouse, suggesting that these structures are important *in vivo*, although they have only been observed in cultured host cells [41–44].

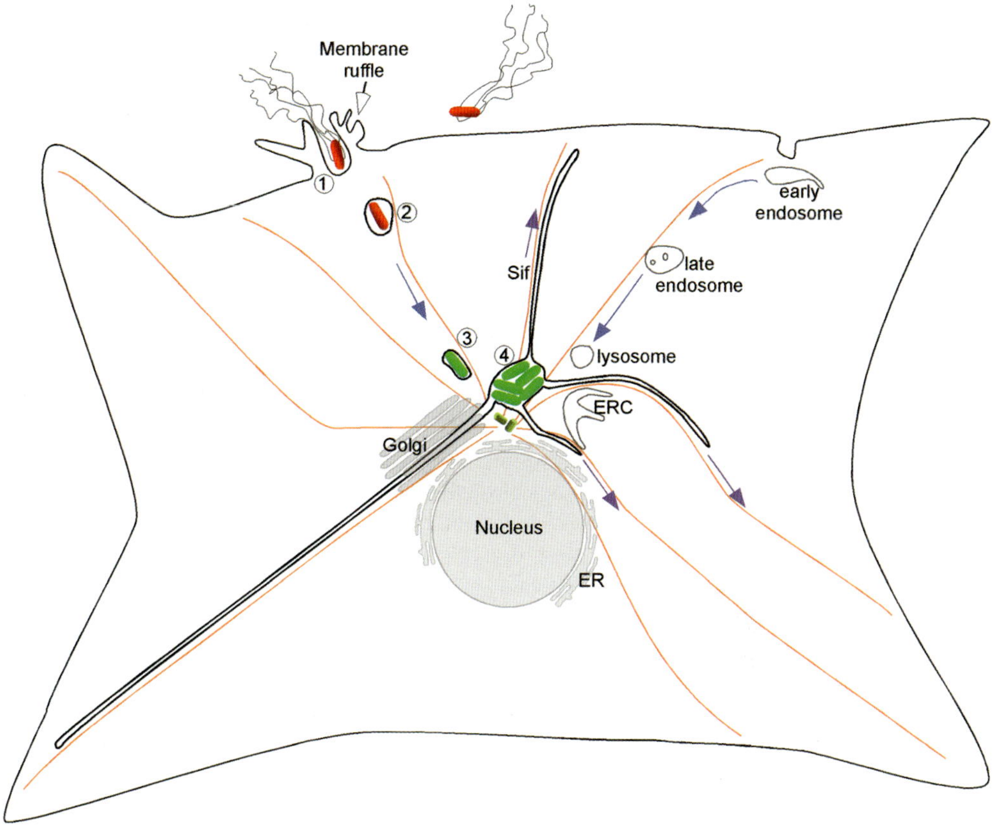

Figure 21.1 Model of *Salmonella*-containing vacuole (SCV) maturation. (1) SPI1-induced invasive bacteria translocate T3SS1 effectors across the plasma membrane, leading to localized membrane ruffles and engulfment of the bacteria into the SCV. (2) Newly formed SCVs are distinguished by the presence of early endosome markers such as early endosomal antigen 1 (EEA1) and Rab5. (3) Within 1–2 h post infection early endosomal markers are replaced by late endosome/lysosome (LE/Lys) markers, such as LAMP1 and vacuolar ATPase, and SCVs become acidified. (4) By 4–5 h post infection the majority of SCVs are localized in a juxtanuclear position, close to the microtubule organizing center (MTOC), the SPI2 regulon is induced and replication is initiated. *Salmonella*-induced filament (SIF) tubules start to extend from the surface of the SCV and radiate through the cell.

A number of lipids have critical roles in phagosome biogenesis, where they are involved in essential steps such as actin assembly, signaling and motor recruitment [45] and there is no doubt that this is also true for the SCV [46, 47]. In particular, *Salmonella* appear to specifically target phospholipids and cholesterol in the SCV membrane. Phosphatidylinositol 3-phosphate (PI(3)P), a phospholipid that mediates sorting of downregulated receptors in endosomes [48], accumulates transiently on the SCV [49–51]. The mechanism of PI(3)P production remains unclear, particularly since some studies show a requirement for host cell phosphatidylinositol 3-kinase (PI3K) whereas others do not [49, 50, 52]. PI(3)P accumulation is dependent on the *Salmonella* effector protein SopB, which is an inositol phosphatase, and mutants lacking SopB have some defects in SCV biogenesis [49, 51]. SopB phosphatase activity is also required for the efficient recruitment of at least two phosphoinositide-binding proteins, sorting nexin-1 (SNX1) and EEA1, to the bacterial entry sites and/or SCV [37, 50]. Another lipid that accumulates in SCVs is cholesterol with up to 30% of total cellular cholesterol accumulating in SCVs at later time points (8 h post infection) [53]. This is particularly interesting given the recently described role of cholesterol in modulating endosome motility [54]. The bacterial factors involved in cholesterol accumulation remain uncertain [53, 55], although the translocated effector protein SseJ is an acyltransferase/lipase that esterifies cholesterol both *in vitro* and *in vivo* [56].

21.3
Factors Affecting SCV Biogenesis

The ability of *Salmonella* to invade and survive within host cells is determined by genes encoded on several discrete regions on the chromosome (see Table 21.2). In particular, *Salmonella* pathogenicity islands 1 and 2 (SPI1 and SPI2) encode T3SS1 and T3SS2 respectively. T3SS1 is induced extracellularly and until recently was believed to be downregulated immediately following internalization, although recently it has been implicated in post-invasion processes including SCV biogenesis [37, 51, 57, 58]. The characteristic signature of *Salmonella* invasion, localized actin rearrangements and plasma membrane ruffling of ruffles, is mediated by three T3SS1 effectors: SopE, SopE2 and SopB/SigD. SopE and SopE2 are potent GTP exchange factors (or GEFs) for

Table 21.2 *Salmonella* pathogenicity islands. (For review see [178]).

Pathogenicity island	Role in pathogenicity	Distribution
SPI1	T3SS1 – Invasion of nonphagocytic cells. Iron uptake	Ubiquitous – except for very few strains for example Senftenberg [179].
SPI2	T3SS2 – Intracellular survival and systemic spread	*S. enterica* (not in *S. bongori*)
SPI3	Mg^{2+} uptake	Ubiquitous
SPI4	T1SS and secreted nonfimbrial adhesin	Ubiquitous
SPI5	T3SS1 and T3SS2 effectors	Ubiquitous (some genes not in *S. bongori* [180])
SPI6 (SCI)	Fimbriae	Subsp. I (partially also in IIIB, IV, VII)
SPI7 (MPI)	Vi antigen, pilus assembly, sopE	Subsp. I
SPI8	?	Typhi
SPI9	Toxin?	Subsp. I
SPI10	Sef fimbriae	Subsp. I
SCI-1	Antibiotic resistance	Subsp. I
HPI	Iron uptake	Subsp. IIIa, IIIb, IV

T3SS, type III secretion system.

the small GTPases Cdc42, RhoG and Rac1, all of which stimulate activation of actin polymerization via WASP family members, N-WASP and WAVE2 and the ARP2/3 complex [59, 60]. In contrast, the inositol phosphatase SopB has an indirect effect on RhoG and Cdc42, not Rac1, by inducing membrane recruitment of SGEF, a GEF that binds to specific phospholipids via a pleckstrin homology (PH) domain [61, 62]. A SopB-dependent increase in PI(3)P also recruits VAMP8 to the nascent phagosome and promotes invasion [52].

Two other T3SS1 effectors that modulate actin polymerization are SipA and SipC, though they are unable to induce ruffling in the absence of SopE, SopE2 and SopB [63]. SipA enhances invasion by locally inhibiting ADF/cofilin- and gelsolin-directed actin disassembly and simultaneously stimulating pathogen-induced actin polymerization [64, 65]. SipC assembles into multimers and causes actin nucleation and bundling and may bind intermediate filaments [66–68]. Actin remodeling is subsequently downregulated by the activity of another T3SS1 effector, SptP, a GTPase specific for Cdc42 and Rac1 [69, 70]. The opposing activities of SptP versus SopE and SopE2 are balanced by temporal regulation; the ubiquitinated effectors, SopE and SopE2, are rapidly degraded after translocation into host cells whereas the non-ubiquitinated SptP has a longer half-life [71]. In this way *Salmonella* insure that ruffling, as well as any other effects of upregulating the small GTPases, is turned off after invasion. Interestingly, SopB is also modified by ubiquitination although this does not seem to play any role in its turnover and this T3SS1 effector can be detected inside infected cells for many hours [72, 73].

Following invasion, expression of the SPI1 regulon is downregulated and the SPI2-encoded T3SS2 system is induced. However, the switch from T3SS1 activity to T3SS2 activity does not occur immediately and it is now apparent that there is temporal overlap, especially for a subset of T3SS1 translocated effectors that can persist in host cells for several hours [71, 73, 74]. Thus SopB regulates the phospholipid content of the SCV and can block normal endosome to lysosome trafficking and SNX1 recruitment [37, 75], it may also affect actin organization via a mechanism separate from its phosphatase activity [76] and can modulate positioning of the SCV via mysosin II activation [58]. The actin-bundling protein SipA [77–81] is translocated very rapidly into the host cell within 10 min [82] but also has been ascribed a long-term effect in juxtanuclear positioning of the SCV [74]. In addition, several T3SS1 effectors have postinvasion activities that are not directly related to SCV biogenesis. The phosphatase activity of SopB induces Akt activation and is an antiapoptotic signal in infected epithelial cells [83, 84] whereas SipB induces cell death in infected macrophages [85, 86].

The SPI2-encoded T3SS2 is required for pathogenesis in the murine and chicken models [87–90] and *in vitro* studies have shown that SPI2 mutants are replication defective in macrophages though not in epithelial cells [91–93]. T3SS2 is not required for initial SCV biogenesis, such as LAMP1 acquisition, but is required for juxtanuclear positioning of the vacuole at the MTOC and also for SIF formation [39, 93, 94]. Almost 20 proteins have been identified as substrates of T3SS2, many of which associate with the SCV/SIF membranes [43, 95–98]. At least three of these, SifA, SseG and SseF, are involved in maintaining the SCV at a juxtanuclear position [39, 94, 99].

Table 21.3 *Salmonella* type III secretion system effector proteins implicated in *Salmonella*-containing vacuole biogenesis.

Effector protein	Invasion or SCV biogenesis	Role	Refs
T3SS1 effectors			
SopE, SopE2	Invasion	Cdc42 and Rac GEF. Induces membrane ruffling of host cells to promote entry Ubiquinated and targeted for rapid degradation (<30 min) post infection	[60, 62, 181–184]
SopB/SigD	Invasion and SCV biogenesis	Inositol phosphatase. Generates PI (3)P and Ins(1,4,5,6)P_4 upon invasion. Promotes indirect Rho-GTPase activation and macropinocytosis	[51, 52, 58, 61, 62, 75, 185–187]
SopD	Membrane fission and macropinosome formation		[185]
SipA	Invasion and SCV biogenesis	Induces actin bundling Inhibits ADF-mediated and gelsolin-mediated actin depolymerization Potentiates SipC activity	[63, 64, 74, 78, 188]
SipC	Invasion and SCV biogenesis	Translocon component Inserts into host cell membrane and promotes actin nucleation and bundling	[64, 66, 67, 189]
SptP	Post invasion recovery of actin.	Cdc42 and Rac GAP Mediates recovery of cytoskeletal architecture by antagonizing SopE/E2 activity Delayed degradation compared with SopE/E2 Facilitates recovery of cellular architecture post invasion	[69, 71, 189, 190]
SopA	Invasion in polarized cells	E3 ligase, targeted to mitochondria	[191–193]
T3SS2 effectors			
SpiC		Interferes with vesicular trafficking Required for translocation of SPI2 effectors	[194–198]
SseF		Interferes with vesicular trafficking, SCV localization, SIF formation	[39, 94, 99, 199, 200]
SseG		Interferes with vesicular trafficking, SCV localization, SIF formation	[38, 94, 99, 200]
SifA		SCV membrane integrity, SIF formation, aggregation/tubulation of endosomes	[41, 42, 44, 74, 100, 101, 110, 201–204]

Table 21.3 (*Continued*)

Effector protein	Invasion or SCV biogenesis	Role	Refs
SspH1		E3 ubiquitin ligase	[205]
SspH2		Actin remodeling	[206]
SseI/SrfH		Actin remodeling	[206]
SseJ		Cholesterol modification, SCV membrane	[43, 44, 56, 104]
PipB2		SIF extension	[108]
SopD2		SIF extension	[106, 107, 201]

T3SS, type III secretion system; PI(3)P, phosphatidylinositol 3-phosphate; SCV, *Salmonella*-containing vacuole; Ins(1,4,5,6)P_4, inositol 1,4,5,6-tetrakisphosphate; ADF, actin depolymerizing factor; SIF, *Salmonella*-induced filament.

Host cell proteins involved in SCV positioning include the minus end directed motor protein dynein, the small GTPase Rab7 and the Rab7 effector RILP [94, 100–102]. Recent studies suggest that SCV positioning may not be solely a function of T3SS2 since two SPI1 effectors, SipA and SopB, may also contribute [58, 74]. Following movement to the juxtanuclear region, approximately 4–6 h post infection, SIF tubule formation is initiated [103]. *Salmonella* mutants unable to translocate T3SS2 effectors, or lacking SifA, do not make SIFs [41, 93]; and transient expression of SifA in HeLa cells causes extensive vacuolation of LAMP1-positive compartments and formation of SIF-like tubules, confirming that SifA activity is required for SIF formation [93].

Other T3SS2 effectors implicated in SIF formation are SSeJ, SopD2 and PipB2, all of which localize to SCV/SIF membranes [104–106]. SseJ is an acyltransferase that esterifies cholesterol and is required for lipid droplet formation in infected host cells [56, 104]. SseJ and SifA seem to work together or on the same pathway, since deletion of SseJ restores the SCV membrane integrity of a SifA deletion [104]. SopD2 can also induce formation of tubular endosomes and is required for maximal SIF extension [106, 107]. PipB2 is required for maximal SIF extension; however, expression of this effector in epithelial cells does not cause tubulation or fusing of endosomes but instead causes a dramatic mislocalization of LE/Lys to the cells periphery [108]. This mislocalization is presumably due to increased anterograde movement of endosomes since PipB2 interacts with the kinesin light chain, a subunit of the kinesin-1 motor complex that drives anterograde transport along microtubules, and mediates recruitment of kinesin-1 to the membrane of the SCV/SIFs [109]. Intriguingly, mutants lacking SifA recruit increased amounts of kinesin-1 to the SCV and ultimately escape from the vacuole into the cytosol [110]. Thus SIF formation depends on a balance between the activities of SifA, SseJ and PipB2 which mediate the precise recruitment of kinesin-1 [103, 110, 111].

Another protein that can be translocated by T3SS2 is the virulence plasmid-encoded SpvB, an actin-specific ADP-ribosyltransferase that, like SseJ, also down-regulates SIF formation [97, 103]. Finally, formation of an actin meshwork around the SCV, although still of unknown function, is mediated by the T3SS2-translocated kinase SteC [112].

21.3.1
Inhibition of Host Response by T3SS2

Phagocytic cell resistance to microbial pathogens is dependent on the production of reactive oxygen species (ROS) and reactive nitrogen species (RNS) by NADPH phagocyte oxidase (Phox) and inducible nitric oxide synthase (iNOS), respectively, and both are required for host resistance to wild-type *Salmonella* [113–115]. In macrophages *Salmonella* are initially killed by Phox activity followed by sustained inhibition of growth by RNS [116]. However, the ability of *Salmonella* to cause systemic disease is dependent on its ability to survive inside macrophages and *Salmonella* limit both accumulation of Phox and RNS production in the SCV via the T3SS2 [117–121]. *Salmonella* also contain several enzymes that can metabolize RNS, of which the flavohemoglobin Hmp seems to be the most important in the intracellular environment [122–124].

Another phagosome membrane protein associated with innate resistance to *Salmonella*, at least in mice, is Slc11a1 (NRAMP1). Slc11a1$^+$ mice are resistant to infection with a number of pathogens, although the mechanism of resistance remains unclear. In macrophages Slc11a1 induces expression of *Salmonella* genes encoding cation transport systems and SPI2 [125, 126] but also counteracts the ability of *Salmonella* to regulate SCV biogenesis [127]. Slc11a1-mediated resistance to *Salmonella* does not appear to involve iNOS or phagocyte oxidase [128] but instead may be associated with the ability of some dendritic cells to respond rapidly to infection [129].

In addition to providing an intracellular niche for bacterial survival, and/or replication, another potential function of the SCV is in the prevention or delay of antigen presentation in infected cells and evidence for the involvement of T3SS2 in this process is accumulating. Mice inoculated with SPI2 mutants develop increased resistance to reinfection compared to mice inoculated with wild-type *Salmonella* and resistance is associated with increased numbers of both CD4$^+$ and CD8$^+$ T cells [130]. More specifically, T3SS2 appears to block antigen processing in infected dendritic cells and the T3SS2 effector SifA prevents cell surface expression of MHC class II [28, 131–133].

21.3.2
Regulation of Virulence Factors in the SCV

The PhoP/PhoQ two-component system, consisting of a sensor kinase, PhoQ and its transcriptional activator PhoP, is used by many Gram-negative pathogens to regulate virulence gene expression [134]. *In vitro* the *Salmonella* PhoP/PhoQ system can be activated by low Mg^{2+} concentrations and low pH, although it is now thought that antimicrobial peptides and acidic pH are major stimuli in the SCV [135–137]. PhoP/-PhoQ regulates some genes indirectly, for example via the PmrA/PmrB two-component system or the DNA-binding protein SlyA [138]. Examples of PhoPQ-regulated intracellular functions include prevention of uncontrolled SCV–lysosome fusion [32], remodeling of the lipid A domain of lipopolysaccharide that may help bacteria withstand the stresses of the intracellular environment [139], downregulation of flagellar expression [140] and expression of the superoxidase dismutase SodCI that is required for surviving the respiratory burst in macrophages [141]. Another two-component regulatory system involved in intracellular virulence gene expression, particularly SPI2 genes, is the SPI2-encoded SsrAB [142–144]. The involvement of PhoP/PhoQ in SsrAB expression is controversial, although similar signals including cation chelation, are involved [143, 145–147]. SlyA and OmpR, both of which regulate SsrAB, also affect response to the intracellular environment and SPI2 induction [145, 148].

21.4 Immunology of the SCV

Virulent *Salmonella* serovars may use a number of strategies to evade the host immune response and the immunology of the SCV is likely to be determined by a variety of factors, particularly the serovar and host species [133]. Two mechanisms by which intracellular bacteria can avoid detection by the host are by restricting the expression of immunogenic molecules inside host cells or by preventing the host cell from processing and displaying these molecules.

In vivo both macrophages and dendritic cells are important targets of *Salmonella*, yet these cells also form an essential component of the host immune system. Both are important antigen-presenting cells (APCs) and *Salmonella* seems to specifically target this process from within the SCV. In particular, recent studies have shown that T3SS2 can prevent antigen presentation to T cells by avoiding lysosomal degradation in dendritic cells [132, 133] and by specifically blocking presentation of peptides by MHC class II at the cell surface [130].

21.4.1 Immunologically Relevant *Salmonella* Antigens

Salmonella have classic inflammatory surface molecules including flagellins and lipopolysaccharide (LPS), the polysaccharide that coats Gram-negative bacteria.

Salmonella swim using flagella that may be required in invasion [149] but are downregulated in the presence of cationic antibiotic peptides [136], presumably following internalization, although expression can be induced intracellularly at late stages post invasion [150]. Flagellins can stimulate an immune response by binding to Toll-like receptor 5 (TLR5) [151], for example, in an *in vitro* model, flagellin is translocated via a T3SS2-dependent mechanism to the basolateral surface of infected epithelial cells where it encounters TLR5 [98, 152]. The two *Salmonella* flagellins, FliC and FljB, can both activate NFκB-dependent gene expression [153]. FliC is an important antigen recognized by $CD4^+$ cells in both humans and mice [154].

One of the really interesting questions in the *Salmonella* field is how systemic versus localized infection is initiated. One molecule that seems to be particularly important in the establishment of typhoid fever is the Vi capsular antigen. Genes necessary for production of Vi are encoded on SPI7, a genetically unstable region that is present in almost all Typhi isolates but not in Typhimurium [155]. Vi can reduce TLR-dependent interleukin 8 (IL-8) expression in the intestinal mucosa [156, 157] and expression of the Vi capsule in Typhimurium leads to suppression of tumor necrosis factor α (TNFα) and IL-6 production by infected murine macrophages in a TLR4-dependent manner, suggesting that Vi prevents recognition of *Salmonella* LPS by TLR4 [158].

Several T3SS-translocated effector proteins are also implicated in immune modulation. In mice infected with serovar Typhimurium the T3SS1-translocated effector SipC is a target antigen for CD4 T cell recognition via a process that requires actin-dependent uptake of the bacteria [159]. The T3SS2 effector SseL suppresses NFκB and SseL-deficient Typhimurium induce stronger inflammatory responses, associated with increased production of NFκB-dependent cytokines [160]. The virulence plasmid-encoded SpvC effector, also translocated into host cells, has phosphothreonine lyase on phospho-ERK, a host mitogen-activated protein kinase and can downregulate cytokine release from infected cells [161]. Finally, the T3SS1 effector SopB has also been shown to have a role in immune escape of *Salmonella* and mutants lacking SopB induce less cytotoxicity in infected dendritic cells [162].

21.5 Summary

The interactions between *Salmonella* and their eukaryotic hosts have evolved over many millennia and we are only beginning to understand the elaborate network of interactions required for pathogenesis. Establishment of the intracellular niche, specifically the SCV wherein *Salmonella* survive and replicate, is a cornerstone in this process. Recent studies are revealing novel nuances in SCV biogenesis that suggest we are entering a new era in *Salmonella* research. A major hurdle now is to connect some of the basic observations, made in simple cultured cell systems, to the complex multidimensional *in vivo* environment that involves interactions with a variety of host cell types. The ability of *Salmonella* to respond to a variety of specialized intracellular environments and to react accordingly is undoubtedly one of the keys to its success as a pathogen.

Acknowledgments

Work in the author's laboratory is supported by the Intramural Research Program of the National Institutes of Allergy and Infectious Disease, NIH.

References

1 Baumler, A.J., Tsolis, R.M., Ficht, T.A. and Adams, L.G. (1998) *Infect Immun*, **66**, 4579–4587.

2 Crump, J.A., Luby, S.P. and Mintz, E.D. (2004) *Bull World Health Organ*, **82**, 346–353.

3 Edsall, G., Gaines, S., Landy, M., Tigertt, W.D. *et al.* (1960) *J Exp Med*, **112**, 143–166.

4 Vugia, D., Cronquist, A., Hadler, J., Tobin-D'Angelo, M., *et al.* (2008) *MMWR Morb Mortal Wkly Rep*, **57**, 366–370.

5 Voetsch, A.C., Van Gilder, T.J., Angulo, F.J., Farley, M.M. *et al.* (2004) *Clin Infect Dis*, **38** (Suppl 3), S127–S134.

6 Nielsen, H., Gradel, K.O. and Schonheyder, H.C. (2006) *APMIS*, **114**, 641–645.

7 Fernandez Guerrero, M.L., Aguado, J.M., Arribas, A., Lumbreras, C. and de Gorgolas, M. (2004) *Medicine (Baltimore)*, **83**, 123–138.

8 Rodriguez, M., de Diego, I., Martinez, N., Rosario Rodicio, M. and Carmen Mendoza, M. (2006) *Int J Med Microbiol*, **296**, 211–222.

9 Foley, S.L., Lynne, A.M. and Nayak, R. (2008) *J Anim Sci*, **86**, E149–E162.

10 Callaway, T.R., Edrington, T.S., Anderson, R.C., Byrd, J.A. and Nisbet, D.J. (2008) *J Anim Sci*, **86**, E163–E172.

11 Barrow, P.A., Huggins, M.B., Lovell, M.A. and Simpson, J.M. (1987) *Res Vet Sci*, **42**, 194–199.

12 Zhang, S., Kingsley, R.A., Santos, R.L., Andrews-Polymenis, H. *et al.* (2003) *Infect Immun*, **71**, 1–12.

13 Monack, D.M., Bouley, D.M. and Falkow, S. (2004) *J Exp Med*, **199**, 231–241.

14 Wigley, P., Berchieri, A. Jr, Page, K.L., Smith, A.L. and Barrow, P.A. (2001) *Infect Immun*, **69**, 7873–7879.

15 Wigley, P., Hulme, S.D., Powers, C., Beal, R.K. *et al.* (2005) *Infect Immun*, **73**, 2986–2990.

16 Brodsky, I.E., Ghori, N., Falkow, S. and Monack, D. (2005) *Mol Microbiol*, **55**, 954–972.

17 Caron, J., Loredo-Osti, J.C., Laroche, L., Skamene, E. *et al.* (2002) *Genes Immun*, **3**, 196–204.

18 Caron, J., Loredo-Osti, J.C., Morgan, K. and Malo, D. (2005) *Genes Immun*, **6**, 500–508.

19 Merrell, D.S. and Falkow, S. (2004) *Nature*, **430**, 250–256.

20 Prouty, A.M., Schwesinger, W.H. and Gunn, J.S. (2002) *Infect Immun*, **70**, 2640–2649.

21 Kingsley, R.A., Humphries, A.D., Weening, E.H., De Zoete, M.R. *et al.* (2003) *Infect Immun*, **71**, 629–640.

22 Kingsley, R.A., Keestra, A.M., de Zoete, M.R. and Baumler, A.J. (2004) *Mol Microbiol*, **52**, 345–355.

23 Weening, E.H., Barker, J.D., Laarakker, M.C., Humphries, A.D. *et al.* (2005) *Infect Immun*, **73**, 3358–3366.

24 Nix, R.N., Altschuler, S.E., Henson, P.M. and Detweiler, C.S. (2007) *PLoS Pathog*, **3**, e193.

25 Jabado, N., Cuellar-Mata, P., Grinstein, S. and Gros, P. (2003) *Proc Natl Acad Sci USA*, **100**, 6127–6132.

26 Drecktrah, D., Knodler, L.A., Ireland, R. and Steele-Mortimer, O. (2006) *Traffic*, **7**, 39–51.

27 Tobar, J.A., Gonzalez, P.A. and Kalergis, A.M. (2004) *J Immunol*, **173**, 4058–4065.

28 Herrada, A.A., Contreras, F.J., Tobar, J.A., Pacheco, R. and Kalergis, A.M. (2007) *Proc Natl Acad Sci USA*, **104**, 13402–13407.

29 Cano, D.A., Martinez-Moya, M., Pucciarelli, M.G., Groisman, E.A. *et al.* (2001) *Infect Immun*, **69**, 6463–6474.

30 Patel, J.C. and Galan, J.E. (2005) *Curr Opin Microbiol*, **8**, 10–15.

31 Steele-Mortimer, O., Meresse, S., Gorvel, J.P., Toh, B.H. and Finlay, B.B. (1999) *Cell Microbiol*, **1**, 33–49.

32 Garvis, S.G., Beuzon, C.R. and Holden, D.W. (2001) *Cell Microbiol*, **3**, 731–744.

33 Hashim, S., Mukherjee, K., Raje, M., Basu, S.K. and Mukhopadhyay, A. (2000) *J Biol Chem*, **275**, 16281–16288.

34 Garcia-del Portillo, F. and Finlay, B.B. (1995) *J Cell Biol*, **129**, 81–97.

35 Oh, Y.K., Alpuche-Aranda, C., Berthiaume, E., Jinks, T. *et al.* (1996) *Infect Immun*, **64**, 3877–3883.

36 Drecktrah, D., Knodler, L.A., Howe, D. and Steele-Mortimer, O. (2007) *Traffic*, **8**, 212–225.

37 Bujny, M.V., Ewels, P.A., Humphrey, S., Attar, N. *et al.* (2008) *J Cell Sci*, **121**, 2027–2036.

38 Salcedo, S.P. and Holden, D.W. (2003) *EMBO J*, **22**, 5003–5014.

39 Ramsden, A.E., Mota, L.J., Munter, S., Shorte, S.L. and Holden, D.W. (2007) *Cell Microbiol*, **9**, 2517–2529.

40 Garcia-del Portillo, F., Zwick, M.B., Leung, K.Y. and Finlay, B.B. (1993) *Proc Natl Acad Sci USA*, **90**, 10544–10548.

41 Stein, M.A., Leung, K.Y., Zwick, M., Garcia-del Portillo, F. and Finlay, B.B. (1996) *Mol Microbiol*, **20**, 151–164.

42 Beuzon, C.R., Meresse, S., Unsworth, K.E., Ruiz-Albert, J. *et al.* (2000) *EMBO J*, **19**, 3235–3249.

43 Freeman, J.A., Ohl, M.E. and Miller, S.I. (2003) *Infect Immun*, **71**, 418–427.

44 Ruiz-Albert, J., Yu, X.J., Beuzon, C.R., Blakey, A.N. *et al.* (2002) *Mol Microbiol*, **44**, 645–661.

45 Yeung, T. and Grinstein, S. (2007) *Immunol Rev*, **219**, 17–36.

46 Brumell, J.H. and Grinstein, S. (2003) *Cell Microbiol*, **5**, 287–297.

47 Hilbi, H. (2006) *Cell Microbiol*, **8**, 1697–1706.

48 Petiot, A., Faure, J., Stenmark, H. and Gruenberg, J. (2003) *J Cell Biol*, **162**, 971–979.

49 Pattni, K., Jepson, M., Stenmark, H. and Banting, G. (2001) *Curr Biol*, **11**, 1636–1642.

50 Scott, C.C., Cuellar-Mata, P., Matsuo, T., Davidson, H.W. and Grinstein, S. (2002) *J Biol Chem*, **277**, 12770–12776.

51 Hernandez, L.D., Hueffer, K., Wenk, M.R. and Galan, J.E. (2004) *Science*, **304**, 1805–1807.

52 Dai, S., Zhang, Y., Weimbs, T., Yaffe, M.B. and Zhou, D. (2007) *Traffic*, **8**, 1365–1374.

53 Catron, D.M., Sylvester, M.D., Lange, Y., Kadekoppala, M. *et al.* (2002) *Cell Microbiol*, **4**, 315–328.

54 Chen, H., Yang, J., Low, P.S. and Cheng, J.X. (2008) *Biophys J*, **94**, 1508–1520.

55 Garner, M.J., Hayward, R.D. and Koronakis, V. (2002) *Cell Microbiol*, **4**, 153–165.

56 Nawabi, P., Catron, D.M. and Haldar, K. (2008) *Mol Microbiol*, **68**, 173–185.

57 Steele-Mortimer, O., Brumell, J.H., Knodler, L.A., Meresse, S. *et al.* (2002) *Cell Microbiol*, **4**, 43–54.

58 Wasylnka, J.A., Bakowski, M.A., Szeto, J., Ohlson, M.B. *et al.* (2008) *Infect Immun.*, **76**, 2722–2735.

59 Ehrbar, K., Mirold, S., Friebel, A., Stender, S. and Hardt, W.D. (2002) *Int J Med Microbiol*, **291**, 479–485.

60 Friebel, A., Ilchmann, H., Aepfelbacher, M., Ehrbar, K. *et al.* (2001) *J Biol Chem*, **276**, 34035–34040.

61 Zhou, D., Chen, L.M., Hernandez, L., Shears, S.B. and Galan, J.E. (2001) *Mol Microbiol*, **39**, 248–259.

62 Patel, J.C. and Galan, J.E. (2006) *J Cell Biol*, **175**, 453–463.

63 Zhou, D., Mooseker, M.S. and Galan, J.E. (1999) *Proc Natl Acad Sci USA*, **96**, 10176–10181.

64 McGhie, E.J., Hayward, R.D. and Koronakis, V. (2001) *EMBO J*, **20**, 2131–2139.

65 McGhie, E.J., Hayward, R.D. and Koronakis, V. (2004) *Mol Cell*, **13**, 497–510.

66 Chang, J., Chen, J. and Zhou, D. (2005) *Mol Microbiol*, **55**, 1379–1389.

67 Chang, J., Myeni, S.K., Lin, T.L., Wu, C.C. *et al.* (2007) *Mol Microbiol*, **66**, 1548–1556.

68 Hayward, R.D. and Koronakis, V. (1999) *EMBO J*, **18**, 4926–4934.

69 Fu, Y. and Galan, J.E. (1998) *Mol Microbiol*, **27**, 359–368.

70 Murli, S., Watson, R.O. and Galan, J.E. (2001) *Cell Microbiol*, **3**, 795–810.

71 Kubori, T. and Galan, J.E. (2003) *Cell*, **115**, 333–342.

72 Marcus, S.L., Knodler, L.A. and Finlay, B.B. (2002) *Cell Microbiol*, **4**, 435–446.

73 Drecktrah, D., Knodler, L.A., Galbraith, K. and Steele-Mortimer, O. (2005) *Cell Microbiol*, **7**, 105–113.

74 Brawn, L.C., Hayward, R.D. and Koronakis, V. (2007) *Cell Host Microbe*, **1**, 63–75.

75 Dukes, J.D., Lee, H., Hagen, R., Reaves, B.J. *et al.* (2006) *Biochem J*, **395**, 239–247.

76 Aleman, A., Rodriguez-Escudero, I., Mallo, G.V., Cid, V.J. *et al.* (2005) *Cell Microbiol*, **7**, 1432–1446.

77 Galkin, V.E., Orlova, A., VanLoock, M.S., Zhou, D. *et al.* (2002) *Nat Struct Biol*, **9**, 518–521.

78 Popp, D., Yamamoto, A., Iwasa, M., Nitanai, Y. and Maeda, Y. (2008) *Cell Motil Cytoskeleton*, **65**, 165–177.

79 Lilic, M., Galkin, V.E., Orlova, A., VanLoock, M.S. *et al.* (2003) *Science*, **301**, 1918–1921.

80 Higashide, W., Dai, S., Hombs, V.P. and Zhou, D. (2002) *Cell Microbiol*, **4**, 357–365.

81 Jepson, M.A., Kenny, B. and Leard, A.D. (2001) *Cell Microbiol*, **3**, 417–426.

82 Schlumberger, M.C., Muller, A.J., Ehrbar, K., Winnen, B. *et al.* (2005) *Proc Natl Acad Sci USA*, **102**, 12548–12553.

83 Steele-Mortimer, O., Knodler, L.A., Marcus, S.L., Scheid, M.P. *et al.* (2000) *J Biol Chem*, **275**, 37718–37724.

84 Knodler, L.A., Finlay, B.B. and Steele-Mortimer, O. (2005) *J Biol Chem*, **280**, 9058–9064.

85 Cook, P., Totemeyer, S., Stevenson, C., Fitzgerald, K.A. *et al.* (2007) *Immunology*, **122**, 222–229.

86 Hersh, D., Monack, D.M., Smith, M.R., Ghori, N. *et al.* (1999) *Proc Natl Acad Sci USA*, **96**, 2396–2401.

87 Coburn, B., Li, Y., Owen, D., Vallance, B.A. and Finlay, B.B. (2005) *Infect Immun*, **73**, 3219–3227.

88 Jones, M.A., Hulme, S.D., Barrow, P.A. and Wigley, P. (2007) *Avian Pathol*, **36**, 199–203.

89 Shea, J.E., Beuzon, C.R., Gleeson, C., Mundy, R. and Holden, D.W. (1999) *Infect Immun*, **67**, 213–219.

90 Shea, J.E., Hensel, M., Gleeson, C. and Holden, D.W. (1996) *Proc Natl Acad Sci USA*, **93**, 2593–2597.

91 Hensel, M., Shea, J.E., Waterman, S.R., Mundy, R. *et al.* (1998) *Mol Microbiol*, **30**, 163–174.

92 Cirillo, D.M., Valdivia, R.H., Monack, D.M. and Falkow, S. (1998) *Mol Microbiol*, **30**, 175–188.

93 Brumell, J.H., Rosenberger, C.M., Gotto, G.T., Marcus, S.L. and Finlay, B.B. (2001) *Cell Microbiol*, **3**, 75–84.

94 Abrahams, G.L., Muller, P. and Hensel, M. (2006) *Traffic*, **7**, 950–965.

95 Abrahams, G.L. and Hensel, M. (2006) *Cell Microbiol*, **8**, 728–737.

96 Coombes, B.K., Lowden, M.J., Bishop, J.L., Wickham, M.E. *et al.* (2007) *Infect Immun*, **75**, 574–580.

97 Browne, S.H., Hasegawa, P., Okamoto, S., Fierer, J. and Guiney, D.G. (2008) *FEMS Immunol Med Microbiol*, **52**, 194–201.

98 Lyons, S., Wang, L., Casanova, J.E., Sitaraman, S.V. *et al.* (2004) *J Cell Sci*, **117**, 5771–5780.

99 Deiwick, J., Salcedo, S.P., Boucrot, E., Gilliland, S.M. *et al.* (2006) *Infect Immun*, **74**, 6965–6972.

100 Harrison, R.E., Brumell, J.H., Khandani, A., Bucci, C. *et al.* (2004) *Mol Biol Cell*, **15**, 3146–3154.

101 Guignot, J., Caron, E., Beuzon, C., Bucci, C. *et al.* (2004) *J Cell Sci*, **117**, 1033–1045.

102 Marsman, M., Jordens, I., Kuijl, C., Janssen, L. and Neefjes, J. (2004) *Mol Biol Cell*, **15**, 2954–2964.

103 Birmingham, C.L., Jiang, X., Ohlson, M.B., Miller, S.I. and Brumell, J.H. (2005) *Infect Immun*, **73**, 1204–1208.

104 Ohlson, M.B., Fluhr, K., Birmingham, C.L., Brumell, J.H. and Miller, S.I. (2005) *Infect Immun*, **73**, 6249–6259.

105 Knodler, L.A., Vallance, B.A., Hensel, M., Jackel, D. *et al.* (2003) *Mol Microbiol*, **49**, 685–704.

106 Jiang, X., Rossanese, O.W., Brown, N.F., Kujat-Choy, S. *et al.* (2004) *Mol Microbiol*, **54**, 1186–1198.

107 Brumell, J.H., Kujat-Choy, S., Brown, N.F., Vallance, B.A. *et al.* (2003) *Traffic*, **4**, 36–48.

108 Knodler, L.A. and Steele-Mortimer, O. (2005) *Mol Biol Cell*, **16**, 4108–4123.

109 Henry, T., Couillault, C., Rockenfeller, P., Boucrot, E. *et al.* (2006) *Proc Natl Acad Sci USA*, **103**, 13497–13502.

110 Boucrot, E., Henry, T., Borg, J.P., Gorvel, J.P. and Meresse, S. (2005) *Science*, **308**, 1174–1178.

111 Henry, T., Gorvel, J.P. and Meresse, S. (2006) *Cell Microbiol*, **8**, 23–32.

112 Poh, J., Odendall, C., Spanos, A., Boyle, C. *et al.* (2007) *Cell Microbiol.*, **10**, 20–30.

113 Mastroeni, P., Vazquez-Torres, A., Fang, F.C., Xu, Y. *et al.* (2000) *J Exp Med*, **192**, 237–248.

114 Alam, M.S., Akaike, T., Okamoto, S., Kubota, T. *et al.* (2002) *Infect Immun*, **70**, 3130–3142.

115 Alam, M.S., Zaki, M.H., Sawa, T., Islam, S. *et al.* (2008) *Microbiol Immunol*, **52**, 197–208.

116 Vazquez-Torres, A., Jones-Carson, J., Mastroeni, P., Ischiropoulos, H. and Fang, F.C. (2000) *J Exp Med*, **192**, 227–236.

117 Gallois, A., Klein, J.R., Allen, L.A., Jones, B.D. and Nauseef, W.M. (2001) *J Immunol*, **166**, 5741–5748.

118 Vazquez-Torres, A., Xu, Y., Jones-Carson, J., Holden, D.W. *et al.* (2000) *Science*, **287**, 1655–1658.

119 Chakravortty, D., Hansen-Wester, I. and Hensel, M. (2002) *J Exp Med*, **195**, 1155–1166.

120 Vazquez-Torres, A., Fantuzzi, G., Edwards, C.K. 3rd, Dinarello, C.A. and Fang, F.C. (2001) *Proc Natl Acad Sci USA*, **98**, 2561–2565.

121 McCollister, B.D., Bourret, T.J., Gill, R., Jones-Carson, J. and Vazquez-Torres, A. (2005) *J Exp Med*, **202**, 625–635.

122 Bang, I.S., Liu, L., Vazquez-Torres, A., Crouch, M.L. *et al.* (2006) *J Biol Chem*, **281**, 28039–28047.

123 Gilberthorpe, N.J., Lee, M.E., Stevanin, T.M., Read, R.C. and Poole, R.K. (2007) *Microbiology*, **153**, 1756–1771.

124 Stevanin, T.M., Poole, R.K., Demoncheaux, E.A. and Read, R.C. (2002) *Infect Immun*, **70**, 4399–4405.

125 Zaharik, M.L., Cullen, V.L., Fung, A.M., Libby, S.J. *et al.* (2004) *Infect Immun*, **72**, 5522–5525.

126 Zaharik, M.L., Vallance, B.A., Puente, J.L., Gros, P. and Finlay, B.B. (2002) *Proc Natl Acad Sci USA*, **99**, 15705–15710.

127 Cuellar-Mata, P., Jabado, N., Liu, J., Furuya, W. *et al.* (2002) *J Biol Chem*, **277**, 2258–2265.

128 White, J.K., Mastroeni, P., Popoff, J.F., Evans, C.A. and Blackwell, J.M. (2005) *J Leukoc Biol*, **77**, 311–320.

129 Valdez, Y., Diehl, G.E., Vallance, B.A., Grassl, G.A. *et al.* (2008) *Cell Microbiol.*, **10**, 1646–1661.

130 Cheminay, C., Mohlenbrink, A. and Hensel, M. (2005) *J Immunol*, **174**, 2892–2899.

131 Mitchell, E.K., Mastroeni, P., Kelly, A.P. and Trowsdale, J. (2004) *Eur J Immunol*, **34**, 2559–2567.

132 Tobar, J.A., Carreno, L.J., Bueno, S.M., Gonzalez, P.A. *et al.* (2006) *Infect Immun*, **74**, 6438–6448.

133 Bueno, S.M., Gonzalez, P.A., Carreno, L.J., Tobar, J.A. *et al.* (2008) *Immunology*, **1242**, 522–533.

134 Prost, L.R. and Miller, S.I. (2008) *Cell Microbiol*, **10**, 576–582.

135 Martin-Orozco, N., Touret, N., Zaharik, M.L., Park, E. *et al.* (2006) *Mol Biol Cell*, **17**, 498–510.

136 Bader, M.W., Navarre, W.W., Shiau, W., Nikaido, H. *et al.* (2003) *Mol Microbiol*, **50**, 219–230.

137 Bader, M.W., Sanowar, S., Daley, M.E., Schneider, A.R. *et al.* (2005) *Cell*, **122**, 461–472.

138 Navarre, W.W., Halsey, T.A., Walthers, D., Frye, J. *et al.* (2005) *Mol Microbiol*, **56**, 492–508.

139 Murata, T., Tseng, W., Guina, T., Miller, S.I. and Nikaido, H. (2007) *J Bacteriol*, **189**, 7213–7222.

140 Adams, P., Fowler, R., Kinsella, N., Howell, G. *et al.* (2001) *Proteomics*, **1**, 597–607.

141 Golubeva, Y.A. and Slauch, J.M. (2006) *J Bacteriol*, **188**, 7853–7861.

142 Bijlsma, J.J. and Groisman, E.A. (2005) *Mol Microbiol*, **57**, 85–96.

143 Deiwick, J., Nikolaus, T., Erdogan, S. and Hensel, M. (1999) *Mol Microbiol*, **31**, 1759–1773.

144 Garmendia, J., Beuzon, C.R., Ruiz-Albert, J. and Holden, D.W. (2003) *Microbiology*, **149**, 2385–2396.

145 Lee, A.K., Detweiler, C.S. and Falkow, S. (2000) *J Bacteriol*, **182**, 771–781.

146 Miao, E.A., Freeman, J.A. and Miller, S.I. (2002) *J Bacteriol*, **184**, 1493–1497.

147 Kim, C.C. and Falkow, S. (2004) *J Bacteriol*, **186**, 4694–4704.

148 Linehan, S.A., Rytkonen, A., Yu, X.J., Liu, M. and Holden, D.W. (2005) *Infect Immun*, **73**, 4354–4362.

149 Tallant, T., Deb, A., Kar, N., Lupica, J. *et al.* (2004) *BMC Microbiol*, **4**, 33.

150 Hautefort, I., Thompson, A., Eriksson-Ygberg, S., Parker, M.L. *et al.* (2008) *Cell Microbiol*, **10**, 958–984.

151 Steiner, T.S. (2007) *Infect Immun*, **75**, 545–552.

152 Gewirtz, A.T., Simon, P.O. Jr, Schmitt, C.K., Taylor, L.J. *et al.* (2001) *J Clin Invest*, **107**, 99–109.

153 Simon, R. and Samuel, C.E. (2007) *Biochem Biophys Res Commun*, **355**, 280–285.

154 Bergman, M.A., Cummings, L.A., Alaniz, R.C., Mayeda, L. *et al.* (2005) *Infect Immun*, **73**, 7226–7235.

155 Pickard, D., Wain, J., Baker, S., Line, A. *et al.* (2003) *J Bacteriol*, **185**, 5055–5065.

156 Raffatellu, M., Chessa, D., Wilson, R.P., Dusold, R. *et al.* (2005) *Infect Immun*, **73**, 3367–3374.

157 Sharma, A. and Qadri, A. (2004) *Proc Natl Acad Sci USA*, **101**, 17492–17497.

158 Wilson, R.P., Raffatellu, M., Chessa, D., Winter, S.E. *et al.* (2008) *Cell Microbiol*, **10**, 876–890.

159 Musson, J.A., Hayward, R.D., Delvig, A.A., Hormaeche, C.E. *et al.* (2002) *Eur J Immunol*, **32**, 2664–2671.

160 Le Negrate, G., Faustin, B., Welsh, K., Loeffler, M. *et al.* (2008) *J Immunol*, **180**, 5045–5056.

161 Mazurkiewicz, P., Thomas, J., Thompson, J.A., Liu, M. *et al.* (2008) *Mol Microbiol*, **67**, 1371–1383.

162 Link, C., Ebensen, T., Standner, L., Dejosez, M. *et al.* (2006) *Microbes Infect*, **8**, 2262–2269.

163 Gordon, M.A. (2008) *J Infect.*, **56**, 413–422.

164 Fica, A.E., Horowitz, H.W., Lior, H. and Cabello, F.C. (1994) *J Clin Microbiol*, **32**, 2327–2330.

165 Layton, A.N. and Galyov, E.E. (2007) *Expert Rev Mol Med*, **9**, 1–17.

166 Monack, D.M., Mueller, A. and Falkow, S. (2004) *Nat Rev Microbiol*, **2**, 747–765.

167 Lawley, T.D., Chan, K., Thompson, L.J., Kim, C.C. *et al.* (2006) *PLoS Pathog*, **2**, e11.

168 Santos, R.L., Zhang, S., Tsolis, R.M., Kingsley, R.A. *et al.* (2001) *Microbes Infect*, **3**, 1335–1344.

169 Pejcic-Karapetrovic, B., Gurnani, K., Russell, M.S., Finlay, B.B. *et al.* (2007) *J Immunol*, **179**, 6088–6096.

170 Giles, N., Hopper, S.A. and Wray, C. (1989) *Epidemiol Infect*, **103**, 235–241.

171 Pullinger, G.D., Paulin, S.M., Charleston, B., Watson, P.R. *et al.* (2007) *Infect Immun*, **75**, 5191–5199.

172 Morgan, E., Campbell, J.D., Rowe, S.C., Bispham, J. *et al.* (2004) *Mol Microbiol*, **54**, 994–1010.

173 Cooper, G.L., Venables, L.M., Woodward, M.J. and Hormaeche, C.E. (1994) *Infect Immun*, **62**, 4739–4746.

174 Huang, D.B. and DuPont, H.L. (2005) *Lancet Infect Dis*, **5**, 341–348.

175 Paulin, S.M., Watson, P.R., Benmore, A.R., Stevens, M.P. *et al.* (2002) *Infect Immun*, **70**, 6788–6797.

176 Segall, T. and Lindberg, A.A. (1991) *Zentralbl Veterinarmed B*, **38**, 169–185.

177 Wallis, T.S., Paulin, S.M., Plested, J.S., Watson, P.R. and Jones, P.W. (1995) *Infect Immun*, **63**, 2755–2761.

178 Gerlach, R.G. and Hensel, M. (2007) *Berl Münch Tierärztl Wochenschr*, **120**, 317–327.

179 Hu, Q., Coburn, B., Deng, W., Li, Y. *et al.* (2008) *J Clin Microbiol*, **46**, 1330–1336.

180 Knodler, L.A., Celli, J., Hardt, W.D., Vallance, B.A. *et al.* (2002) *Mol Microbiol*, **43**, 1089–1103.

181 Friebel, A. and Hardt, W.D. (2000) *Methods Enzymol*, **325**, 82–91.

182 Rudolph, M.G., Weise, C., Mirold, S., Hillenbrand, B. *et al.* (1999) *J Biol Chem*, **274**, 30501–30509.

183 Mukherjee, K., Parashuraman, S., Raje, M. and Mukhopadhyay, A. (2001) *J Biol Chem*, **276**, 23607–23615.

184 Hardt, W.D., Chen, L.M., Schuebel, K.E., Bustelo, X.R. and Galan, J.E. (1998) *Cell*, **93**, 815–826.

185 Bakowski, M.A., Cirulis, J.T., Brown, N.F., Finlay, B.B. and Brumell, J.H. (2007) *Cell Microbiol.*, **9**, 2839–2855.

186 Terebiznik, M.R., Vieira, O.V., Marcus, S.L., Slade, A. *et al.* (2002) *Nat Cell Biol*, **4**, 766–773.

187 Norris, F.A., Wilson, M.P., Wallis, T.S., Galyov, E.E. and Majerus, P.W. (1998) *Proc Natl Acad Sci USA*, **95**, 14057–14059.

188 Schlumberger, M.C., Kappeli, R., Wetter, M., Muller, A.J. *et al.* (2007) *Mol Microbiol*, **65**, 741–760.

189 Cain, R.J., Hayward, R.D. and Koronakis, V. (2008) *PLoS Pathog*, **4**, e1000037.

190 Fu, Y. and Galan, J.E. (1999) *Nature*, **401**, 293–297.

191 Raffatellu, M., Wilson, R.P., Chessa, D., Andrews-Polymenis, H. *et al.* (2005) *Infect Immun*, **73**, 146–154.

192 Layton, A.N., Brown, P.J. and Galyov, E.E. (2005) *J Bacteriol*, **187**, 3565–3571.

193 Zhang, Y., Higashide, W.M., McCormick, B.A., Chen, J. and Zhou, D. (2006) *Mol Microbiol*, **62**, 786–793.

194 Shotland, Y., Kramer, H. and Groisman, E.A. (2003) *Mol Microbiol*, **49**, 1565–1576.

195 Freeman, J.A., Rappl, C., Kuhle, V., Hensel, M. and Miller, S.I. (2002) *J Bacteriol*, **184**, 4971–4980.

196 Uchiya, K., Barbieri, M.A., Funato, K., Shah, A.H. *et al.* (1999) *EMBO J*, **18**, 3924–3933.

197 Yu, X.J., Liu, M. and Holden, D.W. (2004) *Mol Microbiol*, **54**, 604–619.

198 Yu, X.J., Ruiz-Albert, J., Unsworth, K.E., Garvis, S. *et al.* (2002) *Cell Microbiol*, **4**, 531–540.

199 Guy, R.L., Gonias, L.A. and Stein, M.A. (2000) *Mol Microbiol*, **37**, 1417–1435.

200 Kuhle, V., Jackel, D. and Hensel, M. (2004) *Traffic*, **5**, 356–370.

201 Brown, N.F., Szeto, J., Jiang, X., Coombes, B.K. *et al.* (2006) *Microbiology*, **152**, 2323–2343.

202 Dumont, A., Schroeder, N., Gorvel, J.P. and Meresse, S. (2007) *Methods Mol Biol*, **394**, 275–287.

203 Brumell, J.H., Tang, P., Zaharik, M.L. and Finlay, B.B. (2002) *Infect Immun*, **70**, 3264–3270.

204 Brumell, J.H., Tang, P., Mills, S.D. and Finlay, B.B. (2001) *Traffic*, **2**, 643–653.

205 Rohde, J.R., Breitkreutz, A., Chenal, A., Sansonetti, P.J. and Parsot, C. (2007) *Cell Host Microbe*, **1**, 77–83.

206 Miao, E.A., Brittnacher, M., Haraga, A., Jeng, R.L. *et al.* (2003) *Mol Microbiol*, **48**, 401–415.

IIB
Intracytosolic Bacteria

Intracellular Niches of Microbes. Edited by Ulrich E. Schaible and Albert Haas

ISBN: 978-3-527-32207-7

22
Burkholderia pseudomallei

Joanne M. Stevens and Mark P. Stevens

22.1
Burkholderia pseudomallei and Melioidosis

Burkholderia pseudomallei is a Gram-negative facultative intracellular pathogen that causes melioidosis, a severe invasive disease of animals and humans that is endemic in Southeast Asia and northern Australia. Clinical presentation of melioidosis varies from rapidly fatal septicemia and acute pneumonia, through chronic or localized abscess formation to subacute infections that may progress to symptomatic disease depending on the condition of the host (reviewed in [1]). Melioidosis accounts for approximately 20% of community-acquired septicemias in northeastern Thailand, with a mortality rate of about 40% in treated patients. The organism exhibits a striking capacity for latency and relapse is common even in patients treated with appropriate antibiotics. Cutaneous and pulmonary melioidosis has been reported in veterans of Southeast Asian conflicts decades after geographical exposure. Recurring infections do not always involve the same genotypes, indicating that reinfection may occur or that multiple genotypes may form quiescent foci in the host [2].

B. pseudomallei is a saprophyte and transmission is believed to involve contact with contaminated soil or water, especially by inhalation, aspiration and via abraded skin. Risk factors include diabetes mellitus, renal dysfunction, cirrhosis and suppression of immune function as a consequence of disease or drug treatment. Clinical management of melioidosis is reviewed elsewhere [1], and is hindered by the intrinsic resistance of *B. pseudomallei* to penicillin, aminoglycosides, rifamycins, colistin, macrolides and first- and second-generation cephalosporins. Such resistance is conferred by multiple β-lactamases, efflux systems and the ability of the organism to persist in an intracellular niche. Concern exists that *B. pseudomallei* may be used as a biological weapon owing to its aerosol infectivity, severe course of infection and the difficulty of diagnosis and treatment [3].

Intracellular Niches of Microbes. Edited by Ulrich E. Schaible and Albert Haas

ISBN: 978-3-527-32207-7

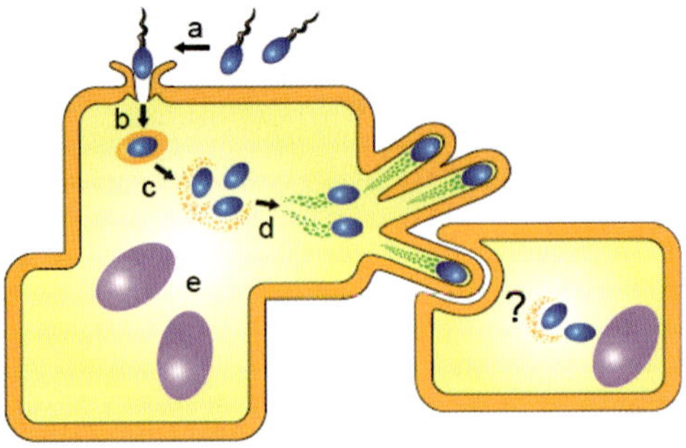

Figure 22.1 Host cell interactions of *Burkholderia pseudomallei* (adapted from [44]). (a) Contact of *B. pseudomallei* with phagocytic and nonphagocytic cells is modulated by bacterial factors, including type IV pili that promote interbacterial interactions and microcolony formation and surface polysaccharides that limit phagocytosis and activation of innate immunity. (b) Invasion involves reorganization of the subcortical actin cytoskeleton and partly requires the Bsa T3SS. (c) Once in endosomes, *B. pseudomallei* resists antimicrobial systems and is able to escape into the cytosol by destruction of the endosomal membrane. (d) In the cytosol *B. pseudomallei* induces the polymerization of actin at a single pole, leading to intracellular motility and the formation of actin-rich membrane protrusions that may project into adjacent cells. Such protrusions may be engulfed and lysed by paracytophagy, a cellular process that may be facilitated by bacterial factors. Bacteria may also spread by cell fusion (e), leading to the formation of multinucleated giant cells.

22.2 An Overview of the Intracellular Life of *B. pseudomallei*

In common with other facultative intracellular pathogens, *B. pseudomallei* is able to promote its uptake by nonphagocytic cells (Figure 22.1a; [4–6]) and to survive and proliferate in phagocytes [4, 7, 8]. *B. pseudomallei* is also able to enter and persist within free-living amoebae of the genus *Acanthamoeba* and arbuscular mycorrhizal fungi of the genus *Gigaspora* and it has been proposed that such interactions may aid survival in the environment, among other factors (reviewed in [9]). Amoebic passage enhances monocyte entry, intracellular replication and virulence of *Legionella pneumophila* [10], and it remains to be determined if such interactions influence the infectivity of *B. pseudomallei*. Following internalization of nonphagocytic or phagocytic cells (Figure 22.1b), *B. pseudomallei* is able to escape from entry vesicles by destruction of the endosomal membrane (Figure 22.1c; [5, 11]). Escape may occur as early as 15 min after uptake [5] with the majority of bacteria being free in the cytosol at 3 h post inoculation of murine macrophage-like cells in the absence of costaining for lysosome-associated membrane glycoprotein 1 (LAMP1) [11] or the lysomotropic agent thorium chloride [5]. In the cytoplasm, the bacteria may then be propelled by continuous polymerization of actin at a single pole, leading to the formation of membrane-bound protrusions tipped with bacteria (Figure 22.1d; [6, 11, 12]). As with other intracytoplasmic pathogens capable of actin-based motility, it is believed that this may facilitate cell-to-cell spread in the absence of immune surveillance (reviewed in [13]). Uniquely among bacterial pathogens described to date, *B. pseudomallei* may also induce the fusion of adjacent cells, leading to the formation of multinucleated giant cells (MNGCs) in infected monolayers (Figure 22.1e; [6, 14]) by a process that requires endosome escape [15]. Foci of *B. pseudomallei*-containing fused cells have been observed in human melioidosis patients [16], however it remains unclear if MNGCs reflect a host response to limit the spread of infection or a programed strategy for bacterial dissemination mediated by specific virulence factors.

22.3
Genome Structure and Diversity of *B. pseudomallei*

Sequencing of the complete genome of *B. pseudomallei* strain K96243 revealed the existence of two chromosomes of 4.07 and 3.17 Mbp, with significant partitioning of core and accessory functions between them [17]. Sixteen genomic islands (GIs) comprising 6.1% of the K96243 genome were identified on the basis of aberrant GC content and other features associated with mobile elements [17]. Such islands were variably conserved in a panel of 40 *B. pseudomallei* isolates [17] and this diversity is reinforced by microarray analysis of the gene content of K96243 relative to the monkey isolate Bp15682 [18] and by analysis of the conservation of *B. pseudomallei*-specific sequences identified by subtractive hybridization [19]. A further level of complexity exists at the level of transcription and translation of shared genes, as only 43% of strain-specific differences in mRNA abundance could be correlated with gene prevalence and about 20% of proteins resolved under specific conditions varied between the two strains, often as a consequence of the presence of isoforms [18].

Plasticity of the *B. pseudomallei* genome is also evident from comparisons with *B. mallei*, the etiological agent of glanders. Glanders shares clinical features of melioidosis and multi-locus sequencing typing (MLST; [20]), comparison of complete genome sequences [17, 21], and microarray analysis of gene content and transcription [22, 23], indicate that *B. mallei* is a clone of *B. pseudomallei* that has undergone gene decay. Such mutational attrition is associated with insertion sequences and recombination between simple sequence repeats and likely dictates why *B. pseudomallei* has a broad host range whereas *B. mallei* is restricted to humans and solipeds. Diversity also exists among environmental isolates of *B. pseudomallei*, which were previously separated into two groups on the basis of their ability to assimilate arabinose, ribotyping and restriction and amplified-fragment profiles. Those which assimilate arabinose (ara^+) were later assigned to species *B. thailandensis*, a distinction supported by MLST [20]. Those which do not assimilate arabinose are associated with melioidosis and are retained in the species *B. pseudomallei*.

Though *B. thailandensis* is capable of invasion, endosome escape, actin-based motility and cell fusion [6], it exhibits a substantially elevated median lethal dose (LD_{50}) relative to *B. pseudomallei* in rodent models of melioidosis [24]. As such, significant effort has been invested to identify differences in the repertoire, sequence and expression of genes that may explain the differential virulence of the two species. Sequencing of the genome of *B. thailandensis* strain E264 [22] and an independently archived version of the same strain [25] has revealed the presence of smaller chromosomes (3.81 and 2.91 Mbp) retaining a high level of synteny and nucleotide identity to those of K96243, but divergence among GIs and loci-encoding fimbriae, surface polysaccharides and type III secretion systems. In addition, though many virulence-associated loci are retained in *B. thailandensis* they exhibit increased sequence diversity [25]. Genomic differences between *B. pseudomallei* and *B. thailandensis* are supported by microarray analysis [22, 23], subtractive hybridization [26, 27] and analysis of the conservation of type III secretion gene clusters [28]. However, differences also exist between *B. pseudomallei* and *B. thailandensis* at the level of the transcriptome [29] and proteome [30] that cannot be attributed to gene prevalence. Variability in both genome content and transcription has also been described following *in vivo* passage of *B. mallei* [31] and further studies are needed to understand the dynamics of genome evolution in *Burkholderia* and its relationship to pathogenicity and long-term persistence in the host.

The advent of complete genome sequences for *B. pseudomallei* and related organisms has primed the application of targeted and genome-wide mutagenesis and other approaches to assign functions to the encoded genes. Insights from such studies are discussed in relation to key stages in the intracellular life of *B. pseudomallei* in the following sections.

22.4
Cell Contact and Adherence

B. pseudomallei is able to adhere to human primary pharyngeal epithelial cells [32] and various immortalized epithelial cell lines (Figure 22.2a; [33, 34]). Adherence of *B. pseudomallei* to A549 human respiratory epithelium cells occurs at significantly greater levels than for *B. thailandensis* [33], and is regulated by temperature, growth phase, media composition and requires bacterial protein synthesis [34]. Adherence is also sensitive to exogenous asialogangliosides GM1 and GM2 and bacteria attach to these tissue glycolipids in solid-phase assays [35], indicating that they may act as receptors. Bacterial appendages associated with cell contact such as fimbriae have yet to be demonstrated by electron microscopy [32, 34], however sequencing of the *B. pseudomallei* K96243 genome identified 13 putative fimbrial loci [17], some of which are absent in *B. mallei* ATCC23344 [21] and *B. thailandensis* E264 [25]. Eight type IV pilus-associated clusters were identified in *B. pseudomallei* K96243, five of which encode one or more predicted pilin subunits. Inactivation of one predicted pilin (BPSL0782, PilA) significantly impaired adherence to human respiratory epithelial cell lines and reduced virulence in nematodes and murine models [36]. However, type IV pili often mediate interbacterial interactions and the role of PilA in temperature-sensitive microcolony formation can largely be explained by its role in bacterial aggregation as opposed to cell contact *per se* [37].

Flagella have been observed to tether *B. pseudomallei* to the surface of *Acanthamoeba* trophozoites and to be required for endocytosis [38], however no defect in invasion or replication of a flagellin (FliC) mutant in A549 cells was found to exist [39].

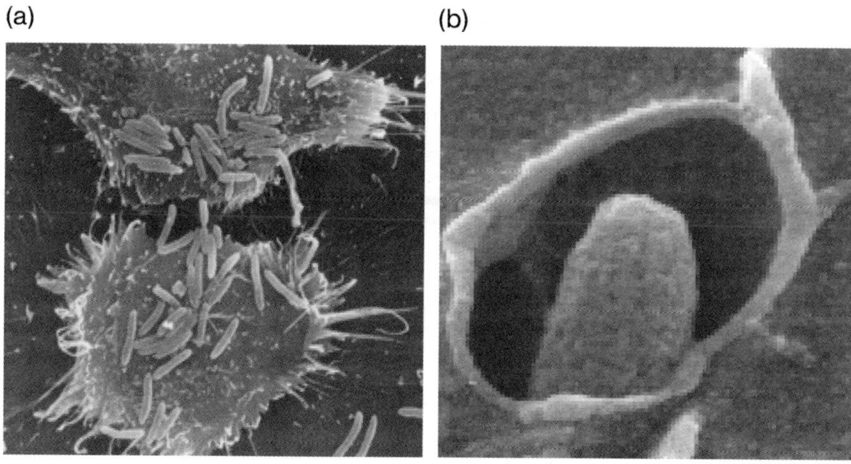

Figure 22.2 (a) Scanning electron micrograph of *Burkholderia pseudomallei* strain 08 adhering to Chang epithelial cells showing both interbacterial interactions and attachment to host cells. From [34]. (b) Scanning electron micrograph of *B. pseudomallei* strain E8 entering a HeLa cell. From [12].

In some bacteria flagellin constituents bind to mucins and it is possible that cultured cell lines fail to reveal roles for flagella in adherence owing to an absence of mucin production. *B. pseudomallei* flagellin has been reported to influence virulence in BALB/c mice [39], but not virulence in *C. elegans* [39], Syrian hamsters or diabetic rats [40]. A requirement exists to separate the role of flagella as adhesins *per se* from their role in motility by disruption of flagella rotation.

Though not considered to act as adhesins, surface-associated polysaccharides modulate the interaction between *B. pseudomallei* and host cells. The -3)-2-*O*-acetyl-6-deoxy-β-D-manno-heptopyranose-(1- capsule (also known as type I O-PS) inhibits C3b deposition and thus opsonophagocytosis [41], consistent with a pivotal role in virulence [26, 42, 43]. Genes required for capsule production are absent in *B. thailandensis* [22, 23, 25, 26]. *B. pseudomallei* also produces an unusual lipopolysaccharide (LPS; type II O-PS), which exhibits weaker pyrogenic activity than enterobacterial LPS and is a relatively poor activator of murine macrophage-like cells and interleukin 8 (IL-8) synthesis (reviewed in [44]). *B. pseudomallei* partially escapes macrophage killing by interfering with the expression of inducible nitric oxide synthase (iNOS) [45], and recent studies indicate that the LPS O-antigen helps prevent the production of phosphorylated Y701-STAT-1 and interferon regulatory factor 1, both of which are essential for iNOS transcription [46]. An O-antigen mutant was internalized by RAW267.4 cells significantly more efficiently than the parent strain, but replicated intracellularly at a lower level, indicating that *B. pseudomallei* may use its LPS to create an intracellular environment favorable to the bacteria [46]. Consistent with such a role, LPS mutations impair virulence in rodent models [47].

Two other polysaccharide biosynthesis gene clusters exist in *B. pseudomallei* (types III and IV O-PS) and mutation of either locus increases the median time to death in mice, though only significantly in the case of type IV O-PS [48]. The mechanism by which exopolysaccharides predicted to be encoded by these loci contribute to virulence remains obscure. Surface-associated lipids may also influence the outcome of host cell contact as purified *B. pseudomallei* rhamnolipid impairs the phagocytic activity of macrophages, likely as a consequence of reorganization of the F-actin network and cell cycle arrest [49]. Further studies with defined mutant strains are required to evaluate the relevance of rhamnolipid-mediated effects on host cells during infection.

22.5 Invasion

B. pseudomallei is able to enter a range of epithelial and fibroblast lines *in vitro* [4–6]. Scanning electron microscopy of *B. pseudomallei*-infected HeLa cells detected membrane invaginations at the bacteria–cell interface similar to those seen during *Streptococcus pyogenes* entry (Figure 22.2b; [12]), however it remains unclear if bacterial endocytosis is preceded by "ruffling" of the membrane as described for other invasive pathogens. Such events are driven by reorganization of the subcortical actin cytoskeleton, which in the case of the enteric pathogens *Salmonella* and *Shigella* is orchestrated by bacterial proteins delivered into host cells via type III secretion systems (T3SSs). Sequencing of the *B. pseudomallei* K96243 genome identified a cluster of genes similar in order and sequence to loci in *S. enterica* serovar Typhimurium (*inv/spa/prg*) and *S. flexneri* (*ipa/mxi/spa*) that encode T3SSs [11, 28]. The function of T3SSs may be likened to that of a molecular syringe as they serve to inject bacterial proteins (effectors) into host cells which then subvert cellular processes (reviewed in [50]). In the case of *Salmonella* a second class of type III secreted proteins (Sip proteins) facilitate translocation of the effectors SopB, SopE and SopE2 which act in concert to remodel subcortical actin. SopB dephosphorylates a key lipid regulator of the actin cytoskeleton (phosphatidylinositol 4,5-bisphosphate) whereas SopE and SopE2 act as guanine nucleotide exchange factors (GEFs) for Rho-family GTPases that regulate actin assembly (reviewed in [51]).

The Inv/Mxi-Spa-like T3SS in *B. pseudomallei* (designated *bsa*, *Burkholderia* secretion apparatus or TTS-3; BPSS1543-1552) encodes homologs of *Salmonella* Sip proteins (BipB, -C and -D) and a homolog of SopE/SopE2 [11]. BopE is secreted in a manner dependent on a predicted structural component of the T3SS (BsaZ) and both BipD and BopE promote invasion of HeLa cells by *B. pseudomallei* [52]. Ectopic expression of BopE stimulates rearrangement of subcortical actin, likely as a result of activation of Cdc42 and Rac1 as purified BopE acts a GEF for these GTPases *in vitro* [52]. Mutation of *bipD* impaired invasion to a greater extent than mutation of *bopE*, implying that BopE may act in concert with other Bsa-secreted effectors to force phagocytosis [52]. Consistent with the key role of T3SSs in the virulence of Gram-negative pathogens, components of the Bsa system are transcribed during acute melioidosis in hamsters [53], recognized by convalescent sera from human melioidosis patients [11] and $CD4^+$ T cells following infection of mice [54], and mutation of structural proteins is highly attenuating in rodent models [53, 55, 56]. Despite this, subunit vaccination with the translocator protein BipD did not confer significant protection in a murine model [55]. Inactivation of *bopE* did not cause a significant delay in median time to death in BALB/c mice [55] or Syrian hamsters [56], indicating that its role in forced phagocytosis may be dispensable *in vivo*, at least upon parenteral infection of highly susceptible hosts. By analogy with other T3SSs it is likely that the Bsa apparatus translocates numerous effectors into host cells and further studies are required to map the repertoire, function and subcellular target of these molecules.

Consistent with the role of actin remodeling in invasion, treatment of cells with an inhibitor of actin polymerization (cytochalasin D) inhibits uptake of *B. pseudomallei* [57, 58]. Invasion can also be reduced by a specific inhibitor of p38 mitogen-activated protein kinase, but not an inhibitor of phosphatidylinositol 3-kinase [57]. Phosphorylation of p38 can be detected early after infection [57], however activated p38 regulates a plethora of cellular processes and further studies are required to dissect the events downstream of p38 that are required for uptake and the molecular mechanism by which *B. pseudomallei* triggers the signal transduction cascade.

The *bsa* locus is conserved in *B. mallei* and *B. thailandensis* and is required for full virulence in experimental models of glanders. Though the *bsa* locus is present in ara$^+$ *B. thailandensis* the genes are repressed relative to their counterparts in *B. pseudomallei* in a manner dependent on the arabinose assimilation operon [29]. Introduction of the *B. thailandensis* E264 *araA-I* operon into *B. pseudomallei* 406e repressed transcription of numerous *bsa* genes and increased the lethal dose for Syrian hamsters [29]. These findings reinforce the role that gene decay may play in the evolution of virulence among *Burkholderia* species and may partly explain why *B. pseudomallei* strains invade cultured cells at higher efficiencies than ara$^+$ *B. thailandensis* [33]. Invasion efficiency may also be influenced by divergence in the sequence of *bsa*-encoded proteins, as comparison of the complete genome sequences of *B. pseudomallei* and *B. thailandensis* has indicated that BPSS1547 (*prgH*), BPSS1549 (*mxiI/prgJ*), BPSS1551 and BPSS1552 (*orgA* and *B*), exhibit the highest level of divergence among the virulence-associated genes [25]. In other species, PrgH and PrgJ orthologs form part of the needle complex that creates a physical interface between the pathogen and host cell.

Two other T3SS-associated gene clusters exist in *B. pseudomallei* (TTS-1, BPSS1390-1408; TTS-2, BPSS1613-1629) which exhibit homology with the *hrp* locus of the plant pathogen *Ralstonia solanacearum* [28, 59]. TTS-1 is absent from *B. mallei* and *B. thailandensis* yet ubiquitous among isolates of *B. pseudomallei*, providing the basis of a diagnostic test (reviewed in [1]). Mutation of predicted structural components of TTS-1 and -2, singly or in combination, did not significantly impair the virulence of *B. pseudomallei* in Syrian hamsters [56], however the intriguing possibility remains that they may mediate symbiotic or pathogenic interactions with plants.

Few other bacterial factors that specifically influence invasion have been identified. A screen of random Tn5-OT182 mutants of *B. pseudomallei* for defects in entry into A549 cells identified a putative two-component sensory system (IrlRS) with homology to those regulating heavy-metal resistance in other organisms [60]. IrlS is predicted to be a sensor kinase that autophosphorylates on receipt of a signal and transfers the phosphate moiety to IrlR, leading to altered expression of genes under its direct or indirect control. Invasion of A459 cells was reduced to about 10% of wild-type levels by an insertion in *irlR* and restored by plasmid-mediated *trans*-complementation, however the insertion did not impair intracellular survival or resistance to human defensin HNP-1 [60]. The genes under the control of IrlRS are unknown, but the locus is not essential for virulence in virulence in hamsters and diabetic rats [60] or the induction of pulmonary proinflammatory responses or lethality in mice [61]. These experiments indicate that while invasion is a key step in the intracellular life of *B. pseudomallei*, it can be markedly reduced without altering inflammatory responses and pathology in the murine pulmonary melioidosis model [61], consistent with the finding that proinflammatory responses can be triggered by cell contact in the absence of internalization [57].

22.6 Endosome Escape

Escape of *B. pseudomallei* from entry vesicles occurs before fusion with the late endocytic pathway in cultured cells, though evidence of phagosome–lysosome fusion can be seen in monocytes isolated from the blood of human melioidosis patients and healthy volunteers [62]. In cultured cells, endosome escape requires the Bsa T3SS, likely as a consequence of the pore-forming activity of translocon components as described for *Shigella* [11]. Inactivation of *bsaZ* or *bipD* confines *B. pseudomallei* to vesicles as determined by transmission electron microscopy and mutant bacteria almost exclusively colocalize with LAMP1 (Figure 22.3; [11]). Confinement of *bsa* mutants to phagosomes results in prolonged exposure to reactive oxygen and nitrogen intermediates, cationic peptides and other antimicrobial systems and may explain why such mutants exhibit impaired intracellular growth [11]. Early studies recorded an absence of endosome escape and thus actin tail formation by *bsa* mutants at up to 6 h post infection [11], however a transposon mutant lacking the hypothetical *bsa*-encoded protein BPSS1539 was reported to proliferate in LAMP1-associated vacuoles in PtK2 cells and escape to form actin tails at later times [63], indicating that other factors may be involved in destruction of the endosomal membrane.

In the case of *Listeria monocytogenes*, phospholipase C enzymes influence endosome escape in human cells, evasion of autophagy, cell-to-cell spread via lysis of membrane protrusions and membrane fusion *in vitro* (reviewed in Chapter 24). The *B. pseudomallei* genome encodes several putative phospholipases [17], including Plc-1 (BPSL2403) and Plc-2 (BPSL0338), which have been proven to hydrolyze phosphatidylcholine and sphingomyelin but lack hemolytic activity [64]. Neither Plc-1 nor -2 was required for *B. pseudomallei*-induced MNGC formation, an event presumed to occur downstream of endosome escape, however they act in concert to promote plaque formation [64]. Transcription of two further putative phospholipases (BPSS0067 and BPSS1876) was observed to be highly upregulated during acute

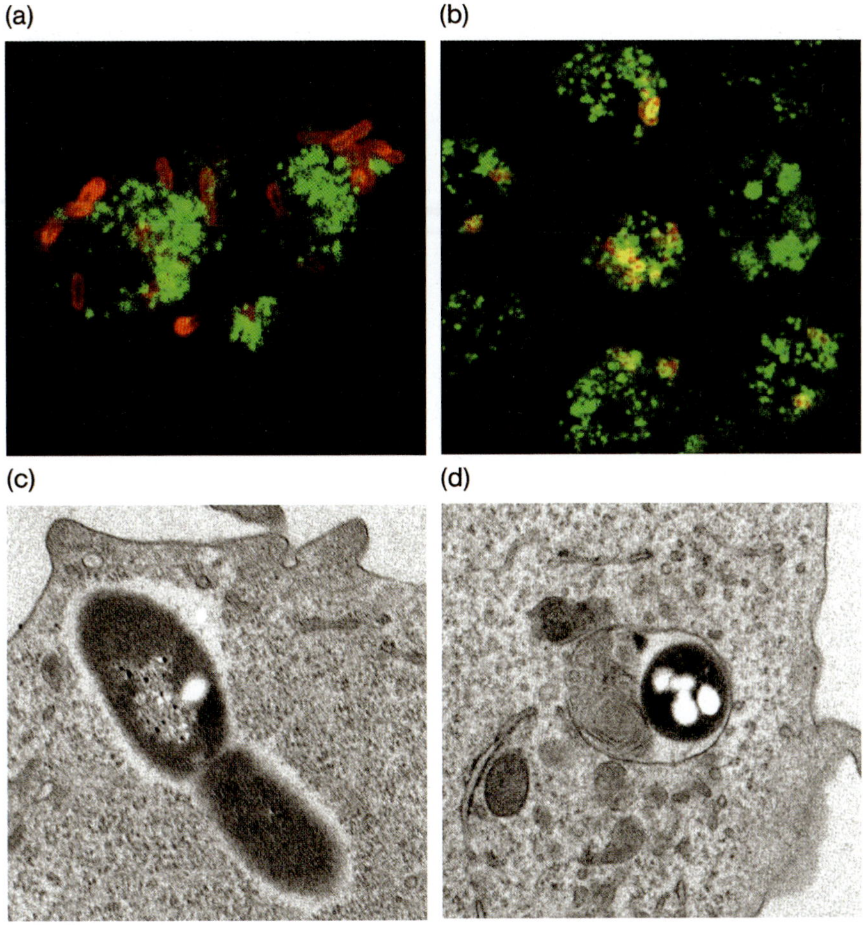

Figure 22.3 Confocal micrographs showing the location of wild-type (a) and *bsaZ* mutant (b) *Burkholderia pseudomallei* (red) relative to LAMP1-containing vesicles 6 h post infection of J774.2 murine macrophage-like cells. The absence of colocalization of wild-type *B. pseudomallei* with LAMP1 indicates that it escapes from entry vesicles prior to fusion with the late endocytic pathway in a manner that requires the Bsa apparatus. Transmission electron microscopy of J774.2 cells 3 h post infection with the wild-type (c) or *bsaZ* mutant (d) confirms the requirement for the Bsa T3SS for destruction of the endosomal membrane. From [11].

melioidosis [53]. Further studies are required to evaluate the role of *B. pseudomallei* phospholipases in lysis of the membranes of primary and spreading vacuoles.

22.7 Actin-Based Motility

In common with selected species of *Listeria*, *Shigella*, *Rickettsia* and *Mycobacterium* (Chapters 19, 24–27, respectively), *B. pseudomallei* shares the ability to induce

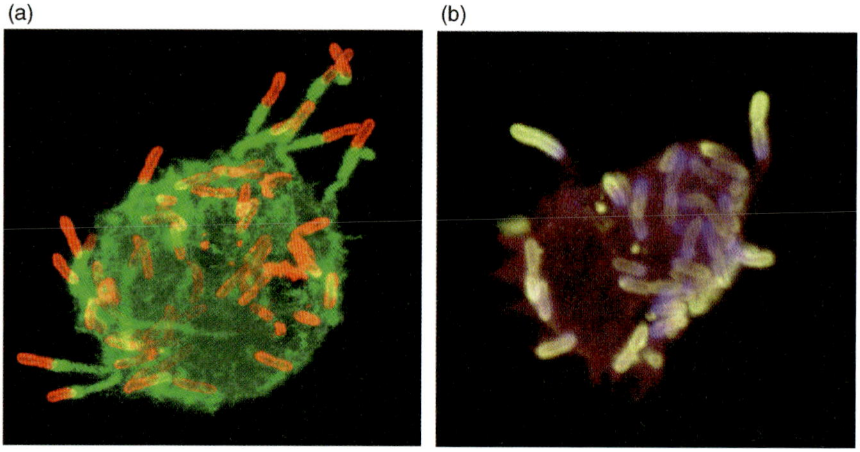

Figure 22.4 Confocal micrographs showing *Burkholderia pseudomallei*-induced actin tails 6 h post infection of J774.2 cells. (a) Bacterial lipopolysaccharide (LPS) is stained red and polymerized (F)-actin is labeled green with fluorophore-conjugated phalloidin. (b) Bacterial LPS is labeled green, F-actin is labeled red and the BimA protein required for actin-based motility of *B. pseudomallei* is labeled blue [66].

polymerization of actin at one pole of the bacterial cell to promote its movement within and between host cells (Figure 22.4a). The formation of actin-rich bacteria-containing membrane protrusions was first described in *B. pseudomallei*-infected HeLa and J774.1 cells [6], but is also a feature of *B. mallei* and *B. thailandensis* infection [65]. Mechanisms of bacterial actin-based motility converge on activation of the cellular Arp (actin-related protein) 2/3 complex. Activation of the Arp2/3 complex requires cellular nucleation-promoting factors (NPFs) such as Wiskott–Aldrich syndrome protein (WASP) family members and pathogens capable of actin-based motility often mimic the activity of NPFs or recruit and activate them at the bacterial pole (reviewed in [13]). By using green fluorescent protein fusions and immunocytochemistry, Arp2/3 components were found to be localized throughout *B. pseudomallei*-induced actin tails [12]. However, the role of the Arp2/3 complex in intracellular motility of *B. pseudomallei* is unclear as overexpression of a dominant negative domain of the cellular NPF Scar1, which inhibits actin tail formation by *S. flexneri*, *L. monocytogenes* and *R. conorii*, does not interfere with actin-based motility of *B. pseudomallei* [12]. Further, in contrast to *Shigella*, N-WASP is not recruited to the surface of intracellular *B. pseudomallei* and normal actin-based motility occurs in N-WASP$^{-/-}$ fibroblasts [12]. Fibroblasts lacking vasodilator-stimulated phosphoprotein (VASP) also support actin-based motility of *B. pseudomallei* and VASP is not recruited to the tails, in contrast to those formed by *Listeria* and *Shigella* [12]. Taken together, these observations indicate that actin-based motility of *B. pseudomallei* occurs by a mechanism distinct to that used by other intracytoplasmic pathogens.

The bacterial factor(s) required for actin-based motility of *B. pseudomallei* have been sought by screening random mutants for a small plaque phenotype [63] and by targeted inactivation of candidate factors. On the basis that *Shigella* uses a Type V secreted protein for actin-based motility (IcsA), the translated genome of *B. pseudomallei* was searched for predicted Type V secreted proteins. This search identified a protein with C-terminal homology to the oligomeric *Yersinia* adhesin YadA, as well as domains associated with actin-binding, including proline-rich motifs and WASP homology-2 (WH2) domains [66]. Actin-based motility of *B. pseudomallei* in J774.2 cells was abolished by mutation of this factor (BimA; BPSS1492) and could be restored by transient expression of the gene *in trans* [66]. BimA is located at the pole of the bacterial cell at which actin polymerization occurs and polar localization is frequently seen in only one daughter cell at division (Figure 22.4b), implying that polar targeting of BimA may rely on distinguishing the new and old poles of the cell [66]. Transient expression of BimA in HeLa cells results in F-actin clustering reminiscent of that seen upon overexpression of cellular NPFs and antibody-mediated clustering at the plasma membrane of a CD32 chimera in which the cytoplasmic domain was replaced with BimA results in localization of the chimaera to the tips of F-actin-enriched membrane protrusions [66]. A purified but truncated BimA protein was reported to bind monomeric actin in a concentration-dependent manner and to weakly stimulate actin polymerization *in vitro* in an Arp2/3-independent manner [66]. These data imply that BimA alone may be sufficient for intracellular motility of *B. pseudomallei*, however a number of post-translational modifications of BimA are possible and it is unclear if these, or bacterial cofactors, are required for full activity.

BimA orthologs exist in *B. mallei* (BMAA0749) and *B. thailandensis* (unassembled genome at www.tigr.org) and can restore actin-based motility of a *B. pseudomallei bimA* mutant [65]. The N-termini of the BimA proteins of the three species differ in the number of WH2 domains, sequence of proline-rich motifs and the presence of a central and acidic (CA) domain, indicating that they may initiate actin assembly by distinct mechanisms [65]. CA domains are required for the ability of WASP family proteins to stimulate the Arp2/3 complex and consistent with the presence of a unique CA domain, only *B. thailandensis* BimA sequesters Arp3 from cell extracts *in vitro* [65].

In *B. mallei*, actin-based motility requires the VirAG two-component sensory system which activates transcription of *bimA*, as well as three linked genes (*bimBCE*) [67]. In *B. mallei bimA* is predicted to be encoded in an operon (*bimBCAD*) next to an operon transcribed in the opposite direction comprising *virAG-bimE*. Mutations in *bimB*, -*C* or -*E* abolished actin tail formation by *B. mallei* [67], raising the possibility that actin-based motility may be multifactorial; however potential exists for polar or second-site effects. Interestingly, whilst *virA* and *virG* mutants are highly attenuated following intraperitoneal infection of Syrian hamsters, mutants harboring defects in *bimA*, -*B*, -*C* and -*E* exhibited comparable virulence to the parent strain [67]. In *B. pseudomallei*, inactivation of a putative exported protein (BPSL1528) results in the formation of shorter actin tails and an absence of membrane protrusions and plaque formation in PtK2 monolayers [63]. The mutant was not impaired in invasion but exhibited reduced intracellular replication in HeLa cells and survival in J774.1 macrophages and was highly attenuated in BALB/c mice [63]. Further studies are required to dissect the role of BPSL1528 and genes encoded proximal to BimA in actin-based motility of *B. pseudomallei* and to unravel the role of domains, predicted posttranslational modifications and protein–protein interactions of BimA in this process. In addition, it remains unclear if double-membrane protrusions caused by BimA-mediated projection of *B. pseudomallei* from one cell to the next are engulfed by paracytophagy and lysed in a manner that requires bacterial factors.

22.8
Cell Fusion

Uniquely among pathogens capable of actin-based motility, *B. pseudomallei* can also fuse adjacent cells in infected monolayers (Figure 22.5) [6, 14]. It is presumed that cell fusion requires endosome escape and intracellular replication owing to the impaired ability of mutants with defects in these processes to form MNGC, however no bacterial factors required for membrane fusion *per se* have been identified at the time of writing. BimA is required for efficient fusion (J.M. Stevens, unpublished observations); however this raises the issue of how *B. pseudomallei*-induced membrane protrusions may elicit fusion whereas those from other pathogens capable of actin-based motility do not.

Fusion occurs naturally among a limited number of mammalian cell types and is vital in development and homeostasis (reviewed in [68]). Macrophages can differentiate and fuse to form two types of multinucleated cells, osteoclasts and giant cells, that respectively influence bone resorption and immunity. Differentiation

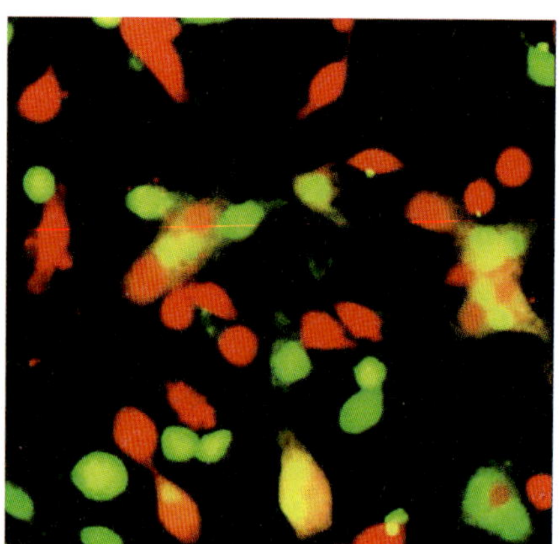

Figure 22.5 Fluorescence micrograph of co-cultured J774A.1 cells labeled separately with the dyes CMTMR (red) and CMFDA (green) prior to infection with *Burkholderia pseudomallei*. Cell fusion induced by the bacteria 6 h post infection results in mixing of the dyes and thus yellow fluorescence. From [6].

of bone marrow-derived macrophages and monocytes into osteoclasts and giant cells is driven by the cytokines RANKL and IL-4, respectively [68]. Infection of RAW264.7 macrophages with *B. pseudomallei* elicits similar effects to treatment with RANKL, including elevated transcription of factors required for osteoclast formation, such as the chemokines MCP-1, MIP-1γ and RANTES as well as increased expression of the osteoclast markers calcitonin receptor, cathepsin K and tartrate-resistant acid phosphatase [69]. The *B. pseudomallei* gene *lfpA* (lactonase family protein A; BPSS2074) is induced in macrophages and required for optimal expression of many of the aforementioned factors, however an *lfpA* mutant still formed MNGCs and production of RANTES was not affected [69]. LfpA exhibits homology with the eukaryotic protein regucalcin and is required for full virulence in Syrian hamsters and BALB/c mice [69], however further studies are required to dissect its role in the intracellular life of *B. pseudomallei* and whether it directly influences the induction of factors required for macrophage fusion. It is noteworthy that LpfA is absent in *B. thailandensis*, even though ara^+ biotypes are able to induce fusion [6, 14].

22.9 Intracellular Replication

The bacterial factors required for intracellular replication of *B. pseudomallei* are relatively ill-defined. Screening of *B. pseudomallei* Tn5-OT182 mutants for impaired plaque formation in PtK2 cell monolayers highlighted the importance of biosynthesis of purine (*purN*, BPSL0908; *purM*, BPSL2818), histidine (*hisF*, BPSL3133) and para-aminobenzoate (*pabB*, BPSL2825), in intracellular proliferation [63]. Mutants with defects in these genes exhibited normal invasion and actin-based motility but were attenuated in mice [63]. Screening of signature-tagged transposon mutants has reinforced the need for selected amino acids and para-aminobenzoate during murine melioidosis [43, 70]. Other genes implicated in delayed plaque formation include BPSL0413 (putative lipoate-protein ligase B, LipB), BPSS1509 (hypothetical protein) and BPSL0395 (putative cytidyltransferase), though their role in intracellular growth requires further study. A role for LipB is supported by the observation that a lipoate-protein ligase (LlpA1) is required for replication of *L. monocytogenes* in the cytosol by enabling the utilization of host-derived lipoate [71] and lipoyl peptides [72].

Microarray analysis of the transcription of *B. pseudomallei* genes in the organs of Syrian hamsters and during culture *in vitro* also supports the role of specific biochemical pathways during infection, for example isomerization of 3-phosphoglycerate, catabolism of inositol and D-glucosamine and biosynthesis of specific amino acids [53]. The same study also indicated that energy production pathways other than those reliant on ATP synthase and NADH dehydrogenase may be used during infection [53]. A recent screen for *B. pseudomallei* promoters induced upon invasion of RAW264.7 relative to culture in broth by use of *in vivo* expression technology (IVET) revealed *in vivo* induction of genes predicted to encode a type VI secretion system [73]. The cluster (designated *tss*-5 owing to the existence of other putative type VI secretion loci) is encoded proximal to the *bsa* locus, but inactivation of a single gene of the cluster (*tssH-5*) did not alter invasion. A system encoded at the same position in *B. mallei* secretes Hcp1 (BMAA0744), a type VI secreted protein produced during natural and experimental glanders that is required for full virulence [67]. The mode of action of type VI secreted proteins and their role in intracellular life remains to be elucidated.

It remains unclear how *B. pseudomallei* senses and adapts to the changing niches it encounters on cell contact, uptake and endosome escape. While two-component sensory systems regulate the expression of genes pertinent to invasion [60] and actin-based motility [67], it is evident that alternative RNA polymerase sigma factors also reprogram bacterial genes during intracellular life. Inactivation of the stationary phase sigma factor RpoS raises the sensitivity of *B. pseudomallei* to oxidative stress and impairs internalization and MNGC formation in RAW264.7 cells [74]. RpoS was not required for intracellular replication in RAW664.7 cells, however in interferon γ (IFNγ)-activated macrophages mutation of RpoS resulted in a failure to suppress iNOS induction and thus reduced survival [74]. The stress response sigma factor RpoE also plays a role in intra-macrophage survival as well as resistance to oxidative and osmotic stress [75], and proteins under its control have been identified that may contribute to these processes [76].

22.10 Immunology of *B. pseudomallei*-Host Cell Interactions

The interaction between *B. pseudomallei* and host immunity is reviewed elsewhere [44, 77], however key features at a cellular level are described below. As documented above, *B. pseudomallei* LPS is a relatively poor agonist of Toll-like receptor 4-mediated innate responses and other surface-associated polysaccharides may mask underlying antigens. As a consequence, *B. pseudomallei* is a relatively weak inducer of proinflammatory cytokines and phagosomal killing mechanisms compared to other Gram-negative pathogens. Indeed, macrophages from human melioidosis patients produce significantly lower levels of nitric oxide and the activation marker 8-iso-PGF(2α) compared to those from the healthy subjects [78]. The activation state of macrophages may further be modulated by interference in IFNγ signaling, as recent studies have shown that *B. pseudomallei* activates the expression of suppressor of cytokine signaling 3 (SOCS3) and cytokine-inducible Src homology 2-containing protein (CIS) in RAW264.7 cells, leading to a reduction in key iNOS regulators [79]. These events require bacterial internalization [80], but the bacterial factors that mediate this effect are unknown. IFNγ and its inducers play pivotal roles in control of experimental melioidosis, therefore further studies are required to determine if (and how) *B. pseudomallei* modulates IFNγ production and signaling. The related pathogen *Pseudomonas aeruginosa* can bind IFNγ via the outer membrane porin OprF and IFNγ activates the expression of its key virulence factor PA/I lectin [81], and it will be exciting to determine if *B. pseudomallei* has also evolved the ability to sense activation of host innate immunity.

Once inside endosomes, suppression of iNOS induction is correlated with improved bacterial survival, however other factors may be deployed to resist the antimicrobial environment. *B. pseudomallei* exhibits resistance to the cationic peptide protamine and purified human defensin HNP-1 [4] and also produces a catalase-peroxidase (KatG) that confers resistance to oxidative stress [82]. The importance of these defenses in intracellular survival in varied cells and over time is ill-defined and their role may be limited once *B. pseudomallei* escapes to the cytosol by destruction of the endosomal membrane.

To date, few studies have addressed how *B. pseudomallei* evades cellular mechanisms for control of bacteria in the cytosol and presentation of its antigens in the context of the MHC complex. In the cytosol pathogens must evade autophagy, a cellular process to capture particles such as organelles, surround them with a new membrane and target the contents for degradation. Some intracytoplasmic pathogens actively inhibit this process, in the case of *Shigella* via the type III secreted effector IcsB (reviewed in [83]). *B. pseudomallei* possesses an IscB homolog that influences virulence [11, 55], however its role in evasion of autophagy remains unclear. Further, it is unknown how *B. pseudomallei* can persist in an intracellular niche for prolonged periods without triggering cytoplasmic sensors of pathogen-associated molecular patterns such as Nod/CARD family proteins that activate inflammatory responses. One immune evasion strategy may be the induction of host cell death. *B. pseudomallei* is able to induce apoptosis of macrophages by a mechanism dependent on caspase-1 [58]. This may result in the depletion of phagocytes and requires the Bsa T3SS proteins BipB [15] and BsaQ [58], likely owing to their role in bacterial invasion and membrane damage.

22.11 Concluding Remarks

B. pseudomallei follows a remarkable path toward the production of melioidosis. Not only is it able to persist in the rhizosphere and free-living organisms, but it is highly adapted to subvert cellular pathways to facilitate invasion, intracellular replication, actin-based motility and long-term persistence in diverse animal hosts and cell types. The advent of complete genome sequences for *Burkholderia* species, together with methods to survey the conservation, expression and function of the encoded genes has rapidly accelerated our understanding of these processes. However, at the time of writing the role of only a fraction of the *B. pseudomallei* genome in pathology and protection is known and the mode of action of the encoded products is ill-defined. Further research is needed to assign temporal and spatial roles to such genes both on a cellular level (during entry, intracellular survival and exit) and during phases of infection (acute, chronic and latent). Even with such knowledge, it will be important to consider the dynamic nature of *Burkholderia* genomes both in the environment and during infection. Rodent models of melioidosis are providing valuable insights into the bacterial and host components that influence the outcome of infection, however vaccines safe for human use remain elusive and an ongoing need exists to identify protective antigens and virulence factors for the rational design of intervention strategies.

Acknowledgments

The authors gratefully acknowledge the support of the Biotechnology & Biological Sciences Research Council, United Kingdom (grants C20021 and E01212).

References

1 Cheng, A.C. and Currie, B.J. (2005) Melioidosis: epidemiology, pathophysiology, and management. *Clin Microbiol Rev*, **18**, 383–416.

2 Pitt, T.L., Trakulsomboon, S. and Dance, D.A. (2007) Recurrent melioidosis: possible role of infection with multiple strains of *Burkholderia pseudomallei*. *J Clin Microbiol*, **45**, 680–681.

3 Rotz, L.D., Khan, A.S., Lillibridge, S.R., Ostroff, S.M. and Hughes, J.M. (2002) Public health assessment of potential biological terrorism agents. *Emerg Infect Dis*, **8**, 225–230.

4 Jones, A.L., Beveridge, T.J. and Woods, D.E. (1996) Intracellular survival of *Burkholderia pseudomallei*. *Infect Immun*, **64**, 782–790.

5 Harley, V.S., Dance, D.A., Tovey, G., McCrossan, M.V. and Drasar, B.S. (1998) An ultrastructural study of the phagocytosis of *Burkholderia pseudomallei*. *Microbiology*, **94**, 35–45.

6 Kespichayawattana, W., Rattanachetku, S., Wanun, T., Utaisincharoen, P. and Sirisinha, S. (2000) *Burkholderia pseudomallei* induces cell fusion and actin-associated membrane protrusion: a possible mechanism for cell-to-cell spreading. *Infect Immun*, **68**, 5377–5384.

7 Egan, A.M. and Gordon, D.L. (1996) *Burkholderia pseudomallei* activates complement and is ingested but not killed by polymorphonuclear leukocytes. *Infect Immun*, **64**, 4952–4959.

8 Pruksachartvuthi, S., Aswapokee, N. and Thankernpol, K. (1990) Survival of *Pseudomonas pseudomallei* in human phagocytes. *Int J Med Microbiol*, **31**, 109–114.

9 Inglis, T.J. and Sagripanti, J.L. (2006) Environmental factors that affect the survival and persistence of *Burkholderia pseudomallei*. *Appl Environ Microb*, **72**, 6865–6875.

10 Cirillo, J.D., Cirillo, S.L., Yan, L., Bermudez, L.E., Falkow, S. and Tompkins, L.S. (1999) Intracellular growth in *Acanthamoeba castellanii* affects monocyte entry mechanisms and enhances virulence of *Legionella pneumophila*. *Infect Immun*, **67**, 4427–4434.

11 Stevens, M.P., Wood, M.W., Taylor, L.A., Monaghan, P., Hawes, P., Jones, P.W., Wallis, T.S. and Galyov, E.E. (2002) An Inv/Mxi-Spa-like type III protein secretion system in *Burkholderia pseudomallei* modulates intracellular behaviour of the pathogen. *Mol Microbiol*, **46**, 649–659.

12 Breitbach, K., Rottner, K., Klocke, S., Rohde, M., Jenzora, A., Wehland, J. and Steinmetz, I. (2003) Actin-based motility of *Burkholderia pseudomallei* involves the Arp 2/3 complex, but not N-WASP and Ena/VASP proteins. *Cell Microbiol*, **5**, 385–393.

13 Stevens, J.M., Galyov, E.E. and Stevens, M.P. (2006) Actin-dependent movement of bacterial pathogens. *Nat Rev Microbiol*, **4**, 91–101.

14 Harley, V.S., Dance, D.A., Drasar, B. and Tovey, G. (1998) Effects of *Burkholderia pseudomallei* and other *Burkholderia* species on eukaryotic cells in tissue culture. *Microbios*, **96**, 71–93.

15 Suparak, S., Kespichayawattana, W., Haque, A., Easton, A., Damnin, S., Lertmemongkolchai, G., Bancroft, G.J. and Korbsrisate, S. (2005) Multinucleated giant cell formation and apoptosis in infected host cells is mediated by *Burkholderia pseudomallei* type III secretion protein BipB. *J Bacteriol*, **187**, 6556–6560.

16 Wong, K.T., Puthucheary, S.D. and Vadivelu, J. (1995) The histopathology of human melioidosis. *Histopathology*, **26**, 51–55.

17 Holden, M.T., Titball, R.W., Peacock, S.J., Cerdeno-Tarraga, A.M., Atkins, T., Crossman, L.C., Pitt, T., Churcher, C. *et al.* (2004) Genomic plasticity of the causative agent of melioidosis, *Burkholderia pseudomallei*. *Proc Natl Acad Sci USA*, **101**, 14240–14245.

18 Ou, K., Ong, C., Koh, S.Y., Rodrigues, F., Sim, S.H., Wong, D., Ooi, C.H., Ng, K.C. *et al.* (2005) Integrative genomic, transcriptional, and proteomic diversity in natural isolates of the human pathogen *Burkholderia pseudomallei*. *J Bacteriol*, **187**, 4276–4285.

19 Monastyrskaya, G., Fushan, A., Abaev, I., Filyukova, O., Kostina, M., Pecherskih, E. and Sverdlov, E. (2004) Genome-wide comparison reveals great inter- and intraspecies variability in *B. pseudomallei* and *B. mallei* pathogens. *Res Microbiol*, **155**, 781–793.

20 Godoy, D., Randle, G., Simpson, A.J., Aanensen, D.M., Pitt, T.L., Kinoshita, R. and Spratt, B.G. (2003) Multilocus sequence typing and evolutionary relationships among the causative agents of melioidosis and glanders, *Burkholderia pseudomallei* and *Burkholderia mallei*. *J Clin Microbiol*, **41**, 2068–2079.

21 Nierman, W.C., DeShazer, D., Kim, H.S., Tettelin, H., Nelson, K.E., Feldblyum, T., Ulrich, R.L., Ronning, C.M. *et al.* (2004) Structural flexibility in the *Burkholderia mallei* genome. *Proc Natl Acad Sci USA*, **101**, 14246–14251.

22 Kim, H.S., Schell, M.A., Yu, Y., Ulrich, R.L., Sarria, S.H., Nierman, W.C. and DeShazer, D. (2005) Bacterial genome adaptation to niches: divergence of the potential virulence genes in three *Burkholderia* species of different survival strategies. *BMC Genomics*, **6**, 174.

23 Ong, C., Ooi, C.H., Wang, D., Chong, H., Ng, K.C., Rodrigues, F., Lee, M.A. and Tan, P. (2004) Patterns of large-scale genomic variation in virulent and avirulent *Burkholderia* species. *Genome Res*, **14**, 2295–2307.

24 Brett, P.J., Deshazer, D. and Woods, D.E. (1997) Characterization of *Burkholderia pseudomallei* and *Burkholderia pseudomallei*-like strains. *Epidemiol Infect*, **118**, 137–148.

25 Yu, Y., Kim, H.S., Chua, H.H., Lin, C.H., Sim, S.H., Lin, D., Derr, A., Engels, R. *et al.* (2006) Genomic patterns of pathogen evolution revealed by comparison of *Burkholderia pseudomallei*, the causative agent of melioidosis, to avirulent *Burkholderia thailandensis*. *BMC Microbiol*, **6**, 46.

26 Reckseidler, S.L., DeShazer, D., Sokol, P.A. and Woods, D.E. (2001) Detection of bacterial virulence genes by subtractive hybridization: identification of capsular polysaccharide of *Burkholderia pseudomallei* as a major virulence determinant. *Infect Immun*, **69**, 34–44.

27 Brown, N.F. and Beacham, I.R. (2000) Cloning and analysis of genomic differences unique to *Burkholderia pseudomallei* by comparison with *B. thailandensis*. *Int J Med Microbiol*, **49**, 993–1001.

28 Rainbow, L., Hart, C.A. and Winstanley, C. (2002) Distribution of type III secretion gene clusters in *Burkholderia pseudomallei*, *B. thailandensis* and *B. mallei*. *Int J Med Microbiol*, **51**, 374–384.

29 Moore, R.A., Reckseidler-Zenteno, S., Kim, H., Nierman, W., Yu, Y., Tuanyok, A.,

Warawa, J., DeShazer, D. *et al.* (2004) Contribution of gene loss to the pathogenic evolution of *Burkholderia pseudomallei* and *Burkholderia mallei*. *Infect Immun*, **72**, 4172–4187.

30 Wongtrakoongate, P., Mongkoldhumrongkul, N., Chaijan, S., Kamchonwongpaisan, S. and Tungpradabkul, S. (2007) Comparative proteomic profiles and the potential markers between *Burkholderia pseudomallei* and *Burkholderia thailandensis*. *Mol Cell Probes*, **21**, 81–91.

31 Romero, C.M., DeShazer, D., Feldblyum, T., Ravel, J., Woods, D., Kim, H.S., Yu, Y., Ronning, C.M. *et al.* (2006) Genome sequence alterations detected upon passage of *Burkholderia mallei* ATCC 23344 in culture and in mammalian hosts. *BMC Genomics*, **7**, 228.

32 Ahmed, K., Enciso, H.D., Masaki, H., Tao, M., Omori, A., Tharavichikul, P. and Nagatake, T. (1999) Attachment of *Burkholderia pseudomallei* to pharyngeal epithelial cells: a highly pathogenic bacteria with low attachment ability. *Am J Trop Med Hyg*, **60**, 90–93.

33 Kespichayawattana, W., Intachote, P., Utaisincharoen, P. and Sirisinha, S. (2004) Virulent *Burkholderia pseudomallei* is more efficient than avirulent *Burkholderia thailandensis* in invasion of and adherence to cultured human epithelial cells. *Microb Pathog*, **36**, 287–292.

34 Brown, N.F., Boddey, J.A., Flegg, C.P. and Beacham, I.R. (2002) Adherence of *Burkholderia pseudomallei* cells to cultured human epithelial cell lines is regulated by growth temperature. *Infect Immun*, **70**, 974–980.

35 Gori, A.H., Ahmed, K., Martinez, G., Masaki, H., Watanabe, K. and Nagatake, T. (1999) Mediation of attachment of *Burkholderia pseudomallei* to human pharyngeal epithelial cells by the asialoganglioside GM1-GM2 receptor complex. *Am J Trop Med Hyg*, **61**, 473–475.

36 Essex-Lopresti, A.E., Boddey, J.A., Thomas, R., Smith, M.P., Hartley, M.G., Atkins, T., Brown, N.F., Tsang, C.H. *et al.* (2005) A type IV pilin, PilA, contributes to adherence of *Burkholderia pseudomallei* and virulence *in vivo*. *Infect Immun*, **73**, 1260–1264.

37 Boddey, J.A., Flegg, C.P., Day, C.J., Beacham, I.R. and Peak, I.R. (2006) Temperature-regulated microcolony formation by *Burkholderia pseudomallei* requires *pilA* and enhances association with cultured human cells. *Infect Immun*, **74**, 5374–5381.

38 Inglis, T.J., Robertson, T., Woods, D.E., Dutton, N. and Chang, B.J. (2003) Flagellum-mediated adhesion by *Burkholderia pseudomallei* precedes invasion of *Acanthamoeba astronyxis*. *Infect Immun*, **71**, 2280–2282.

39 Chua, K.L., Chan, Y.Y. and Gan, Y.H. (2003) Flagella are virulence determinants of *Burkholderia pseudomallei*. *Infect Immun*, **71**, 1622–1629.

40 DeShazer, D., Brett, P.J., Carlyon, R. and Woods, D.E. (1997) Mutagenesis of *Burkholderia pseudomallei* with Tn5-OT182: isolation of motility mutants and molecular characterization of the flagellin structural gene. *J Bacteriol*, **179**, 2116–2125.

41 Reckseidler-Zenteno, S.L., DeVinney, R. and Woods, D.E. (2005) The capsular polysaccharide of *Burkholderia pseudomallei* contributes to survival in serum by reducing complement factor C3b deposition. *Infect Immun*, **73**, 1106–1115.

42 Atkins, T., Prior, R., Mack, K., Russell, P., Nelson, M., Prior, J., Ellis, J., Oyston, P.C. *et al.* (2002) Characterisation of an acapsular mutant of *Burkholderia pseudomallei* identified by signature tagged mutagenesis. *Int J Med Microbiol*, **51**, 539–547.

43 Cuccui, J., Easton, A., Chu, K.K., Bancroft, G.J., Oyston, P.C., Titball, R.W. and Wren, B.W. (2007) Development of signature-tagged mutagenesis in *Burkholderia pseudomallei* to identify genes important in survival and pathogenesis. *Infect Immun*, **75**, 1186–1195.

44 Wiersinga, W.J., van der Poll, T., White, N.J., Day, N.P. and Peacock, S.J. (2006) Melioidosis: insights into the pathogenicity of *Burkholderia pseudomallei*. *Nat Rev Microbiol*, **4**, 272–282.

45 Utaisincharoen, P., Tangthawornchaikul, N., Kespichayawattana, W., Chaisuriya, P. and Sirisinha, S. (2001) *Burkholderia pseudomallei* interferes with inducible nitric oxide synthase (iNOS) production: a possible mechanism of evading macrophage killing. *Microbiol Immunol*, **45**, 307–313.

46 Arjcharoen, S., Wikraiphat, C., Pudla, M., Limposuwan, K., Woods, D.E., Sirisinha, S. and Utaisincharoen, P. (2007) Fate of a *Burkholderia pseudomallei* lipopolysaccharide mutant in the mouse macrophage cell line RAW264.7: possible role for the O-antigenic polysaccharide moiety of lipopolysaccharide in internalization and intracellular survival. *Infect Immun*, **75**, 4298–4304.

47 Deshazer, D., Brett, P. and Woods, D. (1998) The type II O-antigenic polysaccharide moiety of *Burkholderia pseudomallei* lipopolysaccharide is required for serum resistance and virulence. *Mol Microbiol*, **30**, 1081–1100.

48 Sarkar-Tyson, M., Thwaite, J.E., Harding, S.V., Smither, S.J., Oyston, P.C., Atkins, T.P. and Titball, R.W. (2007) Polysaccharides and virulence of *Burkholderia pseudomallei*. *Int J Med Microbiol*, **56**, 1005–1010.

49 Haussler, S., Rohde, M., von Neuhoff, N., Nimtz, M. and Steinmetz, I. (2003) Structural and functional cellular changes induced by *Burkholderia pseudomallei* rhamnolipid. *Infect Immun*, **71**, 2970–2975.

50 Galán, J.E. and Wolf-Watz, H. (2006) Protein delivery into eukaryotic cells by type III secretion machines. *Nature*, **444**, 567–573.

51 Schlumberger, M.C. and Hardt, W.-D. (2005) Triggered phagocytosis by *Salmonella*: bacterial molecular mimicry of RhoGTPase activation/deactivation. *Curr Top Microbiol Immunol*, **291**, 29–42.

52 Stevens, M.P., Friebel, A., Taylor, L.A., Wood, M.W., Brown, P.J., Hardt, W.-D. and Galyov, E.E. (2003) A *Burkholderia pseudomallei* type III secreted protein, BopE, facilitates bacterial invasion of epithelial cells and exhibits guanine nucleotide exchange factor activity. *J Bacteriol*, **185**, 4992–4996.

53 Tuanyok, A., Tom, M., Dunbar, J. and Woods, D.E. (2006) Genome-wide expression analysis of *Burkholderia pseudomallei* infection in a hamster model of acute melioidosis. *Infect Immun*, **74**, 5465–5476.

54 Haque, A., Chu, K., Easton, A., Stevens, M.P., Galyov, E.E., Atkins, T., Titball, R. and Bancroft, G.J. (2006) A live experimental vaccine against *Burkholderia pseudomallei* elicits CD4+ T cell-mediated immunity, priming T cells specific for 2 type III secretion system proteins. *J Infect Dis*, **194**, 1241–1248.

55 Stevens, M.P., Haque, A., Atkins, T., Hill, J., Wood, M.W., Easton, A., Nelson, M., Underwood-Fowler, C. *et al.* (2004) Attenuated virulence and protective efficacy of a *Burkholderia pseudomallei bsa* type III secretion mutant in murine models of melioidosis. *Microbiology*, **150**, 2669–2676.

56 Warawa, J. and Woods, D.E. (2005) Type III secretion system cluster 3 is required for maximal virulence of *Burkholderia pseudomallei* in a hamster infection model. *FEMS Microbiol Lett*, **242**, 101–108.

57 Utaisincharoen, P., Arjcharoen, S., Lengwehasatit, I., Limposuwan, K. and Sirisinha, S. (2005) *Burkholderia pseudomallei* invasion and activation of epithelial cells requires activation of p38 mitogen-activated protein kinase. *Microb Pathog*, **38**, 107–112.

58 Sun, G.W., Lu, J., Pervaiz, S., Cao, W.P. and Gan, Y.H. (2005) Caspase-1 dependent macrophage death induced by *Burkholderia pseudomallei*. *Cell Microbiol*, **7**, 1447–1458.

59 Winstanley, C., Hales, B.A. and Hart, C.A. (1999) Evidence for the presence in *Burkholderia pseudomallei* of a type III secretion system-associated gene cluster. *Int J Med Microbiol*, **48**, 649–656.

60 Jones, A.L., DeShazer, D. and Woods, D.E. (1997) Identification and characterization of a two-component regulatory system involved in invasion of eukaryotic cells and heavy-metal resistance in *Burkholderia pseudomallei*. *Infect Immun*, **65**, 4972–4977.

61 Wiersinga, W.J., de Vos, A.F., de Beer, R., Wieland, C.W., Roelofs, J.J., Woods, D.E. and van der Poll, T. (2008) Inflammation patterns induced by different *Burkholderia* species in mice. *Cell Microbiol*, **10**, 81–87.

62 Puthucheary, S.D. and Nathan, S.A. (2006) Comparison by electron microscopy of intracellular events and survival of *Burkholderia pseudomallei* in monocytes from normal subjects and patients with melioidosis. *Singapore Med J*, **47**, 697–703.

63 Pilatz, S., Breitbach, K., Hein, N., Fehlhaber, B., Schulze, J., Brenneke, B., Eberl, L. and Steinmetz, I. (2006) Identification of *Burkholderia pseudomallei* genes required for the intracellular life cycle and *in vivo* virulence. *Infect Immun*, **74**, 3576–3586.

64 Korbsrisate, S., Tomaras, A.P., Damnin, S., Ckumdee, J., Srinon, V., Lengwehasatit, I., Vasil, M.L. and Suparak, S. (2007) Characterization of two distinct phospholipase C enzymes from *Burkholderia pseudomallei*. *Microbiology*, **153**, 1907–1915.

65 Stevens, J.M., Ulrich, R.L., Taylor, L.A., Wood, M.W., Deshazer, D., Stevens, M.P. and Galyov, E.E. (2005) Actin-binding proteins from *Burkholderia mallei* and *Burkholderia thailandensis* can functionally compensate for the actin-based motility defect of a *Burkholderia pseudomallei bimA* mutant. *J Bacteriol*, **187**, 7857–7862.

66 Stevens, M.P., Stevens, J.M., Jeng, R.L., Taylor, L.A., Wood, M.W., Hawes, P., Monaghan, P., Welch, M.D. *et al.* (2005) Identification of a bacterial factor required for actin-based motility of *Burkholderia pseudomallei*. *Mol Microbiol*, **56**, 40–53.

67 Schell, M.A., Ulrich, R.L., Ribot, W.J., Brueggemann, E.E., Hines, H.B., Chen, D., Lipscomb, L., Kim, H.S. *et al.* (2007) Type VI secretion is a major virulence determinant in *Burkholderia mallei*. *Mol Microbiol*, **64**, 1466–1485.

68 Chen, E.H., Grote, E., Mohler, W. and Vignery, A. (2007) Cell-cell fusion. *FEBS Lett*, **581**, 2181–2193.

69 Boddey, J.A., Day, C.J., Flegg, C.P., Ulrich, R.L., Stephens, S.R., Beacham, I.R., Morrison, N.A. and Peak, I.R. (2007) The bacterial gene *lfpA* influences the potent induction of calcitonin receptor and osteoclast-related genes in *Burkholderia pseudomallei*-induced TRAP-positive multinucleated giant cells. *Cell Microbiol*, **9**, 514–531.

70 Atkins, T., Prior, R.G., Mack, K., Russell, P., Nelson, M., Oyston, P.C., Dougan, G. and Titball, R.W. (2002) A mutant of *Burkholderia pseudomallei*, auxotrophic in the branched chain amino acid biosynthetic pathway, is attenuated and protective in a murine model of melioidosis. *Infect Immun*, **70**, 5290–5294.

71 O'Riordan, M., Moors, M.A. and Portnoy, D.A. (2003) *Listeria* intracellular growth and virulence require host-derived lipoic acid. *Science*, **302**, 462–464.

72 Keeney, K.M., Stuckey, J.A. and O'Riordan, M.X. (2007) LplA1-dependent utilization of host lipoyl peptides enables *Listeria* cytosolic growth and virulence. *Mol Microbiol*, **66**, 758–770.

73 Shalom, G., Shaw, J.G. and Thomas, M.S. (2007) *In vivo* expression technology identifies a type VI secretion system locus in *Burkholderia pseudomallei* that is induced upon invasion of macrophages. *Microbiology*, **153**, 2689–2699.

74 Utaisincharoen, P., Arjcharoen, S., Limposuwan, K., Tungpradabkul, S. and Sirisinha, S. (2006) *Burkholderia pseudomallei* RpoS regulates multinucleated giant cell formation and inducible nitric oxide synthase expression

in mouse macrophage cell line (RAW264.7). *Microb Pathog*, **40**, 184–189.

75 Korbsrisate, S., Vanaporn, M., Kerdsuk, P., Kespichayawattana, W., Vattanaviboon, P., Kiatpapan, P. and Lertmemongkolchai, G. (2005) The *Burkholderia pseudomallei* RpoE operon is involved in environmental stress tolerance and biofilm formation. *FEMS Microbiol Lett*, **252**, 243–249.

76 Thongboonkerd, V., Vanaporn, M., Songtawee, N., Kanlaya, R., Sinchaikul, S., Chen, S.T., Easton, A., Chu, K. *et al.* (2007) Altered proteome in *Burkholderia pseudomallei rpoE* operon knockout mutant: insights into mechanisms of *rpoE* operon in stress tolerance, survival, and virulence. *J Proteome Res*, **6**, 1334–1341.

77 Gan, Y.H. (2005) Interaction between *Burkholderia pseudomallei* and the host immune response: sleeping with the enemy? *J Infect Dis*, **192**, 1845–1850.

78 Nathan, S.A., Qvist, R. and Puthucheary, S.D. (2005) Kinetic studies of bioactive products nitric oxide and 8-iso-PGF(2alpha) in *Burkholderia pseudomallei* infected human macrophages, and their role in the intracellular survival of these organisms. *FEMS Immunol Med Microbiol*, **43**, 177–183.

79 Ekchariyawat, P., Pudla, S., Limposuwan, K., Arjcharoen, S., Sirisinha, S. and Utaisincharoen, P. (2005) *Burkholderia pseudomallei*-induced expression of suppressor of cytokine signaling 3 and cytokine-inducible src homology 2-containing protein in mouse macrophages: a possible mechanism for suppression of the response to gamma interferon stimulation. *Infect Immun*, **73**, 7332–7339.

80 Ekchariyawat, P., Pudla, S., Limposuwan, K., Arjcharoen, S., Sirisinha, S. and Utaisincharoen, P. (2007) Expression of suppressor of cytokine signaling 3 (SOCS3) and cytokine-inducible Src homology 2-containing protein (CIS) induced in *Burkholderia pseudomallei*-infected mouse macrophages requires bacterial internalization. *Microb Pathog*, **42**, 104–110.

81 Wu, L., Estrada, O., Zaborina, O., Bains, M., Shen, L., Kohler, J.E., Patel, N., Musch, M.W. *et al.* (2005) Recognition of host immune activation by *Pseudomonas aeruginosa*. *Science*, **309**, 774–777.

82 Loprasert, S., Whangsuk, W., Sallabhan, R. and Mongkolsuk, S. (2003) Regulation of the *katG-dpsA* operon and the importance of KatG in survival of *Burkholderia pseudomallei* exposed to oxidative stress. *FEBS Lett*, **542**, 17–21.

83 Radtke, A.L. and O'Riordan, M.X. (2006) Intracellular innate resistance to bacterial pathogens. *Cell Microbiol*, **8**, 1720–1729.

23
Francisella tularensis

Lee-Ann H. Allen and Grant S. Schulert

23.1
Francisella tularensis and Tularemia

Francisella tularensis is a Gram-negative coccobacillus and the causative agent of the disease tularemia. This organism was first isolated from small mammals with suspected plague in Tulare County, California, in 1911. The bacterium was further characterized and linked to human disease by Dr. Edward Francis, and was named in honor of his work and the site of original discovery [1]. There are four major subspecies of *F. tularensis*: *tularensis* (also known as *Francisella* type A), *holarctica* (or type B), *novicida*, *mediaciatica*, and a variant of *holarctica* found mainly in Japan [2]. However, of these subspecies only *tularensis* and *holarctica* cause significant disease in immunocompetent humans, with type A strains significantly more virulent and responsible for the vast majority of severe infections.

In nature, tularemia is endemic in much of the Northern Hemisphere. Infections of rodents, rabbits and hares, as well as ticks and other arthropods are common; whereas humans appear to be an accidental host. Type A strains are concentrated in North America. On the other hand, *F. tularensis* subspecies *holarctica* strains are less common in North America and are distributed across Europe, Asia and parts of Scandinavia [1, 2]. There are several clinical forms of tularemia, the severity of which are related to *Francisella* subspecies and the route of exposure [3]. For example inhalation of as few as ten *F. tularensis* subsp. *tularensis* causes rapidly progressing pneumonic infection in humans, and up to 60% of these cases are fatal in the absence of antibiotic treatment. In the lung, bacteria infect macrophages, epithelial cells and neutrophils and then disseminate to the liver and spleen. Death occurs secondary to overwhelming tissue necrosis prior to development of a protective immune response.

Ulceroglandular tularemia is the most common form in nature and can be transmitted by ticks or by direct contact with infected animals. A local papule is present initially at the site of inoculation. Within days this lesion ulcerates and local lymph node enlargement develops which can occasionally suppurate. Infection can also occur by ingestion of contaminated food or water. In humans, infections with

Intracellular Niches of Microbes. Edited by Ulrich E. Schaible and Albert Haas

ISBN: 978-3-527-32207-7

F. tularensis subsp. *holarctica* are rarely fatal. An attenuated live vaccine strain of *F. tularensis* subsp. *holarctica* (called LVS) was developed by the Soviet Union in the 1960s but the mechanism of its attenuation is unknown.

Most of what is known about how *Francisella* causes severe, systemic disease has been elucidated using animal models of infection. Studies of non-human primates defined the basic features of tularemia and noted differences in virulence of type A and type B isolates [4]. Although mice are the most common model in current use, it is noteworthy that these animals are much more susceptible to *F. tularensis*, and both virulent type A and B stains of *Francisella* (including LVS) cause fulminant disease in mice as do organisms of the *novicida* biovar [4–6].

Virulent *Francisella* strains are remarkable in their ability to effectively defeat both the innate and adaptive immune system. Indeed, mice with deficiencies in either innate immune signaling pathways such as interferon γ (IFNγ) or tumor necrosis factor α (TNFα), or lacking effective T cell responses, are no more susceptible to *Francisella* than wild-type mice [7]. This suggests that even immunocompetent hosts are nearly defenseless against even low doses of virulent *Francisella*. In the mouse model of pneumonic tularemia caused by type A strains, there are no remarkable histological changes in the lungs, and only modest changes in the liver and spleen, until day 4 post infection, which is only 24–48 h prior to death [5]. Similarly, extensive bronchopneumonia in monkeys was not observed until day 3 post infection [8]. In addition, transcriptional profiling of mouse lungs found that proinflammatory cytokines were not upregulated until day 4 post infection [9], and recent work has shown that type A *Francisella* actively suppresses the pulmonary immune response [10]. Concordant with these data, there is no evidence of recruitment of inflammatory cells to the lungs of mice in the first 72 h of infection with type A *Francisella* [11, 12].

Of particular note here is the ability of *F. tularensis* to replicate in macrophages, and it is generally believed that mononuclear phagocytes are the major reservoir of this organism *in vivo* [2, 13]. Herein we summarize the results of studies that, in just the last few years, have defined the niche occupied by this organism and have delineated the basic mechanisms by which *Francisella* disrupts macrophage function. Where possible, the effects of this pathogen on neutrophil and dendritic cell (DC) function are also noted.

23.2 *F. tularensis* Inhibits Phagosome Maturation Prior to Egress and Replication in the Cytosol

It has been known for some time that *F. tularensis* is a facultative intracellular pathogen of macrophages, and it is noteworthy that type A and type B strains of *F. tularensis* as well as organisms of the *novicida* biovar are all able to replicate in human and murine macrophages *in vitro*. Thus, what accounts for the pronounced differences in subspecies virulence remains obscure. The basic features of the *Francisella* lifecycle are summarized in Figure 23.1. Herein we review what is known

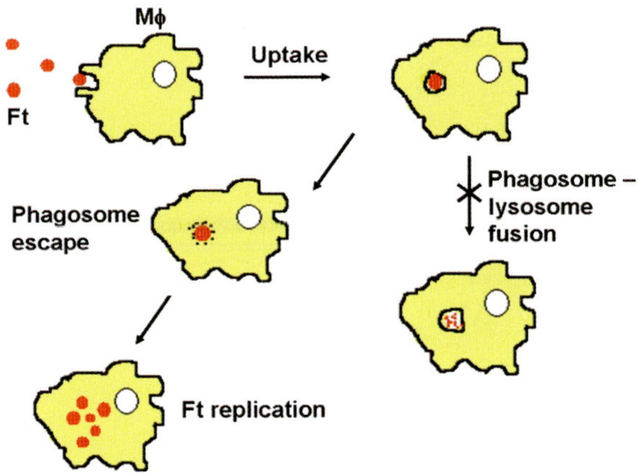

Figure 23.1 Overview of the *Francisella tularensis* (Ft) lifecycle in macrophages. First, bacteria attach to macrophages (Mϕ) and are phagocytosed. Second, whereas killed *F. tularensis* cannot prevent phagosome maturation, live *F. tularensis* inhibits phagosome–lysosome fusion, and bacteria escape the phagosome to replicate in the cytosol.

about this organism's intracellular niche while highlighting current deficits in our understanding.

23.2.1 Receptors for Phagocytosis

Recent data from several research groups provide insight into the receptors utilized by *Francisella* to gain entry into mononuclear phagocytes, as well as neutrophils and DCs. We and others have shown that infection of human monocyte-derived macrophages (MDM) by unopsonized *Francisella* is mediated by the macrophage mannose receptor (MR) [14, 15]. MR accumulates on forming phagosomes, and upregulation of MR (by treatment of MDM with interleukin 4) significantly increases bacterial binding and internalization [14]. Conversely, MR-blocking agents retard infection in a dose-dependent manner [14, 15]. These data are noteworthy since MR is particularly abundant on alveolar macrophages, and as such the ability to bind MR may account in part for the low infectious dose of this organism via the aerosol route. On the other hand, binding of unopsonized *Francisella* to MR-negative monocytes and J774 cells is extremely inefficient [14, 16, 17] and the receptors that mediate infection under these conditions are unknown.

Opsonization of *Francisella* with complement factors in fresh serum is essential for infection of neutrophils, and significantly enhances phagocytosis by monocytes and MDM [14, 18–20]. Macrophage engulfment of serum-opsonized *F. tularensis* is mediated by complement receptor 3 (CR3, CD11b/CD18) as well as MR, and these cells elaborate large asymmetric pseudopodia that appear to ensnare attached bacteria [14, 18]. Thus, the morphology of forming *Francisella* phagosomes is distinct

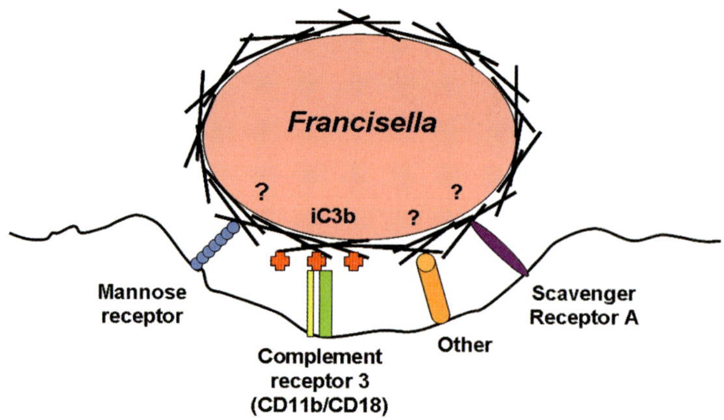

Figure 23.2 Receptors for *Francisella* phagocytosis. Uptake of *F. tularensis* by macrophages is mediated by the mannose receptor (MR) which binds mannose-containing carbohydrates of the organism [14, 15], perhaps lipopolysaccharide (LPS) O-antigen or capsule [3, 14]. Uptake of serum-opsonized bacteria by neutrophils, macrophages and dendritic cells is mediated by complement receptor 3 (CR3, CD11b/CD18) which binds C3bi deposited on the bacterial surface [14, 18]. Scavenger receptor A and CR4 also contribute to phagocytosis of opsonized *Francisella* by macrophages and DCs, respectively [24, 25]. Receptors that mediate entry of nonopsonized *Francisella* into MR-negative cell types are unknown.

from the macropinocytosis triggered by *Salmonella* and the coiling entry characteristic of *Legionella pneumophila* [18, 21, 22]. The significance of "looping phagocytosis" remains to be determined. However, periodate oxidation of *Francisella* lipopolysaccharide (LPS) and capsule prevents looping entry [18], and under these conditions, forming phagosomes exhibit the "sinking" morphology characteristic of uptake via CR3 alone [21, 23]. Other receptors implicated in uptake of opsonized *Francisella* include CR4 on human DCs, and scavenger receptors on J774 cells [24, 25]. These data are summarized in Figure 23.2. To date, receptors that support invasion of epithelial cells by *Francisella* are unknown.

Like *Francisella*, *Leishmania donovani* and virulent strains of *Mycobacterium tuberculosis* use MR and CR3 to enter macrophages [26–28]. The ability of several intracellular pathogens to engage these receptors is of interest as mechanism of uptake can alter microbe fate [29]. MR and CR3 are considered a relatively "safe" portals of entry since their ligation triggers only minimal activation of host microbicidal systems. For example, MR ligation can actively suppress TNFα [30] and this may account for the fact that LVS-infected MDM secrete less TNFα than infected monocytes [31].

23.2.2
Phagosome Composition in Macrophages

Under normal circumstances, the composition of nascent phagosomes is altered by sequential interactions with early endosomes, late endosomes, vesicles derived from

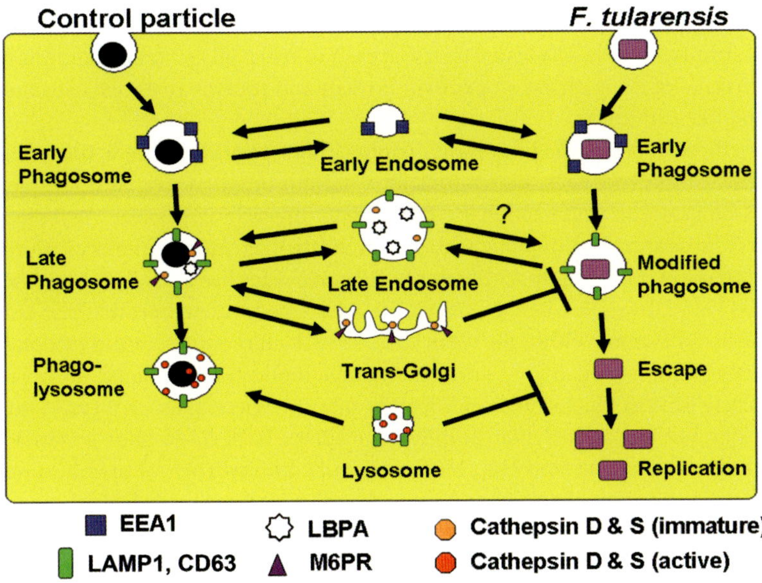

Figure 23.3 *Francisella tularensis* escapes from a phagosome that excludes most late endosomal markers. Control phagosomes mature by sequential interactions with early endosomes, late endosomes and lysosomes. *F. tularensis* phagosomes transiently accumulate the early endosome markers (early endosome antigen 1, EEA1) [33], and thereafter acquire the late endosome glycoproteins LAMP1, LAMP2 and CD63 [33, 34]. However, major late endosome markers, such as lysobisphosphatidic acid (LBPA), Rab7, the mannose 6-phosphate receptor (M6PR) and immature cathepsins are excluded [104]. *Francisella* escapes from this unusual compartment and replicates in the cytosol.

the *trans*-Golgi network (TGN) and lysosomes (Figure 23.3). In the mature phagolysosome, reactive oxygen species, reactive nitrogen intermediates, cationic antimicrobial peptides and lysosomal acid hydrolases collaborate to kill and degrade ingested microbes. Pathogens that survive and replicate in macrophages must either evade or withstand this hostile microenvironment. In 1991, Anthony and colleagues demonstrated for the first time that LVS phagosomes in J774 cells did not fuse with thorium-loaded lysosomes, and that bacteria began to replicate after a lag of approximately 6 h [32]. More recent confocal and immuno-electron microscopy studies have begun to define the composition of the *Francisella* phagosome in primary human MDM and murine bone marrow-derived macrophages [33–35]. These data indicate that nascent phagosomes containing either clinical isolates of type A or type B *F. tularensis*, LVS or *F. tularensis* subsp. *novicida* fuse with early endosomes (as judged by transient accumulation of early endosome antigen 1, EEA1), and thereafter acquire the late endosome membrane glycoproteins LAMP1, LAMP2 and CD63 [33–35]. However, these compartments are not strongly acidified (pH 6.7 at 4 h post infection), and further maturation does not occur as indicated by their failure to acquire the lysosomal enzyme cathepsin D or tracers preloaded into lysosomes [33–35]. Blockade of phagosome maturation is specific for live bacteria and requires genes in the *Francisella* pathogenicity island (discussed below). Of note, all strains of *Francisella* tested to date appear to arrest phagosome maturation at a similar stage of the pathway in primary human and murine macrophages and macrophage-like cell lines.

Although these data provide important insights into the nature of the *F. tularensis* compartment in macrophages, many questions remain unanswered. For example, the role of phagosome acidification in *Francisella*'s lifecycle is unclear. As noted above, direct measurements of phagosome pH indicate that compartments containing live bacteria are not strongly acidified (pH 6.7 and 5.5 for live and killed bacteria, respectively) [33]. However, when macrophages are treated with agents that prevent phagosome acidification – either the weak base NH_4Cl, chloroquine or the vacuolar proton ATPase inhibitor ouabain – intramacrophage replication is ablated, suggesting that partial/limited phagosome acidification may be essential for virulence [36]. Indeed, it has been proposed that acidification allows bacteria to gain access to essential nutrients such as iron [36]. However, an additional role for acidification in regulation or activation of virulence factors that disrupt membrane trafficking to sustain bacterial growth and survival cannot be excluded.

In addition, our understanding of the composition of *Francisella* phagosomes is incomplete, and the fact that these organelles are only moderately acidic suggests that they may not be identical to conventional late endosomes/phagosomes. Indeed, our recent data indicate that MDM phagosomes containing live *Francisella* (LVS, the virulent type A strain SchuS4 or *F. tularensis* subsp. *novicida*) exclude multiple late endosome markers including Rab7, lysobisphosphatidic acid (LBPA), the mannose 6-phosphate receptor (M6PR) and cathepsin S, despite their ability to acquire LAMP1 and CD63 [104] (Figure 23.3). By definition [37], the absence of LBPA and Rab7 demonstrates that *Francisella* does not reside in a normal late phagosome. Thus, one question raised by these data is how *Francisella* alters membrane trafficking to accumulate selectively late endosome membrane glycoproteins. Possible mechanisms include direct acquisition of LAMP1, LAMP2 and CD63 from the biosynthetic pathway via the TGN, fusion with late endosomes followed by rapid remodeling to remove M6PR, LBPA, Rab7 and matrix enzymes, or a variation of "kiss and run" fusion [38].

23.2.3
Phagosome Escape and Replication in the Cytosol

The ability of *Francisella* to grow in macrophages is an essential element of virulence, and bacterial replication begins approximately 6 h after uptake [3]. In 2004 it was discovered that *F. tularensis* replicates in the macrophage cytosol and not inside phagosomes [33, 39]. High-resolution electron micrographs of MDM indicate that beginning 3 h after uptake the phagosome membrane is coated on its cytosolic face with dense fibrillar material, and that the membrane often appears to be forming buds and fragmenting [33, 39]. By 6–9 h post infection, bacteria completely lack a limiting phagosomal membrane, and this correlates with local disappearance of LAMP1 and CD63 as judged by confocal microscopy [33, 34]. Although this general paradigm has been validated by several research groups, it is noteworthy that *Francisella* traverses the endosomal pathway more rapidly, and escapes the phagosome more efficiently in murine macrophages than in with human mononuclear phagocytes. Consequently, within 2 h of uptake at least 50% of ingested *Francisella* appear free in the cytosol of primary murine peritoneal macrophages, BMM or J774 cells, whereas 4–9 h are required for a similar fraction of bacteria to escape the phagosome membrane in human MDM or THP-1 or U937 cells [33–35, 40, 41]. What accounts for this discordance remains to be determined since the data cannot be attributed to differences in phagosome composition, are independent of *Francisella* subspecies and the assay used to assess phagosome disruption, and do not correlate with cultivation of bacteria on solid or liquid media.

23.2.4
Inhibition of the Neutrophil Respiratory Burst

Although most studies of *Francisella* have focused on the fate of this organism in macrophages, neutrophil accumulation in the lung is essential for the development of severe pneumonic tularemia, and studies of monkeys, mice and other animals have demonstrated that infected PMN accumulate in bronchioles and alveoli yet are unable to control infection [42–45]. Additionally, blockade of PMN migration into the lung favors mouse survival even after infection with highly virulent type A organisms [46]. These observations are significant since under normal circumstances neutrophils provide an essential first line of defense against infection. Central to the killing arsenal of neutrophils is the NADPH oxidase, a multicomponent enzyme that catalyzes the conversion of molecular oxygen into superoxide anions. In resting cells the oxidase is inactive, with subunits segregated in the membrane ($gp91^{phox}$, $p22^{phox}$) and cytosol ($p47^{phox}$, $p67^{phox}$, $p40^{phox}$ and the GTPase Rac2). During phagocytosis, NADPH oxidase complexes assemble rapidly on forming phagosomes and oxidants are generated in the phagosome lumen during the first 90 min post infection [47–49].

Pathogens that evade elimination by PMN must either evade or withstand toxic oxidants. Studies performed several years ago demonstrated that neither clinical isolates of *F. tularensis* nor LVS activate human or rhesus monkey neutrophils as indicated by hexose monophosphate shunt activity or production of superoxide, and ~95% of ingested bacteria remained viable [19, 50–52]. We have confirmed and extended these findings to show that *F. tularensis* opsonized with normal serum is rapidly engulfed, yet prevents local NADPH oxidase assembly and activation [20, 49]. Thus, *Francisella* compartments lack membrane and soluble NADPH oxidase subunits and do not accumulate superoxide or other reactive oxygen species [20, 49]. Moreover, within 10 min of infection, the ability of PMN to be activated by heterologous particulate and soluble stimuli, including yeast zymosoan and phorbol esters, is also profoundly impaired [20, 49]. Altogether, these data demonstrate that *Francisella* inhibits overall PMN function, and our unpublished data [105] suggest that macrophage and monocyte NADPH oxidase activity is also curtailed.

Blockade of the respiratory burst allows *Francisella* to evade rapid elimination by resting and activated PMNs. Later in infection, these organisms escape the phagosome and enter the neutrophil cytosol [20]. Viable bacteria persist in this locale for at least 24 h but do not appear to replicate [20]. The ability of *F. tularensis* to escape neutrophil phagosomes and persist in the cytosol may be a distinct element of virulence since other pathogens that escape the phagosome in macrophages and/or epithelial cells, such as *Listeria* and *Shigella*, are eliminated by PMNs. Indeed, neutrophil elastase degrades the *Shigella* virulence factors required for phagosome escape, and *Listeria* is controlled, at least in part, by toxic oxidants [53–55].

23.3
F. tularensis Virulence Factors that Disrupt Phagocyte Function

Despite the fact that *F. tularensis* subsp. *tularensis* is a highly infectious and virulent bacterium, relatively little is known about its mechanisms of pathogenesis. The *F. tularensis* genome does not encode any classical toxins or the type III and type IV secretion systems frequently used by pathogenic bacteria to disrupt host cell function [3]. Nevertheless, the development of genetic tools for manipulation of this organism has provided some insight into mechanisms used to disrupt phagocyte function.

23.3.1
Lipopolysaccharide and Capsule

The lipopolysaccharide (LPS, endotoxin) of most Gram-negative organisms is a potent agonist of host proinflammatory responses. In marked contrast, the endotoxin of type A and type B strains of *F. tularensis* has remarkably little bioactivity [56]. Thus, *F. tularensis* LPS is not a pyrogen and is a relatively poor inducer of TNFα and interleukin 1 (IL-1) in macrophages. The unusual tetraacylated structure of this LPS accounts for its inertness and renders cell activation via Toll-like receptor 4 inefficient [57–62]. Conversely, endotoxin of *F. tularensis* subsp. *novicida* has a different composition and is significantly more bioactive [3, 63, 64].

F. tularensis also expresses a surface exopolysaccharide capsule-like material that is essential for virulence in mice and guinea-pigs [3]. This structure is composed predominantly of mannose but also contains amino acids and fatty acids [3]. LPS and capsule collaborate to trigger activation of the complement cascade [65]; and, as noted above, serum opsonins are essential for infection of neutrophils and enhance phagocytosis by macrophages [14, 18, 33, 50]. At the same time, capsule and LPS collaborate to protect the organism from the lethal effects of late complement effectors, and as such confer serum-resistance [3, 66]. Therefore, mutations affecting LPS biogenesis decrease intramacrophage survival [67, 68], but whether phagosome maturation is affected directly has not been determined, and it is possible that the reduced fitness of these organisms is indirect and reflects membrane perturbations introduced by the membrane attack complex of the complement cascade.

23.3.2
Francsiella Pathogenicity Island

The fact that all *F. tularensis* isolates studied to date escape the phagosome and replicate in the cytosol of human and murine macrophages suggests that this process is controlled by conserved virulence determinants. Indeed, the genome includes a region of DNA called the *Francisella* pathogenicity island (FPI) that contains 17 putative open reading frames arranged in two large operons, and is duplicated in type A and B isolates but not in *F. tularensis* subsp. *novicida* [69]. The significance of this gene dosage difference is unclear, but it has been proposed that duplication of the FPI may enhance virulence of *tularensis* and *holarctica* organisms. Mutagenesis studies have shown that FPI genes *iglA*, *iglC*, *iglD* and *pdpA*, are essential for *Francisella* replication in macrophages and virulence in mice; and microcopy analyses indicate that mutant organisms reside in compartments with lysosomal characteristics and do not reach the cytosol [69–73]. Expression of FPI genes is driven by the transcriptional activator MglA, which is also required for disruption of phagosome maturation, intracellular growth and virulence [74]. Recent data indicate that MglA also controls expression of ~100 genes outside the FPI, one or more of which may also be required to prevent phagosome–lysosome fusion.

Although the mechanisms of action of FPI encoded proteins and their host cell targets are unknown, bioinformatics analyses suggest that genes in the FPI are somewhat similar to a set of genes in *Vibrio cholerae* and *Pseudomonas aeruginosa* that encode a novel type VI secretion system [75, 76]. While this model is attractive, whether *Francisella* contains a functional type VI secretion system remains to be determined.

23.3.3
AcpA

How *Francisella* inhibits PMN activation is unknown. One candidate virulence factor of interest is AcpA, a 57 kDa, soluble, periplasmic acid phosphatase [77]. Purified AcpA catalyzes the hydrolysis of a wide variety of phosphomonoesters *in vitro* and is sufficient to impair activation of porcine neutrophils by formyl peptides or PMA [77]. Nevertheless, the role of this enzyme in *Francisella* virulence is controversial since transposon-mediated disruption of *acpA* in *F. tularensis* subsp. *novicida* has no effect on bacterial survival in J774 cells or virulence in C57Bl/6 mice [78], whereas deletion of *acpA* slows bacterial escape from THP-1 cell phagosomes and has a modest effect on bacterial fitness in BALB/c mice (increasing time to death from 2 to 3 days) [41]. In these studies, the respiratory burst was not measured. Thus, the extent to which AcpA impairs NAPDH oxidase activity of *Francisella*-infected macrophages has not been determined, and the fate of these mutants in neutrophils is also unknown. To complicate matters further, the *Francisella* proteome suggests that virulent *F. tularensis* strains contain large amounts of AcpA but LVS does not [79], and analysis of the SchuS4 genome suggests that *Francisella* expresses three additional acid phosphatases: AcpB, AcpC, AcpD and a histidine acid phosphatase (Hap) [80, 81].

The role of these proteins in virulence is unknown, but Hap exhibits homology to a *Legionella micdadei* acid phosphatase that has also been implicated in inhibition of neutrophil superoxide production [79, 82].

23.4
Immunology of Phagosome and Host Defense

Early in infection, *Francisella* inhibits secretion of TNFα and other proinflammatory cytokines by human mononuclear phagocytes [83–85] and uncouples antigen presentation from cytokine production in murine DCs [86]. These data are in keeping with our demonstration that *Francisella* phagosomes exclude cathepsin S. Determining the mechanism by which *Francisella* alters cytokine production is an area of active investigation. IglC impairs intracellular signaling pathways required for cytokine synthesis [85], and, as noted above, the unusual structure of *F. tularensis* LPS likely accounts for its low bioactivity and prevents cell activation via TLRs [85, 87].

Bacteria in the cytosol of macrophages are detected by members of the NOD-like receptor family, which trigger activation of the inflammasome [88, 89]. This multisubunit complex is required for activation of caspase-1, which in turn induces synthesis and secretion of IL-1β and IL-18 and triggers pyroptosis [88]. This mechanism of proinflammatory cell death plays an essential role in host defense both by initiating an immune response and by depriving intracellular pathogens of their protective niche. This pathway is triggered by *Francisella* in macrophages [89, 90], and is essential for immunity in mice [89]. At the same time, several *Francisella* genes were recently identified that may retard inflammasome activation [91].

Late in infection, IFNγ, TNFα and IL-12 all play important roles in murine resistance to LVS and *F. tularensis* subsp. *novicida* [92–99]. IL-12 is a potent stimulus for production of IFNγ which in turn induces nitric oxide synthase (iNOS) and synthesis of NO by macrophages. Studies of iNOS-null animals confirm a key role for reactive nitrogen intermediates in *Francisella* control [100]. IFNγ also limits bacterial growth inside murine alveolar macrophages by an NO-independent mechanism [101]. In contrast to mouse cells, human macrophages generate NO under a very limited set of conditions [102], and specific mechanisms of control relevant to human tularemia are not well defined. However, recent data indicate that IFNγ enhances maturation of *F. tularensis* subsp. *novicida* phagosomes in human MDMs and impairs phagosome escape [103]. Thus, elimination of bacteria inside mature phagolysosomes may require IFNγ as well as specific antibodies.

23.5
Summary and Outlook

Our understanding of the pathogenesis of tularemia has advanced dramatically in the past few years, and the general features of the *Francisella* lifecycle in macrophages are now established. Collectively, the data support the notion that *F. tularensis* is a stealth pathogen that disrupts many aspects of macrophage function and innate defense. Nevertheless, relative to other pathogens such as *Salmonella* or *M. tuberculosis*, we know very little about *F. tularensis*–phagocyte interactions at the molecular level, and many important questions remain unanswered. For example, the receptors that mediate uptake of unopsonized *F. tularensis* by MR-negative monocytes and J774 cells have not been described. Recent data have begun to define the niche occupied by *Francisella* in macrophages, and bacteria reside in a novel compartment prior to phagosome egress and replication in the cytosol. Nevertheless, how this compartment is generated is unclear, and the extent to which *Francisella* antigens are presented to T cells has not been precisely defined. Also unknown is the composition of the *Francisella* compartment in other cell types, including neutrophils and DCs. Thus, in future studies it will be of interest to ascertain whether inhibition of neutrophil activation by *Francisella* is coupled to defects in granule mobilization.

Although recent genetic screens have expanded the list of putative *Francisella* virulence factors, the function and host cell targets for these proteins remain elusive. Moreover, the hypothesis that the FPI encodes a type VI secretion system is attractive, but this remains to be empirically demonstrated. Identification of putative secreted effector proteins of this system would be a major step towards characterization of the host cell factors targeted by this pathogen to allow intracellular survival. On the other hand, AcpA has well-characterized enzymatic activities but its role in virulence is controversial. The acid phosphatase and putative phospholipase activities of AcpA suggest it could be involved in blockade of the respiratory burst. Nevertheless, although acid phosphatases are present in many pathogens, a role for these enzymes in virulence has never been definitively demonstrated. Given the ability of *F. tularensis* to cause a wide spectrum of disease, we must also consider the possibility that virulence factor expression may be affected by route of infection or bacterial residence in different tissues and cell types. This issue is of particular note since, despite recent advances, we still do not know what accounts for the pronounced difference in virulence of *Francisella* subspecies in humans.

Acknowledgments

Work on this topic on the Allen laboratory is supported by funds from the National Institutes of Health (R01 AI073835 and P01 AI44642 to L.A.). G.S.S. was supported by a predoctoral fellowship (NIH T32-AI07343-17).

References

1 Tarnvik, A. and Berglund, L. (2003) *Eur Respir J*, **21**, 361–373.

2 McLendon, M.K., Apicella, M.A. and Allen, L.-A.H. (2006) *Annu Rev Microbiol*, **60**, 167–185.

3 McLendon, M.K., Apicella, M. and Allen, L.-A.H. (2006) *Annu Rev Microbiol*, **60**, 167–185.

4 Rick Lyons, C. and Wu, T.H. (2007) *Ann N Y Acad Sci*, **1105**, 238–265.

5 Conlan, J.W., Chen, W., Shen, H., Webb, A. and KuoLee, R. (2003) *Microb Pathog*, **34**, 239–248.

6 Elkins, K.L., Cowley, S.C. and Bosio, C.M. (2007) *Ann N Y Acad Sci*, **1105**, 284–324.

7 Chen, W., KuoLee, R., Shen, H. and Conlan, J.W. (2004) *Microb Pathog*, **36**, 311–318.

8 White, J.D., Rooney, J.R., Prickett, P.A., Derrenbacher, E.B., Beard, C.W. and Griffith, W.R. (1964) *J Infect Dis*, **114**, 277–283.

9 Andersson, H., Hartmanova, B., Kuolee, R., Ryden, P., Conlan, W., Chen, W. and Sjostedt, A. (2006) *J Med Microbiol*, **55**, 263–271.

10 Bosio, C.M., Bielefeldt-Ohmann, H. and Belisle, J.T. (2007) *J Immunol*, **178**, 4538–4547.

11 Metzger, D.W., Bakshi, C.S. and Kirimanjeswara, G. (2007) *Ann N Y Acad Sci*, **1105**, 266–283.

12 Kirimanjeswara, G.S., Golden, J.M., Bakshi, C.S. and Metzger, D.W. (2007) *J Immunol*, **179**, 532–539.

13 Santic, M., Molmeret, M., Klose, K.E. and Abu Kwaik, Y. (2006) *Trends Microbiol*, **14**, 37–44.

14 Schulert, G.S. and Allen, L.A.H. (2006) *J Leukoc Biol*, **80**, 563–571.

15 Balagopal, A., MacFarlane, A.S., Mohapatra, N., Soni, S., Gunn, J.S. and Schlesinger, L.S. (2006) *Infect Immun*, **74**, 5114–5125.

16 Lai, X.H., Golovliov, I. and Sjostedt, A. (2004) *Microb Pathog*, **37**, 225–230.

17 Bolger, C.E., Forestal, C.A., Italo, J.K., Benach, J.L. and Furie, M.B. (2005) *J Leukoc Biol*, **77**, 893–897.

18 Clemens, D.L., Lee, B.Y. and Horwitz, M.A. (2005) *Infect Immun*, **73**, 5892–5902.

19 Lofgren, S., Tarnvik, A. and Carlsson, J. (1980) *Infect Immun*, **29**, 329–334.

20 McCaffrey, R.L. and Allen, L.-A.H. (2006) *J Leukoc Biol*, **80**, 1224–1230.

21 Aderem, A. and Underhill, D.M. (1999) *Annu Rev Immunol*, **17**, 593–623.

22 Alpuche-Aranda, C., Racoosin, E.L., Swanson, J.A. and Miller, S.I. (1994) *J Exp Med*, **179**, 601–608.

23 Allen, L.-A.H. and Aderem, A. (1996) *J Exp Med*, **184**, 627–637.

24 Ben Nasr, A., Haithcoat, J., Masterson, J.E., Gunn, J.S., Eaves-Pyles, T. and Klimpel, G.R. (2006) *J Leukoc Biol*, **80**, 774–786.

25 Pierini, L.M. (2006) *Cell Microbiol*, **8**, 1361–1370.

26 Schlesinger, L.S. (1993) *J Immunol*, **150**, 2920–2930.

27 Wilson, M.E. and Pearson, R.D. (1988) *Infect Immun*, **56**, 363–369.

28 Zimmerli, S., Edwards, S. and Ernst, J.D. (1996) *Am J Respir Cell Mol Biol*, **15**, 760–770.

29 Aderem, A. and Underhill, D.M. (1999) *Annu Rev Immunol*, **17**, 593–623.

30 Zhang, J., Tachado, S.D., Patel, N., Zhu, J., Imrich, A., Manfruelli, P., Cushion, M., Kinane, T.B. and Koziel, H. (2005) *J Leukoc Biol*, **78**, 665–674.

31 Bolger, C.E., Forestal, C.A., Italo, J.K., Benach, J.L. and Furie, M.B. (2005) *J Leukoc Biol*, **77**, 893–897.

32 Anthony, L.S.D., Burke, R.D. and Nano, F.E. (1991) *Infect Immun*, **59**, 3291–3296.

33 Clemens, D.L., Lee, B.Y. and Horwitz, M.A. (2004) *Infect Immun*, **72**, 3204–3217.

34 Santic, M., Molmeret, M., Klose, K.E., Jones, S. and Kwaik, Y.A. (2005) *Cell Microbiol*, **7**, 969–979.

35 Checroun, C., Wehrly, T.D., Fischer, E.R., Hayes, S.F. and Celli, J. (2006) *Proc Natl Acad Sci USA*, **103**, 14578–14583.

36 Fortier, A.H., Leiby, D.A., Narayanan, R.B., Asafoadjei, E., Crawford, R.M., Nacy, C.A. and Meltzer, M.S. (1995) *Infect Immun*, **63**, 1478–1483.

37 Gruenberg, J. and van der Goot, F.G. (2006) *Nat Rev Mol Cell Biol*, **7**, 495–504.

38 Desjardins, M. (1995) *Trends Cell Biol*, **5**, 183–186.

39 Golovliov, I., Baranov, V., Krocova, Z., Kovarova, H. and Sjostedt, A. (2003) *Infect Immun*, **71**, 5940–5950.

40 Golovilov, I., Ericsson, M., Sandstrom, G., Tarnvik, A. and Sjostedt, A. (1997) *Infect Immun*, **65**, 2183–2189.

41 Mohapatra, N.P., Balagopal, A., Soni, S., Schlesinger, L.S. and Gunn, J.S. (2007) *Infect Immun*, **75**, 390–396.

42 Schricker, R.L., Eigelsbach, H.T., Mitten, J.O. and Hall, W.C. (1972) *Infect Immun*, **5**, 734–744.

43 Tulis, J.J., Eigelsbach, H.T. and Kerpsack, R.W. (1970) *Am J Pathol*, **58**, 329–336.

44 Hall, W.C., Kovatch, R.M. and Schricker, R.L. (1973) *J Pathol*, **110**, 193–201.

45 Chiavolini, D., Alroy, J., King, C.A., Jorth, P., Weir, S., Madico, G., Murphy, J.R. and Wetzler, L.M. (2008) *Infect Immun*, **76**, 486–496.

46 Malik, M., Bakshi, C.S., McCabe, K., Catlett, S.V., Shah, A., Singh, R., Jackson, P.L., Gaggar, A., Metzger, D.W., Melendez, J.A., Blalock, J.E. and Sellati, T.J. (2007) *J Immunol*, **178**, 1013–1020.

47 DeLeo, F.R., Allen, L.-A.H., Apicella, M. and Nauseef, W.M. (1999) *J Immunol*, **163**, 6732–6740.

48 Nauseef, W.M. (2004) *Histochem Cell Biol*, **122**, 277–291.

49 Allen, L.-A.H. and McCaffrey, R.L. (2007) *Immunol Rev*, **219**, 103–117.

50 Lofgren, S., Tarnvik, A., Bloom, G.D. and Sjoberg, W. (1983) *Infect Immun*, **39**, 715–720.

51 Proctor, R.A., White, J.D., Ayala, E. and Canonico, P.G. (1975) *Infect Immun*, **11**, 146–152.

52 Lofgren, S., Tarnvik, A., Thore, M. and Carlsson, J. (1984) *Infect Immun*, **43**, 730–734.

53 Weinrauch, Y., Drujan, D., Shapiro, S.D., Weiss, J. and Zychlinsky, A. (2002) *Nature*, **417**, 91–94.

54 Mandic-Mulec, I., Weiss, J. and Zychlinsky, A. (1997) *Infect Immun*, **65**, 110–115.

55 Kobayashi, S.D., Braughton, K.R., Whitney, A.R., Voyich, J.M., Schwan, T.G., Musser, J.M. and DeLeo, F.R. (2003) *Proc Natl Acad Sci USA*, **100**, 10948–10953.

56 Ancuta, P., Pedron, T., Girard, R., Sandstrom, G. and Chaby, R. (1996) *Infect Immun*, **64**, 2041–2046.

57 Katz, J., Zhang, P., Martin, M., Vogel, S.N. and Michalek, S.M. (2006) *Infect Immun*, **74**, 2809–2816.

58 Duenas, A.I., Aceves, M., Orduna, A., Diaz, R., Sanchez Crespo, M. and Garcia-Rodriguez, C. (2006) *Int Immunol*, **18**, 785–795.

59 Chen, W., KuoLee, R., Shen, H., Busa, M. and Conlan, J.W. (2004) *Microb Pathog*, **37**, 185–191.

60 Vinogradov, E., Perry, M.B. and Conlan, J.W. (2002) *Eur J Biochem*, **269**, 6112–6118.

61 Phillips, N.J., Schilling, B., McLendon, M.K., Apicella, M. and Gibson, B.W. (2004) *Infect Immun*, **72**, 5340–5348.

62 Barker, J.H., Weiss, J., Apicella, M.A. and Nauseef, W.M. (2006) *Infect Immun*, **74**, 3277–3284.

63 McLendon, M.K., Schilling, B., Hunt, J.R., Apicella, M.A. and Gibson, B.W. (2007) *Infect Immun*, **75**, 5518–5531.

64 Schilling, B., McLendon, M.K., Phillips, N.J., Apicella, M.A. and Gibson, B.W. (2007) *Anal Chem*, **79**, 1034–1042.

65 Sorokin, V.M., Pavlovich, N.V. and Prozorova, L.A. (1996) *FEMS Immunol Med Microbiol*, **13**, 249–252.

66 Sandstrom, G., Lofgren, S. and Tarnvik, A. (1988) *Infect Immun*, **56**, 1194–1202.

67 Thomas, R.M., Titball, R.W., Oyston, P.C., Griffin, K., Waters, E., Hitchen, P.G., Michell, S.L., Grice, I.D., Wilson, J.C. and Prior, J.L. (2007) *Infect Immun*, **75**, 371–378.

68 Raynaud, C., Meibom, K.L., Lety, M.A., Dubail, I., Candela, T., Frapy, E. and Charbit, A. (2007) *Infect Immun*, **75**, 536–541.

69 Nano, F.E., Zhang, N., Cowley, S.C., Klose, K.E., Cheung, K.K., Roberts, M.J., Ludu, J.S., Letendre, G.W., Meierovics, A.I., Stephens, G. and Elkins, K.L. (2004) *J Bacteriol*, **186**, 6430–6436.

70 de Bruin, O.M., Ludu, J.S. and Nano, F.E. (2007) *BMC Microbiol*, **7**, 1.

71 Santic, M., Molmeret, M., Barker, J.R., Klose, K.E., Dekanic, A., Doric, M. and Abu Kwaik, Y. (2007) *Cell Microbiol*, **9**, 2391–2403.

72 Gray, C.G., Cowley, S.C., Cheung, K.K. and Nano, F.E. (2002) *FEMS Microbiol Lett*, **215**, 53–56.

73 Golovliov, I., Sjostedt, A., Mokrievich, A. and Pavlov, V. (2003) *FEMS Microbiol Lett*, **222**, 273–280.

74 Lauriano, C.M., Barker, J.R., Yoon, S.S., Nano, F.E., Arulanandam, B.P., Hassett, D.J. and Klose, K.E. (2004) *Proc Natl Acad Sci USA*, **101**, 4246–4249.

75 Mougous, J.D., Cuff, M.E., Raunser, S., Shen, A., Zhou, M., Gifford, C.A., Goodman, A.L., Joachimiak, G., Ordonez, C.L., Lory, S., Walz, T., Joachimiak, A. and Mekalanos, J.J. (2006) *Science*, **312**, 1526–1530.

76 Pukatzki, S., Ma, A.T., Sturtevant, D., Krastins, B., Sarracino, D., Nelson, W.C., Heidelberg, J.F. and Mekalanos, J.J. (2006) *Proc Natl Acad Sci USA*, **103**, 1528–1533.

77 Reilly, T.J., Baron, G.S., Nano, F.E. and Kuhlenschmidt, M.S. (1996) *J Biol Chem*, **271**, 10973–10983.

78 Baron, G.S., Reilly, T.J. and Nano, F.E. (1999) *FEMS Microbiol Lett*, **176**, 85–90.

79 Hernychova, L., Halada, P., Macela, A., Kroca, M., Johansson, T. and Malina, M. (2001) *Proteomics*, **1**, 508–515.

80 Larsson, P., Oyston, P.C., Chain, P., Chu, M.C., Duffield, M., Fuxelius, H.H., Garcia, E., Halltorp, G., Johansson, D., Isherwood, K.E., Karp, P.D., Larsson, E., Liu, Y., Michell, S., Prior, J., Prior, R., Malfatti, S., Sjostedt, A., Svensson, K., Thompson, N., Vergez, L., Wagg, J.K., Wren, B.W., Lindler, L.E., Andersson, S.G., Forsman, M. and Titball, R.W. (2005) *Nat Genet*, **37**, 153–159.

81 Felts, R.L., Reilly, T.J., Calcutt, M.J. and Tanner, J.J. (2006) *Acta Crystallogr Sect F Struct Biol Cryst Commun*, **62**, 32–35.

82 Saha, A.K., Dowling, J.N., LaMarco, K.L., Das, S., Remaley, A.T., Olomu, N., Pope, M.T. and Glew, R.H. (1985) *Arch Biochem Biophys*, **243**, 150.

83 Bolger, C.E., Forestal, C.A., Italo, J.K., Benach, J.L. and Furie, M.B. (2005) *J Leukoc Biol*, **77**, 893–897.

84 Telepnev, M., Golovilov, I. and Sjostedt, A. (2005) *Microb Pathog*, **38**, 239–247.

85 Telepnev, M., Golovliov, I., Grundstrom, T., Tarnvik, A. and Sjostedt, A. (2003) *Cell Microbiol*, **5**, 41–51.

86 Bosio, C.M. and Dow, S.W. (2005) *J Immunol*, **175**, 6792–6801.

87 Sandstrom, G., Sjostedt, A., Johansson, T., Kuoppa, K. and Williams, J.C. (1992) *FEMS Microbiol Immunol*, **5**, 201–210.

88 Mariathasan, S. and Monack, D.M. (2007) *Nat Rev Immunol*, **7**, 31–40.

89 Mariathasan, S., Weiss, D.S., Dixit, V.M. and Monack, D.M. (2005) *J Exp Med*, **202**, 1043–1049.

90 Gavrilin, M.A., Bouakl, I.J., Knatz, N.L., Duncan, M.D., Hall, M.W., Gunn, J.S. and Wewers, M.D. (2006) *Proc Natl Acad Sci USA*, **103**, 141–146.

91 Weiss, D.S., Brotcke, A., Henry, T., Margolis, J.J., Chan, K. and Monack, D.M. (2007) *Proc Natl Acad Sci USA*, **104**, 6037–6042.

92 Golovliov, I., Sandstrom, G., Ericsson, M., Sjostedt, A. and Tarnvik, A. (1995) *Infect Immun*, **63**, 534–538.

93 Elkins, K.L., Rhinehart-Jones, T.R., Culkin, S.J., Yee, D. and Winegar, R.K. (1996) *Infect Immun*, **64**, 3288–3293.

94 Leiby, D.A., Fortier, A.H., Crawford, R.M., Schreiber, R.D. and Nacy, C.A. (1992) *Infect Immun*, **60**, 84–89.

95 Elkins, K.L., Cooper, A., Colombini, S.M., Cowley, S.C. and Kieffer, T.L. (2002) *Infect Immun*, **70**, 1936–1948.

96 Duckett, N.S., Olmos, S., Durrant, D.M. and Metzger, D.W. (2005) *Infect Immun*, **73**, 2306–2311.

97 Fortier, A.H., Polsinelli, T., Green, S.J. and Nacy, C.A. (1992) *Infect Immun*, **60**, 817–825.

98 Anthony, L.S., Ghadirian, E., Nestel, F.P. and Kongshavn, P.A. (1989) *Microb Pathog*, **7**, 421–428.

99 Lindgren, H., Golovliov, I., Baranov, V., Ernst, R.K., Telepnev, M. and Sjostedt, A. (2004) *J Med Microbiol*, **53**, 953–958.

100 Lindgren, H., Stenmark, S., Chen, W., Tarnvik, A. and Sjostedt, A. (2004) *Infect Immun*, **72**, 7172–7182.

101 Polsinelli, T., Meltzer, M.S. and Fortier, A.H. (1994) *J Immunol*, **153**, 1238–1245.

102 Thomassen, M.J. and Kavuru, M.S. (2001) *Int Immunol*, **1**, 1479–1490.

103 Santic, M., Molmeret, M. and Abu Kwail, Y. (2005) *Cell Microbiol*, **7**, 957–967.

104 Schulert, G.S. and Allen, L.-A.H. (2008) *J Leukoc Biol*, **84**, Supplement 38.

105 Schulert, G.S., McCaffrey, R.L., Buchan, B.W., Lindemann, S.R., Jones, B.D. and Allen, L.-A.H. (2009) *Infect Immun*, **77**, 1324–1336.

24
Listeria monocytogenes

Damien Balestrino and Pascale Cossart

24.1
Listeria and Listeriosis

24.1.1
The Genus *Listeria*

Listeria monocytogenes was first discovered in 1926 by E.G.D. Murray who identified a Gram-positive bacterium as the cause of a lethal disease affecting rabbits and guinea-pigs in Cambridge (UK) [1]. Since these animals developed marked monocytosis, the microorganism was named *Bacterium monocytogenes*. A second report from South Africa described a Gram-positive bacterium as the cause of a lethal necrotizing hepatic infection affecting wild gerbils [2]. In honor of the British surgeon Joseph Lister, the organism was named *Listerella hepatolytica*. These organisms were then discovered to be identical, and as a compromise, the name *Listeria monocytogenes* was settled on.

L. monocytogenes is a facultative intracellular pathogen, facultative anaerobic, nonspore-forming Gram-positive bacillus that grows readily on blood agar, and produces incomplete β-hemolysis. The bacterium possesses flagella which confer a characteristic tumbling motility at 10–25 °C. Optimal growth ranges from 30 to 37 °C but, unlike most bacteria, *L. monocytogenes* grows well at refrigerator temperature (4–10 °C) and is very resistant to extreme conditions (pH, osmotic pressure). The high resistance of the bacterium to a broad range of conditions accounts for its wide distribution in nature. *L. monocytogenes* thrives in diverse environments such as soil, water, various food products and is part of the flora of many mammals. It can also be silently present in the gastrointestinal tract of healthy human adults [3].

Among the six listerial species (*L. monocytogenes*, *L. seeligeri*, *L. welshimeri*, *L. innocua*, *L. ivanovii* and *L. grayi*), only *L. monocytogenes* and *L. ivanovii* are pathogenic, *L. ivanovii* being strictly an animal pathogen. The infectious disease caused by these bacteria is known as listeriosis. *L. monocytogenes* causes serious localized and generalized infections in humans and a variety of other vertebrates.

Intracellular Niches of Microbes. Edited by Ulrich E. Schaible and Albert Haas

ISBN: 978-3-527-32207-7

Listeriosis is usually a severe disease with a mortality rate in humans of 20–30% or higher, despite early antibiotic treatment [3].

24.1.2
Listeriosis and Pathophysiology

The first human cases of listeriosis were reported in Denmark by Nyfeldt in 1929 [4]. Nevertheless, listeriosis was long considered as a zoonosis, and the fact that *L. monocytogenes* is also a human food-borne pathogen was not formally demonstrated until the 1980s [5]. *L. monocytogenes* is now unambiguously recognized as a human food-borne pathogen, but is an infrequent cause of illness in the general population. Some groups, however, including elderly patients, pregnant women, patients with chronic diseases and patients with innate or acquired immunodeficiencies, are particularly vulnerable to infection. *L. monocytogenes* causes a range of clinical manifestations from asymptomatic intestinal carriage and gastroenteritis to invasive and disseminated disease. Septicemia, meningoencephalitis and infection of the fetus in pregnant women are the most serious manifestations of listeriosis [3].

Pathogenesis of listeriosis results mainly from the ability of *L. monocytogenes* to invade, replicate and survive within several different cell types in mammalian tissues, including nonphagocytic cells and macrophages. Central for the pathophysiology is the ability of *L. monocytogenes* to cross three tissue barriers – the intestinal barrier, the hemato-encephalic barrier and the feto-placental barrier – by subversion of cellular effectors and functions that will allow bacterial internalization, replication and survival within different host cell types [6].

The gastrointestinal tract is the primary natural site of entry for pathogenic *Listeria*, and contaminated food is the major source of infection in both epidemic and sporadic cases (Figure 24.1) [7]. The enteric phase of the infection was previously considered to be almost always clinically silent. Nevertheless, it may lead to the development of a non-invasive form of listeriosis which may be presented as febrile gastroenteritis. This presentation may affect people with no known immunocompromising conditions following ingestion of high number of organisms (higher than 10^9 organisms), after which the patients develop fever and diarrhea within 24 h; the illness usually resolves spontaneously. In the population at risk, an invasive listeriosis can appear and the signs accounting for this generalized infection occur after an incubation period that can be as long as 10–70 days, with a mean incubation period of 31 days [3, 8].

At the intestinal level, the way by which *Listeria* precisely breaches the intestinal barrier is still under investigation. In humans, *L. monocytogenes* crosses it by an active invasion of epithelial cells and by a passive process through the Peyer's patches, suggesting a role for specialized phagocytic M cells in bacterial uptake [9–11]. The *L. monocytogenes* translocation across the intestinal barrier is followed by bacterial multiplication in the intestine lamina propria, and then by dissemination via the mesenteric lymphatics and the portal circulation to the liver and the spleen (Figure 24.1) [10]. There, bacteria are rapidly cleared from the bloodstream by resident macrophages (Kupffer cells), resulting in a decrease in the number of viable bacterial in the liver during the first 6 h after infection [10]. Resident macrophages produce

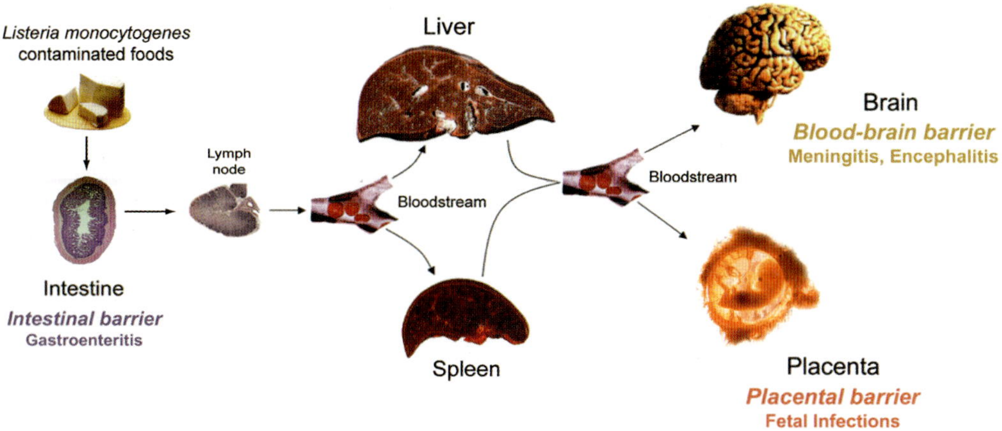

Figure 24.1 Schematic steps of human listeriosis. The gastrointestinal tract is the primary natural site of entry for *Listeria monocytogenes*. Bacterial translocation across the intestinal barrier is followed by their dissemination via the mesenteric lymphatics and the portal circulation to the liver and the spleen. Then, hematogenous dissemination of bacteria to the materno-fetal unit and the central nervous system may occur.

cytokines interleukin 6 (IL-6), IL-12 and tumor necrosis factor α (TNFα), which initiate proinflammatory responses that are important for recruiting neutrophils to the liver to kill *L. monocytogenes*. However, if not controlled properly by the immune system, infection may cause bacteremia which might be clinically silent but might also be responsible for an influenza-like febrile illness, accompanied by muscle pain, joint pain, headache and backache. Hematogenous dissemination of *L. monocytogenes* may then occur at any site, although the particular predilection sites are the materno-fetal unit and the central nervous system (CNS) (Figure 24.1) [10].

During pregnancy, listeriosis can occur at any time but it usually occurs in the third trimester, when pregnancy-associated immunosuppression is thought to be the most intense [3]. *Listeria*, present in the bloodstream, has access to the highly vascularized uterine decidua and can then colonize the trophoblast layer. Subsequently, the translocation of *L. monocytogenes* across the endothelial barrier causes placental abscesses, chorioamniotitis and enables the bacteria to reach the fetal bloodstream, leading to generalized infection and subsequent death of the fetus *in utero* or to premature birth of a severely infected neonate [10].

In addition to its ability to cross the intestinal barrier and the materno-fetal barrier, *Listeria* has the capacity to breach the blood–brain barrier and then reach the CNS. In humans, CNS infection by *L. monocytogenes* presents primarily in the form of meningitis and encephalitis, often associated with the presence of infectious foci in the brain parenchyma, especially in the brainstem [10]. How and where *Listeria* crosses the blood–brain barrier remains largely unknown. *L. monocytogenes* is thought to infect the CNS by a hematogenous route, but an intraaxonal bacterial transport from peripheral sites to CNS may also occur *in vivo* [12, 13].

Since its first identification, *L. monocytogenes* has attracted the attention of a wide range of investigators, including clinicians, food microbiologists, medical microbiologists and cellular biologists. In particular, due to its properties, *L. monocytogenes* is one of the very few intracellular organisms systematically used to study the mechanism underlying the induction and establishment of a T cell response. Since the pioneering studies of Mackaness in the 1960s which demonstrated that cellular immunity was critical for control of *L. monocytogenes* infection in mice, *L. monocytogenes* has become a useful and informative model for elucidating important mechanisms of the immune response to intracellular bacterial pathogens [14].

Both the innate and adaptive immune systems are crucial to the recognition and elimination of *L. monocytogenes* from the host [15]. Infection is first recognized by the innate immune system, which plays an essential role in controlling bacterial growth and dissemination, preventing the spread into systemic, lethal infection. Neutrophils and macrophages, especially Kupffer cells, are the principal mediators of initial killing of the majority of bacteria. Moreover, secretion of TNFα and IL-12 by macrophages triggers the production of interferon γ (INFγ) by natural killer (NK) cells, which in turn leads to activation of macrophages and increases their bactericidal activity. INFγ and TNFα are essential for controlling a primary *L. monocytogenes* infection, since mice that lack these cytokines or their cognate receptors rapidly succumb to infection [16, 17]. Chemokines and cytokines also aid in initiation of the *L. monocytogenes*-specific adaptive immune response, which follows the innate immune response.

Dendritic cells (DCs) are an important link between innate and adaptive immunity. DCs present *L. monocytogenes* antigens and, by this way, drive strong CD4 and CD8 T cell responses that result in a stable population of *L. monocytogenes*-specific memory T cells. The development of a *Listeria*-specific T cell response is important to confer sterilizing immunity since severe combined immunodeficiency (SCID) mice develop a chronic infection, and memory T cells provide lifelong protective immunity [16, 17]. The generation of this protective immunity to rechallenge is dependent on immunization with live *L. monocytogenes*, capable of escaping into the cytosol. It is believed that live infection provides a stimulus that induces T cell differentiation, whereas immunization with heat-killed bacteria provides a distinct stimulus that induces neither the full differentiation of primed T cells nor the protective immunity.

While T cell responses are essential to control infection, *L. monocytogenes*-specific antibodies play no critical role in recovery from infection and protection against a secondary infection since the majority of the bacteria remain intracellular and spread intercellularly during infection. Thus, mice that lack both T cell and humoral immunity are relatively resistant to infection with *L. monocytogenes* at early time points but are unable to clear bacteria in the long term [16, 17].

Most of the *L. monocytogenes* determinants involved in virulence have first been identified by classical genetics (search for a mutant with a specific phenotype followed by gene identification and in-depth analysis) [18]. The first *L. monocytogenes* virulence factor identified by the genetic approach was the hemolysin (listeriolysin O or LLO). In 1986, a transposon mutagenesis was efficiently used in *L. monocytogenes* to isolate hemolysin-negative mutants [19]. Using similar approaches, two papers published soon after referred to the isolation of *L. monocytogenes* hemolysin-negative mutants [20, 21]. In 1987, Mengaud *et al.* sequenced the LLO open reading frame and revealed that LLO is homologous to streptolysin O and pneumolysin [22]. Since its first description, LLO has been extensively studied and is now believed to be one of the main virulence factors of *Listeria*. In 2001, the publication of the complete genomes of *L. monocytogenes* and that of the nonpathogenic species *L. innocua* paved the way towards identification of other virulence factors, foremost among them, surface and secreted proteins [23]. A result of the analysis of the complete genome sequences is that it became possible to describe a large number of genes encoding proteins with unknown functions or similar to poorly characterized proteins. Postgenomic analyses (i.e., systematic gene disruption) are then necessary to identify function for these proteins [24]. Access to the complete genome sequence also made it possible to start large-scale transcriptome and proteome studies. Recent advances in the understanding of the cellular and molecular aspects of the *L. monocytogenes* infection are presented here. They illustrate strikingly how a pathogen can exploit cell functions to its own advantage.

24.2 The Intracellular Lifestyle of *Listeria* and its Relationship to Virulence

L. monocytogenes is a facultative intracellular pathogen which has evolved multiple strategies to invade a large panel of mammalian cells. Host cell invasion, critical for listeriosis pathology, is a multifactorial process which requires several key bacterial factors. After entry inside the cells, *L. monocytogenes* lyses the internalization vacuole and escapes into the host cell cytoplasm, where bacterial replication takes place (Figure 24.2). Inside cells, *L. monocytogenes* then develops an intracellular lifecycle which does not lead to the death of the host cell. Indeed, unlike many other facultative intracellular pathogens, *L. monocytogenes* has evolved mechanisms to avoid killing of its host cell, even until late infection. In the cytosol, free *Listeria* are then able to move using an actin-based motility process to propel themselves (Figure 24.2). This movement is random. Bacteria which come into contact with cytoplasmic membrane stimulate the formation of finger-like protrusions with a bacterium at the tip. In this way, *L. monocytogenes* can penetrate uninfected neighboring cells by becoming engulfed in a double-membrane protrusion, resulting in the formation of a so-called secondary vacuole (Figure 24.2). This compartment is then lysed, allowing a new intracellular infection cycle in adjacent cells without the bacterium coming into contact with the extracellular environment, enabling it to evade the host immune response [18, 24].

24.2.1 Entry in Host Cell

L. monocytogenes can invade and replicate in a wide variety of cells *in vivo* and in tissue culture. It is phagocytosed by professional phagocytic cells, such as macrophages or dendritic cells, and it induces its own internalization in various cell types that are not

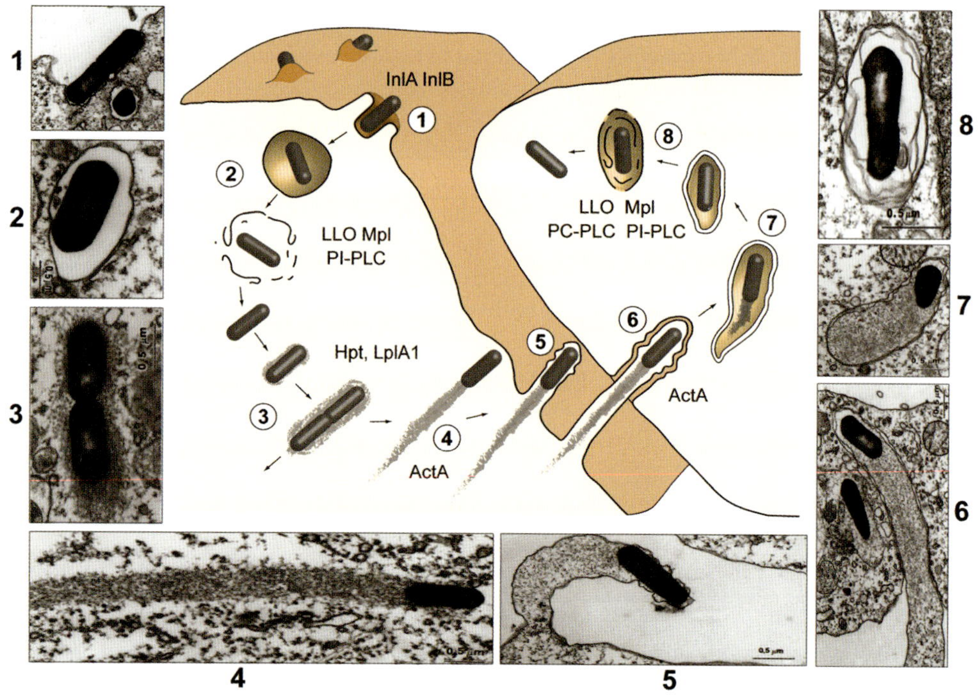

Figure 24.2 Schematic representation and electron micrographs of the *Listeria monocytogenes* intracellular lifecycle stages. The main bacterial factors involved, all described in the text, are indicated. (1) *L. monocytogenes* induces its entry into nonprofessional phagocytes by a dynamic process that requires local cytoskeletal rearrangement, actin polymerization and membrane remodeling. (2) After engulfment, bacteria are present an in internalization vacuole. The vacuole is then disrupted by the action of the secreted bacterial factors. (3) Inside the cytoplasm, bacteria multiply and are rapidly surrounded by a cloud of host actin filaments, which rearrange to form characteristic comet tails with a bacterium at the apex of the tail (Figure 24.3). (4–6) The actin-based motility processes allow bacteria to penetrate into a neighboring cell by the formation of protrusions in the plasma membrane. (7–8) Once inside the neighboring cell, *L. monocytogenes* is engulfed in a double-membrane vacuole, which is then lysed to to perpetuate the cycle.

normally phagocytic, such as enterocytes, hepatocytes, fibroblasts and endothelial cells [18, 25, 26].

Similar to the situation for most pathogens, the invasion of macrophages by *L. monocytogenes* is a passive process, whereas the entry into nonprofessional phagocytes cells is a dynamic process that requires local cytoskeletal rearrangement, actin polymerization and membrane remodeling. It involves a zipper-type mechanism, in that the bacterium gradually sinks into dip-like structures of the host cell surface until it is finally engulfed [25, 27]. This mechanism is common to *Listeria* and *Yersinia*, and differs from the "trigger" mechanism described for the entry of *Salmonella* and *Shigella* into cells, characterized by the formation of the spectacular local processes or membrane ruffles. *L. monocytogenes* entry is induced by binding of the bacterial surface proteins InlA and InlB to receptors on the host cell. Both of these invasins are necessary and sufficient for bacterial entry into nonphagocytic host cell (Figure 24.2). However, InlA-mediated entry is restricted to a smaller number of cell types that express its receptor [25, 26]. Few other bacterial factors involved in *L. monocytogenes* adhesion and/or entry in host cell have been described. Only InlA- and InlB-mediated entries, the two major entry pathways, are discussed in this section.

Internalin (InlA) and InlB, encoded by the *inlAB* operon, are proteins characterized by the presence of a N-terminal domain-containing leucine-rich repeats (LRRs) that interact with their respective host cell receptors, E-cadherin and the hepatocyte growth factor (HGF) receptor Met. Beside InlA and InlB, the *L. monocytogenes* strains possess other LRR-containing proteins referred to as "internalins" [28]. To date, only InlA and InlB are involved in internalization *per se*, and the exact roles of the other internalins are still elusive. Since LRRs provide versatile recognition units for protein–protein interactions in a variety of prokaryotic and eukaryotic proteins, it is tempting to speculate that proteins of the internalin family may play a key role in establishing host–*Listeria* molecular interactions during the infectious process [28]. The invasion processes mediated by InlA and InlB are apparently similar from a mechanistic point of view (zipper type), but the two proteins follow different signaling pathways to achieve entry.

InlA is a covalently linked bacterial cell wall protein that mediates recognition and invasion of epithelial cells through specific interaction with the host epithelial cell protein E-cadherin [25, 26]. E-cadherin is a transmembrane adhesion protein required for adherent junction formation in epithelial cells. The E-cadherin N-terminal extracellular domain mediates homophilic binding with E-cadherin molecules on neighboring cells, while the intracellular domain bridges cadherin to the actin cytoskeleton via catenins. InlA–E-cadherin interactions engage the cellular machinery required for cell junction formation and induce local cytoskeletal rearrangements necessary for uptake of the bacterium into epithelial cells [25, 26]. The recognition of E-cadherin by InlA is species-specific as it requires a proline at position 16 (present in humans and guinea-pigs) whereas the presence of a glutamic acid at this location, such as in the mouse or rat, is not permissive to either adhesion or invasion [29]. Therefore, the mouse model is inadequate to study the crossing of intestinal barrier that is characteristic of listeriosis. Consequently, a novel animal system was developed to study the crossing of the intestinal barrier, i.e. a transgenic mouse that expresses human E-cadherin on the surface of enterocytes. In this model system, InlA can interact with E-cadherin, rendering mice susceptible to InlA-dependent oral infection with *L. monocytogenes* [30].

InlB is associated with the bacterial surface, but it can also be detached from the bacterial cell wall and be released into the culture supernatant. Binding of InlB to its main cellular receptor Met results in the entry of *L. monocytogenes* into different cell types [25, 26]. Met is a protein tyrosine kinase. Upon binding of its endogenous ligand, Met activates survival and proliferative signals, and induces cytoskeletal rearrangements that function in homeostasis, cellular motility and differentiation.

Met also controls growth, invasion and metastasis in cancer cells [31]. Recent studies have revealed that the molecular mechanism underlying InlB-mediated entry into cells unexpectedly exploits the clathrin-mediated endocytosis machinery, a process that is used by mammalian cells to take up nutrients and to recycle membrane proteins [32]. This result is highly surprising since it was thought that the endocytic machinery supported internalization of vesicles smaller than 100 nm. However, it seems that the potential use of the endocytic machinery for bacterial internalization is a broad phenomenon also observed with other bacterial pathogens [33]. Besides the endocytosis mechanism, binding of InlB also activates a signaling cascade that ultimately leads to cytoskeletal rearrangements and entry of *L. monocytogenes*. The proteins crucial for internalization include the small GTPases Rac and/or Cdc42,N-WASP/WAVE proteins, the Arp2/3 complex, the Ena/vasodilator-stimulated phosphoprotein (Ena/VASP) proteins, cofilin and its downregulator LIM kinase [34].

Whereas the effects of InlB on cellular processes are well documented, its function *in vivo* has long remained elusive. A study confirmed the importance of InlB for liver and spleen colonization, but its contribution to crossing the intestinal barrier is insignificant [35]. Interestingly, InlB does not recognize guinea-pig and rabbit Met, revealing a second *L. monocytogenes* species-specificity. Recent results using a novel humanized E-cadherin knock-in mouse model have shown the inter-dependent roles of InlA and InlB for the crossing of the fetoplacental barrier [36].

24.2.2
L. monocytogenes Intracellular Life

After entry, *Listeria* becomes engulfed within a phagocytic vacuole. The first event necessary to allow intracellular survival and multiplication is the rapid lysis of this vacuole, which occurs within 30 min [37]. However, there is evidence that *L. monocytogenes* ensures its intravacuolar viability by preventing phagosome maturation. These aspects of *Listeria* phagosomal life are detailed in this chapter.

Following escape from the phagosome, *L. monocytogenes* is released into the cytosol where it encounters conditions of nutrient and ion starvation. In order to replicate and survive as long as possible within the infected host cells, the bacterium has evolved strategies to adapt its metabolism. It does this without withdrawing nutrients essential for the host cells, thereby allowing host survey and bacterial growth. It seems that *L. monocytogenes* uses excess, storage or even catabolic products that are more or less dispensable for the host cell. Of major importance for replication in the host cell cytosol is the Hpt transporter for phosphorylated hexoses, which is needed for the use of hexose phosphates as a carbon and energy source. The supply of glucose 6-phosphate, a phosphate sugar available in the cytoplasm, may derive from degradation of cellular glycogen, an alternative carbon source abundant in the host cell. An *hpt* deletion mutant showed significantly reduced replication in the cytosol of host cells and attenuated virulence [38]. The expression of *hpt* is dependent on the major virulence regulator PrfA (positive regulatory factor A).

PrfA is a transcriptional activator that belongs to the Crp/Fnr family, and regulates the expression of key virulence determinants of *L. monocytogenes* (i.e., *plcA, hly, mpl, actA, plcB, inlA, inlB, inlC* and *hpt*) [39]. It activates transcription by binding as a homodimer to a palindromic promoter element, the PrfA box. PrfA ensures the correct spatio-temporal and niche-adapted expression of the target genes, with maximum induction inside the host and repression in the environmental habitat. In particular, the PrfA-regulated virulence factors are produced at higher levels as temperature increases to 37 °C. At low temperatures, the untranslated leader region of the *prfA* messenger RNA forms a stable secondary structure masking the Shine–Dalgarno sequence. Consequently, PrfA is not translated and virulence factors depending on PrfA are not expressed [40, 41]. Moreover, PrfA is highly active when *L. monocytogenes* replicates in the cytosol of mammalian host cells, whereas it is repressed when glucose (or other phosphotransferase system sugars) is used as carbon source. Therefore, utilization of an alternative source of carbon by *L. monocytogenes* upregulates PrfA, indispensable for intracellular survival, but also preserves the supply of energy for the host cell. During intracellular growth, *L. monocytogenes* has also developed the ability to use alternative nitrogen sources in its metabolism [38].

Other bacterial factors of major importance for replication in the host cell cytosol have been identified. *L. monocytogenes* lacking the lipoate protein ligase LplA1, encoded by *lplA1*, is defective for growth in the host cytosol and is less virulent in animals. Further analysis suggests that under infection a major target for LplA1, the E2 subunit of pyruvate dehydrogenase (PDH), lacks a critical lipoyl modification when the *lplA1* mutant was grown intracellularly. The abortive growth of the mutant is due to the loss of PDH function [42]. Reduced proliferation in the cytosol of mammalian host cells was also observed with a mutant defective in *oppA*, encoding an oligopeptide-binding protein, suggesting that efficient uptake of oligopeptides may favor intracellular growth of *L. monocytogenes* in macrophages.

Various further aspects of the life of *Listeria* inside the host cell as well as the bacterial functions required have recently been reviewed and are therefore not extensively detailed in this section [38].

24.2.3
Intracellular Motility

L. monocytogenes replicates in the cytosol. Moreover, intracytoplasmic bacteria are rapidly surrounded by a cloud of host actin filaments, which later (around 2 h post infection) rearranges to form long comet tails of short, highly cross-linked actin filaments, with the bacterium at the apex of the tail (Figure 24.3). This actin polymerization into a tail provides the driving force for motility. Bacteria are then propelled randomly into the cytoplasm with an average bacterial speed of 0.05–0.25 $\mu m\,s^{-1}$ [43]. The model of actin polymerization by *L. monocytogenes* has been initially revealed by the unusual phenotype of an isogenic *actA* mutant, which was capable of entering into cell, escaping into the host cytoplasm, but could not move intracellularly and accumulated as microcolonies inside the cell. This mutant,

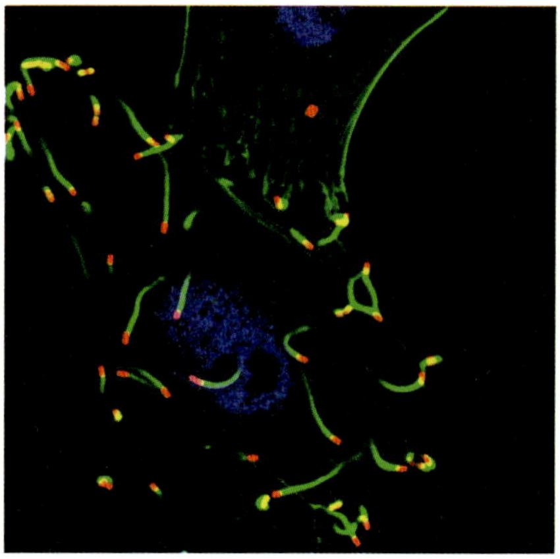

Figure 24.3 Immunofluorescence staining of mammalian cells infected with *Listeria monocytogenes*. *L. monocytogenes* can polymerize actin to form actin tails and protrusions at the edges of cells (bacteria in red; actin in green; cell nucleus in blue).

unable to recruit actin, was defective for cell-to-cell spreading and highly attenuated in a mouse infection model [43].

The *actA* gene encodes the protein ActA. ActA is present on the *L. monocytogenes* surface and asymmetrically distributed; this asymmetry allows a polar polymerization of actin and is correlated with efficient comet tail formation [43]. ActA is sufficient to promote the actin recruitment and polymerization events. Thus, once ActA is displayed on their surface, bacteria can be passive participants while the host cell expends energy to move them within and between cells. The ActA protein was shown to bind members of the enabled/vasodilator-stimulated phosphoprotein (Ena/VASP) family of proteins, which modulate bacterial speed and directionality [43]. Moreover, ActA acts as a nucleation-promoting complex by binding to and activating the Arp2/3 complex, which normally functions in host cells to nucleate actin assembly and to generate dendritic arrays of actin filaments. Indeed, ActA and the mammalian nucleation-promoting factors WASP (Wiskott–Aldrich syndrome protein) share sequence and functional identity, allowing ActA to mimic proteins of the WASP family [43].

Using this actin-based motility system, *L. monocytogenes* can form protrusions that are engulfed by neighboring cells. Very little is known about this critical process of *Listeria* infection, but it has been suggested that bacteria could take advantage of an endogenous process of paracytophagy (engulfment of portions of neighboring cells in non-infected monolayers) to favor the intercellular passage of bacteria [44]. The uptake of these protrusions leaves the bacterium enclosed in a spherical secondary vacuole, distinct from the primary vacuole in that it is surrounded by a double membrane, one from the donor cell and one from the recipient. After escape from this vacuole, the bacterium is believed to acquire motility and restart its intracellular lifecycle. Cell-to-cell spread *in vitro* may occur from *L. monocytogenes*-infected macrophages to other cells, including neurones [45]. Such a mechanism possibly occurs during infection and provides the bacteria with a convenient way to disseminate into several host tissues, particularly at host barriers [46].

24.3
Bacterial Factors Involved in Escape From the Vacuole

L. monocytogenes can reside in macrophages and nonphagocytic cells, and subsequently escape from the membrane-bound phagosomes to replicate in the cytosol of infected cells. The ability of this bacterium to escape from the vacuole is a critical step in the infection process and one of the features that distinguishes *L. monocytogenes* from almost all other facultative intracellular pathogens. Various bacterial factors are required for efficient vacuole escape.

24.3.1
Listeriolysin O (LLO)

LLO, a 60 kDa secreted protein, is the primary *L. monocytogenes* determinant responsible for vacuole escape from primary and secondary intracellular vacuoles of most cultured cell types. Expression of LLO by *Bacillus subtilis* and *L. innocua* confers to these extracellular bacteria the capacity to escape from the vacuole into the cytosol of macrophages [37]. Due to these properties, LLO is a major virulence determinant of *L. monocytogenes*. Indeed, mutants lacking LLO fail to reach the cytoplasm and are five orders of magnitude less virulent than wild-type bacteria *in vivo* [37]. Transcription of *hly*, the gene encoding LLO, is largely controlled by PrfA, and is consequently upregulated dramatically during intracellular infection. Moreover, the presence of full-length 5′UTR is required for maximal expression during intracellular infection, probably by increasing *hly* transcript stability [47].

24.3.1.1 Role in Vacuole Escape
LLO is a member of a large family of pore-forming cholesterol-dependent cytolysins (CDCs), which includes streptolysin O, perfringolysin O and pneumolysin O produced, respectively, by the extracellular pathogens *Streptococcus pyogenes*, *Clostridium perfringens* and *Streptococcus pneumoniae* [37]. Like other members of the pore-forming CDC family, LLO monomers bind to the cholesterol-containing eukaryotic cell membranes and then oligomerize into large complexes that penetrate the membrane by extending transmembrane β-hairpins to form pores of 25–30 nm in diameter. The β-barrel pore is formed by the contribution of twin transmembrane β-hairpins from each CDC monomer in the oligomeric pore complex (Figure 24.4). LLO is an atypical member of this CDC family because it is produced by an intracellular rather than extracellular pathogen. Both the vacuolar and plasma membranes of the host cell contain cholesterol; thus, although LLO is required for phagosomal escape, it is also potentially toxic to the host cell. Yet, once inside the cell,

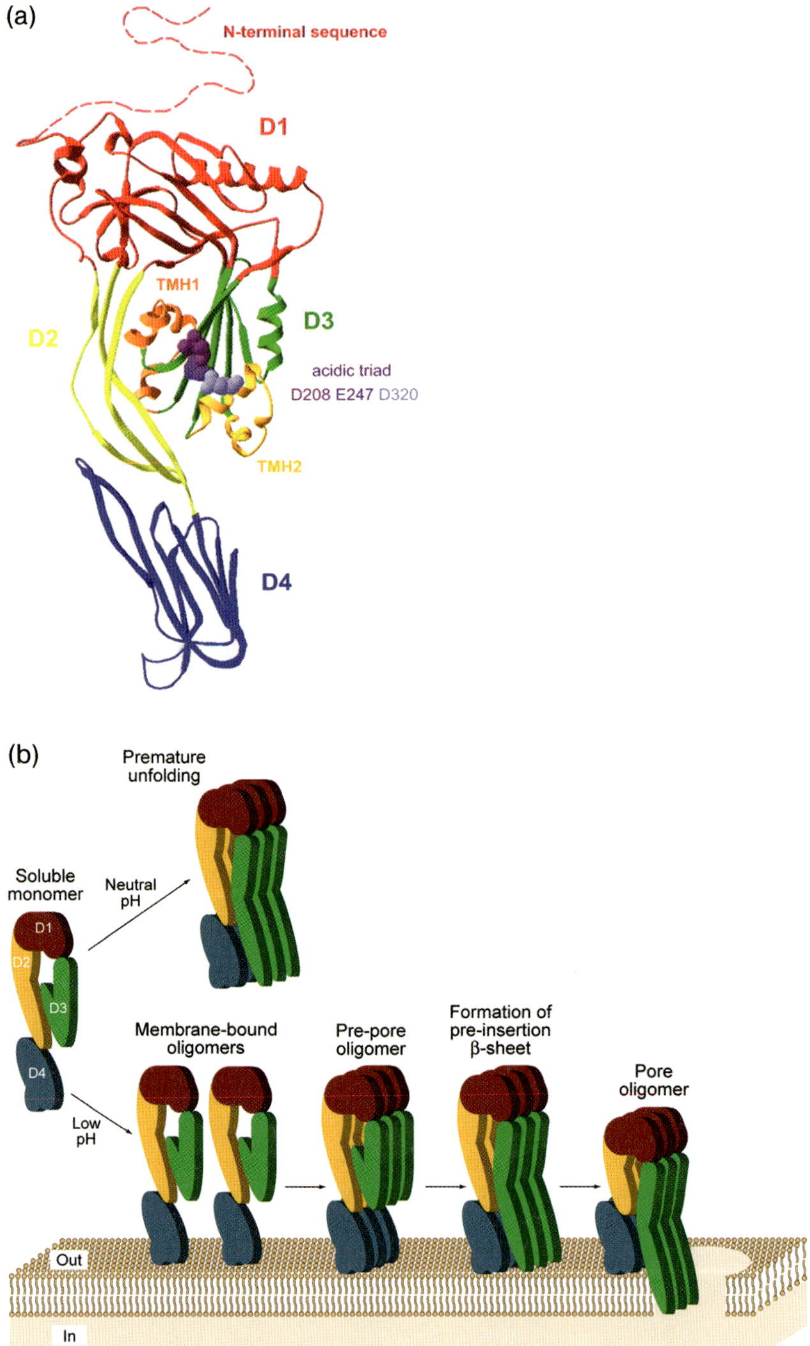

Figure 24.4 (caption see p. 443)

the activity of LLO must be tightly regulated to ensure efficient escape from the vacuole but prevent degradation of host plasma membrane. As shown below, LLO activity is modulated at different levels, including the levels of synthesis (translation) and pore-forming activity (degradation and optimum pH activity).

24.3.1.1.1 **Regulation of Synthesis** During the bacterial intracellular cycle, the correct spatial compartmentalization of LLO has been attributed to the 26-amino-acid sequence located in the extreme N-terminus of the protein. A *L. monocytogenes* mutant lacking this sequence escapes from the host vacuole as efficiently as wild-type bacteria, but it permeabilizes the host plasma membrane of the infected cell shortly after the bacterium gains access to the host cell cytosol. This cytotoxic phenotype was correlated with an increase amount of LLO inside the cytosol of the infected cell [48, 49]. This region of LLO contains a PEST-like sequence, a motif present in eukaryotic proteins. Because PEST sequences in eukaryotic proteins often function as degradation tags, it had originally been hypothesized that this part of LLO limits pore-forming activity in the cytosol by directing LLO for rapid degradation in this cellular compartment [48]. However, neither the amino acid changes nor deletion of this sequence affected the half-life of LLO in the cytosol of infected macrophages [50]. It has now been proposed that during cytosolic growth, the 5′ coding sequence of *hly* functions to repress LLO translation. Thus, *L. monocytogenes* controls LLO activity by negatively regulating its translation in order to minimize level of this protein in the host cytosol [51]. The exact mechanism underlying this regulation remains to be elucidated [51].

24.3.1.1.2 **Regulation of Activity** Several mechanisms for LLO activity regulation during the growth of *L. monocytogenes* in the host cytosol have been described. Inside the cytosol, LLO is the substrate for an eukaryotic protein degradation pathway called the "N-end rule pathway" [52]. In eukaryotes, the N-end rule pathway is a ubiquitin-dependent proteolytic pathway that ultimately allows substrate delivery to the 26S proteasome; it therefore relates the half-life of a protein to the identity of its N-terminal residue. Secreted LLO bears an N-terminal lysine residue (Lys), owing to the removal of the LLO signal sequence by a bacterial signal peptidase during LLO secretion from *L. monocytogenes*. N-terminal Lys is a destabilizing residue in the N-end rule. The result is that LLO is a physiological N-end rule substrate in the cytosol of infected mammalian cells. Replacing the N-terminal

Figure 24.4 (a) Structural model of the listeriolysin O (LLO) monomer. The N-terminal sequence of LLO is drawn with a dashed line. The four domains (D1–D4) of the LLO monomer and the two sets of α-helices that form the transmembrane β-hairpins (TMHs) are mentioned, as well as the triad of acidic residues in domain 3 (D208, E247 and D320) that function as the pH sensor. Figure adapted from Schnupf and Portnoy (2007) [37]. (b) Mechanism of pore formation. At low pH, the LLO monomer interacts with the host cell plasma membrane. Subsquent structural rearrangements allow monomers to form a pre-pore complex, which then leads to the formation of transmembrane β-hairpins that punch through the membrane. At neutral pH, domain 3 of the monomer prematurely unfolds, rendering the protein unable to form pores.

lysine of LLO with a stabilizing residue, such as valine, increased the *in vivo* half-life of LLO, but did not strongly affect the virulence of *L. monocytogenes* [52]. Degradation of LLO by the N-end rule pathway could limit the toxicity of overproduced LLO, and it is therefore believed that this property of LLO has been selected for its positive effects on fitness during the evolution of *L. monocytogenes* [50]. LLO is effectively targeted to the proteasome during cytosolic growth, but degradation of LLO during *L. monocytogenes* infection may not be a major regulatory mechanism that controls LLO activity inside the cell. Indeed, blocking host proteasome function during infection with wild-type *L. monocytogenes* does not lead to substantial host cell toxicity. Altogether, these findings highlight the fine adjustments that modulate the potential cytotoxicity of LLO in the cytosol; both by repression of its synthesis and by its proteasomal degradation.

Another regulating mechanism for the activity of LLO has been identified, and corresponds to an optimum pH pore-forming activity. LLO exhibits little cytolytic activity at neutral pH but has substantial activity at pH <6 [53]. So, LLO is fully active in the acidic environment of the phagosome, whereas it is less active at the neutral pH of the host cell cytoplasm. Within the large family of CDCs, the pH-dependence of pore-forming activity is unique to LLO, and this characteristic appears to be important for the successful intracellular life of *L. monocytogenes*. Indeed, substitution of perfringolysin O, a pH-insensitive CDC, for LLO permits escape of *L. monocytogenes* from the vacuole but leads to cell membrane permeabilization and consequently impairs the intracellular replication of *L. monocytogenes* [54, 55]. The molecular mechanism of the pH optimum of LLO involves a rapid and irreversible denaturation of its structure by a premature unfolding of the transmembrane β-hairpins at neutral pH at temperatures $>30\,^{\circ}\mathrm{C}$. A triad of acidic residues in domain 3 (D208, E247 and D320) functions as the pH sensor and initiates the denaturation of LLO by destabilizing the structure of domain 3, resulting in protein aggregation and loss of hemolytic activity (Figure 24.4) [53].

24.3.1.2 **Role in Host Cell Signaling**

In addition to its major role in vacuole escape, LLO also has the surprising ability to trigger several signaling pathways in host cells that strongly influence the course of infection. These events include activation of NFκB, mitogen-activated protein kinase (MAPK) pathways, host phospholipases and protein kinase C. In this way, LLO modulates bacterial uptake, induction of proinflammatory cytokines and chemokines in macrophages, induction of mucus exocytosis in intestinal cells and induction of apoptosis [52, 56]. Recently, it has been shown that LLO also modulates host gene expression through histone modifications [57]. This modification correlates with transcriptional downregulation of a subset of host genes involved in immune responses, reflecting a mechanism by which *L. monocytogenes* targets the immune response thereby preventing pro-inflammatory response [57]. The mechanisms by which LLO trigger such multiple and diverse cellular responses are still poorly understood. Although some of these LLO signals are due to the direct Ca^{2+} influx through pores, LLO can trigger signaling via mechanisms that are independent of pore formation [56]. Indeed, due to its ability to aggregate membrane cholesterol via oligomerization, LLO is a potent aggregator of lipid rafts and can therefore induce co-clustering of raft-associated kinases and adapter molecules, thus inducing signals [58].

24.3.2
Bacterial Phospholipases PI-PLC and PC-PLC

LLO is largely responsible for mediating internalization and vacuole escape. However, in some human cell lines, LLO has been shown to be dispensable for primary vacuolar lysis and a major role for the two secreted bacterial phospholipases C (PLCs) PI-PLC and PC-PLC has been recognized [59–61]. PI-PLC and PC-PLC are encoded by the genes *plcA* and *plcB* respectively, whose expression is highly expressed during host cell infection because it is controlled by PrfA.

PI-PLC may hydrolyze phosphatidylinositol (PI) and glycosyl-PI-anchored proteins present in the phagolysosomal membrane. PC-PLC is a broad-range PLC and hydrolyzes a wide variety of phospholipids including phosphatidylcholine, phosphatidylserine and sphingomyelin. It was initially named PC-PLC because of its similarity with other bacterial PLCs for which phosphatidylcholine is the preferred substrate. During intracellular growth, the PC-PLC is synthesized as a proenzyme (pro-PC-PLC) which is retained within the bacterial cell wall [61, 62]. During cell-to-cell spread and upon vacuolar acidification, *L. monocytogenes* metalloprotease Mpl contributes to the regulation of PC-PLC activity by two independent mechanisms: it regulates the PC-PLC bacterial cell wall translocation and the PC-PLC activation by proteolytic cleavage of its N-terminal propeptide. Both of these mechanisms are pH-sensitive and require acidic pH [63]. In this way, Mlp tightly regulates the PC-PLC activity during bacterial growth inside the cell and specifies its action in the phagosome. Indeed, a constitutive secretion of active PC-PLC during intracellular infection causes host membrane damage, and results in attenuation for virulence in the mouse [64].

Both these bacterial phospholipases have an effective role in *L. monocytogenes* virulence, since PI-PLC and PC-PLC deletion mutants are recovered in much lower numbers from the livers and spleens of infected mice. In the absence of PI-PLC, bacteria escape from the primary vacuole less efficiently than the wild-type and have a small defect in the ability to spread from cell to cell. Deletion of PC-PLC does not affect escape from primary vacuoles but significantly reduces the bacterium's ability to escape secondary vacuoles formed upon cell-to-cell spread and bacteria accumulate in double-membrane vesicles [61, 65]. However, the ability of a double phospholipase mutant to escape from the primary phagosome and to spread from cell to cell is compromised to a greater extent than with either mutation alone, suggesting that the two phospholipases have overlapping functions in the primary and secondary phagosome [66].

24.3.3
Bacterial Lipoproteins

Bacterial lipoproteins form a subclass of exported proteins that are involved in a broad range of functions; they are key components of many transport systems, such as ABC transporters, but are also known to be involved in host–pathogen interactions. The key step for the subcellular localization of lipoproteins, lipidation, is catalyzed by the prolipoprotein diacylglyceryl transferase (Lgt). It was recently shown that deletion of the *lgt* gene in *L. monocytogenes* impairs intracellular growth of the bacterium. Moreover, comparative extracellular proteome analyses of a *L. monocytogenes* wild-type strain and an *lgt* mutant showed that various lipoproteins are regulated by PrfA, suggesting that several lipoproteins contribute to the pathogenicity of *L. monocytogenes* [67].

The lipoprotein modification by Lgt is a prerequisite for lipoprotein-specific signal peptidase II (Lsp) activity, which cleaves its N-terminal signal peptide to form the mature lipoprotein. A *L. monocytogenes* Lsp-deficient mutant is affected in escape from the phagosomal compartment and its virulence is attenuated *in vivo*, but it does not have defects in intracytosolic multiplication and cell-to-cell spread [42]. Moreover, Lsp production is temporally controlled as the expression of the *lsp* gene, encoding Lsp, is strongly induced when bacteria are still entrapped inside the phagosomal compartments of infected macrophages [42]. Hence, the maturation of lipoproteins seems to be important for virulence, particularly for efficient phagosomal escape of *L. monocytogenes*. However, the molecular mechanism remains unknown.

24.4
Survey Inside the Phagosome and Phagosome Maturation

24.4.1
Factors Involved in Survey Inside the Phagosome

The phagosome constitutes the critical compartment where *L. monocytogenes* infection is confined, and *L. monocytogenes* has developed strategies that enable it to resist in the harsh environment of phagosomes. A listerial peptidoglycan *N*-deacetylase (PgdA), was recently shown to be critical for *L. monocytogenes* persistence inside macrophage vacuoles. In *L. monocytogenes*, PgdA partially deacetylates the peptidoglycan (PG) *N*-acetylglucosamine residues, thus conferring a bacterial resistance to the bacteriolytic activity of lysozyme [68]. A *L. monocytogenes* PgdA-deficient mutant is consequently rapidly destroyed within macrophage vacuoles, and is found frequently in large vacuoles with multiple bacteria or bacterial debris. Thus, PG *N*-deacetylation is a highly efficient mechanism used by *L. monocytogenes* to evade innate host defenses, and therefore plays a central role in *Listeria* virulence [68].

Cells also use reactive oxygen species (ROS) as major mediators of phagocytes microbicidal activity. To counter the phagosomal oxidative burst, *L. monocytogenes* possesses a functional manganese-superoxide dismutase (MnSOD) which catalyzes the conversion of superoxide radical anions to hydrogen peroxide. It has been shown that this bacterial SOD is posttranslationally controlled by phosphorylation [69]. The most active nonphosphorylated form of MnSOD is secreted in infected cells, where it becomes phosphorylated to be less active in the cytoplasm.

A mutant affected in MnSOD is impaired in survival within macrophages and is dramatically attenuated in mice, suggesting that the capacity to counteract ROS is an essential component of *L. monocytogenes* virulence [69].

24.4.2
Vacuole Maturation Delay

Inside the vacuole, the fate of bacteria depends also on host cell functions that modify the phagosomal lumen, including acidification, fusion with endosomes and lysosomes, and delivery of microbicidal molecules. Sequential interactions of the phagosome with the endocytic pathway lead to changes in the lipid and protein composition of its membrane and generation of molecules important for host defense [70]. To successfully invade and replicate within the cell, some intracellular pathogens employ mechanisms to avoid destruction in the harsh environment of lysosomes. *L. monocytogenes* escapes from phagosomal degradation by entering in the cytosol, but there is now evidence that the bacterium also influences the dynamics of phagosome maturation to ensure its intravacuolar viability and promote its escape. Two models of delaying the phagosomal maturation by *L. monocytogenes* have been described.

24.4.2.1 LLO-Independent Vacuole Maturation Delay

L. monocytogenes can delay vacuole maturation by targeting the host protein Rab5a. Rab5a possesses a guanosine triphosphatase (GTPase) activity and cycles between the GDP-inactive and GTP-active forms. Rab5a is involved in the regulation of phagocytic transport, phagosome–endosome fusion and phagosomal maturation [71]. Rab5a GDP/GTP exchange, and therefore Rab5a activity, is one of the critical phagocytic functions determining *L. monocytogenes* survival or clearance within *Listeria* phagosomes. Indeed, increase of the activation rate of Rab5a in epithelial-like cells mimics the listericidal signal induced by IFNγ, i.e. increased transport of bactericidal lysosomal proteases (i.e., cathepsin D) to the phagocytic compartments [72]. Conversely, in the absence of Rab5a, intracellular growth of *L. monocytogenes* in epithelial-like cells was promoted and survival increased considerably as fusion with lysosomes was almost completely avoided [73]. Thus, *L. monocytogenes* has evolved strategy to recruit and inhibit Rab5a exchange activity, and consequently modify the phagocytic compartment by impairing its maturation [73].

Listeria glyceraldehyde-3-phosphate dehydrogenase (GAPDH) has been recently identified as a Rab5a-binding protein responsible for this GTPase interference [74]. GAPDH binds to Rab5a and causes Rab5a-specific ADP ribosylation. GAPDH is consequently responsible for Rab5a recruitment to phagosomes and inhibition by impairment of the GDP/GTP exchange activity. *L. monocytogenes* GAPDH can be considered as a virulence factor as it favors viability of the bacteria within the phagosome as well as the exclusion of putative listericidal lysosomal proteases such as cathepsin D. The inhibition of Rab5a exchange activity constitutes a new *L. monocytogenes* virulence strategy to impede the phagolysosome fusion and subsequent degradation of the bacteria [74].

24.4.2.2 LLO-Dependent Vacuole Maturation Delay

Henry *et al.* suggested another possible mechanism to explain the phagosome maturation delay during *L. monocytogenes* infection in macrophages [75]. They characterized maturation and perforation of *L. monocytogenes*-containing phagosomes, and showed that *L. monocytogenes* delays the vacuole maturation by inhibiting vacuole fusion with lysosomes [75]. Immediately prior to rupture, phagosomes containing wild-type *L. monocytogenes* are Rab5a- and lysosome-associated membrane protein 1 (LAMP1)-negative but phosphatidylinositol 3-phosphate (PI(3)P)- and Rab7-positive [75]. Rab5a is an early endosome marker, whereas Rab7 is a late endosome marker. Thus, it is suggested that *L. monocytogenes* escapes from vacuoles corresponding to a late endosome before fusion of vacuoles with LAMP1-positive lysosomes. Moreover, wild-type *L. monocytogenes* delayed vacuole fusion with lysosomes compared to an LLO-deficient mutant, indicating that LLO impairs vacuole maturation and alters vacuole chemistry prior to release of *L. monocytogenes* into cytoplasm [75, 76]. The delay in phagosome maturation and lysosomal fusion was proposed to occur via small LLO-dependent perforation of the vacuolar membrane early in infection that results in increased pH and decreased Ca^{2+} levels within the vacuole, thereby disrupting the proton and calcium gradients required for vacuolar maturation [76]. Indeed, acidic pH and high concentrations of calcium inside endocytic compartments are essential for the phagosome maturation and fusion with lysosomes. This maturation delay seems to be necessary for efficient *L. monocytogenes* vacuole escape, since escape from LAMP1-positive compartments was determined to be inefficient, probably due to the action of the lysosomal protease cathepsin D which targets LLO and inhibits its function [75, 77].

In agreement with these observations, RNA interference screens in *Drosophila* S2 cells identified vesicular trafficking components affecting escape of *L. monocytogenes* from the phagosomal compartment [78, 79]. Specifically, host knockdowns that bypassed the requirement of LLO for escape were in components involved in fusion of late endosomes/multivesicular bodies with lysosomes. Under this condition, the triple-mutant LLO and PLCs was not able to escape, suggesting that knockdowns of vesicular and trafficking proteins likely led to vacuolar conditions that resembled those produced by LLO [79]. A proposed model is that LLO could form pores at late endosome stage, thereby aborting further maturation and promoting the activity of PLCs to mediate disruption of the vacuole. Indeed, the knockdown affecting vacuolar ATPase subunits blocked *L. monocytogenes* escape, underscoring the need for vacuolar maturation past the early endosome stage before LLO-mediated vacuolar escape at the late stage, where conditions are optimal for its pore-forming toxin LLO [79]. These data correlate well with the fact that *L. monocytogenes* escaped from macrophage endosomes labeled with Rab7, a late endosome marker, but devoid of Rab5 [75]. However, the absence of the early endosome marker Rab5a from *L. monocytogenes* vacuoles in macrophages is at odds with the studies indicating that *L. monocytogenes* modulates vesicular trafficking by targeting Rab5a needed to direct membrane fusion events at the early endosome stage [73, 74]. To date, not much is known about the vesicular trafficking stage that is targeted by *L. monocytogenes*. However, depending on the cell type, both models may be not exclusive and may cooperate to enhance bacterial escape.

24.4.3
Signaling and Vacuole Disruption

Escape of *L. monocytogenes* requires acidification of the maturing phagosome, since blocking acidification inhibits phagosomal escape. The exact requirement of acidification for vacuole escape is still unknown. Surprisingly, the requirement of acidification for LLO-mediated escape does not appear to be due to the higher hemolytic potential of LLO at acidic pH. Indeed, mutated LLO with an increased activity at neutral pH does not lead to escape from the phagosomal compartment quicker or at less acidic pH than wild-type [80]. The precise way in which *L. monocytogenes* escapes from a vacuole is unclear but two mechanisms, not mutually exclusive, have been proposed.

In macrophages, one putative mechanism is that *L. monocytogenes* activates host functions subsequently required for vacuole dissolution. LLO along with PLCs participates in the modulation of host intracellular signaling, including opening of calcium channels and release of Ca^{2+} from stores, activation of host polyphosphoinositide-specific PLC, phospholipase D and protein kinase C (PKC) isoforms [81]. Host PKCs regulate a variety of cellular processes, and require cofactors for their activation, particularly diacylglycerol (DAG) and Ca^{2+}. *L. monocytogenes* modulates intracellular levels of second messengers upon infection; it affects intracellular Ca^{2+} levels and generates DAG by hydrolysis of phospholipids in host membranes. Altogether, these data suggest that LLO activates host functions that could be needed for disruption of the macrophage phagosome.

Upon early *L. monocytogenes* infection of macrophages, PI-PLC may enter the host cell's cytosol via pores formed by LLO in the vacuole membrane. Once inside the cell, PI-PLC reaches the inner leaflet of the host cell plasma membrane and cleaves substrate phospholipids in host membranes, producing DAG and phosphoinositide [81]. DAG subsequently activates recruitment of PKCδ to the plasma membrane, which leads to the opening of plasma membrane calcium channels and subsequent elevation of intracellular Ca^{2+} levels [81, 82]. The generation of the second messenger DAG and the calcium influx are proposed to lead to the activation and translocation of other PKCs to early endosome, thereby facilitating phagosomal escape. At present time there is no available information on how PKCs influence phagosomal perforation and disruption, but several data are in agreement with this model. First, inhibition of calcium elevation and/or inhibition of PLD or PKCβ lead to strong inhibition of escape from the primary vacuole of a macrophage [81, 82]. Second, PKCδ-deficient mice are highly susceptible to *L. monocytogenes* infection and die before the onset of a protective adaptive immune response. This heightened susceptibility was driven by increased phagosomal escape of *Listeria* in macrophages. PKCδ appears therefore to be a critical factor for confinement and killing of *L. monocytogenes* within macrophage phagosomes, although the mechanism whereby PKCδ mediates its action remains unknown [83].

Finally, during *L. monocytogenes* infection of macrophages, an LLO-dependent mobilization of PKCε on vacuoles follows perforation of the vacuole membrane. The LLO-dependent activation of host phosphoinositide-specific PLCs is presumably implicated in PKCε activation. The exact role of PKCε recruitment to vacuoles remains unknown, but it could be part of a cytoplasmic signaling mechanism involved in the breakdown and recycling of damaged organelles, especially damaged *L. monocytogenes* vacuoles [84].

The second possible mechanism to explain phagosomal escape is supported by studies on LLO-mediated delay in phagosome maturation. Henry *et al.* suggested that *L. monocytogenes* delays macrophage phagosome maturation and lysosomal fusion via small LLO-dependent perturbations of the vacuolar membrane that result in disruption of the proton and calcium gradients required for vacuolar maturation [75]. *L. monocytogenes* escapes from vacuoles associated with late endosome markers but devoid of the LAMP1 lysosome marker. Moreover, escape of *L. monocytogenes* from LAMP1-positive compartments was also determined to be inefficient, probably because of the LLO function inhibition by the action of lysosomal proteases such as cathepsin D [73]. Therefore, the phagosomal escape could be that LLO pores formed at the late endosome stage prevent further maturation of the phagosome and provide a favorable environment for the PLCs to mediate disruption of the vacuole [75, 76].

The specific mechanism in which *L. monocytogenes* escapes from secondary vacuole remains elusive. It is suggested that in murine macrophages membrane dissolution could occur in two steps and bacterial escape could occur upon acidification of the vacuolar compartment. PLCs could participate in the dissolution of the inner membrane of the vacuole and LLO could perforate the outer membrane. Then, the vacuole escape mechanism could be the same as in the case of the primary phagosome [66]. Both LLO and PC-PLC require a decrease in vacuolar pH to function. However, even if vacuolar maturation leads to acidification of the compartment between the two membranes of the secondary vacuole, it is not clear how the internal compartment of the vacuole becomes acidified for PC-PLC activation [66].

As described above, LLO has a critical role in *L. monocytogenes* vacuole escape. Recently, it has been shown that *L. monocytogenes* with inefficient LLO activity effectively fails to escape efficiently from the phagosome, but forms "spacious *Listeria*-containing phagosomes" (SLAPs). SLAPs are autophagosome-like compartments, allowing slow bacterial growth within enlarged vesicles [85]. Consistent with the fact that *L. monocytogenes* can be targeted by autophagy early in infection, SLAPs colocalize with the autophagy markers LC3 [85, 86]. Moreover, SLAPs are non-mature phagosomes as a result of LLO-dependent delayed vacuole maturation.

SLAPs may represent a stalemate between bacterial LLO action and the host defense system, resulting in persistent infection. Indeed, a high LLO concentration allows bacterial escape from the phagosome. However, under conditions where LLO activity is not sufficient to drive escape, autophagy could maintain bacteria within nondegradative vacuoles (SLAPs) that allow slow bacterial growth. The formation of SLAPs could represent a delicate balance between virulence factors of the pathogen and innate immune mechanisms of the infected cell. Indeed, the host cell is able to prevent bacterial colonization of the cytosol without eradication of the pathogen. At the same time, the pathogen is unable to escape from the vacuole but replicates in SLAPs at a reduced rate [85].

24.5 Summary and Outlook

The ability of *Listeria monocytogenes* to survive in both phagosomal and cytoplasmic compartments of host cells, as well as its capacity to stimulate the host immune response, has propelled this intracellular bacterium as a useful model for microbiologists, cell biologists and immunologists. The different steps of the *L. monocytogenes* infectious process *in vitro* are clearly established. The main bacterial factors involved in entry, escape from the vacuole and cell-cell spreading are well described. Futhermore, by comparing the genome of pathogenic *L. monocytogenes* to non-pathogenic *L. innocua*, new putative virulence genes have been identified in *L. monocytogenes*. Extensive functional analyses of these bacterial factors are now needed to understand the specific pathogenic traits they confer, and determine their precise role in the interplay between the bacterium and the host.

Although our understanding of the *L. monocytogenes* infectious process is vast, many important questions still remain. *L. monocytogenes* is capable of manipulating the host cell at several levels (i.e. gene transcription, signaling, post-transcriptional modifications,. . .). Many bacterial and host factors involved in these processes have been identified, but the exact mechanism underlying their interaction is only partially understood. One such example is the delayed maturation of the vacuole caused by *Listeria* manipulating the host cell machinery. In doing so, the bacteria assure their survival and trigger their subsequent escape to the cytoplasm. To date several models have been proposed but the exact mechanism of this step remains elusive. Many scientists are intrigued by questions raised by the interactions between the bacteria and the host. The development of new technologies, such as large-scale transcriptome, proteome or metabolome analyses, as well as powerful microscopy techniques, promise to reveal new highlights on the molecular basis of listeriosis.

Acknowledgments

We would like to thanks members of the Cossart's Lab for helpful discussions. We are very grateful to Alejandro Toledo-Arana for his help for the figures and to Serawit Bruck and Serge Mostowy for critical reading of this manuscript. We would like to apologize to all authors whose work we may not have adequately presented or even omitted due to space limitation.

References

1 Murray, E.G.D., Webb, R.A. and Swann, M.B.R. (1926) *J Pathol Bacteriol*, **29**, 407–439.
2 Pirie, J.H.H. (1927) *Publ S Afr Inst Med Res*, **3**, 163–186.
3 Lecuit, M. (2007) *Microbes Infect*, **9**, 1216–1225.
4 Nyfelt, A. (1929) *C R Seances Soc Biol Fil*, **101**, 590–591.
5 Schlech, W.F. 3rd, Lavigne, P.M., Bortolussi, R.A., Allen, A.C., Haldane, E.V., Wort, A.J., Hightower, A.W., Johnson, S.E., King, S.H., Nicholls, E.S. and Broome, C.V. (1983) *N Engl J Med*, **308**, 203–206.
6 Lecuit, M. and Cossart, P. (2002) *Trends Mol Med*, **8**, 537–542.
7 Farber, J.M. and Peterkin, P.I. (1991) *Microbiol Rev*, **55**, 476–511.
8 Swaminathan, B. and Gerner-Smidt, P. (2007) *Microbes Infect*, **9**, 1236–1243.
9 Jensen, V.B., Harty, J.T. and Jones, B.D. (1998) *Infect Immun*, **66**, 3758–3766.
10 Lecuit, M., Sonnenburg, J.L., Cossart, P. and Gordon, J.I. (2007) *J Biol Chem*, **282**, 15065–15072.
11 Pron, B., Boumaila, C., Jaubert, F., Sarnacki, S., Monnet, J.P., Berche, P. and Gaillard, J.L. (1998) *Infect Immun*, **66**, 747–755.
12 Antal, E.A., Loberg, E.M., Bracht, P., Melby, K.K. and Maehlen, J. (2001) *Brain Pathol*, **11**, 432–438.
13 Drevets, D.A., Leenen, P.J. and Greenfield, R.A. (2004) *Clin Microbiol Rev*, **17**, 323–347.
14 Mackaness, G.B. (1962) *J Exp Med*, **116**, 381–406.
15 Cossart, P. (2007) *Microbes Infect*, **9**, 1143–1146.
16 Pamer, E.G. (2004) *Nat Rev Immunol*, **4**, 812–823.
17 Zenewicz, L.A. and Shen, H. (2007) *Microbes Infect*, **9**, 1208–1215.
18 Dussurget, O., Pizarro-Cerda, J. and Cossart, P. (2004) *Annu Rev Microbiol*, **58**, 587–610.
19 Gaillard, J.L., Berche, P. and Sansonetti, P. (1986) *Infect Immun*, **52**, 50–55.
20 Kathariou, S., Metz, P., Hof, H. and Goebel, W. (1987) *J Bacteriol*, **169**, 1291–1297.
21 Portnoy, D.A., Jacks, P.S. and Hinrichs, D.J. (1988) *J Exp Med*, **167**, 1459–1471.
22 Mengaud, J., Chenevert, J., Geoffroy, C., Gaillard, J.L. and Cossart, P. (1987) *Infect Immun*, **55**, 3225–3227.
23 Glaser, P., Frangeul, L., Buchrieser, C., Rusniok, C., Amend, A., Baquero, F., Berche, P., Bloecker, H., Brandt, P., Chakraborty, T., Charbit, A., Chetouani, F., Couve, E., de Daruvar, A., Dehoux, P., Domann, E., Dominguez-Bernal, G., Duchaud, E., Durant, L., Dussurget, O., Entian, K.D., Fsihi, H., Garcia-del Portillo, F., Garrido, P., Gautier, L., Goebel, W., Gomez-Lopez, N., Hain, T., Hauf, J., Jackson, D., Jones, L.M., Kaerst, U., Kreft, J., Kuhn, M., Kunst, F., Kurapkat, G., Madueno, E., Maitournam, A., Vicente, J.M., Ng, E., Nedjari, H., Nordsiek, G., Novella, S., de Pablos, B., Perez-Diaz, J.C., Purcell, R., Remmel, B., Rose, M., Schlueter, T., Simoes, N., Tierrez, A., Vazquez-Boland, J.A., Voss, H., Wehland, J. and Cossart, P. (2001) *Science*, **294**, 849–852.
24 Hamon, M., Bierne, H. and Cossart, P. (2006) *Nat Rev Microbiol*, **4**, 423–434.
25 Pizarro-Cerda, J. and Cossart, P. (2006) *Cell*, **124**, 715–727.
26 Seveau, S., Pizarro-Cerda, J. and Cossart, P. (2007) *Microbes Infect*, **9**, 1167–1175.
27 Cossart, P. and Sansonetti, P.J. (2004) *Science*, **304**, 242–248.
28 Bierne, H., Sabet, C., Personnic, N. and Cossart, P. (2007) *Microbes Infect*, **9**, 1156–1166.
29 Lecuit, M., Dramsi, S., Gottardi, C., Fedor-Chaiken, M., Gumbiner, B. and Cossart, P. (1999) *EMBO J*, **18**, 3956–3963.
30 Lecuit, M., Vandormael-Pournin, S., Lefort, J., Huerre, M., Gounon, P., Dupuy,

C., Babinet, C. and Cossart, P. (2001) *Science*, **292**, 1722–1725.

31 Birchmeier, C., Birchmeier, W., Gherardi, E. and Vande Woude, G.F. (2003) *Nat Rev Mol Cell Biol*, **4**, 915–925.

32 Veiga, E. and Cossart, P. (2005) *Nat Cell Biol*, **7**, 894–900.

33 Veiga, E., Guttman, J.A., Bonazzi, M., Boucrot, E., Toledo-Arana, A., Lin, A.E., Enninga, J., Pizarro-Cerda, J., Finlay, B.B., Kirchhausen, T. and Cossart, P. (2007) *Cell Host Microbe*, **2**, 340–351.

34 Bierne, H., Miki, H., Innocenti, M., Scita, G., Gertler, F.B., Takenawa, T. and Cossart, P. (2005) *J Cell Sci*, **118**, 1537–1547.

35 Khelef, N., Lecuit, M., Bierne, H. and Cossart, P. (2006) *Cell Microbiol*, **8**, 457–470.

36 Disson, O., Grayo, S., Huillet, E., Nikitas, G., Langa Vives, F., Dussurget, O., Ragon, M., Le Monnier, A., Babinet, C., Cossart, P. and Lecuit, M. (2008) *Nature*, **445**, 1114–1118.

37 Schnupf, P. and Portnoy, D.A. (2007) *Microbes Infect*, **9**, 1176–1187.

38 Joseph, B. and Goebel, W. (2007) *Microbes Infect*, **9**, 1188–1195.

39 Milohanic, E., Glaser, P., Coppee, J.Y., Frangeul, L., Vega, Y., Vazquez-Boland, J.A., Kunst, F., Cossart, P. and Buchrieser, C. (2003) *Mol Microbiol*, **47**, 1613–1625.

40 Johansson, J., Mandin, P., Renzoni, A., Chiaruttini, C., Springer, M. and Cossart, P. (2002) *Cell*, **110**, 551–561.

41 Renzoni, A., Klarsfeld, A., Dramsi, S. and Cossart, P. (1997) *Infect Immun*, **65**, 1515–1518.

42 Reglier-Poupet, H., Frehel, C., Dubail, I., Beretti, J.L., Berche, P., Charbit, A. and Raynaud, C. (2003) *J Biol Chem*, **278**, 49469–49477.

43 Gouin, E., Welch, M.D. and Cossart, P. (2005) *Curr Opin Microbiol*, **8**, 35–45.

44 Robbins, J.R., Barth, A.I., Marquis, H., de Hostos, E.L., Nelson, W.J. and Theriot, J.A. (1999) *J Cell Biol*, **146**, 1333–1350.

45 Dramsi, S., Levi, S., Triller, A. and Cossart, P. (1998) *Infect Immun*, **66**, 4461–4468.

46 Drevets, D.A. (1999) *Infect Immun*, **67**, 3512–3517.

47 Shen, A. and Higgins, D.E. (2005) *Mol Microbiol*, **57**, 1460–1473.

48 Decatur, A.L. and Portnoy, D.A. (2000) *Science*, **290**, 992–995.

49 Glomski, I.J., Decatur, A.L. and Portnoy, D.A. (2003) *Infect Immun*, **71**, 6754–6765.

50 Schnupf, P., Portnoy, D.A. and Decatur, A.L. (2006) *Cell Microbiol*, **8**, 353–364.

51 Schnupf, P., Hofmann, J., Norseen, J., Glomski, I.J., Schwartzstein, H. and Decatur, A.L. (2006) *Mol Microbiol*, **61**, 999–1012.

52 Schnupf, P., Zhou, J., Varshavsky, A. and Portnoy, D.A. (2007) *Infect Immun*, **75**, 5135–5147.

53 Schuerch, D.W., Wilson-Kubalek, E.M. and Tweten, R.K. (2005) *Proc Natl Acad Sci USA*, **102**, 12537–12542.

54 Jones, S. and Portnoy, D.A. (1994) *Infect Immun*, **62**, 5608–5613.

55 Pizarro-Cerda, J. and Cossart, P. (2006) *J Pathol*, **208**, 215–223.

56 Kayal, S. and Charbit, A. (2006) *FEMS Microbiol Rev*, **30**, 514–529.

57 Hamon, M.A., Batsche, E., Regnault, B., Tham, T.N., Seveau, S., Muchardt, C. and Cossart, P. (2007) *Proc Natl Acad Sci USA*, **104**, 13467–13472.

58 Gekara, N.O., Jacobs, T., Chakraborty, T. and Weiss, S. (2005) *Cell Microbiol*, **7**, 1345–1356.

59 Grundling, A., Gonzalez, M.D. and Higgins, D.E. (2003) *J Bacteriol*, **185**, 6295–6307.

60 Marquis, H., Doshi, V. and Portnoy, D.A. (1995) *Infect Immun*, **63**, 4531–4534.

61 Vazquez-Boland, J.A., Kocks, C., Dramsi, S., Ohayon, H., Geoffroy, C., Mengaud, J. and Cossart, P. (1992) *Infect Immun*, **60**, 219–230.

62 Snyder, A. and Marquis, H. (2003) *J Bacteriol*, **185**, 5953–5958.

63 Yeung, P.S., Zagorski, N. and Marquis, H. (2005) *J Bacteriol*, **187**, 2601–2608.

64 Yeung, P.S., Na, Y., Kreuder, A.J. and Marquis, H. (2007) *Infect Immun*, **75**, 44–51.

65 Smith, G.A., Marquis, H., Jones, S., Johnston, N.C., Portnoy, D.A. and Goldfine, H. (1995) *Infect Immun*, **63**, 4231–4237.

66 Alberti-Segui, C., Goeden, K.R. and Higgins, D.E. (2007) *Cell Microbiol*, **9**, 179–195.

67 Baumgartner, M., Karst, U., Gerstel, B., Loessner, M., Wehland, J. and Jansch, L. (2007) *J Bacteriol*, **189**, 313–324.

68 Boneca, I.G., Dussurget, O., Cabanes, D., Nahori, M.A., Sousa, S., Lecuit, M., Psylinakis, E., Bouriotis, V., Hugot, J.P., Giovannini, M., Coyle, A., Bertin, J., Namane, A., Rousselle, J.C., Cayet, N., Prevost, M.C., Balloy, V., Chignard, M., Philpott, D.J., Cossart, P. and Girardin, S.E. (2007) *Proc Natl Acad Sci USA*, **104**, 997–1002.

69 Archambaud, C., Nahori, M.A., Pizarro-Cerda, J., Cossart, P. and Dussurget, O. (2006) *J Biol Chem*, **281**, 31812–31822.

70 Desjardins, M., Huber, L.A., Parton, R.G. and Griffiths, G. (1994) *J Cell Biol*, **124**, 677–688.

71 Vieira, O.V., Bucci, C., Harrison, R.E., Trimble, W.S., Lanzetti, L., Gruenberg, J., Schreiber, A.D., Stahl, P.D. and Grinstein, S. (2003) *Mol Cell Biol*, **23**, 2501–2514.

72 Prada-Delgado, A., Carrasco-Marin, E., Bokoch, G.M. and Alvarez-Dominguez, C. (2001) *J Biol Chem*, **276**, 19059–19065.

73 Prada-Delgado, A., Carrasco-Marin, E., Pena-Macarro, C., Del Cerro-Vadillo, E., Fresno-Escudero, M., Leyva-Cobian, F. and Alvarez-Dominguez, C. (2005) *Traffic*, **6**, 252–265.

74 Alvarez-Dominguez, C., Madrazo-Toca, F., Fernandez-Prieto, L., Vandekerckhove, J., Pareja, E., Tobes, R., Gomez-Lopez, M.T., Del Cerro-Vadillo, E., Fresno, M., Leyva-Cobian, F. and Carrasco-Marin, E. (2008) *Traffic*, **9**, 325–337.

75 Henry, R., Shaughnessy, L., Loessner, M.J., Alberti-Segui, C., Higgins, D.E. and Swanson, J.A. (2006) *Cell Microbiol*, **8**, 107–119.

76 Shaughnessy, L.M., Hoppe, A.D., Christensen, K.A. and Swanson, J.A. (2006) *Cell Microbiol*, **8**, 781–792.

77 del Cerro-Vadillo, E., Madrazo-Toca, F., Carrasco-Marin, E., Fernandez-Prieto, L., Beck, C., Leyva-Cobian, F., Saftig, P. and Alvarez-Dominguez, C. (2006) *J Immunol*, **176**, 1321–1325.

78 Agaisse, H., Burrack, L.S., Philips, J.A., Rubin, E.J., Perrimon, N. and Higgins, D.E. (2005) *Science*, **309**, 1248–1251.

79 Cheng, L.W., Viala, J.P., Stuurman, N., Wiedemann, U., Vale, R.D. and Portnoy, D.A. (2005) *Proc Natl Acad Sci USA*, **102**, 13646–13651.

80 Glomski, I.J., Gedde, M.M., Tsang, A.W., Swanson, J.A. and Portnoy, D.A. (2002) *J Cell Biol*, **156**, 1029–1038.

81 Goldfine, H. and Wadsworth, S.J. (2002) *Microbes Infect*, **4**, 1335–1343.

82 Poussin, M.A. and Goldfine, H. (2005) *Infect Immun*, **73**, 4410–4413.

83 Schwegmann, A., Guler, R., Cutler, A.J., Arendse, B., Horsnell, W.G., Flemming, A., Kottmann, A.H., Ryan, G., Hide, W., Leitges, M., Seoighe, C. and Brombacher, F. (2007) *Proc Natl Acad Sci USA*, **104**, 16251–16256.

84 Shaughnessy, L.M., Lipp, P., Lee, K.D. and Swanson, J.A. (2007) *Cell Microbiol*, **9**, 1695–1704.

85 Birmingham, C.L., Canadien, V., Kaniuk, N.A., Steinberg, B.E., Higgins, D.E. and Brumell, J.H. (2008) *Nature*, **451**, 350–354.

86 Birmingham, C.L., Canadien, V., Gouin, E., Troy, E.B., Yoshimori, T., Cossart, P., Higgins, D.E. and Brumell, J.H. (2007) *Autophagy*, **3**, 442–451.

25
Mycobacterium marinum

Monica Hagedorn and Thierry Soldati

25.1
An Introduction to *Mycobacterium marinum* and the Illness Caused by it

Mycobacteria infections, such as tuberculosis and leprosy, cause devastation throughout the world, and even though decades of research have been focused on these diseases, the mechanisms of mycobacteria pathogenicity in general, and especially of *Mycobacterium tuberculosis*, are still not fully understood. The very long antibiotic treatment (between 3 and 6 months or even longer) necessary to cure mycobacteria infections has increasingly led to the appearance of strains resistant to multiple antibiotics. This trend makes the need to understand the course of mycobacterial diseases even more pressing. However, the complexity, difficult accessibility and ethical concerns linked to the use of mammalian models, such as non-human primates and mice, have hampered advancement in understanding the underlying molecular mechanisms. *Mycobacterium marinum* belongs to the genus of mycobacteria and is an established model to study mycobacteria pathogenicity in alternative host organisms [1–7]. In this chapter we focus especially on these "model" features of *M. marinum* and the experimental advantages it brings with it, especially in synergism with genetically tractable host organisms, both at systemic and cellular levels. For more exhaustive overviews on infections caused by *M. marinum*, including its immunology and histopathological aspects, we would refer the reader to recent reviews [1, 3, 8].

25.1.1
Course of Infection and Illness in Animals and Humans

M. marinum is a water-borne Gram-positive, acid-fast, rod bacterium that naturally infects fish and amphibians and is an opportunistic pathogen for humans [8]. In addition to posing a health risk to humans and animals, *M. marinum* infections cause large economic losses in industries that rely on healthy fish breeding and are a serious problem in research facilities that maintain and grow zebrafish. *M. marinum* dissemination via contaminated water is difficult to control and infections lead to wasting and death of fish.

Intracellular Niches of Microbes. Edited by Ulrich E. Schaible and Albert Haas

ISBN: 978-3-527-32207-7

Infections caused by *M. marinum* in its natural hosts follow a course very similar to the ones resulting from *M. tuberculosis*, as has been documented in numerous studies [1, 9]. Also, based on complete genome comparison and 16S rRNA sequence analyses, it has been found that *M. marinum* is closely related to the *M. tuberculosis* complex [10, 11]. Whereas *M. tuberculosis* is an obligatory human pathogen, *M. marinum* has a broad spectrum of hosts ranging from fish and frogs to aquatic mammals. It is not clear how *M. marinum* naturally enters its host to cause an infection. Based on experimental infections of zebrafish, an uptake route from contaminated water via the gastrointestinal tract is favored over uptake through the gill epithelium [12]. Accordingly, in research facilities, force-feeding fish with contaminated food seems to be the most efficient "natural" way to cause an infection. Alternatively, for experimental infections direct injection of the pathogen into the caudal vein of fish embryos is often used [9, 13].

In general, the main cell types pathogenic mycobacteria infect are macrophages or other phagocytic cells. After escaping killing by the cell-intrinsic immune machinery of macrophages, the bacteria proliferate within these cells. Interestingly, elegant infection studies in zebrafish showed that these cells do not offer optimal proliferation conditions for mycobacteria [14]. The authors demonstrated, using a mutant zebrafish strain that was lacking macrophages, that these cells control bacteria growth during the early stages of infection and are critical for tissue dissemination of bacteria, a later step in infection. The latent stage of infection is initiated by the migration of infected macrophages into deeper tissues where they form granulomas, often with a caseating core [9]. In its natural host, the granulomas induced by *M. marinum* are very similar to those induced by *M. tuberculosis* in humans (mammals). There are reports that the infection of other cell types, such as epithelial cells, is experimentally possible [15, 16]. The mechanism by which the pathogen enters these nonphagocytic cells is not clear, however for *M. tuberculosis* a macropinocytic pathway has been proposed [17].

As already mentioned in the introduction, *M. marinum* is capable of causing a nonsystemic infection in humans. The optimal growth temperature for *M. marinum* is adapted to its natural host and therefore in the range of 28–32 °C. Accordingly, *M. marinum* causes ulceration only on cooler surfaces in humans, namely granulomatous skin lesions (variously described as swimming pool or fish tank granulomas) and no systemic infections. Very few cases of systemic spreading have been observed in immunosuppressed patients [8]. The path the pathogen follows to invade human organisms is not known, but most likely it enters the skin from contaminated water via small cuts or punctures. This puts people who handle fish on a regular basis, for example in the fishing industry but also aquarists, at high risk of infection. The infection can be cured with long-term application of combinations of antibiotics (e.g., tetracycline or clarithromycin [18]) in the same way as they are applied to cure other mycobacterial infections, such as tuberculosis and *M. avium* infections. The short, 4–6 h, generation time of *M. marinum* lowers its persistence and increases the efficiency of antibiotic treatment. However, if cutaneous *M. marinum* infections are not identified or treated over the years, the infection can spread to tendons and bone material, causing serious damage [8].

Sequencing of the 6 636 827 bp *M. marinum* genome has been completed and published [10]. It can be searched online using a BLAST engine at the Sanger Institute web site (http://www.sanger.ac.uk/Projects/M_marinum/). Comparison of the *M. marinum* genome (6.6 Mbp) to *M. tuberculosis* (4 Mbp) [19] shows 99.4% sequence homology [1] and an average amino acid identity of 85% [10]. In contrast to *M. tuberculosis* that has evolved into a pathogen specialized to humans, *M. marinum* has maintained a large genome, which is thought to be due to the different lifestyles and specific host adaptations [1, 3, 10]. Interestingly a gene family, encoding so-called PE/PE-PGRS proteins that are known to be present on the bacterial surface, and their associated ESX-secretion systems, are significantly expanded. As was mentioned above and will be emphasized further in Section 25.2, *M. tuberculosis* infections occur exclusively in humans, in contrast to *M. marinum* that has a widespread spectrum of host organisms.

Interestingly, recent evidence indicates that *M. marinum* and *M. ulcerans*, the causative agent of the human skin disease Buruli ulcer, seem to have evolved from a common progenitor [20]. Based on a variety of genetic approaches and analyses it became apparent that mycolactone-producing mycobacteria might originate from a common ancestor and have evolved into several lineages to occupy distinct ecological niches, such as ectotherms for *M. marinum* [10].

25.1.2
Establishment of *M. marinum* Host Model Systems to Study Mycobacteria Virulence

The rigid host restriction of *M. tuberculosis* has necessitated the search and establishment of alternative pathogen and host models to investigate mycobacterial pathogenicity. Since the mid 1980s, *M. marinum* has been established as a model mycobacterial pathogen [4, 21–24]. Its lower safety risk, faster growth and course of infection render it an attractive alternative pathogen to study the general concepts of mycobacterial pathogenicity and, specifically, the conserved mechanisms of *M. tuberculosis* virulence. The combination of *M. marinum* as a mycobacterial pathogen model with genetically tractable host models, such as zebrafish, *Drosophila* and *Dictyostelium discoideum* [5, 6, 9, 25–27] produces a positive synergism and offers a wide spectrum of new approaches. As a result over the past years it has significantly advanced our knowledge of mycobacteria virulence.

The group around Stanley Falkow was a driving force in the establishment of alternative hosts for *M. marinum*. In 1997, they showed that *M. marinum* infection of the leopard frog, *Rana pipiens*, results in a chronic granulomatous infection with all the characteristics of a mycobacterial infection, importantly also including a latent phase [24]. The establishment and characterization of this first host model system was the stepping-stone to identify novel virulence factors by examining the fate of transposon-generated *M. marinum* mutants [28, 29]. Arguably, the most successful model hosts to advance the deciphering of mycobacterial pathogenicity are fish, another class of natural hosts for *M. marinum*. Trucksis and colleagues [30] showed that *M. marinum* infection of the gold-fish, *Carassius auratus*, results in a systemic mycobacterial infection with granuloma formation. However, up to now, the zebrafish, *Danio rerio*, has proven to be the most versatile and successful host model system to study *M. marinum* [6, 9]. Genome sequencing is ongoing and progress can be monitored at: http://www.sanger.ac.uk/Projects/D_rerio/.

The zebrafish provides several unique advantages. It is a genetically tractable organism with forward genetic screens, reverse genetics and established transgenic lines. Combined with *M. marinum* as a genetically tractable pathogen, this offers enormous potential to study host–pathogen interactions by genetic approaches [13, 14, 31–35]. The zebrafish model also offers unique opportunities to study the immune response to pathogenic mycobacteria. The animals acquire adaptive immunity only in adulthood, which allows researchers to focus on innate immunity by investigating embryos [14]. In addition, the embryos are optically transparent, offering the unique opportunity to follow an *in vivo* infection and granuloma formation by microscopy within the same animal [13, 33]. *M. marinum* infection of zebrafish mimics certain aspects of tuberculosis [25] and results in caseating granulomatous lesions [9]. However, as already mentioned, at a systemic level the two pathogens show distinct differences, probably due to adaptations to diverse ecological niches.

Naturally, the fruit fly, *Drosophila melanogaster*, is infected by *M. fortuitum*, but it is also another genetically tractable model host for *M. marinum* infection [5, 36] (reviewed in [6]). Direct injection of bacteria into the flies induces a lethal infection at a relatively low MOI (multiplicity of infection). At a cellular level it presents similarities to the early stages of infection in the natural hosts of *M. marinum* [5]. Interestingly, infected flies show a "wasting" phenotype [36], as was so characteristic for humans suffering from tuberculosis in the eighteenth century. *Drosophila* has a haploid genome, which has been fully sequenced and annotated (search available under http://www.flybase.bio.indiana.edu/), facilitating genetic approaches to uncover novel host defense strategies and pathogen virulence factors.

The host model organisms described above offer new opportunities to study the effects of mycobacteria infections at a systemic level. Clearly, the investigation of very early events in the infection, such as uptake of the pathogen and manipulation/evasion of cell-intrinsic host defense, requires easily accessible cellular host models. Therefore, single-cell systems, including animal cell lines and the social amoeba *Dictyostelium discoideum* [26, 27], have been successfully established as model hosts.

25.2
Characterization of Intra-Host Cell Compartmentation of *M. marinum* and its Relationship to Virulence

Like most pathogenic mycobacteria, *M. marinum* is a facultative intracellular pathogen. It is thought that, *in vivo*, macrophages or macrophage-like phagocytic cells are the main cell type that is infected by pathogenic mycobacteria. It is a cynical circumstance that pathogenic mycobacteria exploit phagocytosis by those cells, whose sole task it is to protect the organism from infections. The infected macrophages will migrate into tissues where, with the recruitment of other cell types, mycobacteria-specific granulomas will be formed [1, 3, 31].

To study the early establishment of infection, the use of cell lines and single-cell model hosts is essential. The infection of a carp leukocyte cell line (CLC) allowed Cirillo and colleagues to dissect how the pathogen avoids killing by leukocytes and establishes a replication niche [4]. Also, macrophage cell lines from mice, such as J774 and RAW are routinely used to investigate infection and the cellular fate of the pathogen. In addition, macrophage-like cell lines have also been established from genetically tractable host organisms. For example the *Drosophila* cell line S2 can be infected with *M. marinum* [37], with a pathogenic mycobacterium of insects, *M. fortuitum* [38], and with the "nonpathogenic" *M. smegmatis* [39].

As briefly mentioned above, the social amoeba *Dictyostelium discoideum* has been recently established as a novel and promising alternative single-cell host model system [26, 27]. It is thought that mycobacteria pathogenicity evolved from the interactions of environmental pathogens with early protozoans. In other words, the pathogens were "training" their virulence in single-cell organisms before expanding their host spectrum to multicellular organisms. This prior adaptation means that the concept of virulence is already an integral part of the amoeba–pathogen relationship. Therefore, the study of genetically and biochemically tractable single-cell organisms as host models provides a unique opportunity to elucidate pathogen virulence tactics in the context of rudimentary, cell-intrinsic host defenses. It should also be noted that evidence is arising that the concept of innate immunity emerged in the social amoeba [40]. However, the amoeba's role as a natural host for pathogens also has a downside. Especially in hospitals, the amoebae provide an environmental breeding ground for intracellular pathogens, which is problematic in these surroundings [41, 42]. Another twist to the amoeba–pathogen story is given by the observation that pathogen growth/passage within the amoebae can increase its virulence [43].

In addition to its role as an experimental host, *Dictyostelium* is an easy model system that is cheap to cultivate and store. Its genome has been sequenced (search and resources available at http://www.dictybase.org/) and both targeted and random mutagenesis approaches are established as routine methods [44]. It has a long tradition as a model for the study of cellular processes, such as cell motility, chemotaxis [45, 46] and membrane trafficking [47]. More recently, its model character extended to cover mechanisms of intracellular bacteria killing and cell-intrinsic immunity [40, 48–50]. *Dictyostelium* is now a well-established host for a number of pathogens [49, 51, 52] including *M. marinum* [26, 27].

In general, it seems that the fate of *M. marinum* in macrophages is very similar to that described for *M. tuberculosis*. The pathogen is engulfed by the cells via phagocytosis. During and after formation of a pathogen-containing phagosome, the bacterium actively induces a phagosomal maturation arrest (or bifurcation) in order to prevent normal maturation into a bactericidal compartment. This course of establishment of infection has been extensively studied for *M. tuberculosis*, and

M. marinum has been shown to undergo a similar fate in a number of host systems. Barker and colleagues [24] showed that viable *M. marinum* resides in a non-acidic vacuole that does not colocalize with markers that usually indicate phagosome maturation, such as cathepsin D. In contrast, phagosomes containing heat-killed bacteria appear to fully acidify and mature into bactericidal compartments, a similar maturation program as for latex bead phagosomes [5, 53]. The small effector protein Rab5, an early delivered small GTPase, is present on *M. tuberculosis* phagosomes, but the later Rab7 is absent [54, 55]. Such detailed studies have not been performed for *M. marinum*, but it is likely comparable, as the arrest or bifurcation from phagosomal maturation seems to be similar.

The fate of *M. marinum* in *Dictyostelium* appears to be similar to the one observed in macrophages [27]. Soon after formation of the pathogen-containing phagosome the vacuolar H^+-ATPase is transiently present on the compartment, which might corroborate the transient acidification of *M. tuberculosis*-containing phagosomes [56]. Cathepsin D is not delivered and finally the pathogen proliferation vacuole evolves in a compartment that bears markers of the late endocytic/phagocytic pathway and is thus likely non-acidic [27].

Interestingly, even though *M. marinum* and *M. tuberculosis* both appear to prevent being killed by phagolysosome formation, later stages of infection might be more divergent for the two pathogens. *M. marinum* has been described to break its replication niche and escape into the cytosol [15, 27], where it exhibits actin-driven motility [15, 57]. Remarkably, in the amoeba *Dictyostelium*, the breaking of the niche has been observed directly by immunofluorescence. Breakage of the replication vacuole was a frequent event and the broken compartment appeared to be surprisingly stable [27]. The frequency of actin tail formation and associated cytosolic motility is disputed. While it has been beautifully described in a fish leukocyte cell line and epithelial cell lines [15, 57], it is not clear if this event occurs in all host model systems. For example, in *Dictyostelium*, no persistent actin tails can be observed, maybe due to the unusually high rate of actin depolymerization in amoebae [58]. However, when it occurs, actin tails and motility can be observed in about 20% of cells, suggesting that niche breakage might be either an infrequent or asynchronous event. The molecular mechanisms of breaking the replication niche are unknown, but potentially involve pathogen factors that will be discussed in Section 25.3. Escape into the cytosol has recently been documented for *M. tuberculosis* and *M. leprae* in myeloid cells [59], but the findings are still controversial.

Virulence and pathogenic mechanisms of *M. tuberculosis* and *M. marinum* are largely conserved, but these recent findings might indicate differences that were selected to overcome various host organisms or to reflect various entry routes in their respective hosts. *M. tuberculosis* is a strictly human pathogen and directly infects alveolar macrophages via uptake of aerosols. In contrast, *M. marinum* infects a broad range of ectotherms and is proposed to reach macrophages after passage through the gut epithelium. Therefore, it is plausible that escape into the cytosol and actin-mediated ejection is essential for *M. marinum* virulence but is a less frequent, accessory strategy for *M. tuberculosis*.

25.3 Genes Involved in Diverting Phagosome Maturation and in Compartment Maintenance

Completion of the *M. marinum* genome sequence has strengthened its close relationship to the tuberculosis complex (extensively discussed in [1, 10]). In general, it is thought that, compared to *M. tuberculosis*, which strictly colonizes humans, the increase in genome size (1.5-fold the tuberculosis genome) results from adaptations to the various environmental niches. The underlying virulence mechanism, and therefore genes involved in pathogenicity, appear to be conserved among those two pathogenic strains [1, 10], and studies describing interspecies complementation and gene interchange between *M. marinum* and *M. tuberculosis* will be described below.

It is a common understanding that, in order to initiate an infection (establishment of the niche, arresting phagolysosome fusion) the pathogen has to activate an "infection program", which is reflected in a specific gene expression pattern. Falkow and colleagues [60] used a fluorescent-activated cell sorting (FACS)-based approach (differential fluorescence induction, DFI) to identify mycobacterial promoters activated upon infection of macrophages (*map*, macrophage-activated promoter) in leopard frogs using a promoter-trap library. In addition, with the same approach, they were able to identify mycobacteria promoters specifically activated in frog granuloma (*gap*, granuloma activated promoter) [60]. Most of the identified genes have homologs in *M. tuberculosis*, but, interestingly, many of them appeared to be constitutively expressed [60]. This probably highlights the fact that *M. marinum* requires more regulation and genetic versatility to successfully occupy its broad spectrum of environmental niches. The role played by some of the activated genes in *M. marinum* virulence is discussed below.

A number of *M. marinum* virulence determinants have been identified using genetic screening approaches, such as for example using an *M. marinum* genomic cosmid library [61] or transposon-based mutagenesis [29, 62]. Cell wall properties of mycobacteria, like permeability and cording growth morphology, are traits seemingly associated with virulence (reviewed in [1]). Accordingly, a mutant strain, *M. marinum kasB*, that produces shorter mycolic acids and shows a defective cording phenotype, is attenuated *in vivo* and in zebrafish [63]. In addition, the *kasB* strain seems to exhibit an altered permeability that increases the susceptibility to lipophilic compounds [32, 63]. Another mutant generated by targeted deletion of the *erp* locus also shows altered permeability properties, such as increased susceptibility to lipophilic antibiotics [32, 64]. The *erp* locus encodes a secreted cell surface protein [65] and has been first identified and characterized as a virulence determinant in *M. tuberculosis* [64]. However, both *M. marinum* and *M. tuberculosis* lacking the Erp protein are attenuated in cultured macrophages and animal host models [32, 64]. In an elegant study (reviewed in [1]), Clay and colleagues made use of the genetic tractability of zebrafish to show that virulence of the *erp* mutant is restored in fish lacking macrophages [14]. This study is a very nice example showing that alternative host models offer new tools for the dissection of complex host–pathogen relationships at both a cellular and a systemic level.

Knockout of another gene, *iipA*, affects cell permeability and virulence in both *M. marinum* and *M. tuberculosis* [66]. The *iipA* mutant shows an attenuated phenotype in adult zebrafish, which can be complemented by expression of the *M. tuberculosis* ortholog [66]. The cell wall virulence lipids characterized in *M. tuberculosis* have not been investigated as extensively in *M. marinum*. Based on genomic analysis of the components of the biosynthetic pathways of *M. tuberculosis* lipids, they are all predicted to be present in *M. marinum* (mentioned in [1]). Interestingly, a study published by Bitter and colleagues challenges the paradigm that mycobacterial glycolipids function as major virulence factors. Instead of using purified glycolipids, they used *M. marinum* and *M. bovis* BCG mutants that are unable to add a mannose cap on lipoarabinomannan. The conclusion of these experiments is that absence of the mannose cap does not lead to a decrease in bacterial survival during infection [67].

A whole new range of virulence determinants were discovered with the identification of the ESX or type 7 secretion system, especially ESX-1 (reviewed in [68]). The ESX secretion systems are present in all pathogenic mycobacteria strains. The absence of the so-called RD1 region (region of difference), which encodes components of the ESX-1 secretion system, seems to be the major locus responsible for the virulence attenuation of the vaccination strain *M. bovis* variant BCG. *M. marinum*, as well as *M. tuberculosis* lacking the RD1 region, show multiple defects in virulence. The strains show decreased efficiency in preventing phagolysosome fusion in cultured cells and a defect in cell-to-cell spreading [32, 33, 69, 70] leading to reduced virulence at the cellular and systemic levels. Few effector proteins exported via the ESX-1 secretion system are known. ESAT-6 and its putative chaperone CFP10 are encoded within the RD1 locus and are proposed to be secreted as a complex [71], which might be destabilized by exposure to an acidic environment [72]. Their molecular functions are unknown, but, *in vitro*, ESAT-6 exhibits hemolytic activity [33].

Van der Wel and colleagues [59] proposed that the RD1 locus plays a role in the translocation of *M. tuberculosis* into the cytosol in myeloid cells. ESX loci have multiple duplications in the mycobacteria genomes and seem to be functionally linked (extensively reviewed in [68]). Interestingly, the expansion of the ESX loci has been linked to the expansion of PE and PPE gene clusters, a large gene family that comprises up to 10% of the genome of pathogenic mycobacteria.

The *ESX-5* locus has been implicated in virulence. *M. marinum* mutants lacking *ESX-5* do not secrete PPE41 and are compromised in cell-to-cell spreading [34]. Another member of the PE-PGRS family has been shown to be a *M. marinum* virulence determinant in various host models. The gene *mag-24* had been identified by Ramakrishnan and colleagues [60] to be activated upon infection of macrophages. *M. marinum* lacking Mag-24 (L1D) are less efficient at establishing a replication niche in *Dictyostelium* [27]. Also, in frog and zebrafish, the mutant was compromised in virulence [60].

In general, the identification of host factors that modulate cell-intrinsic immunity against mycobacteria is relatively difficult. However, the combination of genetically and biochemically tractable host organisms with genetically tractable pathogens, such as *M. marinum*, significantly facilitates and advances the search for important host factors. The application of genome-wide RNA interference screens in *Drosophila*

S2 cells has led to the identification of several host factors that play a role in mycobacteria infection (e.g., [39]), some of which were subsequently confirmed in mammalian host models. Brown and colleagues [37] identified the lysosomal enzyme β-hexoseaminidase, a peptidoglycan hydrolase, as being important to control a mycobacterial infection. Conservation of this function was demonstrated by monitoring infections in macrophages isolated from β-hexoseaminidase knockout mice [37]. Interestingly, a similar *Drosophila* genome-wide RNA interference screen using *M. fortuitum* identified factors that are needed for phagocytosis and specifically for mycobacterial infection [73]. Most importantly, a member of the scavenger receptor family CD36 was found to be necessary for uptake of the pathogen [73].

The *M. marinum*–zebrafish host–pathogen model system has also been used successfully to identify important host factors based on screenings. In a transcriptome profiling approach using whole adult zebrafish, a long list of genes was shown to be affected by infection [74]. Among them CD36-scavenger receptors and other previously suggested factors, as well as components of the inflammatory response were identified.

The use of the easily genetically and biochemically tractable social amoeba as a host organism has also led to the identification of host factors that modulate infection. For example, vacuolin (a *Dictyostelium* flotillin-like protein) was shown to accumulate at the membrane of the mycobacterial replication niche, allowing the growth and rupture of this vacuole and the release of bacteria into the cytosol to be directly observed [75]. The generality of the finding was confirmed in infected human blood monocytes by the selective accumulation of flotillin-1 around *M. marinum* followed by vacuole rupture. Genetic disruption of the vacuolin gene rendered *Dictyostelium* more immune against infection, indicating that presence of vacuolin is important for *M. marinum* to successfully arrest phagosome maturation [27]. The study also revealed a host resistance factor. The RacH-GTPase is reported to play a role in the regulation of the actin cytoskeleton and is important for early endosomal acidification and the morphology and function of the vacuolin-positive compartments [76]. Absence of RacH rendered *Dictyostelium* more susceptible to *M. marinum* proliferation [27]. This illustrates how the *Dictyostelium* system can be used for the identification and study of both host defenses and pathogen virulence strategies.

Another host protein suggested to play a role in intracellular survival and replication of mycobacteria [77] is coronin, an actin-binding protein [78]. However, a direct role for coronin has been disputed in numerous studies that failed to indicate a role for coronin in mycobacteria infection both in macrophages [79] and in *Dictyostelium* [26]. In that respect this highlights the necessity for a continuous and constructive interplay between studies performed in alternative and mammalian host models, as well as the use of *M. marinum* as a model to study mycobacteria pathogenicity.

25.4
Immunological Aspects of *M. marinum* Infection

In contrast to *M. tuberculosis*, little is known about the immunological host response induced by *M. marinum* infection. As in an *M. tuberculosis* infection, macrophages, the first line of defense in the immune system, are the major cell type infected by *M. marinum*. But as was elegantly shown by Ramakrishnan and colleagues [14], the absence of macrophages in zebrafish rapidly results in a higher burden of pathogens. This indicates that macrophages do not provide an optimal breeding ground for the pathogen, but keep the infection "under control". As for *M. tuberculosis*, infection with *M. marinum* results in the formation of granulomas in a time- and dose-dependent fashion [9] as was shown in zebrafish. Those structures formed by the migration of infected macrophages into deeper tissues are aggregates of differentiated macrophages and other immune cells. In zebrafish, granulomas show a necrotic core, which is similar to the granulomas induced by *M. tuberculosis* in humans; however, they show a paucity in lymphocytes which may be host-specific [9]. Recently, using an enhancer trap screening in zebrafish, a subpopulation of neutrophils was identified that is involved in the innate immune response towards granuloma formation [35].

25.5 What's Next? (Outlook and Summary)

The natural fish and frog pathogen *M. marinum* is genetically closely related to *M. tuberculosis*, a major threat to human health. The infection it causes in its natural hosts is in its course and latency phase very similar to the infection caused by *M. tuberculosis*, a strictly human pathogen. Therefore, *M. marinum* offers the unique opportunity to study the mechanism of mycobacteria pathogenicity in its natural host and other biochemically and genetically tractable organisms. Organisms such as zebrafish and *Drosophila* and single-cell systems, like S2-cells (*Drosophila* phagocyte-like), carp leukocytes and the social amoeba *Dictyostelium discoideum*, have been successfully established as surrogate hosts. The application of targeted mutagenesis and random genome-wide RNA interference approaches has already led to the identification of a whole new list of virulence factors, as well as conserved host factors that modulate intrinsic immunity against mycobacteria infections.

A series of investigations in zebrafish and *Drosophila* especially highlight the advantages and versatility of an "alternative" model host that is easily accessible to genetic manipulations. It is the goal not to substitute conventional host model systems (such as mice), which are closer to the natural host of the human pathogens, but to use the simplest systems to initially identify and characterize pathogen factors and drug targets rapidly. This then leads to subsequent testing in higher organisms and clinically relevant host–pathogen systems, thereby either confirming general virulence functions or indicating host-specific mechanisms.

Acknowledgments

Work in the Soldati lab is supported by the Swiss National Science Foundation, and the group participates in the NEMO (non-mammalian experimental models for the study of bacterial infections) network supported by the Swiss 3R Foundation.

References

1 Tobin, D.M. and Ramakrishnan, L. (2008) *Cell Microbiol*, **10**, 1027–1039.

2 Stamm, L.M. and Brown, E.J. (2004) *Microbes Infect*, **6**, 1418–1428.

3 Cosma, C.L., Sherman, D.R. and Ramakrishnan, L. (2003) *Annu Rev Microbiol*, **57**, 641–676.

4 El-Etr, S.H., Yan, L. and Cirillo, J.D. (2001) *Infect Immun*, **69**, 7310–7317.

5 Dionne, M.S., Ghori, N. and Schneider, D.S. (2003) *Infect Immun*, **71**, 3540–3550.

6 Pozos, T.C. and Ramakrishnan, L. (2004) *Curr Opin Immunol*, **16**, 499–505.

7 Cosma, C.L., Swaim, L.E., Volkman, H., Ramakrishnan, L. and Davis, J.M. (2006) *Curr Protoc Microbiol, Chapter 10, Unit 10B*.

8 Petrini, B. (2006) *Eur J Clin Microbiol Infect Dis*, **25**, 609–613.

9 Swaim, L.E., Connolly, L.E., Volkman, H.E., Humbert, O. *et al.* (2006) *Infect Immun*, **74**, 6108–6117.

10 Stinear, T.P., Seemann, T., Harrison, P.F., Jenkin, G.A. *et al.* (2008) *Genome Res*, **18**, 729–741.

11 Rogall, T., Flohr, T. and Bottger, E.C. (1990) *J Gen Microbiol*, **136**, 1915–1920.

12 Harriff, M.J., Bermudez, L.E. and Kent, M.L. (2007) *J Fish Dis*, **30**, 587–600.

13 Davis, J.M., Clay, H., Lewis, J.L., Ghori, N. *et al.* (2002) *Immunity*, **17**, 693–702.

14 Clay, H., Davis, J.M., Beery, D., Huttenlocher, A. *et al.* (2007) *Cell Host Microbe*, **2**, 29–39.

15 Stamm, L.M., Morisaki, J.H., Gao, L.Y., Jeng, R.L. *et al.* (2003) *J Exp Med*, **198**, 1361–1368.

16 El-Etr, S.H. and Cirillo, J.D. (2001) *Front Biosci*, **6**, D737–747.

17 Garcia-Perez, B.E., Mondragon-Flores, R. and Luna-Herrera, J. (2003) *Microb Pathog*, **35**, 49–55.

18 Rallis, E. and Koumantaki-Mathioudaki, E. (2007) *Expert Opin Pharmacother*, **8**, 2965–2978.

19 Cole, S.T. and Barrell, B.G. (1998) *Novartis Found Symp*, **217**, 160–172; discussion 172-167.

20 Yip, M.J., Porter, J.L., Fyfe, J.A., Lavender, C.J. *et al.* (2007) *J Baltic Stud*, **189**, 2021–2029.

21 Mor, N. (1985) *Infect Immun*, **48**, 850–852.

22 Ramakrishnan, L. and Falkow, S. (1994) *Infect Immun*, **62**, 3222–3229.

23 Valdivia, R.H., Hromockyj, A.E., Monack, D., Ramakrishnan, L. and Falkow, S. (1996) *Gene*, **173**, 47–52.

24 Barker, L.P., George, K.M., Falkow, S. and Small, P.L. (1997) *Infect Immun*, **65**, 1497–1504.

25 Prouty, M.G., Correa, N.E., Barker, L.P., Jagadeeswaran, P. and Klose, K.E. (2003) *FEMS Microbiol Lett*, **225**, 177–182.

26 Solomon, J.M., Leung, G.S. and Isberg, R.R. (2003) *Infect Immun*, **71**, 3578–3586.

27 Hagedorn, M. and Soldati, T. (2007) *Cell Microbiol*, **9**, 2716–2733.

28 Bouley, D.M., Ghori, N., Mercer, K.L., Falkow, S. and Ramakrishnan, L. (2001) *Infect Immun*, **69**, 7820–7831.

29 Mehta, P.K., Pandey, A.K., Subbian, S., El-Etr, S.H. *et al.* (2006) *Microb Pathog*, **40**, 139–151.

30 Talaat, A.M., Reimschuessel, R., Wasserman, S.S. and Trucksis, M. (1998) *Infect Immun*, **66**, 2938–2942.

31 Cosma, C.L., Humbert, O. and Ramakrishnan, L. (2004) *Nat Immunol*, **5**, 828–835.

32 Cosma, C.L., Klein, K., Kim, R., Beery, D. and Ramakrishnan, L. (2006) *Infect Immun*, **74**, 3125–3133.

33 Gao, L.Y., Guo, S., McLaughlin, B., Morisaki, H. *et al.* (2004) *Mol Microbiol*, **53**, 1677–1693.

34 Abdallah, A.M., Verboom, T., Hannes, F., Safi, M. *et al.* (2006) *Mol Microbiol*, **62**, 667–679.

35 Meijer, A.H., van der Sar, A.M., Cunha, C., Lamers, G.E. *et al.* (2008) *Dev Comp Immunol*, **32**, 36–49.

36 Dionne, M.S., Pham, L.N., Shirasu-Hiza, M. and Schneider, D.S. (2006) *Curr Biol*, **16**, 1977–1985.

37 Koo, I.C., Ohol, Y.M., Wu, P., Morisaki, J.H. *et al.* (2008) *Proc Natl Acad Sci USA*, **105**, 710–715.
38 Agaisse, H., Burrack, L.S., Philips, J.A., Rubin, E.J. *et al.* (2005) *Science*, **309**, 1248–1251.
39 Philips, J.A., Porto, M.C., Wang, H., Rubin, E.J. and Perrimon, N. (2008) *Proc Natl Acad Sci USA*, **105**, 3070–3075.
40 Chen, G., Zhuchenko, O. and Kuspa, A. (2007) *Science*, **317**, 678–681.
41 Goy, G., Thomas, V., Rimann, K., Jaton, K. *et al.* (2007) *Res Microbiol*, **158**, 393–397.
42 Thomas, V., Herrera-Rimann, K., Blanc, D.S. and Greub, G. (2006) *Appl Environ Microbiol*, **72**, 2428–2438.
43 Cirillo, J.D., Falkow, S., Tompkins, L.S. and Bermudez, L.E. (1997) *Infect Immun*, **65**, 3759–3767.
44 Eichinger, L. and Rivero, F. (2006) *Dictyostelium Discoideum Protocols*, Humana Press, Totowa, NJ.
45 Franca-Koh, J., Kamimura, Y. and Devreotes, P. (2006) *Curr Opin Genet Dev*, **16**, 333–338.
46 Dormann, D. and Weijer, C.J. (2006) *Curr Opin Genet Dev*, **16**, 367–373.
47 Maniak, M. (2003) *Traffic*, **4**, 1–5.
48 Benghezal, M., Fauvarque, M.O., Tournebize, R., Froquet, R. *et al.* (2006) *Cell Microbiol*, **8**, 139–148.
49 Cosson, P., Zulianello, L., Join-Lambert, O., Faurisson, F. *et al.* (2002) *J Bacteriol*, **184**, 3027–3033.
50 Gebbie, L., Benghezal, M., Cornillon, S., Froquet, R. *et al.* (2004) *Mol Biol Cell*, **15**, 3915–3925.
51 Hilbi, H., Segal, G. and Shuman, H.A. (2001) *Mol Microbiol*, **42**, 603–617.
52 Otto, G.P., Wu, M.Y., Clarke, M., Lu, H. *et al.* (2004) *Mol Microbiol*, **51**, 63–72.
53 Via, L.E., Fratti, R.A., McFalone, M., Pagan-Ramos, E. *et al.* (1998) *J Cell Sci*, **111**, 897–905.
54 Deretic, V., Via, L.E., Fratti, R.A. and Deretic, D. (1997) *Electrophoresis*, **18**, 2542–2547.
55 Via, L.E., Deretic, D., Ulmer, R.J., Hibler, N.S. *et al.* (1997) *J Biol Chem*, **272**, 13326–13331.
56 Sturgill-Koszycki, S., Schlesinger, P.H., Chakraborty, P., Haddix, P.L. *et al.* (1994) *Science*, **263**, 678–681.
57 Stamm, L.M., Pak, M.A., Morisaki, J.H., Snapper, S.B. *et al.* (2005) *Proc Natl Acad Sci USA*, **102**, 14837–14842.
58 Zigmond, S.H. (1993) *Cell Motil Cytoskeleton*, **25**, 309–316.
59 van der Wel, N., Hava, D., Houben, D., Fluitsma, D. *et al.* (2007) *Cell*, **129**, 1287–1298.
60 Ramakrishnan, L., Federspiel, N.A. and Falkow, S. (2000) *Science*, **288**, 1436–1439.
61 El-Etr, S.H., Subbian, S., Cirillo, S.L. and Cirillo, J.D. (2004) *Infect Immun*, **72**, 6902–6913.
62 Robinson, N., Wolke, M., Ernestus, K. and Plum, G. (2007) *Infect Immun*, **75**, 581–591.
63 Gao, L.Y., Laval, F., Lawson, E.H., Groger, R.K. *et al.* (2003) *Mol Microbiol*, **49**, 1547–1563.
64 Berthet, F.X., Lagranderie, M., Gounon, P., Laurent-Winter, C. *et al.* (1998) *Science*, **282**, 759–762.
65 Berthet, F.X., Rauzier, J., Lim, E.M., Philipp, W. *et al.* (1995) *Microbiology*, **141** (Pt 9), 2123–2130.
66 Gao, L.Y., Pak, M., Kish, R., Kajihara, K. and Brown, E.J. (2006) *Infect Immun*, **74**, 1757–1767.
67 Appelmelk, B.J., den Dunnen, J., Driessen, N.N., Ummels, R. *et al.* (2008) *Cell Microbiol*, **10**, 930–944.
68 Abdallah, A.M., Gey van Pittius, N.C., Champion, P.A., Cox, J. *et al.* (2007) *Nat Rev Microbiol*, **5**, 883–891.
69 Tan, T., Lee, W.L., Alexander, D.C., Grinstein, S. and Liu, J. (2006) *Cell Microbiol*, **8**, 1417–1429.
70 Xu, J., Laine, O., Masciocchi, M., Manoranjan, J. *et al.* (2007) *Mol Microbiol*, **66**, 787–800.
71 Brodin, P., Majlessi, L., Marsollier, L., de Jonge, M.I. *et al.* (2006) *Infect Immun*, **74**, 88–98.
72 de Jonge, M.I., Pehau-Arnaudet, G., Fretz, M.M., Romain, F. *et al.* (2007) *J Bacteriol*, **189**, 6028–6034.

73 Philips, J.A., Rubin, E.J. and Perrimon, N. (2005) *Science*, **309**, 1251–1253.

74 Meijer, A.H., Verbeek, F.J., Salas-Vidal, E., Corredor-Adamez, M. *et al.* (2005) *Mol Immunol*, **42**, 1185–1203.

75 Wienke, D., Drengk, A., Schmauch, C., Jenne, N. and Maniak, M. (2006) *Eur J Cell Biol*, **85**, 991–1000.

76 Somesh, B.P., Neffgen, C., Iijima, M., Devreotes, P. and Rivero, F. (2006) *Traffic*, **7**, 1194–1212.

77 Ferrari, G., Langen, H., Naito, M. and Pieters, J. (1999) *Cell*, **97**, 435–447.

78 de Hostos, E.L., Bradtke, B., Lottspeich, F., Guggenheim, R. and Gerisch, G. (1991) *EMBO J*, **10**, 4097–4104.

79 Schuller, S., Neefjes, J., Ottenhoff, T., Thole, J. and Young, D. (2001) *Cell Microbiol*, **3**, 785–793.

26
Rickettsia

Sanjeev K. Sahni, Elena Rydkina, and David J. Silverman

26.1
Introduction to Rickettsiae and Rickettsial Diseases

Rickettsial diseases have been a scourge of humankind throughout history, leading to death and disability. Rickettsiae are small, aerobic, coccobacillary α-proteobacteria characterized by their strict intracellular location, fastidious growth requirements, association with arthropods and unique lifecycles, involving both vertebrate and invertebrate hosts. Obligate intracellular bacteria of the genus *Rickettsia*, which were first described as Gram-negative bacillary microorganisms able to retain basic fuchsin dye when subjected to the Giménez staining technique, belong to the family *Rickettsiaceae* in the order *Rickettsiales* of class *Rhodospirilli*. At present, the genus *Rickettsia* contains more than 20 validated species classified into three groups [1]: (i) the ancestral group (AG: associated with ticks), including *Rickettsia bellii* and *R. canadensis*; (ii) the typhus group (TG: associated with fleas and lice), which includes *R. prowazekii* and *R. typhi*; and (iii) the spotted fever group (SFG: a vast majority associated with ticks, but some with fleas and mites), which consists of *R. rickettsii*, *R. conorii*, *R. akari*, *R. africae*, *R. australis*, *R. honei*, *R. sibirica*, *R. slovaca*, *R. japonica*, *R. felis*, *R. aeschlimanii*, *R. helvetica*, *R. massiliae*, *R. rhipicephali*, *R. montanensis*, *R. parkeri*, *R. peacockii* and *R. heilongjiangensis* (Figures 26.1 and 26.2).

Caused by *Rickettsia rickettsii*, Rocky Mountain spotted fever is considered to be the most severe spotted fever group rickettsiosis with more than 20% mortality without adequate antibiotic treatment, while the mortality rate for Mediterranean spotted fever due to *R. conorii* is estimated at 2.5% even in diagnosed cases [2]. With the advent of molecular tools enabling precise detection, an increasing number of emerging rickettsial pathogens are being described throughout the world, as evidenced by identification of at least 11 additional rickettsial species or subspecies during the last two decades (e.g., *R. japonica* in Japan, *R. conorii* subspecies *caspia* as the agent of Astrakhan fever in the Caspian region of Russia, Africa and Kosovo, *R. africae* in sub-Saharan Africa, *R. honei* and *R. marmionii* in Australia, and *R. sibirica* subspecies *mongoltimonae* in China, Europe and Africa. Considered to be nonpathogenic for decades in their accidental hosts, that is, humans, the ability of several

Intracellular Niches of Microbes. Edited by Ulrich E. Schaible and Albert Haas

ISBN: 978-3-527-32207-7

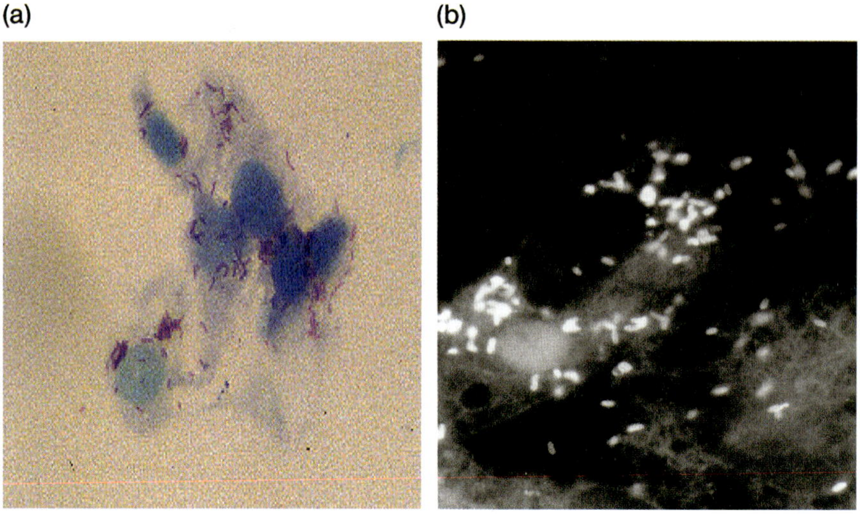

Figure 26.1 Photomicrographs showing (a) Giménez staining of *Rickettsia conorii* in Vero cells and (b) infection of cultured human umbilical vein-derived endothelial cells with *Rickettsia rickettsii* as seen by indirect immunofluorescent staining.

spotted fever species, including *R. aeschlimanii*, *R. massiliae*, *R. helvetica* and *R. slovaca*, to cause human disease is also being recognized. An excellent representative of this group is *R. parkeri*, which was first isolated in 1937 from Gulf Coast ticks *Amblyomma maculatum* in Texas, although the first confirmed human case of infection was not reported until 2004. Transmission of rickettsial diseases by previously unknown and unexpected tick vectors in areas where the disease is

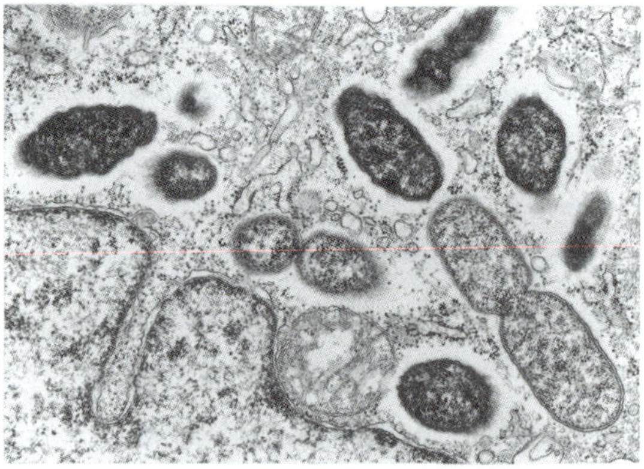

Figure 26.2 Transmission electron micrograph of a chicken embryo fibroblast infected with *Rickettsia prowazekii*. Note the predominantly intracytoplasmic localization of rickettsiae and an intracellular *Rickettsia* organism in the process of replication by simple binary fission.

otherwise rarely reported indicates the pathogen's ability to adapt to new ecological niches and maintain its virulence [3].

Due to its historic impact, recently reported outbreaks and potential for quick spread in socio-economic conditions leading to compromised hygiene, for example massive refugee movements, prison populations and war-associated situations, epidemic typhus caused by *R. prowazekii* also remains a cause for concern. Available estimates attribute more than 3 million human deaths during the last century to louse-vectored *R. prowazekii* infection. Although the majority of these deaths occurred during or immediately after World Wars I and II, epidemic typhus outbreaks in refugee camps of Burundi and evidence of reemergence in Algeria and Russia have been reported fairly recently [4]. With a mortality rate of 10–60%, epidemic typhus is one of the most severe infectious diseases of humans, the differential diagnosis of which can often be difficult due to nonspecific flu-like early symptoms such as fever, headache, malaise and myalgia. Among well-known species of spotted fever as well as typhus groups of rickettsiae, *R. prowazekii* is the only pathogen with recognized capacity to maintain persistent subclinical infection in convalescent patients that can later manifest as Brill–Zinsser disease or recrudescent typhus.

Interestingly, arthropods may act as vectors, reservoirs and/or amplifiers in the lifecycles of various *Rickettsia* species. With the exception of *R. prowazekii*, the role of humans in the natural lifecycles of rickettsiae appears to be secondary. Spotted fever rickettsiae possess the ability to infect and multiply in almost all organs of tick vectors and maintain themselves in nature by transovarial and transstadial transmission. Sexual transmission from male to female ticks has also been described for several species. Dormant rickettsiae present in the salivary glands of infested ticks can be transmitted to vertebrates during blood meal, a process that requires several hours of tick attachment to the host. In general, an eschar at the site of tick bite, development of "centripetal" rash (characterized by initial involvement of extremities followed by spread to the trunk), and dissemination through the vascular endothelium are characteristic features of SFG rickettsioses. Conversely, transmission of epidemic typhus disease occurs by contamination of bite sites by the feces or crushing of rubbed, infected lice. *R. typhi* (the etiologic agent of endemic typhus) has a similar mode of transmission through infected fleas. In further contrast to SFG rickettsiae, which are not harmful to harboring tick vectors, lice infected with *R. prowazekii* cannot sustain the infection and usually die within a couple of weeks.

A unique feature of rickettsiae is the tropism for vascular endothelial cells of small- and medium-sized vessels of their vertebrate hosts. The vascular endothelium serves as the key barrier between the intravascular compartment and extravascular tissues and is integrally involved in regulating blood flow, coagulation, leukocyte trafficking, edema formation and wound healing. Consequently, a common denominator for major features of pathogenesis and most of the complications associated with rickettsial diseases is endothelial cell injury and/or dysfunction, leading to systemic vascular collapse, development of vascular leak syndromes including acute respiratory distress syndrome, multi-organ failure and disseminated intravascular coagulopathies in severe, complicated cases. A number of rickettsial species within spotted fever and typhus subgroups cause the diseases with similar clinical manifestations, but varying severity. The prototype pathogens for both of these groups and the descriptions of the diseases caused by them are listed in Table 26.1.

The potential for deliberate use of pathogenic *Rickettsia* species is abundantly clear from the development of *R. prowazekii* as a battlefield weapon and successful trials of biobombs containing *R. prowazekii* in the past [5]. The possibility that antibiotic-resistant strains may have been or can be developed also remains a matter of grave concern. Thus, considering the potential for illegitimate use to inflict severe disease and associated morbidity/mortality, *R. prowazekii* and *R. rickettsii* have been placed on the list of "select agents" (Centers for Disease Control and Prevention) and categorized, respectively, as class B and C priority pathogens for biodefense research program. It has been recommended recently, however, that criteria such as aerosol infectivity with low doses, a high case-fatality rate, a high ratio of disease to asymptomatic infection, availability in nature, and the potential for genetic engineering of complete antibiotic resistance warrant classification of *R. rickettsii* and *R. prowazekii* as category A agents, and *R. typhi* and *R. conorii* as category B select agents [6].

26.2
Intracytoplasmic Behavior of Rickettsiae

As mentioned above, some of the most common phenotypic characteristics of bacteria classified to genus *Rickettsia* include their strict intracellular location and lifecycle association with arthropods. Competitive inhibition of *R. rickettsii* adhesion to host cells by *R. conorii*, another closely related spotted fever species, implicates a specific interaction between rickettsial ligands and host cell receptors. Two major immunodominant surface-exposed proteins, outer membrane proteins OmpA and OmpB, of which OmpA is present only on SFG organisms, play an important role as rickettsial adhesins. Ku 70, a DNA-dependent serine/threonine kinase and subunit of highly versatile regulatory protein Ku implicated in multiple nuclear processes such as apoptosis, DNA repair and telomere maintenance, serves as the mammalian host cell receptor for rickettsial OmpB [7]. On the other hand, the identity of potential receptor(s) for OmpA and possibly other as yet unidentified rickettsial ligand(s) and their roles in host cell invasion remain to be determined.

The process of host cell invasion involves active participation of metabolically viable host cell and *de novo* protein synthesis by invading rickettsiae. Such an "induced phagocytosis" involves a zipper mechanism, wherein rearrangement of actin cytoskeleton occurs beneath the site of rickettsial attachment, culminating in progressive apposition of host cell membrane over the invading bacterium and resultant engulfment [8]. Evidence from transmission electron microscopy further suggests that the process of internalization into host cells occurs within a few minutes after initial contact. Likely mediated by membranolytic functions of rickettsial hemolysin C and phospholipase D, encoded respectively by *tlyC* and *pldA* genes, rapid release from phagocytic vacuole occurs, allowing internalized rickettsiae to remain "free" in host cytosol and achieve their primary intracellular niche. Yet, there

Table 26.1 Salient clinical features and prognostic outcomes of major spotted fever and typhus group rickettsioses.

Etiologic agent	Disease and synonym	Natural host	Vector	Geographic distribution	Incubation period (days)	Clinical presentation	Prognosis
Rickettsia prowazekii	Epidemic typhus	Humans, flying squirrels	Human body louse, Flying squirrel's (FS) louse	Worldwide USA	6–15	History of louse infestation, onset sudden with fever, chills, malaise, headache. 40 °C. Various neurological manifestations. Maculopapular rash appear first on trunk then to extremities, including palms and soles	Untreated: severe
						Duration: 10–18 d	Treated: excellent
Rickettsia prowazekii	Recrudescent typhus	Humans		Worldwide	–	History of epidemic typhus 10–40 yr previously. Onset mild, flu-like, proceeding to epidemic typhus. Rash may not be present. Duration: 7–11 d	Untreated or treated: good
	Brill–Zinsser disease						

Table 26.1 (*Continued*)

Etiologic agent	Disease and synonym	Natural host	Vector	Geographic distribution	Incubation period (days)	Clinical presentation	Prognosis
Rickettsia typhi	Endemic typhus	Rats	Rat flea	Worldwide	6–14	History of exposure to rats and/or rat fleas. Onset sudden with fever, chills, malaise, headache. 40 °C. Maculopapular rash without petechiae appear on trunk and extremities, including palms and soles	Untreated or treated: good
	Murine typhus					Duration: 8–16 d.	
	Flea-borne typhus						
Rickettsia rickettsii	Rocky Mountain spotted fever (RMSF)	Rodents, rabbits, dogs	Tick	North and South America	2–14	History of tick association. Eschar not typical. Onset sudden with fever, chills, malaise, headache. 39–40 °C. Various neurological manifestations. Maculopapular rash and petechiae appearing first on extremities	Untreated: severe
	Brazilian spotted fever					Duration: 10–20 d	Early treated: good

Rickettsia conorii	Mediterranean spotted fever	Rodents, dogs	Tick	Europe, Asia, Africa	5–7	History of tick association. Eschar typical. Onset sudden with fever, chills, malaise, headache. 39–40 °C. Maculopapular rash and pronounced petechiae	Untreated: mild to poor
	Boutenneuse fever					Duration: 10–20 d	Treated: good
Rickettsia akari	Rickettsialpox	House mice	Mite	Northern Europe, northeastern USA, Asia	7–10	History of mite association. Typical red papulae turning vesicular on the inoculation site. Onset sudden with fever, chills, malaise, headache. 39 °C. Limited maculopapular rash developing into vesicular form Duration: 3–11 d	Untreated or treated: excellent

also exist significant variations in the intracellular growth and motility patterns of SFG versus TG rickettsiae. While formation of a polar actin tail and its exploitation for both intra- and intercellular movements likely serve as a major determinant for establishment, dissemination and pathogenesis of SFG rickettsioses, the typhus organisms either display erratic actin-based motility (*R. typhi*) or do not display this behavior at all (*R. prowazekii*). Surprisingly, despite much shorter actin tail lengths, *R. typhi* is able to move at approximately the same rate, but with reduced cell-to-cell spread in comparison to *R. rickettsii* [9].

Recent investigations of the molecular mechanisms of actin polymerization in SFG rickettsiae have revealed the presence of a Wiskott–Aldrich syndrome protein (WASP) termed RickA, which is capable of activation of actin-nucleating and Y-branching activities of the Arp2/3 complex, and is sufficient to direct actin-based motility. As expected, the single gene encoding RickA (RC0909) is present in *R. conorii*, but not found in the genome sequence of *R. prowazekii* [10]. Interestingly, the evidence for erratic actin-based motility and absence of gene coding for RickA in *R. typhi* indicates the possibility of existence of multiple actin-polymerization mechanisms in pathogenic rickettsiae. Consequently, the primary mechanism for cell-to-cell spread of typhus species involves the necrotic lysis of a heavily infected cell, which occurs as a consequence of constant multiplication by simple binary fission (replication time varies from 9 to 12 h) leading to significant intracellular accumulation and consequent cell lysis [11].

Considered to be "energy parasites" that grow in the ATP-rich cytoplasm of host cell, rickettsiae are typical aerobic, Gram-negative bacilli containing DNA, RNA, enzymes for Krebs cycle and ribosomes for protein synthesis. Their cell wall, surrounded by a slime layer, is composed of a peptidoglycan layer and weakly endotoxic lipopolysaccharide. Although intracellular rickettsiae are able to generate their own metabolic energy by phosphorylating ADP to ATP coupled with oxidation of glutamate in tricarboxylic acid cycle, they also exploit efficient energy exchange through ATP/ADP translocases to utilize the cytoplasmic ATP of the host cell. As a remarkable adaptation to intracellular growth habitat, the expression of the rickettsial ATP/ADP translocase gene is downregulated in heavily infected host cells to prevent sequestering of ATP back into the host cell cytoplasm, when ATP concentration in rickettsiae may be higher than that of the host cell cytoplasm. In contrast, the expression level of the citrate synthase gene *gltA* is high during these later phases of infection [12]. Recent metabolic studies focusing on substrate specificity and/or requirements further reveal that, of the five annotated ATP/ADP translocases (Tlc1, Tlc2, Tlc3, Tlc4 and Tlc5) in its genome, *R. prowazekii* utilizes Tlc1 for energy exchange while Tlc4 and Tlc5 maintain the nucleotide pool for rickettsial nucleic acid synthesis [13].

Included among major families of transporters capable of exporting virulence factors across the membranes of pathogenic bacteria are type IV transporters. These secretion systems use a complex of transmembrane proteins and a pilus to export protein or DNA substrates across bacterial and eukaryotic cell membranes into target cells. Several components of a type IV secretion system, including orthologs of inner membrane components (VirB4 and VirB11), periplasmic components (VirB6, VirB8 and VirB10), and outer membrane components (VirB3 and VirB9) are present in *R. conorii, R. prowazekii* and *R. typhi* [14]. Notably, an ortholog of VirB7, whose function is to bridge the pilus to the transmembrane complex, and both major and minor pilus components (VirB2 or VirB5) are not found. A potential protein substrate for the type IV secretion system, similar to RalF in *Legionella pneumophila*, has also been recognized, but its function(s) remain unknown.

26.3
Host Cell Interactions with Pathogenic *Rickettsia* Species

26.3.1
Spotted Fever Group

Although capable of infecting a variety of host cell types *in vitro*, a majority of sequelae associated with the human rickettsial diseases Rocky Mountain spotted fever (RMSF) and Mediterranean spotted fever (MSF) are attributed to widespread infection of endothelium leading to vessel damage, altered vascular permeability and vascular inflammation/dysfunction, collectively referred to as "rickettsial vasculitis." The clinical complications in severe, untreated cases often manifest as pulmonary edema, hypovolemic hypotension and multiorgan failure and occasionally as disseminated intravascular coagulation. Not surprisingly, rickettsial diseases are generally associated with significant hemostatic/fibrinolytic protein changes due mainly to endothelial perturbations, and infection with *Rickettsia* organisms causes coagulation abnormalities as evidenced by pronounced alterations in plasma concentrations of coagulation factors, natural anticoagulants and components of the fibrinolytic system in animal models, and in human subjects infected with *R. conorii* and *R. rickettsii*.

It is important to keep in mind, however, that activation of procoagulant pathways and microthrombus formation only represent potential life-threatening complications in severe/untreated cases of RMSF and MSF. Indeed, endothelial cells infected *in vitro* with *R. rickettsii* or *R. conorii* display an "activated" phenotype as evidenced by release of endothelial-specific markers from Weibel–Palade bodies, increased expression of surface adhesion molecules and adaptive alterations in antioxidant enzyme systems to cope with infection-induced oxidative stress. Infection of host endothelial cells also results in increased expression and secretion of interleukin 1α (IL-1α) and IL-6, and the chemokines IL-8, monocyte chemoattractant protein 1 (MCP-1) and fractalkine [15,16]. Induced expression of a number of these proinflammatory genes also occurs through signaling mechanisms that require activation of nuclear transcription factor-kappa B (NFκB), a ubiquitous nuclear transcription factor involved in the control of apoptosis, inflammation and stress response, and p38 kinase, the downstream effector component of a stress-activated mitogen-activated protein (MAP) kinase cascade [17]. Studies from authors' laboratories have further implicated a critical role for infection-induced NFκB in maintaining mitochondrial integrity and shifting the intricate balance between pro- and antiapoptotic proteins to prevent host cell death and, in turn, ensure survival of the replication site for intracellular rickettsiae [18].

Clinical case reports suggest that severe cases of human rickettsial infections are often associated with acute respiratory distress syndrome and renal failure and the prognosis of spotted fever and typhus in patients with genetic glucose 6-phosphate dehydrogenase deficiency is poor due to enhanced severity, indicating the potential importance of redox systems in combating the disease [19]. The injury to endothelial cells following *in vitro* infection with *R. rickettsii* results in extensive membrane damage and striking changes in the structural organization of intracellular organelles and cytoskeleton. The accumulation of intracellular peroxides, superoxide anion and extracellular H_2O_2, depletion of protective thiols and reduction in the activities of antioxidant enzymes has also been demonstrated [20]. Endothelial cell infection with SFG rickettsiae induces expression of heme oxygenase 1 (HO-1), the initial rate-limiting enzyme of the heme catabolism pathway known to protect against oxidative damage in a variety of stress situations. Modulation of ferritin, an iron-storage protein with potent antioxidant and antiapoptotic functions in the vasculature further suggests an important role for HO-1 in cellular responses to infection, possibly as a protective mechanism against oxidative injury [21]. In addition, *R. rickettsii* infection of human umbilical vein-derived endothelial cells causes robust induction of cyclooxygenase 2 (COX-2) mRNA and protein expression, but has no apparent effect on the constitutive COX-1 isoform.

Consistently, cells infected with viable rickettsiae display significantly increased secretion of prostaglandins 6-keto $PGF_{1\alpha}$ and PGE_2. Induction of the endothelial COX-2 system and resultant enhanced release of vasoactive prostaglandins may contribute to the regulation of inflammatory responses and vascular permeability changes during spotted fever rickettsioses [22]. Recent evidence implicates dissociated adherens junctions between human cerebral endothelial cells experimentally infected with *R. rickettsii* in the loss of monolayer permeability, reflected by decreased transendothelial electrical resistance and further suggests that proinflammatory stimuli such as tumor-necrosis factor α (TNFα), IL-1β and interferon γ (IFNγ) exacerbate this effect [23].

The ability to effectively control rickettsiae proliferating in the cytoplasm of nonphagocytic host cells, where they are protected from host defense mechanisms such as antibodies and exposure to phagocytes, is another important question. Studies in an established "endothelial target" mouse model of Rocky Mountain spotted fever have suggested a crucial role for synergistic, paracrine effects of IFNγ (secreted by T lymphocytes and natural killer cells) and TNFα (secreted by macrophages) in stimulating infected endothelial cells and other cell types such as hepatocytes and macrophages to produce nitric oxide, which has antirickettsial activity [24]. In macrophages, limited availability of tryptophan due to enhanced degradation by indoleamine-2,3-dioxygenase also mediates cytotoxic effects against rickettsiae, whereas hydrogen peroxide-dependent killing of intracellular rickettsiae also occurs in cultured human endothelial cells, macrophages and peripheral blood monocytes primed with cytokines prior to infection [25]. Identified so far, major mechanisms employed by Fc-dependent antibody-mediated protection against infection of target host cells also include prevention of rickettsial escape from the phagosome and phagolysosomal killing involving nitric oxide, reactive oxygen species and starvation for L-tryptophan [26].

26.3.2 Typhus Group

As for spotted fever rickettsiae, vascular endothelium is the major target of *R. prowazekii* during human infections. However, very little is known about specific host cell responses and no information is available regarding signaling mechanisms that determine the interplay between target endothelial cells and the pathogen. It is important to emphasize here that most of the initial studies aimed at establishing intracellular growth requirements were conducted using *R. prowazekii* infection of either L929 or Vero cells, both of which are fibroblastic in nature. *R. prowazekii* interactions with fibroblast-like L929 cells *in vitro* demonstrated the production of IFNα and IFNβ, but not IFNγ. Interestingly, secreted IFNα as well as IFNβ were found to have inhibitory activity against replication of vesicular stomatitis virus and growth of *R. prowazekii*. However, IFNγ also suppresses the growth of *R. prowazekii* in mouse L929 cells, macrophage-like RAW264.7 cells and human fibroblasts. These findings led to the isolation and characterization of IFN-sensitive and IFN-resistant strains [27].

R. prowazekii-infected human endothelial cells as well as mouse leukocytes/macrophages display increased platelet-activating factor synthesis and prostaglandin secretion [28,29]. *R. prowazekii* infection of both murine and human endothelial cells *in vitro* results in increased transmigration of leukocytes, likely through an exacerbated inflammatory response [30].

26.4 Animal Models and Immunology of Rickettsioses

Although experimental *R. prowazekii* infection in small laboratory animals such as cotton rats, gerbils and guinea-pigs has been attempted, the only well-established models for endothelial target model of rickettsioses exploit the infection of C3H/HeN mice with *R. conorii* and *R. typhi* and infection of BALB/c mice with *R. australis*, the etiologic agent of Queensland tick typhus. Prior to this development, however, disease characteristics and mechanisms of immunity were investigated by rickettsial infection of either guinea-pigs or BALB/c mice. Capable of producing signs and pathologic features similar to human disease, two direct models of epidemic typhus describe the infection of BALB/c mice and cynomolgus monkeys with virulent Breinl strain of *R. prowazekii* [31,32]. To enhance our understanding of the relationship between the host and pathogen, use of *in vivo* models (infection of selective, susceptible strains of laboratory mice) represents an attractive and useful approach. Such model systems, which yield disseminated endothelial-target infection and closely parallel the pathological features of human disease, have been extensively used to investigate the immune responses during SFG as well as TG rickettsioses.

Depletion or adoptive transfer of CD4 and CD8 T lymphocyte subsets during *in vivo* infection with *R. conorii* have revealed a crucial role for immune CD8 T lymphocytes in clearance of rickettsiae from infected endothelial cells. Knockout mice for class 1 major histocompatibility complex (MHC) are extremely susceptible (>50 000-fold) to a lethal infection with *R. australis*, when compared with the wild-type C57BL/6 mice. Mice lacking genes for IFNγ and perforin, on the other hand, are >100-fold more susceptible, further indicating that the activity of cytotoxic T lymphocytes is apparently more critical than the effects of IFNγ and that perforin-mediated activity contributes only partly to anti-rickettsial effect of cytotoxic T lymphocytes [33].

Although expression studies for CXCL9 (Mig: monokine induced by IFNγ) and CXCL10 (IP-10: 10 kDa IFNγ-induced protein), known CXCR3 ligands and chemokines targeting $CD8^+$ T cells and $CD4^+$ T helper cells, have yielded evidence of their increased expression in tissues of experimentally infected mice and in biopsy specimens of patients with RMSF, neutralization by antibodies has no effect on survival of infected hosts or on rickettsial titers in target tissues [34]. Available evidence from established animal models further suggests the occurrence of reciprocal immunological cross-protection between spotted fever and typhus groups of rickettsiae, which is also mediated by T cells. In addition, natural killer cells are also implicated in early anti-rickettsial immune response, likely via an IFNγ-dependent mechanism [35,36].

Some intriguing aspects of rickettsial biology that are starting to emerge from *in vitro* infection of bone marrow-derived dendritic cells include localization of intracellular rickettsiae in both the cytosol and phagosomes and differential interactions with dendritic cells during *in vivo* infection as the basis for stimulating a protective response in resistant C57BL6 mice and suppressive adaptive immune response in susceptible C3H mice. It has been proposed that the presence of rickettsiae in both vacuolar and cytosolic compartments of dendritic cells may serve to enhance the T cell priming function, respectively, through MHC class II and class I pathway-mediated antigen presentation to $CD4^+$ and $CD8^+$ T cells. Suggested by robust upregulation of expression of cell surface molecules such as CD40, CD80, CD86 and MHC class II, and production of proinflammatory cytokines, infection with *R. conorii* induces maturation of dendritic cells *in vitro*. Transfer of such stimulated cells augments immune response through increased $CD4^+$, $CD8^+$ and NK cell activities and confers protection against lethal *in vivo* challenge by inhibiting proliferation and dissemination of rickettsiae into target tissues [37,38].

26.5
Genomics and Proteomics of *Rickettsia* Species

The entire genome of *R. prowazekii* was the first to be sequenced, annotated and published [39], followed by that of *R. conorii* and *R. typhi* [40,41]. The genome sequences for a number of other *Rickettsia* species, including *R. rickettsii* (accession no. AADJ 01000001) and *R. sibirica* (NZ_AABW 01000001) are also available in accessible databases. The genome sizes for rickettsial species range from approximately 1.1 to 1.5 Mb (close to 1.1 Mb for TG and from 1.3 to 1.5 Mb for SFG). The average G + C content for the entire genome is close to 29% for TG and 32–33% for SFG and AG.

Some unique features that emerged towards the end of last century from the description of *R. prowazekii* and *R. conorii* genomes include the tendency of reductive convergent evolution (supported by evidence for a significant loss of genes governing biosynthetic functions), close phylogenetic relationship with mitochondria and the presence of a very high proportion (as much as 24%) of noncoding "junk" DNA or pseudogenes in rickettsial genomes, compared to other microbes. A close comparison of *R. conorii* and *R. prowazekii* genomes reveals the presence of more than 500 genes unique to *R. conorii*, but very few in *R. prowazekii*. *R. conorii* also has as many as 400 orphan genes with no homologs outside of *Rickettsia*. Comparative analyses reveal apparently similar genomic order and content in typhus species *R. prowazekii* and *R. typhi* and absence of *rickA* gene responsible for actin-based motility in *R. conorii* of the spotted fever group. A plethora of split genes have also been recognized – among genes present in *R. conorii* but not in *R. prowazekii*, a significant number have remnant sequences in the *R. prowazekii* genome. On the other hand, the colinearity of the *R. felis* genome, one of the largest among those sequenced so far, is more frequently interrupted by inversions or translocations than smaller rickettsial genomes; it also includes a remarkably high number of paralogs for transposases and proteins containing protein–protein interaction motifs (ankyrin and tetratricopeptide repeats) [42].

While an overall emerging theme suggests that about 775 shared core genes encode proteins governing major metabolic and cellular processes, certain species/strain-specific genes are likely responsible for cell surface proteins (*Sca* genes) and/or potential virulence factors (*Vir* and *pld* genes). Overall, only a small number of rickettsial genes, which can be annotated confidently by their extensive identity to proteins of known function in other organisms, have been characterized in some detail.

In spite of restrictions imposed by the obligate intracellular parasitic nature of the rickettsiae, the last few years have witnessed significant progress in their genetic manipulation. Using selection based on the *arr-2* gene encoding an enzyme that ADP-ribosylates rifampin and transposon mutagenesis, stable *R. prowazekii* insertional knockout mutants have been isolated [43]. More recently, a versatile mariner family-based, Himar1 transposon system has also been employed to generate insertional mutants of *R. prowazekii* [44]. The availability of complete genome sequences for a number of pathogenic species has further facilitated accurate phylogenetic analysis and species/strain characterization via comparative genomics and proteomics. A frameshift mutation in the Rp028/Rp027 gene encoding a methyltransferase enzyme in the avirulent Madrid E strain of *R. prowazekii* has been shown to revert to wild-type in virulent E strain [45]. Comparative evaluation of two strains of *R. rickettsii* (virulent Sheila Smith and avirulent Iowa strain) has revealed deletions and single-nucleotide polymorphisms in *ompA* and *ompB* genes in the genome for Iowa strain, resulting in loss of OmpA protein synthesis and modifications in OmpB protein sequence [46]. Similar studies with avirulent Madrid E and virulent Breinl strains of *R. prowazekii* have shown higher expression levels of five potential virulence genes (*virB4, ftsK, lpxA, rbfE, rpoH*) in the virulent strain [47]. Attempts for the detailed proteomic characterization of *R. prowazekii* and *R. conorii* have also resulted in identification of adhesion proteins, actin-based motility regulators and putative virulence factors (discussed in Section 26.5), and may lead to the discovery of novel pathogenicity regulators.

One of the most striking recent discoveries in the field of rickettsiology is the presence of plasmids in different *Rickettsia* species. Following first identification of short and long forms of plasmid pRF in *R. felis* [42], the presence of the pRM/PRF family of plasmids has been established in diverse species of rickettsiae [48]. The pRF and pRFδ (truncated pRF) of *R. felis* are about 63 and 39 kb long and full-length pRF contains 68 open reading frames, of which 44 may be associated with functional attributes. The pRF plasmid encodes not only homologs of transposases, integrases, recombinases and proteins with ankyrin and tetratricopeptide repeats, but also contains genes for a homolog for patatin and hyaluronidase activity, enzymes possibly involved in supporting spread through the host. The presence of pili-like appendages in *R. felis, R. bellii* and *R. canadensis* supports the conclusions drawn from genomic analyses [49,50]. The existence of plasmids and potential mobile genetic elements in the rickettsiae suggests an even greater potential for genetic diversity and adaptation to variable niches in the hosts and vectors.

26.6
Summary and Outlook

Transmitted to humans in nature by an arthropod vector (fleas, lice, ticks or mites), α-proteobacteria belonging to the *Rickettsia* genus are Gram-negative, obligate intracellular pathogens, which include two major antigenically defined groups: the spotted fever group and the typhus group. Rocky Mountain spotted fever due to *R. rickettsii* infection is one of the most severe rickettsioses, capable of fatal outcome in severe cases. Available estimates attribute more than 3 million human deaths during the last century to louse-vectored epidemic typhus due to *R. prowazekii* infection. Since *Rickettsia* infections are characterized by disseminated infection of vascular endothelium in humans and in animal models of infection, hemostatic/physiologic disturbances such as compromised vascular permeability, hypovolemia, thrombocytopenia, activation of coagulation cascades and the fibrinolytic system, and consumptive coagulopathy are integral features of rickettsial diseases. The advent of new and improved molecular tools, identification and description of previously unrecognized species/subspecies and realization of pathogenic potential of previously known spotted fever rickettsiae has established rickettsioses as emerging infectious diseases. Known for their relatively smaller genome size and tendency for reductive evolution, intracellular rickettsiae occupy a predominantly "intracytoplasmic niche" and primarily utilize host cell energy transported through ATP–ADP exchange systems to meet their metabolic requirements.

Spotted fever rickettsiae display polar actin tail-based movements for intracellular motility and intercellular spread, whereas typhus organisms either lack this property or are known to perform erratic movements within the host cell. Endothelial cell infection *in vitro* with both spotted fever and typhus rickettsiae triggers the activation of ubiquitous NFκB and p38 MAP kinase signal transduction pathway, resulting in transcriptional regulation of a variety of proinflammatory and procoagulant molecules. NFκB, by virtue of its antiapoptotic functions, protects the host cell from undergoing apoptotic death early during the infection, allowing it to remain as a site of replication. Oxidative stress mechanisms play a major role in rickettsial pathogenesis, whereas immune clearance is mediated by cytotoxic T cells and natural killer cells.

Impressive recent advances, including genome sequencing of various species and strains of varying virulence, successful molecular transformation, and characterization of plasmids, should allow for identification of virulence factors, invasion and intracellular survival mechanisms, novel therapeutic strategies and effective vaccine candidates.

Acknowledgments

Work in the authors' laboratories was supported in part by research project grants from the National Institute of Allergy and Infectious Diseases and the National Heart, Lung, and Blood Institute of the National Institutes of Health, Bethesda, Maryland, USA.

References

1 Labruna, M.B., Camargo, L.M., Camargo, E.P. and Walker, D.H. (2005) *Vet Parasitol*, **127**, 169–174.

2 Parola, P. and Raoult, D. (2006) *Clin Dermatol*, **24**, 191–200.

3 Dumler, J.S. and Walker, D.H. (2005) *N Engl J Med*, **353**, 551–553.

4 Roux, V. and Raoult, D. (1999) *J Clin Microbiol*, **37**, 596–599.

5 Azad, A.F. (2007) *Clin Infect Dis*, **45** (Suppl 1), S52–S55.

6 Walker, D.H. (2007) *Clin Infect Dis*, **45** (Suppl 1), S39–S44.

7 Walker, D.H. (2006) *N Engl J Med*, **354**, 1418–1420.

8 Olano, J.P. (2005) *Ann N Y Acad Sci*, **1063**, 187–196.

9 Heinzen, R.A. (2003) *Ann N Y Acad Sci*, **990**, 535–547.

10 Jeng, R.L., Goley, E.D., D'Alessio, J.A., Chaga, O.Y. *et al.* (2004) *Cell Microbiol*, **6**, 761–769.

11 Hackstadt, T. (1996) *Infect Agents Dis*, **5**, 127–143.

12 Andersson, S.G. (1998) *Biochim Biophys Acta*, **1365**, 105–111.

13 Audia, J.P. and Winkler, H.H. (2006) *J Bacteriol*, **188**, 6261–6268.

14 Walker, D.H. and Yu, X.-J. (2005) *Ann N Y Acad Sci*, **1063**, 13–25.

15 Valbuena, G. and Walker, D.H. (2005) *Virchows Arch*, **446**, 21–27.

16 Rydkina, E., Sahni, A., Silverman, D.J. and Sahni, S.K. (2007) *J Med Microbiol*, **56**, 896–906.

17 Rydkina, E., Silverman, D.J. and Sahni, S.K. (2005) *Cell Microbiol*, **7**, 1519–1530.

18 Joshi, S.G., Francis, C.W., Silverman, D.J. and Sahni, S.K. (2004) *FEMS Microbiol Lett*, **234**, 333–341.

19 Walker, D.H. and Kirkman, N.H. (1980) *J Infect Dis*, **142**, 771.

20 Eremeeva, M.E., Dasch, G.A. and Silverman, D.J. (2000) *Subcell Biochem*, **33**, 479–516.

21 Sahni, S.K., Rydkina, E., Sahni, A., Joshi, S.G. and Silverman, D.J. (2005) *Ann N Y Acad Sci*, **1063**, 207–214.

22 Sahni, S.K. (2007) *Thromb Res*, **119**, 531–549.

23 Woods, M.E. and Olano, J.P. (2008) *J Clin Immunol*, **28**, 174–185.

24 Feng, H.M., Popov, V.L. and Walker, D.H. (1994) *Infect Immun*, **62**, 1952–1960.

25 Feng, H.M. and Walker, D.H. (2000) *Infect Immun*, **68**, 6729–6736.

26 Feng, H.M., Whitworth, T., Popov, V. and Walker, D.H. (2004) *Infect Immun*, **72**, 3524–3530.

27 Turco, J. and Winkler, H.H. (1991) *Infect Immun*, **59**, 1647–1655.

28 Walker, T.S., Brown, J.S., Hoover, C.S. and Morgan, D.A. (1990) *J Infect Dis*, **162**, 1136–1144.

29 Walker, T.S. and Mellott, G.E. (1993) *Infect Immun*, **61**, 2024–2029.

30 Bechah, Y., Capo, C., Raoult, D. and Mege, J.L. (2008) *J Infect Dis*, **197**, 142–147.

31 Gonder, J.C., Kenyon, R.H. and Pedersen, C.E. Jr, (1980) *Infect Immun*, **30**, 219–223.

32 Bechah, Y., Capo, C., Mege, J.L. and Raoult, D. (2008) *Future Microbiol*, **3**, 223–236.

33 Valbuena, G., Feng, H.M. and Walker, D.H. (2002) *Microbes Infect*, **4**, 625–633.

34 Valbuena, G., Feng, H.M. and Walker, D.H. (2004) *Am J Trop Med Hyg*, **71**, 393–399.

35 Billings, A.N., Feng, H.M., Olano, J.P. and Walker, D.H. (2001) *Am J Trop Med Hyg*, **65**, 52–56.

36 Valbuena, G., Jordan, J.M. and Walker, D.H. (2004) *J Infect Dis*, **190**, 1221–1227.

37 Fang, R., Ismail, N., Soong, L., Popov, V.L. *et al.* (2007) *Infect Immun*, **75**, 3112–3123.

38 Jordan, J.M., Woods, M.E., Feng, H.M., Soong, L. and Walker, D.H. (2007) *J Infect Dis*, **196**, 629–638.

39 Andersson, S.G., Zomorodipour, A., Andersson, J.O., Sicheritz-Pontén, T. *et al.* (1998) *Nature*, **396**, 133–140.

40 Ogata, H., Audic, S., Renesto-Audiffren, P., Fournier, P.E. *et al.* (2001) *Science*, **293**, 2093–2098.

41 McLeod, M.P., Qin, X., Karpathy, S.E. and Gioia, J. (2004) *J Bacteriol*, **186**, 5842–5855.

42 Ogata, H., Renesto, P., Audic, S., Robert, S. *et al.* (2005) *PLoS Biol*, **3**, e248.

43 Tucker, A.M., Pannell, L.K. and Wood, D.O. (2005) *Ann N Y Acad Sci*, **1063**, 35–46.

44 Liu, Z.M., Tucker, A.M., Driskell, L.O. and Wood, D.O. (2007) *Appl Environ Microbiol*, **73**, 6644–6649.

45 Zhang, J.Z., Hao, J.F., Walker, D.H. and Yu, X.J. (2006) *Vaccine*, **24**, 2317–2323.

46 Ellison, D.W., Clark, T.R., Sturdevant, D.E., Virtaneva, K. *et al.* (2008) *Infect Immun*, **76**, 542–550.

47 Ge, H., Chuang, Y.-Y.E., Zhao, S., Tong, M. *et al.* (2004) *J Bacteriol*, **186**, 556–565.

48 Baldridge, G.D., Burkhardt, N.Y., Felsheim, R.F., Kurtti, T.J. and Munderloh, U.G. (2008) *Appl Environ Microbiol*, **74**, 645–652.

49 Ogata, H., La Scola, B., Audic, S., Renesto, P. *et al.* (2006) *PLoS Genet*, **2**, e76.

50 Fuxelius, H.-H., Darby, A., Min, C.-K., Cho, N.-H. and Andersson, S.G.E. (2007) *Res Microbiol*, **158**, 745–753.

27
Shigella

Guy Tran Van Nhieu and Philippe Sansonetti

27.1
Introduction

27.1.1
Shigellosis, the Disease

The bacillus responsible for dysentery was first identified in 1897 but bacterial dysentery was probably one of the earliest infectious diseases in humans. From the recommendation of Hippocrates to boil water to prevent dysentery, we know that people had known in ancient times that the disease was transmitted by contaminated foodstuff or water. Throughout history, dysentery has been associated with wars of the pre-antibiotic era and profoundly impacted on the destiny of nations, since it has been estimated to kill more soldiers than human weaponries in major conflicts. The poor hygiene conditions and the large concentrations of human beings during war campaigns are key to such an association. Because contamination occurs through feco-oral transmission, the lack of drinkable water supplies or proper latrines and sanitation in camps favors the rapid spread of the disease.

Shigella remains an important pathogen today; its estimated toll from the latest World Health Organization studies report 164.7 million cases with about 1.1 million deaths associated with the disease, mainly in developing countries among children under the age of five [1].

Shigella is responsible for a spectrum of intestinal diseases, ranging from self-contained mild watery diarrhea to the dysenteric syndrome. After an incubation period, usually lasting from 24 to 72 h, the onset of shigellosis is characterized by abdominal cramps, fever and anorexia, accompanied with watery diarrhea. In many cases, and especially in healthy adults, the disease does not progress further and remission occurs spontaneously. In more severe forms, occurring most frequently in malnourished infants, this first phase is followed by acute diarrhea with the emission of bloody and purulent stools. The fatal outcome is often associated with dehydration, hypoglycemia or intestinal complications including toxic megacolon or intestinal perforation. Shigellosis has also been associated with a hemolytic uremic

Intracellular Niches of Microbes. Edited by Ulrich E. Schaible and Albert Haas

ISBN: 978-3-527-32207-7

syndrome often associated with *S. dysenteriae* strains expressing the Shiga toxin. It has been estimated that as little as 100 bacteria are sufficient to induce the disease, underlining the efficient adaptation of *Shigella* to humans. Upon ingestion, *Shigella* invades the colonic mucosa where it elicits an intense inflammatory reaction responsible for mucosal destruction.

Shigella are Gram-negative bacteria that are classified as a genus distinct from *Escherichia coli* mostly for historical reasons, since at the time that Kyoshi Shiga isolated the first *S. dysenteria* strain, it was not believed that pathogenic forms of *E. coli* commensals could exist. Based on modern phylogenetic analysis and genome sequencing, it can be seen that *Shigella* actually belongs to the widely diverse *E. coli* family [3] and constitutes a group of *E. coli* pathovars based on acquisition of a virulence plasmid that bears the invasive capacities of a few chromosomal pathogenicity islands (PAI) and extinction of a certain number of chromosomal genes. Phylogenetic analyses indicate that various *Shigella* strains emanate from the independent acquisition of the large virulence plasmid by horizontal transfer, suggesting a convergent evolution and a selective adjustment to specific requirements linked to infection and replication in the host [2].

Four *Shigella* species are reported. *S. dysenteriae* is usually associated with most severe cases and with bacterial epidemics, whereas *S. flexneri* is responsible for endemics in various parts of the developing world. *S. boydii* is encountered mostly in India, while *S. sonnei* is more prevalent in developed countries. A major feature of *Shigella* is the presence of a large plasmid that carries most of the virulence determinants. This large virulence plasmid has been sequenced in various *Shigella* strains and was shown to contain a very high content of IS sequences [4, 5]. Critical for bacterial virulence, about 25 genes encoded by the *mxi-spa* locus are involved in the expression of a type III secretion system (T3SS) dedicated to the injection of virulence effectors inside host cells. Another 27 genes correspond to T3SS substrates that can be roughly classified in three groups: (i) genes that are constitutively transcribed, such as the *ipa* genes involved in bacterial invasion; (ii) genes for which transcription is upregulated upon induction of type III secretion; (iii) genes for which transcription is only turned on upon induction of type III secretion. For these two last categories of genes, there is evidence that their products are involved in the control of inflammation. Other genes include *icsA*, encoding a surface protein localized at one bacterial pole and responsible for actin-based motility [6].

27.1.2
Inflammation and Bacterial Invasion of the Colonic Mucosa

There is evidence that *Shigella* can downregulate tight junctions of epithelial cells *in vitro*, and that this property may allow bacterial breaching of the epithelium through the paracellular spaces [7]. However, biopsy analysis from experimental shigellosis in macaque monkeys using an *icsA* mutant strain that is deficient for cell-to-cell spread indicated that inflammation does not occur diffusely over mucosal surface but primarily at the levels of lymphoid solitary nodules of the rectal and colonic mucosa [8]. Because *Shigella* invasion does not occur efficiently at the apical surface of intestinal polarized cells, it is assumed that *in vivo*, invasion primarily takes place at the levels of the follicle-associated epithelium (FAE) that overlays the colonic mucosal lymph nodes [9]. This FAE does not form a defined brush border and in particular contains M cells, which are cells with basal endocytic activity that sample and present antigens from the intestinal lumen to the mucosal immune system. In line with this function, M cells are in tight interaction with macrophages that control the incoming flux of antigens and particles. M cell-associated macrophages prevent bacterial access to underlying tissues, and bacterial pathogens that breach the epithelium lining through M cells need to develop means to resist macrophage-mediated killing. *Shigella* was shown to kill macrophages by pyroptosis, a proinflammatory apoptotic process implicating Ipaf, a NOD-like cytosolic pattern-recognition receptor, and activation of caspase 1 leading to the release of interleukin 1β (IL-1β) [10, 11].

Because a hallmark of shigellosis is the acute inflammation and since *Shigella* effectors were shown to induce the release of proinflammatory cytokines, bacterial-induced inflammation was proposed as a strategy used by *Shigella* to breach the colonic epithelium. According to this scheme, following initial infection of M cells and *Shigella*-induced macrophage pyroptosis, the release of IL-1β combined with that of IL-8 from infected epithelial cells leads to the massive recruitment of neutrophils and to disruption of the epithelial layer integrity, thus allowing bacterial access to the baso-lateral surface of polarized epithelial cells and amplifying bacterial invasion (Figure 27.1). Along with the release of proinflammatory cytokines as a consequence of *Shigella* invasion, bacterial products were also shown to contribute to inflammation. As for other Gram-negative bacteria, the lipid A core of the *Shigella*lipopolysaccharide (LPS) can activate cells such as monocytes and neutrophils that are recruited at the site of infection and that constitutively express CD14 and the accessory protein MD-2 through binding to the toll-like receptor 4 (TLR4) and the activation of the proinflammatory transcriptional activator NFκB [12]. In addition, the intracellular location of *Shigella* also triggers signals through the sensing of cell wall components by cytosolic NOD receptors, leading to the activation of NFκB [13]. In the case of *Shigella*, the sensing by these extra- and intracellular pattern recognition molecules amplified by bacterial replication and access to deeper tissues is presumed to account for sustained stimulation leading to massive inflammation (Figure 27.1). *Shigella* invasion of epithelial cells was also reported to induce Ca^{2+} signaling, which may contribute to inflammatory responses [14]. Finally, specific type III secretion effectors, such as OspZ, stimulates the recruitment of neutrophils at the site of infection [15]. The notion that *Shigella* triggers inflammation to invade the colonic mucosa received support from experiments in which the IL-1 receptor antagonist that inhibits IL-1-mediated signaling prevented bacterial colonization of rabbit ileal loops [16].

Shigella, however, also downregulates inflammation through the action of late T3SS effectors. For example, OspF is a phosphothreonine lyase that prevents the activation of the p38 and ERK-2 MAP kinases [17, 18]. OspG prevents NFκB activation by binding to UbCH5, a E2 ubiquitin conjugating enzyme involved in the proteasomal degradation of the NFκB inhibitor I-κB [19]. Consistently, *Shigella ospF* or *ospG* mutants induce an inflammation in rabbit ileal loops that is more massive than that induced by wild-type *Shigella*. Also, the chromosomally encoded factor ShiA was

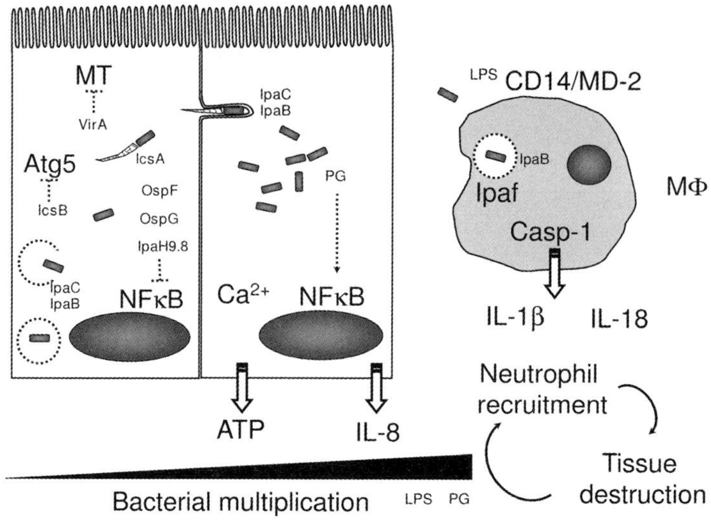

Figure 27.1 Cellular and tissular strategies of *Shigella* invasion. In enterocytes, following vacuolar escape, *Shigella* replicates freely in the cytosol. IcsB, through binding to Atg5, protects bacteria against autophagy. IcsA allows actin-based motility, which is favored by inhibition of microtubule (MT) polymerization by VirA. The late T3SS effector OspF, OspG, and possibly IpaH9.8 downregulate pathways leading to the activation of the proinflammatory transcriptional activator NFκB. *Shigella* cell-to-cell spreading requires cadherin-dependent junctional structures. As for vacuolar escape, lysis of the double-membrane-containing protrusion requires large amounts of IpaB and IpaC. *Shigella* invasion is also associated with the NFκB-dependent release of IL-8 following recognition of bacterial peptidoglycan products (PG) by NOD receptors and with the Ca^{2+}-dependent release of ATP through connexin hemichannels. Proinflammatory signals in the infected macrophage are induced by the recognition of the bacterial LPS by CD14/MD-2 and signaling through TLR4, as well as by the activation of IpaB-induced pyroptosis through the activation of IpaF and caspase-1, leading to the release of IL-1β and IL-18. The release of these proinflammatory cytokines leads to the recruitment of neutrophils at the site of infection that contribute to tissue destruction. The persistent stimulation linked to bacterial replication and LPS and PG production further amplifies neutrophil recruitment and tissue destruction.

reported to inhibit inflammation and neutrophil recruitment at the site of infection, although its mode of action is unclear and probably indirect since this factor is not secreted [20]. Finally, inactivation of the chromosomally encoded *ipaH* alleles points to a role for this family of genes in the downregulation of inflammation [21]. These findings suggest that *Shigella* modulation of inflammation is finely tuned during the various stages of the infectious process.

27.2
Escape into the Cytoplasm and Intracytoplasmic Replication

During *in vitro* challenge of epithelial cells, only a small fraction of bacteria establish a stable contact with the cell surface. Within minutes, localized ruffles form at the site of bacterial interaction that engulf the bacteria in a macropinoytic-like process. This process of bacteria-induced internalization has also been referred to as a "trigger mode" as opposed to the "zipper mode" because it implicates the reorganization of the host cell cytoskeleton, not only at the immediate site of bacterium–cell contact, but also in the neighboring vicinity leading to the formation of membrane ruffles [22]. Although the ruffles induced by *Shigella* during invasion are reminiscent of those induced during *Salmonella* invasion of epithelial cells, the molecular determinants are different. Nevertheless, both "triggering processes" are dependent on a T3SS.

27.2.1
The *Shigella* Type III Secretion System

T3SS are widespread in pathogenic Gram-negative bacteria and have been subdivided into five types according to the levels of conservation in the primary sequences of their structural components [23]. The structural organization of these sophisticated secretion machines appears to allow the optimal presentation of the T3SS tip at the bacterial surface, while burying the remaining part of the apparatus within the LPS layer and bacterial membranes [23, 24]. As a consequence, T3SS structural components and substrates are not targeted by the host humoral responses and antibodies directed against major T3SS effectors fail to induce protection. Consistent with the "LPS shielding" of T3SS, only antibodies directed against the tip complex component IpaD were shown to inhibit *Shigella* invasion *in vitro* [25]. All T3SS share similar features. They are flagella-related structures and the *Shigella* T3SS, as well as T3SS from other species, have been purified and visualized by electron microscopy. T3SS consist of a basal body that spans the inner and outer membranes, prolonged by a needle that, in the case of *Shigella*, has been estimated to measure about 40 nm in length with an external diameter of about 12 nm (Figure 27.2). Single particle reconstruction analysis suggests that the needle is hollow with an internal diameter of 2–3 nm. Since this diameter fits with the estimated size of the pore formed by the T3SS translocator based on osmoprotection studies in erythrocyte lysis assays, it is believed that the translocator is connected to the tip of the T3SS needle to insure a continuum between the bacteria and the host cell [26].

In this model, during injection, T3SS substrates are channeled in an unfolded or partially unfolded configuration through the needle and the translocator. T3SS are activated upon cell contact and the two translocator components as well as IpaD, an additional T3SS substrate, are believed to play a major role in the sensing of host cell membranes and regulation of the secretion [27]. Sensing presumably occurs by the tip of the T3SS, at the distal end of the needle, the needle acting as a spacer that provides the exact length to present this sensor at the bacterial surface [24].

Converging data have led to the proposal of a model, supported by the analysis of T3SS by ultrastructural microscopy, biochemical fractionation and secretion phenotypes in bacterial mutants. From the phenotype of bacterial mutants showing constitutive secretion, it has been known for a long time that IpaB and IpaD regulate T3SS activity, leading to the hypothesis that these two components act like a "cork"

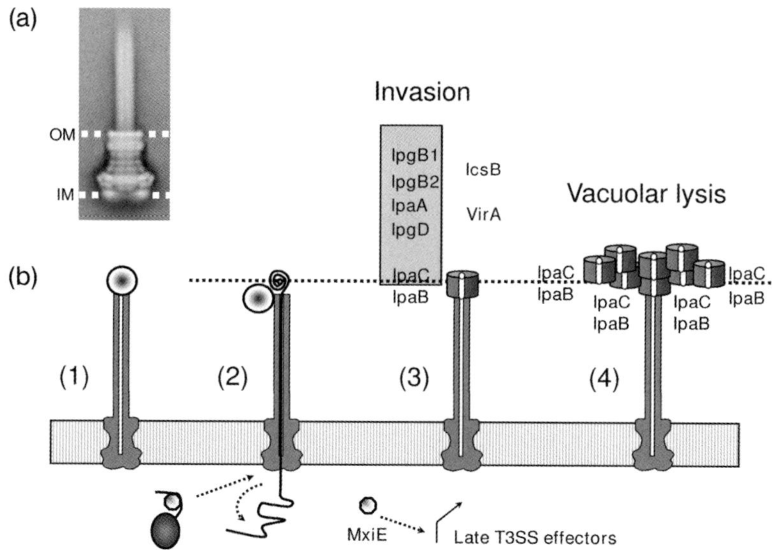

Figure 27.2 The role of the *Shigella* T3SS during the early phases of cell invasion. (a) Single particle reconstruction of transmission electron micrographs of a purified type III secretion system (T3SS). The dashed lines represent the predicted spanning sites of the inner membrane (IM) and outer membrane (OM). (Adapted from Sani *et al.* [25].) (b) Model of T3SS activation. Under nonsecretion conditions, the tip complex (white circle) prevents T3S (1); upon contact with host cell membranes, the tip complex triggers the secretion of IpaB and IpaC that transit through the needle in an unfolded configuration and refold in host cell membranes to form the translocator (gray cylinder) (2). Early T3SS effectors are injected in the cell cytsol. Effectors of invasion are boxed in gray. The release of the IpgC chaperone upregulates the MxiE-dependent transcription of late T3SS effectors (3). Vacuolar lysis is associated with secretion of large amounts of the IpaB and IpaC translocator components (4).

at the tip of the T3SS needle [28]. This view was further supported by the visualization of an IpaD-containing globular structure at the tip of the T3SS needle by electron microscopy analysis of negatively stained bacteria (Figure 27.2) [25].

The visualization of this tip structure is dependent on prior cross-linking of the samples, suggesting that it interacts in a labile manner with the tip of the needle and as expected, was not detected in an *ipaD* mutant. While IpaD presence at the tip does not require IpaB, IpaB association with T3SS does require IpaD [27]. Furthermore, structural analysis of IpaD indicates that its N-terminus presents the characteristics of a chaperone, either involved in self-chaperoning or in the chaperoning of IpaB in the complex [29]. The modeling of IpaD oligomers from dimeric structures and the analysis of the protein composition of T3SS in fractionation experiments are consistent with its fit as a hetero-pentameric complex containing four molecules of IpaD for one molecule of IpaB, closing the tip of the T3SS needle. Since IpaC also appears to associate with the T3SS under nonsecretion conditions, this has led to the suggestion that a ready-to-insert translocator is assembled at the tip of the needle [27].

This view is challenged, however, by the observation that the majority of the IpaB and IpaC pool is not present on the bacterial surface and that secretion of this pool is only induced upon cell contact or in the presence of bile salts [30]. It is possible that two pools of translocator components exist, a minor fraction of which are associated with the T3SS tip, while the majority of the pool are stored in the bacterial cytoplasm and induced upon cell contact.

27.2.2 The Type III Secretion System Effectors of Invasion

Several T3SS substrates have been involved in *Shigella*-induced cytoskeletal reorganization leading to invasion of epithelial cells. The translocator component IpaB was shown to interact with CD44, the hyaluronic acid receptor, suggesting that perhaps along with IpaD, IpaB could also participate in the sensing of host cell membranes [31]. Binding of IpaB to CD44 induced its clustering and recruitment at the levels of lipid microdomains required for *Shigella* invasion [32]. Whether IpaB binding to CD44 assists the proper folding of translocator components or mostly serves the purpose to target T3SS at the levels of microdomains is unclear. Because cholesterol vesicles directly bind to IpaB and promote efficient secretion through the T3SS *in vitro*, targeting microdomains may represent an efficient way to trigger T3SS [33, 34]. Alternatively, signaling leading to cytoskeletal rearrangements may be favored at the levels of microdomains.

The precise contribution of outside-in receptor-mediated signaling in *Shigella*-induced cytoskeletal reorganization is difficult to assess, since various T3SS effectors also directly regulate the actin cytoskeleton. IpaC, in addition to its role in translocating T3SS effectors, was also shown to directly induce actin polymerization through its C-terminal effector domain, either directly or through the activation of the Cdc42- and Rac-GTPases [35, 36]. IpgB1 acts as an activated RhoG mimic by recruiting the ELM0/Dock180 complex at sites of bacterial invasion [37]. Since Dock180 is a well-characterized GEF for the Rac-GTPase, IpgB1 indirectly triggers Rac activation and actin polymerization dependent on the Arp2/3 complex [37]. IpgD a phosphoinositol 4,5-phosphatase, has also been implicated in *Shigella* invasion because its hydrolytic activity allows the disconnection of the cortical actin cytoskeleton from the membranes, thereby allowing free actin monomers for *de novo* polymerization at entry sites [38].

More recent works, however, indicate that IpgD, through the generation of phosphoinositol-5-phosphate, could activate the PI3K/Akt pathway, leading to the transcription of genes involved in cell survival [39]. Finally, IpaA was shown to bind to the focal adhesion protein vinculin, by acting as a talin super-mimic [40]. IpaA binding to vinculin, however, leads to the depolymerization of actin filaments, required for the completion of the invasion process [41]. As opposed to the translocator components IpaB and IpaC that are essential for *Shigella* invasion, the injected T3SS effectors are not critically required for bacterial uptake and rather appear to optimize the invasion process, optimization that may be critical during *in vivo* infection.

27.2.3
Vacuolar Escape and Intracellular Replication

Shortly after internalization, vacuolar lysis occurs and the bacteria are released into the cell cytosol where they multiply. Consistent with a high adaptation to a specific niche, *Shigella* has acquired salient features that favor an intracellular lifestyle. These include: (i) major deletions of loci in the *Shigella* genome that when reintroduced, act as antivirulence factors [42, 43]; (ii) the inactivation of genes involved in flagella biosynthesis required for motility in the environment [44]; and (iii) the acquisition of the polar surface protein IcsA, responsible for actin-based motility [6]. Consistent with important physiological adjustment required for intracellular growth, transcriptomic analysis indicates a significant variation in the gene expression profile during *Shigella* intracellular growth compared with *in vitro* grown bacteria, with approximately one-quarter of the genes showing significant differential regulation [45]. Of interest, these include the upregulation of determinants of transport systems such as the *sitABCD* or the *phoRB* regulon.

Iron acquisition is clearly a critical requirement during host infection, and *Shigella* expresses multiple iron-uptake systems, including the aerobactin (*iual*) system allowing the transport of ferric iron, the *fhu* system for the transport of fungal ferrichrome, and the *sit* and *feo* systems for the transport of ferrous iron [45]. While it is known from the analysis of bacterial mutants that these systems are important for intracellular replication and bacterial dissemination, it appears that they play different roles during growth in aerobic versus anaerobic conditions. Under anaerobic conditions presumably corresponding to the environment in the colonic mucosa, the *feo* ferrous transport system was shown to play a predominant role during intracellular replication of *Shigella*, while the *sit*, *fur* and *iual* operons were downregulated [46]. These findings suggest that these various iron-transport systems are differentially regulated in various tissue locations during infection.

Interestingly, in addition to these iron-uptake systems and as will be developed further, T3SS substrates are also required for *Shigella* intracellular multiplication.

27.2.4
Intracellular Actin-Based Motility and Cell-to-Cell Spreading

During this intracellular replication phase, *Shigella* polymerizes actin at one bacterial pole through IcsA. *In vitro*, IcsA was shown to directly bind to and to activate N-WASP, thus inducing Arp2/3 actin nucleation in a manner similar to the activated GTP-bound form of the Cdc42-GTPase [47]. Inside host cells, *Shigella* actin-based motility is also dependent on the activity of the T3SS, through the recruitment of Toca-1, which is not required for IcsA-mediated N-WASP recruitment at the bacterial pole, but to relieve N-WASP from its auto-inhibitory conformation [48].

Interestingly, it has been observed that actin comet tails do not form efficiently in all cell types. Although these structures are readily observed in nonpolarized epithelial cells, in chicken embryo fibroblasts, *Shigella* was described to move in an organelle-like movement (olm) along actin fibers by a mechanism that still remains to be identified [49]. In polarized intestinal epithelial cells, *Shigella* appears to express both types of actin-dependent motility since actin comet tails are visualized essentially at the cortical actin belt junction at the levels of the zonula adherens, while elsewhere in the cell, intracellular bacteria were observed in association with actin filaments [50]. Since polarized epithelial cells are the most relevant cell system to study *Shigella* dissemination, it thus appears that bacterial dissemination implicates a sequential regulation of cytoskeletal processes, where the bacterium senses the local intracellular environment to express different types of motility.

Besides the recruitment of Toca-1, VirA, another T3SS effector whose expression is upregulated during intracellular replication, can also indirectly regulate actin-based motility. VirA was shown to inhibit microtubule polymerization by binding to tubulin hetero-oligomers [51]. Because *virA* mutants are deficient at disseminating in epithelial cells, it is thought that the microtubular network hinders actin-based bacterial motility that is favored by VirA-mediated disruption of microtubules [51]. Although critical, IcsA-mediated actin-based motility is not sufficient for bacterial cell-to-cell spread. In the absence of cell-to-cell contacts, IcsA mediates the formation of bacteria-containing cell protrusions that do not allow cell-to-cell spread. In particular, transfection of E-cadherin, a cell-to-cell adhesion receptor involved in homotypic interactions, in cadherin-deficient fibroblastic cells was shown to restore the bacterial ability to disseminate [50]. Although these findings indicate that cadherin-dependent cell junctional structures are required, they do not allow a particular junctional organization required for *Shigella* dissemination to be incriminated precisely. This is because by initiating and strengthening cell-to-cell contacts cadherins also allow the establishment of other junctional structures.

Connexin-dependent intercellular communication has also been implicated in *Shigella* cell-to-cell spread. Strikingly, connexin-mediated signals did not appear to diffuse through Gap intercellular communicating junctions, but though the bacterial-induced opening of connexin hemichannels [14]. Hemichannel opening is linked to the increase of intracellular calcium induced by *Shigella* invasion, and leads to the release of ATP in the extracellular milieu. Extracellular ATP, in turn, through the stimulation of purinergic receptors, stimulates *Shigella* invasion and dissemination in neighboring cells in a paracrine manner [14]. The significance of such signaling during *in vivo* infection remains to be established since these findings have been obtained using cultured cell lines. However, since ATP is a well-established modulator of inflammation either at the high concentrations observed at inflammatory sites or at agonist concentrations in combination with other stimuli, it is possible that hemichannel-mediated ATP release also significantly contributes to the regulation of inflammation during various stages of the infection process [52].

27.3
Genes Involved in Phagosome Escape and Intracytoplasmic Life

Histological analysis of rabbit ileal loops infected by *Shigella* indicates that bacteria are found in association with the intestinal mucosa, extracellularly or inside various cell types including enterocytes or monocytes. In M cells, bacteria were essentially detected within a membrane-bound vacuole, suggesting that during infection, these cells allow the transcytosis of *Shigella* to underlying tissues and that vacuolar lysis and bacterial intracellular replication does not occur [8]. The reason why *Shigella* does not escape the vacuole in all cell types is unclear. Since, presumably, M cells are initial sites of uptake, it is possible that the environment of the intestinal lumen that the bacteria face prior to invasion is not proficient for the expression of T3SS substrates in sufficient amounts for efficient vacuolar lysis. Also, it is not surprising that vacuolar lysis is not observed as frequently in cells that are inherently phagocytic compared with epithelial cells that are nonphagocytic. In phagocytic cells, a significant number of internalization events are likely to be driven by the cell phagocytic machinery and not by the bacterial T3SS-dependent invasion machinery.

Since *Shigella*'s escape into the host cell cytosol is linked in many aspects to bacterial invasion, the route of entry may also be important for vacuolar lysis. For example, previous works have shown that cholesterol microdomains that are targeted during *Shigella* invasion favor T3S [32]. Depletion of cholesterol using β-methylcyclodextrin also inhibited vacuolar lysis. Therefore, the composition of the vacuolar membrane, which probably differs in bacterial-induced and "classical" phagosomes, could also impact on the efficiency of vacuolar lysis. Finally, the oxidative response or the delivery of lytic enzymes to the vacuole that accompany uptake by professional phagocytes may also inactivate T3S. In neutrophils, the inability of *Shigella* to lyse the phagocytic vacuole has been associated with the proteolytic activity of the neutrophil elastase that targets *Shigella* determinants of invasion [53].

In epithelial cells, rupture of the vacuole occurs within minutes following bacterial internalization, as a continuum of the entry process. *Shigella* determinants involved in phagosomal escape are also dependent on the T3SS and are major determinants of bacterial invasion. Perhaps supporting the hypothesis that there exist two pools of the translocator components, *Shigella* only needs to secrete a minor fraction of the IpaB and IpaC to invade cells. Evidence for this was provided from polar mutations upstream of the *ipa* operon that led to an important decrease in IpaB and IpaC expression levels, while bacterial invasion remained unaffected [54]. The major defect associated with decreased levels of IpaB and IpaC was the unability of *Shigella* to lyse intracellular host cell membranes, and bacteria found in membrane-trapped vacuoles were deficient for cell-to-cell spread. This suggests that secretion of large amounts of translocator components are required for vacuolar membrane lysis, and that IpaB and IpaC are directly involved in this process (Figure 27.2).

In contrast to *Listeria*, *Shigella* does not express known lipases that would account for the lysis of the vacuolar membrane and escape in the cell cytosol. *Shigella*, however, is known to induce red blood cell hemolysis in a T3SS-dependent manner if bacterial contact is established with host cells. Since it is assumed that the cell contact-dependent hemolysis reflects the ability of *Shigella* to lyse the phagocytic vacuole, this implicates that vacuolar lysis induced by *Shigella* occurs rapidly. This is supported by experimental evidence indicating that *Shigella* escapes from the phagocytic vacuole within minutes after invasion, in a continuum with the entry process (Enninga, Clair *et al.*, unpublished). How, precisely, this occurs is unknown, but because it indicates that constitutively expressed T3SS effectors are involved in lysis, again we can point at translocator components as prime candidates.

Two sets of approaches have been undertaken to identify determinants of vacuolar lysis, while circumventing the problem caused by the fact that both vacuolar escape and bacterial invasion require a functional T3SS. In some studies, vacuolar escape was investigated in a macrophage cell line in which bacteria are internalized because of the host cell inherent phagocytic properties. Since an *ipaB* mutant was deficient for vacuolar escape it was suggested that IpaB is directly involved in membrane lysis [55]. However, since an *ipaB* mutant shows deregulated T3SS activity, an indirect role of IpaB cannot be excluded from these studies. Using this type of approach, the T3SS late effector IpaH 7.8 was also implicated in release from the vacuolar compartment [56]. In other studies, an exchange of functional orthologs showed that the *Salmonella* SipC translocator component could complement a *Shigella ipaC* mutant for invasion. Reciprocally, IpaC was shown to complement a *sipC Salmonella* mutant strain [57]. Similar complementation was observed when IpaB and SipB were exchanged. This was in spite of the modest homologies in the primary sequences of IpaB and SipB or IpaC and SipC, suggesting conserved structural features between these *Shigella* and *Salmonella* components [57].

Interestingly, in these cross-complementation experiments an *ipaB Shigella* mutant strain complemented with SipB, as opposed to a *Shigella ipaC* mutant complemented with SipC, did replicate intracellularly and formed actin comet tails. Although it is possible that these differences may be due to variations in the secretion efficiencies of these T3SS substrates in heterologous systems, these results suggests that IpaC plays a direct role in vacuolar lysis.

Other heterologous complementation studies, however, have suggested a role for chromosomal determinants independent of the *Shigella* invasion determinants in vacuolar escape. When a *Shigella* mutant deleted for its invasion locus was complemented by the related Samonella SPI-1 entry region, invasion and vacuolar escape still occurred [58]. These apparently conflicting reports may be linked to variations in the strains studied or in the read-out used to assess vacuolar escape. It is also possible that there are several pathways for vacuolar escape and that late events of vacuolar escape, instead of corresponding to a direct ability of the pathogen to lyse host cell membranes, rather relates to the inability of the pathogen to maintain the integrity of the phagosome.

This notion has been suggested by studies on a *Salmonella* mutant deficient for SifA, a T3SS that hijacks intracellular trafficking by allowing the recruitment of the microtubule motor kinesin at *Salmonella*-containing vacuole [59]. As opposed to wild-type *Salmonella*, a *sifA* mutant failed to form tubular vesicles and escaped in the cell cytosol, suggesting that vesicle fusion with the *Salmonella*-containing vacuole is important for the maintenance of its integrity during bacterial replication [60]. A critical role of the *Shigella* T3SS could be to allow rapid escape from the vacuole, while later events of vacuolar escape may just correspond to a cell default pathway following phagosome trafficking.

Along these lines, *Burkholderia pseudomallei*, a Gram-negative opportunistic pathogen responsible for melioidosis, a septicemic disease with clinical symptoms related to glanders, expresses three T3SS. Two T3SS are related to plant T3SS, while the T3SS encoded by the *bsa* locus is related to the *Shigella* T3SS. Inactivation of the *bsa* locus altered the kinetics of *B. pseudomallei* in the host cell cytosol [61]. Rapid vacuolar escape may be particularly critical in phagocytic cells to limit killing linked to the oxidative response.

27.4 Host Defense Against Cytoplasmic *Shigella*

27.4.1 Escaping Autophagy

Described in virtually all eukaryotic cells, autophagy is a process that removes intracellular membranes or proteins in a process involving the enclosure by several membrane layers enriched in specific markers such as LC3 and targeting to lysosomal compartments [62]. The detailed mechanism underlying the recognition of intracellular components by the autophagic machinery, by and large ill-defined, implicates the Atg family products. Autophagocytic processes are induced in response to amino acid starvation and are also involved in the physiological removal or turnover of cellular compounds. Autophagy has also been implicated as a cellular defense against intracellular pathogens that get access to the cytosol.

A mutant for IcsB, a T3SS substrate that is not required for *Shigella* invasion, was reported to exhibit an intracellular replication defect and was therefore deficient at spreading from cell to cell [63]. This defect was not linked to a bacterial deficiency in escaping the vacuole or moving intracellularly. Rather, following vacuolar escape, an *icsB* mutant appears to undergo the first round of multiplication in the cell cytosol, but stops replicating after a few hours. Microscopy analysis indicated that the stall in intracellular replication of the *icsB* mutant was linked to its enclosure by LC3-positive membranes, indicating that the bacteria were held in check by the autophagocytic machinery. Consistent with this, the *icsB* mutant did not recruit LC3 and did not show intracellular replication defects in $atg5^{-/-}$ mouse fibroblasts [63].

In the case of *Shigella*, the IcsA/VirG protein that is responsible for actin-based motility also associates with IcsB. In the absence of IcsB, however, IcsA is recognized by the autophagocytic machinery and participates in "seed nucleation," leading to the formation of isolation membranes. IcsB has been shown to bind to Atg5 in a mutually exclusive manner with IcsA, thus indicating that it could prevent *Shigella* recognition by the autophagic machinery. Interestingly, the *icsB* mutant does not show a intracellular replication defect in all cell lines, suggesting that its role in protection against autophagy is restricted to some cellular types *in vivo*.

27.4.1.1 The Role of Cellular Innate Immunity

Monocytes and neutrophils recruited in response to IL-8 production by infected enterocytes, or through the release of IL-1β and IL-18 by macrophage undergoing pyroptosis, are key to the control of bacterial infection because of their bactericidal activities. Thus, cytokines that amplify the recruitment of these cells at the site of infection will play a major role in host resistance. In addition to monocytes and neutrophils recruited at inflammatory sites of the colonic mucosa during *Shigella* infection, T cells are also recruited and activated at the site of infection. Although the role of these cells in controlling the infection remains unclear in humans, there is evidence pointing at such a role in mice. *rag2γC* mice that are deficient for B, T and NK cells are highly sensitive to *Shigella* in pulmonary infections. Resistance to infection can be conferred by the phenotypic engrafting of αβT cells [64]. When engrafting was performed using αβT cells from IFNγ$^{-/-}$ donors, however, *rag2γC* mice were still sensitive to *Shigella* infection, indicating a role in IFN production for the engrafted cells [64]. Similar experiments using *rag2* mice deficient for B and T cells but harboring NK cells further showed the prominent role of NK cells in the control of infection. Since along with T cells, NK cells are known to be major source of IFNγ production, the picture emerging from these experiments is that following recruitment and activation at the site of infection, T and NK cells amplify the immune response through the local production of IFNγ, thus boosting the macrophage bactericidal properties and favoring the resitance of epithelial cells to *Shigella* invasion.

In support of this model, cytokinic profiles of *Shigella*-infected patients indicate a tight correlation between the levels of IFNγ and remission [65]. Strikingly, however, acute shigellosis is accompanied with a drastic decrease in the levels of IFNγ, suggesting that *Shigella* has developed means to divert a T_H1 bactericidal response. This escape from a T_H1 response allows progression of the disease in its most severe forms.

Shigella infections elicit a humoral response with the mucosal production of secretory IgA and IgG directed against the bacterial LPS and other major bacterial secreted proteins [66]. In particular, these responses in HLA-B27 patients can lead to reactive arthritis as a sequalae of *Shigella* infections [67]. Although this is indicative that antigen presentation is functional during *Shigella* infection, these humoral responses do not build up into a long-term memory protection. Repeated infections can lead to a short-term protective immunity that does not allow infected individuals to become resistant to subsequent *Shigella* infections. The modulation of the cytokine production during the initial phases of the infectious process may be responsible for the evolution towards an immunosuppressive response, explaining the lack of long-term immunity against *Shigella* infections [66]. Using mice pulmonary infections, this modulation of the immune response was shown to correlate with a decrease in the levels of IFNγ/IL-1 and an increase in the levels of IL-10/transforming growth factor β (TGFβ) with anti-inflammatory properties [66].

The precise mechanism underlying the control of the immune response by *Shigella* is largely unknown and represents a challenging, but fascinating field of investigation addressing the link between innate and adaptive immunity. Such modulation could have its basis at the cellular level through the action of late T3SS effectors such as OspF, OspG or IpaH9.8 that may interfere with proinflammatory programs. Alternatively, this modulation may be the reflection of the killing of specific immune cells by *Shigella*, unbalancing the mounting of the response. Along these lines, the massive inflammation accompanied by the severe depletion of macrophages, DCs, B and T cells as a hallmark of *Shigella* infection may favor signaling towards immunotolerance.

27.4.1.2 **Resistance to Shigellosis**

As opposed to other infections by enteroinvasive bacteria, *Shigella* infection is limited to the colonic mucosa and, with the exception of septicemia, there are only rare cases of extraintestinal infections. Shigellosis is also restricted to humans and primates. The reasons for this restricted host and tissue tropism are unclear since although the *Shigella* genome sequences indicate putative adhesin-related genes, no bacterial adhesin has been identified to date that could account for such a specificity [68]. *In vitro*, *Shigella* can readily invade and disseminate in cells that are not of primate origin, suggesting that other factors limit bacterial colonization of the epithelium *in vivo*. It is possible that this specificity is determined by differences in the mucosal expression of antimicrobial peptides that prevent bacterial multiplication at the levels of the intestinal mucosa. For example, oral administration of butyrate that leads to upregulation of the antimicrobial peptide CAP 38/LL-37 could be successfully used to eliminate bacteria and has been shown to reduce inflammation in rabbits [69]. Also, lactoferrin, a glycoprotein present in mucosal secretions, was shown to limit *Shigella* invasion of epithelial cells *in vitro*, by degrading the translocator components IpaB and IpaC [70]. Whether such peptides play a role in the control of infection during shigellosis is unclear, since the expression of antimicrobial peptides appears to be downregulated during infection in humans [71]. This particular tropism may also reflect a remarkable adaptation to a specific niche and underscores the diversion of inflammatory processes that are specific to humans.

From transcriptomic analyses of *Shigella*-infected polarized intestinal epithelial cells *in vitro*, IL-8 stands out as the major cytokine induced by invasive bacteria [72]. This IL-8 response is largely superior to the type of response induced by the extracellular addition of LPS or by adherent but noninvasive bacteria, suggesting that the recognition of intracellular motives such as peptidoglycan derivatives by NOD receptors contribute to the sustained activation of the proinflammatory transcriptional activator NFκB. The importance of IL-8 is highlighted by its massive production at inflammatory sites in histological sections of infected rabbit ileal loops, and by the observation that the injection of anti-IL-8-neutralizing antibodies prevents bacterial-induced recruitment of neutrophils and inflammation [73]. Although mice express one copy of the IL-8 receptor (IL-8Rb), the sequence of the mouse genome indicates that they do not express IL-8, presumably because the function of this cytokine is performed by a different pathway. While mice are resistant to *Shigella* intestinal infection, the injection of recombinant IL-8 in combination with invasive *Shigella* in a mouse model of rectal administration resulted in a massive inflammatory response and tissue destruction, suggesting that the absence of IL-8 could explain the lack of mice sensitivity to *Shigella* infections [74].

27.5 Summary and Outlook

Since the advent of the cellular microbiology of *Shigella*, the characterization of pathways leading to bacterial invasion has proceeded to unveil a detailed, yet still incomplete picture of the complex mechanisms involved in the regulation of the actin cytoskeleton. Keys to these developments were the pioneering genetic analysis of bacterial determinants of virulence, followed by the discovery and the on-going dissection of the T3SS function. Advances in transcriptomic analysis have provided us with a general view of the program of genes involved in virulence. The characterization of late T3SS effectors clearly point at a role in the downregulation of proinflammatory signals. We are therefore facing a major shift from the initial concepts needed to trigger inflammation and to efficiently colonize and destroy the colonic mucosa. The apparent paradoxical functions of *Shigella* effectors on the regulation of inflammation highlight the limitations of the current models used to study experimental shigellosis and stress the need to develop models that allow the analysis of the early steps of invasion *in vivo*.

As discussed in this chapter, *Shigella* may very well adopt various cellular strategies according to the cell types and site of infection. *Shigella* does not appear to replicate and is killed in monocytes and macrophages. All these observations support the notion that *Shigella* cannot survive in phagocytic cells, consistent with its inability to spread to other tissues or to persist in the host under a latent state. *In vitro* evidence suggests that macrophage killing does not require vacuolar lysis and bacterial replication. Instead of using phagocytic cells as a replicating niche, it rather appears that bacteria attempt to exploit the proinflammatory potential of these cells to destabilize the integrity of the epithelium. This is in sharp contrast with *Shigella*'s ability to invade and replicate within epithelial cells, where it would be of interest for dampening inflammatory signals, at least during the early phases of bacterial replication. A corollary to this hypothesis is that the action of these effectors would also evolve, depending on the inflammatory context and during the course of infection.

Clearly, an in-depth understanding of this infectious process will implicate the challenging task of deciphering at the molecular level the pathways diverted by virulence effectors, while reintegrating these pathways during the course of infection in the tissue locations and cell types that are targeted by *Shigella*.

References

1 Kotloff, K.L. *et al.* (1999) Global burden of *Shigella* infections: implications for vaccine development and implementation of control strategies. *Bull World Health Organ*, **77** (8), 651–666.

2 Lan, R. and Reeves, P.R. (2002) *Escherichia coli* in disguise: molecular origins of Shigella. *Microbes Infect*, **4** (11), 1125–1132.

3 Wei, J. *et al.* (2003) Complete genome sequence and comparative genomics of *Shigella flexneri* serotype 2a strain 2457T. *Infect Immun*, **71** (5), 2775–2786.

4 Buchrieser, C. *et al.* (2000) The virulence plasmid pWR100 and the repertoire of proteins secreted by the type III secretion apparatus of *Shigella flexneri*. *Mol Microbiol*, **38** (4), 760–771.

5 Venkatesan, M.M. *et al.* (2001) Complete DNA sequence and analysis of the large virulence plasmid of *Shigella flexneri*. *Infect Immun*, **69** (5), 3271–3285.

6 Bernardini, M.L. *et al.* (1989) Identification of icsA, a plasmid locus of *Shigella flexneri* that governs bacterial intra- and intercellular spread through interaction with F-actin. *Proc Natl Acad Sci USA*, **86** (10), 3867–3871.

7 Sakaguchi, T. *et al.* (2002) *Shigella flexneri* regulates tight junction-associated proteins in human intestinal epithelial cells. *Cell Microbiol*, **4** (6), 367–381.

8 Sansonetti, P.J. and Phalipon, A. (1999) M cells as ports of entry for enteroinvasive pathogens: mechanisms of interaction, consequences for the disease process. *Semin Immunol*, **11** (3), 193–203.

9 Mounier, J. *et al.* (1992) *Shigella flexneri* enters human colonic Caco-2 epithelial cells through the basolateral pole. *Infect Immun*, **60** (1), 237–248.

10 Zychlinsky, A., Prevost, M.C. and Sansonetti, P.J. (1992) *Shigella flexneri* induces apoptosis in infected macrophages. *Nature*, **358** (6382), 167–169.

11 Suzuki, T. *et al.* (2007) Differential regulation of caspase-1 activation, pyroptosis, and autophagy via Ipaf and ASC in *Shigella*-infected macrophages. *PLoS Pathog*, **3** (8), e111.

12 Elewaut, D. *et al.* (1999) NF-kappa B is a central regulator of the intestinal epithelial cell innate immune response induced by infection with enteroinvasive bacteria. *J Immunol*, **163** (3), 1457–1466.

13 Girardin, S.E. and Philpott, D.J. (2004) Mini-review: the role of peptidoglycan recognition in innate immunity. *Eur J Immunol*, **34** (7), 1777–1782.

14 Tran Van Nhieu, G. *et al.* (2003) Connexin-dependent inter-cellular communication increases invasion and dissemination of Shigella in epithelial cells. *Nat Cell Biol*, **5** (8), 720–726.

15 Zurawski, D.V. *et al.* (2008) The NleE/OspZ family of effector proteins is required for polymorphonuclear transepithelial migration, a characteristic shared by enteropathogenic *Escherichia coli* and *Shigella flexneri* infections. *Infect Immun*, **76** (1), 369–379.

16 Sansonetti, P.J. *et al.* (1995) Role of interleukin-1 in the pathogenesis of experimental shigellosis. *J Clin Invest*, **96** (2), 884–892.

17 Arbibe, L. *et al.* (2007) An injected bacterial effector targets chromatin access for transcription factor NF-kappaB to alter transcription of host genes involved in immune responses. *Nat Immunol*, **8** (1), 47–56.

18 Li, H. *et al.* (2007) The phosphothreonine lyase activity of a bacterial type III effector family. *Science*, **315** (5814), 1000–1003.

19 Kim, D.W. *et al.* (2005) The *Shigella flexneri* effector OspG interferes with innate immune responses by targeting ubiquitin-conjugating enzymes. *Proc Natl Acad Sci USA*, **102** (39), 14046–14051.

20 Ingersoll, M.A. *et al.* (2003) The ShiA protein encoded by the *Shigella flexneri* SHI-2 pathogenicity island attenuates inflammation. *Cell Microbiol*, **5** (11), 797–807.

21 Ashida, H. *et al.* (2007) *Shigella* chromosomal IpaH proteins are secreted via the type III secretion system and act as effectors. *Mol Microbiol*, **63** (3), 680–693.

22 Cossart, P. and Sansonetti, P.J. (2004) Bacterial invasion: the paradigms of enteroinvasive pathogens. *Science*, **304** (5668), 242–248.

23 Cornelis, G.R. (2006) The type III secretion injectisome. *Nat Rev Microbiol*, **4** (11), 811–825.

24 West, N.P. *et al.* (2005) Optimization of virulence functions through glucosylation of *Shigella* LPS. *Science*, **307** (5713), 1313–1317.

25 Sani, M. *et al.* (2007) IpaD is localized at the tip of the *Shigella flexneri* type III secretion apparatus. *Biochim Biophys Acta*, **1770** (2), 307–311.

26 Blocker, A.J. *et al.* (2008) What's the point of the type III secretion system needle?

Proc Natl Acad Sci USA, **105** (18), 6507–6513.

27 Veenendaal, A.K. *et al.* (2007) The type III secretion system needle tip complex mediates host cell sensing and translocon insertion. *Mol Microbiol*, **63** (6), 1719–1730.

28 Menard, R., Sansonetti, P. and Parsot, C. (1994) The secretion of the *Shigella flexneri* Ipa invasins is activated by epithelial cells and controlled by IpaB and IpaD. *EMBO J*, **13** (22), 5293–5302.

29 Johnson, S. *et al.* (2007) Self-chaperoning of the type III secretion system needle tip proteins IpaD and BipD. *J Biol Chem*, **282** (6), 4035–4044.

30 Olive, A.J. *et al.* (2007) Bile salts stimulate recruitment of IpaB to the *Shigella flexneri* surface, where it colocalizes with IpaD at the tip of the type III secretion needle. *Infect Immun*, **75** (5), 2626–2629.

31 Skoudy, A. *et al.* (2000) CD44 binds to the *Shigella* IpaB protein and participates in bacterial invasion of epithelial cells. *Cell Microbiol*, **2** (1), 19–33.

32 Lafont, F. *et al.* (2002) Initial steps of *Shigella* infection depend on the cholesterol/sphingolipid raft-mediated CD44-IpaB interaction. *EMBO J*, **21** (17), 4449–4457.

33 van der Goot, F.G. *et al.* (2004) Rafts can trigger contact-mediated secretion of bacterial effectors via a lipid-based mechanism. *J Biol Chem*, **279** (46), 47792–47798.

34 Hayward, R.D. *et al.* (2005) Cholesterol binding by the bacterial type III translocon is essential for virulence effector delivery into mammalian cells. *Mol Microbiol*, **56** (3), 590–603.

35 Tran Van Nhieu, G. *et al.* (1999) IpaC induces actin polymerization and filopodia formation during *Shigella* entry into epithelial cells. *EMBO J*, **18** (12), 3249–3262.

36 Kueltzo, L.A. *et al.* (2003) Structure-function analysis of invasion plasmid antigen C (IpaC) from *Shigella flexneri*. *J Biol Chem*, **278** (5), 2792–2798.

37 Handa, Y. *et al.* (2007) *Shigella* IpgB1 promotes bacterial entry through the ELMO-Dock180 machinery. *Nat Cell Biol*, **9** (1), 121–128.

38 Niebuhr, K. *et al.* (2002) Conversion of PtdIns(4,5)P(2) into PtdIns(5)P by the *S. flexneri* effector IpgD reorganizes host cell morphology. *EMBO J*, **21** (19), 5069–5078.

39 Pendaries, C. *et al.* (2006) PtdIns5P activates the host cell PI3-kinase/Akt pathway during *Shigella flexneri* infection. *EMBO J*, **25** (5), 1024–1034.

40 Izard, T., Tran Van Nhieu, G. and Bois, P.R. (2006) *Shigella* applies molecular mimicry to subvert vinculin and invade host cells. *J Cell Biol*, **175** (3), 465–475.

41 Bourdet-Sicard, R. *et al.* (1999) Binding of the *Shigella* protein IpaA to vinculin induces F-actin depolymerization. *EMBO J*, **18** (21), 5853–5862.

42 Maurelli, A.T. *et al.* (1998) "Black holes" and bacterial pathogenicity: a large genomic deletion that enhances the virulence of *Shigella* spp. and enteroinvasive *Escherichia coli*. *Proc Natl Acad Sci USA*, **95** (7), 3943–3948.

43 Prunier, A.L. *et al.* (2007) nadA and nadB of *Shigella flexneri* 5a are antivirulence loci responsible for the synthesis of quinolinate, a small molecule inhibitor of *Shigella* pathogenicity. *Microbiology*, **153** (Pt 7), 2363–2372.

44 Tominaga, A. *et al.* (1994) Molecular characterization of intact, but cryptic, flagellin genes in the genus *Shigella*. *Mol Microbiol*, **12** (2), 277–285.

45 Lucchini, S. *et al.* (2005) Transcriptional adaptation of *Shigella flexneri* during infection of macrophages and epithelial cells: insights into the strategies of a cytosolic bacterial pathogen. *Infect Immun*, **73** (1), 88–102.

46 Boulette, M.L. and Payne, S.M. (2007) Anaerobic regulation of *Shigella flexneri* virulence: ArcA regulates Fur and iron acquisition genes. *J Bacteriol*, **189** (19), 6957–6967.

47 Egile, C. *et al.* (1999) Activation of the CDC42 effector N-WASP by the *Shigella flexneri* IcsA protein promotes actin nucleation by Arp2/3 complex and bacterial actin-based motility. *J Cell Biol,* **146** (6), 1319–1332.

48 Leung, Y., Ally, S. and Goldberg, M.B. (2008) Bacterial actin assembly requires toca-1 to relieve N-wasp autoinhibition. *Cell Host Microbe,* **3** (1), 39–47.

49 Vasselon, T. *et al.* (1992) Movement along actin filaments of the perijunctional area and de novo polymerization of cellular actin are required for *Shigella flexneri* colonization of epithelial Caco-2 cell monolayers. *Infect Immun,* **60** (3), 1031–1040.

50 Sansonetti, P.J. *et al.* (1994) Cadherin expression is required for the spread of *Shigella flexneri* between epithelial cells. *Cell,* **76** (5), 829–839.

51 Yoshida, S. *et al.* (2006) Microtubule-severing activity of *Shigella* is pivotal for intercellular spreading. *Science,* **314** (5801), 985–989.

52 Bours, M.J. *et al.* (2006) Adenosine 5′-triphosphate and adenosine as endogenous signaling molecules in immunity and inflammation. *Pharmacol Ther,* **112** (2), 358–404.

53 Weinrauch, Y. *et al.* (2002) Neutrophil elastase targets virulence factors of enterobacteria. *Nature,* **417** (6884), 91–94.

54 Rathman, M. *et al.* (2000) The development of a FACS-based strategy for the isolation of *Shigella flexneri* mutants that are deficient in intercellular spread. *Mol Microbiol,* **35** (5), 974–990.

55 High, N. *et al.* (1992) IpaB of *Shigella flexneri* causes entry into epithelial cells and escape from the phagocytic vacuole. *EMBO J,* **11** (5), 1991–1999.

56 Fernandez-Prada, C.M. *et al.* (2000) *Shigella flexneri* IpaH(7.8) facilitates escape of virulent bacteria from the endocytic vacuoles of mouse and human macrophages. *Infect Immun,* **68** (6), 3608–3619.

57 Hermant, D. *et al.* (1995) Functional conservation of the *Salmonella* and *Shigella* effectors of entry into epithelial cells. *Mol Microbiol,* **17** (4), 781–789.

58 Paetzold, S. *et al.* (2007) *Shigella flexneri* phagosomal escape is independent of invasion. *Infect Immun,* **75** (10), 4826–4830.

59 Boucrot, E. *et al.* (2005) The intracellular fate of *Salmonella* depends on the recruitment of kinesin. *Science,* **308** (5725), 1174–1178.

60 Beuzon, C.R. *et al.* (2000) *Salmonella* maintains the integrity of its intracellular vacuole through the action of SifA. *EMBO J,* **19** (13), 3235–3249.

61 Burtnick, M.N. *et al.* (2008) *Burkholderia pseudomallei* type III secretion system mutants exhibit delayed vacuolar escape phenotypes in RAW 264.7 murine macrophages. *Infect Immun,* **76** (7), 2991–3000.

62 Kirkegaard, K., Taylor, M.P. and Jackson, W.T. (2004) Cellular autophagy: surrender, avoidance and subversion by microorganisms. *Nat Rev Microbiol,* **2** (4), 301–314.

63 Ogawa, M. *et al.* (2005) Escape of intracellular *Shigella* from autophagy. *Science,* **307** (5710), 727–731.

64 Le-Barillec, K. *et al.* (2005) Roles for T and NK cells in the innate immune response to *Shigella flexneri. J Immunol,* **175** (3), 1735–1740.

65 Raqib, R. *et al.* (1997) A systemic downregulation of gamma interferon production is associated with acute shigellosis. *Infect Immun,* **65** (12), 5338–5341.

66 Phalipon, A. and Sansonetti, P.J. (2007) *Shigella*'s ways of manipulating the host intestinal innate and adaptive immune system: a tool box for survival? *Immunol Cell Biol,* **85** (2), 119–129.

67 Girschick, H.J. *et al.* (2008) Bacterial triggers and autoimmune rheumatic diseases. *Clin Exp Rheumatol,* **26** (1 Suppl 48), S12–S17.

68 Nie, H. *et al.* (2006) Complete genome sequence of *Shigella flexneri* 5b and comparison with *Shigella flexneri* 2a. *BMC Genomics*, **7**, 173.

69 Raqib, R. *et al.* (2006) Improved outcome in shigellosis associated with butyrate induction of an endogenous peptide antibiotic. *Proc Natl Acad Sci USA*, **103** (24), 9178–9183.

70 Gomez, H.F. *et al.* (2001) Protective role of human lactoferrin against invasion of *Shigella flexneri* M90T. *Adv Exp Med Biol*, **501**, 457–467.

71 Islam, D. *et al.* (2001) Downregulation of bactericidal peptides in enteric infections: a novel immune escape mechanism with bacterial DNA as a potential regulator. *Nat Med*, **7** (2), 180–185.

72 Pedron, T., Thibault, C. and Sansonetti, P.J. (2003) The invasive phenotype of *Shigella flexneri* directs a distinct gene expression pattern in the human intestinal epithelial cell line Caco-2. *J Biol Chem*, **278** (36), 33878–33886.

73 Sansonetti, P.J. *et al.* (1999) Interleukin-8 controls bacterial transepithelial translocation at the cost of epithelial destruction in experimental shigellosis. *Infect Immun*, **67** (3), 1471–1480.

74 Singer, M. and Sansonetti, P.J. (2004) IL-8 is a key chemokine regulating neutrophil recruitment in a new mouse model of *Shigella*-induced colitis. *J Immunol*, **173** (6), 4197–4206.

IIC
Intracellular Symbionts

Intracellular Niches of Microbes. Edited by Ulrich E. Schaible and Albert Haas

ISBN: 978-3-527-32207-7

28
Bacterial Symbionts of Plants

Kumiko Kambara, William J. Broughton, and William J. Deakin

28.1
Introduction to Rhizobia and Symbioses

28.1.1
Symbiosis

A symbiotic interaction between two organisms that are widely separated phylogenetically is an intimate association and can be either prolonged or temporary. In symbiosis, the growth, survival and/or reproduction of both the organisms benefit [1]. For plants, associations between fungi and bacteria are thought to have been key innovations in the colonization of land and subsequent habitats. Plant-associated microbes act as metabolic partners accessing limiting nutrients and also as protectors, producing toxins that ward off herbivores or pathogens. The symbiotic interactions of legumes and rhizobia, as well as the widespread mutualistic symbiosis between arbuscular mycorrhizal fungi and vascular flowering plants, have been extensively studied as these symbioses contribute a significant agricultural benefit.

28.1.2
Rhizobia

Soil bacteria belonging to the genera *Azorhizobium, Bradyrhizobium, Mesorhizobium, Rhizobium* and *Sinorhizobium* of the order *Rhizobiales* (collectively called rhizobia) are able to have a symbiotic interaction with the plant family *Leguminosae*. Rhizobia invade legume roots (or occasionally shoots), forming a highly specialized organ called the nodule, and have the ability to fix atmospheric nitrogen to ammonia [2]. In nitrogen-scarce environments, this is important for biological productivity and soil fertility, and thus for agriculture. Within nodules, rhizobia differentiate into bacteroids capable of fixing atmospheric nitrogen to ammonia, which the host legume plant then uses for growth. In return, the rhizobia obtain photosynthetic products and other nutrients from their hosts. Infection of legumes by rhizobia and

Intracellular Niches of Microbes. Edited by Ulrich E. Schaible and Albert Haas

ISBN: 978-3-527-32207-7

thus nodule development is highly restricted in a process termed host specificity. Host plants only interact with a particular species or strains of rhizobia [3]. Host specificity is variable and also depends on the rhizobia strain. Some strains have a very narrow host range, such as *Sinorhizobium meliloti* or *Rhizobium leguminosarum* biovar *trifolii*, which nodulate only a few legume genera, whereas others have a broad host range as exemplified by *Rhizobium* sp. NGR234 (hereafter NGR234), which can nodulate more than 112 genera of legumes as well as the non-legume *Parasponia andersonii* [4, 5].

28.1.3 Arbuscular Mycorrhizae

Arbuscular mycorrhizae (AM) form a symbiotic association with the plant roots supporting vascular plant development under nutrient-limiting and various stress conditions [6]. The AM–root interaction is an ancient symbiosis: fossil evidence shows that it has existed in the roots of the earliest land plants for at least 460 million years [7] and it may have played a key role in facilitating the movement of plants onto land [7–9]. Within angiosperms, more than 80% of species are able to form AM symbioses.

28.2 Invasion of Plant Cells by Rhizobia and Arbuscular Mycorrhizae

Nodulation begins with an exchange of specific molecular signals between the host plant and rhizobia [10–12]. A cocktail of plant-produced flavonoid (2-phenyl-1,4-benzopyrone derivatives) compounds [13] are released from legume roots, attracting rhizobia to the root and inducing the expression of rhizobial nodulation-related (*nod*) genes. Some of the *nod* gene products synthesize and secrete specific chitin-like lipochitooligosaccharides, known as Nod factors. Recognition of Nod factors by the plant causes a series of host responses, including the activation of host gene expression, calcium spiking, root hair deformation and curling, as well as the replication of root cortical cells [14–17]. These physiological and morphological changes ultimately lead to the formation of the nodule, in which rhizobia find an ideal environment to fix atmospheric nitrogen. Thus, Nod factors play a key role during initiation of nodule development and bacterial invasion [18, 19].

Invasion of plant cells by rhizobia is extremely complex (Figure 28.1). In response to Nod factor secretion, root hairs are stimulated and cell wall growth reoriented [20], resulting in curled root hairs. Nod factors also promote the formation of infection threads, which are plant-derived tubular structures. Thus rhizobia enter a pocket within a curled root hair, from which they are taken up into a developing infection thread and begin to travel towards the root cortex where the nodule primordium is developing [21, 22]. At the tip of the infection thread, rhizobia are released into the cytosol of a subset of nodule primordium cells and enveloped in a plant-derived

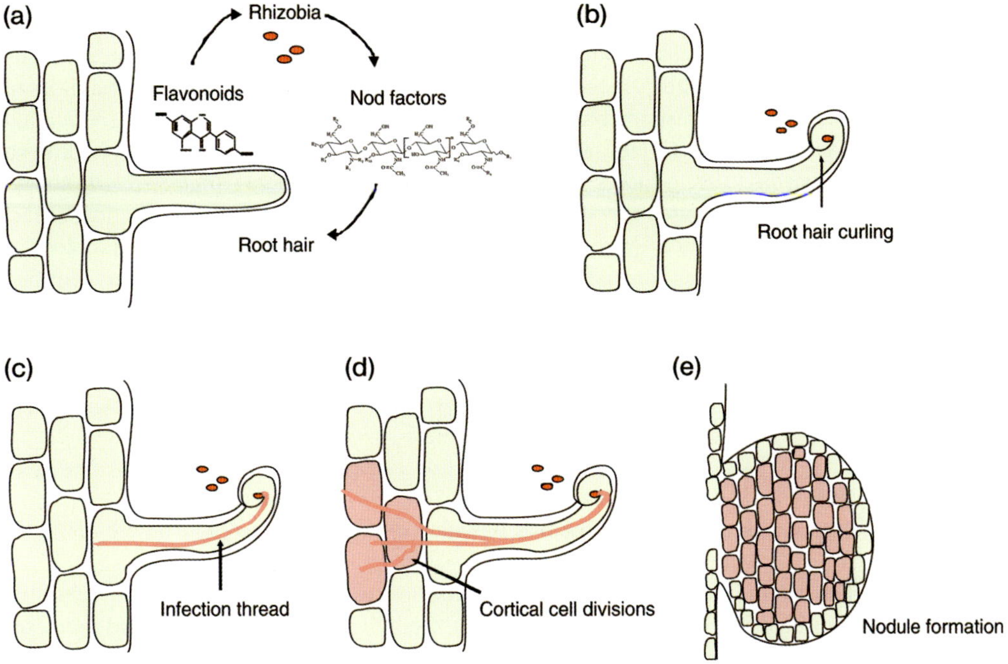

Figure 28.1 Invasion of legume root hairs by *Rhizobium*.(a) Rhizobia naturally colonize the rhizosphere metabolizing organic compounds secreted by root cells. Flavonoids released by host legume roots further attract rhizobia leading to their attachment to root hairs. (b) Elevated flavonoid concentrations at close proximity trigger the synthesis of Nod factors by rhizobia which induce root hair curling and bacterial penetration at the center of the infection pocket. (c) Infection threads develop within the root hair towards the cortical cells of the root. (d) A developing infection thread ramifies near the nodule primordia formed by dividing cortical cells and rhizobia are released from the infection thread to form symbiosomes within nodule cells (shown in pink). (e) Numerous release events and subsequent cortical cell divisions lead to the development of the new root organ, the nodule.

membrane, to form a symbiosome. Subsequent cell divisions and the differentiation of rhizobia into bacteroids, leads to the formation of fully functional nitrogen-fixing root nodules [23].

To initiate an AM symbiosis, following spore germination the hyphal germ tube grows through the soil in search of a host root. Once contact between the symbionts has been established, the fungus forms an appressorium on the root surface through which it enters the root [24]. Then, inside the root, fungal hyphae continue growing until they penetrate the cell wall of an inner cortical cell, where highly ramified fungal hyphae form tree-like structures, termed arbuscules [25]. At the same time, AM also develop extensive hyphae outside of the plant root, and this extraradical hyphal development allows the fungus to supply important nutrients, including phosphate, from the greater area of the soil to the plant. In return, AM receive carbohydrates from the plant [26, 27]. The AM symbiosis also confers resistance to the plant against biotic and abiotic stresses.

The molecular signaling mechanisms between AM and host plants is not as well understood as those for the legume/rhizobia symbiosis. Although a recent study discovered that the strigolactone 5-deoxystrigol is a signal factor in root exudates of *Lotus japonicus* responsible for the induction of hyphal branching in germinating mycorrhizal spores [28]. Prior to this work, strigolactones had only been known as germination inducers of seeds of the parasitic plants *Striga* and *Orobanche*. Whether plants produce further molecules to trigger AM spore germination, attract hyphae or to induce AM root colonization is not known. The identity of any molecular signals emanating from AM that signal to the host plant have also not been identified to date, although the presence of a so-called Myc factors has been postulated [29, 30]. Myc factors have a function conceptually analogous to those of rhizobial Nod factors; that is, they are essential symbiotic signals that activate the host plant's symbiotic program.

28.3 Plant Genes Involved in Symbiosis Initiation

Genes required for the development of the host plant's symbiotic program have been identified by screening mutants of the model legumes *L. japonicus* and *Medicago truncatula* unable to establish efficient symbioses with rhizobia and/or AM. Mutants defective in one type of symbiosis are subsequently checked for their ability to form the other symbiosis and whether the biochemical signal of calcium spiking (see below) can be observed. This has led to the establishment of a signaling cascade of legume genes, with several genes required for both types of symbiosis, but for the initial detection events and subsequent root redevelopment stages specific plant genes are required (Figure 28.2). Despite the extensive morphological differences between the rhizobial and AM symbioses, it is remarkable that they share a number of common signaling components in legumes (Figure 28.2) [31–33]. These so-called common symbiosis (*SYM*) genes [34] are also universally conserved in other legumes and in non-legumes [35]. These observations support the hypothesis that the rhizobial symbiosis in legumes may have evolved from the more ancient AM symbiosis [36].

28.3.1 Plant Perception of Nod and Myc Factors

Potential Nod factor receptor mutants should be blocked at all stages of Nod factor signaling (i.e., Nod factor-induced root hair deformation, calcium influx and spiking and nodule formation) but should potentially still be capable of mycorrhization [37, 38]. Using these phenotypic criteria in mutant screens led to the identification of putative Nod factor receptor mutants and thus genes in *L. japonicus* (*Lj-NFR1* and *Lj-NFR5*) [39, 40], *Pisum sativum* (*Ps-SYM10*) [41, 42] and *M. truncatula* (*Mt-NFP*) [37]. Sequence comparisons of *Lj-NFR5*, *Ps-SYM10* and *Mt-NFP* show that they are orthologs encoding LysM-type receptor kinases (LysM-RKs), and may

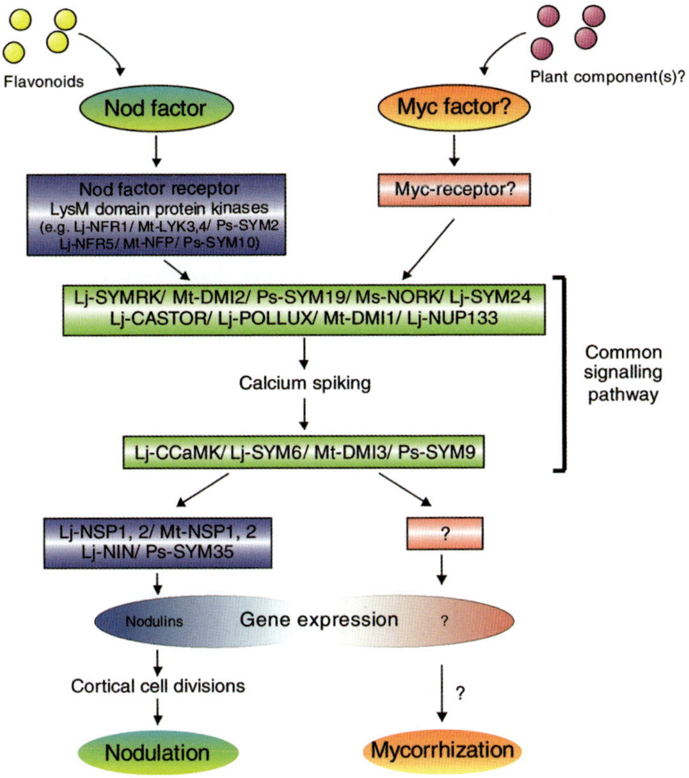

Figure 28.2 Nodulation and endomycorrhization signaling pathways. Specific components are shown in blue (nodulation) and pink (mycorrhization). Nod factor production is induced by plant produced flavonoids and then perceived by plant LysM receptor kinases. Putative Myc factors are also proposed to be perceived by unknown receptors. After the initial recognition events, a common signaling pathway (genes in green) is mediated by at least: seven loci in *Lotus japonicus* (*SYMRK*, *CASTOR*, *POLLUX*, *NUP133*, *SYM24*, *SYM6* and *CCaMK*); three loci in *Medicago truncatula* (*DMI1*, *DMI2*, and *DMI3*); two loci in *Pisum sativum* (SYM19 and SYM9) and one locus in *Melilotus alba* (*NORK*). These are required for both nodulation and mycorrhization. Downstream of the common signaling pathway are specific regulators controlling gene expression for each type of symbiosis.

be located in the plasma membrane. Previously LysM domains have been found in proteins that bind peptidoglycans [43] which are not structurally dissimilar to Nod factors. Therefore these LysM-RKs are good candidates to bind to Nod factors, although direct binding evidence is still lacking. Another potential receptor is *Ps-SYM2* from *P. sativum*, mutation of which has been shown to disrupt perception of specific chemical modifications to Nod factors [41]. Two other candidate receptor genes in *M. truncatula*, the LysM-RKs *Mt-LYK3* and *Mt-LYK4*, have homology to *Lj-NFR1* and *Ps-SYM2*, and mediate Nod factor-induced infection [44].

For AM symbioses, as for the hypothetical Myc factors, no potential host receptors have been identified as yet. Furthermore, no AM mutants blocked in the calcium-signaling response have been identified, which would be indicative of an upstream function (Figure 28.2).

28.3.2 Common Symbiosis Pathway

Genes common to both symbioses have been identified in *M. sativa, P. sativum, L. japonicus, M. truncatula, Phaseolus vulgaris, Vicia faba* and *Melilotus alba*. Mutants of these genes are blocked at an early stage of both the *Rhizobium*–plant and fungus–plant symbiotic interactions. Examples from *L. japonicus* include the symbiosis receptor kinase *SYMRK* [45], two transmembrane ion channel-like proteins *CASTOR & POLLUX* [46–49], the nucleoporin *NUP133* [50] and *SYM24* [38]. As well as being incapable of an AM interaction, mutants of these genes still exhibit root hair deformation in response to Nod factors, but subsequent root hair curling, infection thread formation and calcium spiking are abolished [49]. Therefore, these genes act somewhere downstream of NFR1 and NFR5 and upstream of intracellular calcium spiking (Figure 28.2). *Lj-SYMRK* orthologs were found in *M. truncatula* (*DMI2*), *M. sativa* (*NORK*), and *P. sativum* (*SYM19*) [45, 51]. The *M. truncatula DMI1* gene [52] is a *POLLUX* ortholog, and as expected mutants of *Mt-DMI1* and *Mt-DMI2* block calcium spiking but not root hair deformation [38, 53, 54], indicating that they act upstream of calcium spiking at an early stage of both symbiotic interactions.

Although the *M. truncatula Mt-DMI3* mutant is also blocked at an early stage of both symbiotic interactions [53] its phenotype is subtly different as it is still capable of calcium spiking. *Mt-DMI3* encodes a calcium and calmodulin-dependent protein kinase (CCaMK) [55, 56], and has been placed downstream of the calcium spiking response [15]. Similarly the *Lj-CCaMK, Lj-SYM6* [34, 46, 57] and *Ps-SYM9* [55, 56] are *Mt-DMI3* orthologs, and their mutants cause phenotypes similar to that of the *Mt-DMI3* mutant [58].

28.3.3 Downstream of the Common Signaling Pathway – Specificity of the Symbiosis

Downstream of the common signaling pathway there must be a divergence in the signaling cascade to initiate the transcriptional changes required for the distinct morphological and developmental changes in each symbiosis. Genes involved at this stage identified from screens of legume mutants were still able to perform the early signaling steps such as root hair deformation and calcium spiking, but unable to form nodules or induce Nodulin expression (Figure 28.2). In *M. truncatula*, mutants in two genes, *Mt-NSP1* and *Mt-NSP2* [53, 59], showed that they were required for nodule morphogenesis but acted downstream of *Mt-DMI3* [55, 56]. These mutants exhibit root hair deformation [53, 60, 61] a normal Nod factor-induced calcium influx and spiking [59], but they completely lack infection threads, any sign of cortical cell division and there is no induction of Nodulin genes [53, 59, 62]. NSP1 and NSP2 are predicted to be GRAS domain transcriptional regulators [63]. The mutant phenotypes and the similarity to GRAS domain proteins suggests that they could be Nod factor-activated transcription regulators, possibly controlling key genes in nodule development [60, 61]. Homologs are also present in *L. japonicus*, *Lj-NSP1* and *Lj-NSP2* [63]. Another potential transcriptional regulator has been identified in *L. japonicus*; *Lj-NIN* (orthologous to *Ps-SYM35*) encodes a transmembrane protein with a potential nuclear localization signal and a predicted DNA-binding domain and may also mediate symbiotic gene expression. *Lj-NIN* is thought to act downstream of calcium spiking and is not required for mycorrhization [46, 64].

28.3.4
The Symbiosome

The process by which rhizobia are released from infection threads into nodule primordia is poorly understood. Entire rhizobial cells are engulfed in the plant cell into a compartment termed the symbiosome, which is separated from the plant cell cytoplasm by a plant-derived membrane called the symbiosome membrane (SM). This process resembles phagocytosis in animal cells [65]. Usually bacteria taken up in this way end up in phagosomes and are digested by the fusion of lysosomes with the subsequent acidification of the compartment. Although rhizobia will eventually be digested within the symbiosomes, in a compatible symbiotic interaction, this can be delayed for several weeks. This is despite the fact that the region of the compartment between the bacteroid (the symbiotic form of rhizobia) and the SM, termed the peribacteroid space (PBS), has an acidic pH and contains numerous hydrolytic enzymes, and thus resembles the phagosome. Thus the bacteroids must have some (as yet unknown) mechanisms to avoid their digestion and permit the symbiotic interaction [21].

In eukaryotic cells the vesicular transport pathways underpinning the processes of endo- and exocytosis depend upon the functioning of numerous GTP-binding proteins. A study specifically monitoring the expression of GTP-binding proteins during nodulation found several belonging to the Rab subfamily with elevated expression levels in nodules [66]. One of these, Rab7, had earlier been implicated in nodule formation, as expression of an antisense construct of Rab7 in infected cells of soybean nodules caused symbiosomes to fuse to the cell vacuole and bacteroids to be degraded [67]. Subsequent studies showed that the soybean *rab7* gene was expressed at very early stages of nodule development, possibly even before rhizobial entry to the plant cells [68]. Whether Rab7 is directly involved in the phagocytosis of rhizobia or another critical aspect of early symbiosome functioning remains to be determined.

The SM itself also has some unique properties. Unsurprisingly, as the SM initiates as an invagination of the plasma membrane (PM), marker proteins of the PM can be detected. The SM contains proteins derived from rhizobia as well as the plant, and is particularly enriched in numerous transport systems [69], presumably to facilitate the exchange of nutrients during symbiosis. Furthermore, the SM protein composition changes during nodule development in response to signals emanating from the bacteroid and the plant host [21]. The importance of protein transporters within the SM was demonstrated by the characterization of a plant mutant in a symbiotic sulfate transporter. Under normal growth conditions the plants were unaffected, but in conditions where symbiosis with rhizobia was necessary for plant growth, the mutant was severely compromised. The mutated gene was subsequently shown to encode a sulfate transporter specifically localized to the SM [70].

Rhizobia are phylogenetically related to pathogenic *Brucella* species [71], and there are parallels in their strategies to associate with eukaryotes as both persist for relatively long periods within host-derived membrane compartments. Like rhizobia, *Brucella* spp. must also avoid degradation within a potentially lytic compartment. For both rhizobia and *Brucella* spp. surface polysaccharides (particularly LPS and cyclic glucans) have been shown to be essential for survival of the bacteria, presumably a compromised outer membrane making them more susceptible to lysis [72].

28.4 Symbiotic Signals Produced by Rhizobia

Upon detection of flavonoids released by potential host plants, rhizobia initiate a complex regulatory network for the production of symbiotic signaling molecules. Nod factors are essential for nodule formation and thus can be considered as the primary symbiotic signals. Recent studies have focused not only on the effects of Nod factors (as discussed earlier) but also on the nature and effects of numerous other (secondary) symbiotic signals, such as surface polysaccharides or secreted proteins. These secondary signals are not universally essential for successful nodulation, but can be critical for the interaction with a specific legume and are thus considered as host range determinants. In the case of NGR234 its broad host range on legumes can be partially explained by the variety of secondary signaling molecules it produces, as summarized in Figure 28.3.

The exact stage of action of these rhizobial secondary symbiotic signals is generally unknown, but it may well be at rhizobial release into the plant cell or in the maintenance of bacteroids within the symbiosomes. The role of surface polysaccharides is generally considered to be protective, and may prevent premature digestion of bacteroids within the symbiosomes. The identification of protein secretion systems previously considered characteristic of pathogenic bacteria within rhizobia raises the prospect that the proteins secreted manipulate host pathways for endocytosis or permit chronic infection of eukaryotic cells. As a parallel, the two type III secretion systems of *Salmonella* spp. have been shown to function in exactly this manner [73].

28.4.1 Nod Factors

As described earlier, Nod factors are the first rhizobial signal molecule produced and are essential for nodule formation [61, 74, 75]. Nod factors consist of a β-1,4-linked *N*-acetyl-D-glucosamine backbone of three to six residues of which the non-reducing terminal residue is substituted at the C2 position with an acyl chain. The structure of this acyl chain can vary depending on the rhizobial species. The structural variation of

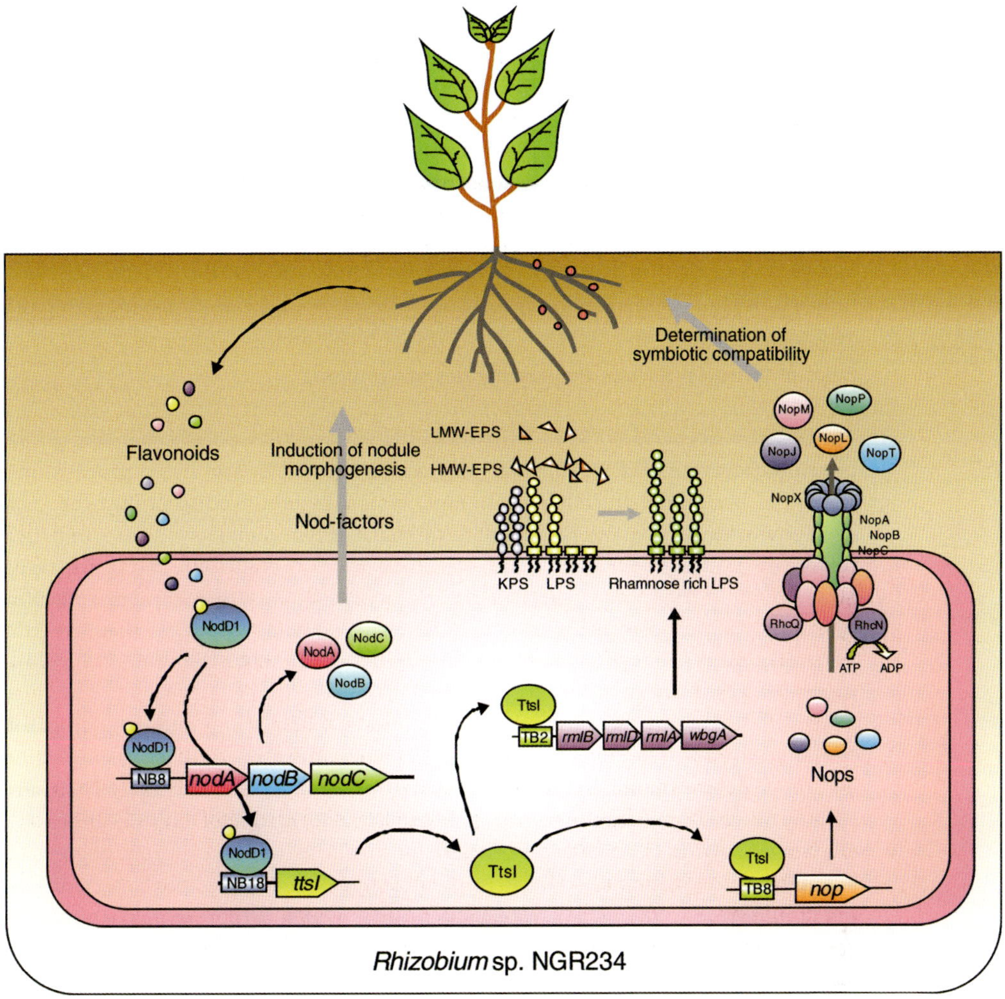

Figure 28.3 Summary of the symbiotic signal molecules produced by NGR234.Flavonoid compounds from legume roots trigger gene expression and synthesis of signals by NodD1 binding to *nod* boxes (NBs). The primary symbiotic signals, the Nod factors, are secreted and allow bacterial entry into root hairs and initiate plant programs for rhizobial infection and root nodule development. NodD1 consequently activates TtsI which activates T3SS and synthesis of rhamnose-rich LPS by binding to *tts* boxes (TBs). LMW EPS, KPS, rhamnose-rich LPS and Nops can all be considered as secondary symbiotic signals, although not essential, they can be extremely important for the nodulation of certain legumes.

a given rhizobial Nod factor can help determine its host specificity. Nod factors are synthesized and exported from the bacteria by the products of *nod* genes. The common nodulation genes *nodABC* are required for basic Nod factor synthesis and are found in most rhizobia that invoke nodules [76]. The only known exception was recently reported in two group II photosynthetic *Bradyrhizobium* strains, BTAi1 and ORS278, which lack *nod* gene homologs [77]. The enzymes encoded by the *nodABC* genes link the individual *N*-acetylglucosamines together, and attach an acyl group to them [78–83]. In addition to the *nodABC* genes, rhizobia also possess species-specific *nod* genes, which modify the basic Nod factor. These host-specific modifications include the addition of sulfuryl, methyl, carbamoyl, acetyl, fucosyl, arabinosyl and other groups to different positions on the backbone, as well as modifications to the structure of the acyl chain. As an example, the *nodSU* genes control the ability of NGR234 to nodulate *Leucaena leucocephala* through *N*-methylation and 6-*O*-carbamoylation of the nonreducing terminus of its Nod factors [84].

28.4.2
Roles of Rhizobial Surface Polysaccharides in Symbiosis

Surface polysaccharides (SPS) which include extracellular polysaccharides (EPS), lipopolysaccharides (LPS), capsular polysaccharides (KPS) and cyclic glucans, are important bacterial extracellular components usually produced to protect the cells from environmental stress. Studies with rhizobial SPS mutants have shown that they can be very important for a successful symbiosis [85]. SPS contribute to various stages of symbiotic development such as root colonization, host recognition, infection thread formation, nodule invasion and host specificity, although they are not normally under the control of NodD proteins as for other symbiotic signaling molecules [12, 86].

EPS has a role in the early stage of symbiosis, establishing and extending the infection thread. The major symbiotically active form of EPS in *S. meliloti* is succinoglycan; mutations in genes required for its synthesis cannot fully invade the root to establish infection threads and lead to the formation of empty nodules [87, 88]. Succinoglycan is produced in two major forms, reflecting different degrees of subunit polymerization: the HMW (high molecular weight) succinoglycan (consisting of hundreds to thousands of repeating units), which is representative of typical bacterial EPS, and a symbiotically active LMW (low molecular weight) succinoglycan of monomers, dimers and trimers produced by digestion of the HMW form by extracellular glycanases [89, 90]. A number of genes involved in synthesis of EPS have been identified in NGR234 [91]. NGR234 synthesizes a HMW form of EPS (similar in structure to that of *S. meliloti*) which is a host range determinant, as an EPS mutant cannot induce nitrogen-fixing nodules on *Leucaena leucocephala*. Although it has been shown that it was the (HMW EPS-derived) LMW EPS produced after glycanase action that was the actual critical factor [92].

KPS are tightly associated with the rhizobial outer membrane, and often play a role in the early stage of symbiosis. Rhizobial KPS are often strain-specific, with structures analogous to the group II K-antigens found in *Escherichia coli*. *S. meliloti* Rm41 produces a form of K-antigen that is symbiotically active but only when EPS is absent [93]. In NGR234, by identifying and deleting the genes responsible for the synthesis of KPS, the resulting mutant had a reduced ability to initiate symbiotic infection [94]. However, the precise role of KPS in root infection and the regulation of KPS expression remain unclear.

LPS are major components of outer membranes of Gram-negative bacteria and are generally thought of as protective molecules. Rhizobial LPS can also play various roles at different stages of symbiosis such as in the initial recognition, infection thread development, root tissue invasion, bacterial release into plant cells and even formation of symbiosomes. LPS molecules consist of a lipid A anchor which maintains the molecule in the hydrophobic outer membrane. Lipid A is associated with a core polysaccharide, which can be substituted by an O-antigen domain. LPS are attached to the membrane by the lipidic part being inserted into the bacterial phospholipid monolayer and thus the saccharidic part is oriented to the exterior of the cell [95–97]. Two forms of LPS are often synthesized, rough LPS (R-LPS) and smooth LPS (S-LPS). The low molecular weight form of LPS (R-LPS) contains only the lipid A and core oligosaccharide, whereas the high molecular weight form of LPS (S-LPS) includes an additional O-antigen [98]. In NGR234 a symbiotic form of S-LPS is produced in the presence of flavonoids, and its absence adversely affected nodulation of several host legume species [99]. This S-LPS molecule is noteworthy as it is predominantly composed of rhamnose residues, and the biosynthetic enzymes required are under the control of TtsI [100].

28.4.3
Secreted Proteins

Although a few secreted proteins that influence nodule formation were identified in early studies, rhizobia has been found to use protein secretion systems usually associated with pathogenic bacteria, such as type III or type IV secretion systems. Type III secretion systems (T3SS) are highly conserved multiprotein complexes and important virulence factors in pathogen–eukaryote interactions. T3SS are found in many Gram-negative bacteria infecting humans, animals and plants [101]. Previously T3SS had been thought to be unique to pathogenic bacteria, but these systems have now been identified in rhizobia [102] such as NGR234 [103], *M. loti* MAFF303099 [104], *B. japonicum* USDA110 [105], *S. fredii* USDA257 [106], USDA191 [107] and HH103 [102, 107]. In contrast, *S. meliloti* 1021 [108] and *M. loti* R7A [109] do not contain T3SS. In these strains, a type IV secretion system may serve a similar function [110].

Mutations in rhizobial T3SS cause symbiotic phenotypes dependent on the host plant [111–116]. For instance, in NGR234 a functional T3SS is required for efficient nodulation of some plant species, such as *Tephrosia vogelii*, *Flemingia congesta* and *Lablab purpureus*, however for *Pachyrhizus tuberosus* and *Crotalaria juncea* the T3SS appears to be extremely deleterious [111, 115].

T3SS of pathogens have been shown to inject proteins (called effectors) into eukaryotic cells and to change normal functioning of eukaryotic cells [117]. Rhizobial T3SS secrete nodulation outer proteins (Nops). NGR234 secretes at least nine Nops, which are classified into two groups: translocatory proteins that are external components of the secretion machine and effector proteins injected into legume cells. For example, NopA, NopB, NopC and NopX are required for the transit of proteins from rhizobia to the plant cell cytoplasm [111, 115, 118, 119]. In NGR234, NopL and NopP have been characterized as rhizobial-specific effector proteins that can be phosphorylated by plant kinases [111, 120–122]. A double mutant of NGRΩ*nopL*Δ*nopP* was shown to have a more pronounced phenotype than either single mutant, suggesting that the effector Nops may also function cooperatively [122]. Additional effector proteins in NGR234 – *nopM*, *nopJ*, and *nopT* (formerly y4fR, y4lO, and y4zC, respectively) – have been identified from homology searches [102, 103].

S. fredii strains USDA257 and HH103 possess very similar T3SS to NGR234, and have also been shown to secrete Nops. In USDA257, NopX, NopB, Nop38 and Nop7 have been detected [113, 114, 123], and in HH103, NopA, NopC, NopL, NopP and NopX have been confirmed to be T3SS-secreted proteins [124]. Furthermore, in HH103, NopM and NopD were also identified as putative secreted effector proteins [124], these Nops are homologous to NopM of NGR234 and XopD of *X. campestris* pv. *vesicatoria*, respectively [125]. XopD in *X. campestris* pv. *vesicatoria* targets SUMO (small ubiquitin-like modifier) conjugated proteins in plants, suggesting that the XopD protease mimics a host protease that removes SUMO modifications. In *B. japonicum* USDA110, genes with homology to *nopA*, *nopB*, *nopL* and *nopP* have been found in the TT3S gene cluster [112]. Using mass spectrometry techniques, eight different flavonoid-inducible secreted protein spots have been identified [126]. Mutation of the *S. fredii* and *B. japonicum* T3SS also leads to host plant-specific phenotypes and a similar result was observed after mutation of the T3SS of *M. loti* MAFF303099. Although protein secretion was not observed by the T3SS of MAFF303099, its mutation enabled nodulation of *Leucaena leucocephala* [110].

28.4.4
Rhizobial Regulation of Symbiotic Signal Production

The rhizobial regulation cascade induced by flavonoids is intricate and many factors participate [18, 19]. The initial signals of nodulation, flavonoids, accumulate in the cytoplasmic membrane of rhizobia [127] and interact with NodD proteins, members of the LysR family of transcriptional regulators. NodD binds to highly conserved DNA sequences called *nod*-boxes (NBs found in the promoter regions of most (nodulation) *nod* genes, inducing a bend in the DNA at the binding site [128]. The chaperonins GroESL modulate the binding activity of NodD and are known to be necessary for the correct folding of NodD in *S. meliloti* [129, 130]. There is no direct evidence for a direct interaction between NodD proteins and flavonoids yet, but it has been suggested that a NodD–flavonoid complex is formed at the NB [131]. Even in the absence of flavonoids, binding of NodD to NB can occur [132] regardless of whether the actual flavonoid can actually activate the downstream *nod* loci [128]. Interactions between flavonoids and NodD proteins do not always lead to transcription; several flavonoids can bind to NodD1 from *S. meliloti*, but only luteolin was capable of activating *nod* gene expression [131]. NodD proteins from different rhizobia respond to different classes of flavonoids, and the spectrum of flavonoids secreted by a legume is considered a determinant of host specificity. Rhizobia usually possess between one and five *nodD* homologs depending on the species. In addition to activation of genes preceded by NBs, some NodD proteins repress the expression of promoters containing NBs. In USDA110 and NGR234, NodD2 is known as a repressor of *nod* genes [133–135], whereas the single *nodD* gene from *R. leguminosarum* bv. *viciae* is autorepressed by its own product [136].

28.4.5
NodD Proteins Initiate a Signaling Cascade

NBs are found upstream of all the *nod* genes responsible for Nod factor synthesis, but they are also found upstream of genes encoding other transcriptional regulators controlling the synthesis of further symbiotic signals. Some rhizobial strains possess one or two copies of another LysR-type regulator *syrM* (for symbiotic regulator) [135, 137–140]. SyrM proteins are NodD homologs and can also act as activators of *nod* genes. In NGR234, there are two copies of *syrM*: *syrM*1 and *syrM*2 [103]. SyrM1 is involved in activation of a number of genes and controls the level of sulfated Nod factors [137]. Transcription of *syrM*2 (unlike *syrM*1) is under the control of a NB and is necessary for the expression of *nodD*2 [141]. In *S. meliloti* SyrM proteins are thought to activate genes by binding to another cis-element, the SyrM motif (*syr* box), found upstream of *nodD*3 and *syrA* [142–145]. A putative *syr* box was found in the promoter region of *nodD*2 of NGR234, suggesting a similar regulatory mechanism may exist [141].

28.4.6
TtsI and tts-Boxes

Certain rhizobia possess another NB-controlled transcriptional activator called *ttsI*. TtsI activates genes by binding to conserved cis-elements termed *tts* boxes (TB) [99, 112, 146]. TtsI controls a number of host-specific symbiotic signals thought to be required for nodule formation later than Nod factors. TtsI has homology to the regulator proteins of the two-component sensor-regulator family [99, 112, 115, 116]. Such regulatory systems usually consist of a sensor histidine protein kinase and a response regulator protein (TtsI). The sensor kinase autophosphorylates at a histidine residue upon detection of a specific signal and subsequently the phosphate group is transferred to an aspartate residue in the response regulator, leading to its activation [147]. No partner sensor has been found for TtsI, however, which poses the question as to how TtsI is activated. Notably, all the TtsI homologs so far identified contain a glutamate residue instead of the conserved (phosphorylated) aspartate residue [99]. In other bacteria such a change from aspartate to glutamate leads to the constitutive activation of the response regulator [148, 149]. Thus TtsI may not require activation by phosphorylation and consequently a sensor kinase partner. Instead only the expression of TtsI is required, which is known to be dependent on flavonoids and NodD [99, 141].

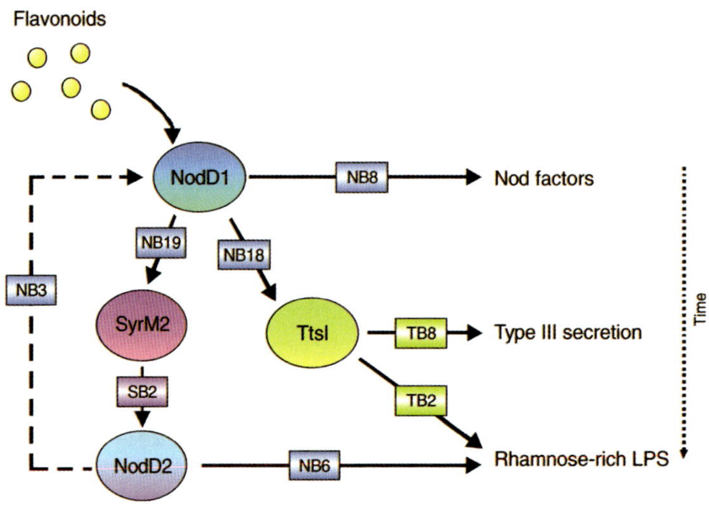

Figure 28.4 Proposed model for the flavonoid- and NodD1-dependent regulatory cascade in NGR234. Modified from Kobayashi and associates [141].In this model, flavonoids are sensed by NodD1 to trigger the regulatory cascade. Activation is shown with solid black arrows, whereas repression by NodD2 of NodD1 expression is marked with a dashed line. Following flavonoid induction, NodD1 rapidly activates the transcription of operons responsible for the synthesis of Nod factors. NodD1 also activates synthesis of TtsI and SyrM2 via NB18 and NB19 respectively. This triggers additional functions that are probably required when more intimate contact between the bacteria and their hosts has occurred. In turn, SyrM2 activates transcription of *nodD2*. At this third regulatory level, NodD2 triggers late flavonoid-inducible loci such as *fixF* (controlled by NB6), which is involved in the synthesis of a rhamnose-rich LPS. TtsI also activates the synthesis of rhamnose-rich LPS as well as T3SS via TB. Concomitantly, NodD2 also represses the expression of *nodD1*.

TtsI homologs are found in several rhizobia, such as USDA110 [105, 112], *M. loti* MAFF303099 [110, 150], *S. fredii* USDA257 [114] and NGR234 [103, 116] which also all possess a T3SS. The genes/operons encoding the rhizobial T3SS machine as well as Nops contain TBs and are thus directly controlled by TtsI. Bioinformatic searches have revealed that numerous TBs are present in T3SS-possessing rhizobial genomes; for example in USDA110 up to 30 TBs were found, suggesting the TtsI-regulon encompasses more than just activation of the T3SS [126]. In NGR234, sequence analysis revealed the presence of 11 TB elements on its symbiotic plasmid alone [99]. The majority of the TBs were found upstream of genes encoding the T3SS machine or possible secreted proteins. Two TBs, however, were located in a cluster of genes involved in rhamnose synthesis, one of which, TB2, activates genes essential for the production of a rhamnose-rich LPS known to be important for successful nodulation of many legumes [99, 100, 151]. Thus, TtsI regulates not only T3SS but also other symbiosis factors. Recently, transcriptional assays have shown that the expression of 10 of the 11 TBs was flavonoid- and TtsI-dependent and that TtsI can bind to TB-containing promoters *in vitro* [146].

28.5
Summary and Outlook

Undoubtedly the flavonoid-induced and NodD-dependent regulation of the synthesis of symbiotic signals has to be carefully controlled. In NGR234 NodD1 heads a signaling pathway composed of several regulators to ensure a temporal gradation in symbiotic signal production and also its own downregulation (Figure 28.4). In this way the expression of a symbiotic gene can be coincided with the requirement of its product at a particular stage of root infection or nodule development [141].

Although a number of rhizobial symbiotic signaling molecules have been identified and there have been recent major advances in how legumes perceive them, infection thread release and the subsequent symbiosome formation remains an understudied area of the rhizobial–legume symbiotic interaction. Few plant proteins have been identified that could play a role in symbiosome development and although suspected, it remains to be seen whether rhizobial factors (such as secreted proteins) can influence this process.

References

1 Odum, E.P. and Smalley, A.E. (1959) *Proc Natl Acad Sci USA*, **45**, 617–622.

2 Mylona, P., Pawlowski, K. and Bisseling, T. (1995) *Plant Cell*, **7**, 869–885.

3 Dénarié, J., Debellé, F. and Rosenberg, C. (1992) *Annu Rev Microbiol*, **46**, 497–531.

4 Pueppke, S.G. and Broughton, W.J. (1999) *Mol Plant Microbe Interact*, **12**, 293–318.

5 Trinick, M.J. (1980) *J Appl Bacteriol*, **49**, 39–53.

6 Graham, J.H. and Miller, M. (2005) in *Plant Ecophysiology*, Vol. 4, Springer, Berlin, pp. 79–100.

7 Remy, W., Taylor, T.N., Hass, H. and Kerp, H. (1994) *Proc Natl Acad Sci USA*, **91**, 11841–11843.

8 Redecker, D., Morton, J.B. and Bruns, T.D. (2000) *Mol Phylogenet Evol*, **14**, 276–284.

9 Heckman, D.S., Geiser, D.M., Eidell, B.R., Stauffer, R.L., Kardos, N.L. and Hedges, S.B. (2001) *Science*, **293**, 1129–1133.

10 Dénarié, J., Debellé, F. and Promé, J. C. (1996) *Annu Rev Biochem*, **65**, 503–535.

11 Ehrhardt, D.W., Wais, R. and Long, S.R. (1996) *Cell*, **85**, 673–681.

12 Spaink, H.P. (2000) *Annu Rev Microbiol*, **54**, 257–288.

13 Reddy, P.M., Rendón-Anaya, M., Río, M. de los Doloros Soto del and Khandual, S. (2007) Flavonoids as Signaling Molecules and Regulators of Root Nodule Development, in Vol. 1, Global Science Books, México, pp. 83–94.

14 Downie, J.A. and Walker, S.A. (1999) *Curr Opin Plant Biol*, **2**, 483–489.

15 Oldroyd, G.E. and Downie, J.A. (2004) *Nat Rev Mol Cell Biol*, **5**, 566–576.

16 Oldroyd, G.E. and Downie, J.A. (2006) *Curr Opin Plant Biol*, **9**, 351–357.

17 Geurts, R., Fedorova, E. and Bisseling, T. (2005) *Curr Opin Plant Biol*, **8**, 346–352.

18 Broughton, W.J., Jabbouri, S. and Perret, X. (2000) *J Bacteriol*, **182**, 5641–5652.

19 Perret, X., Staehelin, C. and Broughton, W.J. (2000) *Microbiol Mol Biol Rev*, **64**, 180–201.

20 Smit, G., Swart, S., Lugtenberg, B.J. and Kijne, J.W. (1992) *Mol Microbiol*, **6**, 2897–2903.

21 Parniske, M. (2000) *Curr Opin Plant Biol*, **3**, 320–328.

22 Cullimore, J.V., Ranjeva, R. and Bono, J.J. (2001) *Trends Plant Sci*, **6**, 24–30.

23 Oldroyd, G.E., Harrison, M.J. and Udvardi, M. (2005) *Plant Physiol*, **137**, 1205–1210.

24 Strack, D., Fester, T., Hause, B., Schliemann, W. and Walter, M.H. (2003) *J Chem Ecol*, **29**, 1955–1979.

25 Harrison, M.J. (1997) *Trends Plant Sci*, **2**, 54–60.

26 Shachar-Hill, Y., Pfeffer, P.E., Douds, D., Osman, S.F., Doner, L.W. and Ratcliffe, R.G. (1995) *Plant Physiol*, **108**, 7–15.

27 Smith, S., Dickson, S. and Smith, F. (2001) *J Plant Physiol*, **28**, 683–694.

28 Akiyama, K., Matsuzaki, K. and Hayashi, H. (2005) *Nature*, **435**, 824–827.

29 Genre, A., Chabaud, M., Timmers, T., Bonfante, P. and Barker, D.G. (2005) *Plant Cell*, **17**, 3489–3499.

30 Harrison, M.J. (2005) *Annu Rev Microbiol*, **59**, 19–42.

31 Albrecht, C., Geurts, R. and Bisseling, T. (1999) *EMBO J*, **18**, 281–288.

32 Hirsch, A.M., Lum, M.R. and Downie, J.A. (2001) *Plant Physiol*, **127**, 1484–1492.

33 Kistner, C. and Parniske, M. (2002) *Trends Plant Sci*, **7**, 511–518.

34 Kistner, C. *et al.* (2005) *Plant Cell*, **17**, 2217–2229.

35 Zhu, H., Riely, B.K., Burns, N.J. and Ane, J.M. (2006) *Genetics*, **172**, 2491–2499.

36 Gianinazzi-Pearson, V. (1996) *Plant Cell*, **8**, 1871–1883.

37 Amor, B.B. *et al.* (2003) *Plant J*, **34**, 495–506.

38 Miwa, H., Sun, J., Oldroyd, G.E. and Downie, J.A. (2006) *Mol Plant Microbe Interact*, **19**, 914–923.

39 Madsen, E.B. *et al.* (2003) *Nature*, **425**, 637–640.

40 Radutoiu, S. *et al.* (2003) *Nature*, **425**, 585–592.

41 Geurts, R., Heidstra, R., Hadri, A.E., Downie, J.A., Franssen, H., Van Kammen, A. and Bisseling, T. (1997) *Plant Physiol*, **115**, 351–359.

42 Walker, S.A., Viprey, V. and Downie, J.A. (2000) *Proc Natl Acad Sci USA*, **97**, 13413–13418.

43 Bateman, A. and Bycroft, M. (2000) *J Mol Biol*, **299**, 1113–1119.

44 Limpens, E., Franken, C., Smit, P., Willemse, J., Bisseling, T. and Geurts, R. (2003) *Science*, **302**, 630–633.

45 Stracke, S. *et al.* (2002) *Nature*, **417**, 959–962.

46 Schauser, L., Handberg, K., Sandal, N., Stiller, J., Thykjaer, T., Pajuelo, E., Nielsen, A. and Stougaard, J. (1998) *Mol Gen Genet*, **259**, 414–423.

47 Szczyglowski, K., Kapranov, P., Hamburger, D. and de Bruijn, F.J. (1998) *Plant Mol Biol*, **37**, 651–661.

48 Kawaguchi, M., Imaizumi-Anraku, H., Koiwa, H., Niwa, S., Ikuta, A., Syono, K. and Akao, S. (2002) *Mol Plant Microbe Interact*, **15**, 17–26.

49 Imaizumi-Anraku, H. *et al.* (2005) *Nature*, **433**, 527–531.

50 Kanamori, N. *et al.* (2006) *Proc Natl Acad Sci USA*, **103**, 359–364.

51 Endre, G., Kereszt, A., Kevei, Z., Mihacea, S., Kalo, P. and Kiss, G.B. (2002) *Nature*, **417**, 962–966.

52 Ane, J.M. *et al.* (2004) *Science*, **303**, 1364–1367.

53 Catoira, R. *et al.* (2000) *Plant Cell*, **12**, 1647–1666.

54 Shaw, S.L. and Long, S.R. (2003) *Plant Physiol*, **131**, 976–984.

55 Levy, J. *et al.* (2004) *Science*, **303**, 1361–1364.

56 Mitra, R.M., Gleason, C.A., Edwards, A., Hadfield, J., Downie, J.A., Oldroyd, G.E. and Long, S.R. (2004) *Proc Natl Acad Sci USA*, **101**, 4701–4705.

57 Harris, J.M., Wais, R. and Long, S.R. (2003) *Mol Plant Microbe Interact*, **16**, 335–341.

58 Schneider, A., Walker, A., Sagan, M., Duc, G., Ellis, N. and Downie, A. (2002) *Theor Appl Genet*, **104**, 1312–1316.

59 Oldroyd, G.E. and Long, S.R. (2003) *Plant Physiol*, **131**, 1027–1032.

60 Kalo, P. *et al.* (2005) *Science*, **308**, 1786–1789.

61 Smit, P., Raedts, J., Portyanko, V., Debelle, F., Gough, C., Bisseling, T. and Geurts, R. (2005) *Science*, **308**, 1789–1791.

62 Mitra, R.M., Shaw, S.L. and Long, S.R. (2004) *Proc Natl Acad Sci USA*, **101**, 10217–10222.

63 Heckmann, A.B., Lombardo, F., Miwa, H., Perry, J.A., Bunnewell, S., Parniske, M., Wang, T.L. and Downie, J.A. (2006) *Plant Physiol*, **142**, 1739–1750.

64 Borisov, A.Y. *et al.* (2003) *Plant Physiol*, **131**, 1009–1017.

65 Brewin, N.J. (1998) in *The Rhizobiaceae* (eds H.P. Spaink, A. Kondorosi and P.J. Hooykaas), Kluwer Academic Publishers, Dordrecht, pp. 417–429.

66 Borg, S., Brandstrup, B., Jensen, T.J. and Poulsen, C. (1997) *The Plant Journal*, **11**, 237–250.

67 Cheon, C.I., Lee, N.G., Siddique, A.B., Bal, A.K. and Verma, D.P. (1993) *EMBO J*, **12**, 4125–4135.

68 Son, O. *et al.* (2003) *Plant Sci*, **165**, 1239–1244.

69 Wienkoop, S. and Saalbach, G. (2003) *Plant Physiol*, **131**, 1080–1090.

70 Krusell, L. *et al.* (2005) *Plant Cell*, **17**, 1625–1636.

71 Tsolis, R.M. (2002) *Proc Natl Acad Sci USA*, **99**, 12503–12505.

72 Jones, K.M., Kobayashi, H., Davies, B.W., Taga, M.E. and Walker, G.C. (2007) *Nat Rev Microbiol*, **5**, 619–633.

73 Schlumberger, M.C. and Hardt, W.D. (2006) *Curr Opin Microbiol*, **9**, 46–54.

74 Downie, J.A. (1998) in *The Rhizobiaceae* (eds H.P. Spaink, A. Kondorosi and P.J.J. Hooykaas), Kluwer Academic Publishers, Dordrecht, pp. 387–402.

75 Lerouge, P., Roche, P., Faucher, C., Maillet, F., Truchet, G., Promé, J. C. and Dénarié, J. (1990) *Nature*, **344**, 781–784.

76 Moulin, L., Munive, A., Dreyfus, B. and Boivin-Masson, C. (2001) *Nature*, **411**, 948–950.

77 Giraud, E. *et al.* (2007) *Science*, **316**, 1307–1312.

78 Atkinson, E.M., Palcic, M.M., Hindsgaul, O. and Long, S.R. (1994) *Proc Natl Acad Sci USA*, **91**, 8418–8422.

79 Geremia, R.A., Mergaert, P., Geelen, D., Van Montagu, M. and Holsters, M. (1994) *Proc Natl Acad Sci USA*, **91**, 2669–2673.

80 John, M., Röhrig, H., Schmidt, J., Wieneke, U. and Schell, J. (1993) *Proc Natl Acad Sci USA*, **90**, 625–629.

81 Kafetzopoulos, D., Thireos, G., Vournakis, J.N. and Bouriotis, V. (1993) *Proc Natl Acad Sci USA*, **90**, 8005–8008.

82 Röhrig, H., Schmidt, J., Wieneke, U., Kondorosi, E., Barlier, I., Schell, J. and John, M. (1994) *Proc Natl Acad Sci USA*, **91**, 3122–3126.

83 Spaink, H.P., Wijfjes, A.H., van der Drift, K.M., Haverkamp, J., Thomas-Oates, J.E. and Lugtenberg, B.J. (1994) *Mol Microbiol*, **13**, 821–831.

84 Jabbouri, S., Fellay, R., Talmont, F., Kamalaprija, P., Burger, U., Relić, B., Promé, J. C. and Broughton, W.J. (1995) *J Biol Chem*, **270**, 22968–22973.

85 Fraysse, N., Couderc, F. and Poinsot, V. (2003) *Eur J Biochem*, **270**, 1365–1380.

86 Becker, A., Fraysse, N. and Sharypova, L. (2005) *Mol Plant Microbe Interact*, **18**, 899–905.

87 Leigh, J.A., Signer, E.R. and Walker, G.C. (1985) *Proc Natl Acad Sci USA*, **82**, 6231–6235.

88 Finan, T.M., Hirsch, A.M., Leigh, J.A., Johansen, E., Kuldau, G.A., Deegan, S., Walker, G.C. and Signer, E.R. (1985) *Cell*, **40**, 869–877.

89 Gonzalez, J.E., Reuhs, B.L. and Walker, G.C. (1996) *Proc Natl Acad Sci USA*, **93**, 8636–8641.

90 Wang, L.X., Wang, Y., Pellock, B. and Walker, G.C. (1999) *J Bacteriol*, **181**, 6788–6796.

91 Streit, W.R., Schmitz, R.A., Perret, X., Staehelin, C., Deakin, W.J., Raasch, C., Liesegang, H. and Broughton, W.J. (2004) *J Bacteriol*, **186**, 535–542.

92 Staehelin, C. *et al.* (2006) *J Bacteriol*, **188**, 6168–6178.

93 Reuhs, B.L., Carlson, R.W. and Kim, J.S. (1993) *J Bacteriol*, **175**, 3570–3580.

94 Le Quéré, A.J., Deakin, W.J., Schmeisser, C., Carlson, R.W., Streit, W.R., Broughton, W.J. and Forsberg, L.S. (2006) *J Biol Chem*, **281**, 28981–28992.

95 Noel, K.D. and Duelli, D.M. (2000) *Rhizobium* lipopolysaccharide and its role in symbiosis. In Prokaryotic Nitrogen Fixation: a Model System for Analysis of Biological Process, Wymondham, UK, pp. 415–431.

96 Carlson, R.W., Reuhs, B.L., Forsberg, L.S. and Kannenberg, E.L. (1999) Rhizobial cell surface carbohydrates: their structures, synthesis, biosynthesis, and functions in Genetics of Bacterial Polysaccharides. (ed. J.B. Goldberg), pp. 53–90.

97 Price, N.P. (1999) *Carbohydr Res*, **317**, 1–9.

98 Reuhs, B.L., Geller, D.P., Kim, J.S., Fox, J.E., Kolli, V.S. and Pueppke, S.G. (1998) *Appl Environ Microbiol*, **64**, 4930–4938.

99 Marie, C., Deakin, W.J., Ojanen-Reuhs, T., Diallo, E., Reuhs, B., Broughton, W.J. and Perret, X. (2004) *Mol Plant Microbe Interact*, **17**, 958–966.

100 Broughton, W.J. *et al.* (2006) *J Bacteriol*, **188**, 3654–3663.

101 Hueck, C.J. (1998) *Microbiol Mol Biol Rev*, **62**, 379–433.

102 Marie, C., Broughton, W.J. and Deakin, W.J. (2001) *Curr Opin Plant Biol*, **4**, 336–342.

103 Freiberg, C., Fellay, R., Bairoch, A., Broughton, W.J., Rosenthal, A. and Perret, X. (1997) *Nature*, **387**, 394–401.

104 Kaneko, T. *et al.* (2000) *DNA Res*, **7**, 331–338.

105 Göttfert, M., Rothlisberger, S., Kundig, C., Beck, C., Marty, R. and Hennecke, H. (2001) *J Bacteriol*, **183**, 1405–1412.

106 Meinhardt, L.W., Krishnan, H.B., Balatti, P.A. and Pueppke, S.G. (1993) *Mol Microbiol*, **9**, 17–29.

107 Bellato, C., Krishnan, H.B., Cubo, T., Temprano, F. and Pueppke, S.G. (1997) *Microbiology*, **143**, 1381–1388.

108 Galibert, F. *et al.* (2001) *Science*, **293**, 668–672.

109 Sullivan, J.T. *et al.* (2002) *J Bacteriol*, **184**, 3086–3095.

110 Hubber, A., Vergunst, A.C., Sullivan, J.T., Hooykaas, P.J. and Ronson, C.W. (2004) *Mol Microbiol*, **54**, 561–574.

111 Ausmees, N., Kobayashi, H., Deakin, W.J., Marie, C., Krishnan, H.B., Broughton, W.J. and Perret, X. (2004) *J Bacteriol*, **186**, 4774–4780.

112 Krause, A., Doerfel, A. and Göttfert, M. (2002) *Mol Plant Microbe Interact*, **15**, 1228–1235.

113 Lorio, J.C., Kim, W.S. and Krishnan, H.B. (2004) *Mol Plant Microbe Interact*, **17**, 1259–1268.

114 Krishnan, H.B., Lorio, J., Kim, W.S., Jiang, G., Kim, K.Y., DeBoer, M. and Pueppke, S.G. (2003) *Mol Plant Microbe Interact*, **16**, 617–625.

115 Marie, C., Deakin, W.J., Viprey, V., Kopcinska, J., Golinowski, W., Krishnan, H.B., Perret, X. and Broughton, W.J. (2003) *Mol Plant Microbe Interact*, **16**, 743–751.

116 Viprey, V., Del Greco, A., Golinowski, W., Broughton, W.J. and Perret, X. (1998) *Mol Microbiol*, **28**, 1381–1389.

117 Galan, J.E. and Collmer, A. (1999) *Science*, **284**, 1322–1328.

118 Deakin, W.J., Marie, C., Saad, M.M., Krishnan, H.B. and Broughton, W.J. (2005) *Mol Plant Microbe Interact*, **18**, 499–507.

119 Saad, M.M., Kobayashi, H., Marie, C., Brown, I.R., Mansfield, J.W., Broughton, W.J. and Deakin, W.J. (2005) *J Bacteriol*, **187**, 1173–1181.

120 Bartsev, A.V., Boukli, N.M., Deakin, W.J., Staehelin, C. and Broughton, W.J. (2003) *FEBS Lett*, **554**, 271–274.

121 Bartsev, A.V., Deakin, W.J., Boukli, N.M., McAlvin, C.B., Stacey, G., Malnoé, P., Broughton, W.J. and Staehelin, C. (2004) *Plant Physiol*, **134**, 871–879.

122 Skorpil, P., Saad, M.M., Boukli, N.M., Kobayashi, H., Ares-Orpel, F., Broughton, W.J. and Deakin, W.J. (2005) *Mol Microbiol*, **57**, 1304–1317.

123 Krishnan, H.B. (2002) *J Bacteriol*, **184**, 831–839.

124 Rodrigues, J.A. *et al.* (2007) *J Proteome Res*, **6**, 1029–1037.

125 Hotson, A., Chosed, R., Shu, H., Orth, K. and Mudgett, M.B. (2003) *Mol Microbiol*, **50**, 377–389.

126 Süβ, C., Hempel, J., Zehner, S., Krause, A., Patschkowski, T. and Göttfert, M. (2006) *J Biotechnol*, **126**, 69–77.

127 Hubac, C., Ferran, J., Guerrier, D., Trémolières, A. and Kondorosi, A. (1993) *J Gen Microbiol*, **139**, 1571–1578.

128 Fisher, R.F. and Long, S.R. (1993) *J Mol Biol*, **233**, 336–348.

129 Yeh, K.C., Peck, M.C. and Long, S.R. (2002) *J Bacteriol*, **184**, 525–530.

130 Fisher, R.F. and Long, S.R. (1989) *J Bacteriol*, **171**, 5492–5502.

131 Peck, M.C., Fisher, R.F. and Long, S.R. (2006) *J Bacteriol*, **188**, 5417–5427.

132 Feng, J., Li, Q., Hu, H.L., Chen, X.C. and Hong, G.F. (2003) *Nucleic Acids Res*, **31**, 3143–3156.

133 Göttfert, M., Holzhauser, D., Bani, D. and Hennecke, H. (1992) *Mol Plant Microbe Interact*, **5**, 257–265.

134 Garcia, M., Dunlap, J., Loh, J. and Stacey, G. (1996) *Mol Plant Microbe Interact*, **9**, 625–636.

135 Fellay, R. *et al.* (1998) *Mol Microbiol*, **27**, 1039–1050.

136 Hu, H., Liu, S., Yang, Y., Chang, W. and Hong, G. (2000) *Nucleic Acids Res*, **28**, 2784–2793.

137 Hanin, M., Jabbouri, S., Broughton, W.J. and Fellay, R. (1998) *Mol Plant Microbe Interact*, **11**, 343–350.

138 Michiels, J., De Wilde, P. and Vanderleyden, J. (1993) *Nucleic Acids Res*, **21**, 3893.

139 Mulligan, J.T. and Long, S.R. (1989) *Genetics*, **122**, 7–18.

140 Swanson, J.A., Mulligan, J.T. and Long, S.R. (1993) *Genetics*, **134**, 435–444.

141 Kobayashi, H., Naciri-Graven, Y., Broughton, W.J. and Perret, X. (2004) *Mol Microbiol*, **51**, 335–347.

142 Barnett, M.J., Rushing, B.G., Fisher, R.F. and Long, S.R. (1996) *J Bacteriol*, **178**, 1782–1787.

143 Barnett, M.J., Swanson, J.A. and Long, S.R. (1998) *Genetics*, **148**, 19–32.

144 Perret, X., Freiberg, C., Rosenthal, A., Broughton, W.J. and Fellay, R. (1999) *Mol Microbiol*, **32**, 415–425.

145 Xiao, H., Shen, S.J. and Zhu, J.B. (1998) *Sci China C Life Sci*, **41**, 157–162.

146 Wassem, R., Kobayashi, H., Kambara, K., Le Quéré, A., Walker, G.C., Broughton, W.J. and Deakin, W.J. (2008) *Mol Microbiol*, **68**, 736–748.

147 Stock, A.M., Robinson, V.L. and Goudreau, P.N. (2000) *Annu Rev Biochem*, **69**, 183–215.

148 Klose, K.E., Weiss, D.S. and Kustu, S. (1993) *J Mol Biol*, **232**, 67–78.

149 Lan, C.Y. and Igo, M.M. (1998) *J Bacteriol*, **180**, 171–174.

150 Kaneko, T. *et al.* (2000) *DNA Research*, **7**, 331–338.

151 Reuhs, B.L. *et al.* (2005) *J Bacteriol*, **187**, 6479–6487.

29
Cyanobacterial and Algal Symbioses

Goran Kovacevic, Jürgen M. Steiner, and Wolfgang Löffelhardt

29.1
Introduction

Symbioses in which cyanobacteria and algae are involved as the "junior partner" are enormously widespread, with animals, fungi, protists and plants as the "senior partners." The relationships between the symbiotic partners range from mutualistic to parasitic ones. In this article we will deal only with strictly intracellular symbioses with emphasis on those cases where (at least) the host cell profits from nutritional (mostly photosynthate) contributions of the symbiont. Symbiosis not only leads to competitive advantages in certain habitats but must be envisaged as a major driving force in evolution [1], resulting in the continual generation of new species. Its most prominent achievement was the creation of eukaryotes around one and a half billion years ago. The ultimate goal of all symbiotic associations must be to reach a level of integration where the endosymbiont ends up as an organelle. This is, of course, a very complicated and lengthy process. Most of the systems dealt with in this chapter can be considered to be at the very beginning or in the first quarter of this "tour de force." In the case of endosymbiotic cyanobacteria the range is from facultative associations to primitive plastids (cyanelles), which show clear vestiges of their cyanobacterial origin (Figure 29.1).

In the case of endosymbiotic algae we will discuss borrowed, that is, ephemeral autotrophy, as well as full integration, finally resulting in secondary or complex plastids (Figure 29.2). Furthermore, some peculiar systems in which only the plastids are stolen from the prey will be discussed, eventually culminating in so-called tertiary endosymbioses.

29.2
Intracellular Cyanobacteria

The enormous impact of intracellular cyanobacteria in the course of evolution is exemplified by their involvement in the origin of the plant kingdom through a single

Intracellular Niches of Microbes. Edited by Ulrich E. Schaible and Albert Haas

ISBN: 978-3-527-32207-7

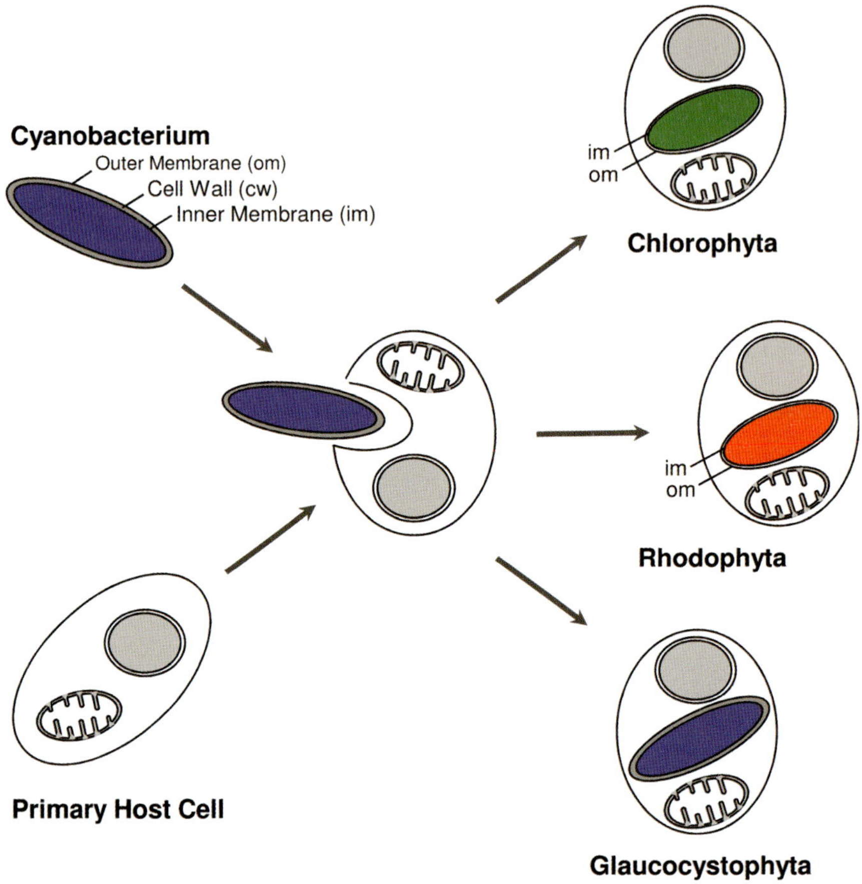

Figure 29.1 Primary endosymbiosis. A single event led to primary plastids which are surrounded by two membranes. Not only primary plastids but also the plant kingdom as a whole can be envisaged to be monophyletic. Modified from [40].

primary endosymbiotic event [2] between a filamentous, heterocyst-forming cyanobacterium [3] and a heterotrophic protist about 1.5 billion years ago [4] (Figure 29.1). Since then, cyanobacteria have continued to contribute, in numerous symbiotic associations, to the fitness and speciation of their eukaryotic hosts, ranging from protists, algae, and plants [5] to invertebrates [6].

29.2.1 Plant Symbioses

The genus *Nostoc* is especially prone to undergo symbioses with a wide variety of plants and fungi. One cluster comprises strains isolated from host plants such as *Gunnera*, *Cycas* and *Anthoceros*, the other cluster evolved to form (extracellular)

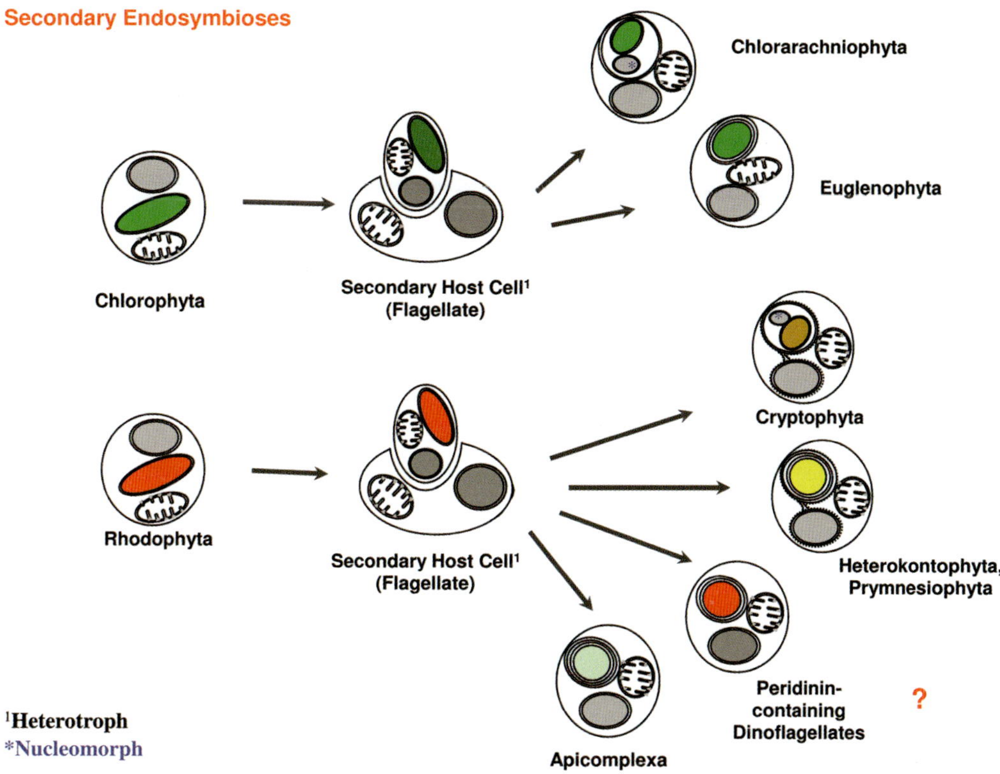

Figure 29.2 Secondary endosymbioses. These were multiple and polyphyletic. Red algae and green algae as well as several heterotrophic protists were involved. The resulting complex plastids are surrounded by three or four membranes. Modified from [40].

symbioses with the water fern *Azolla* [7]. Here, the cyanobiont is invader rather than prey. The filamentous, nitrogen-fixing cyanobacteria differentiate in response to environmental cues into heterocysts, akinetes and hormogonia (especially frequent among symbiotic strains), the latter being the motile form responsible for chemotaxis towards plant-derived substances and ingression into the host tissue involving type IV pili formation [8]. Thus far, discrimination between symbiotic and free-living strains based on either molecular or morphological approaches has not proved possible [7].

Gunnera is the only group of land plants that harbor cyanobacteria intracellularly. The plant seems to need the cyanobiont more than vice versa: all *Gunnera* inspected to date contain *Nostoc* spp., which also thrive when isolated from the host. Infection occurs through peculiar stem glands that develop under nitrogen deprivation [9]. In symbiotic *Nostoc* spp. changes in protein expression are observed, for example, upregulation of nitrogenase [10] and downregulation of photosystem II and of Calvin cycle enzymes [11, 12]. The obvious function of the cyanobiont in the photoautotrophic host is nitrogen fixation, which is supported by the dark, microaerophilic conditions inside the gland cells, whereas its carbon metabolism seems to shift from autotrophy towards heterotrophy [1, 12].

29.2.2
Invertebrate Symbioses

Coral reefs offer an important habitat for intracellular cyanobacteria in sponge and cnidarian hosts [6]. The compartmentalized structure of sponges seems to promote symbiotic associations [13]. Cyanobacteria meet optimal conditions for photosynthesis, rendering some tropical sponges (containing large populations of symbionts) autotrophic. These predominate in the oligotrophic outer reef area. The flattened shapes of sponges have evolved to facilitate light harvesting through the endosymbionts. Host control of endosymbiont number is achieved via nitrogen availability to the symbiont. Diverting of photosynthate (in the form of glycerol) also serves to delimit the number of cyanobacteria. Interestingly, 79 out of 119 cyanobacterial 16S rRNA sequences from sponge endosymbionts proved to be sponge-specific [14]. When vertical transmission is observed, dependence upon the host and thus gene transfer from the cyanobacterium must be envisaged. However, to what extent this occurs is not known for the sponge/symbiont system [13]. From comparison of the 16S rRNA sequences in the largest sponge-specific cluster candidates *Synechococcus spongiarum* (very widespread, found in sponges from all over the world) a combination of vertical (predominantly) and horizontal transmission of the cyanobacterial symbionts is assumed. There are also indications of sponge–cyanobacterium coevolution (also implying vertical transmission). Host-specific, likely more specialized symbionts are retained even under unfavorable conditions whereas generalist symbionts such as *S. spongiarum* might be digested or escape [15].

The fossil record supports at least coexistence of sponges and cyanobacterial mats since the early Cambrian. It can be assumed that there was an ancient association but it should be kept in mind that the degree of integration between symbiont and host and the mode of symbiont transmission is not necessarily an indication of the age of the symbiosis [5]. Concerning the acquisition methods of the endosymbionts, nonspecific enrichment due to the high filtering capacity (24 000 L per day per kilogram of sponge) seems less likely since abundant species such as *Prochlorococcus* are not overrepresented in sponge-derived libraries [13]. Sponge research profits from the medical and biotechnological interest in the vast number of secondary products produced by the symbiotic association although in most cases the spongal or microbial origin of the respective substance is not known [13].

In the other important groups of the coral reef ecosystem, cnidarians and foraminifers, cyanobacterial symbionts are common [6, 16] but eukaryotic phototrophs dominate.

29.2.3 Protist (Algal) Symbioses

Among the numerous cyanobacterial symbioses observed today, some display features as synchronized host–endosymbiont division, inability of the symbionts to be cultivated outside the host cell and pronounced gene loss compared with free-living relatives. All this is indicative of the onset of processes that previously have led to the formation of an organelle. However, it is uncertain if in these cases the organelle (plastid) state ever will be reached or if this lengthy and immensely complicated mutual adaptation will never be finished as a dead end of evolution. Nevertheless, such organisms are very interesting objects to study and we present the two paradigms, *Paulinella chromatophora* and *Rhopalodia gibba*. In the former the cyanobacterial contribution is photoautotrophy, whereas in the latter it is the ability to fix nitrogen.

29.2.3.1 *Paulinella chromatophora*: Result of a Second Independent Primary Endosymbiotic Event?

This photoautotrophic thecamoeba was first grouped among the glaucocystophytes and the sausage-like "chromatophores" were named cyanelles (Figure 29.3). While 18S rRNA-based phylogenetic analysis led to a shift to the cercozoa, the term "cyanelle" (in fact describing an obligate cyanobacterial endosymbiont) is more appropriate here than in the case of the glaucocystophyte plastids. 16S rRNA sequence comparison pointed to a α-cyanobacterial invader/prey from the *Prochlorococcus/Synechococcus* clade. The recently completed genome of the symbiont amounts to 1.02 Mb and comprises 867 protein-coding genes [17]. This genetic information exceeds that of plastids by a factor of four to ten. While the genome

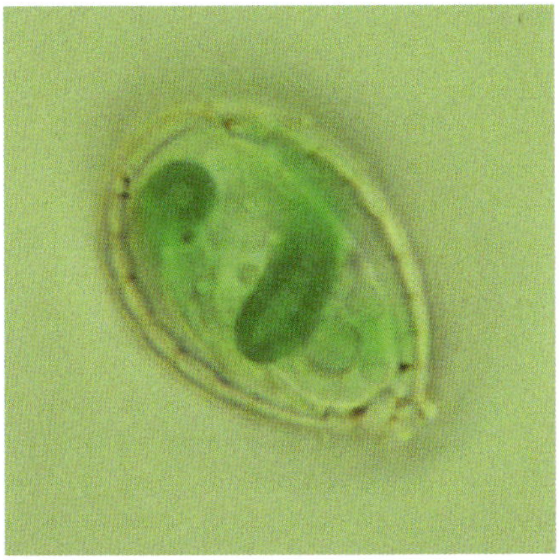

Figure 29.3 *Paulinella chromatophora*. This thecamoeba owes its autotrophic lifestyle to two sausage-like cyanobacterial endosymbionts that are shared between the two daughter cells upon division. Magnification 630×.

appeared to be reduced to one-third as compared with the close relative *Synechococcus* WH5701, it is just 25% smaller than that of some unicellular α-cyanobacteria, for example, *Prochlorococcus* MED4. Does this situation justify calling the endosymbiont a plastid? Not surprisingly, this is a matter of debate [18–20].

Reductive genome evolution typical for bacterial endosymbionts manifests itself here in the lack of genes essential for some pathways of amino acid and cofactor synthesis, of the TCA cycle and so on, obviously with the purpose of rendering the symbiont host-dependent. Keeping in mind a potential compensation at the metabolic level, the complement of transporters in the symbiont genome is unexpectedly low. However, it has been hypothesized that one of the first steps during the onset of an intracellular symbiosis is the implantation of host cell-derived transporters into the endosymbiont plasma membrane [21, 22]. Even a contribution of transporters from a third player, coinfecting chlamydiae, during the establishment of the primary endosymbiosis between protist and cyanobacterium, is discussed [23]. On the other hand, the symbiont is obligatory for the now autotrophic host cell. In this context, it is surprising that the complete set of photosynthetic genes is found. In our opinion, this argues against a plastid nature and the authors were cautious to use the historic denomination "chromatophore" [17] rather than symbiotic cyanobacterium or plastid. Genes for lipopolysaccharide-BS are missing, whereas with the exception of D-Ala-D-Ala ligase (which might have been replaced by an analogous enzyme with deviating sequence) the complete set of genes involved in peptidoglycan biosynthesis is present.

A carboxysomal CO_2-concentrating mechanism (CCM) in the symbiont [24] should result in accumulation of bicarbonate and will thus necessitate osmotic protection [25]. In glaucocystophyte "cyanelles", which in fact are primitive plastids derived from the single primary endosymbiotic event, the genes for peptidoglycan biosynthesis have been completely transferred to the nucleus as well as 50% of the thylakoid protein genes and all Calvin cycle genes except Rubisco large and small subunits [26]. The single "carboxysomes" in glaucocystophyte algae are of pyrenoid size, with rounded or polyhedric contours, and lack a clearly recognizable shell in some species. The multiple, polyhedral and small carboxysomes of *Paulinella* [24] correspond exactly to the CCM-microcompartments of cyanobacteria. A number of pseudogenes indicate that reduction of the A/T-rich genome is an ongoing process while the synteny to *Synechococcus* WH5701 is well preserved in many but not all regions. Interestingly, the closely related marine thecamoeba *Paulinella ovalis* is a heterotroph and predator of cyanobacteria which are digested completely [27].

The minimum age of the endosymbiosis leading to *P. chromatophora* was estimated from the tempo of genome reduction to 60 million years [17]. This likely represents a second independent attempt at establishing a hereditary endosymbiotic association between a protist and a cyanobacterium resulting in photoautotrophy: evidence for a single origin of *P. chromatophora* isolates from Germany and Japan was recently obtained (T. Takayama and K. Ishida, personal communication). In our opinion, classification of the *Paulinella* chromatophore as a plastid is premature. Gene loss is proven, potential gene transfer from endosymbiont to the host cell nucleus must

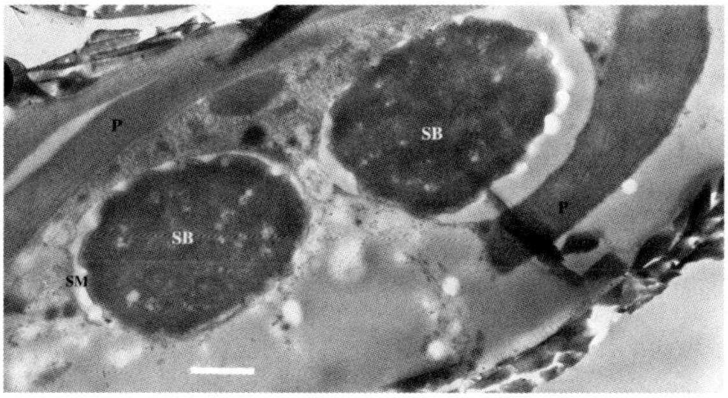

Figure 29.4 *Rhopalodia gibba*. A pinnate diatom containing nonphotosynthetic N_2-fixing endosymbiotic cyanobacteria. SB, sphaeroid body; P, plastid; SM, symbiosomal membrane. Bar: 1 μm.

await the sequencing of the latter genome. However, the critical issue will be the demonstration of a specific protein import apparatus at the chromatophore envelope [19]. Only then, a renaming of the endosymbiont to a (very peculiar, cyanelle-like) plastid is justified.

29.2.3.2 ***Rhopalodia gibba*: Birth of a Novel Organelle (Nitrogenosome)?**

In the second recently investigated case, the "spheroid bodies" of cyanobacterial origin in the diatom *Rhopalodia gibba* (Figure 29.4), the age of the association presumably is lower and the genome reduction less pronounced. Another difference is the autotrophic host cell harboring chlorophyll *a*/*c* plastids and the nitrogen-fixing capacity of the endosymbiont [28, 29]. Therefore the question whether this is plastid or symbiont is not as burning here. Common to *Paulinella* and *Rhopalodia* is the obligate character of the intracellular symbiosis and the vertical transmission of the symbiont, rather the exception than the rule among the mostly facultative cyanobacterial symbioses [5].

The *Rhopalodia* endosymbiont genome size is estimated to be approximately 2.6 Mb, that of the closest free-living relative, *Cyanothece* sp. ATTC 51142, is 5.7 Mb. Thus, genome reduction amounted to more than 50%, and 51 kb of contiguous genomic sequences (the *nif* gene region, paralleled by 63 kb from the corresponding region of *Cyanothece*) and 140 kb of noncontiguous sequences were created from the endosymbiont. All *nif* genes in the spheroid body genome (with the exception of the non-essential *fdxN* and the truncated *nifU*) were found to be intact, in accordance with the assumption that nitrogen fixation constituted the driving force for the onset of the symbiosis [30]. Photosynthetic genes are either missing from the spheroid body genome as *petJ* and *petE* or have been converted into pseudogenes (e.g., *psbC* and *psbD*) which is consistent with the lack of pigmentation and the absence of photosynthetic activity observed [31]. Notable is the higher number of noncoding DNA stretches (>500 bp and AT-rich) in the spheroid body's genome compared to *Cyanothece*. Several transposase genes on the endosymbiont genome were found to be disrupted. This is indicative of a stable and long-term interaction which eventually can be traced back as far as 5–23 million years ago [31]. However, the fact that many pseudogenes still are clearly identifiable and that the major recombination/repair enzymes RecA and RecF seem to be intact, points towards a relatively recent association (Figure 29.4). It remains to be determined whether the cyanobacterial endosymbiont really is contained within a symbiosomal membrane or if the "hyaline space" in Figure 29.4 represents an artefactual enlargement of the periplasmic space. In any event, there is no such membrane in *P. chromatophora* [32] and in primary plastids. Accordingly, the outer envelope membrane of chloroplasts is equivalent to the outer envelope membrane of cyanobacteria, except for the exchange of lipopolysaccharide in the outer leaflet against phosphatidylcholine [33]. This seems to be a characteristic of advanced cyanobacterial symbioses.

A *Cyanothece*-like N_2-fixing endosymbiont was also found in another diatom, *Climacodiun frauenfeldianum* [34]. Here the cyanobiont is not obligatory and its presence or absence is dependent upon water temperature. The photosynthetic apparatus seems to be intact: obviously this association represents an earlier stage than *R. gibba* (Figure 29.5). The micrographs available do not allow a decision about the presence or absence of a symbiosomal membrane.

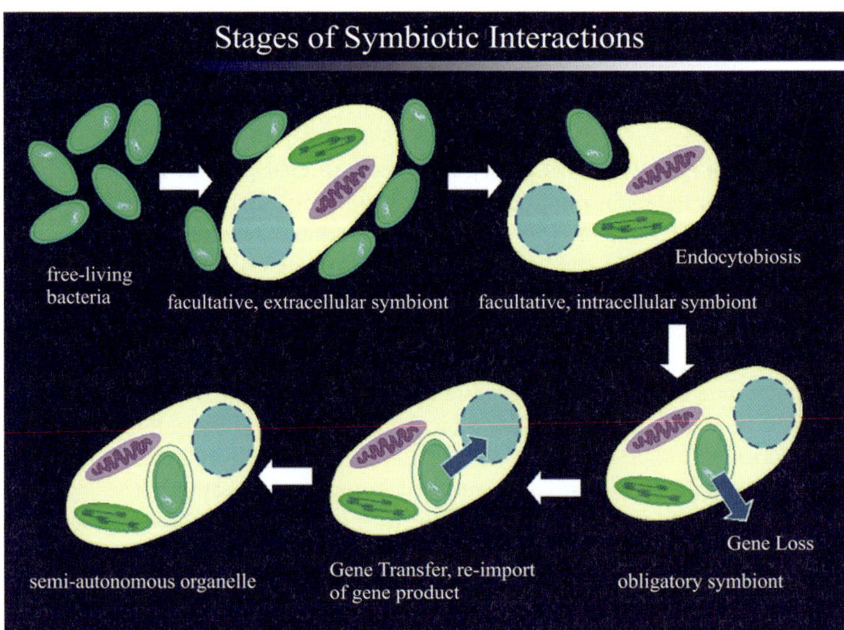

Figure 29.5 Stages of intracellular symbioses. Here a autotrophic host cell is depicted in increasingly tight interactions with a (cyano)bacterial symbiont – a situation resembling that of *Rhopalodia*.

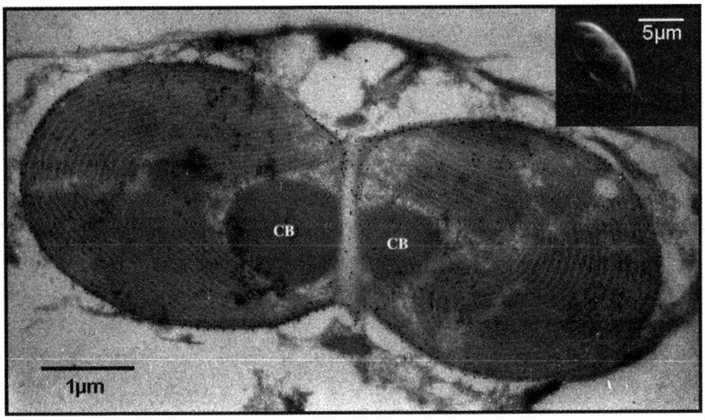

Figure 29.6 *Cyanophora paradoxa*. The living fossil containing peptidoglycan-armored plastids as demonstrated by immuno-electron microscopy. Primary antibodies are directed against peptidoglycan from *Escherichia coli*. Gold particles decorate mainly the envelope and the newly formed septum of the dividing cyanelle. CB, rubisco-containing central body (putative carboxysome).

29.2.3.3 **Glaucocystophyte Cyanelles: The "Cyanobacterial" Plastids**

After the common origin from the "protoplastid", the primordial photosynthetic organelle resulting from the single primary endosymbiotic event, evolution of the glaucocystophyte slowed down, rendering their cyanelles the closest cousins – among plastids – to free-living cyanobacteria [26]. Today they still mimic early stages of plastid evolution and model organisms such as *Cyanophora paradoxa* can be considered as living fossils. The peptidoglycan-armored plastids and the carboxysome-like Rubisco microcompartments [35] – unique features not encountered in any other plastid type – are the extant proof of the endosymbiotic theory (Figure 29.6). Their storage carbohydrate biosynthetic pathways mark the transition from cyanobacterial glycogen to chloroplast starch [36]. The development of a specific protein import apparatus, of utmost importance for organelle evolution, can best be studied in cyanelles. The cyanelle Toc/Tic translocon though primordial nevertheless is homologous to the chloroplast protein import machinery [37] as heterologous *in vitro* import experiments work in both directions (in the case of higher plant precursors a slight modification is necessary).

The key component of the Toc translocon, the Toc75 pore, was recruited among the outer membrane proteins of the cyanobacterial endosymbiont as the starting point of the protein import machinery. Changes in function and directionality of transport were involved [38] but the "substrate specificity" of the primitive receptor/pore, that is, the requirement for terminal phenylalanine, was retained from the prokaryotic Omp85 protein [39]. With the development of the "chlorophytes" a number of receptors were added to the Toc75 pore and it was reduced to a translocation channel. These receptors conferred less stringent and overlapping specificities to the Toc translocon, resulting in loss of the phenylalanine requirement.

29.3 Intracellular Algae

Special emphasis will be given to classical systems such as *Chlorella/Hydra* and *Chlorella/Paramecium* and also some space devoted to the symbiotically versatile dinoflagellate *Symbiodinium*. In general, as in cyanobacteria/eukaryote symbioses, the range expands from facultative to obligate associations, ultimately leading to so-called secondary or complex plastids surrounded by three or four membranes [40], which will be our starting point this time (Figure 29.2). The heterotrophic secondary host cell gained a photosynthetic organelle by enslaving and reducing a phototrophic eukaryote. The phagosome/symbiosome membrane is found throughout the intracellular algal symbioses, ending up as the outermost membrane in complex plastids, in obvious contrast to the primary symbioses mentioned above. The amount of reduction of the primary host cell is variable: In cryptomonads and chlorarachnids, the periplastidic space between the two outermost and the two innermost membranes contains a vestigial nucleus consisting of three mini-chromosomes totaling from 350 to 600 kb [41–43] and 80S ribosomes different from those in the cytosol. In diatoms or euglenids just one or two additional membranes are reminders of the secondary endosymbiotic event.

Gene transfer mainly occurred between the nuclei of the primary and secondary host cells. Protein targeting to the plastid is a lot more complicated. Bipartite presequences lead to a combination of secretory pathway and plastid targeting. In the red line, phenylalanine plays a crucial role for plastid import as the first amino acid of the transit sequence [44]. The flagellate *Hatena arenicola* (*Kathaplepharida*, related to cryptomonads) can be viewed as a green algal endosymbiosis "en route" to becoming a secondary plastid [45]. The *Nephroselmis* symbiont shows an enlarged chloroplast, whereas mitochondria, Golgi and endomembrane system are partially degraded. However, the nucleus exceeds the nucleomorphs mentioned above in size and, likely, DNA content. The lifestyle of *H. arenicola* shows cyclic changes from autotroph to predator as only one of the daughter cells receives the endosymbiont [45].

Still another case is the presence of "kleptoplastids" as remains from algal prey in protists, ciliates or even invertebrates. These perform photosynthesis for their predator for some time (typically a few weeks) and then are degraded and replaced by plastids from new prey. However, in the sea slug *Elysia chlorotica*, the lifespan of the kleptoplastids from the chromophyte *Vaucheria* extends from 9 to 14 months equalling the lifespan of *Elysia*. Gene transfer to the invertebrate genome and – eventually – even protein import into the kleptoplasts that are reduced from four to two surrounding membranes is envisaged: a scenario equivalent to a tertiary endosymbiosis [46]. Another example of "stolen autotrophy" is given by the ciliate *Myrionecta rubra* [47]. Plastids and mitochondria of cryptophycean origin are kept as membrane-delineated complexes (CMCs) including cytoplasm, endoplasmic reticulum and lipid droplets. The plastids are active for about eight weeks (much longer than in the case of most kleptoplasts, with the exception of the *Elysia/Vaucheria* system), divide within the ciliate and perform chlorophyll synthesis. After that time autotrophy is lost unless new prey is taken in. Recently, it became clear that transcriptionally active cryptophyte nuclei are also retained in *M. rubra* (but not included into CMCs) for up to 30 days and are responsible for plastid performance and replication [48]. Karyoklepty thus represents a new strategy for prolonging acquired temporary autotrophy.

29.3.1 Classic Symbioses

The variability of algal symbionts – rhodophytes, chlorophytes, diatoms, cryptomonads, dinoflagellates, and so on – is highest among (marine) porifera and foraminifera. In the latter, the choice of the preferred algal endosymbionts depends on the position of the host in the water column (Maria Holzmann, personal communication).

29.3.1.1 *Symbiodinium* sp., The "Multisymbiont"

This dinoflagellate genus is hosted by many invertebrates from the coral reef ecosystems, such as corals, sponges, jellyfish, sea anemones and molluscs [6]. Coral bleaching caused by global warming is considered a major issue here, although the reasons behind it are not clearly understood [6, 13]. Recently, heat stress-induced photobleaching of cultured *Symbiodinium* was ascribed to decreased *de novo* synthesis of the major chlorophyll *a*/*c*-peridinin antenna protein [49]. The dinoflagellates are contained in symbiosomes and offer a multitude of photosynthetic export products (Table 29.1) to the corals: glycerol, glucose, amino acids, lipids and C_4 acids [6].

Furthermore, *Symbiodinium* endosymbionts have been reported in ciliates and large soritid foraminifers. The latter show high diversity (61 symbiont types in three soritid genera) and, at the same time, pronounced host–symbiont specificity [50]. Recently, evidence for maternal transmission of dinoflagellate endosymbionts was reported for the worm *Waminoa brickeri* [51].

Table 29.1 Characteristics of invertebrate–algal endosymbioses.

Endosymbiont	Host	Localization	Algal export products	Host factors	CCM in alga
Dinoflagellate	Coral	Symbiosome	Glycerol, glucose, amino acids, lipids, C_4 acids	+	yes
	Sea anemone	Symbiosome	Glycerol	+	yes
Chlorella	Hydra	Symbiosome	Maltose	pH	yes

CCM, CO_2-concentrating mechanism. Modified from [5].

29.3.1.2 The *Chlorella/Hydra* Endosymbiosis

Hydras are simple cosmopolitan freshwater invertebrates usually found in unpolluted freshwater on leaves of submerged plants. They belong to the phylum *Cnidaria*, class *Hydrozoa*, order *Hydroida*, suborder *Hydrina*, family *Hydridae*. Hydras have a simple, cylindrical body with a budding region, an adhesive foot at one end and a mouth surrounded by four to eight tentacles at the other [52]. The body is composed of two cellular layers, ectoderm and gastroderm, separated by mesoglea, a layer of extracellular material (glycans, fibrillar proteins, etc.). Hydras show a tremendous rate of regeneration and thus can serve as a classical model organism for our understanding of such fundamental processes and the biology of higher animals in general [53, 54]. Green hydra (*Hydra viridissima* Pallas, 1766) is an endosymbiotic species that contains photoautotrophic endosymbiotic algae (Figure 29.6), whereas the heterotrophic brown hydra (*Hydra oligactis* Pallas, 1766) does not.

Green hydra is a typical example of algal endosymbiosis (Figure 29.7). In its gastrodermal myoepithelial cells it harbors up to 20 individuals of the unicellular green alga *Chlorella* [52], occupying about 10% of the cellular volume. A single hydra could contain up to 200 000 endosymbiotic algae [55]. Each alga is enclosed in one individual vacuolar membrane called a symbiosome (Figure 29.8). Inside the oblong gastrodermal myoepithelial cells of green hydra algae are regularly placed to form columns, one alga above the other, in the basal part of the cell, towards mesoglea (Figure 29.8). It has been noted that some algae are placed very close to the host cell nucleus [55]. A constant number of algae in the cell is maintained by hydra in several ways: inhibition of algal mitosis, expulsion of algae or digestion of excess algae [56]. During starvation in the dark, hydra can digest a number of algae, which now become the food source, while the best-adapted algae survive. Symbiotic hydras in the dark grow slower than the aposymbiotic ones because they feed the algae in order to maintain the reduced population of endosymbionts. The maintenance of

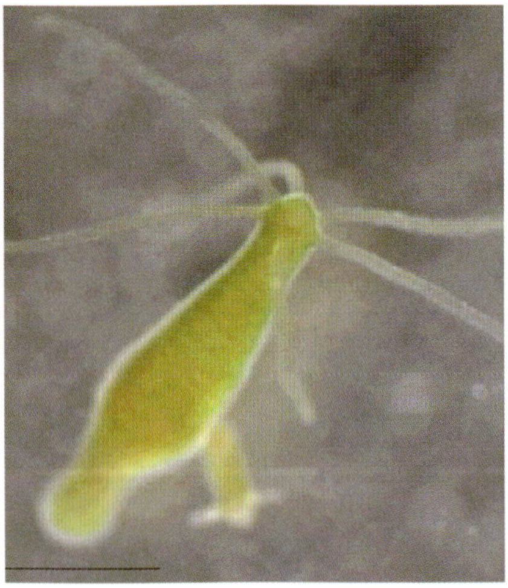

Figure 29.7 *Hydra viridissima* Pallas, 1760. Bar: 2 mm.

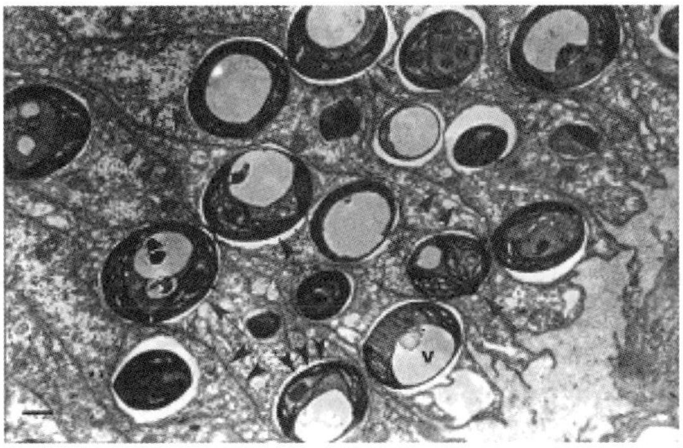

Figure 29.8 Transmission electron microscope image of endosymbiotic algae (arrow) in the gastrodermal myoepithelial cells of green hydra (cell membrane: double arrows). Each alga is surrounded by a symbiosomal membrane, creating the perialgal space (triple arrows). v, vacuole. Bar 1 μm.

symbiosis depends on metabolite flow in both directions. In addition to being protected inside the hydra cells, algae also use the products of host metabolism: CO_2, phosphates, nitrates and sulfates.

Growth and division in the host depend on the inorganic nutrients [57] that the algae receive from hydra. The algae assimilate these compounds, convert them into energy-rich organic molecules and return them back to the host, which represents efficient recycling within this endosymbiotic system. The rate of photosynthesis in symbiotic *Chlorella* is much higher than in their free-living relatives. About 10% of the CO_2 fixed by the algae is incorporated into the protoplasm of hydra in the form of glycogen. Upon nitrogen starvation in the light, algae stimulate the growth of green hydra. Therefore symbiotic hydras can better survive longer periods of starvation than aposymbiotic species [55].

In order to form successful endosymbiosis, hydra and alga have to preadapt. Green hydra possesses a great adaptive potential. As the result of preadaptation of both symbiotic partners, certain changes have been established. During regular phagocytosis green hydra can distinguish the endosymbionts. There are at least two modes of phagocytosis, one for the ingestion of algae and one for all other particles. Phagocytosis of algae into the host cell takes place in the gastroderm [58], which is followed by their placement into the symbiotic vacuoles where they are protected from the digestive enzymes of hydra [59]. After the recognition of both partners, which likely is achieved by the interaction between the phagocytic surface and the algal surface, the algae induce the formation of endocytic microvilli at the ends of gastrodermal myoepithelial cells directed towards the gastric cavity. For hydra the amount and type of the photosynthate released by the alga is important. Releasing of maltose is a characteristic of endosymbiotic *Chlorella* [60]. It is the required quantity of maltose released by algae inside the cells of green hydra that prevents the fusion of symbiosomes with lysosomes [59]. The absence of acid phosphatase in the symbiosomes prevents fusion, that is, prevents the vacuoles from becoming phagolysosomes. Further more, the perialgal space helps in establishing endosymbiosis, maintaining a low pH value that blocks the release of lysosomal enzymes into the phagosome.

Symbiotic algae are acid-tolerant. When the pH drops below 4, the algae start to release a considerable and constant amount of maltose [61]. The fitness of different *Chlorella* strains for symbiosis correlates with the tolerance to low pH. A few hours after infection, the pH of perialgal vacuoles reaches 3.5–4 [62]. Algae are mostly placed into the "quiet" part of the cell (adjacent to mesoglea) where no intensive intracellular decomposition occurs. Nonviable algae are fused to the digestive vacuoles of gastrodermal cells of green hydra. Also, successfully established endosymbiosis with one algal species disqualifies an endosymbiosis with another species [63].

Recently, new ultrastructural features of green hydra symbiosis in unfavorable conditions have been described, including extensive widening of perialgal spaces, degradation and loss of symbiosomes, and symbiosome fusion resulting in up to three algae per vacuole [64] (Figure 29.9). These changes are considered to be defensive and protective mechanisms in green hydra symbiosis and may lead to successful maintenance of the symbiosis.

Algae are transferred vertically from the parent to the new polyps mostly by asexual reproduction via budding [52]. It is known that symbiotic algae exert a strong influence on sexual reproduction of green hydra [55]. There is a chlorophyte-derived gene for ascorbate peroxidase that is found in the genome of hydra and is expressed only during oogenesis. This is the first known gene of this kind in the animal kingdom, and is likely the result of lateral gene transfer from some symbiont, not necessarily from *Chlorella* [65]. All the individual endosymbiotic algae in eggs, zygotes and embryos of green hydra localize to the symbiosomes, only occasionally are two or three algae found inside the same vacuole.

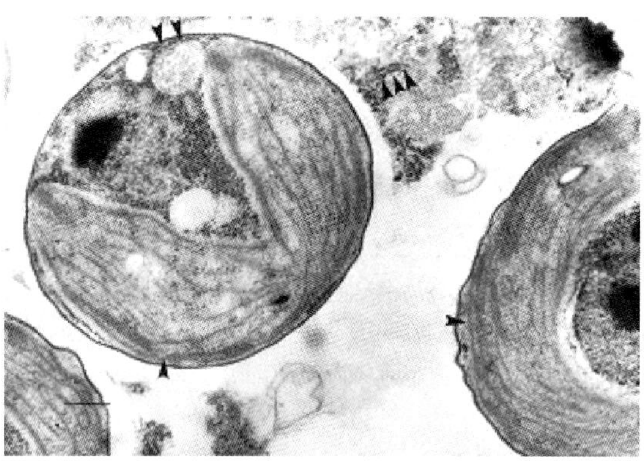

Figure 29.9 Transmission electron microscope image of endosymbiotic algae from green hydra (arrowhead). Algae entering into the perialgal spaces of other algae; joining of their perialgal spaces. Swollen chloroplasts and mitochondria (2 arrowheads). Rough endoplasmic reticulum torn apart (three arrowheads). Bar 0.5 μm.

Until now it has been considered that endosymbiotic algae from green hydra belong to the genus *Chlorella* (*Chlorellales, Trebouxiophyceae*). *Chlorella* is a complex genus of polyphyletic origin [62]. Certain species belong to the class *Chlorophyceae* [66]. It has been known for a long time that algae isolated from hydra cannot be maintained in stable persistent culture [62, 65, 66], while the free-living algae *Chlorella vulgaris* and other *Chlorella* species can usually be grown successfully. The microenvironment seems to have an impact on this feature [55]: “zoochlorellae” from green hydra represent distinct separate species of algae, different from free-living *Chlorella*, based on molecular analyses [62, 65, 66].

29.3.2 Endosymbiotic Algae from *Paramecium*

The ciliate *Paramecium bursaria* Ehrenberg (*Himenostomatida, Peniculina*) is a simple ubiquitous freshwater protoctist. It moves through the rotatory action of several thousand cilia that cover the external surface of the body. Because water is constantly flowing into the cell through osmosis, *Paramecium* has two contractile vacuoles that eliminate excess water and maintain the osmotic equilibrium. *Paramecium* can ingest up to 5000 bacteria per day and therefore plays an important role in cleaning the groundwater. Food is engulfed into a food vacuole by phagocytosis. Food vacuoles merge with the lysosomes and digest the food. Since bacteria tend to create a slightly acidic environment, paramecia are attracted towards acidic regions in the water.

P. bursaria is a well-known object of research in the field of endosymbiosis. In general, it harbors several hundreds of endosymbiotic coccoid unicellular green algae in its cytoplasm and retains them over many generations. The initial establishment of successful symbiosis and its long-term maintenance require that the algae resist digestion by the *Paramecium*. Algae suitable for symbiosis are recognized by a lectin/glycan interaction mechanism whereby specific algal surface groups interact with host membrane-bound proteins [67]. Each alga is enclosed in a perialgal vacuole (symbiosome), which protects it from lysosomal fusion [68]. This membrane structure is different from that of a digestive vacuole [69]. Perialgal vacuoles of green *P. bursaria* are devoid of acid phosphatase activity, unlike digestive vacuoles. Algae-free *P. bursaria* can be obtained by growing them under dark conditions in which algae are digested by lysosomal fusion [68]. A number of *Chlorella*-like algae from *P. bursaria* can be cultured [70].

Chlorella species suitable for symbiosis can be distinguished by the presence of glucosamine in their walls, whereas the walls of nonsymbiotic species contain glucose and mannose [71]. However, other authors indicate that the infectivity of *Chlorella* species for the ciliate *P. bursaria* is not based on certain sugar residues of their cell wall, but rather on their ability to accumulate beneath the host cell membrane after escaping from the host digestive vacuole in the early infection process. Other strains could also escape from the digestive vacuoles but fail to localize beneath the host cell membrane and thus become digested in the end.

Symbiotic algae usually localize at a distance of about 10 μm from each other. The establishment of endosymbiosis seems to be both algal- and host strain-specific [72] and physiological adaptation in symbiotic *Chlorella* is also important [73]. Some authors refer that the usual symbiotic partner in *Paramecium* as *Chlorella vulgaris* [74], others feel that types of endosymbionts depend on the host line(s) [75]. Even the endosymbiotic association of *Paramecium* with two different coccoid algae is reported: one of them is *Chlorella vulgaris,* which cannot be maintained in permanent culture, and the other is *Chlorella*-like *Choricystis minor* [76]. In summary, the analysis of 18S rRNA and ITS2 gene sequences indicate that these endosymbiotic algae are of polyphyletic origin and belong to *C. vulgaris,* "American" and "European" groups of *Chlorella,* and *Coccomyxa* [77].

The algal symbiont has an impact on the growth of *Paramecium* [78]. At low pH, the algae produce a lot of maltose [79]. *Paramecium* receives photosynthate in return for providing a habitat for the algal symbionts. Endosymbiotic alga can also be supplied with nitrogen by the host [80]. This is regarded to be a mutualistic relationship [81]. Since the algae provide sugars and oxygen, *Paramecium* becomes more independent of external food sources [79]. Algae-bearing *Paramecium* grow better than algae-free *Paramecium* and symbiotic *Chlorella* enhances thermotolerance in *P. bursaria* [82]. *P. bursaria* is able to synchronize cell division by regulating the cell cycle of endosymbiotic *Chlorella* and keeping the algal population constant until the host cell enters cytokinesis [83].

P. bursaria demonstrates microtubule-based rotational cytoplasmic streaming, in which some cytoplasmic granules and endosymbionts flow in a constant direction. Cytoplasmic streaming is arrested in dividing green paramecia and the endosymbiotic algae proliferate only during the arrest of cytoplasmic streaming [84]. Equal distribution of endosymbionts to the two daughter cells is also essential to maintain endosymbiosis in *P. bursaria.* However, it is not known how this process is controlled during cytokinesis of *P. bursaria.*

Chlorella can also offer benefits other than photosynthate to ciliate host cells: UV-protecting mycosporine-like amino acids (MLAAs) have been identified as product of the endosymbiont. The same ciliate species, *Askenasia chlorelligera,* may enlist different "zoochlorellae" according to their potential for biosynthesis of MLAAs and the clearness or turbidity of the respective lakes [73]. Similar UV-protection is exerted by cyanobionts in sponges [13].

29.3.3 Algae/Plant Symbioses

One of the strangest associations between an alga and another organism is perhaps that between the immature "precursor" forms of *Coccomyxa*-related green algae and *Ginkgo biloba* [85]. *Ginkgo biloba* is regarded as the oldest living tree species on earth dating back to at least the early Jurassic period, more than 170 million years ago. *Coccomyxa* is a terrestrial coccal alga known as photobiont in some lichens [86]. Within *Ginkgo* host cells, the alga resides in an immature "precursor" state: nucleus and mitochondria are not observed, and diffuse electron-dense areas resemble thylakoid-like membranous structures of a nonfunctional chloroplast [87]. Mature algae with eukaryotic traits and a normal functional chloroplast were observed almost exclusively in host cells undergoing necrosis and could be easily cultured in inorganic, ammonium-containing medium [85]. Algal DNA has been found in reproductive tissues such as microsporophylls, pollen grains, ovules and prothallus, as well as in complete zygotic embryos, while for vegetative tissues, the presence of the endosymbiont could be shown only in cotyledons and young leaves [87].

The symbiotic advantage of two photoautotrophic species with the algal precursors apparently devoid of functional plastids seems unclear. However, since TEM identified numerous storage bodies, probably of lipid nature, in the cytoplasm of the endosymbionts as well as in *G. biloba* host cells, the authors suggest a possible involvement of the endosymbiont in metabolic pathways of its host.

Another explanation comes from paleobotanists studying *Sublagenicula nuda* lycophyte megaspores from the upper Visean of central France, which frequently host a colonial green alga named *Lageniastrum macrosporae* as an endophyte [88]. They considered the adaptive advantages for the algae occupying the interior of megaspores, including the potential effectiveness of the spores in protection during periods of desiccation and against plankton-feeding animals. In addition, the use of the spores as potential vectors for dispersal from one isolated body of water to another by spore-feeding animals is envisaged. The onset of this kind of endosymbiosis could have been facilitated by the unique fertilization process of *Ginkgo* [87]. As a lichen alga, *Coccomyxa* could land upon a pollination droplet, which covers the micropyle of young *Ginkgo* ovules, get inside the pollen chamber together with the pollen, and survive there until fertilization occurs. Since *Coccomyxa* is about 20 times smaller than a *Ginkgo* spermatozoid, the alga could then have entered the egg cell along with the sperm. Once in the egg cell, it had to escape digestion like any other intracellular symbiont, adjust its reproduction rate, and become an inherent part of the host [87].

Acknowledgments

The authors would like to acknowledge the kind and expert help of Prof. dr. sc. Nikola Ljubešić from the Rudjer Boskovic Institute, Zagreb, Croatia, in publishing this work. We are indebted to Prof. Uwe G. Maier and Dr. Christine Voß (Marburg) for Figures 29.1, 29.2, 29.4 and 29.5. We thank Prof. Michael Melkonian and Dr. Eva Nowack (Cologne) for Figure 29.3. Research on *Cyanophora* was made possible through support from the Austrian Research Fund (FWF, grants P19683, P15438 and P14573 to WL).

References

1 Osborne, B. (2007) *New Phytol*, **175**, 602–605.

2 Rodríguez-Ezpeleta, N., Brinkmann, H., Burey, S.C., Roure, B. *et al.* (2005) *Curr Biol*, **15**, 1325–1330.

3 Deusch, O., Landan, G., Roettger, M., Gruenheit, N. *et al.* (2008) *Mol Biol Evol*, **25**, 748–761.

4 Reyes-Prieto, A., Weber, A.P.M. and Bhattacharya, D. (2007) *Annu Rev Genet*, **41**, k147–k168.

5 Usher, K.M., Bergman, B. and Raven, J.A. (2007) *Annu Rev Ecol Evol Syst*, **38**, 255–273.

6 Yellowlees, D., Rees, T.A.V. and Leggat, W. (2008) *Plant Cell Environ*, **31**, 679–694.

7 Papaefthimiou, D., Hrouzek, P., Mugnai, M.A., Lukesova, A. *et al.* (2008) *Int J Syst Evol Microbiol*, **58**, 553–564.

8 Duggan, P.S., Gottardello, P. and Adams, D.G. (2007) *J Bacteriol*, **189**, 4547–4551.

9 Chiu, W.L., Peters, G.A., Levieille, G., Still, P.C. *et al.* (2005) *Plant Physiol*, **139**, 224–230.

10 Wang, C.M., Ekman, M. and Bergman, B. (2004) *Mol Plant Microbe Interact*, **17**, 436–443.

11 Black, K. and Osborne, B. (2004) *New Phytol*, **162**, 125–132.

12 Ekman, M., Tollback, P., Klint, J. and Bergman, B. (2006) *Mol Plant Microbe Interact*, **19**, 1251–1261.

13 Taylor, M.W., Radax, R., Steger, D. and Wagner, M. (2007) *Microbiol Mol Biol Rev*, **71**, 295–347.

14 Hentschel, U., Hopke, J., Horn, M., Friedrich, A.B. *et al.* (2002) *Appl Environ Microbiol*, **68**, 4431–4440.

15 Thacker, R.W. (2005) *Integr Compar Biol*, **45**, 369–376.

16 Lee, J.J. (2006) *Symbiosis*, **42**, 63–76.

17 Nowack, E.C.M., Melkonian, M. and Gloeckner, G. (2008) *Curr Biol*, **18**, 410–418.

18 Yoon, H.S., Reyes-Prieto, A., Melkonian, M. and Bhattacharya, D. (2006) *Curr Biol*, **16**, R670–R672.

19 Theissen, U. and Martin, W.F. (2006) *Curr Biol*, **16**, R1016–R1017.

20 Bodyl, A., Mackiewicz, P. and Stiller, J.W. (2007) *Trends Microbiol*, **15**, 295–296.

21 Weber, A.P.M., Linka, M. and Bhattacharya, D. (2006) *Eukaryot Cell*, **5**, 609–612.

22 Deschamps, P., Colleoni, C., Nakamura, Y., Suzuki, E. *et al.* (2008) *Mol Biol Evol*, **25**, 536–548.

23 Huang, J. and Gogarten, J.P. (2007) *Genome Biol*, **8**, R99.

24 Marin, B., Nowack, E.C.M., Gloeckner, G. and Melkonian, M. (2007) *BMC Evol Biol*, **7**, 85.

25 Raven, J.A. (2003) *Eur J Phycol*, **38**, 47–53.

26 Löffelhardt, W. and Bohnert, H.J. (2001) in *Symbiosis* (ed. J. Seckbach), Kluwer Academic Publishers, Dordrecht, pp. 111–130.

27 Deonarine, S.N., Gobler, C.J., Lonsdale, D.L. and Caron, D.A. (2006) *Aquat Microb Ecol*, **44**, 181–195.

28 Floener, L. and Bothe, H. (1983) in *Endocytobiology II* (eds W. Schwemmler and H.E.A. Schenk), Walter de Gruyter & Co., Berlin, pp. 541–552.

29 Prechtl, J., Kneip, C., Lockhart, P.J., Wenderoth, K. *et al.* (2004) *Mol Biol Evol*, **21**, 1477–1481.

30 Kneip, C., Lockhart, P.J., Voß, C. and Maier, U.G. (2007) *BMC Evol Biol*, **7**, 55.

31 Kneip, C., Voß, C., Lockhart, P.J. and Maier, U.G. (2008) *BMC Evol Biol*, **8**, 30.

32 Kies, L. (1992) in *Algae and Symbioses* (ed. W. Reisser), Biopress, Bristol, pp. 353–377.

33 Block, M.A., Douce, R., Joyard, J. and Rolland, N. (2007) *Photosynth Res*, **92**, 225–244.

34 Carpenter, E. and Janson, S. (2000) *J Phycol*, **36**, 540–544.

35 Fathinejad, S., Steiner, J.M., Reipert, S., Marchetti, M. *et al.* (2008) *Physiol Plant*, **133**, 27–32.

36 Plancke, C., Colleoni, C., Deschamps, P., Dauvillée, D. *et al.* (2008) *Eukaryot Cell*, **7**, 247–257.

37 Steiner, J.M., Yusa, F., Pompe, J.A. and Löffelhardt, W. (2005) *Plant J*, **44**, 646–652.

38 Steiner, J.M. and Löffelhardt, W. (2005) *Mol Membr Biol*, **22**, 123–132.

39 Wunder, T., Martin, R., Löffelhardt, W., Schleiff, E. *et al.* (2007) *BMC Evol Biol*, **7**, 236.

40 Stoebe, B. and Maier, U.G. (2002) *Protoplasma*, **219**, 123–130.

41 Douglas, S., Zauner, S., Fraunholz, M., Beaton, M. *et al.* (2001) *Nature*, **410**, 1091–1096.

42 Gilson, P.R., Su, V., Slamovits, C.H., Reith, M.E. *et al.* (2006) *Proc Natl Acad Sci USA*, **103**, 9566–9571.

43 Archibald, J.M. (2007) *Bioessays*, **29**, 392–402.

44 Gruber, A., Vugrinec, S., Hempel, F., Gould, S.B. *et al.* (2007) *Plant Mol Biol*, **64**, 519–530.

45 Okamoto, N. and Inouye, I. (2006) *Protist*, **157**, 401–419.

46 Rumpho, M.E., Dastoor, F.P., Manhart, J.R. and Lee, J. (2006) in *The Structure and Function of Plastids* (eds R.R. Wise and J.K. Hoober), Springer, Dordrecht, the Netherlands, pp. 452–473.

47 Johnson, M.D., Tengs, D., Oldach, D. and Stoecker, D.K. (2006) *J Phycol*, **42**, 1235–1246.

48 Johnson, M.D., Oldach, D., Delwiche, C.F. and Stoecker, D.K. (2007) *Nature*, **445**, 426–428.

49 Takahashi, S., Whitney, S., Itoh, S., Maruyama, T. *et al.* (2008) *Proc Natl Acad Sci USA*, **105**, 4203–4208.

50 Pochon, X., Garcia-Cuetos, L., Baker, A.C., Castella, E. *et al.* (2007) *Coral Reefs*, **26**, 867–882.

51 Barneah, O., Brickner, I., Hooge, M., Weis, V.M. *et al.* (2007) *Invertebr Biol*, **126**, 113–119.

52 Holstein, T. and Emschermann, P. (1995) *Cnidaria: Hydrozoa, Kamptozoa*, Gustav Fischer Verlag, Stuttgart.

53 Siebert, S., Thomsen, S., Reimer, M.M. and Bosch, T.C.G. (2005) *Mech Development (MOD)*, **122**, 998–1007.

54 Meinhardt, H. (2002) *BioEssays*, **23**, 185–191.

55 Habetha, M., Anton-Erksleben, F., Neumann, K. and Bosch, T.C.G. (2003) *Zoology*, **106**, 1–8.

56 Baghdasarian, G. and Muscatine, L. (2000) *Biol Bull*, **199**, 278–286.

57 McAuley, P.J., Dorling, M. and Hodge, H. (1996) *J Phycol*, **32**, 839–846.

58 Meints, R.H. and Pardy, R.L. (1980) *J Cell Sci*, **43**, 239–251.

59 Hohman, T.C., McNeil, P.L. and Muscatine, L. (1982) *J Cell Biol*, **94**, 56–63.

60 Cernichiari, E., Muscatine, L. and Smith, D.C. (1969) *Proc R Soc Lond B*, **173**, 557–576.

61 Kessler, E., Kauer, G. and Rahat, M. (1991) *Botanica Acta*, **104**, 58–63.

62 Huss, V.A.R., Holweg, C., Seidel, B., Reich, V., Rahat, M. and Kessler, E. (1993/1994) *Endocyt Cell Res*, **10**, 35–46.

63 Rahat, M. (1991) *Oikos*, **62**, 381–388.

64 Kovačević, G., Kalafatić, M. and Ljubešić, N. (2007) *Folia Biol (Krakow)*, **55**, 77–79.

65 Habetha, M. and Bosch, T.C.G. (2005) *J Exp Biol*, **208**, 2157–2165.

66 Friedl, T. (1997) Evolution of the green algae, in *Origins of Algae and their Plastids* (ed. D. Bhattacharya), Springer-Verlag, Berlin.

67 Reisser, W. (1990) Participation of algal cell wall surface structures in the formation of the host-symbiont-interface of endocytobiotic systems, in *Experimental Phycology I* (eds W. Wiessner, D.G. Robinson and R.C. Starr), Springer Verlag, Berlin, Heidelberg, New York.

68 Gu, F.K., Chen, L., Ni, B. and Zhang, X. (2002) *Eur J Protistol*, **38**, 267–278.

69 Reisser, W. and Niess, D. (1983) Photobehavior of ciliate-algae associations: an indicator for symbiosis formation?, in *Endocytobiology II. Intracellular Space as Oligogenetic Ecosystem* (eds H.E.A. Schenk and W. Schwemmler), W de Gruyter, Berlin, New York.

70 Weis, D.S. (1978) *J Protozool*, **25**, 366–370.

71 Takeda, H., Sekiguchi, T., Nunokawa, S. and Usuki, I. (1998) *Eur J Protistol*, **34**, 133–137.

72 Kodama, Y. and Fujishima, M. (2007) *Protoplasma*, **231**, 56–63.

73 Summerer, M., Sonntag, B. and Sommaruga, R. (2008) *J Phycol*, **44**, 77–84.

74 Nakahara, M., Tsobuta, H., Handa, S., Watanabe, S. and Deguchi, H. (2004a) *Hikobia*, **14**, 129–142.

75 Hoshina, R., Kato, Y., Akmako, S. and Imamura, N. (2005) *Plant Biol*, **7**, 526–532.

76 Nakahara, M., Handa, S., Watanabe, S. and Deguchi, H. (2004b) *Symbiosis*, **36**, 127–151.

77 Hoshina, R. and Imamura, N. (2008) *Protist*, **159**, 53–63.

78 Kadono, T., Shiota, K., Tanaka, M., Kawano, T., Kosaka, T. and Hosoya, H. (2004a) *Endocytobiosis Cell Res*, **15**, 63–70.

79 Dorling, M., McAuley, P.J. and Hodge, H. (1996) *Eur J Phycol*, **32**, 19–24.

80 Reisser, W. (1976) *Arch Microbiol*, **107**, 357–360.

81 Karakashian, M.W. (1975) *Symp Soc Exp Biol*, **29**, 145–173.

82 Iwatsuki, K., Nishidoi, M. and Suehiro, K. (1998) *Comp Biochem Physiol A*, **121**, 405–409.

83 Kadono, T., Kawano, T., Hosoya, H. and Kosaka, T. (2004b) *Protoplasma*, **223**, 133–141.

84 Takahashi, T., Shirai, Y., Kosaka, T. and Hosoya, H. (2007) *PLoS ONE*, **2**, e1352.

85 Trémouillaux-Guiller, J., Rohr, T., Rohr, R. and Huss, V.A.R. (2002) *Am J Bot*, **89**, 727–733.

86 Lohtander, K., Oksanen, I. and Rikkinen, J. (2003) *Lichenologist*, **35**, 325–339.

87 Trémouillaux-Guiller, J. and Huss, V.A.R. (2007) *Planta*, **226**, 553–557.

88 Krings, M., Grewing, A., Taylor, T.N., Kerp, H. *et al.* (2005) *Geobios*, **38**, 451–465.

30
Insect Symbionts

Heike Feldhaar and Roy Gross

30.1
Introduction

Insects are among the most successful animal groups on Earth, with regard to both their biomass and their biodiversity. Symbioses with intracellular bacteria are now considered to have been a major driving force in the evolution of this group by enabling insects – as well as numerous other invertebrate hosts – to make use of their metabolic capacities. From vertebrates, in contrast, such symbioses are unknown.

In general, bacterial symbionts can be categorized either by evolutionary age and degree of codependence between host and symbiont or by their effect on the host (i.e., mutualist, commensal or pathogen). The latter characterization may be problematic, however, since the effect on the host may vary during host ontogeny and may be hard to evaluate in the first place. Endosymbionts that are strictly vertically transmitted via the cytoplasm of the egg from one host generation to the next will spread in the population only by two means. Either they contribute directly to host fitness, or they manipulate host reproduction in a way that leads to an increase of number or fitness of infected females in relation to males or uninfected females in the population. Whereas the former are generally viewed as mutualistic associations between host and endosymbiont the latter can mostly be considered as parasitic.

Associations between insect hosts and "primary" endosymbionts are often ancient, with an estimated age of 30 to 250 million years [1], predating the origin of dominant modern groups such as mammals or angiosperms. Following presumably a single infection event, they are transmitted exclusively vertically via the germline, that is, directly from mother to offspring, leading to congruent host and symbiont phylogenies. The majority of primary endosymbionts is considered to be mutualistic since they are usually required for normal host development, often by contributing to the host's nutrition. Associations between primary endosymbionts and their hosts are thus generally obligate for both partners. In addition to their presence in reproductive tissues, primary endosymbionts reside in specialized host cells, the bacteriocytes (Figure 30.1 and 30.2). An estimated 15% of insects harbor this type of endosymbiont [1, 2].

Intracellular Niches of Microbes. Edited by Ulrich E. Schaible and Albert Haas

ISBN: 978-3-527-32207-7

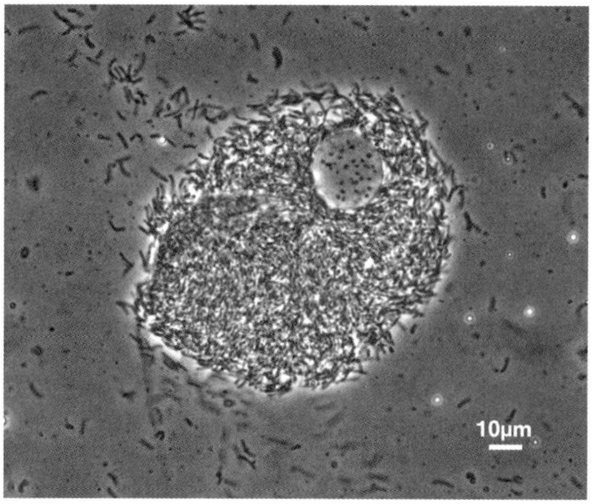

Figure 30.1 An isolated bacteriocyte of the carpenter ant *Camponotus floridanus* entirely filled up with the endosymbiotic bacterium *Blochmannia floridanus*. The bacteriocyte is surrounded by bacteria that have been released from bacteriocytes damaged during the isolation process. Courtesy of Sascha Stoll.

In contrast, "secondary" endosymbionts often have a shorter history of coevolution with a single host and are generally facultative symbionts from the host's perspective. Secondary endosymbionts do not necessarily reside in specialized host tissues but may occur extracellularly in the hemocoel or in other body tissues such as fat

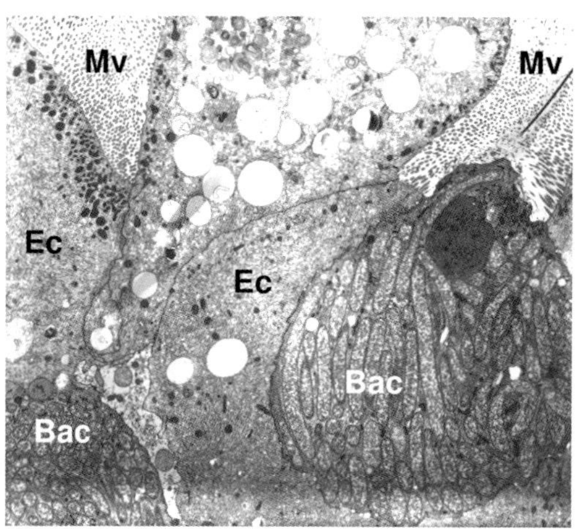

Figure 30.2 Electron micrograph of bacteriocytes (Bac) of *Camponotus floridanus* embedded in midgut tissue comprising enterocytes (Ec) and microvilli protruding from the midgut lumen (Mv). Magnification ~2000-fold. Courtesy of Christina Sauer.

bodies, muscle, nervous tissue or gut. They are primarily transmitted vertically but still have the capacity to invade naïve host cells or tissues on their own account and are able to colonize new hosts by routes other than directly from parent to offspring via horizontal transmission [3]. Secondary enodsymbionts may be transmitted by endoparasites that parasitize several different hosts [4] or among larvae of parasitoids of different species sharing the same host egg [5]. Therefore, related hosts may harbor distantly related strains of an endosymbiont or even different strains within a single host, whereas the same strain may be found in distantly related host species, resulting in incongruencies in the phylogenies of host and endosymbiont [3].

It is thought that some secondary endosymbionts may be on the way to developing into primary endosymbionts. Such secondary endosymbionts may substitute primary endosymbionts whose genomes have already degraded to an extent where some functions in nutritional upgrading for the host can no longer be fulfilled (see below) [6].

Whereas primary endosymbionts have adapted to the host and vice versa, stable associations between secondary endosymbionts and hosts can be formed readily in some cases, implying that no prior host adaptation (e.g., the presence of bacteriocytes) is required for chronic infections – a feature that secondary endosymbionts share with pathogenic bacteria. Relationships of secondary endosymbionts can range from beneficial to detrimental for the host. The most well-known representatives of this group are bacteria of the genera *Wolbachia*, *Spiroplasma*, *Cardinium* and *Rickettsia* that strongly influence host reproduction [7–11]. *Wolbachia* in particular may have been of paramount importance for insect evolution since it is extremely widespread, with an estimated 70% of all insect species being infested with this bacterium and may even have been the cause of speciation events [10, 12].

Here we compare host–symbiont interactions of primary and secondary endosymbiotic bacteria with an emphasis on those whose genomes are known. They comprise roughly two major functional groups: (i) bacteria that contribute to the host needs by providing nutrients that are not accessible or limiting to the host. Such bacteria are considered to be mutualists and are mostly primary endosymbionts and in some cases secondary endosymbionts. (ii) bacteria such as *Wolbachia* that manipulate host reproduction in order to enhance their own spread and are generally considered as parasitic bacteria. The latter group is usually made up of secondary endosymbionts. All these bacteria (i) have a reduced genome size due to a mostly or exclusively intracellular lifestyle, (ii) maintain the association with the host using similar molecular mechanisms and (iii) may not elicit host immune responses so that a chronic infection of the host is facilitated. Ancient primary endosymbionts usually have more strongly reduced genomes in comparison to secondary endosymbionts and have retained few (if any) genes for effector proteins that have a negative impact on the host, and only a limited number of genes coding for bacterial cell wall components [13].

Exceptions are those primary endosymbionts that have established an association with their host more recently, like the primary endosymbiont SOPE of the rice weevil and SZPE of corn weevils [14]. Their genomes resemble those of secondary endosymbionts, still containing large numbers of biosynthetic pathways, numerous pseudogenes as well as insertion elements [3, 15–17]. The genomes of secondary endosymbionts are intermediate in size or only slightly smaller than those of their free-living relatives (1–4.1 Mb) and still exhibit a larger genetic repertoire. They contain mobile elements or phages which together allow recombination within genomes or horizontal gene transfer. Moreover, such bacteria may also encode specific protein secretion systems, in particular type III secretion systems (T3SS) or type IV secretion systems (T4SS). These protein secretion systems are widespread in pathogenic bacteria and enable a direct import of bacterial effector proteins into the host cell cytoplasm. Such effector proteins can modulate the host cells in favor of the bacteria, for example, they may induce uptake of the bacteria into the eukaryotic cell or they are required for their intracellular survival and multiplication [3, 17, 18].

As secondary endosymbionts can also occur in tissues other than the gonads of their hosts and are able to cross several different host tissues, they are expected to get into contact with and be recognized by the host's immune system. In general, insects can mount a sophisticated immune response in response to an infection with pathogens such as bacteria, fungi or other parasites. The insect immune system comprises two parts: (i) the humoral immune response resulting in the production of soluble antimicrobial compounds, melanin formation and clotting, and (ii) a cellular immune response mainly based on hemocytes, involving defense mechanisms such as phagocytosis and encapsulation.

Microorganisms can be detected by hemocyte surface pattern recognition receptors (PRRs). These recognize conserved microbial cell envelope compounds such as peptidoglycan fragments. Once bound, the PRRs activate signaling pathways that are specific for Gram-negative bacteria (Imd pathway) or Gram-positive bacteria and fungi (Toll patheway) which then cause phagocytosis or antimicrobial peptide symthesis [19–21]. Insects carrying endosymbionts – frequently primary and secondary together – thus face the problem of having to mount an immune response that should be effective in eliminating pathogens upon infection while at the same time keeping and controlling endosymbionts in growth and number. The number and variability of specific pathogen-recognition components of the insect immune system has recently been found to be much higher than thought [22, 23], possibly providing a mechanism for specificity in the innate immune system that allows differential control of pathogens and endosymbionts.

30.2 Mutualistic Primary Endosymbionts with a Function in Nutritional Upgrading

One reason for the evolutionary success of insects may be their capability to occupy a vast array of terrestrial habitats, some of which are strongly limited in nutrients or provide only imbalanced food resources. By exploiting the metabolic potential of endosymbionts, insects have been enabled to specialize on such imbalanced food resources such as plant phloem sap or blood [1, 13, 22, 24–27]. Thus incorporation of these mutualistic bacteria can be viewed as a key innovation in the evolution of their hosts.

Representatives of these bacteria include several primary endosymbionts whose genomes have been sequenced, facilitating a better understanding of host–symbiont interactions. These are the endosymbionts of four aphid species (*Buchnera* spp.) [6, 28–30], the endosymbionts of two carpenter ant species (*Blochmannia* spp.) [31, 32], one symbiont each of psyllids (*Carsonella ruddii*) [33] and tsetse flies (*Wigglesworthia glossinidia*) [22] as well as two different endosymbionts coresiding in a xylem-feeding sharpshooter (*Baumannia cicadellinicola* and *Sulcia muelleri*) [26, 27]. All these bacteria belong to the γ-proteobacteria, except for *Sulcia muelleri* that belongs to the *Bacteroidetes*. The best-studied mutualistic secondary symbiont is the vertically transmitted *Sodalis glossinidius* which, together with *Wigglesworthia*, is resident in tsetse flies and whose genome sequence was recently analyzed [16]. This bacterium appears to have a positive effect on host biology, since its selective elimination affected the longevity of the tsetse flies. In addition, genomic data are partially available for the more recently acquired endosymbionts SOPE and SZPE of grain weevils [3, 34] that play a role in nutritional upgrading for their hosts, as well as the secondary endosymbiont *Hamiltonella defensa* that co-resides with *Buchnera* in aphids [35]. *Hamiltonella* confers a benefit to the host by providing protection against eukaryotic parasites to the aphid host [35, 36].

30.2.1
Primary Endosymbionts

The genomes of ancient primary endosymbionts are extremely reduced in size in comparison to their free-living relatives, comprising only 180–800 kb, and are strongly biased in their composition with an AT-content of 70–80%. DNA repair and recombination functions have largely been lost, leading to a highly conserved genome structure within a genus of endosymbionts. As there is no evidence of uptake of foreign genes or homologous recombination among different strains, these bacteria seem to have been excluded from horizontal gene transfer events for millions of years [3, 37]. The metabolic capacity retained in the genome varies among primary endosymbionts [13] in line with specific nutritional requirements of the respective host. The endosymbionts contribute to the host's biology by providing it with metabolites that are required for fertility or survival but are absent or limited in the host's diet such as essential amino acids, vitamins and cofactors. They can also contribute to nitrogen recycling or storage [1, 24, 25].

The extreme reduction of genomic competence reflects the high degree of adaptation to a single ecological niche, the insect host cell. Although genes from all functional categories are lost over time, genes rendered unnecessary in this stable environment are lost faster. Examples are genes involved in the bacterial metabolism that have become redundant because their products are now supplied by the host or that are not required for host or symbiont survival, DNA repair and recombination functions, signal transduction pathways and transcriptional regulatory mechanisms. Thus, in contrast to free-living bacteria that require a high degree of transcriptional regulation in order to be able to react quickly to environmental changes, most genes of primary endosymbionts are constitutively expressed, as shown recently for *Buchnera aphidicola* and *Blochmannia floridanus* in DNA-array experiments [38–40].

Comparative analyses of the sequenced endosymbiont genomes suggest that the molecular mechanisms required by the bacteria for survival in the host cell lead to a subset of genes shared by most endosymbionts. In general, genes involved in essential functions such as DNA replication, transcription and translation are retained and may account for up to one-third of the genome. Chaperone systems and all essential components of the Sec protein translocation machinery that are needed for correct folding and membrane transport of proteins are also retained [16]. Chaperonines like GroEL are probably important in posttranslational correction and stabilization of proteins that may be affected by the strong AT-bias of the genes of endosymbionts. GroEL was among the few genes found to be constitutively overexpressed in the aphid *B. aphidicola* and under positive selection [41]. Approximately 10–14% of the coding capacity of each endosymbiont is devoted to the specific nutritional requirements of the respective host [3].

Primary endosymbionts colonize specialized cells of the host, the bacteriocytes. In concert with the strongly reduced genetic repertoire of the bacteria this may allow the hosts to control endosymbiont proliferation, a process that has often been referred to as "domestication" of the bacteria. The bacteriocytes in aphids are characterized by a specific gene expression pattern and are produced during embryogenesis irrespective of the presence of bacteria [42]. The bacteriocytes can be intercalated between midgut cells as in carpenter ants (Figure 30.1). Alternatively they can form a bacteriome, an organ-like structure consisting of cell clusters adjacent to the gut or within the fat body as in the case of aphids or cockroaches, respectively [1]. Within the bacteriocyte the bacteria can either be located free in the cytosol (*Blochmannia, Wigglesworthia, Baumannia*) or enclosed in host cell-derived vacuoles (*Buchnera*) [1, 22, 43]. Interestingly, the gene *dnaA*, which encodes a protein required for the initiation of DNA replication, was lost in primary endosymbionts that reside free in the cytosol but is still present in vacuole-enclosed bacteria such as *Buchnera* [13].

A feature consistent with the assumed symbiotic function of the bacteriocyte was the specific upregulation of genes encoding functions related to metabolism and transport of amino acids recently found by transcriptome analysis of the bacteriocytes of the pea aphid (*Acyrtosiphon pisum*) harboring *B. aphidicola* [44]. In aphids a third group of genes, namely those usually involved in defense response, such as putative lysozyme genes, were also upregulated. Reverse transcriptase polymerase chain reaction (RT-PCR) experiments have indicated that these lysozyme-encoding genes are among the most highly expressed genes in the aphid bacteriocyte and may have a function in the control of the endosymbiont population [44]. Similar expression profiles have been found in the bacteriome of the corn weevil *Sitophilus zeamais*, which contains the more recently acquired endosymbiont *Sitophilus zeamais* primary endosymbiont SZPE. Genes involved in sugar uptake and metabolism were upregulated in the bacteriocytes as well as various antistress systems that control the oxidative stress resulting from higher metabolic rate of the cells [45].

Factors known to be relevant for bacteria–host cell interactions such as tissue and cell invasion are generally missing from the genomes of the primary endosymbionts.

However, in *Buchnera* and *Wigglesworthia* several genes required for flagellar biosynthesis have been retained, although these bacteria do not express functional flagellae. The remnants of the flagellar apparatus may be used by these bacteria as transport machineries to exchange proteins with the host cell. Flagellar secretion machineries are known to export proteins additional to flagellar components such as virulence factors in pathogenic bacteria like *Yersinia entrocolitica* [46].

Due to the permanent life of primary endosymbionts free in the cytosol of the host cell, their cell envelope has become simplified. This is even more pronounced in endosymbionts living in host-derived vesicles like *Buchnera aphidicola* [13]. The strongly reduced genetic repertoire that has led to the loss of virulence genes and the simplified structure of the bacterial cell envelope may have rendered ancient primary endosymbionts less visible for the insect innate immune system.

30.2.2 Recently Acquired and Secondary Mutualistic Endosymbionts

While endosymbionts are exclusively localized in bacteriocytes, secondary or recently acquired endosymbionts may also be found extracellularly in the hemolymph at least for part of the host's lifecycle. Therefore, these secondary endosymbionts should be more visible to the host immune system than primary endosymbionts.

30.2.2.1 Immune Responses

During metamorphosis of the corn weevil the larval bacteriome dissociates, leading to a release of its endosymbiont SZPE into the hemolymph. Most likely cells of the bacteriome are then reinfected along the intestine at a later stage during metamorphosis. The larval bacteriome is situated at the foregut periphery whereas the bacteriome of the adult weevil is in the proximity of the hindgut [2, 47]. In parallel to this release of bacteria, a peptidoglycan-recognition protein (PGRP) is strongly upregulated in the adjacent tissues, presumably mounting an adaptive immune response that prevents bacterial invasion into insect tissues or organs other than the bacteriome [47]. In the bacteriocytes *w*PGRP, a homolog of the *Drosophila* PGRP-LB with catalytic function, is strongly upregulated when bacteria are present. In *Drosophila*, the PGRP-LB degrades peptidoglycan via its amidase activity and was shown to thereby downregulate the immune response against Gram-negative bacteria by preventing an activation of the Imd-signaling pathway. An upregulation of this gene may decrease the host's immune defense locally and may thus be an adaptive response of the host allowing a long-term interaction with the endosymbiotic bacteria [47].

The permanent presence of *Sodalis glossinidius* in the hemolymph, in addition to the localization in the bacteriocytes, may have led to a constitutive expression of the antimicrobial peptide diptericin in tsetse flies to prevent uncontrolled proliferation of this secondary endosymbiont. *Sodalis* was found to be 10 times more resistant to the antimicrobial activity of diptericin in comparison to its close relative *E. coli*. Thus either the endosymbiont has adapted to the antimicrobial activity or control of symbiont number by the host is facilitated by this constitutive expression [48].

An additionally induced immune response in tsetse flies as a response to trypanosome infection does not seem to have an impact on the primary endosymbiont *Wigglesworthia*, possibly due to its intracellular niche. To date it is unknown however, whether the secondary endosymbiont *Sodalis* is affected by induced antimicrobial peptides other than diptericin [49].

30.2.2.2 **Host Cell Invasion**

Molecular mechanisms that facilitate invasion of host cells are still retained in the genomes of more recently acquired or secondary mutualistic endosymbionts. Thus, the *Enterobacteriaceae* SZPE and SOPE and their close relative, the secondary endosymbiont of the tsetse fly *Sodalis glossinidius*, contain homologs of the *inv/spa* genes that encode a T3SS in *Shigella* and *Salmonella* [34]. It has been shown, however, that the genes encoding a functional T3SS export apparatus are still complete, whereas many genes for effector proteins found in the mammalian pathogens are lacking. Presumably only effector proteins that are specifically involved in the cytoskeletal rearrangements necessary for the invasion of bacterial cells are retained in the genome of the mutualistic secondary endosymbionts [3]. In addition, these bacteria contain many mobile genetic elements [16, 17]. It is estimated that about 25% of the genomes of SZPE and SOPE are bacterial insertion sequences [16].

Another member of the *Enterobacteriaceae*, the secondary endosymbiont *Hamiltonella defensa*, which is found in aphids as well as other insect families [50] harbors two T3SS as well as a bacteriophage [35] in spite of its advanced genome reduction (genome size is approx. 1.7 Mb). The phage APSE-2 was found to be a close homolog of the lambdoid phage P22 of *Salmonella enterica*. Interestingly, the phage encodes an intact homolog of the cytolethal distending toxin (*cdtB*) that interrupts the eukaryotic cell cycle and is present in several mammalian pathogens. Since the presence of *Hamiltonella defensa* carrying the phage correlates with enhanced defense of the aphid hosts against eukaryotic parasites, such as the eggs of parasitoid wasps, it is assumed that the phage-born toxins mediate this observed protection. Experimental infection of aphid hosts with different *Hamiltonella* strains that all carried the phage-conferred parasite resistance irrespective of the host genotype [35, 51, 52]. It is not known by what mechanism eukaryotic parasitoid eggs can be targeted by the toxin without damaging the eukaryotic aphid host.

30.3
Endosymbionts that Interfere with Reproduction

In recent years, inherited arthropod symbionts that interfere specifically with the reproduction of their insect hosts have been discovered in various groups of bacteria like the *Proteobacteria*, *Spiroplasma* (*Mollicutes* within the *Firmicutes*), *Chlamydia* and *Bacteroidetes* (CFB: *Cytophaga–Flexibacter–Bacteroides*) [7]. Four major types of reproductive manipulation have been documented in insect hosts: (i) cytoplasmic incompatibility, whereby uninfected females are reproductively incompatible with infected males; (ii) male-killing, where male offspring die during development; (iii) phenotypic feminization of genetic males; (iv) induced parthenogenesis, where genetic males are turned into genetic females [7, 10, 53]. While host phenotypes resulting from manipulation by bacteria are described well, the underlying mechanisms are less well understood. The availability of genome sequences of *Wolbachia* [18, 54] and related bacterial pathogens will likely provide new insights into the mechanisms that endosymbionts use to manipulate their hosts.

30.3.1 *Wolbachia*

Like other *Rickettsiales*, the Gram-negative α-proteobacterium *Wolbachia* is an obligatory intracellular endosymbiont of eukaroytes. Its closest relatives, *Ehrlichia* and *Rickettsia*, both include parasites that infect mammals and use insects as vectors [9]. In contrast, *Wolbachia* exclusively infects arthropods and nematodes, both members of the "ecdysozoa," a clade of metazoans comprising all molting animals [55]. Currently all *Wolbachia* are considered to belong to a single formally described species, *Wolbachia pipientis*, which is classified into eight supergroups (A–H) based on a multilocus genotype of up to five genes conserved in all groups (*gatB*, *coxA*, *hcpA*, *fbpA* and *ftsZ*) [56–59]. *Wolbachia* belonging to the groups A and B infect a wide range of insect hosts and are often found together in a single host. These *Wolbachia* have not specialized on a particular host and frequent horizontal transfer of the endosymbiont leads to incongruent phylogenies of insect hosts and *Wolbachia* [11]. The bacterium has also been transferred successfully by embryonic injection into closely and distantly related naïve or new hosts, resulting in stable infections and expression of expected host phenotypes [60, 61]. In contrast, *Wolbachia* of group F are less common but have a very broad host range comprising insects, arachnids but also in filarial nematodes [57]. C and D *Wolbachia* infect a single family of nematodes, the *Onchocercidae* or filarial nematodes that are causative agents of filariasis. *Wolbachia* belonging to these groups are considered to be mutualistic as they have been shown to be essential for host development and reproduction, thus making the endosymbiont a possible target for treatment of filariasis [55].

In insects, *Wolbachia* is mostly localized in the host's reproductive tissue but presence in somatic tissue has also been shown [62]. Its reliance on maternal transmission has selected for a number of mechanisms leading to a female-biased sex ratio distortion, such as (i) feminization of genetic males, or (ii) male-killing, (iii) parthenogenesis or (iv) spread of the bacterium in the population through cytoplasmic incompatibility (see below). Cytoplasmic incompatibility leads to a spread of the infection by decreasing the fitness of uninfected females in contrast to those carrying *Wolbachia* [10, 53]. The magnitude of these effects may vary with host genotype, host age, temperature or bacterial genotype [63–65]. A wide range of more subtle effects on host biology, such as decreased host locomotor activity, body size and competitive ability have also been detected [66, 67]. The availability of *Wolbachia* genome data in concert with an increasing number of cytological studies now facilitates an understanding of the mechanisms *Wolbachia* utilizes to maintain itself within a broad range of hosts and to manipulate host biology in its own favor.

30.3.1.1 **Host Phenotypes Caused by *Wolbachia***

Cytological studies have contributed greatly to an understanding of the processes leading to cytoplasmic incompatibility, the most common phenotype of *Wolbachia* in insects. The cytoplasmic incompatibility phenotype results in aberrant offspring production between hosts carrying strains of *Wolbachia* which differ in cytoplasmic factors [10, 68, 69]. Unidirectional cytoplasmic incompatibility is usually expressed when an infected male mates with an uninfected female, which is lacking cytoplasmic factors of *Wolbachia*. The reciprocal cross of an infected female with an uninfected male is fully compatible, as well as crosses between both infected males and females carrying the same cytoplasmic factor. Thus infected females can successfully produce offspring when mating with either an uninfected male or males carrying the same *Wolbachia* strain. In contrast uninfected females will only produce normal offspring when mating with an uninfected male [53].

Werren and colleagues proposed a modification–rescue model for cytoplasmic incompatibility induction [53], assuming two distinct bacterial functions. First, *Wolbachia* modifies the sperm within the testes, possibly by alteration of the chromatin-binding proteins, and second, a rescue factor is expressed in the eggs [53, 63]. The modification of the paternal chromosomes during spermatogenesis alters their behavior during the first mitotic divisions. In insects, the nuclear envelopes containing the maternal and paternal chromosomes do not break down completely when the pronuclei meet in embryogenesis. As a result, a unique spindle apparatus is formed for the first cell cycle that is composed of two physically distinct bundles of microtubules – one attaching to the maternal and the other the paternal chromosomes [70]. In cytoplasmic incompatibility embryos, sperm nuclei that have been modified by *Wolbachia* lag behind their female counterparts during mitosis due to a delay in nuclear envelope breakdown in the male pronucleus. This loss of synchrony in mitosis leads to a complete or partial loss of parental chromosomes, rendering the developing embryo haploid [68, 69]. In diploid insect species the majority of embryos subsequently die, whereas in haplodiploid species such as hymenoptera the offspring develop into normal haploid males. Thus two different phenotypes are produced by cytoplasmic incompatibility depending on host:mortality or male-biased sex ratio. Infection with the same strain of *Wolbachia* restores synchrony of mitosis in the first cell cycle and embryos develop normally [10]. When insects carrying such strains of *Wolbachia* are cured from their infection, embryogenesis proceeds normally.

Other *Wolbachia* have become obligatory symbionts for their insect host, as in the parasitoid wasp *Asobara tabida* [71, 72]. Here, the presence of *Wolbachia* is required for apoptosis during oogenesis, which is an essential process for successful maturation of the eggs [72]. However, to date the mechanism preventing apoptosis in nurse cells of the wasp is not known. Other intracellular pathogens are known to be able to influence disease progression by controlling apoptotic pathways in their hosts, either inducing or blocking them [73].

30.3.1.2 How Does *Wolbachia* Manipulate its Host's Reproductive Biology?

Three major groups of genes have been identified in *Wolbachia* that may play a role in its interactions with the host: (i) those coding for a T4SS and its effector proteins; (ii) those coding for ankyrin-repeat motif proteins and (iii) those present in mobile elements such as bacteriophages [55], including several ankyrin-domain containing genes [18]. Complete T4SS were found in the genome of *Wolbachia* isolated from *Drosophila melanogaster* (*w*Mel) as well as *Wolbachia* from the nematode *Brugia malayi* (*w*Bm), arranged in two separate operons [18, 54, 55]. Interestingly, a homolog of *wsp*, encoding the *Wolbachia* surface protein, lies within one of the operons. This protein may be exported to the vacuole in which the bacterium resides [18]. Whereas it is not known whether it is a virulence factor in insects, WSP of *Wolbachia* in nematodes may interact with signaling pathways of animals and humans that are infected by nematodes. In humans, WSP has been shown to be involved in the inhibition of apoptosis in neutrophils [74] and to induce an immune reaction via Toll-like receptor 2 (TLR2) and TLR4 [75]. That the WSP protein may play an important role in host–*Wolbachia* interactions is supported by the fact that the *wsp* gene is under strong positive selection in parasitic strains but not in mutualistic strains, suggesting a parasite–host arms race in the former [76].

Other proteins that may be exported via T4SS are proteins with ankyrin domains. The ankyrin-repeat (ANK), a tandem motif of ~33 amino acids, is commonly found in eukaryotic proteins, often in those mediating protein–protein interactions, such as linkage of membrane proteins to the cytoskeleton, acting as transcription factors and playing a role in modification of cell cycle-regulating proteins [77]. ANK-containing proteins produced by bacteria have been shown to play a role in host–pathogen interactions by influencing the gene expression of the host cell [78]. In spite of the reduced genome size of *w*Mel, 23 ANK proteins are present in the genome, an unusually high number for a prokaryote [18, 55]. In comparison, its relatives *Rickettsia*, *Anaplasma* or *Ehrlichia* contain only one or two ANK proteins [78]. Since *Wolbachia* has been shown to interact with early events of cell cycling in the zygote of insects [69] it is likely that the ankyrin motif-containing proteins interact with the same cell cycle proteins. Correlative evidence that ANK proteins are involved in cytoplasmic incompatibility comes from different lines of evidence. Comparison of orthologs of these genes in different *Wolbachia* strains from *Drosophila* has shown that they are variable in respect to ANK repeat number, presence/absence of transmembrane domains, disruption by insertion elements (IS) or complete absence of particular orthologs in strains that are known to be incapable of inducing cytoplasmic incompatibility [79]. Such changes in repeat number or disruption by IS are likely altering the strength and specificity of the interaction with host target proteins.

Bacteriophages present in the genome of *Wolbachia* may also be important for the manipulation of the host's biology. Several of the *w*Mel ANK genes are located within integrated prophage segments in the genome. In *Culex pipiens* it was shown that allelic variation of two ANK genes that form part of a prophage region was correlated with cytoplasmic incompatibility in two different mosquito strains [65]. In addition, there is evidence that the *Wolbachia*-specific WO phages can move both within the bacterial genome as well as between bacteria that co-reside in the same host.

In concert with high recombination rates found within the phage, this possibly generates diversity within arthropod *Wolbachia* [80]. Lateral gene transfer via WO phages may also be responsible for the observation that distantly related strains of *Wolbachia* produce the same host phenotype, whereas closely related strains may lead to different phenotypes. Interestingly, the mutualistic *Wolbachia* in nematodes *w*Bm contains only a small number of ANK genes and entirely lacks prophages [18].

30.3.1.3 *Wolbachia* and Host Immune Response

In spite of the partly extracellular occurrence of *Wolbachia*, the bacterium is able to persist in a vast array of arthropod hosts. Unlike other bacterial pathogens of insects, *Wolbachia* infection does not seem to activate an immune response during any phase of the host's lifecycle, although it occurs and moves between different host tissues. Both, *Drosophila simulans* and *Aedes albopinctus* that were heavily infected with *Wolbachia* did not produce more diptericin or cecropin, two known antimicrobial peptides against Gram-negative bacteria. In contrast, these antimicrobial peptides showed increased expression levels when *Wolbachia*-infected flies were challenged with *E. coli* [81]. One possible explanation for the invisibility of *Wolbachia* to host immune detection may be its mostly intracellular localization. Inside host cells *Wolbachia* resides in a host-derived vacuole. It has been suggested that this vacuole resembles a phagosome and that *Wolbachia* is able to modify this vacuole, allowing the bacterium to escape lysis and degradation [82].

30.3.2 Male-Killing by *Spiroplasma*

Male-killing is a widespread phenotype among endosymbiotic bacteria [8]. Male-killing bacteria are passed vertically to both male and female progeny but then selectively kill male offspring during embyrogenesis. Recently Veneti and colleagues were able to show that male-killing by *Spiroplasma* requires a functional dosage compensation complex in the male *Drosophila* embryos [83]. Dosage compensation is the process by which the gene expression titer on the X chromosome is equalized between the two sexes despite the difference in X chromosome number. In males this complex comprises five proteins that bind to and modify the male X chromosome and lead to its hypertranscription [83]. It was recently shown that male-killed embryos are characterized by massive apoptotic death of cells within the embryo preceding complete arrest in development shortly after the formation of the dosage compensation complex. Since this *Spiroplasma*-induced apoptosis occurs at the same point of embryogenesis as normal apoptotic processes begin, it is not clear whether the bacterium directly induces apoptosis or whether apoptosis is merely a response to the effects of the infection [83, 84].

30.4 Endosymbiont Communities

Numerous insect hosts have recently been described that harbor several different endosymbionts coexisting in the bacteriocytes. Co-residence of mutualistic endosymbionts may have been required in ancient symbiotic associations due to advanced genome reduction. In the process of genome reduction an ancient primary endosymbiont may eventually lose gene functions or complete genes whose metabolites are required by the host. By stably maintaining a second endosymbiont, this loss-of-function may be compensated. For the co-residing primary endosymbionts of xylem-feeding sharpshooters, the γ-proteobacterium *Baumannia cicadellinicola* and the *Bacteroidetes* species *Sulcia muelleri*, genomic analysis has revealed that the two symbionts have complementary biosynthetic capabilities [26]. While *Baumannia*'s genome encodes pathways for the biosynthesis of vitamins and cofactors and lacks most amino acid biosynthetic pathways, the latter seem to be contributed by *S. muelleri*. Thus the predicted metabolic capabilities of the two endosymbionts are both essential to the host and presumably the two endosymbionts themselves are also dependent on exchange of metabolites between each other for survival and growth [26]. Similarly, the aphid primary endosymbiont with the most strongly reduced genome, *Buchnera aphidicola* BCc, may not provide all metabolic functions required by its host. Consequently, it may be replaced by the endosymbiont *Serratia symbiotica*, which was formerly assumed to be a facultative symbiont but was found to be present in the bacteriocytes in all cedar aphid populations studied [6].

Frequently one or more secondary endosymbionts co-reside with primary endosymbionts in the bacteriocytes of insect hosts, with up to six different bacterial strains or species found to infect the same host [85]. Recently Gottlieb *et al.* [86] studied the spatial interaction of the primary endosymbiont *Portiera* and five secondary endosymbionts (*Wolbachia*, *Rickettsia*, *Cardinium* and either *Hamiltonella* or *Arsenophonus*) that were found to co-reside in the bacteriocytes of the whitefly *Bemisia tabaci*. The whitefly bacteriocyte is enclosed as a whole by the egg during oogenesis, providing an efficient vehicle for vertical transmission of all symbiotic bacteria together. Coexistence of different bacteria in the same cell may be made possible by niche partitioning. Bacteria occupying a similar niche seem to be mutually exclusive in the same individual. For example *Cardinium* and *Rickettsia* were never found together. Since both bacteria usually invade the same tissues during whitefly development, competition outside the bacteriocyte may be responsible for this exclusion.

Interactions in such communities may be complex and comprise bacteria–bacteria interactions as well as bacteriophages. The latter may facilitate horizontal gene transfer among secondary endosymbionts that contain a large number of mobile genetic elements at the beginning of the process of genome reduction [35, 87]. In addition they may mediate interactions with the host [52].

30.5 Conclusion and Future Directions

With more endosymbionts being studied at the genomic and organismal levels categorizing them becomes increasingly difficult. Genome reduction is a continuous process and the largest genomes of recently acquired primary endosymbionts may be larger than the genomes of secondary endosymbionts. In addition, the line between mutualism and parasitism is becoming increasingly blurred as costs or benefits for symbiont and host may vary with life stage as well as biotic and abiotic environments and are often difficult to measure – especially in insect hosts that harbor consortia of endosymbionts. In general, bacteria with less reduced genomes have retained a higher flexibility that provides them with more tools for interaction with their hosts and possibly also other endosymbionts present in the same host.

Many questions remain open. In particular, the secondary endosymbionts have been more in the focus of research lately, due to their greater resemblance to pathogenic bacteria. In the future, comparative work on the mechanisms used to induce similar host phenotypes (e.g., male-killing or cytoplasmic incompatibility) by different secondary endosymbionts will hopefully yield valuable insights into symbiont–host interactions. Little is known about the regulation of primary and secondary endosymbionts by the respective hosts. Here transcriptomic studies of bacteriocytes but also comparative work of insect immune systems is vital. In addition, finding out how endosymbionts evade the host immune system could yield valuable insights into bacteria–host interactions not only of endosymbionts but also in respect to the vast array of emerging mammalian pathogens that use insects as vectors. Finally, virtually nothing is known about the interaction of the symbiotic bacteria with their hosts on the cellular level, which is a broad field for future research.

Secondary endosymbionts that do not elicit an immune response in the insect hosts such as *Wolbachia* and others can now be cultured *ex vivo*. Thus these bacteria may be relevant models to investigate those factors, for example, alterations in the cell wall structure, *in vivo* that may lead to the evasion of the host immune system.

Research on the evolution of endosymbiont consortia is only at the beginning. Although theoretical work has shown that evolution towards cooperation is possible when multiple infections are stably maintained [88], hardly anything is known of their impact on host biology. A better knowledge of endosymbiont–host interactions or bacteria–bacteria interactions inside the insect host can be an important contribution to the development of symbiont-based pest control strategies [89]. These could either target the endosymbiont itself, as exemplified in new approaches in the treatment of filariasis [90], or transgenic endosymbionts could be used to deliver factors into vector populations detrimental for the insects.

References

1 Baumann, P. (2005) *Annu Rev Microbiol*, **59**, 155–189.

2 Buchner, P. (1965) *Endosymbiosis of Animals with Plant Microorganisms*, Wiley, New York.

3 Dale, C. and Moran, N.A. (2006) *Cell*, **126**, 453–465.

4 Noda, H., Miyoshi, T., Zhang, Q., Watanabe, K. *et al.* (2001) *Mol Ecol*, **10**, 2101–2106.

5 Huigens, M.E., de Almeida, R.P., Boons, P.A.H., Luck, R.F. and Stouthamer, R. (2004) *Proc Roy Soc Lond B Biol Sci*, **271**, 509–515.

6 Pérez-Brocal, V., Gil, R., Ramos, S., Lamelas, A. *et al.* (2006) *Science*, **314**, 312–313.

7 Hunter, M.S. and Zchori-Fein, E. (2006) in *Insect Symbiosis*, Vol. 2 (eds K. Bourtzis and T.A. Miller), CRC Press, Boca Raton, FL, pp. 39–56.

8 Hurst, G.D.D., Jiggins, F.M. and Majerus, M.E.N. (2003) *Insect Symbiosis* (eds K. Bourtzis and T.A. Miller), CRC Press, Boca Raton, FL, pp. 177–198.

9 Perlman, S.J., Hunter, M.S. and Zchori-Fein, E. (2006) *Proc Roy Soc Lond B Biol Sci*, **273**, 2097–2106.

10 Stouthamer, R., Breeuwer, J.A.J. and Hurst, G.D.D. (1999) *Annu Rev Microbiol*, **53**, 71–102.

11 Werren, J.H., Zhang, W. and Guo, L.R. (1995) *Proc Roy Soc Lond B Biol Sci*, **261**, 55–63.

12 Hilgenboecker, K., Hammerstein, P., Schlattmann, P., Telschow, A. and Werren, J.H. (2008) *FEMS Microbiol Lett*, **281**, 215–220.

13 Zientz, E., Dandekar, T. and Gross, R. (2004) *Microbiol Mol Biol Rev*, **68**, 745–770.

14 Lefevre, C., Charles, H., Vallier, A., Delobel, B. *et al.* (2004) *Mol Biol Evol*, **21**, 965–973.

15 Charles, H., Condemine, G., Nardon, C. and Nardon, P. (1997) *Insect Biochem Mol Biol*, **27**, 345–350.

16 Moya, A., Pereto, J., Gil, R. and Latorre, A. (2008) *Nat Rev Genet*, **9**, 218–229.

17 Toh, H., Weiss, B.L., Perkin, S.A.H., Yamashita, A. *et al.* (2006) *Genome Res*, **16**, 149–156.

18 Wu, M., Sun, L.V., Vamathevan, J., Riegler, M. *et al.* (2004) *PLoS Biol*, **2**, 327–341.

19 Feldhaar, H. and Gross, R. (2008) *Microbes Infect*, **10**, 1082–1088.

20 Ferrandon, D., Imler, J.L., Hetru, C. and Hoffmann, J.A. (2007) *Nat Rev Immunol*, **7**, 862–874.

21 Lemaitre, B. and Hoffmann, J. (2007) *Annu Rev Immunol*, **25**, 697–743.

22 Akman, L., Yamashita, A., Watanabe, H., Oshima, K. *et al.* (2002) *Nat Genet*, **32**, 402–407.

23 Dong, Y., Taylor, H.E. and Dimopoulos, G. (2006) *PLoS Biol*, **4**, e229.

24 Douglas, A.E. (1998) *Annu Rev Entomol*, **43**, 17–37.

25 Feldhaar, H., Straka, J., Krischke, M., Berthold, K. *et al.* (2007) *BMC Biol*, **5**, 48.

26 McCutcheon, J.P. and Moran, N.A. (2007) *Proc Natl Acad Sci USA*, **104**, 19392–19397.

27 Nakabachi, A., Yamashita, A., Toh, H., Ishikawa, H. *et al.* (2006) *Science*, **314**, 267–267.

28 Shigenobu, S., Watanabe, H., Hattori, M., Sakaki, Y. and Ishikawa, H. (2000) *Nature*, **407**, 81–86.

29 Tamas, I., Klasson, L., Canback, B., Naslund, A.K. *et al.* (2002) *Science*, **296**, 2376–2379.

30 van Ham, R.C.H.J., Kamerbeek, J., Palacios, C., Rausell, C. *et al.* (2003) *Proc Natl Acad Sci USA*, **100**, 581–586.

31 Degnan, P.H., Lazarus, A.B. and Wernegreen, J.J. (2005) *Genome Res*, **15**, 1023–1033.

32 Gil, R., Silva, F.J., Zientz, E., Delmotte, F. *et al.* (2003) *Proc Natl Acad Sci USA*, **100**, 9388–9393.

33 Wu, D., Daugherty, S.C., Van Aken, S.E., Pai, G.H. *et al.* (2006) *PLoS Biol*, **4**, 1079–1092.

34 Dale, C., Plague, G.R., Wang, B., Ochman, H. and Moran, N.A. (2002) *Proc Natl Acad Sci USA*, **99**, 12397–12402.

35 Moran, N.A., Degnan, P.H., Santos, S.R., Dunbar, H.E. and Ochman, H. (2005) *Proc Natl Acad Sci USA*, **102**, 16919–16926.

36 Oliver, K.M., Russell, J.A., Moran, N.A. and Hunter, M.S. (2003) *Proc Natl Acad Sci USA*, **100**, 1803–1807.

37 Wernegreen, J.J. (2002) *Nat Rev Genet*, **3**, 850–861.

38 Stoll, S., Feldhaar, H. and Gross, R. *Environ Microbiol*, **11**, 877–888.

39 Wilcox, J.L., Dunbar, H.E., Wolfinger, R.D. and Moran, N.A. (2003) *Mol Microbiol*, **48**, 1491–1500.

40 Wilson, A.C.C., Dunbar, H.E., Davis, G.K., Hunter, W.B. *et al.* (2006) *BMC Genomics*, **7**, 50.

41 Fares, M.A., Barrio, E., Sabater-Munoz, B. and Moya, A. (2002) *Mol Biol Evol*, **19**, 1162–1170.

42 Braendle, C., Miura, T., Bickel, R., Shingleton, A.W. *et al.* (2003) *PLoS Biol*, **1**, E21.

43 Sauer, C., Dudaczek, D., Hölldobler, B. and Gross, R. (2002) *Appl Environ Microbiol*, **68**, 4187–4193.

44 Nakabachi, A., Shigenobu, S., Sakazume, N., Shiraki, T. *et al.* (2005) *Proc Natl Acad Sci USA*, **102**, 5477–5482.

45 Heddi, A., Vallier, A., Anselme, C., Xin, H. *et al.* (2005) *Cell Microbiol*, **7**, 293–305.

46 Young, G.M., Schmiel, D.H. and Miller, V.L. (1999) *Proc Natl Acad Sci USA*, **96**, 6456–6461.

47 Anselme, C., Vallier, A., Balmand, S., Fauvarque, M.O. and Heddi, A. (2006) *Appl Environ Microbiol*, **72**, 6766–6772.

48 Hao, Z.R., Kasumba, I., Lehane, M.J., Gibson, W.C. *et al.* (2001) *Proc Natl Acad Sci USA*, **98**, 12648–12653.

49 Lehane, M.J., Aksoy, S. and Levashina, E. (2004) *Trends Parasitol*, **20**, 433–439.

50 Moran, N.A., Russell, J.A., Koga, R. and Fukatsu, T. (2005) *Appl Environ Microbiol*, **71**, 3302–3310.

51 Degnan, P.H. and Moran, N.A. (2008) *Appl Environ Microbiol*, **74**, 6782–6791.

52 Oliver, K.M., Moran, N.A. and Hunter, M.S. (2005) *Proc Natl Acad Sci USA*, **102**, 12795–12800.

53 Werren, J.H., Baldo, L. and Clark, M.E. (2008) *Nat Rev Microbiol*, **6**, 741–751.

54 Foster, J., Ganatra, M., Kamal, I., Ware, J. *et al.* (2005) *PLoS Biol*, **3**, 599–614.

55 Fenn, K. and Blaxter, M. (2006) *Trends Parasitol*, **22**, 60–65.

56 Baldo, L., Hotopp, J.C.D., Jolley, K.A., Bordenstein, S.R. *et al.* (2006) *Appl Environ Microbiol*, **72**, 7098–7110.

57 Baldo, L., Prendini, L., Corthals, A. and Werren, J.H. (2007) *Curr Microbiol*, **55**, 367–373.

58 Casiraghi, M., Bordenstein, S.R., Baldo, L., Lo, N. *et al.* (2005) *Microbiology-SGM*, **151**, 4015–4022.

59 Lo, N., Paraskevopoulos, C., Bourtzis, K., O'Neill, S.L. *et al.* (2007) *Int J Syst Evol Microbiol*, **57**, 654–657.

60 Boyle, L., O'Neill, S.L., Robertson, H.M. and Karr, T.L. (1993) *Science*, **260**, 1796–1799.

61 Braig, H.R., Guzman, H., Tesh, R.B. and O'Neill, S.L. (1994) *Nature*, **367**, 453–455.

62 Dobson, S.L., Bourtzis, K., Braig, H.R., Jones, B.F. *et al.* (1999) *Insect Biochem Mol Biol*, **29**, 153–160.

63 Bourtzis, K., Braig, H.R. and Karr, T.L. (2003) *Insect Symbiosis* (eds K. Bourtzis and T.A. Miller), CRC Press, Boca Raton, FL, pp. 217–246.

64 Reynolds, K.T., Thomson, L.J. and Hoffmann, A.A. (2003) *Genetics*, **164**, 1027–1034.

65 Sinkins, S.P., Walker, T., Lynd, A.R., Steven, A.R. *et al.* (2005) *Nature*, **436**, 257–260.

66 Fleury, F., Vavre, F., Ris, N., Fouillet, P. and Bouletreau, M. (2000) *Parasitology*, **121**, 493–500.

67 Huigens, M.E., Hohmann, C.L., Luck, R.F., Gort, G. and Stouthamer, R. (2004) *Entomol Exp Appl*, **110**, 115–123.

68 Callaini, G., Dallai, R. and Riparbelli, M.G. (1997) *J Cell Sci*, **110**, 271–280.

69 Tram, U., Ferree, P.A. and Sullivan, W. (2003) *Microbes Infect*, **5**, 999–1011.

70 Kawamura, N. (2001) *Dev Growth Differ*, **43**, 343–349.

71 Dedeine, F., Bandi, C., Bouletreau, M. and Kramer, L.H. (2003) *Insect Symbiosis* (eds K. Bourtzis and T.A. Miller), CRC Press, Boca Raton, FL, pp. 267–282.

72 Pannebakker, B.A., Loppin, B., Elemans, C.P.H., Humblot, L. and Vavre, F. (2007) *Proc Natl Acad Sci USA*, **104**, 213–215.

73 Faherty, C.S. and Maurelli, A.T. (2008) *Trends Microbiol*, **16**, 173–180.

74 Bazzocchi, C., Comazzi, S., Santoni, R., Bandi, C. *et al.* (2007) *Parasite Immunol*, **29**, 73–79.

75 Brattig, N.W., Bazzocchi, C., Kirschning, C.J., Reiling, N. *et al.* (2004) *J Immunol*, **173**, 437–445.

76 Jiggins, F.M., Hurst, G.D.D. and Yang, Z.H. (2002) *Mol Biol Evol*, **19**, 1341–1349.

77 Elfring, L.K., Axton, J.M., Fenger, D.D., Page, A.W. *et al.* (1997) *Mol Biol Cell*, **8**, 583–593.

78 Caturegli, P., Asanovich, K.M., Walls, J.J., Bakken, J.S. *et al.* (2000) *Infect Immun*, **68**, 5277–5283.

79 Iturbe-Ormaetxe, I., Burke, G.R., Riegler, M. and O'Neill, S.L. (2005) *J Bacteriol*, **187**, 5136–5145.

80 Bordenstein, S.R. and Wernegreen, J.J. (2004) *Mol Biol Evol*, **21**, 1981–1991.

81 Bourtzis, K., Pettigrew, M.M. and O'Neill, S.L. (2000) *Insect Mol Biol*, **9**, 635–639.

82 Siozios, S., Sapountzis, P., Ioannidis, P. and Bourtzis, K. (2008) *Insect Sci*, **15**, 89–100.

83 Veneti, Z., Bentley, J.K., Koana, T., Braig, H.R. and Hurst, G.D.D. (2005) *Science*, **307**, 1461–1463.

84 Bentley, J.K., Veneti, Z., Heraty, J. and Hurst, G.D.D. (2007) *BMC Biol*, **5**, 9.

85 Tsuchida, T., Koga, R., Shibao, H., Matsumoto, T. and Fukatsu, T. (2002) *Mol Ecol*, **11**, 2123–2135.

86 Gottlieb, Y., Ghanim, M., Gueguen, G., Kontsedalov, S. *et al.* (2008) *FASEB J*, **22**, 2591–2599.

87 Bordenstein, S.R. and Reznikoff, W.S. (2005) *Nat Rev Microbiol*, **3**, 688–699.

88 Vautrin, E., Genieys, S., Charles, S. and Vavre, F. (2008) *J Evol Biol*, **21**, 145–161.

89 Douglas, A.E. (2007) *Trends Biotechnol*, **25**, 338–342.

90 Hoerauf, A., Mand, S., Adjei, O., Fleischer, B. and Buttner, D.W. (2001) *Lancet*, **357**, 1415–1416.

IID
Parasites and Fungi

Intracellular Niches of Microbes. Edited by Ulrich E. Schaible and Albert Haas

ISBN: 978-3-527-32207-7

31
Histoplasma capsulatum

S. L. Newman

31.1
Histoplasma capsulatum and Histoplasmosis

The dimorphic fungal pathogen *Histoplasma capsulatum* was originally described in 1906 by Samuel Darling [1], and eventually documented as the etiologic agent of histoplasmosis in 1934 by de Monbreun [2]. Although *H. capsulatum* has a worldwide distribution, it is particularly endemic to the Ohio and Mississippi river valleys, and the southeastern, central and mid-atlantic States, where it infects about 500 000 individuals yearly [3].

Several varieties of *H. capsulatum* have been described based on their morphology, geography and pathology. Two of these are pathogenic to humans: *H. capsulatum* var. *capsulatum* that causes disease in the United States, and *H. capsulatum* var. *duboisii* that causes disease in Africa [4]. The US form of *H. capsulatum* tends to cause pulmonary lesions and a high fungal burden in secondary lymphoid organs, whereas the African variety manifests mainly as skin and bone infections. A third variety, *H. capsulatum* var. *farciminosum*, causes infection in horses. This review will focus on *H. capsulatum* var. *capsulatum*.

Clinically, infection with *H. capsulatum* causes a spectrum of illnesses ranging from acute or chronic pulmonary disease, to a progressive disseminated form. Both acute and chronic pulmonary histoplasmosis are characterized by discrete granuloma formation, whereas in acute disseminated disease, granulomas are formed poorly, if at all. Instead, infected tissues are heavily infiltrated by macrophages engorged with yeasts. In immunocompetent individuals, the course of infection may be asymptomatic or cause a mild self-limiting influenza-like syndrome. However, *H. capsulatum* may produce progressive disseminated infections in individuals immunocompromised by hematologic malignancies, cytotoxic therapy or HIV infection [5]. Recently, rheumatoid arthritis patients that were treated with commercial preparations of anti-tumor necrosis factor α (TNFα) were reported to have developed disseminated histoplasmosis [6], a finding that fits the importance of this cytokine in the development and maintenance of cell-mediated immunity to *H. capsulatum* [7].

Intracellular Niches of Microbes. Edited by Ulrich E. Schaible and Albert Haas

ISBN: 978-3-527-32207-7

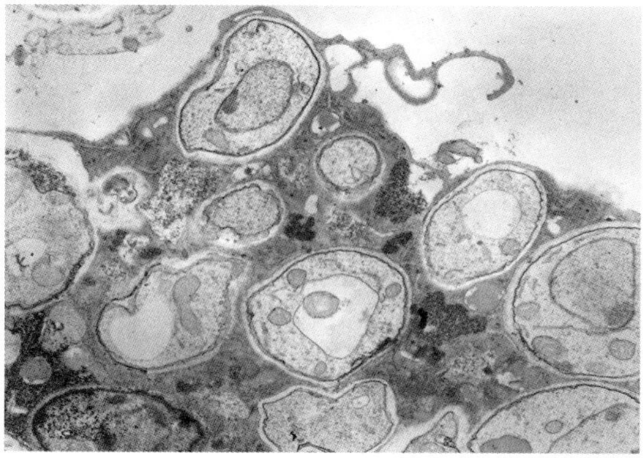

Figure 31.1 Human macrophage infected with *Histoplasma capsulatum* yeasts. Each yeast appears to be in a single phagosome. Note the thick electron translucent cell wall from which it derives its species name.

H. capsulatum is an ascomycete that grows as multicellular mycelia in the environment, being found in rich soil and areas contaminated with bird or bat guano. The mycelia bear spores or conidia (termed macroconidia or microconidia depending on their size and morphology) that are the infectious form of the microorganism. Infection is initiated by incidental inhalation of conidia (2–4 μm), which then become deposited in the terminal bronchioles and alveoli of the lung. Within a period of hours to days, the conidia transform into oval budding yeasts, and it is these yeasts that are responsible for all of the immunopathologic and clinical manifestations of histoplasmosis. The yeast is not encapsulated, although in tissues and macrophages the thick cell wall appears translucent (Figure 31.1).

In vivo, it is not clear whether conidia convert into the yeast form within alveolar macrophages or extracellularly, as a temperature of 37 °C is sufficient for transformation of conidia into yeasts [8]. Although *H. capsulatum* conidia are phagocytosed as rapidly as yeasts by alveolar macrophages [9], *in vitro*, the conidia convert into yeasts at the same rate whether they are confined in human or murine macrophages, or in tissue culture medium (Newman, unpublished observations). Thus, in the lung, conidia may convert into yeasts both intracellularly and extracellularly.

Regardless, once conidia convert into yeasts, the yeasts are phagocytosed by alveolar macrophages, within which they multiply [5, 10]. Replicating yeasts destroy the alveolar macrophages, and subsequently yeasts are ingested by other resident alveolar macrophages and by inflammatory phagocytes recruited into the lung. Repetition of this cycle leads to lympho-hematogenous spread of infection to lymph nodes and other organs rich in mononuclear phagocytes. Eventual development of specific cell-mediated immunity by the host results in activation of macrophages and resolution of the infection [7, 11]. As with other intracellular pathogens, the interaction of *H. capsulatum* with macrophages is the key event in the pathogenesis of histoplasmosis. Macrophages first provide an environment for fungal replication and dissemination, and then subsequently act as the final effector cells to remove the organism from the host. The details of this interaction are discussed below.

31.2 Characterization of the Macrophage Phagosomes Containing *H. capsulatum* Yeasts

31.2.1 Recognition and Phagocytosis

During the early phase of infection, macrophages recognize and phagocytose *H. capsulatum* yeasts thereby providing them access to a permissive intracellular environment for replication. This initial contact between *H. capsulatum* and macrophages is crucial in the pathogenesis of histoplasmosis. Human monocyte-derived macrophages recognize *H. capsulatum* yeasts via the CD18 integrin family of adhesins (LFA-1 (CD11a), CR3 (CD11b), p150,95 (CD11c)) [9, 12]. Efficient attachment requires the CD18 receptors to be mobile within the membrane, and mobility requires intact actin microfilaments of the cellular cytoskeleton [9]. No other macrophage receptors appear to be required for recognition of *H. capsulatum* yeasts, and macrophage-derived complement components also are not involved [12]. Although similar studies have not been performed with murine macrophages, it is presumed, but not proven, that since these cells possess CD18 receptors [13], they are required for recognition of *H. capsulatum* yeasts. Once bound, phagocytosis of *H. capsulatum* yeasts by human macrophages and murine peritoneal macrophages is rapid [9, 14], and opsonization with complement component C3 is not required.

Remarkably, the ligand on the surface of *H. capsulatum* yeasts that is recognized by macrophage CD18 receptors is heat shock protein 60 (hsp60) [15]. Hsp60 sits on the cell wall of *H. capsulatum* yeasts in discrete clusters, and there are about 250 hsp60 antigenic sites per yeast cell. This low number of sites explains why macrophage CD18 receptors must be mobile to efficiently bind *H. capsulatum* yeasts. Further, when hsp60 is depleted from a freeze/thaw extract (F/TE) of yeasts that contains *H. capsulatum* surface proteins, the F/TE no longer inhibits the binding of yeasts to macrophages, suggesting that hsp60 is the major, if not the sole ligand, by which macrophages recognize *H. capsulatum*. Interestingly, vaccination of mice with *H. capsulatum* recombinant hsp60 is protective against pulmonary histoplasmosis [16]. Thus, *H. capsulatum* hsp60 serves a dual role in the immunopathology of histoplasmosis.

31.2.2 Intracellular Survival and Replication

Based on electron micrographs, it appears that each yeast resides in a singular phagosome (Figure 31.1). *In vitro*, *H. capsulatum* yeasts multiply readily within human alveolar macrophages and monocyte-derived macrophages with an intracellular generation time of about 20 h [17]. The intracellular generation time of yeasts in murine peritoneal macrophages is about 10 h [18]. Opsonization of yeasts with normal or *H. capsulatum*-immune serum does not alter the ability of *H. capsulatum* yeasts to replicate in either mouse macrophages [19] or human macrophages [17].

Table 31.1 *In vitro* interaction of *Histoplasma capsulatum* yeasts with human and murine macrophages.

	Human macrophages	**Mouse macrophages**
Recognition by CD18	Yes	?
Respiratory burst stimulated	Yes	No
PL-fusion	No	Yes
Intraphagosomal pH 6.5	Yes	Yes
Activated by IFNγ	No	Yes
Activated by IL-3, GM-CSF, M-CSF	Yes	?
Activated by adherence to collagen matrix	Yes	?

Once inside the phagosome, *H. capsulatum* must avoid the microbicidal armamentarium of the macrophage that consists of the generation of toxic oxygen molecules (respiratory burst) and lysosomal hydrolases (phagolysosomal fusion, PL-fusion). Interestingly, there are some significant differences in the way in which *H. capsulatum* yeasts interact with human vs. murine macrophages, and these differences are highlighted where appropriate (Table 31.1).

H. capsulatum yeasts have a thick cell wall, but it is not sufficient to protect the yeasts from destruction by macrophages, and heat-killed or fixed yeasts are easily digested by human [20] and mouse macrophages [21]. *H. capsulatum* yeasts also must rapidly synthesize new proteins to survive within macrophages, as the addition of cycloheximide to infected human macrophages leads to killing and digestion of the fungus [20]. As cycloheximide inhibits protein synthesis in both yeasts [22] and mammalian cells [23], it is clear that macrophages do not need to synthesize new proteins to kill and digest the yeasts, but that this is accomplished by the fusion of preformed lysosomal hydrolases with yeast-containing phagosomes. Although *H. capsulatum* yeasts synthesize new proteins within macrophages, the yeasts do not secrete a product(s) that globally paralyzes macrophage fungicidal mechanisms [20, 24]. Thus, the intracellular survival strategy of *H. capsulatum* yeasts is focused on regulating the phagosomal compartment.

Phagocytosis of *H. capsulatum* yeasts by human macrophages stimulates the respiratory burst that apparently has no deleterious effect on the yeasts as they replicate readily in these cells [12]. The fact that the potent fungistatic activity of human neutrophils is completely unrelated to the production of toxic oxygen radicals [25] also lends credence to the idea that *H. capsulatum* yeasts are resistant to respiratory burst products. Thus, the yeasts apparently focus all of their survival strategy on avoiding the contents of lysosomes. In human macrophages, *H. capsulatum* yeasts inhibit PL-fusion [26], whereas normal PL-fusion occurs in mouse P388D1 macrophages [27] and resident peritoneal macrophages [28]. Mouse macrophages avoid the destructive effects of lysosomal hydrolases by regulating intraphagosomal pH to 6.5. At this pH, the destructive capacity of the lysosomal contents are neutralized as they require an acid pH for optimal activity [26]. In addition, at a pH of 6.5, iron transferrin would be half-saturated [29] leaving some free iron available for growth.

Interestingly, yeast-containing phagosomes in human macrophages also are maintained at pH 6.5, in addition to inhibiting PL-fusion [26]. Why the dual strategy for human macrophages? Apparently, this is necessary for two reasons. First, it turns out that unlike the lysosomal enzymes of murine macrophages that require an acid pH for activity, the contents of the lysosomes of human macrophages are active at relatively neutral pH [26]. Second, the yeasts must maintain an intraphagosomal pH of 6.5 to obtain iron for initial growth and replication. Proof of this latter idea comes from experiments designed to disrupt this strategy. Thus, raising the intraphagosomal pH by addition of the weak base chloroquine [30] to infected human macrophages, leads to the complete destruction of *H. capsulatum* yeasts [31]. However, no killing of the yeasts occurs when iron nitriloacetate (FeNTA), an iron compound that is soluble at neutral to alkaline pH [32], is included with the chloroquine.

The importance of iron acquisition for the pathogenesis of *H. capsulatum* yeasts is further emphasized by the fact that chloroquine given intraperitoneally for 6 days to *H. capsulatum*-infected mice, significantly reduces the growth of yeasts in the spleen and liver. Moreover, treatment of mice with chloroquine for 10 days following a lethal inoculum of *H. capsulatum* protects 67% of the mice, whereas all control mice die by day 11 [31]. Currently, the mechanism(s) by which *H. capsulatum* yeasts acquire iron from macrophages is unknown. However, there are some data (described below) suggesting that iron is acquired, at least in part, by the production of low molecular weight siderophores that bind ferric iron.

31.2.3 Activation of Macrophage Anti-Histoplasma Activity

It has been known since the 1950s that the induction of cell-mediated immunity is the hallmark for host defense against *H. capsulatum* [33]. Thus, experiments in mice and guinea-pigs demonstrated that resistance to a lethal inoculum of *H. capsulatum* yeasts could be induced by prior injection of dead yeasts, or by a sublethal infection with viable yeasts. Resistance to *H. capsulatum* yeasts did not correlate with the titer of complement-fixing antibodies, and was not transferred by homologous immune serum. It now is clear that, in murine models of histoplasmosis, a successful immune response comes from the induction of T_H1 type helper T cells and the subsequent production of the cytokines interleukin 12 (IL-12), interferon γ (IFNγ) and tumor necrosis factor α (TNFα) [34–39]. The secretion of IL-12 is required for the production of IFNγ [34, 38], which presumably is required for activation of macrophages. TNFα production is required for a successful host response in both naïve and *H. capsulatum*-immune animals [35–39], but the mechanism of action of TNFα is unknown. However, TNFα does *not* appear to directly activate macrophages [35, 40]. Interestingly, granulocyte-macrophage colony stimulating factor (GM-CSF) is necessary for survival in naïve, but not immune mice, and appears to act via suppression of the T_H2 cytokines IL-4 and IL-10 [11].

31.2.3.1 **Activation of Mouse Macrophage Anti-Histoplasma Activity**

Observations in murine models of histoplasmosis are confirmed by *in vitro* experiments, at least for murine macrophages. Thus, *in vitro*, resident mouse peritoneal macrophages are activated by IFNγ to inhibit the intracellular growth of yeasts [41]. Remarkably, the yeasts remain viable but do not divide. Confounding this observation is the fact that, *in vitro*, splenic macrophages require both IFNγ and LPS to be activated to a fungistatic state [40]. Thus, at least one other cytokine(s) besides IFNγ is required to induce fungicidal activity in murine macrophages.

Activated murine macrophages inhibit the growth of *H. capsulatum* by restricting the availability of intracellular iron [40, 42], and by the production of NO [43, 44]. Thus, macrophages fungistatic activity can be reversed by the addition of holotransferrin or ferrous sulfate [40, 42], or by the addition of *N*-monomethyl-L-arginine, a competitive inhibitor of nitric oxide synthase [44]. Further, only murine macrophages that produce NO have anti-histoplasma activity [43]. Together, these data suggest that IFNγ/LPS activates macrophages to synthesize NO, and NO nitrosylates iron transferrin so that iron is no longer available to the yeasts. *In vivo* studies in mice also demonstrate that the production of NO is required for successful clearance of *H. capsulatum* yeasts in the primary immune response. Remarkably, however, NO does not appear to be essential for clearance of yeasts in the secondary immune response to *H. capsulatum* [45].

31.2.3.2 **Activation of Human Macrophage Anti-Histoplasma Activity**

In contrast to our knowledge of murine immunology, the cytokines required to activate human macrophage fungistatic/fungicidal activity against *H. capsulatum* remain obscure. Neither IFNγ nor TNFα, alone or in combination, activate human macrophages to inhibit the growth of *H. capsulatum* yeasts [46, 47]. Human macrophages are activated by IFNγ to kill and/or inhibit the intracellular growth of intracellular pathogens such as *Trypanosoma cruzi* [48], *Leishmania donovani* [49], *Toxoplasma gondii* [50] and *Chlamydia psittaci* [51]. The most likely explanation for this lack of effectiveness by IFNγ against *H. capsulatum* is that the mechanism(s) of killing/growth inhibition of the aforementioned pathogens is mediated, at least in part, through the production of toxic oxygen metabolites, and, as discussed earlier, *H. capsulatum* yeasts appear to be impervious to toxic oxygen radicals.

Interestingly, culture of human macrophages with IFNγ downregulates transferrin receptors [52], which should result in decreased uptake of iron by macrophages. As *H. capsulatum* is extremely sensitive to the intracellular iron concentration, one might predict that downregulation of transferrin receptors would be sufficient to inhibit the growth of *H. capsulatum* yeasts. However, after 5–7 days of culture in serum-containing medium, human macrophages probably have sufficient iron stores to support the intracellular replication of yeasts, and, therefore, downregulation of transferrin receptors by IFNγ may not reduce the intraphagosomal iron concentration enough to affect the growth of yeasts.

Although IFNγ is ineffective, the presence of IL-3, GM-CSF or monocyte CSF (M-CSF) during the differentiation of monocytes into macrophages does activate macrophage fungistatic (but not fungicidal) activity against *H. capsulatum* yeasts. Addition of IFNγ or TNFα to CSF-activated macrophages during the last 24 h prior to infection with *H. capsulatum* yeasts, does not further enhance macrophage fungistatic activity [47]. The mechanism(s) by which CSFs stimulate human macrophage fungistatic activity against *H. capsulatum* yeasts is unknown.

Despite the dearth of knowledge concerning cytokines, it has been demonstrated that human macrophage fungicidal activity can be activated simply by adherence to type 1 collagen matrices [53]. Other extracellular matrix proteins, such as fibronectin, laminin, and vitronectin, or nongelled collagen do not induce macrophage anti-histoplasma activity. Macrophages are activated immediately upon adherence to the collagen matrices, and fungicidal activity is maintained for up to 4 days post infection. Culture of collagen-adherent macrophages with IFNγ or TNFα, or IL-3, GM-CSF or M-CSF does not augment macrophage anti-histoplasma activity [53]. The mechanism (s) of collagen-induced macrophage-mediated fungicidal activity is through PL-fusion, directly countering the yeast's survival strategy. Thus, the yeasts are destroyed by preformed lysosomal hydrolases. There is no evidence that toxic oxygen radicals, NO, or the restriction of intracellular iron participates in this macrophage antifungal activity [53]. We hypothesize that, *in vivo*, some combination of cytokines and extracellular matrix may be required to activate macrophage fungicidal activity against *H. capsulatum* yeasts.

31.3 Genes Involved in the Intracellular Survival of *H. capsulatum* Yeasts

As discussed above, in murine macrophages *H. capsulatum* yeasts reside in a mature phagosome that has fused with lysosomes. In contrast, in human macrophages, the yeast resides in an immature phagosome that does not fuse with lysosomes. To date, no *H. capsulatum* genes have been described that are required for survival in murine macrophages vs. human macrophages. In fact, it has only been in the last 10 years that molecular genetic techniques have been developed that allow for the disruption of specific genes in *H. capsulatum*, and, therefore, the ability to perform experiments to identify genes required for virulence. Indeed, only five genes have been characterized that appear to be involved in the survival of *H. capsulatum* yeasts in macrophages, and that affect virulence in a mouse model of histoplasmosis. These genes are discussed in the following paragraphs, and summarized in Table 31.2.

31.3.1 Calcium-Binding Protein

The first reported gene required for the virulence of *H. capsulatum* was reported in 2000. This gene, *CBP1*, codes for a calcium-binding protein, CBP1 [54]. CBP1 is a low molecular weight protein produced in copious quantities by *H. capsulatum* yeasts, but not mycelia, during exponential growth in liquid culture media [55–57]. The gene

Table 31.2 *Histoplasma capsulatum* genes required for intracellular growth in macrophages and/or virulence in mice.

Gene	Protein	Mechanism
CBP1	Calcium-binding protein	Calcium homeostasis? Other?
YPS3	Yeast phase-specific protein 3	Unknown
AGS1	α-1,3-Glucan synthase	Surface α-glucan prevents macrophage TNFα secretion?
VMA1	v-ATPase subunit A	Iron homeostasis; other metabolic processes?
SIDA	L-ornithine-N^5-monooxygenase	Siderophore production and iron acquisition

Functional depletion of all genes except *YPS3* leads to deficient growth in macrophages, and attenuated virulence in mice. *YPS3*-silenced mutants grow normally in macrophages, but are attenuated in mice.

CBP1 is yeast-phase specific, and its expression is transcriptionally regulated [57, 58]. *CBP1* is expressed when *H. capsulatum* yeasts replicate within P388D1 macrophages and hamster tracheal epithelial cells, and *CBP1* is a secreted *in vivo* in a murine model of histoplasmosis [55, 58]. Most importantly, CBP1 is a virulence gene for *H. capsulatum*. A *cpb1*-null strain grows slowly in liquid culture when calcium is limited, fails to replicate in and kill P388D1 macrophages *in vitro*, and is unable to infect the lungs of mice. Complementation of the null mutant with a full-length copy of *CBP1* restores the wild-type phenotype, including virulence, thereby fulfilling Koch's molecular postulates [54].

Interestingly, the requirement for *CBP1* expression for virulence does not appear to be due to CBP1 acting in a nutritional capacity as a scavenger of calcium. *H. capsulatum* mycelia, which do not produce CBP1, do not grow well under calcium-limiting conditions, whereas the yeasts replicate relatively well [56]. Thus, resistance to calcium limitation directly correlates with the expression of *CBP1*. However, in yeasts, the expression of *CBP1* is not regulated by the availability of calcium, and this fact seems inconsistent with a simple nutritional role. Thus, the exact role of CBP1 in regulating the growth of *H. capsulatum* yeasts in macrophages, and the virulence of yeasts in mice probably is more complex than the mere scavenging of calcium.

31.3.2
α-1,3-Glucan

A second virulence factor identified for *H. capsulatum* yeasts is the yeast phase-specific cell wall polysaccharide α-1,3-glucan [59]. The initial recognition that α-1,3-glucan might be a virulence factor came from the fact that spontaneous *H. capsulatum* variants of the G184A strain that lack α-1,3-glucan in their cell wall had reduced virulence with respect to their capacity to kill mice after intravenous infection [60, 61]. Recently this observation was confirmed using RNA interference (RNAi) targeting of *AGS1* (encoding α-1,3-glucan synthase) to deplete the cell wall of α-1,3-glucan [62]. In these studies the silenced strain was impaired in its ability to lyse monolayers of P388D1 murine macrophages over 6 days of culture, and impaired in its ability to colonize the lungs of mice. The specificity of the RNA-silencing data were confirmed by generating a deletion of the *AGS1* gene in the chromosome [62]. α-1,3-Glucan comprises the outermost layer of the cell wall, and it has been suggested that the mechanism(s) by which α-1,3-glucan affects the pathogenesis of *H. capsulatum* yeasts is by blocking binding to the macrophage β-glucan receptor dectin-1, and, thereby reducing the production of TNFα [63].

Interestingly, the presence or absence of α-1,3-glucan in the cell wall does not affect phagocytosis of the yeasts by P388D1 macrophages. Further, α-1,3-glucan-deficient yeasts are able to persist in macrophages for up to 7 days, suggesting that α-1,3-glucan is not required for intracellular survival [64]. Finally, chemotype 1 strains such as G217B, which are fully virulent, do not contain α-1,3-glucan in their cell wall [65, 66]. Thus, the role of α-1,3-glucan as a virulence factor may only be specific for the G186AR strain of *H. capsulatum*.

31.3.3 Yeast Phase-Specific Protein 3 (YPS3)

A third virulence factor, YPS3, was originally identified in a differential hybridization screen as one of five unique yeast phase transcripts [67]. YPS3 is a 20 kDa protein that is secreted and localized on the surface of *H. capsulatum* yeasts, and is a homolog of the adhesion molecule BAD1 from *Blastomyces dermatitidis* [68, 69]. YPS3 is expressed mainly on North American isolates such as G217B that possess the highest virulence levels [67, 70]. Although RNA silencing of *YPS3* demonstrates that gene expression is required for optimum virulence in mice, the silenced strains replicate as well as the parent strain in RAW264.7 murine macrophages [71]. Thus, there is an interesting disconnect between the ability of the yeasts to replicate in macrophages, and their virulence in a mouse model of histoplasmosis. The mechanism by which YPS3 promotes virulence in mice is unknown.

31.3.4 Vacuolar ATPase (v-ATPase)

Historically, identification of genes associated with microbial virulence has been through the reverse genetics approach. This strategy is useful when promising virulence genes have been identified by one means or another. While molecular genetic techniques have been successful in "knocking out" genes in the G186AR strain [54], the use of these techniques in strain G217B has been problematic. Fortunately, the development of an efficient insertional mutation technique for *H. capsulatum* using *Agrobacterium tumefasciens*-mediated transformation has made it possible to use a forward genetic screen to identify genes required for a specific phenotype. Characterization of Agrobacterium-mediated transformants in *H. capsulatum* yeasts demonstrate that approximately 90% have a single unique insertion with no chromosomal rearrangements or deletions [72]. A forward genetic screen of 50 000 *H. capsulatum* insertional mutants was used successfully to identify novel genes required for the synthesis of α-1,3-glucan [73].

Using this mutagenesis technique, we screened insertional mutants for an inability to replicate in human macrophages, and identified two mutants with insertions within the gene encoding for the vacuolar ATPase (v-ATPase) catalytic subunit A that were deficient in *VMA1* expression. One of these mutants was characterized in detail, and found to exhibit properties similar to other *vma* mutants in fungi [103]. Eukaryotic v-ATPases are proton pumps, and their primary role is the ATP-driven transport of protons from the cytosol into acidic organelles, including lysosomes, early and late endosomes, and the late golgi apparatus [74]. Organelle acidification appears to be involved in protein sorting in biosynthetic and endocytic pathways, proteolytic activation of zymogen precursors, and transmembrane transport of viral contents and toxins [74–76]. In plants and fungi, the lysosome-like vacuole is involved in the storage of metabolic building blocks, calcium and metal ion homeostasis, and osmotic control [77], and vacuolar acidification is critical for these functions. Thus, *vma1* mutants are disrupted in multiple metabolic functions, and exhibit several interesting phenotypes.

Our mutant, designated *vma1* :: *HPH*, grew normally at pH 5.5 and 6.5, but grew moderately slower at pH 7.5. This phenotype is not as severe as that reported for *Saccharomyces cerevisiae*, in which loss of v-ATPase function leads to a pH-dependent lethality in which mutants can not grow at pH 7 or above, but are able to grow at pH 5.0–5.5 [78]. Other fungi, including *Neurospora crassa*, *Candida albicans* and *Schizosaccharomyces pombe* also show a pH-dependent growth phenotype upon disruption of v-ATPase activity [79–81].

A second characteristic of the *vma1* :: *HPH* mutant is that it is unable to form mycelia at 28 °C. This inability to germinate also is found in *vma* mutants of *N. crassa*, *Aspergillus nidulans* and *C. albicans* [79, 80, 82]. In *S. cerevisiae vma* mutants the germination defect can be suppressed at low pH [83], but culture of the *H. capsulatum vma1* mutant at pH 5.5 did not result in the production of mycelia.

Of particular interest with regards to intracellular survival in macrophages, fungal *vma* mutants are disrupted in iron acquisition [84]. Indeed, the *H. capsulatum vma1* :: *HPH* mutant yeasts are unable to grow on iron-poor medium, demonstrating that the *H. capsulatum* v-ATPase is involved in iron homeostasis. This idea is reinforced by the demonstration that bafilomycin, an inhibitor of the v-ATPase [85], partially blocked the growth of wild-type G217B on iron replete medium, and completely blocked the growth of wild-type yeasts on iron-poor medium. Thus, treatment of wild-type yeasts with bafilomycin recapitulates the *vma1* :: *HPH* mutant phenotype. As iron acquisition is critical for *H. capsulatum* intracellular survival and replication [31, 86], this function of the v-ATPase may be the most critical with regards to the inability of the *vma1* mutant to survive in macrophages.

Although the inability of the *vma1* :: *HPH* mutants to grow on iron-poor medium could be reversed by the addition of the siderophore rhodotorulic acid, the mutant was found to be able to produce siderophores. We hypothesize that under iron starvation conditions, *vma1* :: *HPH* mutant yeasts expend all their effort making and secreting siderophores to acquire iron, but because of overwhelming metabolic stress the mutant becomes nonviable, and is unable to subsequently use the siderophores.

In *vma1*:: *HPH* mutant-infected macrophages, the addition of exogenous free ferric, but not ferric bound to transferrin, partially restores intracellular growth. However, the additional growth is minor compared to the intracellular growth of wild-type or *vma1/VMA1* complemented yeasts. The *vma1* mutant also is unable to inhibit PL-fusion, presumably because of metabolic stress. As yeasts that lack a functional v-ATPase also are sensitive to oxidative stress [87, 88], we hypothesize that the occurrence of PL-fusion and the production H_2O_2 upon ingestion of *H. capsulatum vma1* mutant yeasts [12], is sufficient to lead to their being killed by human macrophages.

These data, and previous data on the iron sensitivity of intracellular *H. capsulatum* yeasts [31, 86] suggest a model for iron acquisition and intracellular survival in macrophages. Thus, when *H. capsulatum* yeasts are initially ingested by macrophages, they require a small amount of free iron to enable them to rapidly transcribe and translate genes necessary for survival and growth. Once these initial gene products are made, the *H. capsulatum* yeasts utilize siderophores (see below) and possibly ferric reductases to obtain iron from transferrin.

In a murine model of pulmonary histoplasmosis, the *vma1* mutant was completely unable to colonize the lungs or disseminate to the spleen. Further, a normally lethal inoculum of the mutant did not kill any of the mice, whereas the wild-type and complemented strain killed all the mice in 12 days. Thus, a functional v-ATPase is an absolute requirement for *H. capsulatum* yeasts to survive and replicate in macrophages, and for virulence in mice.

31.3.5 *SIDA* (*SID1*) (L-Ornithine-N^5-Monooxygenase)

The acquisition of iron from the host is essential for all pathogenic microbes [89], as over 100 metabolic enzymes utilize iron as a cofactor for their catalytic activity [90]. As mentioned earlier, restriction of iron availability in *H. capsulatum*-infected macrophages leads to killing and digestion of the yeasts [31, 86]. Several iron-acquisition systems have been identified in *H. capsulatum*, but their roles in intracellular survival and virulence are unknown. *H. capsulatum* secretes several types of ferric reductases and also has surface ferric reductase activity. The *H. capsulatum* ferric reductases can reduce iron from transferrin and from hydroxamate siderophores [91, 92]. *H. capsulatum* also secretes the hydroxamate siderophores dimerum acid, coprogen B, and fusarine when grown in low-iron media [93]. In addition, *H. capsulatum* yeasts are able to bind hemin through protein-based receptors [94].

A third iron acquisition system, found in *S. cerevisiae*, utilizes a complex of a multicopper oxidase (FET3) and a ferric permease (FTR1) for high-affinity transport of ferric iron [95, 96]. A BLAST search of the G186AR *H. capsulatum* genome reveals sequence homologies of 61% for *FTR1* and 65% for *FET3*. This FET3/FTR1 high-affinity iron uptake complex is required for the virulence and the iron-limited growth of *C. albicans* [97]. However, the genome of *H. capsulatum* strain G217B does not contain *FET3* and *FTR1* homologs (Genome Sequencing Center, Washington University, St. Louis, USA; http://www.genome.wustl.edu/projects/hcapsulatum).

To determine if siderophores were required for intracellular growth in macrophages, we disrupted the expression of the *SIDA* gene, that codes for L-ornithine-N^5-monooxygenase, in *H. capsulatum* strains G217B and G186AR using RNAi silencing and gene knockout strategies, respectively (Hilty *et al.*, manuscript in revision). This enzyme catalyzes the first step in siderophore biosynthesis, and, therefore, loss of expression would result in no siderophore production. Neither the *SIDA*-silenced G217B yeasts, nor the Δ*sidA* G186AR yeasts grow on iron-restricted medium, although both strains grow as well as wild-type and complemented yeasts on iron-rich medium. Further, it was confirmed that neither of the mutants synthesized and secreted siderophores.

When tested for their ability to replicate in human and murine macrophages, the knockout and silenced strains gave somewhat different results. The replication of G186AR Δ*sidA* was inhibited by 52% in human macrophages, and by 47% in murine peritoneal macrophages, compared to the growth of wild-type yeasts. Intracellular growth of the Δ*sidA* yeasts was restored to the level of wild-type yeasts by the addition of FeNTA. G186AR Δ*sidA/SIDA* complemented yeasts grew as well as wild-type yeasts in both human and murine macrophages. In contrast, the intracellular growth of G217B *SIDA*-silenced yeasts was inhibited by 77 and 76%, in human macrophages and murine macrophages, respectively, compared with wild-type yeasts. Again, the addition of FeNTA restored the growth of the *SIDA*-silenced yeasts to the level of wild-type yeasts.

The difference in growth in macrophages between the Δ*sidA* mutant and *SIDA*-silenced mutant may be because the G186AR strain also contains the FET3/FTR1 iron uptake pathway, which could compensate for the loss of siderophore production. However, further experiments are required to verify this hypothesis.

When tested in a murine model of pulmonary histoplasmosis, the G217B *SIDA*-silenced yeasts and the G186AR Δ*sidA* yeasts gave similar results. Thus, one week after infection, the fungal burden of G217B *SIDA*-silenced yeasts was reduced by 1.5 logs in both the lung and spleen compared to wild-type yeasts, and the fungal burden of the G186AR Δ*sidA* yeasts was reduced by 1 log in both organs. The complemented G186AR Δ*sidA/SIDA* grew the same as the wild-type strain. In an additional experiment, a lethal inoculum of G217B wild-type yeasts caused all of the mice to die by day 8, whereas mice infected with G217B *SIDA*-silenced yeasts demonstrated delayed morbidity and mortality with 40% of the mice surviving out to 20 days. A similar experiment could not be performed with the G186AR Δ*sidA* yeasts as the hygromycin-resistance gene, *hph*, that is used to disrupt the *SIDA* gene is uniformly lethal to the mice [98].

Using microarray analysis of *H. capsulatum* strain G217B, Hwang *et al.* [99] identified a discrete set of co-regulated genes whose transcription was upregulated under iron-poor conditions. As these genes were involved in the synthesis and secretion of siderophores, they disrupted the *SID1* (*SIDA*) gene by allelic replacement in *H. capsulatum* strain G186AR. The strains deficient in *SID1* demonstrated a significant growth defect in murine bone-marrow-derived macrophages (BMDMs), that could be reversed by the addition of $FeSO_4$. The Δ*sid1* strain also was attenuated in a murine model of histoplasmosis. Complementation of the Δ*sid1* strain restored normal replication in BMDM and virulence in mice. Although our data and the data of Hwang *et al.* are essentially identical, they cannot be compared directly because of the different techniques used to quantify intracellular growth in macrophages, and the different methods used to expressed the data regarding colonization of mice.

31.4
Immunology of Phagosomes Containing *H. capsulatum* Yeasts

Currently, nothing is known about how antigens are processed or presented by either murine or human macrophages, or which immunologically relevant molecules are present in the phagosomes. Indeed, although murine macrophages are capable of presenting antigens to T cells [100], human monocyte-derived macrophages are unable to stimulate T cell proliferation after phagocytosis of viable or heat-killed yeasts (Newman, unpublished observations). Interestingly, if monocytes are allowed to differentiate in the presence of lymphocytes (as mixed mononuclear cells) and heat-killed *H. capsulatum* yeasts, T cell proliferation is stimulated [101].

Although human macrophages are not good antigen-presenting cells (APCs), human dendritic cells (DCs) are excellent APC. Unlike macrophages, phagocytosis of *H. capsulatum* yeasts by human DCs results in killing and digestion of the yeasts through massive PL-fusion [101, 102]. Further, phagocytosis of viable and heat-killed yeasts leads to antigen presentation to T cells, as quantified by T cell proliferation. Interestingly, antigen presentation is more efficient when DCs ingest viable yeasts [101]. We hypothesize that *in vivo*, the DC is the preferred APC and is required for the generation of cell-mediated immunity.

31.5
Summary and Outlook

Clearly the amount of information available concerning the intracellular survival strategies, and the genes involved, in *H. capsulatum* parasitism of the macrophage phagosome are only just beginning to be discovered. Unlike bacteria, whose genetic manipulation has been available for many years, molecular genetic techniques for *H. capsulatum* have been developed only over the last 10 years. Further, the number of laboratories that study fungal pathogens is tiny compared with the number of labs studying bacterial pathogens. However, with molecular genetic techniques being constantly improved, the next 10 years of research on *H. capsulatum* should reveal many of its secrets to surviving intracellularly in macrophages.

References

1 Darling, S.T. (1906) *JAMA*, **46**, 1283–1285.
2 de Monbreun, W.A. (1934) *Am J Trop Med*, **14**, 93–135.
3 Ajello, L. (1971) in *Histoplasmosis* (eds L. Ajello, E.W. Chick and M.F. Furcolow), CC Thomas, Springfield, IL, pp. 103–122.
4 Vincent, R.D., Goewert, R., Goldman, W.E., Kobayashi, G.S. *et al.* (1986) *J Bacteriol*, **165**, 813–818.
5 Deepe, G.S. and Bullock, W.E. (1992) in *Inflammation: Basic Principles and Clinical Correlates* (eds J.I. Gallin, I.M. Goldstein and R. Snyderman), Raven Press, New York, pp. 943–958.
6 Wood, K.L., Hage, C.A., Knox, K.S., Kleiman, M.B. *et al.* (2003) *Am J Respir Crit Care Med*, **167**, 1279–1282.
7 Newman, S.L. (2001) *Semin Respir Infect*, **16**, 102–108.
8 Maresca, B. and Kobayashi, G.S. (1989) *Microbiol Rev*, **53**, 186–209.
9 Newman, S.L., Bucher, C., Rhodes, J. and Bullock, W.E. (1990) *J Clin Invest*, **85**, 223–230.
10 Baughman, R.P., Kim, C.K., Vinegar, A., Hendricks, D.E. *et al.* (1986) *Am Rev Respir Dis*, **134**, 771–776.
11 Deepe, G.S. Jr., and Seder, R.A. (1998) *Res Immunol*, **149**, 397–406; discussion 509–310.
12 Bullock, W.E. and Wright, S.D. (1987) *J Exp Med*, **165**, 195–210.
13 Sanchez-Madrid, F., Nagy, J.A., Robbins, E., Simon, P. and Springer, T.A. (1983) *J Exp Med*, **158**, 1785–1803.
14 Kimberlin, C.L., Hariri, A.R., Hempel, H.O. and Goodman, N.L. (1981) *Infect Immun*, **34**, 6–10.
15 Long, K.H., Gomez, F.J., Morris, R.E. and Newman, S.L. (2003) *J Immunol*, **170**, 487–494.
16 Gomez, F.J., Allendoerfer, R. and Deepe, G.S. Jr., (1995) *Infect Immun*, **63**, 2587–2595.
17 Newman, S.L., Gootee, L., Bucher, C. and Bullock, W.E. (1991) *Infect Immun*, **59**, 737–741.
18 Howard, D.H. (1964) *J Bacteriol*, **87**, 33–38.
19 Howard, D.H. (1965) *J Bacteriol*, **89**, 518–523.
20 Newman, S.L., Gootee, L., Morris, R. and Bullock, W.E. (1992) *J Immunol*, **149**, 574–580.
21 Howard, D.H. (1959) *J Bacteriol*, **78**, 69–78.
22 Siegel, M.R. and Sisler, H.D. (1964) *Biochim Biophys Acta*, **87**, 83–89.
23 Ennis, H.L. and Lubin, M. (1964) *Science*, **146**, 1474–1476.
24 Eissenberg, L.G., Goldman, W.E. and Schlesinger, P.H. (1993) *J Exp Med*, **177**, 1605–1611.
25 Newman, S.L., Gootee, L. and Gabay, J.E. (1993) *J Clin Invest*, **92**, 624–631.
26 Newman, S.L., Gootee, L., Hilty, J. and Morris, R.E. (2006) *J Immunol*, **176**, 1806–1813.
27 Eissenberg, L.G., Schlesinger, P.H. and Goldman, W.E. (1988) *J Leukoc Biol*, **43**, 483–491.
28 Taylor, M.L., Espinosa-Schoelly, M.E., Iturbe, R., Rico, B. *et al.* (1989) *Clin Exp Immunol*, **75**, 466–470.
29 Princiotto, J.V. and Zapolski, E.J. (1975) *Nature*, **255**, 87–88.
30 Krogstad, D.J. and Schlesinger, P.H. (1987) *N Engl J Med*, **317**, 542–549.
31 Newman, S.L., Gootee, L., Brunner, G. and Deepe, G.S. Jr., (1994) *J Clin Invest*, **93**, 1422–1429.
32 Bates, G.W. and Wernicke, J. (1971) *J Biol Chem*, **246**, 3679–3686.
33 Salvin, S.B. (1955/56) *Trans NY Acad Sci*, **18**, 462–468.
34 Allendoerfer, R., Biovin, G.P. and Deepe, G.S. Jr. (1997) *J Infect Dis*, **175**, 905–914.
35 Allendoerfer, R. and Deepe, G.S. Jr. (1998) *J Immunol*, **160**, 6072–6082.
36 Smith, J.G., Magee, D.M., Williams, D.M. and Graybill, J.R. (1990) *J Infect Dis*, **162**, 1349–1353.
37 Wu-Hsieh, B.A., Lee, G.S., Franco, M. and Hofman, F.M. (1992) *Infect Immun*, **60**, 4230–4238.

38 Zhou, P., Sieve, M.C., Bennett, J., Kwon-Chung, K.J. *et al.* (1995) *J Immunol*, **155**, 785–795.

39 Zhou, P., Sieve, M.C., Tewari, R.P. and Seder, R.A. (1997) *Infect Immun*, **65**, 936–942.

40 Lane, T.E., Wu-Hsieh, B.A. and Howard, D.H. (1993) *Infect Immun*, **61**, 1468–1473.

41 Wu-Hsieh, B.A. and Howard, D.H. (1987) *Infect Immun*, **55**, 1014–1016.

42 Lane, T.E., Wu-Hsieh, B.A. and Howard, D.H. (1991) *Infect Immun*, **59**, 2274–2278.

43 Lane, T.E., Otero, G.C., Wu-Hsieh, B.A. and Howard, D.H. (1994) *Infect Immun*, **62**, 1478–1479.

44 Lane, T.E., Wu-Hsieh, B.A. and Howard, D.H. (1994) *Infect Immun*, **62**, 1940–1945.

45 Zhou, P., Miller, G. and Seder, R.A. (1998) *J Immunol*, **160**, 1359–1368.

46 Fleischmann, J., Wu-Hsieh, B. and Howard, D.H. (1990) *J Infect Dis*, **161**, 143–145.

47 Newman, S.L. and Gootee, L. (1992) *Infect Immun*, **60**, 4593–4597.

48 Nogueira, N., Chaplan, S., Reesink, M., Tydings, J. and Cohn, Z.A. (1982) *J Immunol*, **128**, 2142–2146.

49 Murray, H.W. and Cartelli, D.M. (1983) *J Clin Invest*, **72**, 32–44.

50 Murray, H.W., Rubin, B.Y., Carriero, S.M., Harris, A.M. and Jaffee, E.A. (1985) *J Immunol*, **134**, 1982–1988.

51 Rothermel, C.D., Rubin, B.Y., Jaffe, E.A. and Murray, H.W. (1986) *J Immunol*, **137**, 689–692.

52 Byrd, T.F. and Horwitz, M.A. (1989) *J Clin Invest*, **83**, 1457–1465.

53 Newman, S.L., Gootee, L., Kidd, C., Ciraolo, G.M. and Morris, R. (1997) *J Immunol*, **158**, 1779–1786.

54 Sebghati, T.S., Engle, J.T. and Goldman, W.E. (2000) *Science*, **290**, 1368–1372.

55 Batanghari, J.W., Deepe, G.S. Jr., Di Cera, E. and Goldman, W.E. (1998) *Mol Microbiol*, **27**, 531–539.

56 Batanghari, J.W. and Goldman, W.E. (1997) *Infect Immun*, **65**, 5257–5261.

57 Patel, J.B., Batanghari, J.W. and Goldman, W.E. (1998) *J Bacteriol*, **180**, 1786–1792.

58 Kugler, S., Young, B., Miller, V.L. and Goldman, W.E. (2000) *Cell Microbiol*, **2**, 537–547.

59 Kanetsuna, F., Carbonell, L.M., Gil, F. and Azuma, I. (1974) *Mycopathol Mycol Appl*, **54**, 1–13.

60 Klimpel, K.R. and Goldman, W.E. (1987) *Infect Immun*, **55**, 528–533.

61 Klimpel, K.R. and Goldman, W.E. (1988) *Infect Immun*, **56**, 2997–3000.

62 Rappleye, C.A., Engle, J.T. and Goldman, W.E. (2004) *Mol Microbiol*, **53**, 153–165.

63 Rappleye, C.A., Eissenberg, L.G. and Goldman, W.E. (2007) *Proc Natl Acad Sci USA*, **104**, 1366–1370.

64 Eissenberg, L.G., West, J.L., Woods, J.P. and Goldman, W.E. (1991) *Infect Immun*, **59**, 1639–1646.

65 Domer, J.E. (1971) *J Bacteriol*, **107**, 870–877.

66 Reiss, E., Miller, S.E., Kaplan, W. and Kaufman, L. (1977) *Infect Immun*, **16**, 690–700.

67 Keath, E.J., Painter, A.A., Kobayashi, G.S. and Medoff, G. (1989) *Infect Immun*, **57**, 1384–1390.

68 Bohse, M.L. and Woods, J.P. (2005) *Eukaryot Cell*, **4**, 685–693.

69 Brandhorst, T.T., Wuthrich, M., Warner, T. and Klein, B. (1999) *J Exp Med*, **189**, 1207–1216.

70 Bohse, M.L. and Woods, J.P. (2007) *Eukaryot Cell*, **6**, 609–615.

71 Bohse, M.L. and Woods, J.P. (2007) *Infect Immun*, **75**, 2811–2817.

72 Sullivan, T.D., Rooney, P.J. and Klein, B.S. (2002) *Eukaryot Cell*, **1**, 895–905.

73 Marion, C.L., Rappleye, C.A., Engle, J.T. and Goldman, W.E. (2006) *Mol Microbiol*, **62**, 970–983.

74 Mellman, I., Fuchs, R. and Helenius, A. (1986) *Annu Rev Biochem*, **55**, 663–700.

75 Stevens, T.H. and Forgac, M. (1997) *Annu Rev Cell Dev Biol*, **13**, 779–808.

76 Nishi, T. and Forgac, M. (2002) *Nat Rev Mol Cell Biol*, **3**, 94–103.

77 Klionsky, D.J., Herman, P.K. and Emr, S.D. (1990) *Microbiol Rev*, **54**, 266–292.

78 Nelson, H. and Nelson, N. (1990) *Proc Natl Acad Sci USA*, **87**, 3503–3507.

79 Bowman, E.J., Kendle, R. and Bowman, B.J. (2000) *J Biol Chem*, **275**, 167–176.

80 Poltermann, S., Nguyen, M., Gunther, J., Wendland, J. *et al.* (2005) *Microbiology*, **151**, 1645–1655.

81 Iwaki, T., Goa, T., Tanaka, N. and Takegawa, K. (2004) *Mol Genet Genomics*, **271**, 197–207.

82 Melin, P., Schnurer, J. and Wagner, E.G. (2004) *Microbiology*, **150**, 743–748.

83 Enyenihi, A.H. and Saunders, W.S. (2003) *Genetics*, **163**, 47–54.

84 Davis-Kaplan, S.R., Compton, M.A., Flannery, A.R., Ward, D.M. *et al.* (2006) *J Biol Chem*, **281**, 32025–32035.

85 Bowman, E.J., Siebers, A. and Altendorf, K. (1988) *Proc Natl Acad Sci USA*, **85**, 7972–7976.

86 Newman, S.L., Gootee, L., Stroobant, V., van der Goot, H. and Boelaert, J.R. (1995) *Antimicrob Agents Chemother*, **39**, 1824–1829.

87 Thorpe, G.W., Fong, C.S., Alic, N., Higgins, V.J. and Dawes, I.W. (2004) *Proc Natl Acad Sci USA*, **101**, 6564–6569.

88 Milgrom, E., Diab, H., Middleton, F. and Kane, P.M. (2007) *J Biol Chem*, **282**, 7125–7136.

89 Weinberg, E.D. and Weinberg, G.A. (1995) *Curr Opin Infect Dis*, **8**, 164–169.

90 Miethke, M. and Marahiel, M.A. (2007) *Microbiol Mol Biol Rev*, **71**, 413–451.

91 Timmerman, M.M. and Woods, J.P. (1999) *Infect Immun*, **67**, 6403–6408.

92 Timmerman, M.M. and Woods, J.P. (2001) *Infect Immun*, **69**, 7671–7678.

93 Howard, D.H., Rafie, R., Tiwari, A. and Faull, K.F. (2000) *Infect Immun*, **68**, 2338–2343.

94 Foster, L.A. (2002) *Can J Microbiol*, **48**, 437–442.

95 Askwith, C., Eide, D., Van Ho, A. Bernard, P.S. *et al.* (1994) *Cell*, **76**, 403–410.

96 Stearman, R., Yuan, D.S., Yamaguchi-Iwai, Y., Klausner, R.D. and Dancis, A. (1996) *Science*, **271**, 1552–1557.

97 Ramanan, N. and Wang, Y. (2000) *Science*, **288**, 1062–1064.

98 Smulian, A.G., Gibbons, R.S., Demland, J.A., Spaulding, D.T. and Deepe, G.S. Jr. (2007) *Eukaryot Cell*, **6**, 2066–2071.

99 Hwang, L.H., Mayfield, J.A., Rine, J. and Sil, A. (2008) *PLoS Pathog*, **4**, e1000044.

100 Deepe, G.S. Jr., Smith, J.G., Sonnenfeld, G., Denman, D. and Bullock, W.E. (1986) *Infect Immun*, **54**, 714–722.

101 Gildea, L.A., Morris, R.E. and Newman, S.L. (2001) *J Immunol*, **166**, 1049–1056.

102 Gildea, L.A., Ciraolo, G.M., Morris, R.E. and Newman, S.L. (2005) *Infect Immun*, **73**, 6803–6811.

103 Hilty, J., Smulian, A.G. and Newman, S.L. (2008) *Mol. Micro*, **70**, 127–139.

32
Leishmania: *L. mexicana* vs. *donovani* vs. *major*; Amastigotes vs. Promastigotes

Christine Matte, Julia Mallégol, and Albert Descoteaux

32.1
Leishmania and Leishmaniases

Protozoan parasites of the genus *Leishmania* are unicellular eukaryotic organisms belonging to the trypanosomatid family, characterized by the presence of a single flagellum and a DNA-rich organelle, the kinetoplast. *Leishmania* is endemic in 88 countries on four continents and remains a serious health problem in those countries, with over 12 million cases worldwide. The various *Leishmania* species are the causative agents of a spectrum of human diseases ranging from a confined cutaneous disease to a progressive visceral disease that can be fatal. Cutaneous leishmaniasis is the most common form of leishmaniasis, characterized by the apparition of ulcerative lesions of the skin (Baghdad ulcer, Delhi boil), and is caused by *L. major*, *L. tropica* and *L. mexicana*. Lesions generally heal spontaneously, leaving disfigurating scars, but some forms will persist and disseminate. Mucocutaneous leishmaniasis (espundia) is a variant form of cutaneous leishmaniasis caused by *L. braziliensis braziliensis*. Years after the initial skin ulcers, the infection spreads and causes massive tissue destruction of the oro-naso-pharyngeal region. Finally, *L. donovani*, the causative agent of visceral leishmaniasis (Kala-azar), disseminates and infects macrophages of the liver, the spleen and the bone marrow. This infection is chronic and may be fatal in untreated cases. Current treatment is based on chemotherapy, which is difficult to administer, expensive and becoming ineffective due to the emergence of drug resistance. Although there are currently no effective vaccines, leishmaniasis is considered one of a few parasitic diseases potentially controllable by vaccination.

Leishmania exists under two distinct developmental stages. In mammals, the parasite proliferates intracellularly as non-motile amastigotes, within phagolysosomes of macrophages [1]. In the insect vector, the blood-sucking sand fly of either the genus *Phlebotomus* or the genus *Lutzomyia*, *Leishmania* develops as motile promastigotes. When feeding on an infected mammal, the sand fly takes up amastigote-containing macrophages/monocytes or free amastigotes. As digestion of the blood meal progresses, amastigotes differentiate into the motile promastigote form, which

Intracellular Niches of Microbes. Edited by Ulrich E. Schaible and Albert Haas

ISBN: 978-3-527-32207-7

must attach to the sand fly midgut epithelium to avoid being excreted together with the digested blood meal. Attachment is mediated by the interaction between the parasite surface glycolipid lipophosphoglycan (LPG) and lectins expressed by sand fly midgut epithelial cells [2]. Dividing, noninfective promastigotes (procyclics) undergo metacyclogenesis to become nondividing virulent metacyclic promastigotes [3]. Owing to compositional and structural modifications of LPG, metacyclic promastigotes detach from the midgut epithelial cells and migrate towards the anterior end of the digestive tract [4]. Upon the next blood meal of the infected sand fly, metacyclic promastigotes are regurgitated into the mammal where they are phagocytosed by macrophages. There, they differentiate into amastigotes to replicate inside a vacuole with lysosomal features.

Throughout its lifecycle, *Leishmania* is confronted to hostile environments where it must avoid destruction, including the sand fly midgut and the phagolysosomes of host macrophages. To successfully persevere into both hosts, *Leishmania* uses a panoply of strategies. In this chapter, we provide an overview of our current knowledge on the nature of the intramacrophage compartment(s) in which *Leishmania* species establish and maintain infection.

32.2
Characterization of Intra-Host Cell Compartmentation of *Leishmania* and its Relationships to Virulence

32.2.1
Amastigotes

Internalization of amastigotes is a classical receptor-mediated phagocytic process requiring cytoskeletal reorganization in which the parasite plays no or little active role. As for other particles, the opsonization status determines the mode of entry. Hence, internalization in CHO cells of non-opsonized axenic *L. amazonensis* amastigotes involves the guanosine triphosphatases (GTPases) Rho and Cdc42, whereas parasite internalization is Rac1-dependent when these amastigotes are opsonized with immunoglobulin G [5]. In the case of *L. donovani* amastigotes isolated from the spleen of infected hamsters, their internalization by macrophages is mediated by Rac1 and Arf6, and occurs in the absence of significant phosphorylation of cytosolic $p47^{phox}$, thereby preventing assembly of the NADPH oxidase and exposure to oxidants [6].

When amastigotes of *L. donovani* and *L. major* enter macrophages, no obvious signs of inhibition of phagosome maturation are observed, as the parasitophorous vacuoles (PVs) interact with endosomes and lysosomes, leading to the acquisition of lysosomal features [1, 7–9]. However, *L. donovani* amastigotes disrupt phagosome membrane lipid raft integrity, as determined by the alteration of GM1 distribution and the impairment of flotillin recruitment [6]. The mechanisms by which *L. donovani* amastigotes disrupt lipid rafts at the phagosome membrane and the consequences on pathogenesis are not known, but this may interfere with assembly of an active

NADPH oxidase enzyme. Regardless of the *Leishmania* species involved, vacuoles harboring amastigotes are strongly acidic (pH 4.7–5.2) [10], contain numerous lysosomal hydrolases and their membranes are enriched with late endosomal/lysosomal proteins, such as Rab7, macrosialin, LAMP1 and LAMP2, and the vacuolar H^+-ATPase. Interestingly, whereas amastigotes from most *Leishmania* species (*L. major, L. donovani*) reside in tight individual PVs from where daughter cells segregate into their own PVs, amastigotes of the *L. mexicana* complex reside in communal PVs that become rapidly and increasingly distended. The molecular basis of PV enlargement and the consequences for the intracellular survival of these parasites are poorly understood. That amastigotes reside in an acidic environment is consistent with the optimal metabolism of *L. donovani* amastigotes (respiration, catabolism of energy substrates and incorporation of precursors into macromolecules) at acidic pH (pH 4.0 and 5.5), while these activities are optimal at neutral pH for promastigotes [11].

32.2.2
Promastigotes

The predominant route of entry of serum-opsonized promastigotes is via the complement receptor 3 (CR3) and is mainly dependent on the GTPase RhoA [12]. In contrast to amastigotes, promastigotes exist only transiently inside mammalian hosts. Thus, to establish infection and differentiate into amastigotes, they must first avoid being killed by macrophages following their internalization. One mechanism used by promastigotes to evade the microbicidal consequences of phagocytosis is the inhibition of phagolysosome biogenesis. Hence, in contrast to the amastigotes, *L. donovani* and *L. major* promastigotes are internalized in phagosomes that poorly interact with late endosomes and lysosomes, and as a consequence, recruitment of Rab7 is impaired and, in the case of LAMP1, recruitment is delayed [13–15]. This effect of *Leishmania* promastigotes on phagosome–lysosome fusion is confined to parasite-containing phagosomes, as the fusion machinery remains operational in infected macrophages [13].

This *L. donovani*-induced phagosome maturation arrest is characterized by a progressive accumulation of periphagosomal F-actin [16]. In addition, components of the actin polymerization machinery, including Arp2/3, Wiskott–Aldrich syndrome protein (WASP), α-actinin, myosin II and Nck, accumulate with similar kinetics on *L. donovani* promastigote-containing phagosomes. The Rho-family GTPases, Cdc42, Rac1 and RhoA, are also present on promastigote-containing phagosomes [17]. Accumulating periphagosomal F-actin might interfere with the recruitment of signal transducers and vesicles trafficking to the forming phagolysosome, thus contributing to the arrest of phagolysosome maturation. However, periphagosomal F-actin accumulation does not seem to occur with all *Leishmania* species. Indeed, F-actin rapidly disassembles from newly formed phagosomes harboring *L. amazonensis* metacyclic promastigotes and at 30 min post infection no more than 10–20% of the phagosomes are still surrounded by F-actin [18].

Similar to amastigotes, promastigotes disrupt lipid rafts on the phagosomal membrane [9]. Flotillin-1, a component of lipid rafts, is recruited to phagosomes during the maturation process from late endocytic organelles. Remarkably, *L. donovani* promastigotes inhibit the acquisition of flotillin-1-enriched lipid rafts by phagosomes [9]. As these membrane microdomains play key roles in various cellular functions, including signal transduction [19], targeting of these structures in phagosomes by pathogens might contribute to their ability to establish infection.

An important consequence of the *L. donovani* promastigote-induced phagosome remodeling may be an impairment of the assembly of the NADPH oxidase complex at the phagosome, resulting in the creation of an environment devoid of oxidants [20]. Assembly of the NADPH oxidase requires that cytosolic phosphorylated $p47^{phox}$ and $p40/p67^{phox}$ heterodimers associate to form $p47/p67/p40^{phox}$ heterotrimers prior to their membrane translocation, where they interact with membrane-associated flavocytochrome *b*558 [21, 22]. Internalization of *L. donovani* promastigotes triggers the phosphorylation of $p47^{phox}$ and the formation of complexes containing both $p47^{phox}$ and $p67^{phox}$. However, these cytosolic components of the NADPH oxidase complex fail to associate to the promastigote-containing phagosome membrane. This impairment is not the consequence of a general defect in the NADPH oxidase activation process but is restricted to promastigote-containing phagosomes. Thus, as one consequence of phagosome remodeling, *L. donovani* promastigotes establish infection in an environment devoid of oxidants, which may be propitious to their differentiation into amastigotes.

32.3
Leishmania Molecules Involved in Diverted Phagosome Maturation/Compartment Maintenance

Leishmania produces a variety of glycoconjugates, the most abundant belonging to the phosphoglycan (PG) family. These glycoconjugates all share a unique structure not found in mammals, namely the disaccharide-phosphate Gal(β1,4)Man(α1-PO_4 → 6) unit [23]. These PGs are either secreted (phosphoglycan, proteophosphoglycan, acid phosphatase), or membrane-bound (lipophosphoglycan or LPG). With 1–5 million copies per cell, LPG covers the entire promastigote surface including the flagellum and forms a dense glycocalyx. LPG is a stage-specific molecule as it is either strongly downregulated or absent in the amastigote stage. The abundance, location and uniqueness of these glycoconjugates are consistent with the functions that LPG and other PGs play throughout the *Leishmania* lifecycle and establishment of infection within macrophages. The present section describes the role of LPG in phagosome remodeling induced by promastigotes.

32.3.1
Molecular Structure

Due to its abundance, imposing size and structure, LPG is the dominant molecule of the promastigote glycocalyx. The backbone of LPG consists of a polymer of the repeating Gal(β1,4)Man(α1-PO_4 → 6) unit, linked to a 1-*O*-alkyl-2-*lyso*-phosphatidyl (*myo*)inositol (PI) anchor via a glycan core (Figure 32.1). The number of LPG-repeating

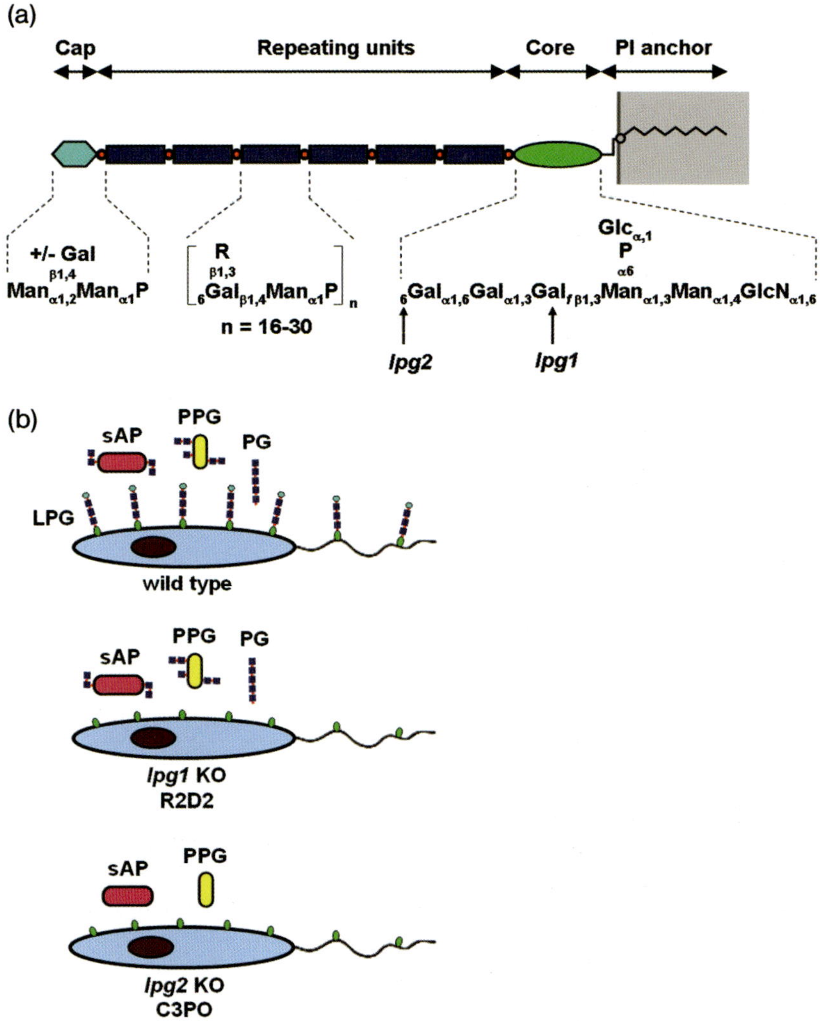

Figure 32.1 Structure and diversity of the *Leishmania* surface phosphoglycans. (a) The sugar composition of the three *Leishmania* lipophosphoglycan distal domains is depicted, along with the membrane-embedded 1-*O*-alkyl-2-*lyso*-phosphatidyl(*myo*)inositol (PI) anchor. The *lpg1* and *lpg2* genes function at the indicated steps of lipophosphoglycan (LPG) biosynthesis. (b) Disaccharide phosphate repeating units are found on all phosphoglycans (PGs) of wild-type promastigotes, while they are either partially or completely absent in LPG-deficient (*lpg1* KO) and repeating unit-deficient (*lpg2* KO) promastigotes, respectively. Man, mannose; Gal, galactose; Gal$_f$, galactofuranosyl; GlcN, glucosamine; P, phosphate; R, saccharide side-chain.

units varies from 16 to 30, depending on the promastigote developmental stage (procyclic vs. metacyclic) and species. At the nonreducing end of the repeating unit moiety is a small cap composed of neutral oligosaccharides, mostly galactose and mannose residues.

Structural analyses of LPG from several *Leishmania* species have revealed complete conservation of the PI anchor and extensive conservation of the glycan core, the only exception being that 40% of *L. major* cores are not substituted with Glcα1-PO_4 [24]. Conversely, the oligosaccharide cap displays some degree of variability among *Leishmania* species, in both sugar composition and sequence. *L. major* LPG is capped with the simplest structure, Manα1-2Man [24], while *L. donovani* and *L. mexicana* LPGs are predominantly capped with the branched trisaccharide Manα1-2[Galβ1-4] Man [25, 26]. The most striking differences in LPG structure amongst *Leishmania* species pertain to the repeating unit moiety. While *L. donovani* LPG follows the basic Gal(β1,4)Man(α1-PO_4 → 6) repeat sequence [25], LPG molecules from other species have additional saccharide side-chains branching off the C3 position of the galactose residue (Figure 32.1, R). In *L. mexicana* for instance, 25% of the galactose residues are substituted with βGlc [26]. *L. major* LPG presents the most complex repeating unit moiety, with approximately 87% substitution for small chains containing one to four galactose, glucose or pentose arabinose residues [24]. Additionally, a small portion (<20%) of *L. major* LPG may have two saccharide chains branching off the glycan core, forming a Y-shaped molecule [24].

32.3.2
A Role for LPG in the Inhibition of Phagosome Maturation

Chemical mutagenesis experiments performed during the 1980s and 1990s led to the isolation of four *L. donovani* LPG-defective mutants, affectionately named the Star Wars mutants R2D2, C3PO, OB1 and JEDI, that display different alterations in LPG structure often accompanied by defects in the biosynthesis of other PGs [27–30]. A functional genetic complementation approach was developed to identify the genes affected by these mutations. Thus the *LPG1* gene, discovered upon complementation of the R2D2 mutant, encodes a putative galactofuranosyltransferase responsible for the addition of the unique galactofuranosyl (Gal_f) residue present within the LPG glycan core [31, 32]. Mutations in *LPG1* therefore produce mutants expressing a truncated LPG core to which repeating units cannot be added, without affecting the biosynthesis of secreted PGs. Similarly, complementation of the C3PO mutant led to the identification of the *LPG2* gene, which codes for a membrane protein required for the transport of GDP-mannose into the Golgi lumen, where repeating units are assembled on LPG and all other PGs [33, 34]. Hence, mutations in *LPG2* have a global impact on the biosynthesis of all PGs.

R2D2, C3PO and the corresponding $lpg1^{-/-}$ and $lpg2^{-/-}$ deletion mutants created thereafter are essential tools in the investigation of the role played by PGs in the interactions between *Leishmania* promastigotes with both the sand fly and the mammalian host. Their use first helped to shed some light on the role played by the unique disaccharide-phosphate structure of PGs. Repeating unit-deficient *L. donovani* or *L. major* promastigotes ($lpg2^{-/-}$) are rapidly destroyed upon phagocytosis by macrophages [20, 35, 36]. Consistently, these mutants fail to impede with phagolysosome biogenesis [13, 15], and recruitment of the late endocytic and lysosomal markers Rab7 and LAMP1 occurs [14]. Biochemical and genetic approaches identified LPG as the sole PG responsible for inhibition of phagosome maturation. Hence, opsonization of the PG-deficient C3PO with purified *L. donovani* LPG conferred the capacity to inhibit phagosome–lysosome fusion, indicating that LPG repeating units are necessary and sufficient to mediate this effect [13]. In addition, phagosomes containing repeating unit-deficient (*lpg2*$^{-/-}$) or LPG-deficient (*lpg1*$^{-/-}$) *L. donovani* promastigotes displayed comparably high levels of fusion with endosomes, indicating that the sole absence of LPG prevents promastigotes from blocking phagolysosomal biogenesis [15]. These conclusions correlated well with the observation that amastigotes, which do not make LPG, are internalized into compartments that readily fuse with late endosomes and lysosomes and rapidly acquire late endosome/lysosome markers [15, 18]. This *L. donovani* promastigote-induced phagosome remodeling is associated with the exclusion of cytosolic components of the NADPH oxidase complex (p47^{phox} and p67^{phox}) from the phagosome membrane and the impairment of superoxide production at the phagosome, a process strictly dependent on the presence of LPG [20]. Hence, instead of a system based on effector proteins injected into the host cells through a specialized secretion apparatus, *Leishmania* promastigotes use a single molecule, LPG, to arrest phagosome maturation and establish infection in an environment devoid of oxidants.

Further studies attributed importance to the length of the LPG repeating unit moiety. Hence, a *L. donovani* mutant expressing truncated forms of LPG with only three to five repeating units was unable to inhibit phagosome–endosome fusion [13]. However, the level of structural complexity of this moiety, increased for instance by the presence of side-chains, was ruled out as a potential contributing factor upon the observation that the *L. major* Spock mutant, defective in LPG oligosaccharide side-chain biosynthesis, impairs phagosome–endosome fusion to the same extent as wild-type *L. major* promastigotes [15].

32.3.3
Putative Modes of Action

Several hypotheses have been put forward in recent years regarding the mechanism (s) by which LPG may exert its impact on phagosome remodeling. The periphagosomal accumulation of F-actin is strictly dependent on the presence of LPG [16] and is caused by the LPG-mediated retention of the Rho family GTPase Cdc42 at the phagosome. Indeed, expression of the dominant-negative Cdc42N17 mutant inhibited LPG-mediated periphagosomal F-actin accumulation in RAW264.7 macrophages [17, 37]. Likewise, components of the host cell machinery involved in actin polymerization and cytoskeleton rearrangement, such as WASP, Arp2/3, Nck, α-actinin and myosin II, are retained at the phagosome in an LPG-dependent manner (Figure 32.2) [17]. LPG also interferes with the association of protein kinase Cα (PKCα) to the phagosome membrane [16]. PKCα normally translocates to nascent phagosomes and participates in periphagosomal F-actin breakdown [38] and in the regulation of phagosome maturation [39–41]. LPG-mediated exclusion of PKCα from *L. donovani* promastigote-containing phagosomes may therefore contribute to the periphagosomal accumulation of F-actin. Another function of LPG is

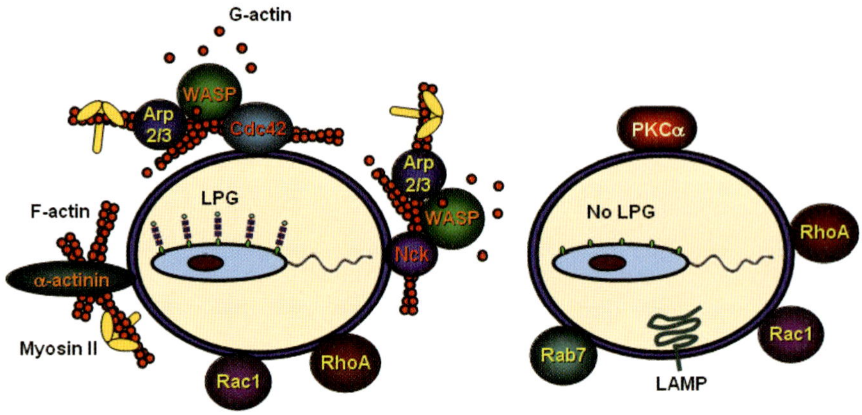

Figure 32.2 Lipophosphoglycan (LPG)-mediated retention of the actin polymerization machinery at the phagosome membrane. *Leishmania* promastigotes induce periphagosomal accumulation of F-actin through LPG-mediated abnormal retention of Cdc42 and Nck at the phagosome membrane, as well as retention of the actin polymerization, crosslinking and remodeling machinery. In the absence of LPG, no accumulation of F-actin occurs and recruitment of PKCα, Rab7 and LAMP1 to the phagosome membrane takes place.

the disruption of lipid rafts observed on phagosomes harboring *L. donovani* promastigotes [9], and this may involve the insertion of the LPG lipid anchor into these membrane microdomains [42].

It is puzzling to imagine how a glycolipid such as LPG can exert such pleiotropic effects on phagosome maturation-related activities. One interesting hypothesis is derived from the observation that when promastigotes attach to the macrophage surface, LPG is transferred from the parasite to the outer leaflet of the plasma membrane and, upon internalization, to the inner leaflet of the phagosome membrane [43, 44]. This insertion may disturb the normal occurrence of events at the phagosome membrane. For instance, it appears to be the membrane insertion of LPG that prevents PKCα association to the membrane [45]. Alternatively, LPG transfer to the inner leaflet of the phagosome membrane may play a direct role in the inhibition of phagosome maturation. LPG insertion alters the biophysical properties of the entire phagosome membrane, which becomes stabilized and refractory to the formation of fusion intermediates [46]. Regardless, the phenomenon of LPG insertion provides an explanation for the restriction of promastigote-mediated effects on phagosome maturation to *Leishmania*-containing phagosomes.

32.3.4
The Peculiar Case of *Leishmania mexicana*

For both *L. donovani* and *L. major*, LPG is recognized as a key promastigote virulence factor, and its role in mediating phagosome remodeling is well established. However, the contribution of LPG in the establishment of *L. mexicana* infection is less obvious.

Hence, *L. mexicana lpg1*$^{-/-}$ promastigotes fail to show decreased virulence upon mouse infections, cause lethal disseminated disease, and display normal replication within macrophages [47]. The species difference with regard to virulence factors appears to include all other phosphoglycans. Whereas *L. major lpg2*$^{-/-}$ mutants show a much more severe loss of virulence than their *lpg1*$^{-/-}$ counterparts [48], *L. mexicana lpg2*$^{-/-}$ mutants remain just as virulent in both mice and macrophages [49]. Thus *Leishmania* species may have evolved an alternative strategy for establishment of infection and survival within mammalian macrophages. Whether these differences are attributable to the fact that *L. mexicana* resides in a large communal vacuole compared to the tight individual vacuole that characterizes *L. donovani* and *L. major* is a hypothesis that will deserve further investigation.

32.3.5 Compartment Maintenance by Amastigotes

Although successful establishment of an intracellular infection by promastigotes is a crucial step in the *Leishmania* lifecycle, replication and long-term survival as an amastigote are thought to be the most important requirements for virulence for *Leishmania* parasites. While studies focusing on the amastigote stage may be clinically more relevant, little attention has been devoted to the impact of amastigotes on their intracellular niche, the phagolysosome.

32.3.6 LIT1 and the Battle for Iron

An adequate supply of iron is essential for the survival of *Leishmania* parasites within macrophages, a requirement somewhat problematic as only low amounts of iron are found in late endosomes and lysosomes. When iron is internalized by mammalian cells, it exists as the poorly soluble oxidized ferric (Fe^{3+}) form complexed to transferrin. Upon endosome acidification, Fe^{3+} is released from transferrin, reduced to the soluble ferrous (Fe^{2+}) form and ultimately transported to the cytoplasm, a function executed in large part by the NRAMP proteins. NRAMP1 (*Slc11a1*) is a pH-dependent efflux pump located on late endosome/lysosome membranes of macrophages that extrudes both Fe^{2+} and Mn^{2+} from these compartments to the cytoplasm. Numerous studies have reported a crucial role for this transporter in innate resistance to intracellular pathogens, probably through its role in iron deprivation from the endocytic system. Hence, inactivation of NRAMP1 renders macrophages permissive to *L. donovani* replication [50].

LIT1, a ZIP family ferrous iron transporter expressed in *L. amazonensis* and *L. major*, may counterbalance the action of NRAMP1 in favor of parasite virulence [51]. LIT1 is solely expressed by amastigotes and upregulated by the intracellular milieu, and is essential for intracellular replication of amastigotes in macrophages. In *L. amazonensis*, LIT1 plays a role in vacuole expansion and maintenance, as 48 h post infection the overall size of the compartments containing *lit1*$^{-/-}$ parasites was much smaller and, generally, contained only one amastigote per vacuole. By 72 h post infection, these compartments appeared shrunken, and some *lit1*$^{-/-}$ parasites showed signs of degeneration. Moreover, overexpression of LIT1 leads to collapse of the *L. amazonensis*-containing vacuole, suggesting that the balance between Fe^{2+} translocation to the parasite by LIT1 and Fe^{2+} efflux to the cytoplasm by NRAMP1 has consequences on vacuole size and maintenance.

32.3.7
Phagolysosome Enlargement

A proteophosphoglucan (PPG) molecule secreted by *L. mexicana* amastigotes seems to cause vacuolization in macrophages, and has therefore been suggested to play a role in the formation of the large communal PVs observed in macrophages infected by this species [52]. However, the biological significance of this observation for parasite virulence is still unknown.

32.4
Immunology of Phagosomes Containing *Leishmania*

Antigen presentation to T lymphocytes is key to the development of an effective immune response against *Leishmania*, and phagosomes play a pivotal role in this process as these organelles are fully competent to process antigens and generate peptide : MHC class II complexes that are transported to the cell surface and presented to T cells [53]. Therefore, the late endosomal/lysosomal features of *Leishmania* amastigote-containing phagosomes suggest that they should be competent to process and generate peptide : MHC class II complexes. On the other hand, alteration of phagosome composition and function by *Leishmania* promastigotes may potentially reduce their ability to process antigens. As MHC class I molecules have not been observed to be associated with *Leishmania*-containing vacuoles [63, 64], the issue of cross-presentation will not be addressed here.

Several studies indicated that antigen presentation in the context of MHC class II is deficient in *Leishmania*-infected macrophages [54–59]. This deficiency is not due to the absence of MHC class II and H-2M molecules in phagosomes formed after internalization of infectious promastigotes and amastigotes, since these molecules are rapidly found in the phagosomes after infection [60]. The observation that *L. amazonensis* and *L. mexicana* amastigotes capture by endocytosis and degrade MHC class II and H-2M [60] may contribute to the inhibition of antigen presentation by these *Leishmania* species. Furthermore, amastigote antigens can be sequestered, and thus unavailable for presentation in the MHC class II context [57]. Alternatively, amastigotes may secrete molecules such as complex polysaccharides, which block antigen presentation [61]. One such molecule is the high molecular weight proteophosphoglycan, a polyanionic molecule released by *L. amazonensis* and *L. mexicana* amastigotes in the phagolysosomal lumen [62].

Efficiency of presentation of the internal LACK antigen of *L. donovani*, *L. major* and *L. amazonensis* closely correlates with the developmental stage and infectivity of these parasites. Hence, whereas a transient presentation of LACK occurs when macrophages are infected with non-infectious procyclic promastigotes, there is no detectable presentation of this antigen when macrophages are infected with either infectious metacyclic promastigotes or amastigotes [56]. A similar observation was made with two other protective endogenous antigens (P-8 and GP46) [57], indicating that MHC class II-mediated antigen presentation is inhibited during *Leishmania* infection. The mechanisms involved remain to be understood, and functional characterization of *Leishmania*-harboring vacuoles may provide some clues. In this regard, disruption of lipid raft integrity of the *Leishmania*-containing phagosome membrane may contribute to the defective MHC class II-mediated antigen presentation.

32.5 Summary and Outlook

The features and properties of the *Leishmania*-containing vacuoles depend on whether macrophages internalize promastigotes or amastigotes. To survive phagocytosis, promastigotes must block the maturation process of their vacuole and create an environment propitious to promastigote-to-amastigote differentiation. This process is strictly dependent on the presence of the surface glycolipid LPG. On the other hand, amastigotes proliferate with apparent impunity within an hydrolytic compartment with late endosomal/lysosomal features. Future issues to be addressed include the mechanisms involved in the enlargement of the parasitophorous vacuole harboring amastigotes of the *L. mexicana* complex, as well as the origin of the membrane required for the formation of new parasitophorous vacuoles when amastigotes of *L. donovani* or *L. major* replicate.

References

1. Chang, K.P. and Dwyer, D.M. (1976) *Science*, **193**, 678–680.
2. Sacks, D.L. (2001) *Cell Microbiol*, **3**, 189–196.
3. Sacks, D.L. (1989) *Exp Parasitol*, **69**, 100–103.
4. Sacks, D.L. (1992) *Infect Agents Dis*, **1**, 200–206.
5. Morehead, J., Coppens, I. and Andrews, N.W. (2002) *Infect Immun*, **70**, 4571–4580.
6. Lodge, R. and Descoteaux, A. (2006) *Eur J Immunol*, **36**, 2735–2744.
7. Berman, J.D., Dwyer, D.M. and Wyler, D.J. (1979) *Infect Immun*, **26**, 375–379.
8. Antoine, J.C., Prina, E., Lang, T. and Courret, N. (1998) *Trends Microbiol*, **6**, 392–401.
9. Dermine, J.F., Duclos, S., Garin, J., St-Louis, F. *et al.* (2001) *J Biol Chem*, **276**, 18507–18512.
10. Antoine, J.C., Prina, E., Jouanne, C. and Bongrand, P. (1990) *Infect Immun*, **58**, 779–787.
11. Mukkada, A.J., Meade, J.C., Glaser, T.A. and Bonventre, P.F. (1985) *Science*, **229**, 1099–1101.
12. Lodge, R. and Descoteaux, A. (2005) *Clin Immunol*, **114**, 256–265.
13. Desjardins, M. and Descoteaux, A. (1997) *J Exp Med*, **185**, 2061–2068.

14 Scianimanico, S., Desrosiers, M., Dermine, J.F., Meresse, S. *et al.* (1999) *Cell Microbiol*, **1**, 19–32.

15 Dermine, J.F., Scianimanico, S., Privé, C., Descoteaux, A. and Desjardins, M. (2000) *Cell Microbiol*, **2**, 115–126.

16 Holm, A., Tejle, K., Magnusson, K.E., Descoteaux, A. and Rasmusson, B. (2001) *Cell Microbiol*, **3**, 439–447.

17 Lodge, R. and Descoteaux, A. (2005) *Cell Microbiol*, **7**, 1647–1658.

18 Courret, N., Frehel, C., Gouhier, N., Pouchelet, M. *et al.* (2002) *J Cell Sci*, **115**, 2303–2316.

19 Simons, K. and Vaz, W.L. (2004) *Annu Rev Biophys Biomol Struct*, **33**, 269–295.

20 Lodge, R., Diallo, T.O. and Descoteaux, A. (2006) *Cell Microbiol*, **8**, 1922–1931.

21 Bokoch, G.M. and Diebold, B.A. (2002) *Blood*, **100**, 2692–2696.

22 Brown, G.E., Stewart, M.Q., Liu, H., Ha, V.L. and Yaffe, M.B. (2003) *Mol Cell*, **11**, 35–47.

23 Descoteaux, A. and Turco, S.J. (1999) *Biochim Biophys Acta*, **1455**, 341–352.

24 McConville, M.J., Thomas-Oates, J.E., Ferguson, M.A. and Homans, S.W. (1990) *J Biol Chem*, **265**, 19611–19623.

25 Thomas, J.R., McConville, M.J., Thomas-Oates, J.E., Homans, S.W. *et al.* (1992) *J Biol Chem*, **267**, 6829–6833.

26 Ilg, T., Etges, R., Overath, P., McConville, M.J. *et al.* (1992) *J Biol Chem*, **267**, 6834–6840.

27 King, D.L. and Turco, S.J. (1988) *Mol Biochem Parasitol*, **28**, 285–293.

28 McNeely, T.B., Tolson, D.L., Pearson, T.W. and Turco, S.J. (1990) *Glycobiology*, **1**, 63–69.

29 Descoteaux, A., Avila, H.A., Zhang, K., Turco, S.J. and Beverley, S.M. (2002) *EMBO J*, **21**, 4458–4469.

30 Descoteaux, A., Mengeling, B.J., Beverley, S.M. and Turco, S.J. (1998) *Mol Biochem Parasitol*, **94**, 27–40.

31 Ryan, K.A., Garraway, L.A., Descoteaux, A., Turco, S.J. and Beverley, S.M. (1993) *Proc Natl Acad Sci USA*, **90**, 8609–8613.

32 Huang, C. and Turco, S.J. (1993) *J Biol Chem*, **268**, 24060–24066.

33 Descoteaux, A., Luo, Y., Turco, S.J. and Beverley, S.M. (1995) *Science*, **269**, 1869–1872.

34 Ma, D., Russell, D.G., Beverley, S.M. and Turco, S.J. (1997) *J Biol Chem*, **272**, 3799–3805.

35 McNeely, T.B. and Turco, S.J. (1990) *J Immunol*, **144**, 2745–2750.

36 Späth, G.F., Lye, L.F., Segawa, H., Sacks, D.L. *et al.* (2003) *Science*, **301**, 1241–1243.

37 Lerm, M., Holm, A., Seiron, A., Sarndahl, E. *et al.* (2006) *Infect Immun*, **74**, 2613–2618.

38 Holm, A., Tejle, K., Gunnarsson, T., Magnusson, K.E. *et al.* (2003) *Biochem Biophys Res Commun*, **302**, 653–658.

39 Allen, L.H. and Aderem, A. (1995) *J Exp Med*, **182**, 829–840.

40 Aderem, A. and Underhill, D.M. (1999) *Annu Rev Immunol*, **17**, 593–623.

41 Ng Yan Hing, J.D., Desjardins, M. and Descoteaux, A. (2004) *Biochem Biophys Res Commun*, **319**, 810–816.

42 Dermine, J.F., Goyette, G., Houde, M., Turco, S.J. and Desjardins, M. (2005) *Cell Microbiol*, **7**, 1263–1270.

43 Tolson, D.L., Turco, S.J. and Pearson, T.W. (1990) *Infect Immun*, **58**, 3500–3507.

44 Duque-de-Mello, M., Ho, J.L., Vannier-Santos, M.A. and Pimenta, P.F.P. (1999) *Mem Inst Oswaldo Cruz*, **94**, 70.

45 Giorgione, J.R., Turco, S.J. and Epand, R.M. (1996) *Proc Natl Acad Sci USA*, **93**, 11634–11639.

46 Miao, L., Stafford, A., Nir, S., Turco, S.J. *et al.* (1995) *Biochemistry*, **34**, 4676–4683.

47 Ilg, T. (2000) *EMBO J*, **19**, 1953–1962.

48 Sacks, D.L., Modi, G., Rowton, E., Spath, G. *et al.* (2000) *Proc Natl Acad Sci USA*, **97**, 406–411.

49 Ilg, T., Demar, M. and Harbecke, D. (2001) *J Biol Chem*, **276**, 4988–4997.

50 Forbes, J.R. and Gros, P. (2001) *Trends Microbiol*, **9**, 397–403.

51 Huynh, C., Sacks, D.L. and Andrews, N.W. (2006) *J Exp Med*, **203**, 2363–2375.

52 Peters, C., Stierhof, Y.D. and Ilg, T. (1997) *Infect Immun*, **65**, 783–786.

53 Ramachandra, L., Song, R. and Harding, C.V. (1999) *J Immunol*, **162**, 3263–3272.

54 Fruth, U., Solioz, N. and Louis, J.A. (1993) *J Immunol*, **150**, 1857–1864.

55 Prina, E., Jouanne, C., de Souza Lao, S., Szabo, A. *et al.* (1993) *J Immunol*, **151**, 2050–2061.

56 Prina, E., Lang, T., Glaichenhaus, N. and Antoine, J.C. (1996) *J Immunol*, **156**, 4318–4327.

57 Kima, P.E., Soong, L., Chicharro, C., Ruddle, N.H. and McMahon-Pratt, D. (1996) *Eur J Immunol*, **26**, 3163–3169.

58 Wolfram, M., Ilg, T., Mottram, J.C. and Overath, P. (1995) *Eur J Immunol*, **25**, 1094–1100.

59 Courret, N., Prina, E., Mougneau, E., Saraiva, E.M. *et al.* (1999) *Eur J Immunol*, **29**, 762–773.

60 Antoine, J.C., Lang, T., Prina, E., Courret, N. and Hellio, R. (1999) *J Cell Sci*, **112** (Pt 15), 2559–2570.

61 Leyva-Cobian, F., Outschoorn, I.M., Carrasco-Marin, E. and Alvarez-Dominguez, C. (1997) *Clin Immunol Immunopathol*, **85**, 1–15.

62 Ilg, T., Stierhof, Y.D., McConville, M.J. and Overath, P. (1995) *Eur J Cell Biol*, **66**, 205–215.

63 Lang, T., de Chastellier, C., Frehel, C., Hellio, R. *et al.* (1994) *J Cell Sci*, **107** (Pt 1), 69–82.

64 Lang, T., Hellio, R., Kaye, P.M. and Antoine, J.C. (1994) *J Cell Sci*, **107** (Pt 8), 2137–2150.

33
Plasmodium and *Babesia*

Markus Winterberg, Jude M. Przyborski, and Klaus Lingelbach

33.1
Introduction

Babesia and *Plasmodium* belong to the phylum of *Apicomplexa*, unicellular eukaryotes characterized by the so-called apical complex, located at one pole of the cell. This complex consists of several compartments (e.g., rhoptries, micronemes and, in *Babesia* spp., the spherical body) which play an important role during the invasion of the respective host cell. Both parasites are obligatory intracellular pathogens with short-lived extracellular stages in their mammalian hosts. Most apicomplexan parasites, including *Plasmodium* and *Babesia*, contain a relict plastid. Initially it was believed that this plastid originated from a green alga, but later results strongly suggested that it was derived from an ancestral red alga which the free-living ancestors of these parasites presumably acquired in a process called "secondary endosymbiosis" [1]. Several plant- and prokaryote-specific biosynthetic pathways, such as isoprenoid synthesis, are localized in this organelle [2–4].

Babesia and *Plasmodium* are transmitted by ticks and by mosquitoes, respectively. Infections are widespread in warm-blooded vertebrates, albeit with a high host specificity. Four species (*P. falciparum, P. vivax, P. malariae* and *P. ovale*) are the causative agents of malaria in humans. In addition, in recent years a growing body of evidence suggest that *P. knowlesi* (a primate parasite) also causes disease in human patients, particularly in Malaysia [5, 6]. Whether these infections can be viewed as a recent zoonosis, or reflect improvements in species-specific diagnosis and monitoring of patients remains to be seen, but demonstrates that our understanding of the epidemiology of malaria infection is still far from complete. *P. falciparum* is responsible for the most severe form of malaria, resulting in 2–3 million deaths annually, mainly children under the age of five. Owing to its impact on public health in endemic countries, *P. falciparum* is at the focus of medical and biomedical research. The typical vertebrate hosts of the various *Babesia* species include wild rodents (*B. microti*), cattle (*B. divergens, B. bovis*) and dogs (*B. gibsoni*). *B. microti, B. bovis* and *B. divergens* occasionally infect humans, in particular immunocompromised patients.

Intracellular Niches of Microbes. Edited by Ulrich E. Schaible and Albert Haas

ISBN: 978-3-527-32207-7

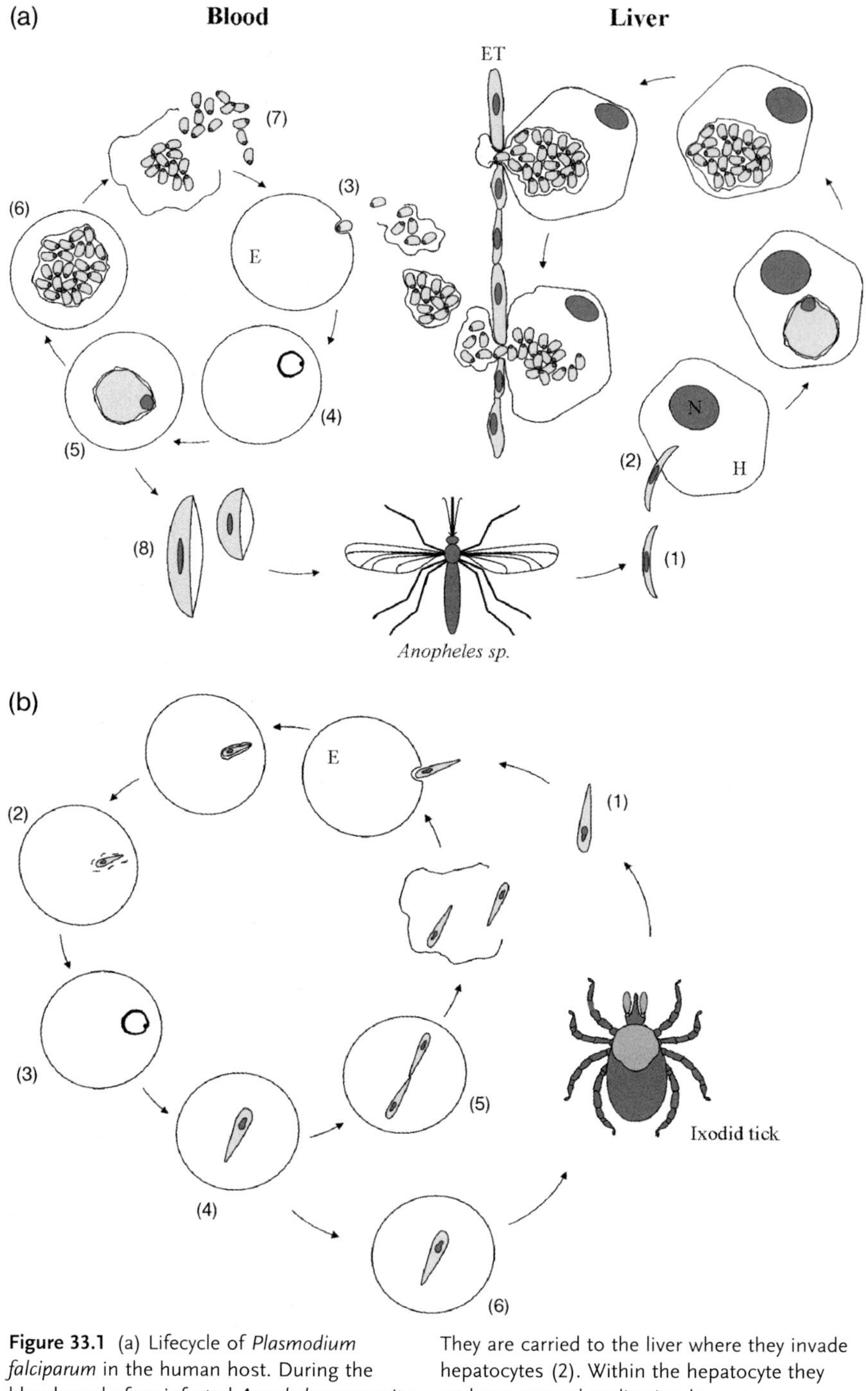

Figure 33.1 (a) Lifecycle of *Plasmodium falciparum* in the human host. During the blood meal of an infected *Anopheles* mosquito infectious sporozoites (1) are transmitted. They are carried to the liver where they invade hepatocytes (2). Within the hepatocyte they undergo asexual replication known as exoerythrocytic schizogony, resulting in the

Within their vertebrate hosts, both parasites display high host cell specificities. While *Babesia* spp., presumably exclusively, invade differentiated red blood cells (RBCs), *Plasmodium* spp. first replicate in liver cells before they are released into the bloodstream where they invade RBCs (Figure 33.1a and b). While *P. falciparum* infects differentiated RBCs, some species, such as *P. vivax*, preferentially invade reticulocytes. The time of development within the RBC varies between species from 24 to 72 h. Intracellular multiplication of the parasite occurs by schizogony, a process that involves multiple nuclear divisions leading to a multinucleated cell (a structure which in biological terms is called "plasmodium") followed by the formation of individual plasma membranes and the differentiation into infectious "merozoites." Upon rupture of the infected cell, 16–32 merozoites are released and invade uninfected cells almost instantaneously. Schizogony in *Babesia* occurs by binary fission (budding), with a lower replication rate than that of *Plasmodium*, and a generation time of approximately 10 h.

33.2 Illness

Malaria is endemic in tropical and subtropical countries, with a third of the world's population being at risk and with the highest prevalence in sub-Saharan Africa. Pathogenesis and virulence of malaria vary between species and are accordingly termed malaria tropica (*P. falciparum*), malaria tertiana (*P. ovale, P. vivax*) and malaria quartana (*P. malariae*), although in endemic regions mixed infections may be frequent.

P. falciparum infections cause acute disease in infants and in non-immune adults, with the erythrocytic stages being responsible for the symptoms. Adults living in endemic areas usually acquire a nonsterile and short-lived, so-called semi-immunity.

formation of merozoites. Infected hepatocytes then detach from the liver and enter the bloodstream as merosomes [59], where the merozoites are released (3) to invade erythrocytes. After entering the erythrocyte the parasites undergo an asexual blood-stage replicative cycle within a parasitophorous vacuole. After invasion the parasite forms the so-called ring-stage (4) which is followed by the metabolically active trophozoite stage (5). The subsequent schizont stage (6) is characterized by the formation of merozoites which are finally released upon rupture of the infected erythrocyte (7). Invasion of erythrocytes reinitiates another round of the blood-stage replicative cycle. In addition to the asexual replication cycle, the parasite occasionally differentiates into sexual forms known as macro- and microgametocytes (8). When ingested by a mosquito, these stages fuse to form a zygote which, after meiosis, undergoes sporogony leading to the formation of infectious merozoites. H, hepatocyte; N, nucleus; E, erythrocyte; ET, endothelium. (b) Lifecycle of *Babesia* spp. in the vertebrate host. After attachment of an infected tick, *Babesia* spp. merozoites are released into the bloodstream (1) and invade erythrocytes. After invasion the parasites are released from the vacuolar membrane (2) and an asexual replication takes place. The erythrocytic stages include a ring-stage (3), a trophozoite stage (4), followed by a schizont stage (5) and finally the rupture of the cell and the release of merozoites which have the ability to reinvade erythrocytes. Alternatively to the asexual replication cycle, some parasite can differentiate into sexual forms (6) that can be taken up by ticks. The sexual forms cannot be distinguished from trophozoite stages at the light microscope level. E, erythrocyte.

In these individuals infections are often seen as a chronic disease, accompanied by anemia and enlargement of the spleen. During first and second pregnancy, infections bear a high risk of premature birth or stillbirth. Acute malaria tropica in a naïve person begins 5–8 days post infection with noncharacteristic symptoms accompanied by extremely high body temperatures (>40 °C). During febrile periods, a parasite-encoded protein, called *Plasmodium falciparum* ring-infested erythrocyte surface antigen (PfRESA), appears to protect the infected host cell from the deleterious effects of febrile temperatures by stabilizing the erythrocyte cytoskeleton [7]. Following the initial fever period, a complex syndrome of various symptoms follows, including neurological disorders and multiorgan dysfunctions. Often, coma is associated with severe malaria, with 15–20% mortality at this stage. Surprisingly, despite the dramatic neurological symptoms, survivors of this stage usually recover without sequelae. Importantly, despite the high mortality in naïve patients, patients living in endemic regions tolerate a high parasite burden of up to 10^7 parasites per ml of blood. In comparison, a single Gram-negative bacterium per ml can cause severe disease [8].

One of the parasite toxins thought to be responsible for the high production of tumor necrosis factor α and the resulting high body temperature is the glycosyl phosphatidylinositol moiety (GPI) of some membrane-bound proteins. The plasmodial GPI differs in structure slightly from its mammalian counterpart and has been shown to induce an inflammatory response [9, 10]. The GPI anchor is also discussed as a possible target for antidisease vaccines [11, 12].

In addition to fever, the severe symptoms of acute malaria coincide with a sequestration of infected erythrocytes in the capillaries and venules of various organs. The ability of infected erythrocytes to adhere to other erythrocytes (called rosetting) and to endothelial cells (called cytoadherence) is the result of an extensive modification of the infected host cell by the intracellular parasite, and this is thought to play a key role in the pathogenesis of the disease [13]. Although antimalarial drugs are available, the spread of parasite strains resistant to affordable medication in endemic areas poses a challenge for the continuous development of novel compounds. Despite considerable research efforts over more than 20 years and several field trials, an effective vaccine is not available.

In general, the symptoms of babesiosis in human patients resemble those of acute malaria with far more dramatic symptoms in splenectomized patients compared with fully immunocompetent patients. In normal patients, fatalities are rare, whereas in splenectomized patients infections may be lethal.

33.3 Characterization of Intra-Host Compartmentation and its Relationship to Virulence

33.3.1 Morphological Description

Plasmodium species can be differentiated by light microscopy of Giemsa-stained blood films. The parasite develops from a ring stage, to the metabolically highly active trophozoite and finally undergoes schizogony. Some parasites develop to male or female gametocytes that are infectious to a biting mosquito where they fuse and, after recombination and mitosis, turn into haploid sporozoites (Figure 33.1). Morphologically, at least at the light microscopical level, the erythrocytic stages of *Babesia* and *Plasmodium* parasites look similar and therefore the two parasites can be mistaken for one another. However, the presence of malaria pigment (hemozoin) and Maurer's clefts/Schuffner's dots are indicative of infection with *Plasmodium*, as *Babesia* do not exhibit these structures.

33.3.2 Choice of Host Cell

To a large extent, the pathogenicity caused by *Plasmodium* and *Babesia* is the result of a rather peculiar host cell–parasite relationship. By infecting the mammalian erythrocyte, these parasites choose a host cell that lacks intracellular compartments and the machineries for antigen processing and presentation. From an immunological standpoint, this may be a "safe haven" because mammalian RBCs have lost their genetic information, and are devoid of *de novo* protein and lipid biosynthesis. Consequently, they have only limited resources to react to a parasite infection, especially as they are unable to present foreign antigens in an MHC context. On the other hand, the parasite's nutritional requirements are met only partially in this rather specialized cell.

33.3.3 Invasion of Host Cells

In order to enter nucleated cells, many intracellular pathogens take advantage of the phagocytic properties of their respective host cells. As a consequence they reside, either permanently or temporarily, in a phagocytic compartment. In contrast, *Plasmodium* and other apicomplexan parasites invade their host cells by an active mechanism, involving the discharge of the contents of organelles located within the apical complex, particularly rhoptries and micronemes. Initial attachment of the merozoite to the host cell plasma membrane can occur at any position of the parasite surface, and it is followed by a so-called reorientation which brings the apical end of the parasite in contact with the target membrane, thus forming a moving junction between the two membranes [14]. The parasite then actively invades the host cell by a process called gliding motility. Many of the proteins involved in this gliding movement have been identified and characterized. There are considerable functional similarities between *Plasmodium* and the taxonomically related parasite *Toxoplasma gondii* where gliding and invasion have been studied in considerable detail [15]. The events following initial attachment, reorientation and discharge of the apical organelles are rapid and poorly understood at a molecular level. Morphologically, the erythrocyte membrane invaginates, the parasite sheds a protein surface coat, and the contents of the single pair of club-shaped rhoptries are released. There is no indication that the parasite actually penetrates the lipid bilayer of the RBC membrane.

Immediately after invasion the "ring-stage" parasite is secluded in a vacuole which, to discriminate it from endocytic vacuoles, is referred to as a "parasitophorous vacuole" (PV). Although the major erythrocyte membrane proteins are excluded from the vacuolar membrane (PVM), recent reports have suggested the presence of some selected proteins derived from the erythrocyte membrane. There is some evidence that proteins recruited from the erythrocyte plasma membrane are organized in lipid rafts [16, 17]. In contrast to endocytic compartments, the pH of the PV is close to neutral.

Although *Babesia* invades the same cell type as *Plasmodium*, the PV rapidly degrades following invasion, possibly involving proteins released from the spherical body [18]. The reason for this different compartmentation is still subject to speculation.

33.3.4
Host Cell Modification

As *P. falciparum* matures from the ring to the trophozoite stage, novel membrane-bound structures appear in the cytoplasm of the infected RBCs. They differ in size and morphology, and have been termed tubulovesicular network and Maurer's clefts, respectively. These structures have not been observed in *Babesia*-infected cells. Approximately 12 h post invasion electron-dense protrusions (called "knobs") appear at the plasma membrane of the *P. falciparum*-infected cell, coinciding with the appearance of the cytoadhesive phenotype (Figure 33.2). Since the differentiated RBC lacks the machineries required for *de novo* protein and lipid biosynthesis, it is probable that these structures are parasite-derived. Indeed, it has been shown that knobs contain various proteins of parasite origin which intercalate into the erythrocyte cytoskeleton that underlies the RBC plasma membrane. While the appearance of novel structures and parasite-derived proteins in erythrocytes is technically easy to detect and therefore evident, the extent of such modifications

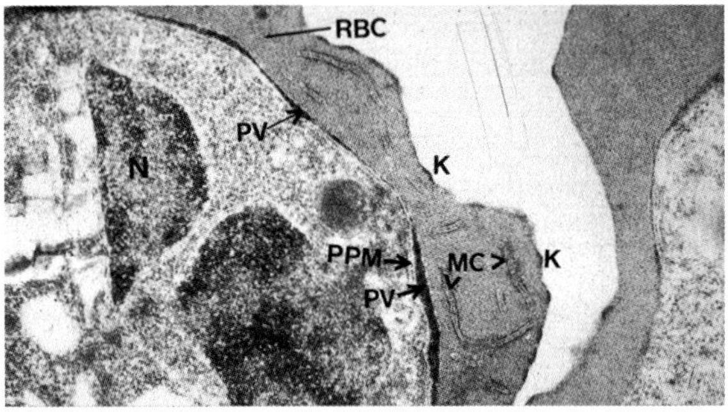

Figure 33.2 *Plasmodium falciparum* infected erythrocyte. Transmission electron micrograph of an erythrocyte infected with *Plasmodium falciparum* in the trophozoite stage. K, knobs; MC, Maurer's clefts and tubulovesicular network; PPM, parasite plasma membrane; PV, parasitophorous vacuole; RBC, red blood cell. From [60].

in liver cells is less clear, which is also partly due to the fact that liver stage parasites are less amenable to *in vitro* cultivation.

Although *Babesia*-infected cells appear to lack intraerythrocytic compartments equivalent to Maurer's clefts or the tubulovesicular network, they nevertheless develop knob structures underneath their surface, although these "knobs" appear to be morphologically different to those in *Plasmodium*-infected cells [19]. For example, the knobs of *P. falciparum*-infected RBCs contain a large amount of electron-dense material that is absent in those of *Babesia*-infected cells. Furthermore there is no evidence that knob-associated histidine-rich protein (KAHRP), a major (possibly only) structural component of *P. falciparum*-derived knobs is found within the *Babesia* knobs.

33.3.5 Relationship to Virulence

The appearance of knobs within the RBC membrane correlates with the ability of *P. falciparum*-infected cells to cytoadhere to endothelial cells of the blood capillaries and to form rosettes with other erythrocytes. Members of a protein family, collectively called the *Plasmodium falciparum* erythrocyte membrane protein-1 (PfEMP-1), are part of the knob structure and are exposed on the outer face of the RBC plasma membrane. The individual members of the PfEMP-1 family range in size from 200 to 350 kDa [20]. They are highly diverse and encoded by the so-called *var* multi-gene family, comprising of approximately 60 different variants. PfEMP-1 proteins are comprised of a highly variable N-terminal region exposed on the outer face of the erythrocyte membrane, and a conserved acidic domain on the cytoplasmic side (Figure 33.3). The N-terminal domain mediates binding of the infected cell to receptors on endothelial cells. Although it appears that there is no strict receptor–ligand specificity, different PfEMP-1 molecules show preferences for certain receptors [21].

A special situation arises in placentas of women during first and second pregnancy because some PfEMP-1 molecules preferentially adhere to chondroitin sulfate A which is expressed on placental tissue. Sequestration of infected cells in the placenta is implicated in low birthweight and stillbirth [22, 23]. Epidemiological

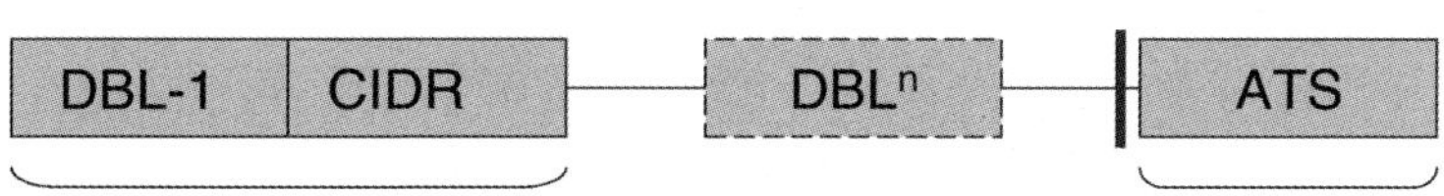

Conserved head structure and multi-adhesive domain — Intracellular domain

Figure 33.3 Primary structure of *Plasmodium falciparum* erythrocyte membrane protein-1 (PfEMP-1). PfEMP-1 proteins are composed of a highly variable N-terminal region exposed on the outer face of the erythrocyte membrane. The Duffy-binding-like 1 (DBL-1) domain is present in all *var* genes sequenced to date. Downstream of the DBL-1 domain is a cysteine-rich interdomain region (CIDR) and a variable number of less conserved DBL structures. Furthermore a highly conserved acidic terminal sequence (ATS) is located on the cytoplasmic side of the erythrocyte membrane. The N-terminal domain mediates binding of the infected cell to receptors on endothelial cells. Modified from [61].

studies indicate that multigravid women develop an immunity that protects from malaria-induced complications during pregnancy. *In vivo*, antibodies directed against specific PfEMP-1 molecules can inhibit cytoadherence of infected RBCs expressing the respective *var* genes. These observations indicate that individual PfEMP-1 molecules may become targets of a humoral immune response. However, the large antigenic repertoire encoded by *var* genes and the controlled but variant expression of these genes (see below) render PfEMP-1 proteins unlikely targets for a general vaccine, although in the case of maternal malaria the possibility of an anti-disease vaccine is being discussed [24].

Recent epidemiological studies indicate a correlation between the expression of certain *var* genes and severity of disease [25–28], although the picture is still incomplete. Thus, the *var* genes contribute to virulence twofold, as mediators of cytoadherence and as a means of immune evasion. *Babesia* also encodes a >100-member family of variant antigens called "variant erythrocyte surface antigen 1 or VESA1" which, like PfEMP-1, are located in knob structures [29]. VESA1, as well as its role in antigenic variation, also appears to mediate cytoadherence, and the interacting ligands on endothelial cells seem to correspond to those involved in PfEMP-1-mediated adherence in *Plasmodium*-infected cells.

33.3.6 Nutritional Aspects

Many pathogens exploit endocytic compartments of nucleated and biosynthetically active host cells as niches that provide an ample supply of soluble low molecular weight nutrients. In contrast, the PV in *Plasmodium* infected cells is a nutrient-poor compartment. The PVM appears to exclude host cell proteins from the parasite surface, but it contains nonselective pores that allow access of molecules smaller than 1300 Da [30]. During its development the parasite ingests approximately 70% of the host cell's hemoglobin, which is proteolytically degraded by a cascade of different *Plasmodium*-specific enzymes in the so-called digestive vacuole [31]. Although hemoglobin is in large supply, it lacks the amino acid isoleucine, which is essential for parasite development. In addition the parasite is reliant on the provitamin pantothenic acid. Thus it must acquire both of these nutrients from the extracellular medium. Within the first 15 h post infection, erythrocytes show an increased permeability for a variety of further solutes that are not taken up by non-infected erythrocytes. These pathways are referred to as "novel permeation pathways" (NPPs) [32, 33].

Despite intense research it is still unclear whether these new pathways are due to altered specificities of erythrocyte membrane transporters, novel transporters synthesized by the parasite and transported to the RBC membrane or by an interaction of both parasite and host cell proteins. Although studied in less detail, recent reports have demonstrated that erythrocytes infected with *Babesia* also show an increased permeability [34]. It has been suggested that NPPs, apart from being important for nutrient acquisition, play a role in the regulation of cell volume under changing osmotic conditions resulting from hemoglobin degradation [35].

33.3.7 A Cell Biological Enigma

The PVM forms an interface between the intracellular malaria parasite and the host cell cytoplasm which prevents the access of host cell proteins to the parasite surface but allows export of proteins into the host cell cytoplasm [36]. From a cell biological standpoint it is a unique phenomenon that a eukaryotic cell (the parasite) controls the biogenesis of compartments (e.g., tubulovesicular network, Maurer's clefts) outside the confines of its own plasma membrane. Likewise, the transport of secreted parasite proteins to defined destinations within a cell that lacks machineries required for a directed protein transport is without precedence and has attracted the interest of cell biologists. The details of these unusual trafficking pathways are still elusive. In general, proteins destined for locations within the host cell are first secreted from the parasite into the PV, and subsequently translocated across the PVM [37–39].

The Maurer's clefts appear to play a role in the translocation of PfEMP-1 and possibly other knob-associated proteins across the erythrocyte cytoskeleton into the plasma membrane [40]. Proteomic analyses of the various compartments (PV, Maurer's clefts) and high-resolution imaging technologies, such as electron tomography, are the current state-of-the-art tools to identify the proteins and to characterize the structural components involved in this fascinating aspect of parasite–host cell interaction [40–42].

33.3.8 The Role of the PV as Host–Parasite Interface

As outlined above, with respect to both biogenesis and maintenance, the PV induced by apicomplexan parasites differs from endocytic or phagocytic compartments. This is particularly evident in parasites that invade mammalian erythrocytes, as RBCs carry out neither endo- nor phagocytosis, and lack the proteins required to mediate these processes. It has been argued for apicomplexan parasites of nucleated cells that one biological role of the PV is to prevent parasite proteins from being released into the host cell cytoplasm and processed for presentation in an MHC class I context. In fact, a cytotoxic T cell response against infected liver cells appears to play an important role in immunity against *P. falciparum*. A reduction of antigen presentation via MHC could thus help to ensure parasite survival by a reduction in the T cell response. However, this phenomenon cannot explain the need for a PV in intraerythrocytic stages. In this context it is noteworthy that *Babesia*, which does not have developmental stages in nucleated antigen-presenting cells, is not secluded in a PV.

As the malaria parasite matures, the PV and the PVM expand considerably. While several secreted parasite proteins are inserted into the PVM, the source of lipids required for this expansion is not entirely clear. Recent reports have suggested that at least a portion of the lipids, presumably organized in lipid rafts, are recruited from the erythrocyte plasma membrane [16, 17]. For a long time it has been argued that the PV, tubovesicular network and Maurer's clefts form an interconnected tubular network within the RBC cytoplasm. Recent proteomic studies of both compartments, the distribution of reporter constructs targeted to the vacuolar space, and studies applying electron tomography suggest, however, that these are distinct compartments, connected by tether-like, nontubular structures [40–42].

Apart from protecting the parasite from potentially harmful host cell proteins, additional roles of the PV are as follows:

- *Nutrient acquisition.* Small solutes taken up by the infected RBCs, either through endogenous transporters or the novel permeation pathways, are thought to gain access to the parasite through nonselective pores within the PVM [30, 43]. These pores have been described functionally, but still await a molecular identification.
- *Protein sorting.* Several secreted parasite proteins remain PV-resident proteins, whereas others are translocated across the PVM and transported to locations within the RBC. It is the current view that this sorting event, and the subsequent translocation processes, are mediated by PV-resident auxiliary machineries [44–46].
- *Protein processing and parasite release.* Upon completion of the intraerythrocytic development 16–32 merozoites are formed which expose on their surface a 196 kDa, GPI-anchored major protein, merozoite surface protein 1 (MSP-1). Immediately prior to or during release of parasites from the infected cell, this precursor is processed to a membrane-resident C-terminal fragment of 19 kDa. This processing occurs in discrete steps and it is essential for merozoite infectivity. Inhibition of the final processing step inhibits invasion [47]. The processing involves a parasite-encoded protease that is secreted into the PV (Mike Blackman, personal communication). Therefore, the PV may provide a subcellular environment maintaining sufficient concentrations of parasite proteases and possibly other enzymes required for parasite maturation and release.

A recent proteome analysis of the PV has revealed a high representation of chaperones and proteases within this compartment, which is consistent with functions related to protein translocation and processing [43]. On the other hand, *Babesia* can survive in erythrocytes without a PV, suggesting that it either uses different mechanisms to carry out these processes, or that the *Plasmodium* PV fulfills another, currently unknown, function.

33.3.9 Virulence Factors

From serological studies using antisera recognizing the surface of the infected RBC it became obvious almost 20 years ago that the parasite undergoes antigenic variation. This variation was found to be mediated by the PfEMP-1 family encoded by *var* genes located mainly in the subtelomeric regions of the parasite's chromosomes. PfEMP-1 is involved in pathogenesis (see above) and is regulated such that the parasite effectively evades a protective host immune response by translation and presentation of a single PfEMP-1 variant per parasite at any time. This variation involves a switching in the expression of *var* genotypes in successive parasite populations and involves a highly complex allelic exclusion mechanism. To date, it appears that the regulated expression (and silencing) of *var* genes is mediated by epigenetic factors affecting chromatin structure by differential acetylation and methylation of histones [48].

Another conspicuous genetic phenomenon is the occurrence of genes encoding proteins with tandemly repeated amino acid sequences. The tandem repeats can vary in size and numbers and, generally, these structures are highly immunogenic. The proteins are composed of small, more or less conserved regions and repetitive structures. Since a cell biological function of such proteins is difficult to envisage, it has been suggested that these polymorphic antigens contribute to immune evasion by inducing a strong but ineffective T cell-independent stimulation of B cells, the so-called "smokescreen effect" [49].

33.4 From Parasite Cell Biology and Genomics to Drug Targets

Attempts to control malaria in endemic areas include vector control as well as vaccine and drug development. It is clearly beyond the scope of this chapter to discuss all approaches being followed by the malaria research community. Therefore we would like to focus on some selected candidate drug targets that were discovered as a result of better understanding of the parasite's cell biology. The complete annotated *P. falciparum* genome sequence was released in 2002, and revealed that the about 23 Mb genome encodes for about 5300 proteins, over 50% of which are hypothetical proteins with no homologs in other organisms sequenced to date [50]. A large proportion of the proteins are predicted to be involved in immune evasion and host–parasite interactions, or involved in several biochemical pathways that could be used for drug development (discussed in more detail in Section 33.4.3). Depending on the algorithms applied, between 150 and 250 proteins are predicted to be transported to the cytosol of the infected host cell (referred to as the secrete- or export-ome). It has been proposed that these proteins are involved in host cell remodeling [45, 46].

The genome of *Babesia bovis* was released in 2007, and showed that the 8.2 Mb genome encodes for approximately 3700 proteins [51]. Analysis of these genes revealed several important differences between the biochemistry of *Babesia* and that of *Plasmodium*, including the lack of specific metabolic pathways in *Babesia* (discussed in more detail below).

Interestingly, comparative analysis of the *Plasmodium falciparum* and *Babesia bovis* genome sequences reveal ± 300 hypothetical proteins that are conserved between these parasites, and thus may be involved in interaction of these parasites with their chosen host cells.

33.4.1 The Apicoplast

The discovery of the apicoplast in many apicomplexan parasites raised questions regarding its cell biological function. The apicoplast is bounded by three or four membranes, and is the relict of an ancient phototrophic eukaryote that has lost all its intracellular compartments [1, 52]. During the course of evolution, almost all plastid genes have been transferred to the nucleus of the parasite, requiring that the protein products then be reimported into the plastid. This transport step is mediated by a so-called bipartite topogenic signal encoded within the primary sequence of the protein. Attempts to purify this organelle in order to identify proteins involved in various metabolic pathways have been unsuccessful, but algorithms have been developed to predict this topogenic signal, and suggest that about 500 proteins are imported into the apicoplast, constituting almost 10% of the parasite's total protein contingent [52]. Based on these predictions it was possible to construct a preliminary map of the metabolic pathways located in this organelle. Even in the absence of a detailed knowledge of the metabolic pathways located in the apicoplast, this organelle was identified as a potential drug target when the parasiticidal effects of furosemide were studied [53]. The *Plasmodium* apicoplast contains various plant-specific metabolic pathways such as type II fatty acid biosynthesis. This observation has opened new avenues for the identification of novel drug targets and agents, such as triclosan [54].

The release of the *Babesia* genome sequence also revealed several biochemical pathways resident within the plastid, however strikingly the type II fatty acid biosynthesis pathway appears to be absent [51]. Thus compounds like triclosan which specifically target this pathway are unlikely to be of use as anti-babesial drugs.

33.4.2
The Food Vacuole

During its growth the parasite digests approximately 70% of the host cell hemoglobin. Phagocytosis occurs at a structure called the cytostome from where the hemoglobin is transported to the "food vacuole" a compartment that can be considered as large phagolysosome. Degradation of the four polypeptide chains involves discrete, parasite-specific proteases. Consequently, proteases involved in hemoglobin degradation are believed to be promising drug targets [55]. As a result of hemoglobin digestion, heme, which has cytotoxic properties is released. This potentially toxic compound is polymerized within the food vacuole to an insoluble product called hemozoin, a process that appears to involve a parasite protein [56]. Several drugs that interfere with parasite growth also affect hemozoin formation [57].

Babesia spp. also appear to metabolize hemoglobin, although the pathway it uses is not well studied. The lack of cytostomes, a digestive vacuole and the formation of hemozoin-like pigment may suggest that this parasite has evolved a different pathway to utilize hemoglobin, and therefore drugs targeting the *Plasmodium* hemoglobin degradation pathway may not function against *Babesia* spp.

33.4.3
Novel Permeation Pathways

The observation that the malaria parasite depends on nutrient acquisition and on a physiological modification of the host cell plasma membrane to allow uptake of nutrients not taken up by the uninfected cell has kindled hopes of developing drugs that block nutrient uptake into the infected cell. In addition, since these pathways also allow passage of a wide variety of compounds not related to parasite growth, they may be used as a Trojan horse to facilitate access of drugs that are excluded from uninfected cells [58]. Similarly, *Babesia* spp. seem to require nutrients from the external medium, and although the uptake mechanism seems to differ from that in *Plasmodium*, these novel pathways may again offer an attractive target for drug development [34].

33.5
Summary and Outlook

The infection of human erythrocytes that are devoid of protein and lipid biosynthesis with eukaryotic, fast replicating parasites is a unique cell biological phenomenon. While life in a metabolically quiescent cell which is unable to process and present antigens may at first appear to be a safe haven, this highly specialized cell does not fulfill all the nutritional requirements for the parasite living inside it. From an experimental standpoint the simple organization of the erythrocyte offers distinct advantages to study the cell biology of intracellular pathogens and their interactions with the host cell. In particular, physiological and biochemical changes of the host cell can easily be attributed to metabolic activities of the parasite. Thus, our knowledge of the cell biology of *Plasmodium falciparum* has advanced considerably during the past two decades and has allowed the identification of novel drug targets. The haploid genotype and the poor genetic amenability of the parasite still limit the use of recombinant DNA technologies to assess the essentiality of specific genes, but this is sure to change with time.

While the *Plasmodium*-infected erythrocyte has been studied in considerable detail, less is known about *Babesia*-infected cells although several similarities are to be noted, including host cell modification, the resulting cytoadherance, antigenic variation and the necessity of nutrient acquisition from the external media. Each parasite appears to have mastered these challenges in its own particular manner, a special case of convergent functional evolution between two related parasites. Study of both of these pathogens in parallel is sure to reveal even more novel cell biological features of these obligate intracellular parasites that choose to invade and replicate within what has been referred to as "just a bag of hemoglobin."

References

1 Köhler, S., Delwiche, C.F., Denny, P.W., Tilney, L.G. *et al.* (1997) *Science*, **275**, 1485–1489.

2 Foth, B.J. and McFadden, G.I. (2003) *Int Rev Cytol*, **224**, 57–110.

3 Fichera, M.E. and Roos, D.S. (1997) *Nature*, **390**, 407–409.

4 Ralph, S.A., D'Ombrain, M.C. and McFadden, G.I. (2001) *Drug Resist Updat*, **4**, 145–151.

5 Cox-Singh, J., Davis, T.M., Lee, K.S., Shamsul, S.S., Matusop *et al.* (2008) *Clin Infect Dis*, **46**, 165–171.

6 McCutchan, T.F. (2008) *Future Microbiol*, **3**, 115–118.

7 Silva, M.D., Cooke, B.M., Guillotte, M., Buckingham, D.W. *et al.* (2005) *Mol Microbiol*, **56**, 990–1003.

8 Smith, T., Genton, B., Baea, K., Gibson, N. *et al.* (1994) *Parasitology*, **109**, 539–549.

9 Naik, R.S., Branch, O.H., Woods, A.S., Vijaykumar, M. *et al.* (2000) *J Exp Med*, **192**, 1563–1576.

10 Gerold, P., Schofield, L., Blackman, M.J., Holder, A.A. *et al.* (1996) *Mol Biochem Parasitol*, **75**, 131–143.

11 Schofield, L. (2007) *Microbes Infect*, **9**, 784–791.

12 Schofield, L., Hewitt, M.C., Evans, K., Siomos, M.A. *et al.* (2002) *Nature*, **15**, 785–789.

13 Miller, L.H., Baruch, D.I., Marsh, K. and Doumbo, O.K. (2002) *Nature*, **415**, 673–679.

14 Baum, J., Richard, D., Healer, J., Rug, M. *et al.* (2006) *J Biol Chem*, **281**, 5197–5208.

15 Soldati, D. and Meissner, M. (2004) *Curr Opin Cell Biol*, **16**, 32–40.

16 Lauer, S., VanWye, J., Harrison, T., McManus, H. *et al.* (2000) *EMBO J*, **19**, 3556–3564.

17 Murphy, S.C., Samuel, B.U., Harrison, T., Speicher, K.D. *et al.* (2004) *Blood*, **103**, 1920–1928.

18 Yokoyama, N., Okamura, M. and Igarashi, I. (2006) *Vet Parasitol*, **138**, 22–32.

19 Hutchings, C.L., Li, A., Fernandez, K.M., Fletcher, T. *et al.* (2007) *Mol Microbiol*, **65**, 1092–1105.

20 Cooke, B., Coppel, R. and Wahlgren, M. (2000) *Parasitol Today*, **16**, 416–420.

21 Flick, K. and Chen, Q. (2004) *Mol Biochem Parasitol*, **134**, 3–9.

22 Nunes, M.C. and Scherf, A. (2007) *Parasitology*, **134**, 1863–1869.

23 Fried, M. and Duffy, P.E. (1996) *Science*, **272**, 1502–1504.

24 Sharma, S. and Pathak, S. (2008) *J Vector Borne Dis*, **45**, 1–20.

25 Montgomery, J., Mphande, F.A., Berriman, M., Pain, A. *et al.* (2007) *Mol Microbiol*, **65**, 959–967.

26 Jensen, A.T., Magistrado, P., Sharp, S., Joergensen, L. *et al.* (2004) *J Exp Med*, **199**, 1179–1190.

27 Trimnell, A.R., Kraemer, S.M., Mukherjee, S., Phippard, D.J. *et al.* (2006) *Mol Biochem Parasitol*, **148**, 169–180.

28 Kaestli, M., Cockburn, I.A., Cortés, A., Baea, K. *et al.* (2006) *J Infect Dis*, **193**, 1567–1574.

29 Allred, D.R. (2001) *Microbes Infect*, **3**, 481–491.

30 Desai, S.A. and Rosenberg, R.L. (1997) *Proc Natl Acad Sci USA*, **94**, 2045–2049.

31 Francis, S.E., Sullivan, D.J. Jr. and Goldberg, D.E. (1997) *Annu Rev Microbiol*, **51**, 97–123.

32 Kirk, K. (2001) *Physiol Rev*, **81**, 495–537.

33 Staines, H.M., Alkhalil, A., Allen, R.J., De Jonge, H.R. *et al.* (2007) *Int J Parasitol*, **37**, 475–482.

34 Alkhalil, A., Hill, D.A. and Desai, S.A. (2007) *Cell Microbiol*, **9**, 851–860.

35 Lew, V.L., Tiffert, T. and Ginsburg, H. (2003) *Blood*, **101**, 4189–4194.

36 Lingelbach, K. and Joiner, K.A. (1998) *J Cell Sci*, **111**, 1467–1475.

37 Ansorge, I., Jeckel, D., Wieland, F. and Lingelbach, K. (1995) *Biochem J*, **308**, 335–341.

38 Marti, M., Baum, J., Rug, M., Tilley, L. *et al.* (2005) *J Cell Biol*, **171**, 587–592.

39 Charpian, S. and Przyborski, J.M. (2008) *Traffic*, **9**, 157–165.

40 Tilley, L., Sougrat, R., Lithgow, T. and Hanssen, E. (2008) *Traffic*, **9**, 187–197.

41 Nyalwidhe, J. and Lingelbach, K. (2006) *Proteomics*, **6**, 1563–1573.

42 Vincensini, L., Fall, G., Berry, L., Blisnick, T. *et al.* (2008) *Mol Biochem Parasitol*, **160**, 81–89.

43 Nyalwidhe, J., Baumeister, S., Hibbs, A.R., Tawill, S. *et al.* (2002) *J Biol Chem*, **277**, 40005–40011.

44 Marti, M., Good, R.T., Rug, M., Knuepfer, E. *et al.* (2004) *Science*, **306**, 1930–1933.

45 Hiller, N.L., Bhattacharjee, S., van Ooij, C., Liolios, K. *et al.* (2004) *Science*, **306**, 1934–1937.

46 Hiss, J.A., Przyborski, J.M., Schwarte, F., Lingelbach, K. *et al.* (2008) *PLoS ONE*, **3**, e1560.

47 Uthaipibull, C., Aufiero, B., Syed, S.E., Hansen, B. *et al.* (2001) *J Mol Biol*, **307**, 1381–1394.

48 Lopez-Rubio, J.J., Riviere, L. and Scherf, A. (2007) *Curr Opin Microbiol*, **10**, 560–568.

49 Anders, R.F., McColl, D.J. and Coppel, R.L. (1993) *Acta Trop*, **53**, 239–253.

50 Gardner, M.J., Shallom, S.J., Carlton, J.M., Salzberg, S.L. *et al.* (2002) *Nature*, **419**, 531–534.

51 Brayton, K.A., Lau, A.O., Herndon, D.R., Hannick, L. *et al.* (2007) *PLoS Pathog*, **19**, 1401–1413.

52 Foth, B.J., Ralph, S.A., Tonkin, C.J., Struck, N.S. *et al.* (2003) *Science*, **299**, 705–708.

53 Jomaa, H., Wiesner, J., Sanderbrand, S., Altincicek, B. *et al.* (1999) *Science*, **285**, 1573–1576.

54 Goodman, C.D. and McFadden, G.I. (2008) *Curr Pharm Des*, **14**, 901–916.

55 Coombs, G.H., Goldberg, D.E., Klemba, M., Berry, C. *et al.* (2001) *Trends Parasitol*, **17**, 532–537.

56 Jani, D., Nagarkatti, R., Beatty, W., Angel, R. *et al.* (2008) *PLoS Pathog*, **4**, e1000053.

57 Kumar, S., Guha, M., Choubey, V., Maity, P. *et al.* (2007) *Life Sci*, **80**, 813–828.

58 Kirk, K. (2004) *Acta Trop*, **89**, 285–298.

59 Sturm, A., Amino, R., van de Sand, C., Regen, T. *et al.* (2006) *Science*, **313**, 1287–1290.

60 Nyalwidhe, J. and Lingelbach, K. (2003) *Zoology*, **106**, 341–348.

61 Chen, Q., Fernandez, V., Sundström, A. and Schlichtherle, M. (1998) *Nature*, **384**, 392–395.

34
Theileria

Dirk Dobbelaere and Martin Baumgartner

34.1
Introduction to *Theileria* and the Disease

Theileriosis is a tick-borne disease complex caused by several members of the genus *Theileria*. *Theileria* belongs to the *Apicomplexa*, a phylum that also encompasses other pathogenic parasites affecting humans and animals such as *Toxoplasma*, *Plasmodium*, *Babesia*, *Neospora* and *Eimeria*. *Theileria* parasites can be divided into "non-transforming" and "transforming" species. Non-transforming *Theileria* spp. predominantly affect red blood cells, causing subclinical infections, but outbreaks of clinical cases have sporadically been reported on several continents. "Transforming" *Theileria* spp., differ from the "non-transforming" *Theileria* by virtue of the fact that they are capable of inducing uncontrolled proliferation of parasitized leukocytes, causing severe disease. In this chapter, we will focus on transforming *Theileria* species that cause East Coast fever (*T. parva*) and tropical theileriosis (*T. annulata*). Details relating to the pathology and diagnostics [1], epidemiology [2] and immunology [3] of *Theileria* infection have been reviewed elsewhere. We will summarize how transforming *Theileria* parasites establish their niche in the host cell cytoplasm and how they interact with the host cell to orchestrate their survival and clonal expansion.

T. parva is found in sub-Saharan East, Central and Southern Africa where it is transmitted by the tick vector *Rhipicephalus appendiculatus*. It causes East Coast fever in cattle, resulting in severe economic losses and posing a considerable hurdle for sustainable livestock production. Under epidemic conditions, high mortality and morbidity, both exceeding 90%, can be observed. Unprotected animals imported into an endemic area are highly susceptible. Animals that live in endemic areas, on the other hand, show a high degree of resistance acquired as a result of early exposure and continuous challenge. African buffalo (*Syncerus caffer*) can also become infected and harbor the parasite, but infections largely remain asymptomatic [4]. It can be expected that a long process of coevolution between *Theileria* and the buffalo resulted in the development of resistance.

Intracellular Niches of Microbes. Edited by Ulrich E. Schaible and Albert Haas

ISBN: 978-3-527-32207-7

T. annulata is a parasite of cattle and domestic buffalo (*Bubalis bubalis* or Asian water buffalo) causing tropical theileriosis or Mediterranean coast fever of cattle. It is transmitted by *Hyalomma* ticks and is prevalent in Northern Africa (extending down through Egypt towards Sudan), the Middle East and large areas of Asia including Pakistan, India and central Asia. It has been speculated that the water buffalo is the original host of *T. annulata* as infection only causes subclinical reactions, whereas cattle are more severely affected. *T. lestoquardi* is a highly pathogenic parasite of sheep and goats that is related to *T. annulata*, causing theileriosis in south-eastern Europe, North Africa, the Middle East and Russia. For a detailed description of the *Theileria* lifecycle the reader is referred to a chapter by Shaw [5].

In the case of East Coast fever, the severity of disease varies depending on the parasite strain, the susceptibility of the host and the size of the sporozoite inoculum. Initially, there is fever and, 5 days after infection, macroschizonts can be observed in the lymph node draining the site of tick attachment. From there, the parasite spreads and parasitized cells can be detected in most lymphoid organs and also in peripheral blood leukocytes. During later stages, parasitized cells are also found proliferating in nonlymphoid tissues. Macroschizonts undergo merogony to form microschizonts that contain numerous merozoites. Upon release from the leukocyte, these invade red blood cells to form piroplasms. These are usually detected 11–12 days after onset of the disease. As the disease progresses, a marked depletion of the lymphoid tissues can be observed, which can be attributed to the elimination of both infected and uninfected lymphoid cells, and potentially involves natural killer cells.

The most striking post mortem lesion is massive pulmonary edema, hyperemia and emphysema. Petechial hemorrhages can be observed in mucosa and serosa of a wide range of organs. Necrotic Peyer's patches depleted of lymphoid cells are also often found. In *T. annulata* infections, both the schizont and the piroplasm stage contribute to pathogenesis. The reticuloendothelial system is severely affected by the schizont, but leukopenia, typical of *T. parva* infection, is not a characteristic of *T. annulata* infection. The number of circulating parasitized leukocytes continues to increase and local hyperplasia due to disseminated infected leukocytes gives rise to lesions, for instance in the kidneys and liver. Classical lesions are pronounced hypertrophy of the spleen and lymph nodes, widespread hemorrhages and ulcers, particularly in the abomasum and small intestine. During piroplasm parasitemia, anemia is often observed and death is usually delayed until parasitemia and anemia are well advanced [6]. In cattle, there is evidence for a certain degree of genetic resistance, with cattle of the *Bos indicus* type (Zebu) showing more resistance than the European (*Bos taurus*) breeds [7, 8].

T. annulata and *T. parva* differ in the spectrum of cells they transform [9]. *T. parva* invades and transforms B cells and T cells expressing CD4 or CD8 and either $\alpha\beta$ or $\gamma\delta$ T cell receptors [10, 11]. Defining the cells targeted by *T. annulata* initially proved to be more difficult because of the extensive alterations in surface marker expression that occur upon infection [12–14]. Infective sporozoites can be harvested from tick salivary glands and used to infect and transform cells *in vitro* [15]. This not only allowed detailed studies of the entry process (see below), but, using cell populations highly purified by cell sorting, also provided better insights into the nature of the target cells [11]. It is now generally assumed that *T. annulata* transforms monocytes/macrophages and B cells [16, 17]. Interestingly, upon transformation by *T. annulata* a parasite-dependent phenotypic and functional dedifferentiation could be observed in macrophages [14]. This was accompanied by the downregulation of important surface proteins such as CD14 and CD11b and failure to express typical macrophage functions such as Fc receptor-mediated phagocytosis, oxidative burst, lipopolysaccharide-induced tumor necrosis factor α (TNFα) secretion, nitric oxide generation and upregulation of procoagulant activity. These functions were restored to a large degree by eliminating the parasite from the cell using a specific theilericidal drug, BW720c. To what extent these changes contribute to the survival of the parasite in the macrophage and play a role in the pathogenesis of *T. annulata* infection, is not known.

T. parva infection is associated with *de novo* expression of T cell markers [18] and changes potentially reflecting dedifferentiation are less pronounced. Nevertheless, in a number of cases, cells expressing CD4 were found also to express CD8 after infection with *T. parva* [10, 19]. On the other hand, bovine cytotoxic T cell clones transformed by *T. parva* were shown to retain their cytolytic activity and antigen specificity [20].

Transformation of dendritic cells by *T. annulata* has also been reported [21]. In the case of *T. parva*, however, sporozoites were found to invade afferent lymph dendritic cells, where they apparently differentiate into schizonts without inducing transformation [22].

34.2 Characterization of the Intra-Host Cell Compartment ... Which Compartment?

34.2.1 Entry

To establish its special niche, *Theileria* has developed an entry mechanism that deviates in several important aspects from that observed for other apicomplexan parasites (reviewed in [5, 23, 24]). Tick attachment to the host and feeding triggers the completion of *Theileria* sporogony, resulting in the rapid production of thousands of mature sporozoites. Between 0.75 and 1.5 μm in size, they are released into the host where they attach to target cells and rapidly complete invasion. Compared to other apicomplexan parasites, *Theileria* sporozoites have a much-reduced apical complex. No conoid is present, and sporozoites appear to contain only two types of secretary organelles: rhoptries, which are located apically, and dispersed microspheres. The typical micronemes, which are usually involved in the release of proteins from the apical tip and contribute to parasite motility and host cell invasion, are missing. The parasite also lacks a subpellicular membrane complex or microtubular basket (typical of motile apicomplexan zoites). These differences help to explain the fact that *Theileria* sporozoites do not depend on motility for host cell entry and can enter cells in any orientation. Entry is quick (3 min) and is accomplished by the progressive circumferential zippering of sporozoite and host cell plasma membranes. The major sporozoite surface proteins for *T. parva* and *T. annulata* are p67 [25, 26] and SPAG-1 [27], respectively. These antigens share sequence homology and are serologically cross-reactive [28]. Antibodies against these proteins prevent sporozoite entry [25, 29] and, for this reason, both have been tested for their potential as vaccines (reviewed in [3]).

The host cell surface molecules engaged by the sporozoite are still not known and no progress has been made since the topic was last reviewed [5, 24]. Based on antibody inhibition studies, it has been proposed that MHC class I and the associated β_2 microglobulin are involved in the initial binding of *T. parva* sporozoites; this could help explain the species specificity [30, 31]. A picture emerging from these studies suggested that sporozoite binding and internalization require different surface molecules. Interestingly, proteolytic cleavage of target cell surface proteins was found to enhance lymphocyte susceptibility to invasion by *T. parva* sporozoites [32]. Furthermore, during entry, part of the sporozoite surface coat is cleaved off [33, 34]. This, and the fact that protease inhibitors block sporozoite entry, point towards an important role for proteolytic processes. For many other intracellular pathogens, the surrounding host cell membrane is modified after entry and constitutes an important part of the surrounding vacuole, thus forming a niche in which the organism will survive, at least temporarily. Once inside the cell, *Theileria* takes a different tack. Rather than modifying the surrounding host cell membrane, in less than 15 min after entry the sporozoite dissolves the surrounding endocytotic membrane, making the host cell cytoplasm its permanent niche. This process coincides with discharge of the rhoptries and microspheres, but the nature of the membranolytic component(s) is not known. In the absence of an endocytotic membrane with which they can fuse, lysosomes, despite rapid recruitment, are incapable of destroying the invading sporozoite [35]. Elimination of the surrounding host cell membrane is thus essential and failure to do so results in death of the young parasite [36].

While many intracellular pathogens tend to remodel the host cell actin cytoskeleton, *Theileria* predominantly interacts with the microtubule network: the freed parasite is surrounded almost immediately by an orderly array of host cell microtubules that converge towards the microtubule organizing center (Figure 34.1); this association is maintained during the entire stay of the schizont in the host cell cytoplasm and is also essential for parasite division over the two daughter cells during mitosis (Figure 34.2). Parasite-associated microtubules are highly stable, but the exact nature of this interaction and the molecules involved are not yet known.

34.2.2
Development in the Cytoplasm

Out of reach of lysosomal destruction, the haploid sporozoite differentiates to the trophozoite and then undergoes multiple rounds of DNA synthesis and nuclear division, without cytokinesis, to form the multinucleate schizont stage. Schizont development is accompanied by the induction of host cell transformation. The exact sequence of early events involved in reprogramming the host cell into a continuously proliferating cell has not yet been established. Studying these early events will require monitoring at the single cell level, as transformation does not take place in all

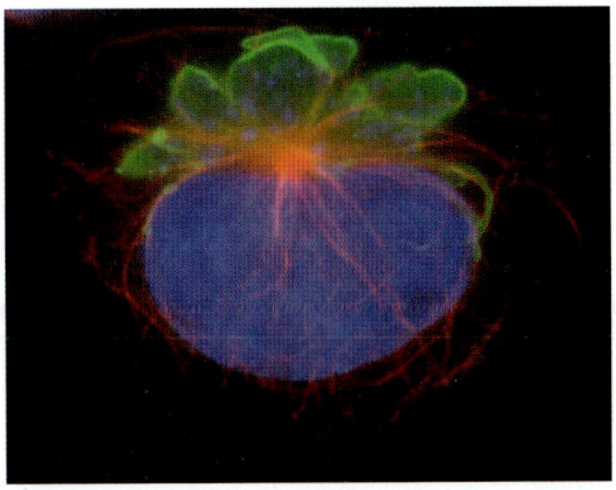

Figure 34.1 At all stages, the schizont is in close contact with host cell microtubules (stained red using anti-α-tubulin), orienting the parasite towards the microtubule-organizing center. The parasite (green) is stained using anti-TaSP (kind gift of Jabbar Ahmed, Borstel); the nuclei are stained blue using DAPI. Image generated by Gongda Xue, Dobbelaere Laboratory.

instances. Instead, specific cell types appear to undergo selection early after infection and only a small proportion of the parasitized cells become established as transformed cell lines [37]. For instance, it could be shown that B cells are at least as permissive for infection as T cells, but in mixed cultures T cells quickly take the upper hand, suggesting the existence of early selection events that may also help to explain the eventual dominance of T cells *in vitro* and *in vivo*. During the early stages of transformation, the parasitized cell is still susceptible to a number of cellular defense mechanisms aimed at eliminating invading pathogens. The initial establishment of macroschizont-infected cell lines is suppressed by low concentrations of nitric oxide (NO) that do not affect the proliferation of established macroschizont-infected cell lines [38]. Higher doses of NO, however, do inhibit established cultures [39]. Likewise, several cytokines, including TNFα, interferons, interleukin 1 (IL-1) and IL-2 significantly inhibit the *in vitro* development of cells harboring *T. annulata* or *T. parva* trophozoites, but fail to inhibit, and in some cases even enhance, the proliferation of macroschizont-infected cell lines of both species [40, 41].

Once established, *Theileria*-transformed cells can be propagated indefinitely, independent of exogenous growth factors or antigenic stimulation [42]. In a number of cell lines, the involvement of an autocrine loop could be demonstrated [43, 44], whereas other cell lines may have acquired total growth-factor independence (reviewed in [45]). Proliferation is also enhanced by cell–cell contact [46, 47], supporting the notion that signaling pathways triggered through surface ligand/receptor interaction promote proliferation.

34.2.3 Once Inside, Always Inside

An important biological property of the transforming *Theileria* schizont is that it can ensure its propagation and cell-to-cell spread without exiting the transformed cell. During host cell mitosis and cytokinesis the schizont associates with the mitotic

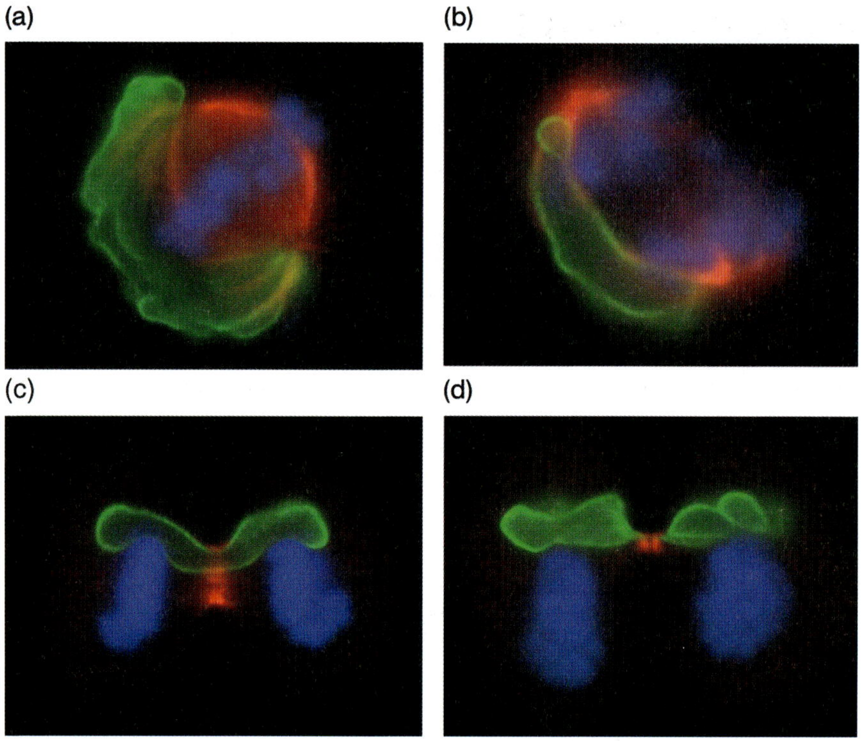

Figure 34.2 (a) By interacting with the spindle pole and mitotic spindles, the parasite positions itself on both sides of the metaphase chromosomes. (b) During anaphase, the schizont remains associated with the mitotic spindle. (c) At telophase, the parasite interacts closely with the central spindle, positioned between the two sets of separated chromosomes. (d) During cytokinesis, the parasite is divided over the two daughter cells and the connection between the daughter cells and the schizonts severed. In all images, the parasite (green) is stained using anti-TaSP (kind gift of Jabbar Ahmed, Borstel) and chromosomes are stained blue (DAPI). In (a) and (b), microtubules (red) are stained using anti-α-tubulin; in (c) and (d), anti-PLK1 (red) was used which localize the central spindle and midbody. Images generated by Conrad von Schubert, Dobbelaere Laboratory.

apparatus and central spindle (see Figure 34.2). DNA replication is synchronized with that of the host cell. Surprisingly, however, parasite DNA replication occurs almost exclusively during late host cell G_2 and early mitosis [48], suggesting the existence of cell cycle-specific molecular cross-talk between the host cell and the parasite. In addition, schizonts also appear to be able to transfer to uninfected cells via a mechanism that is only poorly understood.

When parasitized cell lines are inoculated into a host, the parasite transfers from the donor cells into the cells of the recipient animal [49–51]. This transfer forms the basis of tissue culture vaccines and is essential for the induction of protective immunity in the vaccinated recipient animals [3, 52]. One possibility is that transfer occurs by fusion of infected donor cells with uninfected recipient cells. Alternatively, it is conceivable that uninfected cells with phagocytic properties take up infected cells, with the schizont gaining access to the phagocytosing cell upon destruction of the plasma membrane.

34.2.4
Manipulation of the Host Cell: Live and Let Live

Through evolution, pathogens have acquired an array of mechanisms to counteract host cell death, thus avoiding elimination together with damaged host cells undergoing apoptosis. As is the case for many viruses [53], the activation of antiapoptotic mechanisms in infected host cells is also a common feature of intracellular parasites [54, 55]. Nevertheless, *Theileria* faces the formidable task of having to fight apoptosis on at least two fronts. Inducing host cell transformation is not a trivial event. The parasite needs to carefully unhinge the control mechanisms that normally prevent uncontrolled proliferation, without triggering the safety mechanisms that are activated when the cell cycle is severely perturbed, resulting in programmed cell death. In the case of *T. parva*-transformed T cells, the parasite also needs to counter the gradual increase in susceptibility to death receptor-induced apoptosis that normally contributes to the elimination of activated T cells [56].

Theileria parasites appear to have mobilized the full repertoire of antiapoptotic mechanisms. In *T. parva*-transformed T cells, antiapoptotic proteins such as c-FLIP, c-IAP1, c-IAP2 and X chromosome-linked IAP, which counteract caspase activity, are upregulated in a parasite-dependent manner [57]. This also applies to the antiapoptotic Bcl-2 family member, Mcl-1, which neutralizes proapoptotic proteins such as Bak and Bim (reviewed in [58–60]). Thus, although death receptors such as TNFα or Fas and their corresponding ligands are expressed in several *Theileria*-transformed cell lines, signaling through these receptors does not result in apoptosis [40, 41, 44, 57]. On the contrary, TNFα receptor occupancy has been linked with increased proliferation [40, 44]. Increased resistance to Fas-induced apoptosis might also contribute to the pathogenesis of East Coast fever.

Theileria-transformed cells have been shown to induce a pronounced autologous mixed lymphocyte reaction [61, 62], which is thought to contribute to the dramatic accumulation of uninfected lymphoblastoid cells that temporarily occurs in the animal's lymphoid tissues. It can be expected that in such cells, Fas and FasL will be upregulated as a consequence of activation. The enhanced resistance against Fas-induced apoptosis in *T. parva*-infected cells might impart a selective advantage to parasitized cells over uninfected cells, thus contributing to their clonal expansion and spread through the body. It would also be of interest to determine to what extent Fas-induced apoptosis contributes to the massive lymphocytolysis that follows the initial phase of lymphoproliferation and involves uninfected as well as *T. parva*-infected cells [63].

34.2.4.1 Interference with Hubs and Bottlenecks

NFκB Activation There is mounting evidence from studies on protein–protein interactions that pathogens have evolved to interact with host proteins that function as hubs (those involved in many interactions) or bottlenecks (proteins central to many pathways) [64]. The IκB kinase (IKK) complex is a central regulator of most NFκB activation pathways and, by hijacking IKK, *Theileria* provides at least one striking manifestation of this evolutionary process. To ensure continuous NFκB activity [65], the schizont accumulates large aggregates of IKK on its surface where they become constitutively activated, most probably by proximity-induced *trans*-autophosphorylation [66]. This results in the continuous degradation of the cytoplasmic inhibitors of NFκB, IκBα and IκBβ, and NFκB translocation to the nucleus ([67]. This way, the parasite bypasses the activation signals normally emanating from surface and intracellular receptors and it can be assumed that parasite-induced IKK activation is also not subject to the negative control mechanisms that normally regulate these pathways. In addition, in contrast to normal pathways, parasite-induced IKK activation was found to be independent of functional Hsp90 [68]. Signaling via surface receptors appears not to be blocked, however, as *Theileria*-transformed cells can respond to autocrine signals with increased NFκB activity [44].

Despite intensive efforts, the parasite protein(s) involved in recruiting IKK signalosomes to the parasite surface have not yet been identified, perhaps pointing to the involvement of host cell adapter proteins. At mitosis, when the microtubular network is completely reorganized, IKK signalosomes disappear, potentially suggesting a role of microtubules in IKK recruitment [66].

Whereas constitutive NFκB activity may help protect the cell against different types of apoptosis, it is also likely to contribute to the proliferation potential of the parasitized cell. NFκB contributes to the control of DNA replication by regulating the expression of cyclin D and c-Myc (reviewed in [69]) and recent findings also point towards an important role of IKK mitosis [70, 71].

c-Myc Activation: Walking a Dangerous Tightrope Typically for transformed cells, the proto-oncogene c-Myc is also expressed in *Theileria*-transformed cells [72, 73]. The c-Myc promoter is active and the half-life of the protein is significantly prolonged. Several reports point towards a central role for c-Myc in *Theileria*-induced host cell transformation (reviewed in [59]). Functioning foremost as a transcription factor, c-Myc potentially regulates the transcription of as many as 15% of the genes in species ranging from flies to humans [74]. Target genes include cell cycle genes, protein synthesis components, genes encoding cell adhesion and cytoskeletal proteins, antiapoptotic genes and also microRNAs, small regulatory molecules that regulate the stability or translational efficiency of target mRNAs. This multitude of vital functions may help explain the deleterious effects on host cell survival observed when c-Myc is downregulated in *Theileria*-transformed cells [73]. c-Myc could contribute to the proliferation of *Theileria*-transformed cells by directly regulating genes involved in cell cycle progression or by repressing the cyclin-dependent kinase inhibitors p21 and p15INK4A. c-Myc also represses growth arrest genes such as Gadd45 and gas1 (see review [75] and references therein).

However, cells that rely on c-Myc for uncontrolled proliferation walk a dangerous tightrope, as they also have to contend with specific proapoptotic functions, designed to inhibit c-Myc-dependent tumorigenesis. In several transformed cell lines, rather than promoting survival, c-Myc actually suppresses the cells' antiapoptotic mechanisms, increasing sensitivity to death receptor-induced apoptosis by interfering with the NFκB-dependent survival circuitry that controls the expression of c-IAP2 and Mcl-1 [76] or the caspase-8 inhibitor c-FLIP [77]. Furthermore, c-Myc has also been shown to prime the activation of mitochondrial apoptosis pathways controlled by Bcl-2 family members. Proapoptotic Bak plays an central role in the complex, c-Myc-dependent interplay between pro- and antiapoptotic Bcl-2 family members (reviewed in [78]).

An interesting link also exists between c-Myc and expression of the transferrin receptor (TfR), a major mediator of iron uptake in mammalian cells. The TfR is a downstream target of c-Myc and overexpression is a common feature of human malignancies. In human B cell lymphoma models TfR inhibition results in G_1 arrest, while enforced expression conferred a growth advantage [79]. Overexpression is also observed in *Theileria*-transformed cells, and is accompanied by enhanced transferrin binding, uptake and release, and receptor recycling [80]. The latter was found to be dependent on the ras-related GTPase Rab11, a component of the vesicular trafficking machinery upregulated in *Theileria*-transformed B cells [81]. TfR expression was also found to require activation of phosphatidylinositol 3-kinase (PI3K) and JNK, two additional kinases with central regulatory functions in proliferation and cell survival that are constitutively activated in *Theileria*-transformed cells (see below).

JAK/STAT STATs (signal transducers and activators of transcription) are transcription factors that participate in cytokine and growth factor-mediated signaling. STATs function as messengers that relay signals delivered at the plasma membrane directly to the nucleus where they regulate transcription by binding to specific ISRE/GAS promoter sequences. Whereas STAT activation is normally tightly controlled and functions with finite kinetics, constitutive activation (most notably of STAT3 and STAT5) is frequently observed in malignant transformation. Aberrant STAT3 promotes uncontrolled growth and survival through dysregulation of gene expression, including cyclin D1, c-Myc, Bcl-xL, Mcl-1 and survivin genes. In *T. parva*-transformed B cells, permanent induction of a JAK2/STAT3 signaling pathway was shown to contribute the expression of c-Myc and Mcl-1 gene expression [73]. The fact that JAK2/STAT3 signaling could also be stimulated by recombinant GM-CSF, could point towards the involvement of an autocrine loop [82], but the contribution of other lymphokines and cytokines cannot be excluded [40, 41, 51, 83–85]. Whatever the stimulus, interference with the JAK2/STAT pathway results in reduced c-Myc and Mcl-1 expression, suggesting a synergistic role of the JAK/STAT and NFκB pathways in c-Myc and Mcl-1 expression.

Mitogen-Activated Protein Kinases Mitogen-activated protein kinases (MAPKs) are involved in the cellular responses to almost all stimuli that involve membrane, cytoplasmic and nuclear signaling networks. They all phosphorylate specific serines and threonines on a wide array of target substrates and their biological functions are largely defined by a complex system of spatio-temporal control in the cell (reviewed in [86] and accompanying special review articles). They can be subdivided into the extracellular signal-regulated kinases (ERKs), the c-Jun NH_2-terminal kinases (JNKs), the p38 kinases and ERK5. Given their importance for so many cellular functions, MAPKs were among the earliest to be investigated in *Theileria*-transformed cells. The pattern of induction did not match initial expectations, however. Whereas sustained activation of the ERK1/2 pathway is necessary for proliferation of normal cells, only baseline activity could be detected in *T. parva*-transformed T cells and *T. annulata*-transformed B cells [42, 87]. p38 activity was also low or undetectable [87, 88]. Instead, the JNK kinases were found to be constitutively active in a parasite-dependent manner in all cell lines tested. Recent findings by Seitzer *et al.* [89], however, revealed significant ERK activity in bulk cultures of *Theileria*-transformed cells, whereas ERK activity was much lower in cloned cell lines. The reason for this discrepancy is not known.

JNK was found to contribute directly to the activation of central transcription factors such as AP-1 and ATF-2 [87, 88], which regulate a plethora of cellular functions such as cell adhesion, lymphokine and cytokine expression. Constitutive AP-1 activity also contributes to transferrin receptor expression and – via the induction of Rab11 expression – to receptor recycling [81]. In addition, JNK was also shown to promote the survival and metastatic potential of *T. parva*-transformed B lymphocytes in mice [90]. In *T. annulata*-transformed cells, AP-1 has been shown to regulate the expression of MMP9, an metalloproteinase with an important role in metastasis [91] and the same may apply to *T. parva*-infected cells.

IKK and JNK activation are often coregulated, but contrary to parasite-induced IKK activation, there is no evidence for direct activation of JNK at the schizont. Instead, signals emanating from the surface of the *Theileria*-transformed cell probably contribute to constitutive JNK activity. In this context, it has been proposed that autocrine stimulation of the receptors for TNFα [44] and GM-CSF [82] contribute to AP-1 stimulation. AP-1 activity is also dependent on Src-family kinase activation [92], but whether this pathway also feeds into the JNK activation pathway was not demonstrated.

The Src Family Src family members are proteins tyrosine kinases (PTKs) that share a high degree of homology. They are widely expressed, but differ in their expression patterns depending on the cell type. By phosphorylating downstream components they contribute to the activation of signaling pathways triggered through immune and cytokine receptors. Activated Src-related kinases such as Fyn, Lck and Hck can be detected in different *Theileria*-transformed cell lines [47, 92, 93] and inhibitors block proliferation. Src family PTKs can be found in glycolipid-enriched microdomains (GEMs, also called lipid rafts) and are negatively regulated by the PTK Csk, complexed to tyrosine-phosphorylated transmembrane adapter protein PAG/Cbp (phosphoprotein associated with GEMs, also called Cbp, Csk-binding protein). In the case of *Theileria*-induced Hck activation, it could be shown that this involves the constitutive exclusion of the inhibitor kinase Csk from Hck-containing membrane microdomains, through downmodulation of the transmembrane adapter protein PAG/Cbp [92].

Casein Kinase II Casein kinase II (CK2), an enzyme now known to phosphorylate several hundred cellular substrates, was one of the first kinases found to be upregulated and activated in *T. parva*-transformed cells [94]. Subsequent inhibition experiments confirmed the importance of CK2 in cell survival and proliferation [94–97]. Because of the presence of a predicted signal peptide, it was proposed that the α-subunit of *T. parva* CK2 might be exported into the host cell cytoplasm where it could activate host cell mitogenic pathways. Further studies [98], however, demonstrated that translation is initiated from a downstream initiation site omitting the putative signal sequence, making it unlikely that the protein is secreted via the classical secretory pathway.

Phosphatidylinositol 3-Kinase Optimal proliferation of *Theileria*-transformed cells requires cell–cell contact and a number of different autocrine loops have also been proposed [43, 44, 46, 73, 82]. These processes depend on ligand interactions with receptors that are usually linked to the PI3K signaling pathway. PI3K was found to be permanently active in different *Theileria*-transformed cell lines [82, 99]. By activating PKB/Akt, the arsenal of weapons providing protection against apoptosis is further expanded in the parasitized cell. Essential requirements for the initiation of DNA synthesis are also fulfilled, as the PI3K pathway feeds into the activation of two transcription factors, E2F and c-Myc, with prominent roles in proliferation. Blocking the PI3K pathway clearly affected the proliferation of *Theileria*-transformed cells, but cell viability was not reduced – at least not in the short term-suggesting that antiapoptotic pathways other than those controlled by PKB/Akt, suffice to guarantee survival.

34.3
Genes Involved in Host–Parasite Interaction

The genomes of *T. parva* and *T. annulata* have been sequenced and annotated [100, 101]. When compared to other apicomplexan parasites, interesting differences came to light [102], but obvious "transforming" genes, as have been found in transforming viruses, were not found. There are also no obvious indications that the parasite, at one stage during evolution, might have acquired host cell genes that are now being used to induce transformation. Host cell transformation is accompanied by extensive changes in the pattern of protein phosphorylation [94, 103]. However, in contrast to other apicomplexan parasites such as *Plasmodium* or *Toxoplasma*, no secreted or surface-associated kinases or phosphatases could be found in the *Theileria* proteomes. Nevertheless, data-mining of the genome opens up new opportunities to study parasite genes with important functions in this special form of host–parasite interaction (see review [104]). Of particular interest are proteins that are predicted to be released into the host cell or expressed on the schizont surface, from where they can interact with host cell cytoskeletal or signaling components.

34.3.1
Proteins Released Into the Host Cell

Already before the genomes had been sequenced, a family of *T. annulata* schizont proteins was identified that contain mammalian AT-hook DNA-binding domains. They were designated *TashATs* and shown to be released into the host cell cytoplasm from where they translocate into the host cell nucleus. TashAT expression appears to be linked to the transforming schizont stage and decreases as the parasite differentiates to merozoites [105–107]. As mammalian AT-hook proteins have been implicated in the differentiation, proliferation and apoptosis of myeloid cells, TashATs attracted immediate interest. Intriguingly, however, AT-hook domains are not obvious in the *T. parva* orthologs. Nevertheless, it is tempting to assume that TashATs could modify host cell transcription. For instance, overexpression of TashAT2 in (nonparasitized) BoMac cells induces changes in expression of bUBP43, an ubiquitin-like protease, and its substrate bISG15 [108]. It was hypothesized that these interferon-stimulated genes could be involved in avoiding the host immune defenses and help the transformation process by controlling cell growth and differentiation. SuAT1 is related to the TashATs, but its expression is also maintained as the parasite progresses to merogony. When overexpressed in nonparasitized cells it induces morphological alterations and changes in the expression profile of cytokeratin similar to that found during the infection of leukocytes by the parasite [109].

The subtelomeric regions of all four chromosomes of both the *T. parva* and *T. annulata* harbor a large family of genes encoding subtelomere-encoded variable secreted proteins (SVSPs). Most of the SVSPs have a signal peptide for secretion, a highly variable N-terminal region rich in Q and P residues, and a more conserved C-terminal region. All SVSPs contain so-called FAINT domains (frequently associated in *Theileria*), highly polymorphic domains of unknown function, with an average length of 70 residues. Interestingly, expression analysis suggests that many of the genes are transcribed [110] following a pattern of expression that is independent of the host cell background and determined solely by the parasite genotype (Laboratory Dobbelaere, unpublished). As is the case for TashATs, most of these proteins also have functional nuclear localization signals. Considering the number of genes and range of expression, SVSPs are likely to have important functions, possibly in the process of transformation or perhaps by thwarting host cell immune responses.

34.3.2
The Schizont Surface

Interestingly, as is the case for SVSPs, the major schizont surface protein PIM (polymorphic immunodominant molecule) also contains a QP-rich N-terminal domain [111–113]. The C-terminal domain, which is highly conserved between *T. annulata* (where it is called TaSP) and *T. parva*, is predicted to span the plasma membrane three times, and shows a strong resemblance to copper transporters. However, whether PIM really functions as a transporter has not yet been confirmed.

The extensive polymorphism of the QP-rich domains of both PIM and the SVSPs makes it difficult to predict a specific function and *in vitro* experiments have shown that the QP-rich domain of PIM can undergo unusual processing [114]. In addition to PIM, several other schizont surface proteins have been identified. *11E*, a protein belonging to the glutaredoxin family, mostly resides within the parasite, often in close vicinity to membranous structures, but also appears at the surface membrane [115].

34.3.3 Interactions with the Host Cell Cytoskeleton

During interphase and mitosis, the position of the parasite in the host cell appears to be defined by its interaction with host cell microtubules. Two schizont proteins have been identified that could participate in schizont–host cell microtubule interactions and possibly also distribution of the schizont over the two daughter cells during host cell cytokinesis. A 37 kDa schizont protein, designated TaSE (*T. annulata* secretory protein), was identified that interacts with host cell microtubules and has an affinity for several mitotic structures when overexpressed in mammalian cells [116]. In addition, a GPI-anchored protein of 34 kDa protein, *gp34*, was recently also identified on the surface of *T. parva* and *T. annulata* schizonts. gp34 can interact with different mitotic host cell structures and kinases involved in cytokinesis, and overexpression disrupts cytokinesis (Laboratory Dobbelaere, unpublished observation). In contrast to microtubules, there is no evidence for a direct interaction of the schizont with the host cell actin cytoskeleton (Martin Baumgartner, unpublished). Nevertheless, the organization of the actin cytoskeleton of *Theileria*-transformed cells differs significantly from that in cells from which the parasite has been eliminated, with the number of filopodia being strongly reduced in the latter.

34.4 Immune Responses Directed Against the Schizont

Elimination of the schizont-infected cell by the host immune system poses an obvious, major threat to the parasite. Because of the specific mode of intracellular propagation the schizont is not accessible to antibodies. However, its intracytoplasmic location favors immune recognition by $CD8^+$ cytotoxic T cells stimulated by parasite-derived peptides presented by host cell MHC class I molecules. Such schizont-specific peptides are generated as parasite proteins, released into the host cell cytoplasm, undergo proteasomal degradation and enter the class I MHC antigen processing pathway. To exist, the schizont depends on the expression of hundreds of proteins, at least some of which must find their way into the host cytoplasm. Despite the fact that the T cell receptor repertoire can – at least theoretically – recognize a large number of pathogen epitopes, protective immunity in individual animals seems to be focused on only few polymorphic antigens and cross-protection between strains is often lacking (reviewed in [117]). Intuitively, one would expect the schizont to have evolved ways to limit the repertoire of secreted proteins to an absolute minimum, but genome analysis clearly suggests the opposite. Several members of the TashAT protein family are secreted into the host cell, but apparently not presented. In an elaborate screening exercise for *T. parva* antigens, none of the SVSPs were identified as T cell target antigens, despite the large number of genes expressed [118]. The same applies to PIM, the polymorphic immunodominant molecule, which is abundantly expressed, but only triggers strong B cell responses. Could QP-rich domains conceivably interfere with MHC class I-restricted antigen presentation? N-terminal peptide trimming of protein fragments, imported into the endoplasmic reticulum after proteasomal degradation, is a key step in antigen processing [119]. Interestingly, the aminopeptidases responsible for efficient trimming clip all residues except prolines that flank the N-termini. It would be interesting to test whether an excess of aminopeptidase-resistant QP-rich peptide fragments could render the antigen-presentation process less efficient.

34.5 Summary and Outlook

Rather than trying to survive within the limited confines of a parasitophorous vacuole, transforming *Theileria* parasites opt for a grander lifestyle in a more spacious environment: the cytoplasm. Protected from lysosomal destruction, the schizont establishes a long-lasting interaction with host cell microtubules, thus securing its distribution over the daughter cells at each host cell mitosis.

Intracellular pathogens target predominantly protein hubs and bottlenecks, from where important cellular processes are regulated. True to this tradition, by recruiting the IKK complex to its surface, *Theileria* directly targets IKK, a pivotal regulator of NFκB signaling pathways. Several other pathways, among them important nodes such as the PI3K, JNK and JAK/STAT pathways, are also activated, most likely in an indirect manner (summarized in Figure 34.3). These *Theileria*-induced signaling cascades lead to the activation of a number of central transcription factors, including NFκB, AP-1, ATF-2, E2F and c-Myc, which control many of the genes that regulate cellular replication and survival. Thus, by strategically modulating a network of different host cell signaling pathways, the parasite induces continuous proliferation of the infected host cell and arms the cells against a range of different apoptotic stimuli. These transcription factors also regulate the expression of surface receptors and ligands with important roles in immune and inflammatory responses, which, depending on the cell type, can contribute to transformation-associated phenotypic changes and possibly also to the pathogenesis of the disease.

Compared to other parasites, only little is known about schizont surface proteins, or proteins that are released into the host cell cytoplasm. The intriguing TashAT proteins that can translocate to the host cell nucleus have not yet revealed their secrets and there is more to come: in both *T. parva* and *T. annulata*, large families of SVSPs, again containing nuclear localization signals, have been identified. Just like the major schizont surface protein PIM, they contain N-terminal domains rich in prolines and glutamines. Two candidate proteins have been identified that appear

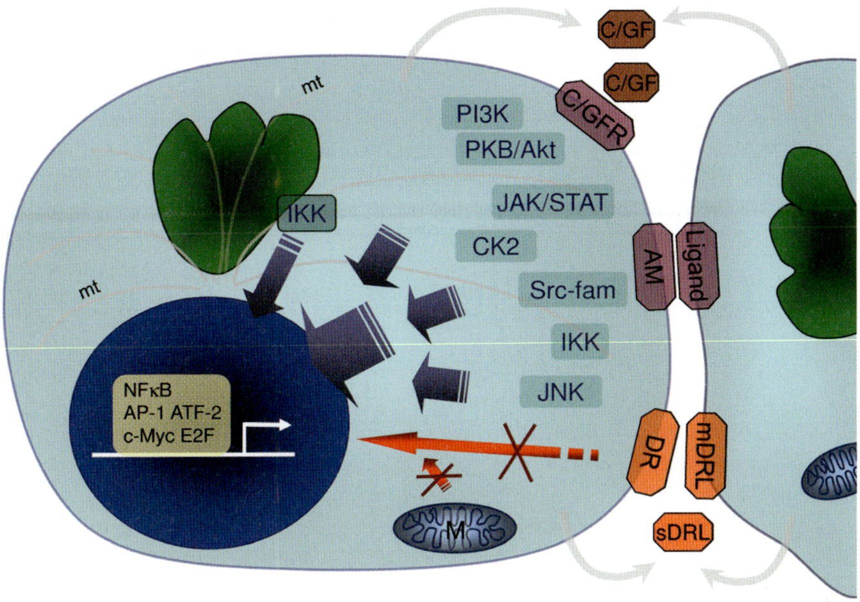

Figure 34.3 Continuous host cell proliferation and protection against apoptosis induced by the intracellular parasite *Theileria* involves a range of different signaling cascades. At all times, the schizont (green) is in close contact with host cell microtubules (mt). By recruiting IKK signalosomes to its surface the parasite directly activates the NFκB pathway, independent of other upstream stimuli. Additional signals, emanating from cytokines or growth factor receptors (C/GFR), death receptors (DR) and possibly also adhesion molecules (AM) contribute to the activation of the JNK, PI3K/PKB, and JAK/STAT pathways. Different Src-family protein tyrosine kinases are also activated, depending on the cell type and, in some cases additional IKK stimulation can also take place. The activity of casein kinase 2 (CK2), which has a broad spectrum of substrates, is constitutively upregulated. Through the combined activity of these pathways, a set of transcription factors is activated, which, often in combination with each other, regulate the expression of a wide range of genes. Some of these genes encode death receptors and their soluble (sDRL) or membrane-bound (mDRL) ligands; others encode cytokines, lymphokines and their receptors. These are likely to contribute to the pathogenesis of theileriosis and, in some cases, also activate autocrine stimulatory loops (gray arrows). Genes essential for initiating DNA replication are activated through E2F, c-Myc and NFκB. Importantly, several of these transcription factors, including NFκB and c-Myc promote the expression of antiapoptotic proteins, including caspase inhibitors and members of the Bcl-2 family; these protect against apoptosis induced through DR and/or mitochondrial (M) pathways (red arrows).

to interact with the host cell mitotic spindle and could participate in facilitating the distribution of the schizont over the two daughter cells at host cell cytokinesis. One of these proteins, gp34, belongs to a rare group of GPI-anchored proteins that, rather than facing the outside world, faces into the cytoplasm of a eukaryotic cell.

Because of the availability of an excellent drug, *Theileria*-transformed cells provide a unique, reversible model to study malignant transformation, and also offer exciting opportunities to study the intricacies of host–pathogen interactions. One of the great challenges ahead will be to define the function TashATs and SVSPs, which are likely to be important players in transformation and potentially also immune evasion processes. The establishment of efficient methods to transfect *Theileria* should greatly facilitate these studies. Several other important questions await answers. What are the host cell receptors required for sporozoite binding and entry? Which parasite enzymes are involved in the elimination of the surrounding host cell membrane after invasion? What is the early sequence of events that leads to transformation? What is the molecular basis that controls the synchronization of host and parasite cell cycles? The genetic factors responsible for increased resistance (*Bos indicus* vs. *Bos taurus*) have not yet been determined, nor is it known how the buffalo controls *Theileria*.

Despite significant efforts spanning several decades, scientists striving to develop cheap, safe and efficient recombinant *Theileria* vaccines finally had to accept that parasites were not about to forfeit the efficient strategies, developed during ages of coevolution, to counter host immune responses aimed at eliminating them. Only live vaccines (in the case of *T. annulata*) or infection and treatment (*T. parva*) appear to be able to induce protective immunity against *Theileria*, underpinning the theory that only live parasites are capable of inducing protective responses [3]. International institutions and laboratories that have focused their research efforts on testing recombinant *Theileria* vaccines are rapidly running out of candidates. The emphasis is therefore shifting back towards the application of live *T. parva* vaccines based on the infection and treatment method developed 30 years ago [120, 121]. Such vaccines are costly and the logistics for delivery problematic. Other problems include the lack of cross-protection and sometimes severe reactions to vaccination. By enhancing our insights into the molecular, cell biological and immunological mechanisms used by the parasite to survive in its cytoplasmic niche, it may be possible to identify parasite genes involved in virulence, pathogenicity and immune-evasion. This knowledge, in turn, might contribute to the development of safer, live vaccines.

Acknowledgments

Isabel Roditi is thanked for corrections and suggestions.

References

1 Irvin, A.D. and Morrison, W.I. (1987) *Immunopathology, Immunology and Immunoprophylaxis of Theileria Infection*, CRC Press, Boca Ratan, Florida.

2 Geysen, D., Bishop, R., Skilton, R., Dolan, T.T. and Morzaria, S. (1999) *Trop Med Int Health*, **4**, A21–A27.

3 Morrison, W.I. and McKeever, D.J. (2006) *Parasitology*, **133** (Suppl), S169–S187.

4 Grootenhuis, J.G., Leitch, B.L., Stagg, D. A., Dolan, T.T. and Young, A.S. (1987) *Parasitology*, **94**, 425–431.

5 Shaw, M.K. (2002) in *Theileria* (eds D.A. E. Dobbelaere and D.J. McKeever),

Kluwer Academic Publishers, Boston, pp. 1–22.

6 Preston, P.M., Brown, C.G., Bell Sakyi, L., Richardson, W. and Sanderson, A. (1992) *Res Vet Sci*, **53**, 230–243.

7 Ndungu, S.G., Brown, C.G. and Dolan, T. T. (2005) *Onderstepoort J Vet Res*, **72**, 13–22.

8 Glass, E.J., Preston, P.M., Springbett, A., Craigmile, S. *et al.* (2005) *Int J Parasitol*, **35**, 337–347.

9 Spooner, R.L., Innes, E.A., Glass, E.J. and Brown, C.G. (1989) *Immunology*, **66**, 284–288.

10 Dobbelaere, D.A.E., Prospero, T.D., Roditi, I.J., Kelke, C. *et al.* (1990) *Infect Immun*, **58**, 3847–3855.

11 Baldwin, C.L., Black, S.J., Brown, W.C., Conrad, P.A. *et al.* (1988) *Infect Immun*, **56**, 462–467.

12 Campbell, J.D., Brown, D.J., Glass, E.J., Hall, F.R. and Spooner, R.L. (1994) *Parasite Immunol*, **16**, 501–505.

13 Howard, C.J., Sopp, P., Preston, P.M., Jackson, L.A. and Brown, C.G. (1993) *Vet Immunol Immunopathol*, **39**, 275–282.

14 Sager, H., Davis, W.C., Dobbelaere, D.A. and Jungi, T.W. (1997) *J Leukoc Biol*, **61**, 459–468.

15 Brown, C.G., Stagg, D.A., Purnell, R.E., Kanhai, G.K. and Payne, R.C. (1973) *Nature*, **245**, 101–103.

16 Sager, H., Bertoni, G. and Jungi, T.W. (1998) *J Immunol*, **161**, 335–341.

17 Moreau, M.F., Thibaud, J.L., Miled, L.B., Chaussepied, M. *et al.* (1999) *Infect Immun*, **67**, 6678–6682.

18 Naessens, J., Newson, J., Bensaid, A., Teale, A.J. *et al.* (1985) *J Immunol*, **135**, 4183–4188.

19 Emery, D.L., MacHugh, N.D. and Morrison, W.I. (1988) *Parasite Immunol*, **10**, 379–391.

20 Baldwin, C.L. and Teale, A.J. (1987) *Eur J Immunol*, **17**, 1859–1862.

21 Stephens, S.A. and Howard, C.J. (2002) *Parasitology*, **124**, 485–493.

22 Wells, W.C., Awino, E. and McKeever, D.J. (1999) *Microsc Soc S Africa Proc*, **29**, 61.

23 Shaw, M.K. (1997) *Int J Parasitol*, **27**, 457–474.

24 Shaw, M.K. (2003) *Trends Parasitol*, **19**, 2–6.

25 Dobbelaere, D.A.E., Shapiro, S.Z. and Webster, P. (1985) *Proc Natl Acad Sci USA*, **82**, 1771–1775.

26 Nene, V., Iams, K.P., Gobright, E. and Musoke, A.J. (1992) *Mol Biochem Parasitol*, **51**, 17–27.

27 Hall, R., Hunt, P.D., Carrington, M., Simmons, D.C *et al.* (1992) *Mol Biochem Parasitol*, **53**, 105–112.

28 Hall, R.N.R. b., Brown, C.G., Wilkie, G.C *et al.* (2000) *Parasite Immunol*, **22**, 223–230.

29 Musoke, A.J., Nantulya, V.M., Rurangirwa, F.R. and Buscher, G. (1984) *Immunology*, **52**, 231–238.

30 Shaw, M.K., Tilney, L.G. and Musoke, A.J. (1991) *J Cell Biol*, **113**, 87–101.

31 Shaw, M.K., Tilney, L.G., Musoke, A.J. and Teale, A.J. (1995) *J Cell Sci*, **108**, 1587–1596.

32 Syfrig, J., Wells, C., Daubenberger, C., Musoke, A.J. and Naessens, J. (1998) *Eur J Cell Biol*, **76**, 125–132.

33 Dobbelaere, D.A., Webster, P., Leitch, B. L., Voigt, W.P. and Irvin, A.D. (1985) *Exp Parasitol*, **60**, 90–100.

34 Webster, P., Dobbelaere, D.A. and Fawcett, D.W. (1985) *Eur J Cell Biol*, **36**, 157–162.

35 Fawcett, D., Musoke, A. and Voigt, W. (1984) *Tissue Cell*, **16**, 873–884.

36 Jura, W.G. (1988) *Vet Parasitol*, **26**, 189–198.

37 Rocchi, M.S., Ballingall, K.T., MacHugh, N.D. and McKeever, D.J. (2006) *Int J Parasitol*, **36**, 771–778.

38 Visser, A.E., Abraham, A., Sakyi, L.J., Brown, C.G. and Preston, P.M. (1995) *Parasite Immunol*, **17**, 91–102.

39 Richardson, J.O., Forsyth, L.M., Brown, C.G. and Preston, P.M. (1998) *Vet Res Commun*, **22**, 31–45.

40 Preston, P.M., Brown, C.G. and Richardson, W. (1992) *Parasite Immunol*, **14**, 125–141.

41 DeMartini, J.C. and Baldwin, C.L. (1991) *Infect Immun*, **59**, 4540–4546.

42 Galley, Y., Hagens, G., Glaser, I., Davis, W. *et al.* (1997) *Proc Natl Acad Sci USA*, **94**, 5119–5124.

43 Dobbelaere, D.A.E., Coquerelle, T.M., Roditi, I.J., Eichhorn, M. and Williams, R. O. (1988) *Proc Natl Acad Sci USA*, **85**, 4730–4734.

44 Guergnon, J., Chaussepied, M., Sopp, P., Lizundia, R. *et al.* (2003) *Cell Microbiol*, **5**, 709–716.

45 Dobbelaere, D.A.E. and Heussler, V.T. (1999) *Annu Rev Microbiol*, **53**, 1–42.

46 Dobbelaere, D.A.E., Roditi, I.J., Coquerelle, T.M., Kelke, C. *et al.* (1991) *Eur J Immunol*, **21**, 89–95.

47 Fich, C., Klauenberg, U., Fleischer, B. and Broker, B.M. (1998) *Parasitology*, **117** (Pt 2), 107–115.

48 Irvin, A.D., Ocama, J.G. and Spooner, P. R. (1982) *Res Vet Sci*, **33**, 298–304.

49 Newson, J., Naessens, J., Stagg, D.A. and Black, S.J. (1986) *Parasite Immunol*, **8**, 149–158.

50 Nichani, A.K., Campbell, J.D., Glass, E.J., Graham, S.P. *et al.* (2003) *Vet Immunol Immunopathol*, **93**, 39–49.

51 Graham, S.P., Brown, D.J., Vatansever, Z., Waddington, D. *et al.* (2001) *Vaccine*, **19**, 2932–2944.

52 Pipano, E. and Shkap, V. (2000) *Ann NY Acad Sci*, **916**, 484–500.

53 Meinl, E., Fickenscher, H., Thome, M., Tschopp, J. and Fleckenstein, B. (1998) *Immunol Today*, **19**, 474–479.

54 Heussler, V.T., Kuenzi, P. and Rottenberg, S. (2001) *Int J Parasitol*, **31**, 1166–1176.

55 Schaumburg, F., Hippe, D., Vutova, P. and Luder, C.G. (2006) *Parasitology*, **132** (Suppl), S69–S85.

56 Thome, M. and Tschopp, J. (2001) *Nat Rev Immunol*, **1**, 50–58.

57 Kuenzi, P., Schneider, P. and Dobbelaere, D.A. (2003) *J Immunol*, **171**, 1224–1231.

58 Dobbelaere, D.A. and Kuenzi, P. (2004) *Curr Opin Immunol*, **16**, 524–530.

59 Dessauge, F., Lizundia, R., Baumgartner, M., Chaussepied, M. and Langsley, G. (2005) *Trends Parasitol*, **21**, 377–385.

60 Heussler, V., Sturm, A. and Langsley, G. (2006) *Parasitology*, **132** (Suppl), S49–S60.

61 Pearson, T.W., Lundin, L.B., Dolan, T.T. and Stagg, D.A. (1979) *Nature*, **281**, 678–680.

62 Goddeeris, B.M. and Morrison, W.I. (1987) *Immunology*, **60**, 63–69.

63 Morrison, W.I., Buscher, G., Murray, M., Emery, D.L. *et al.* (1981) *Exp Parasitol*, **52**, 248–260.

64 Dyer, M.D., Murali, T.M. and Sobral, B.W. (2008) *PLoS Pathog*, **4**, e32.

65 Ivanov, V., Stein, B., Baumann, I., Dobbelaere, D.A. *et al.* (1989) *Mol Cell Biol*, **9**, 4677–4686.

66 Heussler, V.T., Rottenberg, S., Schwab, R., Kuenzi, P. *et al.* (2002) *Science*, **298**, 1033–1036.

67 Palmer, G.H., Machado, J.J., Fernandez, P., Heussler, V. *et al.* (1997) *Proc Natl Acad Sci USA*, **94**, 12527–12532.

68 Hermann, P. and Dobbelaere, D.A. (2006) *FEBS Lett*, **580**, 5023–5028.

69 Joyce, D., Albanese, C., Steer, J., Fu, M. *et al.* (2001) *Cytokine Growth Factor Rev*, **12**, 73–90.

70 Blazkova, H., von Schubert, C., Mikule, K., Schwab, R. *et al.* (2007) *Cell Cycle*, **6**, 2531–2540.

71 Prajapati, S., Tu, Z., Yamamoto, Y. and Gaynor, R.B. (2006) *Cell Cycle*, **5**, 2371–2380.

72 Eichhorn, M., Prospero, T.D., Heussler, V. T. and Dobbelaere, D.A. (1993) *J Exp Med*, **178**, 769–776.

73 Dessauge, F., Hilaly, S., Baumgartner, M., Blumen, B. *et al.* (2005) *Oncogene*, **24**, 1075–1083.

74 Fernandez, P.C., Frank, S.R., Wang, L., Schroeder, M. *et al.* (2003) *Genes Dev*, **17**, 1115–1129.

75 Dang, C.V., O'Donnell, K.A., Zeller, K.I., Nguyen, T. *et al.* (2006) *Semin Cancer Biol*, **16**, 253–264.

76 Ricci, M.S., Kim, S.H., Ogi, K., Plastaras, J.P. *et al.* (2007) *Cancer Cell*, **12**, 66–80.

77 Ricci, M.S., Jin, Z., Dews, M., Yu, D. *et al.* (2004) *Mol Cell Biol*, **24**, 8541–8555.

78 Nieminen, A.I., Partanen, J.I. and Klefstrom, J. (2007) *Cell Cycle*, **6**, 2464–2472.

79 O'Donnell, K.A., Yu, D., Zeller, K.I., Kim, J.W. *et al.* (2006) *Mol Cell Biol*, **26**, 2373–2386.

80 Baumgartner, M., Chaussepied, M., Raposo, G., Goud, B. and Langsley, G. (2005) *Cell Microbiol*, **7**, 637–644.

81 Lizundia, R., Chaussepied, M., Naissant, B., Masse, G.X. *et al.* (2007) *Cell Microbiol*, **9**, 1936–1945.

82 Baumgartner, M., Chaussepied, M., Moreau, M.F., Werling, D. *et al.* (2000) *Cell Microbiol*, **2**, 329–339.

83 McKeever, D.J., Nyanjui, J.K. and Ballingall, K.T. (1997) *Parasite Immunol*, **19**, 319–324.

84 Brown, D.J., Campbell, J.D., Russell, G. C., Hopkins, J. and Glass, E.J. (1995) *Clin Exp Immunol*, **102**, 507–514.

85 Schnittger, L., Hollmann, C., Diemer, U., Boguslawski, K. and Ahmed, J.S. (2000) *Ann NY Acad Sci*, **916**, 676–680.

86 Dhanasekaran, D.N. and Johnson, G.L. (2007) *Oncogene*, **26**, 3097–3099.

87 Chaussepied, M., Lallemand, D., Moreau, M.F., Adamson, R. *et al.* (1998) *Mol Biochem Parasitol*, **94**, 215–226.

88 Botteron, C. and Dobbelaere, D. (1998) *Biochem Biophys Res Commun*, **246**, 418–421.

89 Seitzer, U., Schnittger, L., Boguslawski, K. and Ahmed, J.S. (2006) *Ann NY Acad Sci*, **1081**, 473–475.

90 Lizundia, R., Chaussepied, M., Huerre, M., Werling, D. *et al.* (2006) *Cancer Res*, **66**, 6105–6110.

91 Baylis, H.A., Megson, A. and Hall, R. (1995) *Mol Biochem Parasitol*, **69**, 211–222.

92 Baumgartner, M., Angelisova, P., Setterblad, N., Mooney, N. *et al.* (2003) *Blood*, **101**, 1874–1881.

93 Eichhorn, M. and Dobbelaere, D.A. (1994) *Parasitol Today*, **10**, 469–472.

94 ole MoiYoi, O.K., Brown, W.C., Iams, K. P., Nayar, A. *et al.* (1993) *EMBO J*, **12**, 1621–1631.

95 Dessauge, F., Lizundia, R. and Langsley, G. (2005) *Parasitology*, **130** (Suppl), S37–44.

96 Seldin, D.C. and Leder, P. (1995) *Science*, **267**, 894–897.

97 Shayan, P. and Ahmed, J.S. (1997) *Parasitol Res*, **83**, 526–532.

98 Biermann, R., Schnittger, L., Beyer, D. and Ahmed, J.S. (2003) *J Cell Physiol*, **196**, 444–453.

99 Heussler, V.T., Kuenzi, P., Fraga, F. Schwab, R.A. *et al.* (2001) *Cell Microbiol*, **3**, 537–550.

100 Gardner, M.J., Bishop, R., Shah, T. de Villiers, E.P. *et al.* (2005) *Science*, **309**, 134–137.

101 Pain, A., Renauld, H., Berriman, M., Murphy, L. *et al.* (2005) *Science*, **309**, 131–133.

102 Roos, D.S. (2005) *Science*, **309**, 72–73.

103 Dyer, M., Hall, R., Shiels, B. and Tait, A. (1992) *Exp Parasitol*, **74**, 216–227.

104 Shiels, B., Langsley, G., Weir, W., Pain, A. *et al.* (2006) *Int J Parasitol*, **36**, 9–21.

105 Swan, D.G., Phillips, K., Tait, A. and Shiels, B.R. (1999) *Mol Biochem Parasitol*, **101**, 117–129.

106 Swan, D.G., Stadler, L., Okan, E., Hoffs, M. *et al.* (2003) *Cell Microbiol*, **5**, 947–956.

107 Swan, D.G., Stern, R., McKellar, S., Phillips, K. *et al.* (2001) *J Cell Sci*, **114**, 2747–2754.

108 Oura, C.A., McKellar, S., Swan, D.G., Okan, E. and Shiels, B.R. (2006) *Cell Microbiol*, **8**, 276–288.

109 Shiels, B.R., McKellar, S., Katzer, F., Lyons, K. *et al.* (2004) *Eukaryot Cell*, **3**, 495–505.

110 Bishop, R., Shah, T., Pelle, R., Hoyle, D. *et al.* (2005) *Nucleic Acids Res*, **33**, 5503–5511.

111 Baylis, H.A., Allsopp, B.A., Hall, R. and Carrington, M. (1993) *Mol Biochem Parasitol*, **61**, 171–178.

112 Schnittger, L., Katzer, F., Biermann, R., Shayan, P. *et al.* (2002) *Mol Biochem Parasitol*, **120**, 247–256.

113 Toye, P.G., Goddeeris, B.M., Iams, K., Musoke, A.J. and Morrison, W.I. (1991) *Parasite Immunol*, **13**, 49–62.

114 Casanova, C.L., Xue, G., Taracha, E.L. and Dobbelaere, D.A. (2006) *Mol Biochem Parasitol*, **149**, 144–154.

115 Ebel, T.J.F.S.M., Frisch, A. and Lipp, J. (1997) *J Biol Chem*, **272**, 3042–3048.

116 Schneider, I., Haller, D., Kullmann, B., Beyer, D. *et al.* (2007) *Parasitol Res*, **101**, 1471–1482.

117 Morrison, W.I. (2007) *Vet Parasitol*, **148**, 21–30.

118 Graham, S.P., Pelle, R., Honda, Y., Mwangi, D.M. *et al.* (2006) *Proc Natl Acad Sci USA*, **103**, 3286–3291.

119 Serwold, T., Gaw, S. and Shastri, N. (2001) *Nat Immunol*, **2**, 644–651.

120 Brown, C.G., Radley, D.E., Burridge, M.J. and Cunningham, M.P. (1977) *Tropenmed Parasitol*, **28**, 513–520.

121 Uilenberg, G., Schreuder, B.E., Silayo, R. S. and Mpangala, C. (1976) *Tropenmed Parasitol*, **27**, 329–336.

35
Toxoplasma gondii

L. David Sibley

35.1
Introduction of *Toxoplasma gondii* and Toxoplasmosis

35.1.1
Lifecycle, Transmission and Genetic Tools

Toxoplasma gondii is a member of the phylum *Apicomplexa*, a group of ~5000 obligate intracellular protozoan parasites that are most closely related to dinoflagellates and ciliates [1]. Apicomplexans are parasitic on a wide range of hosts and several cause disease in humans, including *Plasmodium* spp., the causative agents of malaria, and *Cryptosporidium parvum*, an important opportunistic pathogen. *T. gondii* is one of the world's most common protozoan parasites, capable of infecting virtually all types of warm-blooded vertebrates [2]. Serological studies indicate that 25% of the human population of the world may be chronically infected, although prevalence rates in the United States have dropped to ~10% in recent years [3]. *T. gondii* also commonly infects domestic, companion and wild animals, and this remarkable success can be attributed to its efficient spread through the foodchain.

Infections can occur in one of two ways. First, when members of the cat family are infected, the parasite undergoes sexual development and is shed in the feces as a spore-like stage, called the oocyst [2]. Highly resistant, oocysts can contaminate the environment, leading to water-borne infections [2]. Ingestion of the oocyst stage by a wide variety of animals, including humans, leads to acute infections that are characterized by asexual replication of fast-growing forms called tachyzoites [2]. Following the initial acute infection, the parasite develops into a semi-dormant form called the bradyzoite, which resides within cysts in the muscle, internal organs and central nervous system [2]. Ingestion of these cyst stages is the second means by which infections are commonly acquired. Due the high prevalence in animals, *T. gondii* is an important cause of food-borne infection in humans [4].

Intracellular Niches of Microbes. Edited by Ulrich E. Schaible and Albert Haas

ISBN: 978-3-527-32207-7

T. gondii is unique in having excellent forward genetic systems [5], robust *in vitro* culture systems that support both biochemical and cell biological studies [6], and excellent small animal models [7]. *T. gondii* is also equipped with a variety of techniques for reverse genetics including both transient and stable selection that allow for gene tagging, gene disruption and expression of heterologous genes [8]. Forward genetic crosses can be conducted in cats and several interlineage crosses have been used to develop genetic linkage mapping in *T. gondii* [9]. Approximately ~10X genome sequence coverage for several prototypic strains has been generated, annotated and assembled into chromosomes to create a genome map (www.ToxoDB.org) [9].

35.1.2
Strain Diversity and Human Toxoplasmosis

Toxoplasma gondii has a unique population structure consisting of three predominant clonal lineages that are abundant in North America and Europe [10]. The three clonal lineages are remarkably similar, diverging only 1–2% on average for any given gene [11]. Despite their close genetic similarity, there are significant phenotypic differences between the strain types. Type I strains are acutely virulent in mice [12] and several studies indicate they are more likely to cause severe disease in humans [13–15]. Type II strains are commonly found in food animals and also associated with the majority of human infections that have been genotyped, while type III strains are largely confined to animals [10, 16]. While the clonal lineages account for ~95% of all isolates in North America and Europe, a smaller percentage of isolates show evidence of recombination [10]. Recent studies demonstrate that strains of *T. gondii* in South America are genetically distinct from those in the North and that the population structure has been shaped by a series of recent genetic bottlenecks [17, 18]. This unusual population structure indicates that sexual recombination is rare in nature, likely as the result of two aspects of the lifecycle. First, coinfection of animals appears to be uncommon, such that cats are rarely simultaneously infected with two strains. Second, the ability of *T. gondii* to be transmitted via carnivorous–omnivorous feeding bypasses the cat, thus avoiding genetic recombination.

Infection with *T. gondii* in healthy individuals typically results in a mild flu-like syndrome known as toxoplasmosis, which resolves without need for intervention [19]. Most human infections are uncomplicated; however, they persist in a chronic state and can predispose individuals to problems due to subsequent reactivation. Toxoplasmosis is most severe in individuals with compromised immune systems including those undergoing organ transplants, or cancer chemotherapy, individuals infected with HIV, and during pregnancy when infection threatens development of the fetus [19]. Additionally, *T. gondii* has recently been recognized as an important cause of ocular disease, either as a result of congenital infection, or due to new infections in otherwise healthy adults [20, 21]. More severe forms of ocular disease have been associated with unusual genotypes, in particular from regions of South America [22].

35.2
Characterization of the Intracellular Compartment Occupied by *T. gondii*

35.2.1
Intracellular Cycle

Toxoplasma is an obligate intracellular parasite that is able to infect a variety of nucleated cells, including macrophages, and dendritic cells. During acute infection, replication of tachyzoites results in amplification of parasite numbers followed by dissemination throughout the host. This intracellular cycle lasts from 24 to 72 h depending on the host cell and strain of the parasite. Cell invasion is powered by actin-based motility [23], which is also responsible for migration through tissues [24] (Figure 35.1a). Invasion of host cells occurs by active penetration into both phagocytic (macrophages) and nonphagocytic cells (Figure 35.1b). Within the host cell, the

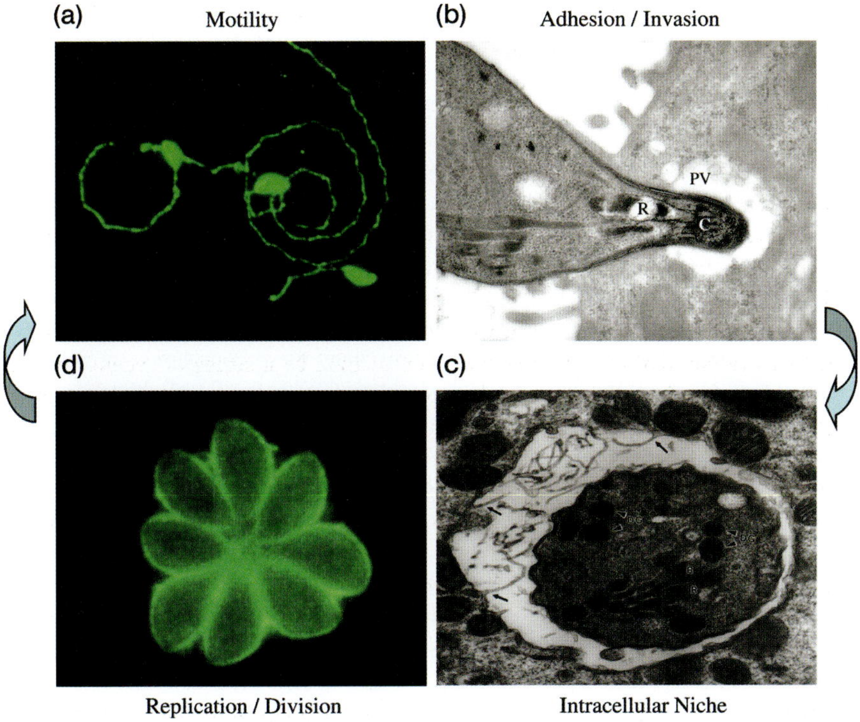

Figure 35.1 Intracellular lifecycle of *Toxoplasma gondii*. (a) Gliding motility on solid substrates is characterized by deposition of trails of the surface protein SAG1. (b) Cell invasion is powered by actin-based motility: the apical end penetrates to the host cell, led by the extended conoid (C). Discharge of rhoptries (R) leads to formation of the parasitophorous vacuole (PV). (c) Intracellular vacuole occupied by *T. gondii* is characterized by recruitment of mitochondria and elaboration of a system of tubular membranes within the lumen (arrows). D, dense granule, R, rhoptry (with permission of [117]). (d) Intracellular replication leading to a rosette of parasites. Stained with antibodies to the surface protein SAG1.

parasite resides in a specially modified compartment called the parasitophorous vacuole (PV) (Figure 35.1c). Replication occurs by a process of binary fission, leading to a rosette of daughter cells [25] (Figure 35.1d). Ultimately the vacuole and host cell lyse, liberating parasites that are motile and infectious for new cells. Each of these steps in the intracellular cycle is considered in more depth below.

35.2.2 Actin-Based Motility

Apicomplexan parasites have an elaborate cytoskeleton characterized by a unique apical microtubular organizing center called the conoid, which connects single microtubules that emanate from the anterior pole and run down the length of the cell [26]. A meshwork of filaments comprised of articullin-like proteins, lies above the microtubules and subtends a system of flattened cisternae that comprise the inner membrane complex (IMC) [26]. The motor complex lies within the space that separates the outer and inner membranes. It consists of a small myosin [27] that is anchored into the IMC through interaction with several accessory proteins [28], actin filaments that form beneath the membrane in motile parasites [29], and aldolase that serves a critical bridging function between the cytoskeleton (via F-actin binding) and the cytoplasmic domains of adhesions [30].

Actin dynamics are highly unusual in apicomplexans, with a majority of actin existing in an unpolymerized state [31]. Rapid assembly of new filaments is essential for motility [29], although the mechanism by which this is controlled remains uncertain. Motility is dependent on the secretion of adhesions from microneme organelles at the apical end of the parasite [32]. Translocation of surface adhesins is coupled to the actin cytoskeleton, and collectively the rearward movement of these complexes drives forward motility [23]. Ultimately, these complexes are released by intramembrane proteolysis through the action of a family of rhomboid proteases [33, 34]. While this system seems energetically inefficient, it provides several advantages for the parasite: (i) adhesins are contained within the parasite until needed for attachment, thus protecting them from immune recognition and neutralization, (ii) motility and cell invasion are polarized due to the gradient of adhesins that develops along the parasite cell surface, and (iii) motility is coupled to adhesion with the substrate, thus conserving energy.

Apicomplexan parasites lack cilia and flagella (with the exception of microgametocytes) and rely on unusual process described as "gliding" to propel the parasite across the substrate. Sporozoites of a variety of apicomplexans undergo gliding, as do more primitive apicomplexans such as gregarines, which are parasites of invertebrates [23]. Gliding in *T. gondii* is characterized by two basic patterns: circular gliding that occurs in a counter clockwise direction, and helical gliding that occurs clockwise and which leads to forward motion and cell invasion [35]. Gliding motility is blocked by agents that prevent actin filament assembly such as cytochalasin D and latrunculins, and is also severely altered by jasplakinolide [29, 36], a compound that stabilizes actin filaments. These studies indicate that gliding motility relies on proper regulation of actin filament turnover. Gliding motility also enables migration across tissue barriers and contributes to dissemination *in vivo* [24, 37]. Migration across the intestinal epithelium, the blood–brain barrier and placental barrier are important in pathogenesis, and active motility of *T, gondii* may contribute to this spread.

35.2.3 Active Entry

T. gondii entry into macrophages is distinct from phagocytosis in a number of features. Invasion is not accompanied by obvious rearrangements of the host actin cytoskeleton, membrane ruffling, or phosphorylation of tyrosine residues on host proteins [38], all of which are dramatically activated during Fc receptor-mediated phagocytosis [39]. Invasion takes approximately 20–30 s, ~5–10 times faster than phagocytic uptake [38]. Parasite entry occurs following apical attachment to cells and the parasite glides into the cell, moving through a visible constriction termed the "moving junction." Recent studies have revealed a complex of proteins released from the neck of secretory organelles called the rhoptries (termed RON proteins), form a ring-like structure that defines the junction [40, 41]. During formation of the junction, RON2 and RON4 are found together in a complex with AMA1, a micronemal protein needed for intimate cell association and invasion [42]. These studies demonstrate that the junction is partly mediated by secretion of parasite proteins; however, they do not resolve how the junction is anchored and what host components participate in its formation.

Direct evidence for the active penetration of host cells by *T. gondii* comes from pharmacological and genetic studies examining the role of host and parasite actins [43]. Agents that disrupt formation of actin filaments, such as cytochalasins and latrunculins, block phagocytosis [44], uptake of many bacterial pathogens [45] and also invasion by parasites such as *T. gondii* [23]. However, the presence of actin-based motility in both the host and eukaryotic parasite complicates interpretation of these latter findings. Using a combination of host cell mutants that were resistant to cytochalasin D and chemically induced mutants in the parasite, it was conclusively shown that parasite actin filaments are essential for cell entry [43]. Cytochalasin resistance was conferred by a point mutation in the single isoform of actin expressed by *T. gondii*, and presence of this mutation allowed gliding and invasion into wild type host cells in the presence of cytochalasin D [43]. These studies do not rule out participation of host actin in this process, however it would appear to play an accessory role, given the fact that parasite actin-based motility is both necessary and sufficient for invasion. The active penetration of cells by apicomplexans is distinct from mechanisms used by other intracellular pathogens, and it has direct consequences for the composition and hence fate of the vacuole that is formed.

35.2.4 Egress

Toxoplasma tachyzoites replicate every 6–9 h and after repeated rounds of binary fission, the vacuole swells to occupy most of the volume of the host cell. Ultimately, this triggers egress, which occurs when the host cell membrane integrity is breached, leading to activation of parasite motility [46]. Induced egress was initially studied following addition of the calcium ionophore A23187, which is a potent but unnatural trigger. Subsequent studies demonstrated that the parasite actually senses the drop in potassium that occurs once the host cell membrane is breached and that this activates calcium-dependent motility in the parasite [47]. Active egress also involves actin-based motility, discharge of micronemes and formation of a "moving junction" during exit from the vacuole [41]. Recent studies have revealed the surprising finding that *T. gondii* contains a plant-like pathway for production of the hormone abscisic acid (ABA), which controls calcium-mediated egress through a cyclic ADP-ribose-mediated pathway [48]. This pathway was likely acquired from an algal endosymbiont that gave rise to the apicoplast [49], and has since been retained to control calcium signaling and perhaps other functions. Suppression of ABA synthesis prevents egress and leads to a developmental switch to slow growing brayzoites, which remain within tissue cysts [48]. Recently, human granulocytes have also been shown to contain ABA [50], albeit at much lower levels than the parasite, suggesting the presence of this metabolite in host cells may also influence intracellular development.

35.3 Parasite Factors Involved in Maturation and Maintenance of the Parasite-Containing Vacuole

35.3.1 Formation of the Parasitophorous Vacuole

During invasion, the parasite propels its way into the host cell using actin-based motility to invaginate the plasma membrane and form the parasitophorous vacuole (PV). Consistent with this, lipid dyes placed in the plasma membrane readily gain entry to the PV [51]. Electrophysiology measurements demonstrate that the vast majority of membrane for the vacuole is derived from the host cell plasma membrane [52]. Remarkably, the composition of the PV membrane is dramatically altered during entry. Host cell single-pass transmembrane proteins are largely excluded, especially those that have cytoplasmic domains that interact with the host cell cytoskeleton [53]. Sorting of proteins from the vacuole likely occurs at the "moving junction," which operates as a molecular sieve. Sorting operates in part based on fluidity within the membrane. Consequently, many proteins that are resident in cholesterol-rich microdomains (i.e., rafts) readily gain entry to the PV, while proteins not found in these domains are generally excluded [51]. However, this association is not absolute, indicating that other factors also influence sorting. As a consequence, the composition of the nascent PV is highly unusual, including many raft components (i.e., GM1, GPI-anchored proteins, cholesterol and CCR5) but excluding many proteins found in the plasma membrane (i.e., β_1 integrin, Na^+/K^+-ATPase and CD44) as well as those that would typically be enriched in phagosomes (i.e., FcR, Src and Lck) [51].

35.3.2
Secretion of Parasite Proteins: Virulence Determinants

Active invasion by *T. gondii* is accompanied by three waves of protein secretion from the parasite: (i) microneme proteins, which mediate attachment; (ii) rhoptry proteins, which contribute to the formation of the moving junction and PV; and (iii) dense granule proteins, which modify the PV [54]. During parasite entry, the integrity of the plasma membrane is temporarily breached, followed by rapid resealing. Transient disruption of the membrane during entry has been detected by electron microscopy [55], electrophysiology recordings of the membrane conductance [52], and by the direct injection of rhoptry proteins into the host cell cytosol during invasion [56]. The mechanism of rupture and resealing is not understood, but may nonetheless be important for delivery of parasite effectors into the host cell cytosol. Recent forward genetics studies identified ROP proteins as key mediators of alteration in host gene transcription and in parasite virulence [57, 58]. Genetic crosses between type I and III strains of *T. gondii* mapped the major determinant of virulence differences between these lineages to a single major locus on chromosome VIIa [59].

Fine mapping revealed that the rhoptry protein ROP18 controls most of the phenotypic variation in virulence between these lineages, which differs in LD_{50} by >4 logs in outbred mice. Like other ROP family members, ROP18 contains a signal peptide and a serine/threonine (S/T) kinase domain (Figure 35.2a). Unlike the majority of ROPs that appear to be pseudokinases, ROP18 is catalytically active and this activity is required for virulence [59, 60]. The virulence-enhancing potential of ROP18 is also dependent on secretion into the host cell where it is targeted to the PV membrane [59].

Previous studies have shown that following secretion, the pseudokinase ROP4 is localized to the PV where it becomes phosphorylated [61]. Hence, potential substrates of ROP18 include both host and parasite proteins that are enriched on the PV (Figure 35.2b). In a separate study comparing progeny from a genetic cross between the less virulent type II and III strains, ROP18 was also detected among a group of loci that contribute to differences in pathogenicity in inbred mice [62]. ROP proteins are highly divergent, suggesting they are under strong selective pressure, and additional family membranes are also likely to be involved in modulating host activities [57, 58].

35.3.3
Avoidance of Fusion

The nonfusigenic nature of the *T. gondii* PV was first demonstrated more than 35 years ago using macrophages labeled with colloidal tracers and examined by electron microscopy. These studies demonstrated that only live parasites are capable of causing the block and that intracellular survival depends on avoidance of fusion [63]. Entry into macrophages also occurs without triggering a respiratory burst [64] and survival is dependent on avoidance of delivery of the proton pump, resulting in a neutral pH in the PV [65]. The intracellular fate of *T. gondii* in macrophages is strongly dependent on the route of entry. Active invasion is accompanied by discharge of parasite secretory proteins within the vacuole lumen and this leads to avoidance of endosome–lysosome

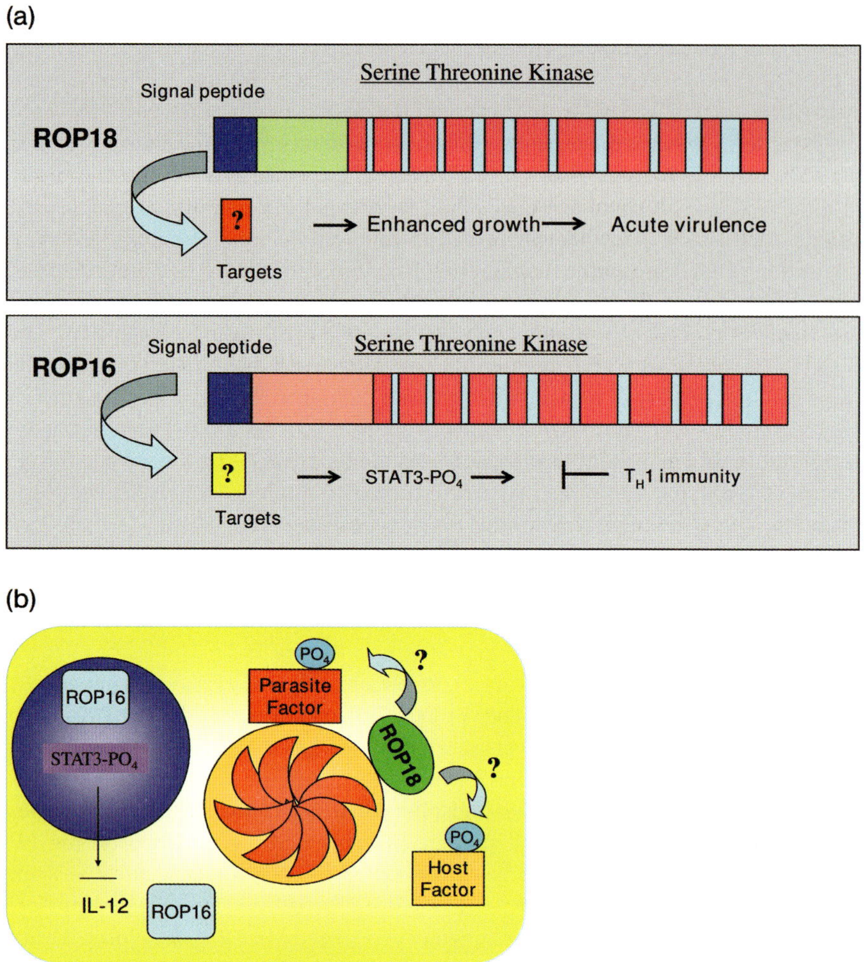

Figure 35.2 Modulation of host cell functions by parasite ROP proteins that are injected into the host cell. (a) ROP18 and ROP16 both contain N-terminal signal peptides followed by a divergent N-terminal domain and a conserved serine/threonine kinase domain. ROP18 enhances parasite growth [60] and leads to enhanced virulence in the mouse model [59, 62]. ROP16 leads to activation of STAT3-PO_4 and inhibition of T_H1 immunity [115]. (b) ROP16 is secreted during invasion and traffics to the nucleus, although it may also be active in the cytosol [115]. Indirectly, ROP16 leads to prolonged phosphorylation of STAT3, resulting in decreased production of IL-12 in type I and III strains. ROP18 is secreted at the time of invasion and traffics to the parasite-containing vacuole [59, 60]. The virulence-enhancing potential of ROP18 relies on its kinase activity and it may act on either host- or parasite-derived substrates.

fusion [66]. Markers of early and late endosomes, as well as lysosomes, are uniformly absent from the PV. However, heat-killed parasites, or live parasites that are opsonized with specific antibody, are engulfed by phagocytosis [38]. Engulfed parasites fail to discharge secretory organelles and are instead delivered to phagosomes that also rapidly acquire markers of the endocytic pathway [66]. The kinetics of acquisition of endocytic markers is similar to the maturation described for Fc receptor-mediated phagocytosis [39], and opsonized parasites are ultimately delivered the lysosomes for degradation [66]. Opsonized parasites are also delivered to lysosomes in CHO cells transfected with Fc receptor [67], indicating that fusion and degradation relies on a conserved pathway and is not a property limited to phagocytic cells.

The ability of *T. gondii* to actively penetrate the host cell, remodel the PV membrane at the time of invasion, and further modify the compartment by secretion of parasite proteins is likely sufficient to explain the avoidance of fusion with endosomes and lysosomes. No specific mediators of fusion resistance have been identified in *T. gondii*, rather this process may depend on the wholesale remodeling of the PV membrane at the time of invasion. The highly unique composition of the PV is unlikely to resemble any other intracellular membrane. Hence, it is neither targeted for degradation, nor does it interact directly with recycling endosomes, or vesicles in the secretory pathway.

35.3.4
Nutrient Acquisition

Sequestered within the PV, the parasite divides by binary fission to create two daughter cells within each maternal cell by a process called endodyogeny [25]. This exclusionary existence leaves the parasite with one major dilemma: acquisition of nutrients. The metabolic requirements of the parasite are likely substantial given their relatively large size (2×5 m), rapid division time (6–9 h), and consumption of the host cell cytosol over a period of several days. Several adaptations may facilitate nutrient acquisition, although the molecular details of these pathways remain unknown. The PV contains pores that allow rapid diffusion of small molecules ($\sim$1300 Da) from the host cell cytosol into the lumen of the vacuole [68]. Following entry, the PV is rapidly surrounded by host endoplasmic reticulum (ER) and mitochondria [63], features that may facilitate acquisition of nutrients, especially lipids. A variety of host lipids are accessible to the parasite [69] and recent evidence demonstrates that *T. gondii* obtains lipoic acid, a mitochondrial product, from its host cell [70]. Toxoplasma lacks the capacity to synthesize sterols, yet its membranes are rich in this lipid and the parasite scavenges cholesterol from host low-density lipoprotein (LDL) via the lysosomal pathway [71].

Recent studies indicate that host lysosomes may become trapped within the lumen of the PV in a process that depends on host cell microtubules [72]. The contents of these trapped vesicles remain segregated from the vacuole by two membrane bilayers, so acquisition of cholesterol or other lipids would still require a separate mechanism for transport.

35.4
Immunology of the Intracellular Compartment Occupied by *T. gondii*

35.4.1
Control of Infection

Although *T. gondii* is able to survive and replicate in normal macrophages, it is rapidly killed by activated cells. Interferon γ (IFNγ) is the key mediator of resistance *in vivo* [73], and inflammatory macrophages provide the critical first line of defense for controlling acute infection [74]. Consistent with this role, macrophages from infected mice are activated and able to control parasite replication [75]. Treatment of naïve macrophages with IFNγ plus tumor necrosis factor α (TNFα) (or LPS as a surrogate) *in vitro* induces potent anti-toxoplasmal activity [76]. Numerous effector mechanisms have been identified, including induction of nitric oxide and production of reactive nitrogen intermediates (RNI) [77], elevated respiratory burst and production reactive oxygen intermediates (ROIs) [78] and degradation of tryptophan [79]. *T. gondii* can survive in variety of cell types, many of which also show the ability to restrict growth when activated, and consistent with this, IFNγ receptors are required in both hematopoietic and nonhematopoietic cells [80].

Immune responses to *T. gondii* rely on both CD8$^+$ and CD4$^+$ T cells that are stimulated following antigen processing and presentation through MHC class I and class II pathways, respectively [81]. This seems at odds with the sequestered intracellular vacuole occupied by *T. gondii*, which is devoid of MHC class I and MHC class II molecules [66]. However, antigen processing may occur in macrophages or dendritic cells that have ingested parasites via phagocytosis, a pathway that shunts parasites for degradation [66]. In contrast to the PV, parasites within fusigenic phagosomes readily acquire MHC class II molecules and digested antigens may also be presented to MHC class I via cross-presentation [82]. An alternative mechanism is suggested by recent studies indicating that leakage of antigens from the PV may provide direct access to presentation via MHC class I within infected cells [83]. Secretory components such as rhoptry proteins also gain direct access to the host cell cytosol and thus are candidates for MHC class I presentation.

Although a variety of cell types respond to infection with *T. gondii* by secreting interleukin 12 (IL-12) and other cytokines, including neutrophils [84, 85], dendritic cells [86, 87] and macrophages [88], recent evidence indicates that inflammatory monocytes are of paramount importance in controlling oral infection [89]. Murine macrophages can be divided into two main categories: resident cells that are responsible for surveillance (Gr1$^-$ CCR2$^-$ CX3CR1^{hi}) and inflammatory monocytes (Gr1$^+$ CCR2$^+$ CX3CR1$^-$) that home to sites of inflammation [90, 91]. Depletion of CCR2 in Ccr2$^{-/-}$ mice leads to an inability of inflammatory moncytes to exit the bone marrow [92], hence they are prevented from migrating to sites of infection. Likewise, ablation of the major chemokine CCL2 (MCP-1) blocks traffic to sites of inflammation, despite normal elevation of circulating CCR2$^+$ cells. CCR2$^+$ monocytes are rapidly recruited to the lamina propria of the small intestine following infection with *T. gondii*, where they line the villus beneath the basement membrane. Infection of either Ccr2$^{-/-}$ or Ccl2$^{-/-}$ mice with *T. gondii* leads to extreme pathology due to an inability to control parasite replication in the gut [89]. CCR2$^+$ monocytes express iNOS, generate TNFα, and secrete IL-12, all important mediators of parasite control. In the absence of these inflammatory monocytes, mice succumb to infection due to extreme pathology in the gut characterized by influx of neutrophils and excessive tissue damage [89]. Finally, adoptive transfer of inflammatory but not resident macrophages, protects Ccr2$^{-/-}$ mice against an otherwise lethal challenge [89]. Importantly, these studies establish inflammatory monocytes as key determinants of mucosal immunity to toxoplasmosis and they are also likely important control of other enteric infections.

35.4.2
Immunity-Related GTPases

Recent studies have revealed another cellular pathway for clearance of intracellular pathogens following activation by IFNγ and upregulation of a class of small GTPases of the p47 family or immunity-related GTPases (Irgs) [93]. Mouse Irg proteins bind to intracellular membranes including the ER and Golgi and following upregulation by IFNγ they are often targeted to pathogen-containing vacuoles [93]. Irg proteins have been implicated in resistance to a number of intracellular bacterial and protozoan pathogens including *T. gondii*. The GTPases Irgm1 (LRG-47) and Irgm3 (IGTP) are required for control of acute toxoplasmosis, while Irgd (IRG-47) plays a role in chronic infection [93]. Recent data suggest that several Irgs lacking a completely conserved GTP-binding site serve as chaperones to sequester active Irgs on intracellular membranes [94]. Irgs are required for macrophages to control *T. gondii* replication in response to IFNγ, as shown by *in vitro* studies [95]. IFNγ-treated astrocytes are also able to control intracellular *T. gondii*: IIGP1 (Irga6) and other Irgs accumulate at the PV, resulting in disruption of the membrane and release of the parasite into the cytosol [96]. While parasites are ultimately degraded, the mechanism of their disposal remains uncertain as they do not acquire lysosomal markers [96]. Nonetheless, targeted deletion of the IIGP1 gene compromises parasite clearance in astrocytes, implicating Irgs in control of intracellular infection [96]. Mouse macrophages require IGTP to control and eliminate *T. gondii* following IFNγ activation [97]. While there is a prominent role for directing the intracellular fate of pathogens in macrophages, lack of Irg genes also leads to lymphopenia in infected animals, which likely contributes to compromised immune function in knockout animals [93]. The diversity of Irgs is much more limited in humans; however, at least one analogous gene product, Irgm, has been implicated in resistance of human monocytes to mycobacteria [98].

Recent studies indicate that the IFNγ clearance pathway involving Irgs also relies on proteins normally involved in autophagy [99]. The autophagy pathway involves formation of curved double-membrane isolation membranes to capture organelles and cytosol, shunting them to the lysosome for degradation [100]. Autophagy requires the action of two Atg5-dependent conjugation systems, and relies on a complex assembly of protein complexes [101]. Using mice that were selectively depleted in Atg5 in myeloid cells using Cre-lox recombination (referred to as "floxed"), we recently demonstrated that Atg5 is required for the IFNγ-mediated clearance of *T. gondii* in activated macrophages [99]. Atg5 was not required for upregulation of IFNγ signaling, or for increased expression of Irgs. However, in the absence of Atg5, IIGP was not recruited to a parasite-containing vacuole, leading to a defect in parasite clearance in IFNγ-activated macrophages [99]. Clearance of *T. gondii*-containing vacuoles does not resemble classical autophagy as the parasite is degraded within the cytosol rather than being engulfed in an isolation membrane. When Atg5-floxed mice were challenged with *T. gondii*, they failed to control parasite growth and rapidly succumbed to an otherwise nonlethal infection [99]. Similar results were obtained when Atg5-deficient monocytes were challenged with BCG or *Listeria* [99]. These studies demonstrate an intricate interplay between proteins required for autophagy during cellular turnover and IFNγ-mediated pathways for resistance to intracellular pathogens.

35.4.2.1 CD40 and Lysosomal Clearance

A distinct pathway for elimination of intracellular *T. gondii* has been described in human and mouse macrophages expressing CD40. Interaction of $CD4^+$ T cells expressing the CD40 ligand CD154 induces killing of intracellular *T. gondii* by macrophages [102]. Parasite killing and degradation are associated with delivery of host markers such as Rab7, cathepsin D and LAMP1, to the parasite-containing vacuole, reflecting fusion with endosomes and lysosomes [103]. Several features are remarkable about this activity. First, it is independent of IFNγ, STAT1 and the p47 GTPases, and it is enhanced by autocrine production of TNFα. Second, it can induce fusion with already established PVs, which are normally nonfusigenic throughout their intracellular existence. Finally, it provides direct evidence for the role of autophagy in clearance. Stimulation of CD40 results in recruitment of LC3 (Atg8) to the PV and disruption of Beclin 1 by RNAi inhibited the clearance of *T. gondii* [103].

Unlike the situation in macrophages activated with IFNγ, the PV membrane does not appear to be disrupted following CD40 ligation [103]. Rather CD40 ligation appears to mediate killing of intracellular *T. gondii* by inducing intact vacuoles to fuse with lysosomes. Presumably the hydrolytic nature of secondary endosomes is responsible for the demise of the parasite, as supported by studies demonstrating that protease inhibitors prevent clearance of the parasite [103]. Whether this pathway also relies on Atg5 has not been addressed.

35.4.3
Alterations in Host Cell Signaling

While infection with *T. gondii* induces a potent T_H1 response that controls the infection, the parasite also has numerous adaptations to thwart immune responses and thereby assure survival, including modulation of cytokine/chemokine secretion, prevention of apoptosis, alteration is cellular maturation, and induction of signaling networks, as reviewed recently [104, 105]. There are striking differences in the pathogenic nature of *T. gondii* strain types *in vivo*, and not surprisingly, some of these are also reflected in differences in alterations of host cell signaling *in vitro*. We will focus here on two important alterations in host cell signaling where information is available on the contribution of strain type and where potential mechanisms are beginning to emerge.

35.4.4
Disruption of IFNγ Signaling

Infection of mouse macrophages with the NTE strain of *T. gondii* (a type II strain) results in suppression of the induction of MHC class II following treatment with

IFNγ [106]. Similarly, infection with either type I or type II strains of *T. gondii* has been reported to delay the maturation of dendritic cells, thus preventing upregulation of MHC class II [107]. These effects would be expected to prevent optimal expansion and activation of $CD4^+$ T cells and result in less effective adaptive immunity. A global microarray analysis of IFNγ-inducible genes revealed that all three strain types suppress this response in human fibroblasts [108]. Infected cells failed to upregulate interferon regulatory factor 1 (IRF1), a transcription factor controlled by STAT1. This and previous studies have shown that STAT1 is normally phosphorylated and trafficked to the nucleus [106, 108], and yet it fails to activate transcription. For some genes, this may be the result of dephosphorylaton of STAT1 in the nucleus, an effect that was more pronounced for type I strains [108]. In the case of other genes, the promoter binding sites may already be activated by additional factors, as infection upregulates a number of genes in the absence of IFNγ. Collectively, repression of IFNγ-inducible gene expression appears to be a common mechanism that underlies the ability of *T. gondii* to survive in a wide range of cell types.

35.4.5 Disruption of Cytokine Production

Infection with *T. gondii in vivo* induces production of proinflammatory cytokines including IL-12 and TNFα, leading to production of IFNγ, which is a critical mediator for resistance [81]. Levels of T_H1 cytokines in the serum are exceedingly high following infection with strains that induce lethal infection even at low inoculum (type I strains). Despite these high levels of cytokines, parasite replication is incompletely controlled and animals succumb due to adverse effects of the cytokine storm [109, 110]. In contrast, less virulent type II strains induce modest levels of T_H1 cytokines, leading to control parasite replication, and survival of the acute phase of infection [109, 110]. Paradoxically, the inverse phenotypes are seen *in vitro* when macrophages are challenged with viable parasites. Infection with type I strains results in very low levels of IL-12 production, while type II strains are potent inducers [88]. Induction of IL-12 by type II strains requires active infection, occurs late in the infection cycle and is not repressed by coinfection with type I stains [88]. Studies on the host cell signaling pathways involved in this response indicate that both strains types initially repress NFκB translocation to the nucleus, which is associated with delayed IL-12 production [111]. Recovery from this refractory period is much greater following infection with type I strains, perhaps due to differences in MAPK and MyD88 signaling pathways that feed into IL-12 production [112].

Importantly, repression of LPS-induced TNFα production persists beyond this time period, indicating that separate mechanisms control these responses [113]. Collectively, these studies indicate that type I strains initially repress the induction of T_H1 cytokines (i.e., IL-12), allowing parasite expansion, whereas type II strains induce early IL-12 that results in more effective control.

Infection with *T. gondii* also results in phosphorylation of STAT3, leading to repression of IL-12 and TNFα in LPS-treated macrophages [114]. The ability of type I strains to suppress LPS-induced TNFα and IL-12 is dependent on infection with live parasites but does not require host IL-10 [114]. Recent genetic studies probing the parasite factors that alter host cell transcription identified a prominent locus on ChrVIIb in *T. gondii* that affects STAT3/STAT6-mediated transcription [115]. Genome-wide analysis of the host cell transcriptome following parasite infection was compared for 19 progeny between type II and type III strains [115]. Analysis of the differences in gene expression led to identification of pathways related to IL-4 and IL-6, which are controlled by the STAT3 and STAT6 transcription factors. Fine mapping was used to identify the parasite protein responsible as ROP16, a rhoptry protein bearing a signal peptide and a S/T kinase domain (Figure 35.2a). ROP16 is released into the host cell and accumulates in the nucleus (Figure 35.2a). Reverse genetic studies demonstrated that ROP16 is capable of inducing prolonged phosphorylation of STA3/STAT6, thus shutting off IL-12 production in type I and III strains [115]. STAT3 is also phosphorylated following infection with type II strains, but this effect is transient, allowing robust production of IL-12 [115].

Although ROP16 does not act directly on STAT3 or STAT6, these studies reveal the exploitation of host signaling networks by parasite secretory proteins that act in a stain-specific manner to alter immune responses and pathogenesis.

35.5 Summary and Outlook

Infection by *T. gondii* results in delivery of a large number of secretory proteins into the host cell, some of which are targeted to the nucleus, others to the PV, and hence the host–parasite interface [57, 58]. The majority of these proteins come from rhoptries, organelles that discharge their contents during invasion, and proteomic studies suggest a very diverse repertoire of components [116]. While the role for several of these is starting to emerge, the function of the majority of ROP proteins remains uncertain. As well, the molecular basis for the large number of changes in host cell signaling networks and gene expression remain unresolved. Defining the parasite proteins that alter these specific host pathways may reveal new mechanisms of pathogenesis and identify new targets for intervention.

Several studies illustrate that cellular immune responses involve autophagy, although the precise mechanisms of killing and clearance of intracellular parasites remain to be defined. It has been suggested that Irgs may function to induce membrane vesiculation [96], analogous to the fission reaction stimulated by dynamin, a GTPases involved in endocytosis. Such an activity might be key in disrupting the vacuole and possibly parasite membranes. If so, there must be specific mechanisms to target Irgs to the vacuole in IFNγ-treated cells, and to regulate their activity such that it does not also compromise host membranes. Recent evidence suggests that degenerate and inactive members of this family function to sequester and thereby regulate active members [94]. However, the mechanisms targeting individual Irgs to different membrane in the cell remains unresolved. Following membrane damage, further degradation would likely involve ubiquitination and proteosomal processing, or activation of macro-autophagy, as suggested by recent reports showing a role for

Atg5 [99]. Unraveling these pathways may identify novel mechanisms for pathogen destruction and may also clarify the role of IRGs and autophagy in delivery of antigens for processing and presentation.

Acknowledgments

I am grateful to many colleagues who have provided useful comments and critical input, as well as current and former members of my laboratory who have contributed to the studies described here. Work in the author's laboratory has been supported by the Burroughs Wellcome Fund and the National Institutes of Health.

References

1 Baldauf, S.L. (2003) The deep roots of eukaryotes. *Science*, **300**, 1703–1706.

2 Dubey, J.P. (2007) The history and life cycle of *Toxoplasma gondii*, in *Toxoplasma gondii the Model Apicomplexan: Perspectives and Methods* (eds L.M. Weiss and K. Kim), Academic Press, Elsevier, New York, pp. 1–17.

3 Jones, J.L., Kruszon-Moran, D., Sanders-Lewis, K. and Wilson, M. (2007) *Toxoplasma gondii* infection in the United States, 1999–2004, Decline from the prior decade. *Am J Trop Med Hyg*, **77**, 405–410.

4 Mead, P.S., Slutsker, L., Dietz, V., McCaig, L.F., Bresee, J.S., Shapiro, C., Griffin, P.M. and Tauxe, R.V. (1999) Food-related illness and death in the United States. *Emerg Infect Dis*, **5**, 607–625.

5 Ajioka, J.W. and Sibley, L.D. (2007) Development and application of classical genetics in *Toxoplasma gondii*, in *Toxoplasma gondii the Model Apicomplexan: Perspectives and Methods* (eds L.M. Weiss and K. Kim), Academic Press, Elsevier, New York, pp. 367–389.

6 Pfefferkorn, E.R. (1981) *Toxoplasma gondii* and the biochemistry of intracellular parasitism. *Trends Biochem Sci*, **6**, 311–313.

7 Reichard, U. and Gross, U. (2007) Toxoplasma animal models and therapeutics, in *Toxoplasma gondii the Model Apicomplexan: Perspectives and Methods* (eds L.M. Weiss and K. Kim), Academic Press, Elsevier, New York, pp. 153–184.

8 Striepen, B. and Soldati, D. (2007) Genetic manipulation of *Toxoplasma gondii*, in *Toxoplasma gondii the Model Apicomplexan: Perspectives and Methods* (eds L.M. Weiss and K. Kim), Academic Press, Elsevier, New York, pp. 391–418.

9 Khan, A., Taylor, S., Su, C., Mackey, A.J., Boyle, J., Cole, R.H., Glover, D., Tang, K., Paulsen, I. and Berriman, M. (2005) Composite genome map and recombination parameters derived from three archetypal lineages of *Toxoplasma gondii*. *Nucleic Acids Res*, **33**, 2980–2992.

10 Howe, D.K. and Sibley, L.D. (1995) *Toxoplasma gondii* comprises three clonal lineages: correlation of parasite genotype with human disease. *J Infect Dis*, **172**, 1561–1566.

11 Grigg, M.E., Bonnefoy, S., Hehl, A.B., Suzuki, Y. and Boothroyd, J.C. (2001) Success and virulence in Toxoplasma as the result of sexual recombination between two distinct ancestries. *Science*, **294**, 161–165.

12 Sibley, L.D. and Boothroyd, J.C. (1992) Virulent strains of *Toxoplasma gondii* comprise a single clonal lineage. *Nature (Lond.)*, **359**, 82–85.

13 Fuentes, I., Rubio, J.M., Ramírez, C. and Alvar, J. (2001) Genotypic characterization of *Toxoplasma gondii* strains associated with human toxoplasmosis in Spain: direct analysis from clinical samples. *J Clin Microbiol*, **39**, 1566–1570.

14 Grigg, M.E., Ganatra, J., Boothroyd, J.C. and Margolis, T.P. (2001) Unusual abundance of atypical strains associated with human ocular toxoplasmosis. *J Infect Dis*, **184**, 633–639.

15 Khan, A., Su, C., German, M., Storch, G.A., Clifford, D. and Sibley, L.D. (2005) Genotyping of *Toxoplasma gondii* strains from immunocompromised patients reveals high prevalence of type I strains. *J Clin Microbiol*, **43**, 5881–5887.

16 Ajzenberg, D., Cogné, N., Paris, L., Bessieres, M.H., Thulliez, P., Fillisetti, D., Pelloux, H., Marty, P. and Dardé, M.L. (2002) Genotype of 86 *Toxoplasma gondii* isolates associated with human congenital toxoplasmosis and correlation with clinical findings. *J Infect Dis*, **186**, 684–689.

17 Khan, A., Fux, B., Su, C., Dubey, J.P., Darde, M.L., Ajioka, J.W., Rosenthal, B.M. and Sibley, L.D. (2007) Recent transcontinental sweep of *Toxoplasma gondii* driven by a single monomorphic chromosome. *Proc Natl Acad Sci USA*, **104**, 14872–14877.

18 Lehmann, T., Marcet, P.L., Graham, D.H. Dahl, E.R. and Dubey, J.P., (2006) Globalization and the population structure of *Toxoplasma gondii. Proc Natl Acad Sci USA*, **103**, 11423–11428.

19 Peterson, E. (2007) Epidemiology, diagnostics, and chemotherapy, in *Toxoplasma: Molecular and Cellular Biology* (eds J.W. Ajioka and D. Soldati), Horizon Bioscience, Norfolk, UK, pp. 37–58.

20 Holland, G.N. (2003) Ocular toxoplasmosis: a global reassessment Part I: epidemiology and course of disease. *Am J Ophthalmol*, **136**, 973–988.

21 Holland, G.N. (2004) Ocular toxoplasmosisi: a global reassessment Part II: disease manifestations and management. *Am J Ophthalmol*, **137**, 1–17.

22 Jones, L.A., Alexander, J. and Roberts, C.W. (2006) Ocular toxoplasmosis: in the storm of the eye. *Parasite Immunol*, **28**, 635–642.

23 Sibley, L.D. (2004) Invasion strategies of intracellular parasites. *Science*, **304**, 248–253.

24 Barragan, A. and Sibley, L.D. (2003) Migration of *Toxoplasma gondii* across biological barriers. *Trends Microbiol*, **11**, 426–430.

25 Striepen, B., Jordan, C.N., Reiff, S. and van, G.G. (2007) Dooren Building the perfect parasite: cell division in Apicomplexa. *PLoS Pathog*, **3**, 691–698.

26 Morrissette, N.S. and Sibley, L.D. (2002) Cytoskeleton of apicomplexan parasites. *Microbiol Mol Biol Rev*, **66**, 21–38.

27 Meissner, M., Schluter, D. and Soldati, D. (2002) Role of *Toxoplasma gondii* myosin A in powering parasite gliding and host cell invasion. *Science*, **298**, 837–840.

28 Gaskins, E., Gilk, S., DeVore, N., Mann, T., Ward, G.E. and Beckers, C. (2004) Identification of the membrane receptor of a class XIV myosin *Toxoplasma gondii. J Cell Biol*, **165**, 383–393.

29 Wetzel, D.M., Håkansson, S., Hu, K., Roos, D.S. and Sibley, L.D. (2003) Actin filament polymerization regulates gliding motility by apicomplexan parasites. *Mol Biol Cell*, **14**, 396–406.

30 Jewett, T.J. and Sibley, L.D. (2003) Aldolase forms a bridge between cell surface adhesins and the actin cytoskeleton in apicomplexan parasites. *Mol Cell*, **11**, 885–894.

31 Sahoo, N., Beatty, W.L., Heuser, J.E., Sept, D. and Sibley, L.D. (2006) Unusual kinetic and structural properties control rapid assembly and turnover of actin in the parasite *Toxoplasma gondii. Mol Biol Cell*, **17**, 895–906.

32 Keeley, A. and Soldati, D. (2004) The glideosome: a molecular machine powering motility and host-cell invasion by Apicomplexa. *Trends Cell Biol*, **14**, 528–532.

33 Dowse, T. and Soldati, D. (2005) Rhomboid-like proteins in Apicomplexa: phylogeny and nomenclature. *Trends Parasitol*, **35**, 747–756.

34 Urban, S. (2006) Rhomboid proteins: conserved membrane proteases with divergent biological functions. *Genes Dev*, **20**, 3054–3068.

35 Håkansson, S., Morisaki, H., Heuser, J.E. and Sibley, L.D. (1999) Time-lapse video microscopy of gliding motility in *Toxoplasma gondii* reveals a novel, biphasic mechanism of cell locomotion. *Mol Biol Cell*, **10**, 3539–3547.

36 Poupel, O. and Tardieux, I. (1999) *Toxoplasma gondii* motility and host cell invasiveness are drastically impaired by jasplakinolide, a cyclic peptide stabilizing F-actin. *Microbes Infect*, **1**, 653–662.

37 Barragan, A. and Sibley, L.D. (2002) Transepithelial migration of *Toxoplasma gondii* is linked to parasite motility and virulence. *J Exp Med*, **195**, 1625–1633.

38 Morisaki, J.H., Heuser, J.E. and Sibley, L.D. (1995) Invasion of *Toxoplasma gondii* occurs by active penetration of the host cell. *J Cell Sci*, **108**, 2457–2464.

39 Swanson, J.A. and Hoppe, A.D. (2004) The coordination of signaling during Fc receptor-mediated phagocytosis. *J Leukoc Biol*, **76**, 1093–1103.

40 Lebrun, M., Michelin, A., El Hajj, H., Poncet, J., Bradley, P.J., Vial, H.J. and Dubremetz, J.F. (2005) The rhoptry neck protein RON4 relocalizes at the moving junction during *Toxoplasma gondii* invasion. *Cell Microbiol*, **7**, 1823–1833.

41 Alexander, D.L., Mital, J., Ward, G.E., Bradley, P.J. and Boothroyd, J.C. (2005) Identification of the moving junction complex of *Toxoplasma gondii*: a collaboration between distinct secretory organelles. *PLoS Pathog*, **1**, 137–149.

42 Mital, J., Meissner, M., Soldati, D. and Ward, G.E. (2005) Conditional expression of *Toxoplasma gondii* apical membrane antigen-1 (TgAMA1) demonstrates that TgAMA1 plays a critical role in host cell invasion. *Mol Biol Cell*, **16**, 4341–4349.

43 Dobrowolski, J.M. and Sibley, L.D. (1996) Toxoplasma invasion of mammalian cells is powered by the actin cytoskeleton of the parasite. *Cell*, **84**, 933–939.

44 Greenberg, S. (1999) Modular components of phagocytosis. *J Leukoc Biol*, **66**, 712–717.

45 Cossart, P. and Sansonetti, P.J. (2004) Bacterial invasion: the paradigms of enteroinvasive pathogens. *Science*, **304**, 242–248.

46 Black, M.W., Arrizabalaga, G. and Boothroyd, J.C. (2000) Ionophore-resistant mutants of *Toxoplasma gondii* reveal host cell permeabilization as an early event in egress. *Mol Cell Biol*, **20**, 9399–9408.

47 Moudy, R., Manning, T.J. and Beckers, C.J. (2001) The loss of cytoplasmic potassium upon host cell breakdown triggers egress of *Toxoplasma gondii*. *J Biol Chem*, **276**, 41492–41501.

48 Nagamune, K., Hicks, L.M., Fux, B., Broissier, F., Chini, E.N. and Sibley, L.D. (2008) Abscisic acid controls calcium-dependent egress and development in *Toxoplasma gondii*. *Nature*, **451**, 207–211.

49 Waller, R.F. and McFadden, G.I. (2005) The apicoplast: a review of the derived plastid of apicomplexan parasites. *Curr Issues Mol Biol*, **7**, 57–79.

50 Bruzzone, S., Moreschi, I., Usai, C., Guida, L., Damonte, G., Salis, A., Scarfi, S., Millo, E., De Flora, A. and Zocchi, E. (2007) Abscisic acid is an endogenous cytokine in human granulocytes with cyclic ADP-ribose as a second messenger. *Proc Natl Acad Sci USA*, **104**, 5759–5764.

51 Charron, A.J. and Sibley, L.D. (2004) Molecular partitioning during host cell penetration by *Toxoplasma gondii*. *Traffic*, **5**, 855–867.

52 Suss-Toby, E., Zimmerberg, J. and Ward, G.E. (1996) *Toxoplasma* invasion: The parasitophorous vacuole is formed from host cell plasma membrane and pinches off via a fusion pore. *Proc Natl Acad Sci USA*, **93**, 8413–8418.

53 Mordue, D.G., Desai, N., Dustin, M. and Sibley, L.D. (1999) Invasion by *Toxoplasma gondii* establishes a moving junction that selectively excludes host cell plasma membrane proteins on the basis of their membrane anchoring. *J Exp Med*, **190**, 1783–1792.

54 Carruthers, V.B. (2002) Host cell invasion by the opportunistic pathogen *Toxoplasma gondii*. *Acta Trop*, **81**, 111–122.

55 Nichols, B.A. and O'Connor, R.G. (1981) Penetration of mouse peritoneal macrophages by the protozoan *Toxoplasma gondii*. *Lab Invest*, **44**, 324–335.

56 Håkansson, S., Charron, A.J. and Sibley, L.D. (2001) *Toxoplasma* evacuoles: a two-step process of secretion and fusion forms the parasitophorous vacuole. *EMBO J*, **20**, 3132–3144.

57 Boothroyd, J.C. and Dubremetz, J.F. (2008) Kiss and spit: the dual roles of *Toxoplasma* rhoptries. *Nat Rev Microbiol*, **6**, 79–88.

58 Bradley, P.J. and Sibley, L.D. (2007) Rhoptries: an arsenal of secreted virulence factors. *Curr Opin Microbiol*, **10**, 582–587.

59 Taylor, S., Barragan, A., Su, C., Fux, B., Fentress, S.J., Tang, K., Beatty, W.L., Haijj, E.L., Jerome, M. and Behnke, M.S. (2006) A secreted serine-threonine kinase determines virulence in the eukaryotic pathogen *Toxoplasma gondii*. *Science*, **314**, 1776–1780.

60 El Hajj, H., Lebrun, M., Arold, S.T., Vial, H., Labesse, G. and Dubremetz, J.F. (2007) ROP18 is a rhoptry kinase controlling the intracellular proliferation of *Toxoplasma gondii*. *PLoS Pathog*, **3**, e14.

61 Carey, K.L., Jongco, A.M., Kim, K. and Ward, G.E. (2004) The *Toxoplasma gondii* rhoptry protein ROP4 is secreted into the parasitophorous vacuole and becomes phosphorylated in infected cells. *Eukaryot Cell*, **3**, 1320–1330.

62 Saeij, J.P.J., Boyle, J.P., Coller, S., Taylor, S., Sibley, L.D., Brooke-Powell, E.T., Ajioka, J.W. and Boothroyd, J.C. (2006) Polymorphic secreted kinases are key virulence factors in toxoplasmosis. *Science*, **314**, 1780–1783.

63 Jones, T.C. and Hirsch, J.G. (1972) The interaction of *Toxoplasma gondii* and mammalian cells II The absence of lysosomal fusion with phagocytic vacuoles containing living parasites. *J Exp Med*, **136**, 1173–1194.

64 Wilson, C.B., Tsai, V. and Remington, J.S. (1980) Failure to trigger the oxidative burst of normal macrophages. *J Exp Med*, **151**, 328–346.

65 Sibley, L.D., Weidner, E. and Krahenbuhl, J.L. (1985) Phagosome acidification blocked by intracellular *Toxoplasma gondii*. *Nature*, **315**, 416–419.

66 Mordue, D.G. and Sibley, L.D. (1997) Intracellular fate of vacuoles containing *Toxoplasma gondii* is determined at the time of formation and depends on the mechanism of entry. *J Immunol*, **159**, 4452–4459.

67 Joiner, K.A., Furhman, S.A., Miettinen, H.M., Kasper, L.H. and Mellman, I. (1990) *Toxoplasma* gondii: Fusion competence of parasitophorous vacuoles in Fc-receptor transfected fibroblasts. *Science*, **249**, 641–646.

68 Schwab, J.C., Beckers, C.J.M. and Joiner, K.A. (1994) The parasitophorous vacuole membrane surrounding intracellular *Toxoplasma gondii* functions as a molecular sieve. *Proc Natl Acad Sci USA*, **91**, 509–513.

69 Charron, A.J. and Sibley, L.D. (2002) Host cells: mobilizable lipid resources for the intracellular parasite *Toxoplasma gondii*. *J Cell Sci*, **115**, 3049–3059.

70 Crawford, M.J., Thomsen-Ziegler, N., Ray, M., Schachtner, J., Roos, D.S. and Seeber, F. (2006) *Toxoplasma gondii* scavenges host-derved lipic acid despite its de novo synthesis in the apicoplast. *EMBO J*, **25**, 3214–3222.

71 Coppens, I. (2006) Contribution of host lipids to Toxoplasma pathogenesis. *Cell Microbiol*, **8**, 1–9.

72 Coppens, I., Dunn, J.D., Romano, J.D., Pypaert, M., Zhang, H., Boothroyd, J.C. and Joiner, K.A. (2006) *Toxoplasma gondii* sequesters lysosomes from mammalian hosts in the vacuolar space. *Cell*, **125**, 261–274.

73 Suzuki, Y., Orellana, M.A., Schreiber, R.D. and Remington, J.S. (1988) Interferon-γ: the major mediator of resistance against *Toxoplasma gondii*. *Science*, **240**, 516–518.

74 Robben, P.R., Laregina, M., Kuziel, W.A. and Sibley, L.D. (2005) Recruitment of Gr-1$^+$ monocytes is essential for control of acute toxoplasmosis. *J Exp Med*, **201**, 1761–1769.

75 Ruskin, I. and Remington, J.S. (1968) Role for the macrophage in acquired immunity to phylogenetically unreleated intracellular organisms. *Antimicrob Agents Chemother*, **8**, 474.

76 Sibley, L.D., Adams, L.B., Fukutomi, Y. and Krahenbuhl, J.L. (1991) Tumor necrosis factor-α tiggers antitoxoplasmal activity of IFN-γ primed macrophages. *J Immunol*, **147**, 2340–2345.

77 Adams, L.B., Hibbs, J.B., Taintor, R.R. and Krahenbuhl, J.L. (1990) Microbiostatic effect of murine-activated macrophages for *Toxoplasma gondii*: Role for synthesis of inorganic nitrogen oxides from L-arginine. *J Immunol*, **144**, 2725–2729.

78 Murray, H.G., Spitalny, G.L. and Nathan, C.F. (1985) Activation of mouse peritoneal macrophages *in vitro* and *in vivo* with interferon-γ. *J Immunol*, **134**, 1619–1622.

79 Pfefferkorn, E.R. (1984) Interferon-γ blocks the growth of *Toxoplasma gondii* in human fibroblasts by inducing the host to degrade tryptophan. *Proc Natl Acad Sci USA*, **81**, 908–912.

80 Yap, G.S. and Sher, A. (1999) Effector cells of both nonhemopoietic and hemopoietic origin are required for interferon (IFN)-γ- and tumor necrosis factor (TNF)-α-dependent host resistance to the intracellular pathogen, *Toxoplasma gondii*. *J Exp Med*, **189**, 1083–1091.

81 Hunter, C.A. and Reichmann, G. (2001) Immunology of *Toxoplasma* infection, in *Toxoplasmosis: A Comprehensive Clinical Guide* (eds D.H. Joynson and T.J. Wreghitt), Cambridge University Press, Cambridge, UK, pp. 43–57.

82 Basta, S. and Alatery, A. (2007) The cross-priming pathway: a protrait of an intricate immune system. *Scand J Immunol*, **65**, 311–319.

83 Gubbels, M.J., Streipen, B., Shastri, N., Turkoz, M. and Robey, E.A. (2005) Class I major histocompatibility complex presentation of antigens that escape from the parasitophorous vacuole of *Toxoplasma gondii*. *Infect Immun*, **73**, 703–711.

84 Bliss, S.K., Marshall, A.J., Zhang, Y. and Denkers, E.Y. (1999) Human polymorphonuclear leukocytes produce IL-12, TNF-α, and the chemokines macrophage-inflammatory protein-1α and -1β in response to *Toxoplasma gondii* antigens. *J Immunol*, **162**, 7369–7375.

85 Bliss, S.K., Zhang, Y. and Denkers, E.Y. (1999) Murine neutrophil stimulation by *Toxoplasma gondii* antigen drives high level production of IFN-γ-independent IL-12. *J Immunol*, **163**, 2081–2088.

86 Sousa, C.R., Yap, G., Schulz, O., Rogers, N., Schito, M., Aliberti, J., Hieny, S. and Sher, A. (1999) Paralysis of dendritic cell IL-12 production by microbial products prevents infection-induced immuno-pathology. *Immunity*, **11**, 637–647.

87 Scharton-Kersten, T., Casper, P., Sher, A. and Denkers, E.Y. (1996) *Toxoplasma gondii*: evidence for interleukin-12-dependent and -independent pathways of interferon-γ production induced by an attenuated parasite strain. *Exp Parasitol*, **84**, 102–114.

88 Robben, P.M., Mordue, D.G., Truscott, S.M., Takeda, K., Akira, S. and Sibley, L.D. (2004) Induction of IL-12 by *Toxoplasma gondii* depends on the parasite genotype. *J Immunol*, **172**, 3686–3694.

89 Dunay, I.R., DaMatta, R.A., Fux, B., Presti, R., Greco, A., Colonna, M. and Sibley, L.D. (2008) Gr1^{+} inflammatory monocytes are required for mucosal resistance to the pathogen *Toxoplasma gondii*. *Immunity*, **29**, 306–317.

90 Geissmann, F., Auffrey, C., Palframan, R., Wirrig, C., Ciocca, A., Campisis, L., Narni-Manichelli, E. and Lauvau, G. (2008) Blood monocytes: distinct subsets, how they relate to dendritic cells, and their possible roles in the regulation of T-cell responses. *Immunol Cell Biol*, **86**, 398–408.

91 Geissmann, F., Jung, S. and Littman, D.R. (2003) Blood monocytes consist of two principal subsets with distinct migratory properties. *Immunity*, **19**, 71–82.

92 Serbina, N.V. and Pamer, E.G. (2006) Monocyte emigration from bone marrow during bacterial infection requires signals mediated by chemokine receptor CCR2. *Nat Immunol*, **7**, 311–317.

93 Taylor, G.A., Feng, C.G. and Sher, A. (2007) Control of IFN-γ mediated host resistance to intracellular pathogens by immunity-related GTPases (p47 GTPases). *Microbes Infect*, **9**, 1644–1651.

94 Hunn, J.P., Koenen-Waisman, S., Papic, N., Schroeder, N., Pawlowski, N., Lange, R., Jaiser, F., Zerrahn, J., Martens, S. and Howard, J.C. (2008) Regulatory interacitons between IRG resistance GTPases in the cellular response to *Toxoplasma gondii*. *EMBO J*, **27**, 2495–2509.

95 Butcher, B.A., Greene, R.I., Henry, S.C., Annecharico, K.L., Weinberg, J.B., Denkers, E.Y., Sher, A. and Taylor, G.A. (2005) p47 GTPases regulate *Toxoplasma gondii* survival in activated macrophages. *Infect Immun*, **73**, 3278–3286.

96 Martens, S., Parvanova, I., Zerrahn, J., Griffiths, G., Schell, G., Reichmann, G. and Howard, J.C. (2005) Disruption of *Toxoplasma gondii* parasitophorous vacuoles by the mouse p47-resistance GTPases. *PLoS Pathog*, **1**, e24.

97 Ling, Y.M., Shaw, M.H., Ayala, C., Coppens, I., Taylor, G.A., Ferguson, D.J.P. and Yap, G.S. (2006) Vacuolar and plasma mebrane stripping and autophagic elimination of *Toxoplasma gondii* in primed effector macropahges. *J Exp Med*, **203**, 2063–2071.

98 Singh, S.B., Davis, A.S., Taylor, G.A. and Deretic, V. (2006) Human IRGM induces autophagy to eliminate intracellular mycobacteria. *Science*, **313**, 1438–1441.

99 Zhao, Z., Fux, B., Goodwin, M., Dunay, I.R., Strong, D., Miller, B.C., Cadwell, K., Delgado, M., Pnopuak, M. and Green, K.G. Atg5 is essential for cellular immunity *in vivo* and recruitment of a p47 GTPase to the *Toxoplasma gondii* parasitophorous vacule in activated macrophages. *Cell Host Microbe*, in print.

100 Levine, B. and Kroemer, G. (2008) Autophagy in the pathogenesis of disease. *Cell*, **132**, 27–42.

101 Mizushima, N., Ohsumi, Y. and Yoshimori, T. (2002) Autophagosome formation in mammalian cells. *Cell Struct Funct*, **27**, 421–429.

102 Andrade, R.M., Wessendarp, M. and Subaste, C.S. (2003) CD154 activates macrophage anti-microbial activity in the absence of IFN-γ through a TNF-α dependent mechanism. *J Immunol*, **171**, 6750–6756.

103 Andrade, R.M., Wessendarp, M., Gubbels, J.M., Striepen, B. and Subaste, C.S. (2006) CD40 induces macrophage anti-*Toxoplasma gondii* activity by triggering autophagy-dependent fusion of pathogen-containing vacuoles and lysosomes. *J Clin Invest*, **116**, 2366–2377.

104 Laliberte, J. and Carruthers, V.B. (2008) Host cell manipulation by the human pathogen *Toxoplasma gondii*. *Cell Mol Life Sci*, **65**, 1900–1915.

105 Lang, C., Griss, U. and Luder, C.G.K. (2007) Subversion of innate and adaptive immune responses by *Toxoplasma gondii*. *Parasitol Res*, **100**, 191–203.

106 Lang, C., Algner, M., Beinert, N., Gross, U. and Luder, C.G.K. (2006) Diverse mechanisms employed by *Toxoplasma gondii* to inhibit IFN-γ induced major histocompatibility complex class II gene expression. *Microbes Infect*, **8**, 1994–2005.

107 McKee, A.S., Dzierszinski, F., Boes, M., Roos, D.S. and Pearce, E.J. (2004) Functional inactivation of immature dendritic cells by the intracellular parasite *Toxoplasma gondii*. *J Immunol*, **173**, 2632–2640.

108 Kim, S.K., Fouts, A.E. and Boothroyd, J.C. (2007) *Toxoplasma gondii* dysregulates IFN-γ inducible gene expression in human fiboblasts: insights from a genome-wide transcriptional profiling. *J Immunol*, **178**, 5154–5165.

109 Mordue, D.G., Monroy, F., La Regina, M., Dinarello, C.A. and Sibley, L.D. (2001) Acute toxoplasmosis leads to lethal overproduction of Th1 cytokines. *J Immunol*, **167**, 4574–4584.

110 Gavrilescu, L.C. and Denkers, E.Y. (2001) IFN-γ overproduction and high level apoptosis are associated with high but not low virulence *Toxoplasma gondii* infection. *J Immunol*, **167**, 902–909.

111 Butcher, B.A., Kim, L., Johnson, P.F. and Denkers, E.Y. (2001) *Toxoplasma gondii* tachyzoites inhibit proinflammatory cytokine induction in infected macrophages by preventing nuclear translocation of the transcription factor NF-κB. *J Immunol*, **167**, 2193–2201.

112 Kim, L., Butcher, B.A., Lee, C.W., Uematsu, S., Akira, S. and Denkers, E.Y. (2006) *Toxoplasma gondii* genotype determines MyD88-dependent signaling in infected macrophages. *J Immunol*, **177**, 2584–2591.

113 Butcher, B.A. and Denkers, E.Y. (2002) Mechanism of entry determines the ability of *Toxoplasma gondii* to inhibit macrophage proinflammatory cytokine production. *Infect Immun*, **70**, 5216–5224.

114 Butcher, B.A., Kim, L., Panopoulos, A.D., Watawich, S.S., Murray, P.J. and Denkers, E.Y. (2005) IL-10-independent STAT3 activation by *Toxoplasma gondii* mediates supression of IL-12 and TNF-alpha in host macrophages. *J Immunol*, **174**, 3148–3152.

115 Saeij, J.P.J., Coller, S., Boyle, J.P., Jerome, M.E., White, M.E. and Boothroyd, J.C. (2007) *Toxoplasma* co-opts host gene expression by injection of a polymorphic kinase homologue. *Nature*, **445**, 324–327.

116 Bradley, P.J., Ward, C., Cheng, S.J., Alexander, D.L., Coller, S., Coombs, G.H., Dunn, J.D., Ferguson, D.J., Sanderson, S.J. and Wastling, J.M. (2005) Proteomic analysis of rhoptry organelles reveals many novel constituents for host-parasite interactions in *T. gondii*. *J Biol Chem*, **280**, 34245–34258.

117 Sibley, L.D., Krahenbuhl, J.L., Adams, G.M.W. and Weidner, E. (1986) Toxoplasma modifies macrophage phagosomes by secretion of a vesicular network rich in surface proteins. *J Cell Biol*, **103**, 867–874.

36
Trypanosoma cruzi

Martin C. Taylor

36.1
Introduction: The Parasite and its Disease

Trypanosoma cruzi is a unicellular protozoan and the causative agent of Chagas disease in humans. It is a member of the order *Kinetoplastida* along with the related parasites *Trypanosoma brucei* and *Leishmania* spp. which cause African sleeping sickness and the leishmaniases, respectively. Within the mammalian host all three of these parasites occupy very different niches, with *T. brucei* living extracellularly in the bloodstream, *Leishmania* in the phagosomal system of macrophages and dendritic cells, and *T. cruzi* principally in the cytoplasm of host cells. *T. cruzi* is unique among the three parasites in being able to infect both phagocytic and nonphagocytic cells. The parasite shows a very high level of genetic diversity between strains and has been subdivided into two major lineages: Tc I and Tc II, with Tc II being further subdivided into five groups TcIIA–TcIIE. *T. cruzi* is transmitted by the insect vector, the triatomine bug or vinchuca. Unlike most insect-transmitted pathogens it is not injected during the bite but is deposited in feces which are then rubbed into the bite wound involuntarily.

Chagas disease is characterized by three distinct clinical stages, the first of which is the acute phase where there is rapid parasite amplification and trypanosomes can be readily seen in the bloodstream. This phase is characterized by general malaise and fever; additional symptoms may include lymphadenopathy, hepatosplenomegaly and myocarditis. In approximately 5–10% of cases death occurs at this stage, particularly among young children, due mainly to meningoencephalitis or cardiomyopathy. The acute phase is self-limiting and lasts between 4 and 8 weeks.

The disease then enters the indeterminate phase where it is difficult to detect parasites and there are few clinical indications. Of those in the indeterminate phase 30% will go on to develop chronic Chagas disease (usually 10–20 years later) with symptoms including cardiomyopathy and/or digestive megasyndromes (mega-esophagus and megacolon). Death at this stage is often due to congestive heart failure although sudden death may also occur. Once chagasic cardiomyopathy becomes apparent the 5-year mortality rate is greater than 50%, though some success

Intracellular Niches of Microbes. Edited by Ulrich E. Schaible and Albert Haas

ISBN: 978-3-527-32207-7

in prolonging life has been achieved through heart transplantation coupled with close monitoring and antitrypanosomal chemotherapy [1, 2]. In HIV/AIDS patients with prior *T. cruzi* infection there may be a reactivation of quiescent disease with parasites being found in large pseudocysts in the central nervous system. Death in these individuals is usually caused by meningoencephalitis.

Chagas disease is both a significant public health problem throughout Latin America and a major economic burden to those countries affected. It is primarily a disease of the rural poor, being transmitted by triatomine bugs (*Triatoma, Rhodnius* and *Panstrongylus* species) which inhabit cracks in the walls of adobe houses and in thatched roofs. Increasingly Chagas disease is occurring in urban settings due to migration and transmission via blood and organ donation. Increased migration from Latin America to the United States and Western Europe has seen the number of cases of Chagas disease in developed countries increase in recent years. The vector is also found in the United States, suggesting the possibility of increased autochthonous Chagas disease in the future. Recent microepidemics in Brazil have been traced to the ingestion of sugarcane juice contaminated with triatomine feces, indicating an oral route of transmission. This seems to result in particularly virulent infection.

At the moment there are only two drugs available for the treatment of Chagas disease, nifurtimox and benznidazole. Both of these have toxicity issues and in many cases may not clear the parasites. Infection with *T. cruzi* is generally a lifelong problem. The role of chemotherapy in indeterminate and chronic Chagas disease is the subject of much current research. Complete parasitological cure has only been demonstrated very recently in an animal model of Chagas disease using benznidazole treatment [3]. However loss of a single gene has been shown to be involved in resistance to both compounds and so there is a strong possibility of significant resistance arising in the wild [4].

The parasite has a complex lifecycle composed of replicating and nonreplicating stages. The nonreplicating forms (metacyclic trypomastigotes and bloodstream trypomastigotes) are preadapted for switching between host and vector while the replicating forms (epimastigote and amastigote) establish infection in the vector and host respectively. The metacyclic trypomastigotes are shed in the feces as the bug feeds on blood. Scratching the wound leads to the ingress of trypomastigotes into the bloodstream where they invade macrophages (see below for details). Once inside the macrophage they briefly inhabit the endosome/lysosome system (12–18 h post infection), where they secrete proteins which allow escape into the cytoplasm. In the cytoplasm of the host cell the parasites transform into nonflagellated amastigotes which replicate by binary fission undergoing nine divisions before differentiating into flagellated trypomastigotes which escape from the host cell and can then infect new host cells or be taken up in the blood meal by the insect vector. In the insect vector they will transform into epimastigotes and replicate, colonizing the gut of the bug. As they migrate to the hindgut they will differentiate into metacyclic trypomastigotes ready to infect a new host.

Since *T. cruzi* consists of two genetically distinct groups, and can invade a variety of mammalian cells, many different proteins and pathways have been implicated in the initiation and maintenance of infection. The current state of knowledge is summarized below.

36.2 Intracellular Compartmentation of *T. cruzi*

36.2.1 Invasion, Lysosome Recruitment and Trypomastigote Retention

Trypanosoma cruzi infects a wide range of cells in the mammalian host, primarily heart, smooth muscle, adipose tissue, CNS and macrophages. This ability to invade numerous different cell types, of necessity, implies that the parasite can utilize multiple host cell receptors to gain entry. Several candidates have been described in the literature including the bradykinin receptors, transforming growth factor β (TGFβ) receptor and most recently the nerve growth factor receptor TrkA [5]. In this latter case it has been shown, by pharmacological and genetic inhibition, that the parasite requires the tyrosine kinase activity of the receptor to mediate enhanced infection via activation of phosphatidylinositol 3-kinases (PI3Ks) (see below). The parasite binds to the TrkA receptor through the GPI-anchored trypomastigote surface protein *trans*-sialidase (TS), this interaction with TrkA promotes neuronal survival and therefore parasite survival within the infected neuron. Parasite binding to neuronal cells is inhibited by soluble TrkA, indicating a direct role for the interaction between anchored TS and the receptor.

Bradykinin receptors have been implicated in both invasion by *T. cruzi* and in the control of the development of immunity via the maturation of dendritic cells [6, 7]. Bradykinin is a peptide which acts as a danger signal. It has been shown that the trypanosome proteinase cruzipain, which is the major cysteine proteinase expressed in parasites, can release lysyl-bradykinin from kininogens, and that cells overexpressing the bradykinin receptor B_2R are more susceptible to invasion. Engagement of B_2R results in Ca^{2+} transients, as do the other receptors implicated in trypanosome invasion. Ca^{2+} signaling in the host cell appears to be a prerequisite for productive infection.

A second parasite proteolytic enzyme has also been shown to be involved in the generation of Ca^{2+} transients in the target cell. This protein, oligopeptidase B, acts on an unidentified parasite product to generate a calcium agonist which enhances invasion. Deletion of the oligopeptidase B genes from the parasite results in a drastic decrease in both *in vitro* invasion of cultured cells and *in vivo* infection of mice [8].

The oligopeptidase B pathway appears to be active in invasion of fibroblasts and myoblasts. The bradykinin pathway seems to be utilized by some strains of *T. cruzi* but not by all, suggesting that at least some of the differences in receptor/signaling usage observed in different studies may reflect both the genetic diversity of the parasite and its ability to invade multiple cell types [6].

The trypanosome actively invades host cells, initially residing in a parasitophorous vacuole but later escaping into the cytoplasm where it undergoes replication. During the infection process the parasite subverts endogenous host signaling pathways to recruit endosomal and lysosomal vesicles to the parasitophorous vacuole. There are two pathways by which the parasite can invade nonphagocytic cells. These are differentiated by the source of the host cell-derived membranes surrounding the incoming trypomastigote. The first pathway to be discovered involves the kinesin-dependent recruitment of lysosomes along microtubules to the cytoplasmic face of the plasma membrane, such that the membrane surrounding the invading trypomastigote is derived from lysosomes [9, 10] (Figure 36.1). Here the parasite is exploiting a host cell mechanism whereby lysosomal membranes are used to repair damage to the plasma membrane in a Ca^{2+}-regulated exocytic process dependent on synaptotagmin VII.

More recently, a lysosome-independent pathway of invasion has been found in which binding of parasites to the cell surface results in a rapid accumulation of phosphoinositide 3-phosphates due to activation of PI3Ks [11]. This pathway involves invagination of the host cell membrane to form the parasitophorous vacuole. The initial sequence involves immediate activation of PI3K with ensuing enrichment of $PtdInsP_3/PtdIns(3,4)P_2$ in the vacuolar membrane, followed by recruitment of the protein kinase Akt to the site of trypomastigote binding. Accumulation of Akt can be observed 2 min after the parasite makes contact. Five minutes after contact the trypanosome is surrounded by plasma membrane derived from the host cell. Rab5 also accumulated around the internalizing trypomastigote prior to any association with early endosomal antigen 1 (EEA1). In the case of metacyclic trypomastigotes invading phagocytic cells it has been shown recently that Rab5 activation is dependent on Toll-like receptor 2 (TLR2) and PI3K activation in murine macrophages. In this system Rab5 inhibition by expression of the dominant negative S34N mutant was shown to reduce both phagosome formation and parasite internalization, indicating a major role for Rab5 in the invasion process [12].

Within 15 min of binding, 50% of internalized trypomastigotes were found in vacuoles positive for plasmamembrane markers (myristoyl-GFP and PLCδ pleckstrin homology domain-GFP) with a further 20–30% positive for EEA1, indicating fusion with early endosomes. The other 20–30% of parasites were present in LAMP1-positive vacuoles and had utilized the lysosome-dependent invasion pathway. However within 60 min almost all parasites are found in LAMP1-positive vesicles and therefore lysosomal fusion with the parasite-containing vacuole occurs no matter which invasion pathway was used. Indeed lysosomal fusion is an absolute requirement for productive infection, in contrast to many other intracellular microbes which seek to avoid lysosomal fusion [13].

The plasma membrane-mediated invasion pathway is enhanced by disruption of the host cell actin cytoskeleton which allows the invagination of the plasmamembrane. The parasite modifies the host cell actin cytoskeleton by specifically downregulating the activity of the GTPase RhoA; no effect is seen on the activity of the other cytoskeletal GTPases, Rac and Cdc42 [14]. The lysosomal entry pathway is exquisitely sensitive to inhibition of PI3K. The observation that the majority of parasites invade via the plasma membrane pathway and not by early lysosome recruitment explains why interference with synaptotagmin VII function has unexpectedly little effect on parasite infectivity, while blocking Ca^{2+}-mediated lysosome exocytosis by 90% [15, 16].

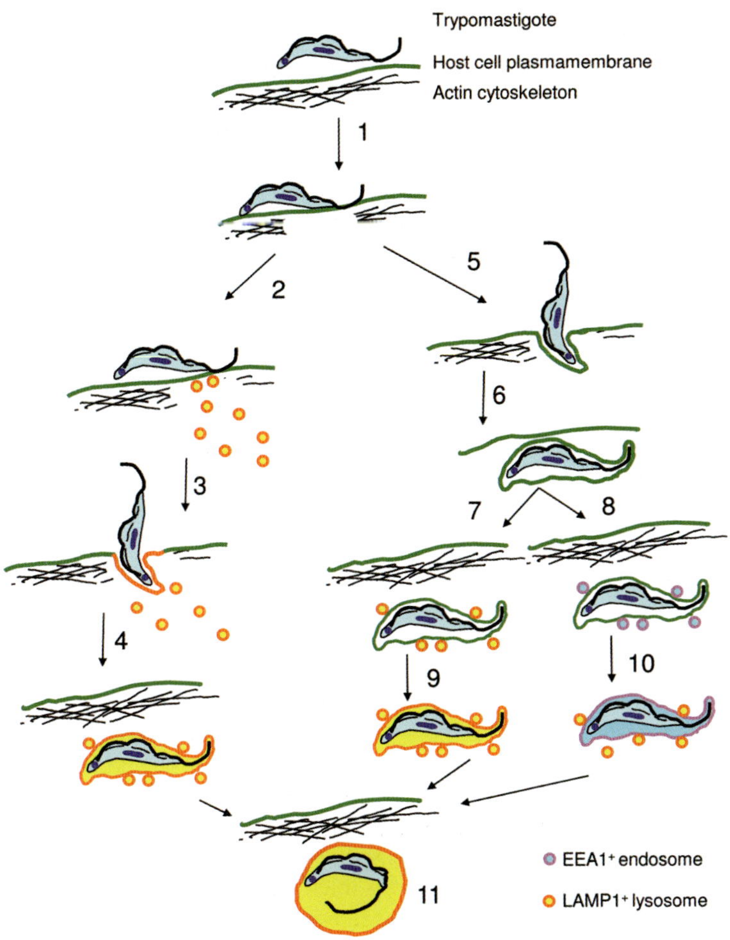

Figure 36.1 Invasion pathways used by *Trypanosoma cruzi*.The trypomastigote binds to the host cell through receptor–ligand interactions triggering phosphatidylinositol 3-kinase activation, Akt recruitment and downregulation of RhoA leading to localized breakdown of the actin cytoskeleton (1). At this stage the invasion process may take different routes. In the lysosome-dependent or "classical" invasion pathway LAMP1$^+$ lysosomes are recruited along microtubules to the site of trypanosome binding in a Ca^{2+}-regulated manner (2). The lysosomal membranes are then utilized to form the invagination surrounding the invading parasite (3). At this stage the parasite is wholly engulfed in lysosome-derived membranes, and the actin cytoskeleton reforms, leaving the parasitophorous vacuole internalized (4). In the lysosome-independent pathway the parasite invades the cell by invagination of the plasmamembrane at the point where the actin skeleton is removed (5). The parasite is then internalized in a vacuole consisting of plasma membrane-derived material (6). At this stage the vacuole may recruit lysosomes (7) or EEA1$^+$ early endosomes (8). Both sets of vesicles fuse with the vacuole (9 and 10). At this point all the pathways converge with the recruitment of further lysosomes and lysosomal fusion to the vacuole (11). Now the parasite is internalized in a LAMP1$^+$ acidified vacuole with lysosomal characteristics regardless of which invasion pathway or receptor was utilized.

Trypomastigotes are highly motile organisms and so they need a mechanism to promote their retention in the newly invaded cell. This process of ensuring a productive infection requires the reassembly of the host cell actin cytoskeleton as interference with actin polymerization by cytochalasin D promotes escape of trypomastigotes after invasion and leads to a lower number of retained parasites in the infected cell [14]. Reassembly of the actin cytoskeleton is also required for lysosomal fusion with the parasitophorous vacuole. Lysosomal fusion is a prerequisite for parasite retention since there is a direct correlation between the rate of lysosomal fusion in infected vacuoles and the ability of the parasite to be retained within the host cell. If lysosomal fusion does not occur the parasite is more likely to escape. The reassembly of the actin cytoskeleton then appears to serve two roles in infection, first the retention of the invading trypomastigote and second promotion of the recruitment of lysosomes to the parasitophorous vacuole to allow the escape of the parasite into the host cell cytoplasm.

The maturation of the *T. cruzi* vacuole differs from the maturation of standard phagosomes in that while phagosomes require the class III PI3K VPS34 for maturation [17] the *T. cruzi* vacuole will acquire lysosomal markers in the presence of the PI3K inhibitor wortmannin and therefore the later phases of vacuole maturation are not dependent on PI3K activity although the initial activation of class I PI3K during invasion appears to be necessary (however some invasion is observed even when PI3K is inhibited). The acquisition of LAMP1 will occur even if PI3K activity is completely blocked if the inhibition occurs more than 10 min after cellular invasion and hence the maturation step rapidly becomes refractory to a requirement for PI3K activity. It has therefore been suggested that the maturation of the *T. cruzi* vacuole is a faster version of standard phagosome maturation [11]. At the moment the molecular details of the interactions between *T. cruzi*, the vacuole, the cytoskeleton and the lysosome remain to be elucidated and so the exact relationship between the *T. cruzi* vacuole and a “standard” phagosome is unclear.

Microarray analysis of infected fibroblasts has shown that *T. cruzi* infection triggers off remarkably few changes in host cell gene expression in the first 2–24 h post invasion [18]. Between 2 and 6 h after infection no genes were induced by more than twofold while six genes were repressed even though the parasite demonstrably triggers several host cell-signaling pathways. At the 24 h timepoint 106 genes were induced, including the type 1 interferon-stimulated genes. This was associated with secretion of interferon β (IFNβ) by the infected cells after 18 h, and this timepoint may coincide with the escape of parasites into the host cytoplasm [18]. No changes in proinflammatory cytokine expression were observed in these experiments although such changes have been demonstrated in other cell types infected by *T. cruzi* and therefore it is likely that different host cells may respond differently to the infection.

36.2.2
Escape from the Vacuole and Intracytoplasmic Replication

The trypomastigote stages of *T. cruzi* are nonreplicative and to re-engage the cell division cycle they must transform into the amastigote stage. However before this occurs, the parasites have to escape from the parasitophorous vacuole. Several mechanisms are involved in this process. First the surface-expressed *trans*-sialidase of the parasite strips lysosomal membrane proteins of their sialic acid (Figure 36.2) [19]. The sialidase activity of intact parasites has been shown to increase threefold upon acidification to pH 5.5. Loss of sialylation of lysosomal membrane glycoproteins incubated with parasite

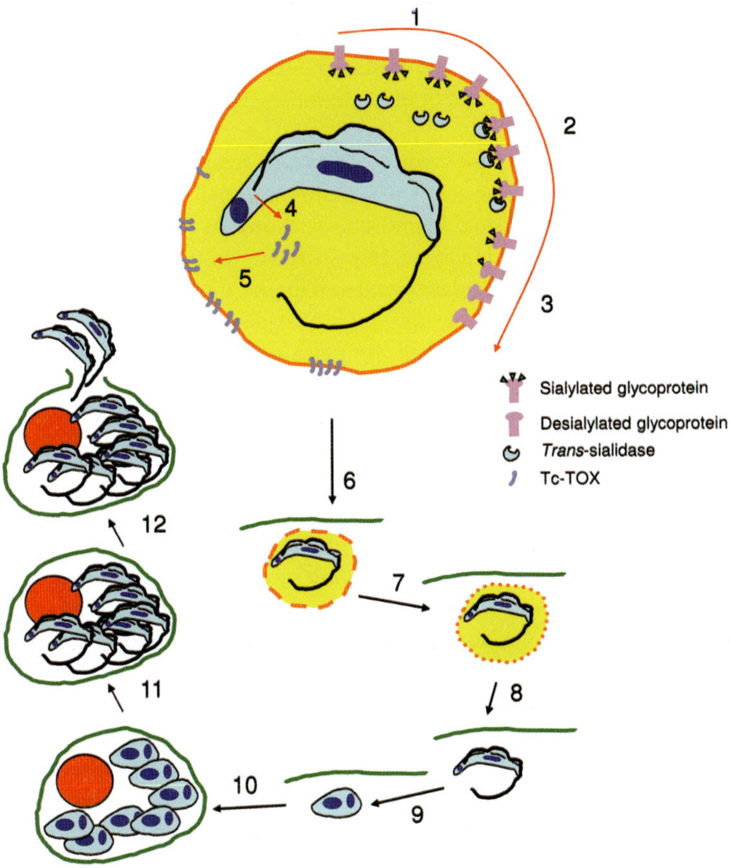

Figure 36.2 Escape from the parasitophorous vacuole by *Trypanosoma cruzi*.The trypomastigote secretes *trans*-sialidase into the lumen or surface *trans*-sialidase may be used (1). The *trans*-sialidase recognizes and binds to sialic acid residues on lysosomal membrane glycoproteins (2). The *trans*-sialidase then strips the sialic acid from the lysosome membrane glycoproteins (3). At the same time the parasite secrets the hemolysin Tc-TOX/Lyt1 from its flagellar pocket (4). The toxin is activated in the acidic pH of the lysosomal lumen and inserts into the vacuolar membrane (5). The concerted action of these two parasite proteins leads to weakening (6) and eventual breakdown of the membrane (7) leaving the trypomastigote free in the cytoplasm (8). The trypomastigote then completes differentiation into the amastigote (9) which undergoes approximately nine rounds of binary fission (10). The amastigotes then differentiate into bloodstream from trypomastigotes (11), the host cell ruptures and the parasites are released to invade more cells (12). After release, if they get into the peripheral blood they may also be taken up by the insect vector to continue the life cycle in another host.

extracts was demonstrated, indicating that these proteins are a substrate for the trypanosome *trans*-sialidase. In addition parasites infecting sialylation-deficient cells escape into the cytoplasm much faster than those in wild-type cells [19]. It has also been shown that metacyclic trypomastigotes express 20-fold lower levels of *trans*-sialidase activity than bloodstream trypomastigotes. The metacyclic forms take longer to exit the vacuole than do the bloodstream trypomastigotes, but metacyclic trypomastigotes engineered to express the bloodstream form of *trans*-sialidase on their surface could also exit the vacuole faster. Both forms of trypomastigote could escape faster in sialic acid-deficient Lec 2 cells than in sialic acid positive CHO wild-type cells [20]. Taken together these findings indicate an important role for *trans*-sialidase in the escape process.

Second the parasite secretes a pore-forming hemolytic toxin Tc-TOX/LYT1 into the vacuole, which is activated by the acidic pH of the lysosomal compartment [21]. This protein inserts into the lysosomal membrane as shown by its ability to induce conductance changes in planar lipid bilayers. It was also shown that pretreatment of erythrocytes with neuraminidase (sialidase) rendered them more susceptible to lysis by Tc-TOX. Thus together the action of these two proteins renders the lysosomal membrane susceptible to rupture, allowing the parasites to escape into the cytoplasm [19]. The escape process is dependent on the acidification of the vesicle as both the Tc-TOX and *trans*-sialidase activities are increased at pH5.5. As corroboration for this, trypanosomes in HeLa cells take longer to escape the vacuole than those in CHO cells. HeLa cell endosomes become acidified later due to the presence of papillomavirus 18 E5 protein, which causes alkalinization of the endosome [22]. Tc-TOX is probably encoded by the LYT1 gene as the two proteins are similar sizes, exhibit pH-dependent hemolytic activity and both cross-react with antibodies against the C9 component of human complement, although the sequence of LYT1 does not share homology with C9, suggesting a structural epitope [23].

LYT1 mRNA is developmentally regulated by stage-specific alternative *trans*-splicing to produce two different proteins [24]. The epimastigote protein kLYT1 has a shorter N-terminal and localizes to the zone between the base of the flagellum and the kinetoplast. This protein appears to be involved in the regulation of differentiation. In the infective mammalian stages of the parasite an upstream splicing site is utilized resulting in a longer N-terminal with a signal peptide for plasma membrane expression. This form of the protein exhibits the Tc-TOX-like activity. Thus the same gene encodes proteins doing two very different jobs, merely by changing the location of the spliced leader addition site in the mRNA. This finding has greater implications in suggesting that the trypanosomatid genomes may encode more diverse protein functions than is predictable from sequence alone.

Once the parasites have entered the cytoplasm they undergo differentiation into the amastigote with a drastic shortening of the flagellum and rounding up of the cell body. The differentiation process appears to be triggered in the vacuole as it can be induced *in vitro* by exposing bloodstream trypomastigotes to acidified medium, although this does not trigger differentiation of metacyclic trypomastigotes suggesting that they require another or a different signal [25]. The amastigotes undergo up to nine rounds of cell division before transforming back into flagellated trypomastigotes. At this stage the host cell undergoes lysis and the highly motile trypomastigotes escape into the surrounding fluid. The exact mechanism of cell lysis is unclear. The released parasites can then infect new cells or reach the peripheral blood and be taken up by a triatomine bug to continue the cycle. The amastigotes themselves also possess the inherent ability to infect new cells should the host cell be lysed before the differentiation process has occurred. Although amastigotes express very low levels of *trans*-sialidase they express high levels of Tc-TOX and it appears that this allows them to escape the vacuole quite effectively after invasion. So each of the three infective parasite stages appears to differ in its requirements, and mechanisms of cell invasion, vacuole formation and cytoplasmic escape.

36.3 Maintenance of the Infected Cell

36.3.1 Subversion of Host Cell Apoptosis by *T. cruzi*

Since *T. cruzi* primarily infects fully differentiated cells such as cardiomyocytes and neurons it is important for the trypanosomes to prevent the infected cell from undergoing apoptosis while they are replicating. It has been shown in two instances that soluble trypanosomal proteins can be involved in inhibition of apotosis. In neurons a noncatalytic domain of the *trans*-sialidase has been shown to bind the nerve growth factor receptor TrkA. Activation of this receptor results in PI3K activation, Akt autophosphorylation and consequent inhibition of apoptosis. Activation of Akt results in multiple prosurvival downstream effects, including the suppression of BAD-induced apoptosis, inhibition of caspase 9-induced cell death, activation of NFκB with consequent transcriptional activation of antiapoptotic/prosurvival genes [26]. Aoki *et al.* (2004) [27] have shown that exposure of neonatal cardiomyocytes to the parasite protease cruzipain results in increased expression of the antiapoptotic protein Bcl-2 and upregulation of arginase activity. However in cardiac myocytes during an established infection, blockage of apoptosis is achieved by the activation of NFκB signaling and was found to be dependent on the presence of intracellular parasites [28]. Conditioned medium from parasites or infected cells failed to prevent TNFα-mediated apoptosis in this model, indicating that soluble parasite-derived molecules were not responsible. Thus it appears that mechanisms for maintenance of infected cells differ according to the stage of parasite which has invaded the cell.

As well as blocking apoptosis it is important for *T. cruzi* to reduce inflammatory damage to its host cells during the early phase of infection. In cardiomyocytes this is achieved by engagement of TLR2 via ligands secreted by trypomastigotes. Activation of TLR2/6 heterodimers in cardiomyocytes results in an upregulation of NFκB leading to expression of proinflammatory cytokines [29]. In particular it has been shown that IL-1β secretion from these cells switches on the hypertrophic response in both the infected cell and surrounding non-infected cells. The hypertrophic response is thought to allow the maintenance of cardiac function when cardiac cell are exposed to stress and this may benefit the parasite by preventing the death of the host during the acute phase. In addition, the exposure of cardiomyocytes to low levels of proinflammatory cytokines appears to engage a survival response which prevents cell death during an ischemic event. Thus it has been hypothesized that this early response preadapts the cardiomyocytes to survival in the face of the inflammatory response to the parasite which otherwise may cause extensive cardiac damage [29]. It is of course in the parasite's interest to keep the host alive long enough for transmission to the vector to occur and so, by inducing protective stress responses, this is achieved.

In addition to blocking endogenous cell death pathways in infected cells the parasite is also capable of inhibiting the extrinsic pathway triggered by the immune response. Intracellular amastigotes but not trypomastigotes inhibit Fas/FasL signaling by prevention of the degradation of c-FLIP. This increased level of C-FLIP blocks activation of procaspase 8, thus preventing the extrinsic apoptotic cascade [30].

Of course the converse of this is that it would also favour parasite growth to stimulate apoptosis of $CD8^+$ Tcells directed against infected cells and trypanosomes also show the ability to effect lymphocyte death. During experimental infection apoptotic cells can be readily detected in the spleen, heart and lymph nodes. Both B and T lymphocytes undergo apoptosis during *T. cruzi* infection. B cells also undergo polyclonal expansion in an antigen-independent manner mediated by a parasite-secreted protein, proline racemase [31]. Tcell apoptosis during infection is associated with activation of caspase 8, indicating that the FAS pathway is responsible. Inhibition of caspase 9 does not block activation-induced cell death and so the intrinsic pathway does not appear to play a significant role. In contrast it has also been shown that the *trans*-sialidase from *T. cruzi* can protect $CD4^+$ T-cells from activation-induced cell death via CD43, but the same protein triggers apoptosis in thymocytes [32]. Enzyme inhibition studies using lactitol have shown that the *trans*-sialidase activity of the protein is required to induce apoptosis in thymocytes. The depletion of thymocytes during the acute phase appears to be a transient effect which may allow the parasite to gain a foothold. Chronic chagasic patient's serum contains neutralizing antibodies targeted at the *trans*-sialidase, indicating that a functional $CD4^+$ response is occurring [32].

The parasite is therefore capable of modulating host apoptotic pathways both to protect its host cell from apoptosis and to trigger apoptosis in cells that may have an adverse effect on its survival.

36.4
Immunology of *T. cruzi* Infection

There is much evidence accumulating that chronic intracellular parasites successfully modulate the host immune response to prevent their elimination and to establish productive infection. *T. cruzi* is no exception to this paradigm and multiple parasite-derived molecules have been shown to interfere with host responses. On initial infection the parasite is recognized by the innate immune response via interaction with the Toll-like receptors. TLRs 2, 4 and 9 have been implicated in the recognition of *T. cruzi*. Intracellular TLR9 binds CpG-rich parasite DNA. TLRs 2 and 4 are activated by parasite-derived lipids, namely the GPI anchor for TLR2 and glycoinositol phospholipid ceramides from epimastigotes for TLR4. TLR4 signaling may play only a very minor role, if any, *in vivo* as the cells from mice lacking the TLR4 pathway still respond to *T. cruzi* as well as TLR4-competent cells. In addition it has been shown that the response directed by TLR2 and TLR9 is equivalent to the entire MyD88-dependent activity, suggesting that TLR4 is of minor importance [33]. The importance of TLR signaling for parasite control in the early stages of infection is underlined by the finding that cells deficient in downstream components of the TLR pathway, namely MyD88 and TRIF, will allow greatly enhanced parasite replication. MyD88-mediated parasite control is thought to involve expression of proinflammatory cytokines, while TRIF exerts its effect by the induction of IFNβ and its downstream effectors. As outlined above, TLR2 is also involved in mediating the hypertrophic response in cardiomyocytes during *T. cruzi* infection [29].

36.4.1
Antigen Presentation by Infected Cells

Successful control of *T. cruzi* infection requires a CD8$^+$ T cell response as the majority of infected cells such as cardiomyocytes express only class I MHC molecules. An elegant series of experiments using trypanosomes transfected with the model T cell antigen ovalbumin demonstrated that the nature of the protein could affect its presentation to the immune system. It was shown that OVA with either a GPI anchor or secreted from the parasite was readily presented via MHC class I, whereas cytoplasmic or transmembrane-anchored OVA was not and therefore only a subset of the endogenous parasite proteins could be expected to be recognized [34]. Indeed it has been shown that by far the majority of T cell responses in both patients and animal models are directed at members of the GPI-anchored *trans*-sialidase family of proteins [35]. The *trans*-sialidase family appear to be extremely immunodominant in *T. cruzi* infections. As this family consists of >1000 hypervariable members it is unlikely to form the basis for a vaccine in the immediate future.

In addition, the upregulation of MHC class I expression on dendritic cells by LPS or IFNγ has been shown to be blocked by *T. cruzi* infection reducing the ability of infected tissues to present antigen [36]. As dendritic cells are the predominant antigen-presenting cells capable of inducing a CD8$^+$ T cell response this inhibition has the capability of reducing the protective response of the host. Recent work has also suggested that infection with *T. cruzi* blocks the IFNγ-mediated switch from constitutive to immunosubunits in the proteasome, thereby further compromising antigen presentation [37]. Thus the parasite can interfere with the generation of a protective immune response at many levels.

Drug-treated mice which achieve sterile parasitological cure under a specific benznidazole treatment regimen have been shown to evolve a protective CD8$^+$ T cell memory response which is stable in the long term. This response occurs regardless of whether the mice are treated at an early stage of infection or during the chronic phase, suggesting that long-term antigen exposure does not lead to T cell "exhaustion" at least in *T. cruzi* infection [3]. This work was also the first well-characterized demonstration that chemotherapy alone could result in total elimination of the parasites. The ability to create a population of $CD8^+$ memory T cells that are protective against *T. cruzi* also implies that a vaccine should be possible.

36.5 Summary and Outlook

It is clear that *T. cruzi* infection involves a particularly complex interplay between host factors and the parasite and that this interplay may vary depending on the particular host cell involved and the stage and strain of parasite.

Firstly the metacyclic trypomastigote must invade naïve cells, recruit lysosomes to allow its escape into the cytoplasm and subsequent differentiation into the amastigote. The amastigote itself has to prevent the infected cell from undergoing apoptosis or being targeted by innate or acquired immune response while it undergoes replication. It then has to differentiate into the bloodstream trypomastigote and escape from the host cell. At this point it has to avoid humoral immune factors and can be taken up by an insect vector or it can invade further host cells, where again it has to recruit lysosomes and escape into the cytoplasm before replication as an amastigote.

The parasite has evolved numerous ways of interfering with the signaling and metabolism of the infected cell, and with the generation of an effective immune response by the host. Encouragingly, the development of an animal model in which parasites have been completely removed by chemotherapy has shown that a protective and stable $CD8^+$ memory T cell population can be generated. This gives impetus to the prospects of developing a vaccine or immunotherapy-based approach to this most intractable infection.

The remarkable ability of *T. cruzi* to infect a wide panoply of host cells is underlined by its genetic variability and its aptitude for modulating multiple host cell regulatory pathways from membrane repair to cytokine transcription and cell death. The recent publication of the parasite genome sequence, coupled with advances in the genetic manipulation toolkit available for host and parasite, live-cell imaging techniques and molecular probes for signaling events will allow the fascinating biology of this cell:pathogen interaction to be elucidated in ever greater detail. As stated above, there are still many gaps in our understanding of exactly how this parasite can establish a chronic life-long infection in immunologically competent hosts.

References

1 Yacoub, S., Mocumbi, A.O. and Yacoub, M.H. (2008) *Heart*, **94**, 244–248.

2 Campos, S.V., Strabelli, T.M., Amato Neto, V., Silva, C.P., Bacal, F., Bocchi, E.A. and Stolf, N.A. (2008) *J Heart Lung Transplant*, **27**, 597–602.

3 Bustamante, J.M., Bixby, L.M. and Tarleton, R.L. (2008) *Nat Med*, **14**, 542–550.

4 Wilkinson, S.R., Taylor, M.C., Horn, D., Kelly, J.M. and Cheeseman, F I. (2008) *Proc Natl Acad Sci USA*, **105**, 5022–5027.

5 de Melo-Jorge, M. and PereiraPerrin, M. (2007) *Cell Host Microbe*, **1**, 251–261.

6 Scharfstein, J., Schmitz, V., Morandi, V., Capella, M.M., Lima, A.P., Morrot, A., Juliano, L. and Muller-Esterl, W. (2000) *J Exp Med*, **192**, 1289–1300.

7 Monteiro, A.C., Schmitz, V., Morrot, A., de Arruda, L.B., Nagajyothi, F., Granato, A., Pesquero, J.B., Muller-Esterl, W., Tanowitz, H.B. and Scharfstein, J. (2007) *PLoS Pathog*, **3**, e185.

8 Caler, E.V., Vaena de Avalos, S., Haynes, P. A., Andrews, N.W. and Burleigh, B.A. (1998) *EMBO J*, **17**, 4975–4986.

9 Tardieux, I., Webster, P., Ravesloot, J., Boron, W., Lunn, J.A., Heuser, J.E. and Andrews, N.W. (1992) *Cell*, **71**, 1117–1130.

10 Rodriguez, A., Samoff, E., Rioult, M.G., Chung, A. and Andrews, N.W. (1996) *J Cell Biol*, **134**, 349–362.

11 Woolsey, A.M., Sunwoo, L., Petersen, C.A., Brachmann, S.M., Cantley, L.C. and Burleigh, B.A. (2003) *J Cell Sci*, **116**, 3611–3622.

12 Maganto-Garcia, E., Punzon, C., Terhorst, C. and Fresno, M. (2008) *Traffic*, **9**, 1299–1315.

13 Andrade, L.O. and Andrews, N.W. (2004) *J Exp Med*, **200**, 1135–1143.

14 Woolsey, A.M. and Burleigh, B.A. (2004) *Cell Microbiol*, **6**, 829–838.

15 Caler, E.V., Chakrabarti, S., Fowler, K.T., Rao, S. and Andrews, N.W. (2001) *J Exp Med*, **193**, 1097–1104.

16 Chakrabarti, S., Andrade, L.O. and Andrews, N.W. (2005) *Mol Biochem Parasitol*, **141**, 125–128.

17 Vieira, O.V., Botelho, R.J., Rameh, L., Brachmann, S.M., Matsuo, T., Davidson, H.W., Schreiber, A., Backer, J.M., Cantley, L.C. and Grinstein, S. (2001) *J Cell Biol*, **155**, 19–25.

18 Vaena de Avalos, S., Blader, I.J., Fisher, M., Boothroyd, J.C. and Burleigh, B.A. (2002) *J Biol Chem*, **277**, 639–644.

19 Hall, B.F., Webster, P., Ma, A.K., Joiner, K. A. and Andrews, N.W. (1992) *J Exp Med*, **176**, 313–325.

20 Rubin-de-Celis, S.S., Uemura, H., Yoshida, N. and Schenkman, S. (2006) *Cell Microbiol*, **8**, 1888–1898.

21 Andrews, N.W., Abrams, C.K., Slatin, S.L. and Griffiths, G. (1990) *Cell*, **61**, 1277–1287.

22 Straight, S.W., Herman, B. and McCance, D.J. (1995) *J Virol*, **69**, 3185–3192.

23 Manning-Cela, R., Cortes, A., Gonzalez-Rey, E., Van Voorhis, W.C., Swindle, J. and Gonzalez, A. (2001) *Infect Immun*, **69**, 3916–3923.

24 Benabdellah, K., Gonzalez-Rey, E. and Gonzalez, A. (2007) *Mol Microbiol*, **65**, 1559–1567.

25 Tomlinson, S., Vandekerckhove, F., Frevert, U. and Nussenzweig, V. (1995) *Parasitology*, **110** (Pt 5), 547–554.

26 Nicholson, K.M. and Anderson, N.G. (2002) *Cell Signal*, **14**, 381–395.

27 Aoki, M.P., Guinazu, N.L., Pellegrini, A.V., Gotoh, T., Masih, D.T. and Gea, S. (2004) *Am J Physiol Cell Physiol.*, **286**, C206–212.

28 Petersen, C.A., Krumholz, K.A., Carmen, J., Sinai, A.P. and Burleigh, B.A. (2006) *Infect Immun*, **74**, 1580–1587.

29 Petersen, C.A., Krumholz, K.A. and Burleigh, B.A. (2005) *Infect Immun*, **73**, 6974–6980.

30 Hashimoto, M., Nakajima-Shimada, J. and Aoki, T. (2005) *Mol Biol Cell*, **16**, 3521–3528.

31 Reina-San-Martin, B., Degrave, W., Rougeot, C., Cosson, A., Chamond, N., Cordeiro-Da-Silva, A., Arala-Chaves, M., Coutinho, A. and Minoprio, P. (2000) *Nat Med*, **6**, 890–897.

32 Mucci, J., Risso, M.G., Leguizamon, M.S., Frasch, A.C. and Campetella, O. (2006) *Cell Microbiol*, **8**, 1086–1095.

33 Bafica, A., Santiago, H.C., Goldszmid, R., Ropert, C., Gazzinelli, R.T. and Sher, A. (2006) *J Immunol*, **177**, 3515–3519.

34 Garg, N., Nunes, M.P. and Tarleton, R.L. (1997) *J Immunol*, **158**, 3293–3302.

35 Martin, D.L., Weatherly, D.B., Laucella, S. A., Cabinian, M.A., Crim, M.T., Sullivan, S., Heiges, M., Craven, S.H., Rosenberg, C.S., Collins, M.H., Sette, A., Postan, M. and Tarleton, R.L. (2006) *PLoS Pathog*, **2**, e77.

36 Van Overtvelt, L., Andrieu, M., Verhasselt, V., Connan, F., Choppin, J., Vercruysse, V., Goldman, M., Hosmalin, A. and Vray, B. (2002) *Int Immunol*, **14**, 1135–1144.

37 Bergeron, M., Blanchette, J., Rouleau, P. and Olivier, M. (2008) *Parasite Immunol*, **30**, 280–292.

37
Trichinella and the Nurse Cell

David B. Guiliano and Yelena Oksov

37.1
Introduction to *Trichinella* and Trichinellosis

Trichinellids are an important group of obligate human and animal nematode pathogens which, until recently, have posed a serious threat to human health and agricultural interests. Despite being relatively large (1–3 mm) multicellular organisms, these nematodes occupy intracellular niches within the host for most of their lifecycle. While this is not unique among the Nematoda [1, 2], Trichinellids distinguish themselves from their peers by orchestrating a complex series of events transforming the parasitized tissue into a novel structure called a "nurse cell." This nurse cell serves as long-term residence for the parasite until subsequent transmission.

37.1.1
Systematics

Trichinellids are enoplid nematodes belonging to a diverse order of obligate vertebrate pathogens called trichocephalids. A number of other important nematode pathogens belong to this order including the trichurids, of which *Trichuris trichuria*, the causative agent of human whipworm, is a member. One important feature of the trichocephalids is their predilection towards intestinal trophic environments and this may represent an ancestral feature of the group. Within the trichinellids eight species and three additional distinct genotypic isolates have been identified and these can be divided into two subgroups based on whether their nurse cells develop thick (encapsulated) or thin (non-encapsulated) collagenous capsules. A striking characteristic of the genus is the relatively broad host specificity displayed by most of its members; in addition to infecting many mammals, some non-encapsulating species infect birds (*T. pseudospiralis*) or reptiles (*T. papuae* and *T. zimbabwensis*) [3]. Analysis of mitochondrial and nuclear gene evolution in the *Trichinellidae* indicates that while separation of the trichinellids from other trichocephalids may have occurred over 250 million years ago, the diversification of the genus occurred relatively recently ~15–20 million years ago [4].

Intracellular Niches of Microbes. Edited by Ulrich E. Schaible and Albert Haas

ISBN: 978-3-527-32207-7

37.1.2
Parasite Lifecycle

Trichinellids are autoheteroxenous, completing their development within a single vertebrate host. There are three phases of the parasite lifecycle: enteral, migratory and parenteral. After ingestion, muscle stage larvae (mL1) contained within contaminated meat are liberated by the actions of host digestive processes. The parasites are carried into the small intestine, invade the columnar epithelium and adopt a para-intracellular lifestyle occupying syncytial tunnels they create by burrowing through enterocytes [5, 6]. Within 30 h post ingestion the nematodes molt four times, develop into adults, mate and within 5 days the females begin shedding hundreds of newborn larvae (NBLs) [7]. The NBLs exit the intestine, burrowing through the smooth muscle until they enter blood or lymphatic vessels. Once in circulation the NBLs become systemic, migrate nonspecifically through the host and penetrate all major anatomical areas including the brain and heart. However, within hours most leave circulation and invade a skeletal muscle fiber [8]. After establishing themselves in the muscle fiber's cytoplasm the NBLs begin a process that within 20 days transforms it into a new structure called the nurse cell. This complex and the parasite it contains are extremely stable and can persist for decades until ingestion by a new host [7].

37.1.3
Trichinella; the Causative Agents of Trichinellosis

Prior to the middle of the twentieth century outbreaks of trichinellosis were regular occurrences and historical records indicate that human populations have continuously coexisted with this disease. Changes in agricultural practices over the past 50 years have drastically reduced human trichinellosis. However, because of risks posed by zoonotic infections agricultural agencies monitor meat to prevent its re-entry into our food supply. This is not a new practice and while the first recorded observations of *Trichinella* in humans were made in the mid-nineteenth century, the knowledge of the disease caused by the parasite dates back to antiquity. Severe illness linked to pork-derived *Trichinella* and tapeworm infections (*Taenia sp.*) may have prompted prohibitions against consumption of this meat contained in Jewish and Islamic law. "And the swine, because he parteth the hoof, and is cloven-footed, but cheweth not the cud, he is unclean unto you. Of their flesh ye shall not eat, and their carcasses ye shall not touch; they are unclean unto you." Leviticus 11: 4,7–8. Thus, it is likely the relationship between the consumption of pork and the acute symptoms of disease have been appreciated for some time.

Tissue damage caused by all lifecycle stages or reactions to dead and dying worms can cause symptoms ranging from nausea, vomiting and diarrhea to muscle pains, fever and conjunctivitis. In severe infections complications arising from migrating NBLs can result in meningitis, encephalitis, neuropathy, brain damage, paralysis, endocarditis, myocarditis and occasionally death.

37.2
Remodeling of the Host Cell Compartment by *Trichinella*

There are a number of characteristics that distinguish the *Trichinella* nurse cell complex from other intracellular pathogens or symbiont associations described in this volume. First, the parasite is not reproductive during its occupation of the muscle fiber and the structure appears to be designed solely for housing and sustaining its resident until transmission. Second, the transformation process completely remodels the myofiber and its local tissue microenvironment. The new structure bears little phenotypic or functional resemblance to its progenitor or any other known host tissue. Third, despite being "foreign" this structure can be sustained for the lifetime of the host, suggesting active modulation/tolerization of the immune system. Finally, the steps undergone to achieve this new structure are phenomena we currently cannot reproduce in muscle *in vivo*. Most studies of nurse cell biology have been carried out using *T. spiralis* infections and initially we will focus on observations of this encapsulating species. However, data comparing the development of encapsulated and non-encapsulated structures will also be discussed.

Before describing what is known about the remodeling of the host cell by *Trichinella* a short description of the most relevant characteristics of skeletal muscle fibers will be given.

37.2.1
Skeletal Muscle; a Unique Multicellular Structure and Compartment for *Trichinella* Infection

Skeletal muscle fibers are large syncytia produced by the fusion of hundreds of muscle stem cells (myoblasts) into a tube-like structure. Their size varies depending on organism and muscle mass examined. However, in humans they range from 10–60 μm in diameter and 1–30 cm in length. Once completed, each fiber contains hundreds of nuclei associated with its plasma membrane (sarcolemma). Within the cytoplasm (sarcoplasm) of each fiber there are highly ordered filaments (myofilaments) composed of myosin, actin and other components of the contractile machinery. These run the length of the fiber and are attached to the sarcolemma. If a muscle fiber is damaged regeneration occurs via mobilization, differentiation and finally fusion of resident myoblasts. Their fusion contributes fresh cytoplasmic and nuclear material to the damaged syncytium and in mammals this constitutes the only known mechanism for fiber regeneration. The molecules and pathways that are involved in differentiation of myoblasts into terminally differentiated muscle fibers have been extensively characterized (see [9, 10] for recent reviews of the field).

From the perspective of *Trichinella* infection biology four facets of this research are particularly interesting. First, muscle stem cells are responsive to factors released after muscle damage or during primary myogenic events. A percentage of the daughter cells produced by proliferation of these stem cells commit to terminal differentiation and subsequently fuse to damaged fibers or form new ones. Second, lineage-specific basic helix-loop-helix (bHLH) and homeobox transcription factors families control various parts of this process. Some of these proteins are involved in maintaining primary stem cell populations while others in response to external factors, commit the cell to terminal differentiation and activate a muscle-specific gene (MSG) expression program. Cell cycle regulatory proteins such as retinoblastoma protein (Rb) interact with these transcription factors modulating their activities and Rb hypophosphorylation is one of the first steps to myogenic commitment [11]. Third, once fused into a syncytium the terminally differentiated cells are refractory to subsequent treatment with growth stimuli. Again regulators such as Rb are implicated in maintaining the postmitotic block and loss of these proteins or modulation of their function can uncouple this process, either directly inducing cell cycle re-entry or by restoring sensitivity to growth stimuli [11, 12]. Finally, other vertebrates such as some urodele species are more developmentally flexible and possess muscle fiber syncytia that can reverse this process, re-enter cell cycle and cellularize into functional mononuclear stem cells [13].

In vitro murine/urodele myotube hybrids (heterokaryon) studies have shown that the mammalian postmitotic block can be reversed by urodele-derived cytoplasmic factors [14]. This suggests that if the mammalian myofibers are supplied with the appropriate intracellular signaling environment they can respond to mitogens.

37.2.2
Invasion of the Fiber

The initial events underlying the invasion of the skeletal myofiber remain poorly studied. Morphological evidence gathered from histological analyses indicates invasion is a purely physical process with the larva using its head to depress the myofiber's sarcolemmal membrane, producing a puncture wound [7]. After penetration the larva "tunnels" to the center of the fiber. Resolution of the puncture appears to occur rapidly (within hours) and inside the myofiber the larva initially lives within the sarcoplasm with no apparent barrier between itself and the fiber's internal components [15]. There is some evidence that different species of *Trichinella* have preferences for colonizing particular subtypes of muscle fibers and one study reported that *T. spiralis* exclusively invades slow-twitch fibers while *T. pseudospiralis* invades both slow and fast-twitch fibers [16].

37.2.3
Unmaking the Muscle: the Initial Stages of the Transformation Process

Transmission electron microscopy (TEM) studies of infected tissue indicate that there is little morphological change in either the muscle or nematode (see Section 37.3.1) until 4 days post fiber invasion (pfi) [15]. Figure 37.1 summarizes these alterations over the course of the infection and shows representative images of hematoxylin and eosin (H&E) stained infected tissue. By day 4 or 5 pfi the nuclei within the infected fiber undergo a number of striking changes, swelling to average volumes of $\sim$500 μm^3, approximately six times the volume of normal myofiber nuclei (see Figure 37.2a). They dissociate from the sarcolemmal membrane, developing extremely prominent nucleoli indicative of high levels of transcriptional activity (see Figure 37.2b). Tunnel staining indicates that the nuclei undergo damage to their DNA but remain transcriptionally active, producing RNA throughout the lifetime of the infection [17–19]. Studies of these structures (referred to as hypertrophic nuclei) provided the first indications that *Trichinella* could reverse the post-mitotic block normally present in this terminally differentiated tissue. Treatment of developing nurse cells with tritiated thymidine (^{3}HT) established that DNA synthesis consistent with cell cycle re-entry was occurring, with incorporation peaking between days 4–8 pfi [20]. *In utero*, whole-animal ^{3}HT labeling experiments indicate that the hypertrophic nuclei develop from the initial preinfection population of myofiber nuclei, and not from the fusion of stem cells to the damaged syncytium [20]. The nuclear alterations are chronic and the hypertrophic nuclei remain suspended in cell cycle at some point after G_1/S containing a 4N DNA content and never complete mitosis [20].

In conjunction with the nuclear alterations a profound suppression of MSG expression occurs with dissolution of myofilament structures and the disappearance of their component from the sarcoplasm by 8 days pfi [15, 21, 22]. Portions of transforming sarcoplasm become basophilic, a feature normally observed in regenerating muscle fibers. This basophilia peaks $\sim$14 days pfi and then rapidly dissipates. The myofibers organelles are completely reorganized and the sarcoplasm fills with polysome-rich rough endoplasmic reticulum (RER) and mitochondria [15]. In addition, the dispersed Golgi elements normally associated with the muscle fiber's plasma membrane return to perinuclear localizations [23, 24]. The inner membrane of the fiber's mitochondria become abnormally vacuolated and ATP synthesis becomes uncoupled [15, 25]. Both the RER and mitochondria collect around the larvae during the early stages of infection but later disperse to the cell periphery [15].

The parasite begins the transformation process with no apparent compartmentalization within the fiber, however, by 10 days pfi a double membrane structure is observed overlaying the nematode's epicuticle [15]. The origins of this membrane structure are unknown, but Despommier and colleagues speculate it is derived from the RER that surrounds the parasites early in the infection process [15]. Alternatively, the epicuticles of mL1 possess a peculiar trilaminar lipid layer as well as an electron-dense surface coat, and the membranes observed could be components of these structures [26, 27].

37.2.4
Making Something New: Parasite-Induced Remodeling of the Muscle Fiber Compartment

Coincidental with the development of the hypertrophic nuclei the proto-nurse cell begins to acquire a number of new features. These secondary characteristics include changes in metabolic activity such as the conversion to an anaerobic metabolism, higher glucose uptake and increased lysosomal acid phosphatase (ACP), aminopeptidase and antioxidant enzyme activities [25, 28–30]. Synthesis of a capsule composed

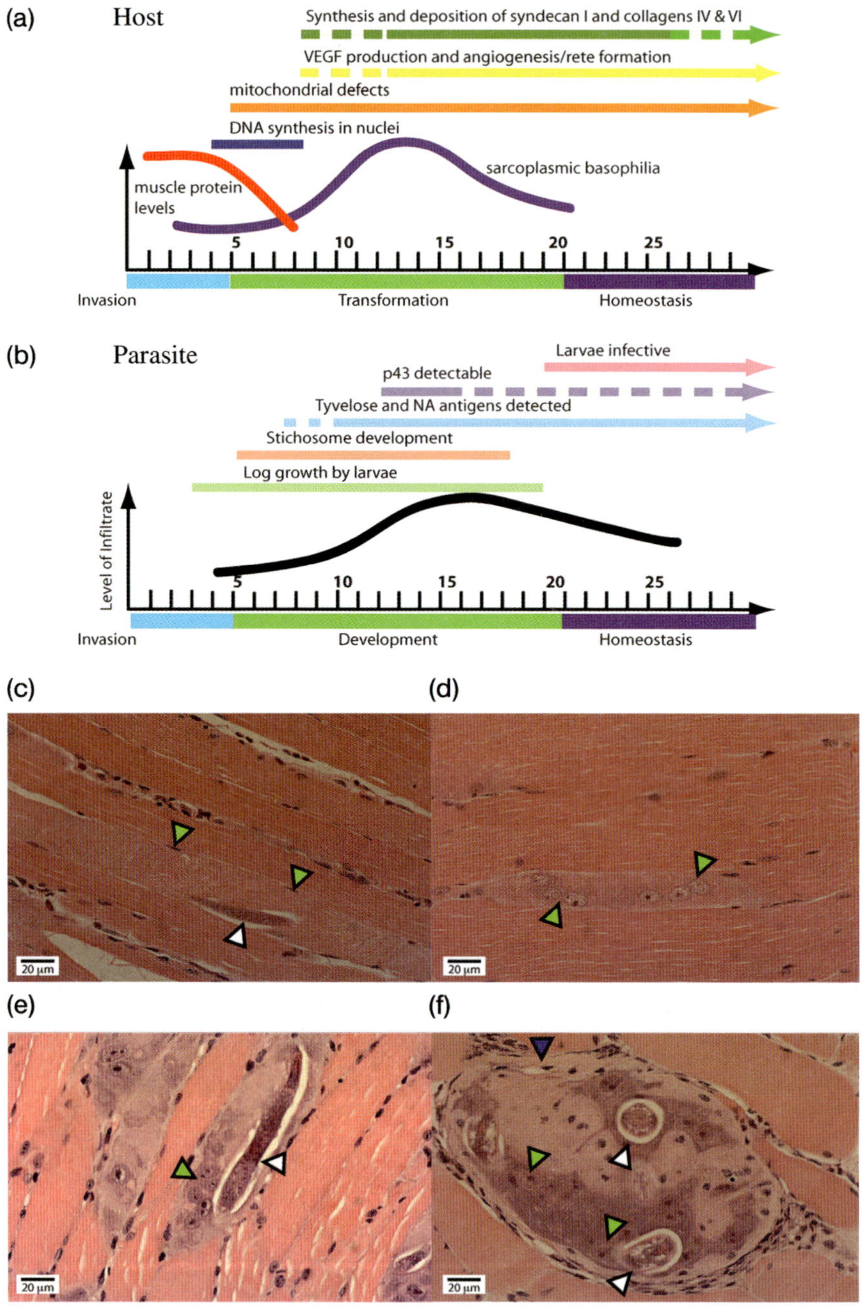

Figure 37.1 (caption see p. 675)

of extracellular matrix, and the recruitment of a new blood vessel network (circulatory rete) around the infected fiber are the most prominent of the new secondary features.

While the function of the capsule is still being debated, the rete is believed to aid in the diffusion of nutrients into and waste products out of the nurse cell. Parasite initiation of a new transcriptional program within the proto-nurse cell appears to be the foundation of all of these alterations and a number of transcripts or proteins associated with these processes have been identified. Seven days pfi transcripts derived from the hypertrophic nuclei encoding collagen types IV and VI and vascular endothelial growth factor (VEGF) can be detected [31, 32]. Development is extremely rapid, with observable capsule condensation by day 10 pfi and angiogenic processes by day 12 pfi. In addition, the heparin-bearing glycoprotein syndecan I, normally found as a component of the basement membrane in epithelia and vascular endothelium, is also expressed during this period [33, 34]. However, unlike the collagens and VEGF, syndecan I never translocates out of the nurse cell and accrues within its cytoplasm. Retention of syndecan I is specific, as other heparinated basement membrane glycoproteins such as perlican are processed and translocated to the plasma membrane normally [33]. Expression of these proteins peaks around day 15 pfi and then gradually diminishes. However, detectable levels of VEGF and collagen VI transcript continue to be synthesized throughout the infection [31, 32].

These observations offer two very important insights into nurse cell biology. First the transcriptional reprogramming is dynamic and the parasites do not "flick" a single transcriptional/developmental switch, turning some genes on and others off. Rather the expression profile of the nurse cell alters as it develops and transcripts no longer required are repressed (i.e., collagen IV). Second, the parasite acquires an exquisite level of control over the fiber's architecture and cellular machinery. This is evidenced by the ultrastructural changes observed in the organelles and the specific retention of some proteins within the nurse cell cytoplasm while other related products are exported normally.

Figure 37.1 The development of the *Trichinella spiralis* nurse cell. Major changes are observed in the infected fiber (a) and parasite (b) during the development of the nurse cell. Days post fiber invasion (pfi) are shown on the x-axis and levels of muscle proteins, sarcoplasmic basophilia or numbers of infiltrating mononuclear cells are shown on the y-axis. VEGF and collagen mRNAs (dashed yellow or dark green line) can be detected several days before protein is detected with the nurse cell cytoplasm (solid lines). Type IV collagen production ceases around day 26 pfi but type VI continues throughout the period of the infection (dashed light green line). Tyvelose containing glycoproteins are initially detected within the nematodes developing stichosome (dashed light blue line) and later within the cytoplasm and hypertrophic nuclei of the nurse cell (solid line). The p43 glycoprotein is synthesized continuously by the parasite from day 12 pfi (dashed light purple line) but is detectable within the cytoplasm of the nurse cell only from days 12–15 pfi (solid light purple line) [48]. (c–f) Images of H&E stained infected tissue containing developing nurse cells approximately 3, 5, 7 and 14 days pfi. The nematodes (white triangles), myofibre nuclei (c) and hypertrophic nuclei (d–f, green triangles), and collagenous capsule (dark blue triangle) are highlighted. Sarcoplasmic basophilia within the transforming fiber peaks around day 14 pfi (f).

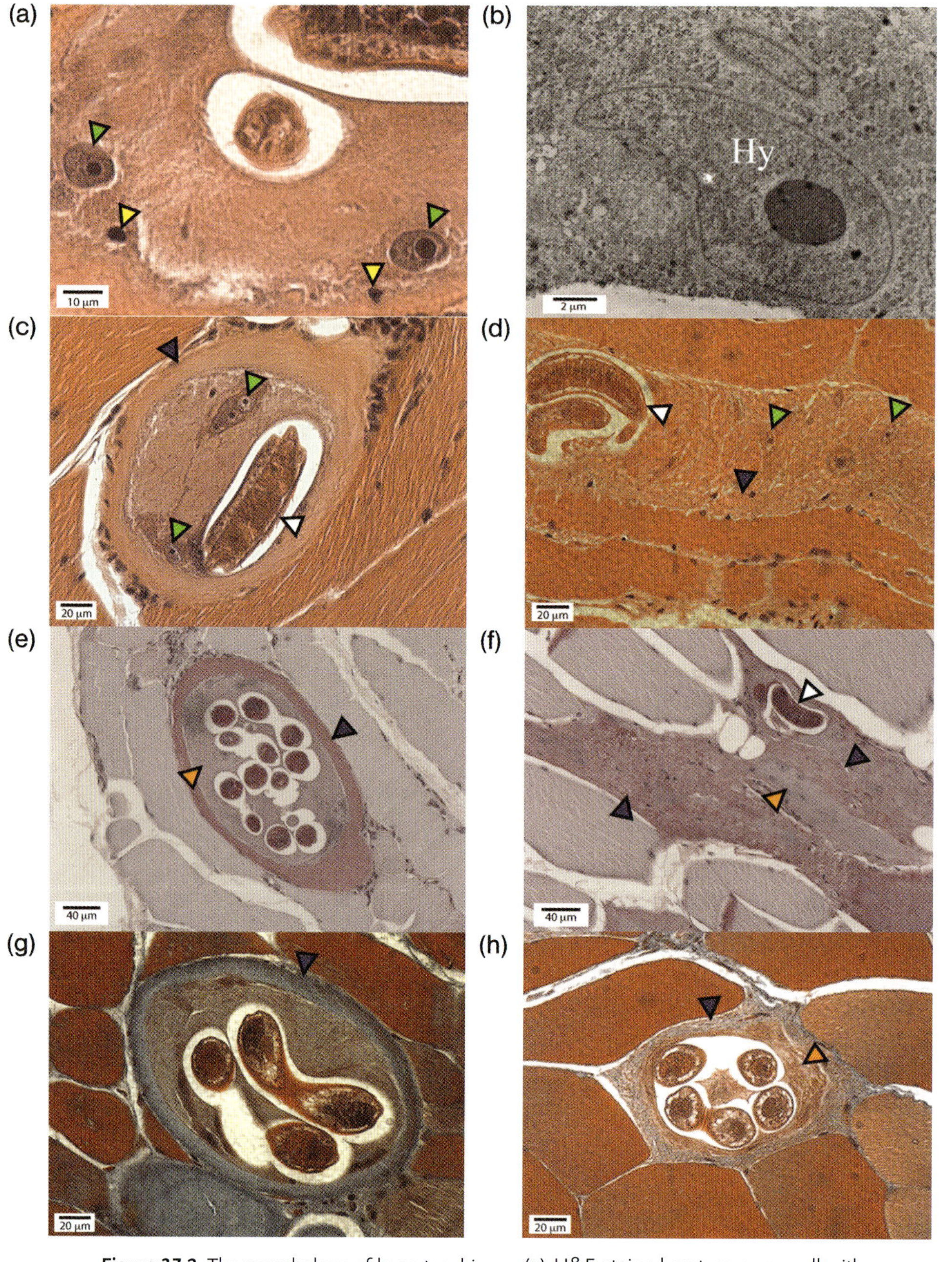

Figure 37.2 The morphology of hypertrophic nuclei and a comparison of nurse cells from encapsulating and non-encapsulating *Trichinella* species. The alterations observed in the nurse cell nuclei persist for the lifetime of the infection. (a) H&E stained mature nurse cell with prominent hypertrophic nuclei (green triangles). Other nuclei possibly derived from infiltrating mononuclear cells can be observed (yellow triangles).

37.2.5
Nurse Cell Polymorphism

Trichinellid species can be divided into encapsulating and non-encapsulating species depending on whether a thick or thin collagen capsule is deposited around the nurse cell. *T. pseudospiralis* is the most extensively studied non-encapsulating species and its comparison with *T. spiralis* has provided some interesting insights into the scope of variation possible in nurse cell morphology (see Figure 37.2c and d). While both species induce myofilament dissolution, cycle re-entry and initiate a new transcriptional program (ACP activity increases) the kinetics of these processes differ with *T. pseudospiralis* nurse cells having longer development periods. Skeletal muscle structural proteins persist within the cytoplasm of *T. pseudospiralis* nurse cells indicating incomplete repression of MSGs by the parasite or initiation of a divergent transcriptional program from its encapsulating sister species [22]. Both species express the necessary components to form a capsule, and both develop PAS-positive components in their cytoplasm and on the surface of their nurse cells (see Figure 37.2e and f). However, levels of collagen expression are lower in *T. pseudospiralis* nurse cells, the infected fiber never separates from adjacent tissue and does not condense into the ovoid structure typical of encapsulating species. Instead two discernable compartments develop within the nurse cell cytoplasm (see Figure 37.2c and d).

Histochemical analysis of *T. pseudospiralis* nurse cells using Masson's trichrome reagent indicates the central region has a different composition to *T. spiralis* nurse cell cytoplasm while the outer cortex has staining properties indicative of loosely packed extracellular matrix (see Figure 37.2g and h). In *T. pseudospiralis*, portions of adjacent non-infected tissue often undergo similar morphological changes as the infected fiber indicating that in the absence of encapsulation, diffusion of "transformation" effectors may occur, resulting in more systemic myopathy.

(b) Transmission electron microscope analysis of a mature nurse cell shows the interior of the hypertrophic nuclei (Hy) contains patchy sections of chromatin, a few dense inclusions and an enlarged nucleolus. The nurse cell complex is extremely stable. (c) H&E stained nurse cell within chronically infected muscle (>6 months post fiber invasion). The nematode (white triangle), hypertrophic nuclei (green triangles), collagenous capsule (dark blue triangle) and infiltrating mononuclear cells are shown. (d) Section of H&E stained muscle chronically infected with the non-encapsulating *Trichinella pseudospiralis*. The nematode (white triangle), myofilaments dissolution, hypertrophic nuclei (green triangles) and outer cytoplasmic compartments (dark blue triangle) can be observed. In both species PAS staining (bright pink color) indicates glycans accumulate within the capsule of *T. spiralis* or the outer cytoplasmic compartment of *T. pseudospiralis* nurse cells (e and f). The capsule and outer cytoplasmic compartment are indicated with dark blue triangles while the cytoplasm/inner cytoplasmic compartments are indicated with orange triangles. Masson's trichrome reagent shows a defined collagen capsule around *T. spiralis* nurse cells (green band indicated by dark blue arrow, g) and differentially stains the inner (orange triangle) and outer (dark blue triangle) compartments of the *T. pseudospiralis* nurse cell cytoplasm (h).

37.2.6
Potential Pathways Used by the Parasite to Achieve the Transformation Process

While we have some information regarding the dysregulation of specific molecules that contribute directly to the infected cell phenotype (i.e., production of extracellular matrix components and angiogenic factors) little is known about the upstream factors that drive their expression. The lack of an *in vitro* invasion/transformation model and the inherent problems associated with studying developing nurse cells *in situ* have made characterization of these upstream processes difficult. Early stages of *Trichinella*-induced changes to the muscle fiber, such as the presence of centralized nuclei and local basophilic transformation of the sarcoplasm, superficially resemble those seen during conventional muscle regeneration. However, the delay between invasion and the appearance of the nuclear abnormalities suggests a secondary event, and not the initial damage to the fiber is driving the process. The temporal association between the nuclear abnormalities, transcriptional reprogramming and data yielded from infection studies using irradiated NBLs (see Section 37.3.1) offers compelling evidence that cell cycle re-entry may be the lynch pin upon which the rest of the transformation process is built.

The transcriptional reprogramming may require the relaxation of the mechanisms that normally restrict muscle fiber gene expression. Cell cycle status is known to play an important role in this process and pushing the myofiber nuclei out of G_0/G_1 may be sufficient to create a new transcriptional environment the parasite can manipulate [11].

The significance of the subsequent loss of MSG expression may simply reflect a bystander effect that is not deliberately produced or required by the parasite. Observations from *T. pseudospiralis* nurse cells support the theory that MSG expression is not directly deleterious to the parasite and does not prevent the induction of secondary features such as increased ACP activity. Whether movement into S phase and DNA synthesis within the hypertrophic nuclei is actively promoted by the parasite or simply occurs as a consequence of entrance into G_1 is unknown. It is tempting to speculate that this is actively promoted by the parasite, perhaps to provide additional transcriptional resources for the remodeling process. However, several *in vitro* studies using myoblastic cell lines and viral oncoproteins such as E1A indicate that relaxation of Rb function is sufficient to allow passage of syncytial nuclei into G_1/S [12]. The mechanisms driving this induction in *Trichinella* infection remain an intriguing question. Manipulation of Rb or upstream factors that modulate its function seems a likely possibility, though analysis of parasite products for molecules that can influence Rb's activities has proved difficult.

Why the nuclei remain suspended after G_1/S is unclear, but intrinsic properties of the muscle syncytia may preclude movement through cell cycle past this point. It is interesting to note that in urodele *in vitro* systems myotube nuclei re-enter cell cycle and traverse S phase and arrest in G_2 while *in vivo* completion of mitosis by these cells is accompanied by cellularization [13]. However, the DNA damage observed within hypertrophic nuclei, and the block this might cause to further cell cycle progression argues that the parasite may actively try to prevent completion of mitosis.

Boonmars and colleagues assessed what role the tumor suppressor p53 plays in the chronic cell cycle suspension observed in the hypertrophic nuclei but found that nurse cells formed normally in p53 knockout mice [18]. This indicates that the DNA damage observed in hypertrophic nuclei is not the source of the observed post-G_1/S block or that p53-independent mechanisms are sufficient to halt cell cycle progression.

A number of recent gene expression studies from the Takahashi Laboratory have shown that components of TGFβ and TNFα signaling pathways are modulated during infection [35, 36]. The exact roles these pathways play in nurse cell genesis, and whether they are primary mediators of the transformation process remains to be determined. However, both are implicated in a variety of biological processes, including muscle differentiation, cell cycle control, apoptotic responses and immunomodulation, making them attractive candidates for further study. A recent microarray study also from the Takahashi lab that compares *T. spiralis* or *T. pseudospiralis*-infected muscle tissue provides the most comprehensive global gene expression analysis to date [37]. A large palette of differentially expressed genes including transcriptional regulators involved in muscle differentiation, cell cycle components and proteins involved in immunological responses were identified. Because the RNA used in this study was derived from whole infected tissue it is impossible to differentiate between alteration occurring within the nurse cell and those derived from the surrounding milieu of host tissue and infiltrating cells [37]. However, this study represents the first attempt to globally measure changes within the infected tissue transcriptome and will undoubtedly serve as a foundation for additional research.

37.3 Parasite Molecules Involved in the Remodeling and Maintenance of the Nurse Cell Complex

The identity of and roles played by *Trichinella*-derived products in the transformation and maintenance of the nurse cell phenotype continues to be of great interest. These molecules will offer vital clues for understanding both host and parasite biology and could also serve as potential reagents for the manipulation of muscle developmental biology. It seems likely the parasite directly modulates gene expression via secretion of its own transcriptional mediators or manipulates upstream signal transduction processes that subsequently alter the transcriptome of the infected fiber. So far there is no evidence that exclusively supports either possibility. Experiments testing the effects of depleting or altering secretion of parasite products offer some direct evidence for their role in the transformation process. Future identification and functional characterization of individual constituents may provide tools for uncovering the details of how interactions between parasite proteins and the host cell lead to its subversion.

37.3.1
Evidence for Parasite Regulation of the Nurse Cell Phenotype and the Role Played by Parasite-Derived Nuclear Antigens

It was established in the mid 1980s that parasite-derived antigens were present in the nurse cell (see Section 37.3.2) [38] and specific manipulation of parasite development or protein secretion has provided evidence for a direct and consistent manipulation of the host cell by the nematode. During nurse cell genesis the parasite undergoes a number of developmental changes (see Figure 37.1b) including a rapid logarithmic growth increasing its length by ~800 μm and volume by ~890 000 μm^3 [39] and development of a highly organized secretory structure called the stichosome (see Figure 37.3a). Completion of growth and stichosome genesis around day 20 pfi coincides with the onset of infectivity. The stichosome is composed of ~50 cells (stichocytes) and occupies about half the parasites final length [40].

TEM and histochemical analysis of this organ have identified a minimum of four cell types that are classified by the characteristics of their morphologically distinct secretory granules (α, β, δ and γ see Figure 37.3b and c) [40, 41]. Each stichocyte cell population produces unique components of the parasite's secreted proteome although the significance of this is as yet unclear (see Figure 37.4a–d). As more stichosome-derived proteins are identified, and their functions uncovered, the significance of their specific compartmentalization will hopefully become more obvious.

Antiparasite antibodies produced during infections or antisera generated against *Trichinella*-secreted products detect protein and glycan antigens in the developing stichosomes from day 6 pfi and within the cytoplasm and nuclei of the nurse cell from day 8 pfi (see Figure 37.4e–g). Their role in the transformation process remains controversial. Studies using irradiated NBLs have shown that while they invade tissue normally their growth pfi is retarded and this coincides with a delay in nurse cell development. These larvae also fail to develop histochemically distinct α or β stichosomes. Despite this defect the major features of the infected cell phenotype still occur [42]. Other studies using mebendazole, which specifically disrupts parasite microtubule networks and secretion processes, are more definitive, showing reduced levels of parasite antigen within the nurse cells of treated animals [43]. The loss of parasite proteins coincided with shrinking of the hypertrophic nuclei and nucleoli, loss of nurse cell RNA, reduced nuclear laminin content and ACP activities. However, these changes did not coincide with any recovery of MSG expression [43].

While these data indicate that the initiation and maintenance of the chronic cell cycle dysregulation phenotype is independent of α and β stichocyte-derived antigens, it is unknown how the δ and γ cells are affected by this treatment and thus a functional role for the stichosome cannot be ruled out. Indeed, it suggests that repression of MSGs is independent of the new transcriptional program and therefore may be linked to chronic cell cycle suspension.

Finally, these experiments have established that maintenance of the infected cell phenotype requires constant input from the parasite presumably in the form of parasite-secreted proteins. However, this does not extend to chronic cell cycle

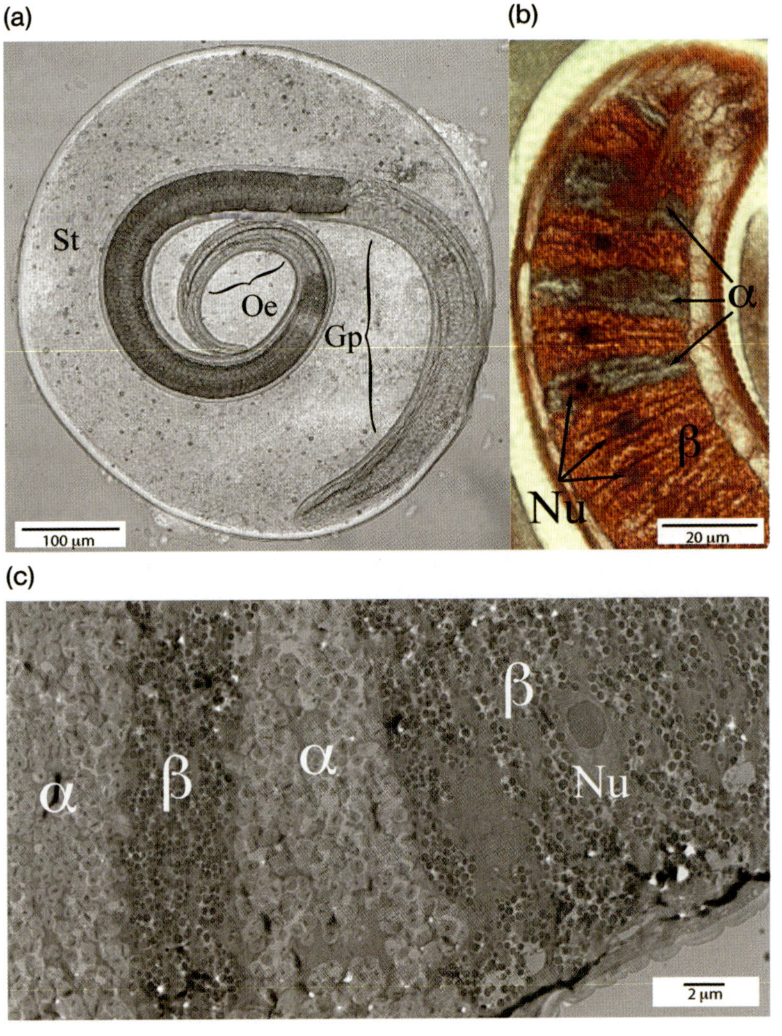

Figure 37.3 The anatomy of *Trichinella spiralis* muscle stage larvae and the stichosome. (a) Phase-contrast image showing an isolated nurse cell containing a mature *T. spiralis* mL1. Three major anatomical features are highlighted: head/oesophageal region (Oe), stichosome (St) and genital primordium (Gp). Immunohistochemical localization studies have established that most of the *T. spiralis* antigens and secreted proteins identified so far are synthesized and stored in the stichosome. (b) Masson's trichrome stained thin section of the mL1 stichosome showing the interdigitating α-stichocyte (green) and β-stichocyte (red) cells. (c) Equivalent transmission electron microscopy section showing the fine structure of the interdigitated stichocytes and morphology of the granules they contain (Nu, stichocyte cell nucleus).

dysregulation, which is maintained after drug treatment. Parasite proteins are found within the insoluble, laminin-containing fraction of the nuclear proteome and microscopic studies indicate they colocalize to the RNA polymerase II containing

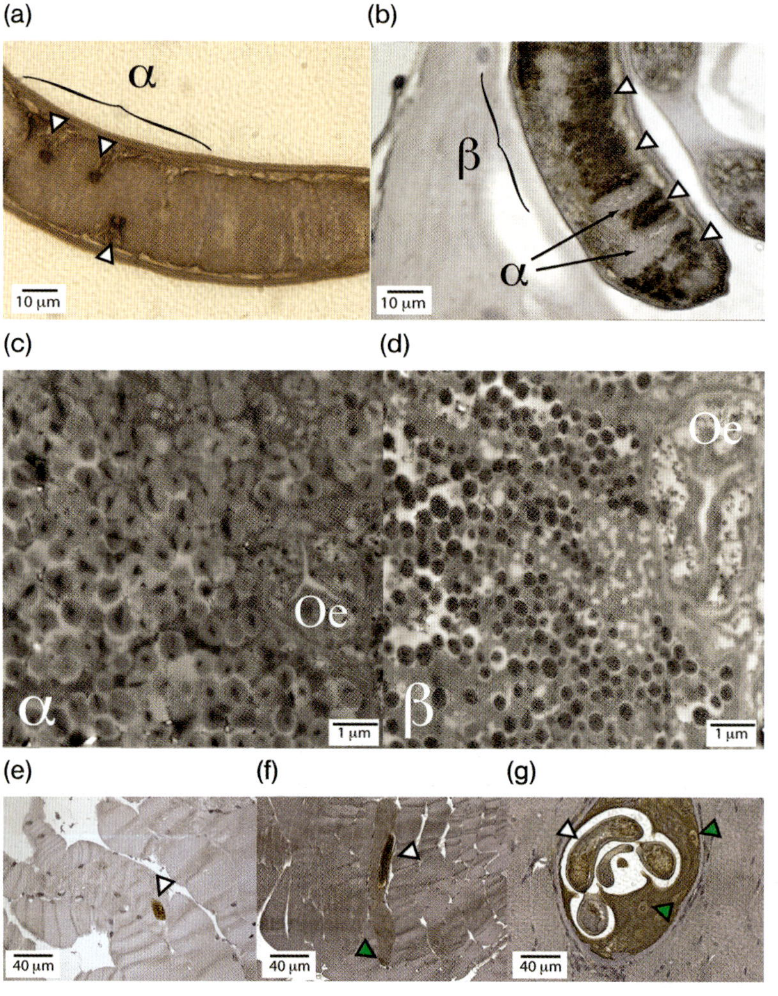

Figure 37.4 Compartmentalization of secreted proteins within the stichosome and localization of *Trichinella*-derived antigens to the developing nurse cell. Characterization of several *Trichinella* secreted proteins has shown that they are synthesized and stored in distinct stichocyte cell populations. (a) Immunohistochemical localization of a novel 7 kDa secreted protein, Ts-SML-5, to a subset of α-stichocyte cells (brown/black color indicated by white triangles). (b) A second secreted protein Ts-SML-2/MCD-1 is localized within the β-stichocyte cells (white triangles) [52]. Fine structural localization of Ts-ANT-2/MCD-1 using immunogold transmission electron microscopy confirms it is synthesized and stored in granules within the β-stichocytes. (c, d) α- and β-granules within the mL1 stichocytes. ANT-2/MCD-1 is detected (black dots) exclusively within the β-granules (Oe, oesophagus). To visualize secretion of parasite proteins into the nurse cell thin sections of paraffin-embedded infected muscle were probed with the anti-tyvelose monoclonal antibody 18 H (brown color) [78]. (e) Cross-reactive glycans can be detected in the nematode (white triangles) at day 6 pfi, while (f) weak cross-reactivity in the nurse cell cytoplasm and hypertrophic nuclei (green triangles) is evident from day 8 pfi. (g) Levels of tyvelosated antigens increase and intense staining is evident in the nematode (white triangle), the nurse cell cytoplasm and hypertrophic nuclei (green triangles) for the lifetime of the infection (day 32 pfi).

speckles [44, 45]. These localization patterns indicate they could act as transcriptional regulators, directly interacting with the transcriptional machinery and DNA of target genes or alter chromatin superstructure within the hypertrophic nuclei in a manner that facilitates the transcriptional reprogramming.

37.3.2
The Composition of the Parasite-Secreted Proteome

Identification of the products secreted by the parasite into the nurse cell remains an important step in elucidating how the nematode achieves the remodeling of the muscle fiber. Like the study of the nurse cell itself, a number of experimental hurdles have made this difficult and the only easily accessible stage of larval development is mature infective mL1. Second, the isolation method used, digestion of infected muscle in an acidified buffer containing pepsin, partially activates the mL1 for their subsequent enteral phase of development. Proteomic analysis of material derived from time courses of cultured mL1s indicates this process results in alterations in the composition of the proteins secreted by the nematode [46]. However, milligrams of secreted proteins (SP) can be collected from mL1 isolated in this manner facilitating identification of a number of its components. One striking feature shared by many mL1 SPs is the presence of a unique family of highly fucosylated tri- and tetra-antennary glycan structures capped with 3,6-dideoxy-*d*-arabinohexose (tyvelose) [47]. These glycans are highly immunodominant and mediate rapid expulsion of intestinal parasites in rat infection models [47]. Glycosylation with the tyvelose moiety is restricted to the larval stages of development, indicating it may have functions either during latter stages of nurse cell development or the initial establishment of the parasite within the intestine.

Depending on the visualization technique used, analysis of *T. spiralis* mL1 SPs by two-dimensional gel electrophoresis indicates it is composed of between 60 and 100 discrete protein species. Proteomic analysis has revealed that many of these species are closely related isoforms or processing intermediates [46]. However, it is anticipated that at least 30–40 unique proteins are present in SP preparations. Currently, the p43 (43 kDa glycoprotein) is the only protein secreted by mL1 that has been shown to localized to the nurse cell. Initially characterized as a tyvelosated antigen, subsequent cloning and sequence analysis established that p43 is member of the DNase II family [48]. However, because p43 lacks several catalytic residues required by other DNase II members it is expected to lack enzyme activity [49]. Immunohistochemical analysis using polyclonal antisera generated against its C-terminus determined that p43 was synthesized within the α-stichocytes and cross-reactive proteins were detected within the nurse cell cytoplasm and hypertrophic nuclei between days 12 and 15 pfi. Subsequently, Jasmer and colleagues established that these antibodies detect a complex family of secreted antigens (eight or more proteins) who share cross-reactive epitopes within the core or C-terminus of p43 [50]. The cross-reactive proteins detected within SP preparations were also distinct from those detected within hypertrophic nuclei preparations. This indicates p43-related nuclear antigen secretion was not sustained after mL1 isolation, or modifications that affect SDS-PAGE migration occur post secretion within the nurse cell.

Unlike p43, these nuclear antigens were secreted independently of the stichosome and p43/α stichocyte-deficient larvae that were capable of inducing the infected cell phenotype also had detectable p43-related antigens within their hypertrophic nuclei [50]. This suggests a secretion process or organ independent of the α and β stichocytes provides the early factors, including p43- related nuclear antigens, which are sufficient to initiate the transformation.

Analysis of other cloned mL1 SPs indicates that p43 may not be unique and several additional examples of enzyme (gp45/TspE1) or enzyme inhibitor (MCD-1) families have been identified that lack conserved residues proven to be required for protein activity [51, 52]. Like p43 these proteins appear to be members of conserved genes families within the trichinellids, indicating that within this group these proteins may have evolved properties that are independent of their previously ascribed functions.

A second parasite molecule implicated in nurse cell genesis, the TsJ5, was identified during a directed screen for parasite-derived myogenic bHLH homologs. The TsJ5 is a novel mL1-specific protein that inhibits the formation of MyoD–DNA complexes *in vitro* [53]. While it is a component of mL1 SPs, unlike p43, it is not localized within the α or β stichosome. However, while polyclonal antibodies failed to detect it within the nurse cell, immunogold TEM analysis revealed that TsJ5 is localized in the cuticle and on the surface of the nematode [54]. This suggests it could be secreted transcuticularly at low levels or that it functions by interacting with factors at the host–parasite interface.

A number of enzymes have been identified within SPs including lipid-activating factors, nucleotide-metabolizing, DNase, RNase, glycolytic, proteolytic, protein kinase and protein phosphatase activities [55–65]. The unusual ecto-protein kinase activities identified in *Trichinella* SPs are particularly intriguing because of their potential modulatory properties on host signal transduction pathways. Arden and colleagues identified at least two distinct kinase activities migrating at 70 and 135 kDa in mL1 SP preparations [61]. As well as acting on endogenous substrates within the SP or on the surface of the parasite, these enzymes are capable of phosphorylating host substrates such as MyoD [63]. Their relative resistance to inhibitors has made it impossible to define which class of kinase the parasite molecules belong to and it is unknown whether they are secreted into the nurse cell. However, once the cognate genes for these enzymes have been cloned, high-throughput substrate screening procedures and ectopic expression in mammalian cell lines should help to identify potential endogenous targets.

37.4 The Immunology of the Nurse Cell

While the mechanisms underlying immunity to the intestinal phases of the parasites are well understood, the immunology of the nurse cell remains poorly defined. In primary infections a relatively high percentage (~70%) of circulating NBLs successfully colonize skeletal muscle [8, 39]. Migration and trapping of NBLs in various capillary-rich organs (liver, spleen, lungs, etc.) and killing by innate immune cells (neutrophils, eosinophils, etc.) is believed to account for some of the pathology observed during the migratory phase [66–68], while larval attrition during the nurse cell development accounts for some of the pathological symptoms observed later in infection. However, once established in the nurse cell parasites can persist for long periods of time (decades) without obvious deleterious effects to the host. This suggests a profound and specific manipulation of the immune system. The exception to this rule, Chinese hamsters are not permissive to muscle infection and mount intense and directed inflammatory response against the developing nurse cells. These inflammatory responses appear to directly target the larvae and treatment with immunosuppressive drugs restores normal nurse cell development [69].

The mechanisms underlying the clearance of larvae are just beginning to be uncovered. Local inflammation and recruitment of infiltrates around developing nurse cells indicates that damage caused by the parasite and the subsequent remodeling process is noticed by the host immune system. Analysis of these cell populations indicates that they alter during the transformation process with early recruitment of neutrophils that are rapidly replaced with a dominant population of $CD45/CD11b^{+}/MHCII^{high}$ macrophages [70]. The activation status of these cells and the roles they play in the infection process is still being determined. The cells appear to freely penetrate the nurse cell plasma membrane and can be readily identified within its cytoplasm at all stages of the infection. This suggests they could interact directly with the parasite. Like other intracellular pathogens T_H1-mediated inflammatory responses, particularly INFγ and the production of toxic mediators such as NO by the local macrophage population is positively associated with the killing of muscle stage larvae [71, 72].

The induction of high levels of antioxidant enzyme expression in the nurse cell may be one of the parasites strategies for dealing with noxious agents produced by activated immune cells [30]. In addition, parasite SPs have been shown to exert direct modulatory properties deactivating or altering inflammatory cell function [73, 74]. The level of parasite SP diffusion into the local tissue microenvironment has not been established. However both parasite antigens and host IgE can be detected on the capsule and interior of the nurse cell, indicating that some degree of permeability exists [75]. This appears to be a double-edged sword and recognition of parasite proteins in nurse cells, followed by production of IgE and tissue eosinophilia is implicated in larval killing [75–77]. T_H2-mediated responses have been shown to be important in the clearance of other tissue-dwelling nematodes. However, while IL-4/IL-13-associated T_H2 responses are linked to the recruitment of infiltrating cell populations around the nurse cell they do not appear to have any protective effect at this stage of the parasites development [71]. While these cytokines may be required for clearance of adults from the intestine the skewing towards T_H2 and the corresponding antagonism of T_H1-associated responses may inadvertently aid larval establishment.

Long-term local immunosuppression through the production of TGFβ and IL-10 also appears to play a role in protecting the complex. Co-depletion studies indicate that these cytokines not only modulate levels of cellular infiltrate around the nurse cell but also specifically suppress T_H1-associated cellular responses, with IL-10 producing $CD4^+/CD25^-$ T-effector cells controlling INFγ production during chronic infection [70, 71]. Recent observations of TGFβ production by nurse cells during the transformation process suggests that *Trichinella*, through modification of its host cell transcriptome, may adjust the local immune response in a manner which facilitates its survival ([36] and David B. Guiliano, unpublished observations). Therefore both host and parasite may contribute to the immunosuppressive phenotype, making the complex highly resistant to clearance mechanisms.

37.5
Future Work and Outlook

The altered phenotype evident in *Trichinella* remodeled muscle fibers bespeaks the work of a sophisticated manipulator of both molecular and cellular physiology. That we still understand very little about how this remodeling is achieved reinforces how important sustained future research in this relatively neglected system will be. Other species of nematodes (hookworm species and *Haycocknema* spp.) occupy intracellular niches within skeletal muscle fibers. However these infections do not induce drastic phenotypic alterations and invariably lead to fiber necrosis. The broad host-specificity and long-term nature of *Trichinella* infection indicates the parasite is manipulating basic aspects of host developmental biology that are likely to be conserved across the *Mammalia*.

Experimental hurdles limiting our manipulation of *Trichinella* or the nurse cell complex have severely hindered progress in the field. However, new micro-isolation techniques such as laser capture microscopy and global gene expression analysis through microarrays or serial analysis of gene expression (SAGE) will allow monitoring of molecular changes in both host and parasite on a level not previously possible. Newly available tools such as muscle promoter-driven lentiviral expression systems will allow us to test the functions of parasite SPs *in vivo* or in *ex vivo* cultures. The advent of more sophisticated techniques may even allow us to genetically manipulate the parasite. The upcoming release of a completed *T. spiralis* genome sequence will also provide researchers with a complete catalogue of potential SPs for testing in functional assays and allow the use of more sensitive micro-proteomic techniques in the analysis of the nurse cell. Thus, we have only just begun deciphering what this unique host parasite system has to tell us about our own biology.

References

1 Stewart, G.L. and Giannini, S.H. (1982) *Exp Parasitol*, **53**, 406–447.

2 Basuroy, R., Pennisi, R., Robertson, T., Norton, R. *et al.* (2008) *Med J Aust*, **188**, 254–256.

3 Pozio, E. and Zarlenga, D.S. (2005) *Int J Parasitol*, **35**, 1191–1204.

4 Zarlenga, D.S., Rosenthal, B.M., La Rosa, G., Pozio, E. and Hoberg, E.P. (2006) *Proc Natl Acad Sci USA*, **103**, 7354–7359.

5 ManWarren, T., Gagliardo, L., Geyer, J., McVay, C. *et al.* (1997) *Infect Immun*, **65**, 4806–4812.

6 Despommier, D.D., Sukhdeo, M. and Meerovitch, E. (1978) *Exp Parasitol*, **44**, 209–215.

7 Despommier, D.D. (1993) *J Parasitol*, **79**, 472–482.

8 Wang, C.H. and Bell, R.G. (1986) *Exp Parasitol*, **62**, 430–441.

9 Zammit, P.S., Partridge, T.A. and Yablonka-Reuveni, Z. (2006) *J Histochem Cytochem*, **54**, 1177–1191.

10 Charge, S.B. and Rudnicki, M.A. (2004) *Physiol Rev*, **84**, 209–238.

11 De Falco, G., Comes, F. and Simone, C. (2006) *Oncogene*, **25**, 5244–5249.

12 Tiainen, M., Spitkovsky, D., Jansen-Durr, P., Sacchi, A. and Crescenzi, M. (1996) *Mol Cell Biol*, **16**, 5302–5312.

13 Brockes, J.P. and Kumar, A. (2002) *Nat Rev Mol Cell Biol*, **3**, 566–574.

14 Velloso, C.P., Simon, A. and Brockes, J.P. (2001) *Curr Biol*, **11**, 855–858.

15 Despommier, D. (1975) *Am J Pathol*, **78**, 477–496.

16 Bagheri, A., Ubelaker, J.E., Stewart, G.L. and Wood, B. (1986) *J Parasitol*, **72**, 277–282.

17 Despommier, D., Symmans, W.F. and Dell, R. (1991) *J Parasitol*, **77**, 290–295.

18 Boonmars, T., Wu, Z., Nagano, I. and Takahashi, Y. (2005) *Parasitology*, **131**, 705–712.

19 Gustowska, L., Gabryel, P., Blotna-Filipiak, M. and Rauhut, W. (1989) *Wiad Parazytol*, **35**, 401–411.

20 Jasmer, D.P. (1993) *J Cell Biol*, **121**, 785–793.

21 Jasmer, D.P. (1990) *Exp Parasitol*, **70**, 452–465.

22 Jasmer, D.P., Bohnet, S. and Prieur, D.J. (1991) *Exp Parasitol*, **72**, 321–331.

23 Purkenson, M. and Despommier, D.D. (1974) *Fine Structure of the Muscle Phase of Trichinella spiralis in the Mouse in Trichinellosis*, Intext Educational Publishers, New York, pp. 7–23.

24 Ralston, E., Lu, Z. and Ploug, T. (1999) *J Neurosci*, **19**, 10694–10705.

25 Michejda, J.W. and Boczon, K. (1972) *Exp Parasitol*, **31**, 161–171.

26 Gounaris, K., Smith, V.P. and Selkirk, M.E. (1996) *Biochim Biophys Acta*, **1281**, 91–100.

27 Modha, J., Roberts, M.C., Robertson, W.M., Sweetman, G. *et al.* (1999) *Parasitology*, **118** (Pt 5), 509–522.

28 Montgomery, J.M., Augostini, P. and Stewart, G.L. (2003) *Int J Parasitol*, **33**, 401–412.

29 Bruce, R.G. (1970) *Parasitology*, **60**, 223–227.

30 Bruschi, F., Saviozzi, M., Piaggi, S., Malvaldi, G. and Casini, A. (2003) *Int J Parasitol*, **33**, 1035–1042.

31 Capo, V.A., Despommier, D.D. and Polvere, R.I. (1998) *J Parasitol*, **84**, 209–214.

32 Polvere, R.I., Kabbash, C.A., Capo, V.A., Kadan, I. and Despommier, D.D. (1997) *Exp Parasitol*, **86**, 191–199.

33 Beiting, D.P., Park, P.W. and Appleton, J.A. (2006) *Infect Immun*, **74**, 1941–1943.

34 Alexopoulou, A.N., Multhaupt, H.A. and Couchman, J.R. (2007) *Int J Biochem Cell Biol*, **39**, 505–528.

35 Wu, Z., Nagano, I., Boonmars, T. and Takahashi, Y. (2005) *Parasitology*, **131**, 373–381.

36 Wu, Z., Nagano, I., Boonmars, T. and Takahashi, Y. (2006) *Int J Parasitol*, **36**, 1159–1166.

37 Wu, Z., Nagano, I. and Takahashi, Y. (2008) *Parasitol Int*, **57**, 368–378.

38 Pritchard, D.I. (1985) *J Helminthol*, **59**, 71–77.

39 Despommier, D., Aron, L. and Turgeon, L. (1975) *Exp Parasitol*, **37**, 108–116.

40 Despommier, D.D. and Muller, M. (1976) *J Parasitol*, **62**, 775–785.

41 Sanmartin, M.L., Iglesias, R., Santamarina, M.T., Leiro, J. and Ubeira, F.M. (1991) *Parasitol Res*, **77**, 301–306.

42 Jasmer, D.P. and Neary, S.M. (1994) *Exp Parasitol*, **78**, 317–325.

43 Yao, C., Bohnet, S. and Jasmer, D.P. (1998) *Mol Biochem Parasitol*, **96**, 1–13.

44 Yao, C. and Jasmer, D.P. (1998) *Mol Biochem Parasitol*, **92**, 207–218.
45 Yao, C. and Jasmer, D.P. (2001) *Infect Immun*, **69**, 4065–4071.
46 Robinson, M.W. and Connolly, B. (2005) *Proteomics*, **5**, 4525–4532.
47 Appleton, J.A. and Romaris, F. (2001) *Vet Parasitol*, **101**, 249–260.
48 Vassilatis, D.K., Despommier, D., Misek, D.E., Polvere, R.I. *et al.* (1992) *J Biol Chem*, **267**, 18459–18465.
49 MacLea, K.S., Krieser, R.J. and Eastman, A. (2003) *Gene*, **305**, 1–12.
50 Jasmer, D.P., Yao, S., Vassilatis, D., Despommier, D. and Neary, S.M. (1994) *Mol Biochem Parasitol*, **67**, 225–234.
51 Arasu, P., Ellis, L.A., Iglesias, R., Ubeira, F.M. and Appleton, J.A. (1994) *Mol Biochem Parasitol*, **65**, 201–211.
52 Robinson, M.W., Massie, D.H. and Connolly, B. (2007) *Mol Biochem Parasitol*, **151**, 9–17.
53 Lindh, J.G., Connolly, B., McGhie, D.L. and Smith, D.F. (1998) *Mol Biochem Parasitol*, **92**, 163–175.
54 Kuratli, S., Hemphill, A., Lindh, J., Smith, D.F. and Connolly, B. (2001) *Mol Biochem Parasitol*, **115**, 199–208.
55 Selkirk, M.E., Hussein, A.S., Chambers, A.E., Goulding, D. *et al.* (2004) *Mol Biochem Parasitol*, **135**, 49–56.
56 Gounaris, K., Thomas, S., Najarro, P. and Selkirk, M.E. (2001) *Infect Immun*, **69**, 3658–3662.
57 Gounaris, K. (2002) *Infect Immun*, **70**, 4917–4924.
58 Gounaris, K., Selkirk, M.E. and Sadeghi, S.J. (2004) *Mol Biochem Parasitol*, **136**, 257–264.
59 Bruce, A.F., Gares, M.P., Selkirk, M.E. and Gounaris, K. (2006) *Mol Biochem Parasitol*, **147**, 224–229.
60 Bruce, A.F. and Gounaris, K. (2006) *Mol Biochem Parasitol*, **145**, 84–93.
61 Arden, S.R., Smith, A.M., Booth, M.J., Tweedie, S. *et al.* (1997) *Mol Biochem Parasitol*, **90**, 111–119.
62 Romaris, F., North, S.J., Gagliardo, L.F., Butcher, B.A. *et al.* (2002) *Mol Biochem Parasitol*, **122**, 149–160.
63 Smith, V.P., Selkirk, M.E. and Gounaris, K. (2000) *FEBS Lett*, **483**, 104–108.
64 Mak, C.H., Chung, Y.Y. and Ko, R.C. (2000) *Parasitology*, **120** (Pt 5), 527–533.
65 Mak, C.H. and Ko, R.C. (1999) *Eur J Biochem*, **260**, 477–481.
66 Buys, J., Wever, R., van Stigt, R. and Ruitenberg, E.J. (1981) *Eur J Immunol*, **11**, 843–845.
67 Ruitenberg, E.J., Buys, J., Teppema, J.S. and Elgersma, A. (1983) *Z Parasitenkd*, **69**, 807–815.
68 Kazura, J.W. (1981) *J Infect Dis*, **143**, 712–718.
69 Ritterson, A.L. (1959) *J Infect Dis*, **105**, 253–266.
70 Beiting, D.P., Bliss, S.K., Schlafer, D.H., Roberts, V.L. and Appleton, J.A. (2004) *Infect Immun*, **72**, 3129–3137.
71 Beiting, D.P., Gagliardo, L.F., Hesse, M., Bliss, S.K. *et al.* (2007) *J Immunol*, **178**, 1039–1047.
72 Helmby, H. and Grencis, R.K. (2003) *Eur J Immunol*, **33**, 2382–2390.
73 Bruschi, F., Carulli, G., Azzara, A., Homan, W. *et al.* (2000) *Int Arch Allergy Immunol*, **122**, 58–65.
74 Gounaris, K. and Selkirk, M.E. (2005) *Trends Parasitol*, **21**, 17–21.
75 Gurish, M.F., Bryce, P.J., Tao, H., Kisselgof, A.B. *et al.* (2004) *J Immunol*, **172**, 1139–1145.
76 Dessein, A.J., Parker, W.L., James, S.L. and David, J.R. (1981) *J Exp Med*, **153**, 423–436.
77 Gurish, M.F., Humbles, A., Tao, H., Finkelstein, S. *et al.* (2002) *J Immunol*, **168**, 5730–5736.
78 Appleton, J.A., Schain, L.R. and McGregor, D.D. (1988) *Immunology*, **65**, 487–492.

Index

Intracellular Niches of Microbes. Edited by Ulrich E. Schaible and Albert Haas

ISBN: 978-3-527-32207-7

j

k